Drug-Drug Interactions

DRUGS AND THE PHARMACEUTICAL SCIENCES

A Series of Textbooks and Monographs

1. Pharmacokinetics, *Milo Gibaldi and Donald Perrier*
2. Good Manufacturing Practices for Pharmaceuticals: A Plan for Total Quality Control, *Sidney H. Willig, Murray M. Tuckerman, and William S. Hitchings IV*
3. Microencapsulation, *edited by J. R. Nixon*
4. Drug Metabolism: Chemical and Biochemical Aspects, *Bernard Testa and Peter Jenner*
5. New Drugs: Discovery and Development, *edited by Alan A. Rubin*
6. Sustained and Controlled Release Drug Delivery Systems, *edited by Joseph R. Robinson*
7. Modern Pharmaceutics, *edited by Gilbert S. Banker and Christopher T. Rhodes*
8. Prescription Drugs in Short Supply: Case Histories, *Michael A. Schwartz*
9. Activated Charcoal: Antidotal and Other Medical Uses, *David O. Cooney*
10. Concepts in Drug Metabolism (in two parts), *edited by Peter Jenner and Bernard Testa*
11. Pharmaceutical Analysis: Modern Methods (in two parts), *edited by James W. Munson*
12. Techniques of Solubilization of Drugs, *edited by Samuel H. Yalkowsky*
13. Orphan Drugs, *edited by Fred E. Karch*
14. Novel Drug Delivery Systems: Fundamentals, Developmental Concepts, Biomedical Assessments, *Yie W. Chien*
15. Pharmacokinetics: Second Edition, Revised and Expanded, *Milo Gibaldi and Donald Perrier*
16. Good Manufacturing Practices for Pharmaceuticals: A Plan for Total Quality Control, Second Edition, Revised and Expanded, *Sidney H. Willig, Murray M. Tuckerman, and William S. Hitchings IV*
17. Formulation of Veterinary Dosage Forms, *edited by Jack Blodinger*
18. Dermatological Formulations: Percutaneous Absorption, *Brian W. Barry*
19. The Clinical Research Process in the Pharmaceutical Industry, *edited by Gary M. Matoren*
20. Microencapsulation and Related Drug Processes, *Patrick B. Deasy*
21. Drugs and Nutrients: The Interactive Effects, *edited by Daphne A. Roe and T. Colin Campbell*
22. Biotechnology of Industrial Antibiotics, *Erick J. Vandamme*
23. Pharmaceutical Process Validation, *edited by Bernard T. Loftus and Robert A. Nash*
24. Anticancer and Interferon Agents: Synthesis and Properties, *edited by Raphael M. Ottenbrite and George B. Butler*

25. Pharmaceutical Statistics: Practical and Clinical Applications, *Sanford Bolton*
26. Drug Dynamics for Analytical, Clinical, and Biological Chemists, *Benjamin J. Gudzinowicz, Burrows T. Younkin, Jr., and Michael J. Gudzinowicz*
27. Modern Analysis of Antibiotics, *edited by Adjoran Aszalos*
28. Solubility and Related Properties, *Kenneth C. James*
29. Controlled Drug Delivery: Fundamentals and Applications, Second Edition, Revised and Expanded, *edited by Joseph R. Robinson and Vincent H. Lee*
30. New Drug Approval Process: Clinical and Regulatory Management, *edited by Richard A. Guarino*
31. Transdermal Controlled Systemic Medications, *edited by Yie W. Chien*
32. Drug Delivery Devices: Fundamentals and Applications, *edited by Praveen Tyle*
33. Pharmacokinetics: Regulatory • Industrial • Academic Perspectives, *edited by Peter G. Welling and Francis L. S. Tse*
34. Clinical Drug Trials and Tribulations, *edited by Allen E. Cato*
35. Transdermal Drug Delivery: Developmental Issues and Research Initiatives, *edited by Jonathan Hadgraft and Richard H. Guy*
36. Aqueous Polymeric Coatings for Pharmaceutical Dosage Forms, *edited by James W. McGinity*
37. Pharmaceutical Pelletization Technology, *edited by Isaac Ghebre-Sellassie*
38. Good Laboratory Practice Regulations, *edited by Allen F. Hirsch*
39. Nasal Systemic Drug Delivery, *Yie W. Chien, Kenneth S. E. Su, and Shyi-Feu Chang*
40. Modern Pharmaceutics: Second Edition, Revised and Expanded, *edited by Gilbert S. Banker and Christopher T. Rhodes*
41. Specialized Drug Delivery Systems: Manufacturing and Production Technology, *edited by Praveen Tyle*
42. Topical Drug Delivery Formulations, *edited by David W. Osborne and Anton H. Amann*
43. Drug Stability: Principles and Practices, *Jens T. Carstensen*
44. Pharmaceutical Statistics: Practical and Clinical Applications, Second Edition, Revised and Expanded, *Sanford Bolton*
45. Biodegradable Polymers as Drug Delivery Systems, *edited by Mark Chasin and Robert Langer*
46. Preclinical Drug Disposition: A Laboratory Handbook, *Francis L. S. Tse and James J. Jaffe*
47. HPLC in the Pharmaceutical Industry, *edited by Godwin W. Fong and Stanley K. Lam*

48. Pharmaceutical Bioequivalence, *edited by Peter G. Welling, Francis L. S. Tse, and Shrikant V. Dinghe*
49. Pharmaceutical Dissolution Testing, *Umesh V. Banakar*
50. Novel Drug Delivery Systems: Second Edition, Revised and Expanded, *Yie W. Chien*
51. Managing the Clinical Drug Development Process, *David M. Cocchetto and Ronald V. Nardi*
52. Good Manufacturing Practices for Pharmaceuticals: A Plan for Total Quality Control, Third Edition, *edited by Sidney H. Willig and James R. Stoker*
53. Prodrugs: Topical and Ocular Drug Delivery, *edited by Kenneth B. Sloan*
54. Pharmaceutical Inhalation Aerosol Technology, *edited by Anthony J. Hickey*
55. Radiopharmaceuticals: Chemistry and Pharmacology, *edited by Adrian D. Nunn*
56. New Drug Approval Process: Second Edition, Revised and Expanded, *edited by Richard A. Guarino*
57. Pharmaceutical Process Validation: Second Edition, Revised and Expanded, *edited by Ira R. Berry and Robert A. Nash*
58. Ophthalmic Drug Delivery Systems, *edited by Ashim K. Mitra*
59. Pharmaceutical Skin Penetration Enhancement, *edited by Kenneth A. Walters and Jonathan Hadgraft*
60. Colonic Drug Absorption and Metabolism, *edited by Peter R. Bieck*
61. Pharmaceutical Particulate Carriers: Therapeutic Applications, *edited by Alain Rolland*
62. Drug Permeation Enhancement: Theory and Applications, *edited by Dean S. Hsieh*
63. Glycopeptide Antibiotics, *edited by Ramakrishnan Nagarajan*
64. Achieving Sterility in Medical and Pharmaceutical Products, *Nigel A. Halls*
65. Multiparticulate Oral Drug Delivery, *edited by Isaac Ghebre-Sellassie*
66. Colloidal Drug Delivery Systems, *edited by Jörg Kreuter*
67. Pharmacokinetics: Regulatory • Industrial • Academic Perspectives, Second Edition, *edited by Peter G. Welling and Francis L. S. Tse*
68. Drug Stability: Principles and Practices, Second Edition, Revised and Expanded, *Jens T. Carstensen*
69. Good Laboratory Practice Regulations: Second Edition, Revised and Expanded, *edited by Sandy Weinberg*

70. Physical Characterization of Pharmaceutical Solids, *edited by Harry G. Brittain*
71. Pharmaceutical Powder Compaction Technology, *edited by Göran Alderborn and Christer Nyström*
72. Modern Pharmaceutics: Third Edition, Revised and Expanded, *edited by Gilbert S. Banker and Christopher T. Rhodes*
73. Microencapsulation: Methods and Industrial Applications, *edited by Simon Benita*
74. Oral Mucosal Drug Delivery, *edited by Michael J. Rathbone*
75. Clinical Research in Pharmaceutical Development, *edited by Barry Bleidt and Michael Montagne*
76. The Drug Development Process: Increasing Efficiency and Cost Effectiveness, *edited by Peter G. Welling, Louis Lasagna, and Umesh V. Banakar*
77. Microparticulate Systems for the Delivery of Proteins and Vaccines, *edited by Smadar Cohen and Howard Bernstein*
78. Good Manufacturing Practices for Pharmaceuticals: A Plan for Total Quality Control, Fourth Edition, Revised and Expanded, *Sidney H. Willig and James R. Stoker*
79. Aqueous Polymeric Coatings for Pharmaceutical Dosage Forms: Second Edition, Revised and Expanded, *edited by James W. McGinity*
80. Pharmaceutical Statistics: Practical and Clinical Applications, Third Edition, *Sanford Bolton*
81. Handbook of Pharmaceutical Granulation Technology, *edited by Dilip M. Parikh*
82. Biotechnology of Antibiotics: Second Edition, Revised and Expanded, *edited by William R. Strohl*
83. Mechanisms of Transdermal Drug Delivery, *edited by Russell O. Potts and Richard H. Guy*
84. Pharmaceutical Enzymes, *edited by Albert Lauwers and Simon Scharpé*
85. Development of Biopharmaceutical Parenteral Dosage Forms, *edited by John A. Bontempo*
86. Pharmaceutical Project Management, *edited by Tony Kennedy*
87. Drug Products for Clinical Trials: An International Guide to Formulation • Production • Quality Control, *edited by Donald C. Monkhouse and Christopher T. Rhodes*
88. Development and Formulation of Veterinary Dosage Forms: Second Edition, Revised and Expanded, *edited by Gregory E. Hardee and J. Desmond Baggot*
89. Receptor-Based Drug Design, *edited by Paul Leff*
90. Automation and Validation of Information in Pharmaceutical Processing, *edited by Joseph F. deSpautz*

91. Dermal Absorption and Toxicity Assessment, *edited by Michael S. Roberts and Kenneth A. Walters*
92. Pharmaceutical Experimental Design, *Gareth A. Lewis, Didier Mathieu, and Roger Phan-Tan-Luu*
93. Preparing for FDA Pre-Approval Inspections, *edited by Martin D. Hynes III*
94. Pharmaceutical Excipients: Characterization by IR, Raman, and NMR Spectroscopy, *David E. Bugay and W. Paul Findlay*
95. Polymorphism in Pharmaceutical Solids, *edited by Harry G. Brittain*
96. Freeze-Drying/Lyophilization of Pharmaceutical and Biological Products, *edited by Louis Rey and Joan C. May*
97. Percutaneous Absorption: Drugs–Cosmetics–Mechanisms–Methodology, Third Edition, Revised and Expanded, *edited by Robert L. Bronaugh and Howard I. Maibach*
98. Bioadhesive Drug Delivery Systems: Fundamentals, Novel Approaches, and Development, *edited by Edith Mathiowitz, Donald E. Chickering III, and Claus- Michael Lehr*
99. Protein Formulation and Delivery, *edited by Eugene J. McNally*
100. New Drug Approval Process: Third Edition, The Global Challenge, *edited by Richard A. Guarino*
101. Peptide and Protein Drug Analysis, *edited by Ronald E. Reid*
102. Transport Processes in Pharmaceutical Systems, *edited by Gordon L. Amidon, Ping I. Lee, and Elizabeth M. Topp*
103. Excipient Toxicity and Safety, *edited by Myra L. Weiner and Lois A. Kotkoskie*
104. The Clinical Audit in Pharmaceutical Development, *edited by Michael R. Hamrell*
105. Pharmaceutical Emulsions and Suspensions, *edited by Francoise Nielloud and Gilberte Marti-Mestres*
106. Oral Drug Absorption: Prediction and Assessment, *edited by Jennifer B. Dressman and Hans Lennernäs*
107. Drug Stability: Principles and Practices, Third Edition, Revised and Expanded, *edited by Jens T. Carstensen and C. T. Rhodes*
108. Containment in the Pharmaceutical Industry, *edited by James P. Wood*
109. Good Manufacturing Practices for Pharmaceuticals: A Plan for Total Quality Control from Manufacturer to Consumer, Fifth Edition, Revised and Expanded, *Sidney H. Willig*
110. Advanced Pharmaceutical Solids, *Jens T. Carstensen*
111. Endotoxins: Pyrogens, LAL Testing, and Depyrogenation, Second Edition, Revised and Expanded, *Kevin L. Williams*
112. Pharmaceutical Process Engineering, *Anthony J. Hickey and David Ganderton*

113. Pharmacogenomics, *edited by Werner Kalow, Urs A. Meyer and Rachel F. Tyndale*
114. Handbook of Drug Screening, *edited by Ramakrishna Seethala and Prabhavathi B. Fernandes*
115. Drug Targeting Technology: Physical • Chemical • Biological Methods, *edited by Hans Schreier*
116. Drug–Drug Interactions, *edited by A. David Rodrigues*
117. Handbook of Pharmaceutical Analysis, *edited by Lena Ohannesian and Anthony J. Streeter*
118. Pharmaceutical Process Scale-Up, *edited by Michael Levin*
119. Dermatological and Transdermal Formulations, *edited by Kenneth A. Walters*
120. Clinical Drug Trials and Tribulations: Second Edition, Revised and Expanded, *edited by Allen Cato, Lynda Sutton, and Allen Cato III*
121. Modern Pharmaceutics: Fourth Edition, Revised and Expanded, *edited by Gilbert S. Banker and Christopher T. Rhodes*
122. Surfactants and Polymers in Drug Delivery, *Martin Malmsten*
123. Transdermal Drug Delivery: Second Edition, Revised and Expanded, *edited by Richard H. Guy and Jonathan Hadgraft*
124. Good Laboratory Practice Regulations: Second Edition, Revised and Expanded, *edited by Sandy Weinberg*
125. Parenteral Quality Control: Sterility, Pyrogen, Particulate, and Package Integrity Testing: Third Edition, Revised and Expanded, *Michael J. Akers, Daniel S. Larrimore, and Dana Morton Guazzo*
126. Modified-Release Drug Delivery Technology, *edited by Michael J. Rathbone, Jonathan Hadgraft, and Michael S. Roberts*
127. Simulation for Designing Clinical Trials: A Pharmacokinetic-Pharmacodynamic Modeling Perspective, *edited by Hui C. Kimko and Stephen B. Duffull*
128. Affinity Capillary Electrophoresis in Pharmaceutics and Biopharmaceutics, *edited by Reinhard H. H. Neubert and Hans-Hermann Rüttinger*
129. Pharmaceutical Process Validation: An International Third Edition, Revised and Expanded, *edited by Robert A. Nash and Alfred H. Wachter*
130. Ophthalmic Drug Delivery Systems: Second Edition, Revised and Expanded, *edited by Ashim K. Mitra*
131. Pharmaceutical Gene Delivery Systems, *edited by Alain Rolland and Sean M. Sullivan*
132. Biomarkers in Clinical Drug Development, *edited by John C. Bloom and Robert A. Dean*

133. Pharmaceutical Extrusion Technology, *edited by Isaac Ghebre-Sellassie and Charles Martin*
134. Pharmaceutical Inhalation Aerosol Technology: Second Edition, Revised and Expanded, *edited by Anthony J. Hickey*
135. Pharmaceutical Statistics: Practical and Clinical Applications, Fourth Edition, *Sanford Bolton and Charles Bon*
136. Compliance Handbook for Pharmaceuticals, Medical Devices, and Biologics, *edited by Carmen Medina*
137. Freeze-Drying/Lyophilization of Pharmaceutical and Biological Products: Second Edition, Revised and Expanded, *edited by Louis Rey and Joan C. May*
138. Supercritical Fluid Technology for Drug Product Development, *edited by Peter York, Uday B. Kompella, and Boris Y. Shekunov*
139. New Drug Approval Process: Fourth Edition, Accelerating Global Registrations, *edited by Richard A. Guarino*
140. Microbial Contamination Control in Parenteral Manufacturing, *edited by Kevin L. Williams*
141. New Drug Development: Regulatory Paradigms for Clinical Pharmacology and Biopharmaceutics, *edited by Chandrahas G. Sahajwalla*
142. Microbial Contamination Control in the Pharmaceutical Industry, *edited by Luis Jimenez*
143. Generic Drug Product Development: Solid Oral Dosage Forms, *edited by Leon Shargel and Isadore Kanfer*
144. Introduction to the Pharmaceutical Regulatory Process, *edited by Ira R. Berry*
145. Drug Delivery to the Oral Cavity: Molecules to Market, *edited by Tapash K. Ghosh and William R. Pfister*
146. Good Design Practices for GMP Pharmaceutical Facilities, *edited by Andrew Signore and Terry Jacobs*
147. Drug Products for Clinical Trials, Second Edition, *edited by Donald Monkhouse, Charles Carney, and Jim Clark*
148. Polymeric Drug Delivery Systems, *edited by Glen S. Kwon*
149. Injectable Dispersed Systems: Formulation, Processing, and Performance, *edited by Diane J. Burgess*
150. Laboratory Auditing for Quality and Regulatory Compliance, *Donald Singer, Raluca-Ioana Stefan, and Jacobus van Staden*
151. Active Pharmaceutical Ingredients: Development, Manufacturing, and Regulation, *edited by Stanley Nusim*
152. Preclinical Drug Development, *edited by Mark C. Rogge and David R. Taft*
153. Pharmaceutical Stress Testing: Predicting Drug Degradation, *edited by Steven W. Baertschi*

154. Handbook of Pharmaceutical Granulation Technology: Second Edition, *edited by Dilip M. Parikh*
155. Percutaneous Absorption: Drugs–Cosmetics–Mechanisms–Methodology, Fourth Edition, *edited by Robert L. Bronaugh and Howard I. Maibach*
156. Pharmacogenomics: Second Edition, *edited by Werner Kalow, Urs A. Meyer and Rachel F. Tyndale*
157. Pharmaceutical Process Scale-Up, Second Edition, *edited by Michael Levin*
158. Microencapsulation: Methods and Industrial Applications, Second Edition, *edited by Simon Benita*
159. Nanoparticle Technology for Drug Delivery, *edited by Ram B. Gupta and Uday B. Kompella*
160. Spectroscopy of Pharmaceutical Solids, *edited by Harry G. Brittain*
161. Dose Optimization in Drug Development, *edited by Rajesh Krishna*
162. Herbal Supplements-Drug Interactions: Scientific and Regulatory Perspectives, *edited by Y. W. Francis Lam, Shiew-Mei Huang, and Stephen D. Hall*
163. Pharmaceutical Photostability and Stabilization Technology, *edited by Joseph T. Piechocki and Karl Thoma*
164. Environmental Monitoring for Cleanrooms and Controlled Environments, *edited by Anne Marie Dixon*
165. Pharmaceutical Product Development: In Vitro-In Vivo Correlation, *edited by Dakshina Murthy Chilukuri, Gangadhar Sunkara, and David Young*
166. Nanoparticulate Drug Delivery Systems, *edited by Deepak Thassu, Michel Deleers, and Yashwant Pathak*
167. Endotoxins: Pyrogens, LAL Testing and Depyrogenation, Third Edition, *edited by Kevin L. Williams*
168. Good Laboratory Practice Regulations, Fourth Edition, *edited by Anne Sandy Weinberg*
169. Good Manufacturing Practices for Pharmaceuticals, Sixth Edition, *edited by Joseph D. Nally*
170. Oral-Lipid Based Formulations: Enhancing the Bioavailability of Poorly Water-soluble Drugs, *edited by David J. Hauss*
171. Handbook of Bioequivalence Testing, *edited by Sarfaraz K. Niazi*
172. Advanced Drug Formulation Design to Optimize Therapeutic Outcomes, *edited by Robert O. Williams III, David R. Taft, and Jason T. McConville*
173. Clean-in-Place for Biopharmaceutical Processes, *edited by Dale A. Seiberling*
174. Filtration and Purification in the Biopharmaceutical Industry, Second Edition, *edited by Maik W. Jornitz and Theodore H. Meltzer*

175. Protein Formulation and Delivery, Second Edition, *edited by Eugene J. McNally and Jayne E. Hastedt*
176. Aqueous Polymeric Coatings for Pharmaceutical Dosage Forms, Third Edition, *edited by James W. McGinity and Linda A. Felton*
177. Dermal Absorption and Toxicity Assessment, Second Edition, *edited by Michael S. Roberts and Kenneth A. Walters*
178. Preformulation in Solid Dosage Form Development, *edited by Moji Christianah Adeyeye and Harry G. Brittain*
179. Drug-Drug Interactions, Second Edition, *edited by A. David Rodrigues*
180. Generic Drug Product Development: Bioequivalence Issues, *edited by Isadore Kanfer and Leon Shargel*

Drug-Drug Interactions
Second Edition

Edited by

A. David Rodrigues
Bristol-Myers Squibb
Research & Development
Princeton, New Jersey, USA

informa
healthcare

New York London

Informa Healthcare USA, Inc.
52 Vanderbilt Avenue
New York, NY 10017

Printed in the United States of America on acid-free paper
10 9 8 7 6 5 4 3 2 1

International Standard Book Number-10: 0-8493-7593-2 (Hardcover)
International Standard Book Number-13: 978-0-8493-7593-4 (Hardcover)

Library of Congress Cataloging-in-Publication Data

Drug-drug interactions / edited by A. David Rodrigues. — 2nd ed.
p. ; cm. — (Drugs and the pharmaceutical sciences ; v. 179)
Includes bibliographical references and index.
ISBN-13: 978-0-8493-7593-4 (hb : alk. paper)
ISBN-10: 0-8493-7593-2 (hb : alk. paper)
1. Drug interactions. 2. Pharmacokinetics. I. Rodrigues, A. David. II. Series.
[DNLM: 1. Drug Interactions. 2. Pharmaceutical Preparations—metabolism.
3. Pharmacokinetics. W1 DR893B v.179 2008 / QV 38 D79154 2008]

RM302.D784 2008
615′.7045—dc22 2007034623

For Corporate Sales and Reprint Permissions call 212-520-2700 or write to: Sales Department, 52 Vanderbilt Avenue, 16th floor, New York, NY 10017.

Visit the Informa Web site at
www.informa.com

and the Informa Healthcare Web site at
www.informahealthcare.com

Preface

Since the publication of the first edition of *Drug-Drug Interactions* in 2002, our knowledge of the various human drug-metabolizing enzyme systems and drug transporters has continued to grow at a rapid pace. This continued growth in knowledge has been fueled by further advances in molecular biology, the continued availability of human tissue, and the development of additional model systems and sensitive assay methods for studying drug metabolism and transport in vitro and in vivo. Broadly speaking, there has been considerable progress in six major areas: in silico (computer-based) approaches, transgenic animal models, pharmacokinetic-based predictions and modeling, the characterization of additional drug-metabolizing enzymes (e.g., CYP2B6 and CYP2C8), characterization of nuclear hormone receptors (e.g., pregnane X receptor), and the classification, characterization, and study of influx and efflux drug transporters. Consequently, it became necessary to revise the first edition of *Drug-Drug Interactions,* and more than three quarters of the original chapters were updated. In response to the constructive feedback of numerous readers and reviewers, the index was expanded and the sequence of the chapters rearranged. However, the second edition still presents the subject of drug-drug interactions from a preclinical, clinical, toxicological, regulatory, industrial, and marketing perspective.

During the preparation of the second edition, many of us were saddened by the passing of Grant Wilkinson, Ph.D, D.Sc. He contributed extensively to the fields of drug metabolism, drug interactions, pharmacokinetics, and pharmacogenetics. As editor, I appreciate greatly his contributions to *Drug-Drug Interactions* and dedicate the second edition of the book to him.

A. David Rodrigues, Ph.D.

Contents

Contributors

Upendra A. Argikar Novartis Pharmaceuticals, Cambridge, Massachusetts, U.S.A.

Stephen E. Clarke GlaxoSmithKline Pharmaceuticals, Ware, U.K.

K. Anton Feenstra Department of Pharmacochemistry, Vrije Universiteit, Amsterdam, The Netherlands

Liang-Shang Gan Biogen Idec, Inc., Cambridge, Massachusetts, U.S.A.

Chris de Graaf Department of Pharmacochemistry, Vrije Universiteit, Amsterdam, The Netherlands

David J. Greenblatt Tufts University School of Medicine and Tufts-New England Medical Center, Boston, Massachusetts, U.S.A.

Houda Hachad Department of Pharmaceutics, University of Washington, Seattle, Washington, U.S.A.

Stephen D. Hall Indiana University School of Medicine, Indianapolis, Indiana, U.S.A.

Shiew-Mei Huang U.S. Food and Drug Administration, Silver Spring, Maryland, U.S.A.

Nina Isoherranen University of Washington, Seattle, Washington, U.S.A.

Barry C. Jones Pfizer Global Research & Development, Kent, U.K.

David R. Jones Indiana University School of Medicine, Indianapolis, Indiana, U.S.A.

Beverly M. Knight The University of North Carolina at Chapel Hill, Chapel Hill, North Carolina, U.S.A.

Kenneth R. Korzekwa Preclinical Research and Development, AllChemie Inc., Wayne, Pennsylvania, U.S.A.

Hiroyuki Kusuhara Department of Molecular Pharmacokinetics, Graduate School of Pharmaceutical Sciences, The University of Tokyo, Tokyo, Japan

Lawrence J. Lesko U.S. Food and Drug Administration, Silver Spring, Maryland, U.S.A.

René H. Levy Department of Pharmaceutics, University of Washington, Seattle, Washington, U.S.A.

Jiunn H. Lin Department of Drug Metabolism, Merck Research Laboratories, West Point, Pennsylvania, U.S.A.

Gang Luo Covance Laboratories Inc., Madison, Wisconsin, U.S.A.

Sidney D. Nelson Department of Medicinal Chemistry, School of Pharmacy, University of Washington, Seattle, Washington, U.S.A.

Deborah A. Nicoll-Griffith Merck Research Laboratories, Rahway, New Jersey, U.S.A.

Brian W. Ogilvie XenoTech LLC, Lenexa, Kansas, U.S.A.

Andrew Parkinson XenoTech LLC, Lenexa, Kansas, U.S.A.

Kevin J. Petty Johnson and Johnson, Raritan, New Jersey, U.S.A.

Isabelle Ragueneau-Majlessi Department of Pharmaceutics, University of Washington, Seattle, Washington, U.S.A.

Rory P. Remmel Department of Medicinal Chemistry, College of Pharmacy, University of Minnesota, Minneapolis, Minnesota, U.S.A.

Malcolm Rowland University of Manchester, Manchester, U.K.

Danny D. Shen University of Washington, Seattle, Washington, U.S.A.

Jose M. Silva Johnson and Johnson, Raritan, New Jersey, U.S.A.

Yuichi Sugiyama Department of Molecular Pharmacokinetics, Graduate School of Pharmaceutical Sciences, The University of Tokyo, Tokyo, Japan

Robert Temple U.S. Food and Drug Administration, Silver Spring, Maryland, U.S.A.

Dhiren R. Thakker The University of North Carolina at Chapel Hill, Chapel Hill, North Carolina, U.S.A.

Kenneth E. Thummel University of Washington, Seattle, Washington, U.S.A.

Matthew D. Troutman Pfizer Inc., Groton, Connecticut, U.S.A.

Etsuko Usuki XenoTech LLC, Lenexa, Kansas, U.S.A.

Jose M. Vega Amgen, Thousand Oaks, California, U.S.A.

Nico P. E. Vermeulen Department of Pharmacochemistry, Vrije Universiteit, Amsterdam, The Netherlands

Lisa L. von Moltke Tufts University School of Medicine and Tufts-New England Medical Center, Boston, Massachusetts, U.S.A.

Grant R. Wilkinson[†] Vanderbilt University School of Medicine, Nashville, Tennessee, U.S.A.

Phyllis Yerino XenoTech LLC, Lenexa, Kansas, U.S.A.

Xin Zhang Indiana University School of Medicine, Indianapolis, Indiana, U.S.A.

Jin Zhou Department of Medicinal Chemistry, College of Pharmacy, University of Minnesota, Minneapolis, Minnesota, U.S.A.

[†]Deceased.

1

Introducing Pharmacokinetic and Pharmacodynamic Concepts

Malcolm Rowland
University of Manchester, Manchester, U.K.

I. SETTING THE SCENE

All effective drugs have the potential for producing both benefits and risks associated with desired and undesired effects. The particular response to a drug by a patient is driven in one way or another by the concentration of that drug, and sometimes its metabolites, at the effect sites within the body. Accordingly, it is useful to partition the relationship between drug administration and response into two phases, *a pharmacokinetic* phase, which relates drug administration to concentrations within the body produced over time, and *a pharmacodynamic* phase, which relates response (desired and undesired) produced to concentration. In so doing, we can better understand why patients vary in their response to drugs, which includes genetics, age, disease, and the presence of other drugs.

Patients often receive several drugs at the same time. Some diseases, such as cancer and AIDS, demand the need for combination therapy, which works better than can be achieved with any one of the drugs alone. In other cases, the patient is suffering from several conditions, each of which is being treated with one or more drugs. Given this situation and the many potential sites for interaction that exist within the body, it is not surprising that an interaction may occur between them, whereby either the pharmacokinetics or the pharmacodynamics of one drug is altered by another. More often than not, however, the interaction is of no clinical significance, because the response of most systems within the body is graded, with the intensity of response varying continuously with the concentration of the compound producing it. Only when the magnitude of change in response is large enough will an interaction become of clinical significance, which in turn varies with the drug. For a drug with a narrow therapeutic window, only a small change in response may precipitate a clinically significant interaction, whereas for a drug with a wide margin of safety, large changes in, say, its pharmacokinetics will have no clinical consequence. Also, it is well to keep in mind that some interactions are intentional, being designed for benefit, as often arises in combination therapy. Clearly, those of concern are the unintentional ones, which lead to either ineffective therapy through antagonism or lower concentrations of the affected drug or, more worryingly, excessive toxicity, which sometimes is so severe as to limit the use of the offending drug or, if it produces fatality, result in its removal from the market.

This chapter lays down the conceptual framework for understanding the quantitative and temporal aspects of drug-drug interactions, hereafter called drug interactions for simplicity. Emphasis is placed primarily on the pharmacokinetic aspects, partly because pharmacokinetic interactions are the most common cause of undesirable and, to date, unpredictable interactions and also because most of this book is devoted almost exclusively to this aspect and indeed to one of its major components, drug metabolism. Some pharmacodynamic aspects are also covered, however, for there are many similarities between pharmacokinetic and pharmacodynamic interactions at the molecular level and because ultimately one has to place a pharmacokinetic interaction into a pharmacodynamic perspective to appreciate the likely therapeutic impact (1–5).

II. BASIC ELEMENTS OF PHARMACOKINETICS

As depicted in Figure 1, it is useful to divide pharmacokinetic processes in vivo broadly into two parts, absorption and disposition. *Absorption,* which applies to all sites of administration other than direct injection into the bloodstream, comprises all processes between drug administration and appearance in circulating blood. Bioavailability is a measure of the extent to which a drug is absorbed. *Disposition* comprises both the distribution of a drug into tissues within the body and its elimination, itself divided into metabolism and excretion of unchanged drug. Disposition is characterized independently following intravenous administration, when absorption is not involved.

Increasingly, aspects of potential drug interactions are being studied in vitro not only with the aim of providing a mechanistic understanding but also with the hope that the findings can be used to predict quantitatively events in vivo, and thereby avoid or limit undesired clinical interactions. To achieve this aim, we need a holistic approach whereby individual processes are nested within a whole body frame—that is, constructs (models) that allow us to explore the impact, for example, of inhibition or induction of a particular metabolic pathway on, say, the concentration–time profile of a drug in the circulating plasma or blood, which delivers the drug to all parts of the

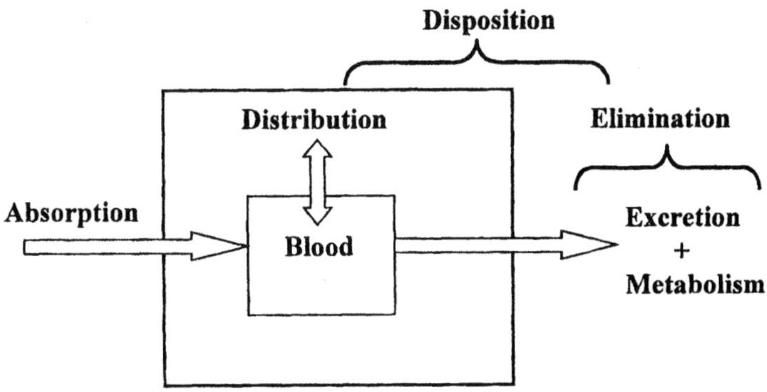

Figure 1 Schematic representation of processes comprising the pharmacokinetics of a compound. Here terms are defined with respect to measurement in blood or plasma. *Absorption* comprises all events between drug administration and appearance at the site of measurement. *Distribution* is the reversible transfer of the drug from and to other parts of the body. *Elimination* is the irreversible loss of the drug either as unchanged compound (excretion) or by metabolism. *Disposition* is the movement of the drug out of blood by distribution and elimination.

body, including sites of action and elimination. This approach also allows us to better interpret the underlying events occurring in vivo following a drug interaction. To appreciate this last statement, consider the events shown in Figures 2 and 3 and the corresponding summary data given in Table 1.

In Figure 2, pretreatment with the antibiotic rifampin shortened the half-life and decreased the area under the plasma concentration–time curve (AUC) profile, but not materially the peak concentration, of the oral anticoagulant warfarin, whether given intravenously or orally. In contrast, pretreatment with the sedative-hypnotic pentobarbital reduced both the peak concentration and AUC of the antihypertensive agent alprenolol following oral administration, while apparently producing no change in alprenolol's pharmacokinetics after intravenous dosing. As can be seen, these clinical studies show clear evidence of an interaction, with both actually involving the same mechanism, enzyme induction, but the effect is clearly expressed in different ways. To understand why this is so, we need to deal first with the intravenous data and then with the oral data—that is, to separate disposition from absorption.

For many purposes, because distribution is often much faster than elimination, as a first approximation the body can be viewed as a single compartment, of volume V, into which drugs enter and leave. This is an apparent volume whose value varies widely among drugs, owing to different distribution patterns within the body. The larger the volume, the lower the plasma concentration for a given amount in the body. The other important parameter controlling the plasma concentration (C)–time profile after an intravenous bolus dose (the disposition

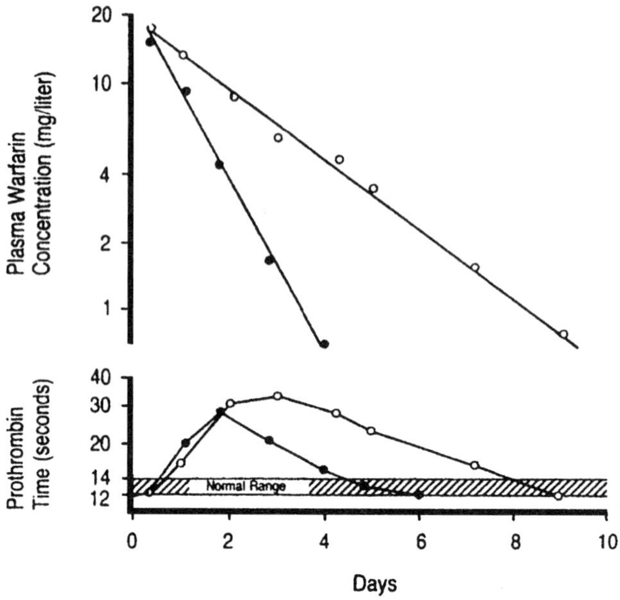

Figure 2 The half-life of the oral anticoagulant warfarin is shortened and its clearance increased when given as a single dose (1.5 mg/kg) before (○) and while (●) subjects have taken the enzyme inducer rifampin 600 mg daily for 3 days prior to and 10 days following warfarin administration. The peak and duration in elevation of the prothrombin time, a measure of the anticoagulant response, are both decreased when rifampin is coadministered. *Source*: From Ref. 6.

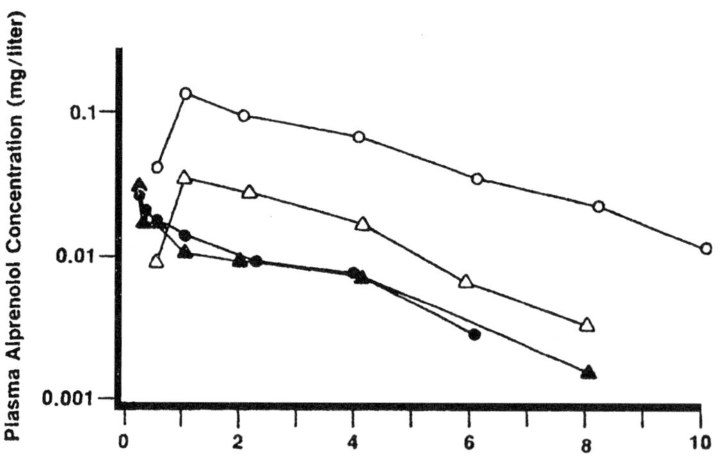

Figure 3 Enzyme induction of alprenolol metabolism following pentobarbital treatment produces minimal changes in events in plasma following intravenous administration of alprenolol 5 mg to subjects (● before, ▲ during pentobarbital) but a marked lowering of the plasma concentrations following oral administration of alprenolol 200 mg (○ before, △ during pentobarbital). *Source*: From Ref. 7.

Table 1 Summary Pharmacokinetic Parameters Before and During Drug Interactions

Warfarin-rifampin interaction[a]

	Warfarin pharmacokinetics				
	Dose (mg/kg)	AUC (mg · hr/L)	CL (L/hr)	$t_{1/2}$ (hr)	V (L)
Warfarin alone	1.5	600	0.18	47	12
Warfarin + rifampin	1.5	258	0.41	18	11

Alprenolol-pentobarbital interaction[b]

	Alprenolol pharmacokinetics							
	Intravenous				Oral			
	Dose (mg)	AUC (mg · hr/L)	CL (L/hr)	$t_{1/2}$ (hr)	Dose (mg)	AUC (mg · hr/L)	$t_{1/2}$ (hr)	F (%)
Alprenolol alone	5	0.067	75	2.3	200	0.71	2.3	26
Alprenolol + pentobarbital	5	0.058	86	1.9	200	0.15	2.4	6.5

[a]Abstracted from Ref. 8.
[b]Abstracted from Ref. 7.

kinetics) is clearance (CL), a measure of the efficiency of the eliminating organs to remove drug, given by

$$\text{Rate of elimination} = \text{CL} \cdot C \tag{1}$$

with units of flow (e.g., mL/min) such that

$$C = \frac{\text{Dose}}{V} e^{-}(\text{CL}/V)t \tag{2}$$

Often, Eq. (2) is recast by substituting k, the fractional rate of elimination of the drug, for CL and V, since

$$k = \frac{\text{Rate of elimination } (\text{CL} \cdot C)}{\text{Amount in body } (V \cdot C)} = \frac{\text{CL}}{V} \tag{3}$$

So

$$C = \frac{\text{Dose}}{V} e^{-kt} \tag{4}$$

It should be noted that k is related to half-life ($t_{1/2}$) by

$$t_{1/2} = \frac{0.693}{k} = \frac{0.693 \cdot V}{\text{CL}} \tag{5}$$

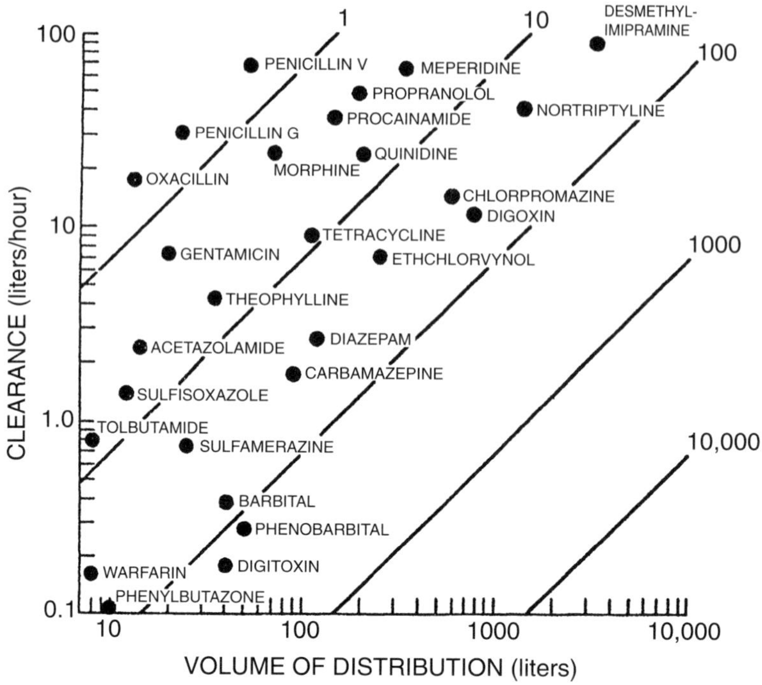

Figure 4 Log-log plot of clearance versus volume of distribution of various drugs in humans illustrating that for a given half-life, clearance and volume of distribution can vary widely. *Source*: Adapted from Ref. 8.

Being independent parameters, one a measure of the extent of distribution of drug within the body and the other a measure of the efficiency of the eliminating organs to remove drug from plasma, V and CL are frequently referred to as *primary* pharmacokinetic parameters, while the dependent ones, k and $t_{1/2}$, are *secondary* parameters, whose values change as a consequence of a change in CL, V, or both. Thus, drugs can have the same half-life but very different values of clearance and volume of distribution, as seen in Figure 4. Also, clearly, once any two parameters are known, the others are readily calculated.

A further important relationship, which follows by summing (integrating) Eq. (1) over all times, when the total amount eliminated equals the dose, is

$$\mathrm{CL} = \left[\frac{\mathrm{Dose}}{\mathrm{AUC}}\right]_{\mathrm{IV}} \tag{6}$$

which allows the estimation of CL from the plasma data. Armed with these relationships, the changes in the disposition kinetics for the two drugs become clear. For alprenolol, because there was no measurable change in either AUC or $t_{1/2}$, there must have been no change in CL or V either. In contrast, the smaller

AUC during rifampin treatment signifies that the clearance of warfarin has increased, although there was no change in *V*, since substitution of the respective values shows that the decrease in $t_{1/2}$ (and increase in *k*) is totally explained by the increase in CL (Table 1). Turning to the oral data, the only other relationship that one needs is

$$F = CL \cdot \left[\frac{AUC}{Dose}\right]_{Oral} \tag{7}$$

Equation (7) follows from the knowledge that the total amount eliminated from the body (CL · AUC) must equal the total amount entering the systemic circulation (*F* · Dose), where *F* is the extent of absorption, or oral bioavailability, of the drug. Note that without the intravenous data to provide an estimate of CL, only the ratio *F*/CL can be assessed following oral dosing, severely limiting the interpretation of events. Returning to the two interaction studies, analysis of the combined oral and intravenous plasma data indicates that whereas there was no change in the oral bioavailability of warfarin (which is totally absorbed) following pretreatment with rifampin, it was reduced from an already low control value of 22% to an even lower value of just 6% for alprenolol after pentobarbital pretreatment (Table 1). To gain further insights into these two interactions, we need to place everything, and particularly clearance, on a more physiological footing. To do so, consider the scheme in Figure 5, which depicts events occurring across an eliminating organ, receiving blood at flow rate *Q* with the drug entering at concentration C_A and leaving at concentration C_v. It follows that

$$\text{Rate of elimination} = Q(C_A - C_v) \tag{8}$$

Often it is useful to express the rate of elimination relative to the rate of presentation ($Q \cdot C_A$) to give the extraction ratio

$$E = \frac{Q(C_A - C_v)}{Q \cdot C_A} = \frac{C_A - C_v}{C_A} \tag{9}$$

And therefore, from the definition of clearance in Eq. (1), it follows that

$$CL = Q \cdot E \tag{10}$$

It is immediately evident from Eq. (10) that clearance depends on both organ blood flow and extraction ratio. The extraction ratio can vary from 0, when no drug is removed, to 1, when all drug within the blood is removed on a single passage though the organ. Then, CL (strictly based on measurements in whole blood to conserve mass balance) is equal to, and cannot exceed, organ blood flow; clearance is then limited by, and is sensitive to, changes in perfusion rate. For both warfarin and alprenolol, essentially all elimination occurs by hepatic metabolism, and comparison of the estimated respective clearance values (0.18 L/hr and 65 L/hr) with the hepatic blood flow of 81 L/hr reveals that warfarin has a low hepatic

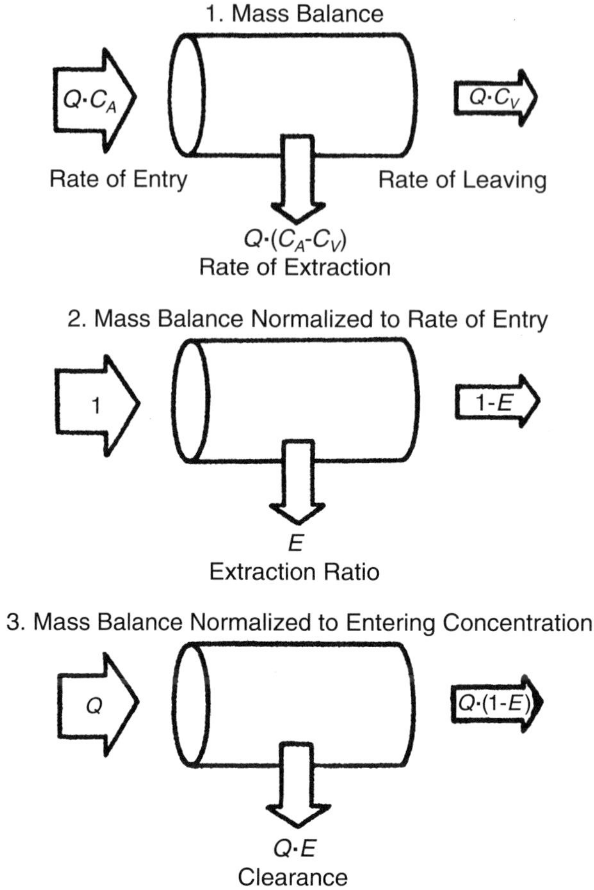

Figure 5 Schematic of the extraction of a drug by an eliminating organ at steady state, illustrating the interrelationships between blood clearance, extraction ratio, and organ blood flow. See the text for appropriate equations. *Source*: From Ref. 1.

extraction ratio (E_H), while for alprenolol it is very high, at 0.80. This difference in extraction ratios has a direct impact on oral bioavailability, since all blood perfusing the gastrointestinal tract drains into the liver via the portal vein before entering the general circulation. Consequently, because only the drug escaping the liver enters the systemic circulation, the oral bioavailability of a high extraction ratio compound, such as alprenolol, is expected to be low because of high *first-pass* hepatic loss. As already mentioned, this is indeed so. Furthermore, its low observed bioavailability (22%) is very close to that predicted, assuming that the liver is the only site of loss of the orally administered compound. Then,

$$\text{Predicted oral bioavailability, } F_H = 1 - E_H \tag{11}$$

that is, 20%. In contrast, on this basis, warfarin, with its very low estimated E_H, is expected to have an oral bioavailability close to 100%. This agrees with the experimental findings, supporting the view that such factors as dissolution of the solid drug (administered as a tablet) and permeation through the intestine wall do not limit the overall absorption of this drug.

A. A Model of Hepatic Clearance

To complete the task of explaining why the effect of induction manifests itself so differently in the pharmacokinetics of warfarin and alprenolol, we need a model that quantitatively relates changes in metabolic enzyme activity to changes in extraction ratio and clearance. Fundamental to all models and indeed to much of both pharmacokinetics and pharmacodynamics is the fact that events are driven by the unbound drug in plasma and tissues, the drug bound to proteins and other macromolecules being too bulky to enter cells and interact with sites of elimination and action. The most widely employed model of hepatic clearance in pharmacokinetics, but not the only one, is the well-stirred model (9–12) depicted in Figure 6. This model assumes that the distribution of a drug is so fast in this highly vascular organ that the concentration of the unbound drug in the blood leaving it is equal to that in it. For this model,

$$E_H = \frac{f_u \cdot CL_{int}}{Q + f_u \cdot CL_{int}} \tag{12}$$

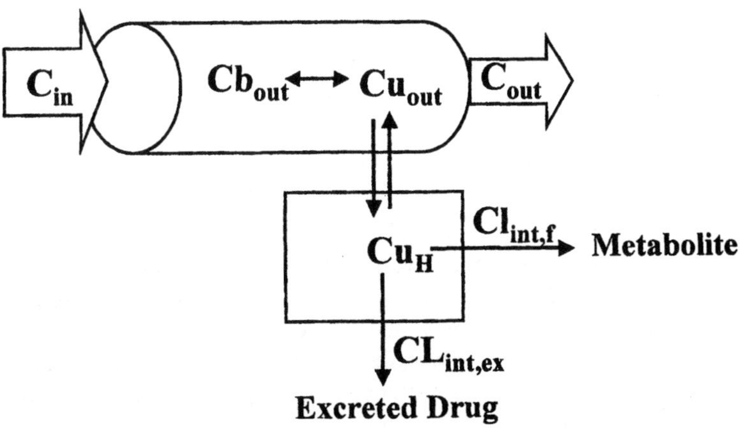

Figure 6 Well-stirred model of hepatic clearance. The exchange of a drug between plasma and hepatocyte and its removal from this cell involves an unbound compound. Intrinsic clearance, CL_{int}, relates the rate of the elimination (by formation of metabolites, $CL_{int,f}$, and secretion of unchanged compound into bile, $CL_{int,ex}$) to the unbound drug in the cell, $C_{u_H} \cdot C_{b_{out}}$ and $C_{u_{out}}$ are the bound and unbound concentrations of the drug leaving the liver at total concentration C_{out}.

and therefore

$$\mathrm{CL} = Q \cdot E_{\mathrm{H}} = \frac{Q \cdot f_{\mathrm{u}} \cdot \mathrm{CL}_{\mathrm{int}}}{Q + f_{\mathrm{u}} \cdot \mathrm{CL}_{\mathrm{int}}} \tag{13}$$

which shows that in addition to blood flow, CL and E_{H} are controlled by f_{u}, the fraction of the unbound drug in plasma (the ratio of unbound concentration in plasma, C_{u}, to the total measured plasma concentration, C, or strictly f_{ub}, the ratio of C_{u} to the whole blood concentration, to maintain mass balance across the liver), and $\mathrm{CL}_{\mathrm{int}}$, the intrinsic clearance.

1. Intrinsic Clearance

Like clearance in general, (hepatic) intrinsic clearance is a proportionality constant, in this case between the rate of elimination and unbound concentration within the liver, $C_{\mathrm{u_H}}$. That is, $\mathrm{CL}_{\mathrm{int}}$ = (Rate of elimination)/$C_{\mathrm{u_H}}$. Conceptually, it is the value of clearance one would obtain if there were no protein binding or perfusion limitation, and is regarded as a measure of the activity within the cell, divorced from any limitations imposed by events in the perfusing blood. As such, the value of intrinsic clearance is often many orders of magnitude greater than for hepatic blood flow. Inferred through the analysis of in vivo data, where one cannot measure events within the cell, and determined experimentally in vitro, the concept of intrinsic clearance is critical not only to the quantitative interpretation and prediction of drug interactions within the liver, but to pharmacokinetics in general. And since elimination can be by both metabolism and excretion, often operating additively within an organ to remove a drug, under nonsaturating conditions,

$$\mathrm{CL}_{\mathrm{int}} = \sum \frac{V_{\mathrm{m}}}{K_{\mathrm{m}}} + \sum \frac{T_{\mathrm{m}}}{K_{\mathrm{d}}} \tag{14}$$

or

$$\mathrm{CL}_{\mathrm{int}} = \sum \mathrm{CL}_{\mathrm{int,f}} + \sum \mathrm{CL}_{\mathrm{int,ex}} \tag{15}$$

where V_{m} and K_{m} are the maximum velocity of metabolism and the Michaelis-Menten constant of each of the enzymes involved, alternatively expressed as their ratio, the intrinsic clearance associated with formation of the metabolite, $\mathrm{CL}_{\mathrm{int,f}}$. Similarly, T_{m} and K_{d} are the transport maximum and dissociation constant of each of the transporters involved in excretion, with their ratio, $\mathrm{CL}_{\mathrm{int,ex}}$, being the intrinsic clearance associated with excretion. Now, recognizing that V_{m} is directly proportional to the total amount of the respective enzyme and that induction involves an increase in its synthesis that increases the amount of the enzyme, it follows that the intrinsic clearance of the affected enzyme, and hence total $\mathrm{CL}_{\mathrm{int}}$, also increases during induction.

Examination of Eqs. (12) and (13) provides an understanding of the conditions determining the extraction ratio and CL of a drug, and hence the influence of induction itself. These relationships between CL, E, Q, f_u, and CL_{int} are displayed graphically in Figure 7. Also, examination of Eq. (12) reveals that plasma protein binding effectively lowers the intrinsic clearance by decreasing the unbound concentration for a given total concentration delivered in blood. However, when the effective intrinsic clearance ($f_u \cdot CL_{int}$) $\gg Q$, it is seen that $E_H \rightarrow 1$ and CL $\rightarrow Q$. Under these circumstances, CL is the perfusion rate that is limited and insensitive to changes in CL_{int}, which explains why induction of the metabolism of alprenolol produced no noticeable increase in its clearance, whereas for a low-extraction drug, such as warfarin (which is both a poor substrate for the metabolic enzymes and very highly protein bound, $f_u = 0.005$),

$$f_u \cdot CL_{int} \ll Q$$

so

$$CL = f_u \cdot CL_{int} \tag{16}$$

which explains why the increase in intrinsic clearance due to enzyme induction is reflected in direct proportion by the measured clearance.

It remains to resolve the oral data, which are achieved as follows. Substituting Eq. (12) in Eq. (11) gives

$$F_H = \frac{Q}{Q + f_u \cdot CL_{int}} \tag{17}$$

which, when further substituted with Eq. (12) in Eq. (7), provides the useful relationship

$$AUC_{oral} = \frac{Dose}{f_u \cdot CL_{int}} \tag{18}$$

From Eq. (18) we see that AUC following an oral dose depends only on f_u and CL_{int} when all of the administered drug reaches the liver essentially intact, as happens with both warfarin and alprenolol. Accordingly, the oral AUC should decrease with enzyme induction, irrespective of whether the drug is of high or low extraction ratio, as was observed.

In summary, changes in (hepatic) intrinsic clearance, whether due to induction or inhibition, are manifest differently in the whole-body pharmacokinetics of a drug, depending on whether it is of high or low clearance when given alone. For drugs of low hepatic extraction ratio, changes in intrinsic clearance produce changes in total clearance and half-life, but minimal changes in oral bioavailability. In contrast, for high extraction ratio drugs, which obviously must be exceptionally good substrates for the (hepatic) metabolic or excretory transport processes, a change in intrinsic clearance is reflected in a noticeable change in oral bioavailability, but not in clearance or half-life.

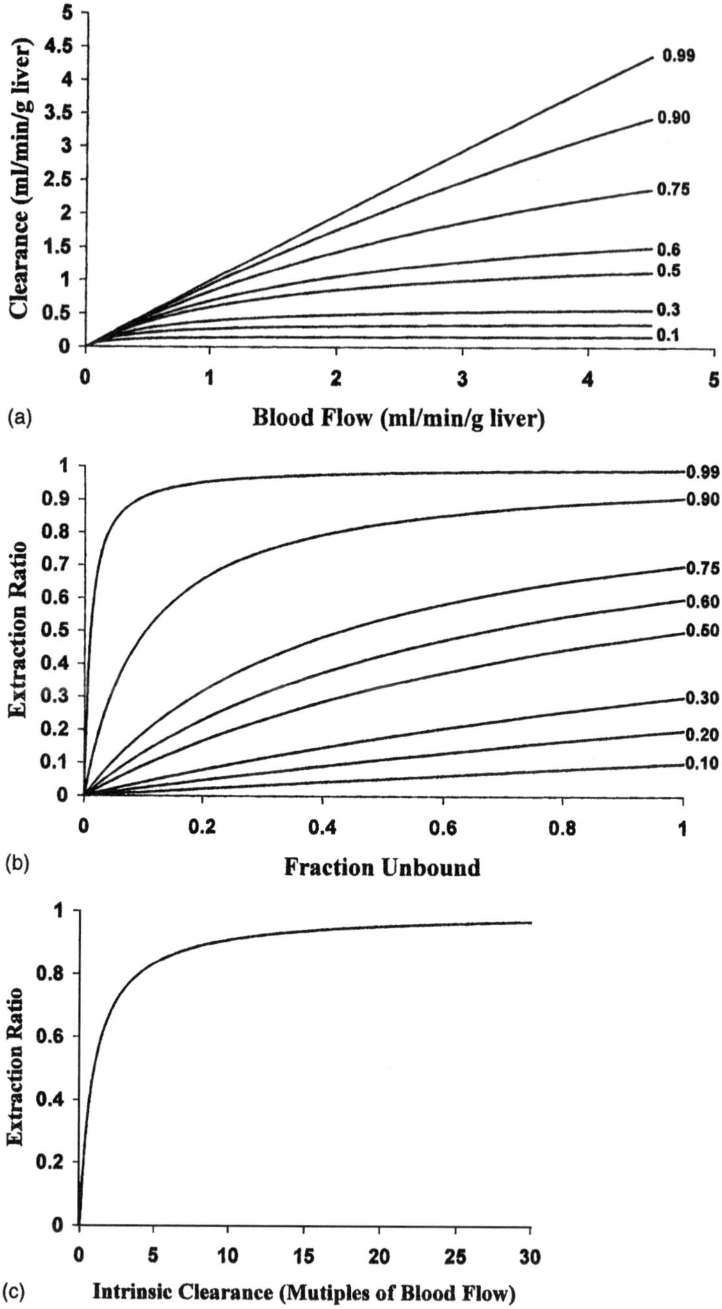

Figure 7 Influence of changes in (a) organ blood flow on clearance, (b) fraction of the drug unbound in plasma (f_u) on extraction ratio, and (c) intrinsic clearance on extraction ratio as predicted by the well-stirred model of hepatic clearance.

2. Plasma Protein Binding

In drug interactions, the most common cause of altered protein binding is displacement, whereby one drug competes with another for one or more binding sites, increasing f_u of the affected drug. This can readily be assessed in vitro in plasma using one of a variety of methods, such as equilibrium dialysis, ultrafiltration, or ultracentrifugation. However, being a competitive process, the degree of displacement depends on the concentrations of the drugs relative to those of the binding sites. Only when the concentration of one of the drugs approaches the molar concentration of the binding sites will substantial displacement occur. In practice, because most drugs are relatively potent, this displacement does not occur as often as one might have supposed, given so relatively few specific binding sites on plasma proteins. Even when substantial displacement does occur, it often is of little to no therapeutic importance.

As seen from Eq. (13) (and Fig. 7) and emphasized in Eq. (16), an increase in f_u will only increase CL of drugs with a low extraction ratio, such as warfarin. When the extraction ratio is high, as with alprenolol, CL is essentially unaffected by a change in f_u, since all drugs, whether initially bound or not, must have been removed on their passage of the drug through the organ. That is, within the contact time of blood in the liver, the bound drug dissociates so rapidly that all of it is available for removal as the unbound drug is cleared. Nevertheless, examination of Eq. (18) shows that for all drugs, the AUC of the pharmacologically important unbound species ($f_u \cdot \text{AUC}$) should be unaffected by displacement following oral administration, which probably explains why no clinically significant pure displacement interactions have been reported to date. Even so, displacement may affect the half-life of a drug. As now examined, much depends on the overall effect of displacement on the volume of distribution as well as on clearance.

B. Model of Distribution

In its simplest form, the body may be viewed as comprising two aqueous spaces, the plasma (volume V_p) and the rest of the body (volume V_T), as depicted in Figure 8, with distribution continuing until at equilibrium the unbound concentrations, C_u and C_{u_T}, respectively, are equal. Then, in each space relating unbound to total drug concentration, and noting that the total amount of drug in the body, $A = V \cdot C = V_p \cdot C + V_T \cdot C_T$, it follows that

$$V = V_p + V_T \cdot \frac{f_u}{f_{u_T}} \tag{19}$$

where f_{u_T} is the fraction of the unbound drug in the tissue. The plasma volume is around 0.05 L/kg. And for drugs that access all the cells, V_T is 0.55 L/kg,

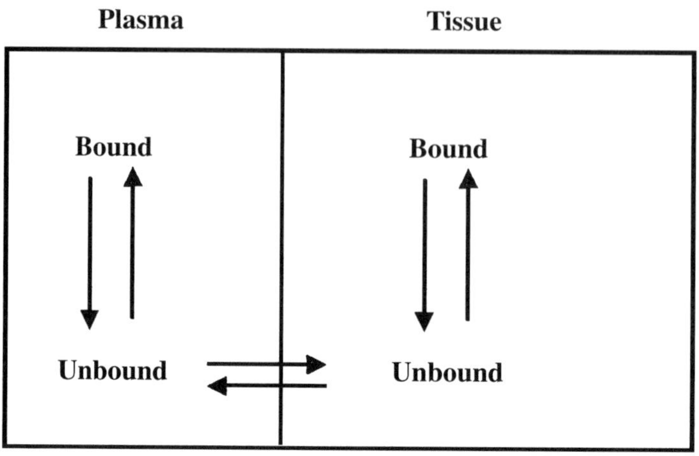

Figure 8 A simple model of drug distribution, with the unbound drug equilibrating between plasma and tissue.

giving a total body water space of 0.6 L/kg. For many drugs, the volume of distribution is quite large, on the order of 1 L/kg or much greater. In these cases, the fraction of drug in the body located in plasma can be ignored, and so V reduces to $V_T \cdot f_u/f_{u_T}$, from which it is apparent that the volume of distribution varies directly with f_u and inversely with f_{u_T}. So displacement in plasma alone will always increase the volume of distribution. For drugs of low volume of distribution (< 0.2 L/kg), because they are predominantly located outside of cells, the situation is complicated by the presence of substantial amounts of drug in the interstitial space, bathing the cells within tissues, where plasma proteins also reside. Dealing with this situation is beyond the scope of this chapter (1).

Combining Eq. (19) with the model for organ clearance [Eq. (13)] facilitates prediction of the effect of displacement on half-life. For low extraction ratio drugs, since $\mathrm{CL} = f_u \cdot \mathrm{CL}_{\mathrm{int}}$ and $V = V_T \cdot f_u/f_{u_T}$, both CL and V will increase to the same extent with displacement within plasma, so $t_{1/2}$ ($= 0.693 \cdot V/\mathrm{CL}$) should remain unchanged. In contrast, half-life is expected to increase with displacement in plasma of high-clearance drugs, since V always increases but CL remains unchanged, being limited by organ blood flow.

III. CHRONIC ADMINISTRATION

Pharmacokinetic information gained following single-dose administration can be used to help predict the likely events following chronic dosing, either as a constant-rate infusion or multiple dosing, which often involves giving a fixed dose at set time intervals.

A. Constant-Rate Infusion

During the infusion, the plasma concentration of the drug continues to rise until a steady state is reached, when the rate of elimination ($\mathrm{CL} \cdot C$) matches the rate of infusion. These relationships, displayed in Figure 9, are defined by

$$C = C_{\mathrm{ss}}(1 - \mathrm{e}^{-kt}) \quad \text{during infusion} \tag{20}$$

and

$$C_{\mathrm{ss}} = \frac{\text{Rate of infusion}}{\mathrm{CL}} \quad \text{at steady state} \tag{21}$$

Clearly, events at steady state depend only on clearance, while the time course on approach to the plateau is governed only by k, and hence by half-life; this information was obtained from a single-dose study. Furthermore, calculations show that 50% of the plateau is reached in 1 half-life and 90% in 3.3 half-lives. Accordingly, drugs with short half-lives will reach steady state quickly, and those with half-lives in the order of days will take over a week. Hence, knowing the $t_{1/2}$ of a drug is important when planning the duration of a study and the frequency of sampling of blood to characterize kinetic events.

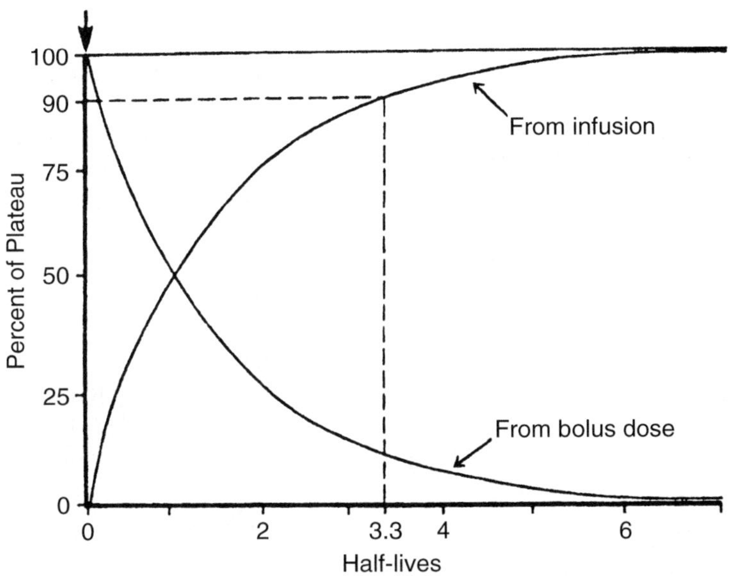

Figure 9 Approach to plateau following a constant rate of input is controlled solely by the half-life of the drug. Depicted is the situation in which a bolus (↓) is immediately followed by an infusion that exactly matches the rate of elimination, thereby maintaining the plasma concentration. As the plasma concentration associated with the bolus falls exponentially, there is a complementary rise in that associated with the infusion. In 3.3 half-lives, the plasma concentration associated with the infusion has reached 90% of the plateau value. *Source*: From Ref. 1.

B. Multiple Dosing

Two additional features are observed on multiple dosing, accumulation and fluctuation (Figure 10). The former arises because there is always drug remaining in the body from preceding doses, and the latter because the rate of input varies throughout each dosing interval. Nonetheless, the rise to the plateau still depends essentially only on the half-life of the drug, while within a dosing interval at plateau, the amount eliminated ($CL \cdot AUC_{ss}$) equals the amount absorbed. That is,

$$F \cdot \text{Dose} = CL \cdot AUC_{ss} \tag{22}$$

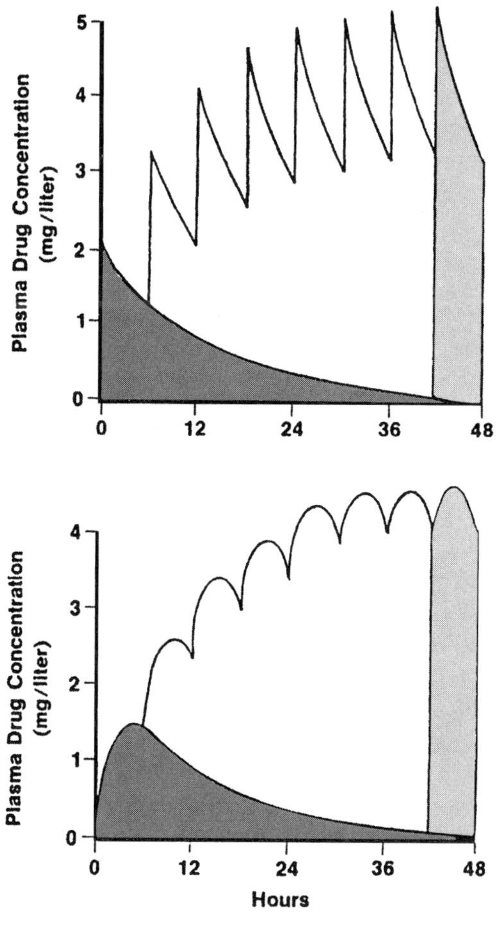

Figure 10 Plasma concentrations of a drug following a multiple-dosing regimen, of fixed dose and interval, intravenously (top) and orally (bottom). Note that in both cases the area under the plasma concentration–time curve within a dosing interval at plateau is equal to the total area following a single dose. *Source*: From Ref. 1.

where AUC_{ss} is the AUC at plateau. Furthermore, comparison of Eq. (22) with Eq. (7) provides a useful expectation when the same-size dose is given on a single occasion and after multiple dosing, namely,

$$AUC_{ss} = AUC_{single} \tag{23}$$

Any deviation from this expectation implies that CL, F, or both must have changed on multiple dosing. If found, the kinetics of the drug are said to be time dependent. An understanding of these kinetic principles helps in the planning and interpretation of in vivo drug interaction studies, which are of many designs. One goal is often to evaluate the full effects of an interaction, which generally requires exposing the affected drug to the highest concentration of the offending drug, which is at its plateau. So the offending drug needs to be administered for at least 3.3 of its half-lives and often for longer to ensure that the exposure is maintained throughout the time course of the affected drug.

IV. A GRADED EFFECT

As already mentioned, practically all drug interactions are graded, being dependent on the concentrations of the interacting drugs and, hence, on their pharmacokinetics as well as manner of administration (1,4). While many scenarios are possible, for illustrative purposes consider the case of competitive inhibition of one pathway (A) of metabolism of a low-clearance drug operating under linear (nonsaturing) conditions in the absence of the inhibitor, all other factors being constant. Then, for the affected pathway,

$$CL_{int,A,inhibited} = \frac{V_m}{K_m(1 + I/K_i)} \quad \text{or} \quad CL_{int,A,inhibited} = \frac{CL_{int,A}}{(1 + I/K_i)} \tag{24}$$

where $CL_{int,A}$ and $CL_{int,A,inhibited}$ are the respective intrinsic clearances of the affected pathway in the absence and presence of the inhibitor, at unbound concentration I. Also characterizing the inhibitor is the inhibitor constant K_i, defined as the unbound concentration of the inhibitor that effectively reduces the value of $CL_{int,A}$ by one-half. Rearrangement of Eq. (24) gives the degree of inhibition of the affected pathway, DI, namely,

$$DI = \frac{I/K_i}{1 + I/K_i} \tag{25}$$

which gives an alternative definition for K_i as the value of I that produces 50% of the maximum degree of inhibition. It is immediately clear from Eqs. (24) and (25) that the important factor is the ratio I/K_i. Thus, a compound may be a potent inhibitor, expressed by a low K_i, but in practice a significant inhibitory effect will arise only if I is high enough so that I/K_i is large. Proceeding further, let f_m be the fraction of the total elimination of drug by the affected pathway in the absence of inhibitor. Then, by reference to previous equations, with appropriate

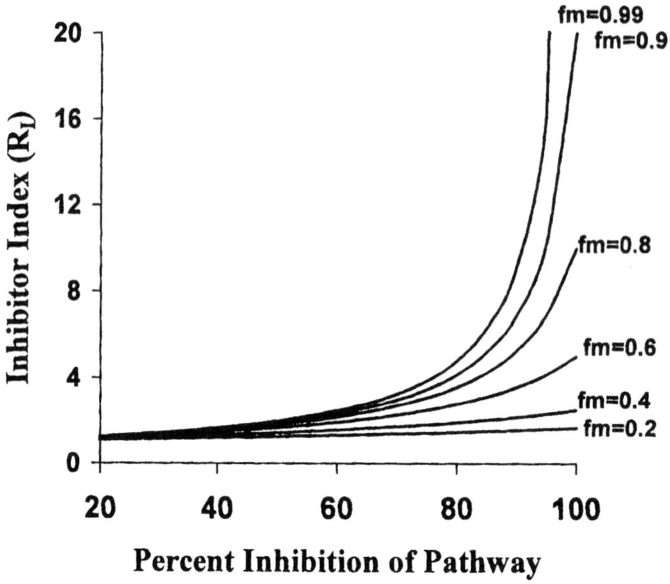

Figure 11 Relationship between the inhibitor index R_I and the degree of inhibition of a metabolic pathway for various values of the fraction of the drug eliminated by that pathway in the absence of the inhibitor, f_m.

rearrangements, one obtains the following generalized equation that permits exploration of the kinetics of this situation:

$$
\begin{aligned}
R_I &= \frac{C_{ss,\text{inhibited}}}{C_{ss,\text{normal}}} = \frac{\text{AUC}_{\text{single,inhibited}}}{\text{AUC}_{\text{single,normal}}} = \frac{\text{AUC}_{ss,\text{inhibited}}}{\text{AUC}_{ss,\text{normal}}} \\
&= \frac{t_{1/2,\text{ inhibited}}}{t_{1/2,\text{ normal}}} = \frac{1}{f_m(1-\text{DI}) + (1-f_m)}
\end{aligned}
\tag{26}
$$

noting that $(I - \text{DI}) = 1/(1 + I/K_i)$. Here R_I is the ratio of C_{ss}, $\text{AUC}_{\text{single}}$, AUC_{ss}, and $t_{1/2}$ in the presence (inhibited) and absence (normal) of the inhibitor. R_I might be thought of as the *inhibitor index,* giving a measure of the severity of the impact of the interaction on whole-body events. Figure 11 shows the relationship between R_I and DI for various values of f_m. It is immediately apparent that the increase in R_I becomes substantial only when $f_m > 0.5$, no matter how extensive the degree of inhibition of the affected pathway. Furthermore, note that the closer DI and f_m both approach 1, R_I increases dramatically to values approaching 10 or greater. In other words, the problem becomes very serious when the affected pathway is the obligatory route for elimination of the drug and is substantially inhibited. Fortunately, this situation does not arise that often in clinical practice.

The other important aspect is the timescale over which the effect of inhibition is seen in plasma, such as on the time to reach plateau following chronic

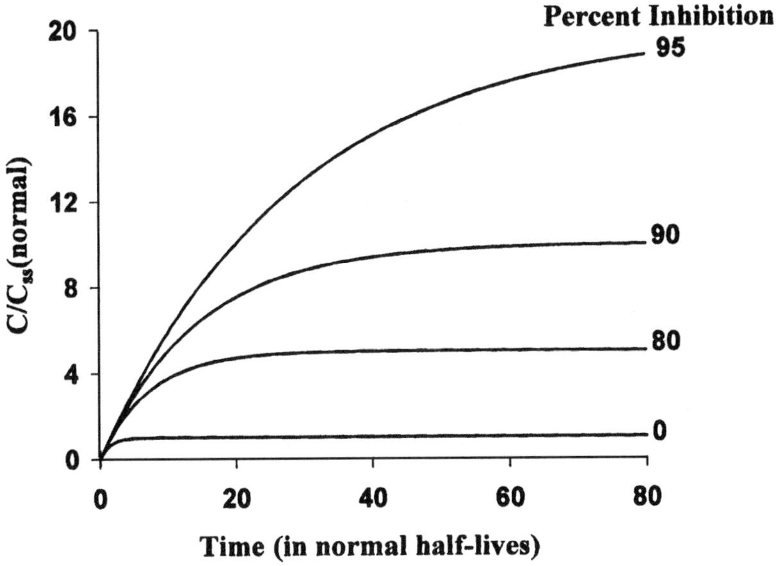

Figure 12 Effect of inhibition on the rate of accumulation of a drug given as a constant-rate infusion when $f_m = 1$. Note that time is expressed in units of normal half-life and concentration in units of the steady-state concentration in the absence of the inhibitor, $C_{ss,normal}$. The greater the degree of inhibition, the longer the half-life and the longer it takes to reach, and the higher is, the plateau.

drug administration, as illustrated in Figure 12 for the extreme case when $f_m = 1$. Recall that it takes approximately four half-lives to reach the plateau. So, although greater inhibition results in a substantial increase in the plateau concentration of the affected drug, because its half-life is also progressively increasing in association with the decrease in clearance, it takes longer and longer to reach the new plateau. This effect has several implications. First, the full effects of an interaction may occur long after the inhibitor has been added to the dosage regimen of the affected drug, with the danger that any resulting toxicity may not be associated with the offending drug by either the patient or the clinician. Second, in planning in vivo interaction studies during development, administration of the affected drug may need to be maintained for much longer in the presence of the potential inhibitor than on the basis of the normal half-life of the drug. On passing, it is worth noting that a possible exception is inhibition of a drug of high hepatic extraction ratio, such as alprenolol. In this case, for moderate degrees of inhibition of intrinsic clearance, the major changes will be in the AUC and peak plasma concentration, with little change in half-life, because, as discussed previously for such drugs, clearance is blood flow limited. Only when inhibition is so severe that the drug is effectively converted from one of high extraction ratio to one of low extraction ratio will half-life also increase.

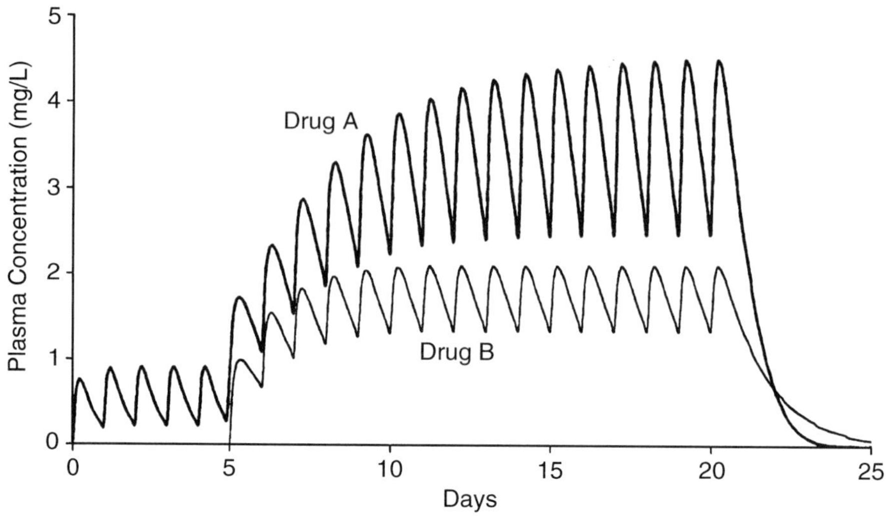

Figure 13 Simulation of drug interaction kinetics involving competitive inhibition. In this scenario, drug A is administered as a fixed oral dosage regimen, first alone until a steady state is reached and then in the presence of a fixed oral dosage regimen of drug B, which inhibits the obligatory pathway for the elimination of drug A, that is, $f_m = 1$. As the plasma concentration of drug B rises, so does the degree of inhibition of drug A, which in turn reduces its clearance and effectively prolongs its half-life. Accordingly, the rise to the new, higher plateau of drug A takes much longer than when it is given alone, being determined by both the pharmacokinetics and dosage regimen of drug B as well as its inhibitory potency. In the current scenario, the clearance of drug A is reduced by an average of 86%, and its half-life increased sevenfold during a dosing interval at the plateau of drug B.

Third, the current scenario corresponds to the clinical situation of the affected drug being added to the regimen of an individual already stabilized on the inhibitor. Another, perhaps more common scenario, especially when the inhibitor has just been introduced into clinical practice, is addition of the inhibitor to the maintenance regimen of the affected drug. Then one needs to consider both the pharmacokinetics and dosage regimen of the inhibitor as well as the changing kinetics of the affected drug. This last scenario is illustrated in Figure 13. On initiating the regimen of the second drug (inhibitor), its plasma concentration rises toward its plateau with a timescale governed by its half-life. And as it rises, so does the degree of inhibition of the affected drug, which in turn decreases its clearance and prolongs its half-life. The net result is that it takes even longer for the plasma concentration of the affected drug to reach its new plateau than anticipated from even its longest half-life, which is at the plateau of the inhibitor. The reason for this is that in essence one has to add on the time it takes for the inhibitor to reach its plateau. Occasionally, the inhibitor has a much longer half-life than the affected drug, even when inhibited. In this case, the rise of the affected drug to its new plateau virtually mirrors in time the approach of the inhibitor to its plateau.

Also shown in Figure 13 is the return of the affected drug to its previous plateau on withdrawing the offending drug. This return is faster than during the rise in the presence of the inhibitor, because as the inhibitor falls, so does the degree of inhibition, which then causes a shortening in the half-life and thus an ever-accelerating decline of the affected drug. However, the speed of decline is strongly determined by the kinetics of the inhibitor. If it has a long half-life, its decline may be the rate-limiting step in the entire process, in which case the decline of the inhibited drug parallels that of the inhibitor itself.

V. ADDITIONAL CONSIDERATIONS

So far, analysis has centered on metabolic drug interactions. But there are many pharmacokinetic interactions other than those occurring at enzymatic sites, such as those involving transporters or altered physiological function.

A. Transporters

The quantitative and kinetic conclusions reached with metabolic drug interactions apply equally well to those involving transporters effecting excretion, which reside in organs connected with the exterior, such as the liver via the bile duct (see Chaps. 5, 8, and 12 for more details). This is readily seen by examination of Eq. (15). Being additive, a given change in either a metabolic or an excretory intrinsic clearance ($CL_{int,f}$ or $CL_{int,ex}$) will produce the same change in the overall intrinsic clearance. Sometimes, a transporter interaction occurs within internal organs, such as the brain, to produce altered drug distribution, not excretion. It occurs, for example, with inhibition of the efflux transporter P-glycoprotein (PGP), located within the blood-brain barrier. For example, normally virtually excluded from the brain by efflux, inhibition of PGP leads to an elevation in brain levels of the substrate cyclosporin (13). Even so, because the brain comprises less than 1% of total body weight, changes in the distribution of a drug within it, even when quite profound and of major therapeutic consequence, will have minimal effect on the volume of distribution of the drug, V, which reflects its overall distribution within the body.

B. Absorption

Many interactions involve a change in either the rate or the extent of drug absorption, particularly following oral administration. There are many potential sites for interaction: within the gastric and intestinal lumen, at or within the gut wall, as well as within the liver (Figure 14). As indicated in Figure 15, the consequences of a change in absorption kinetics depend on whether the affected

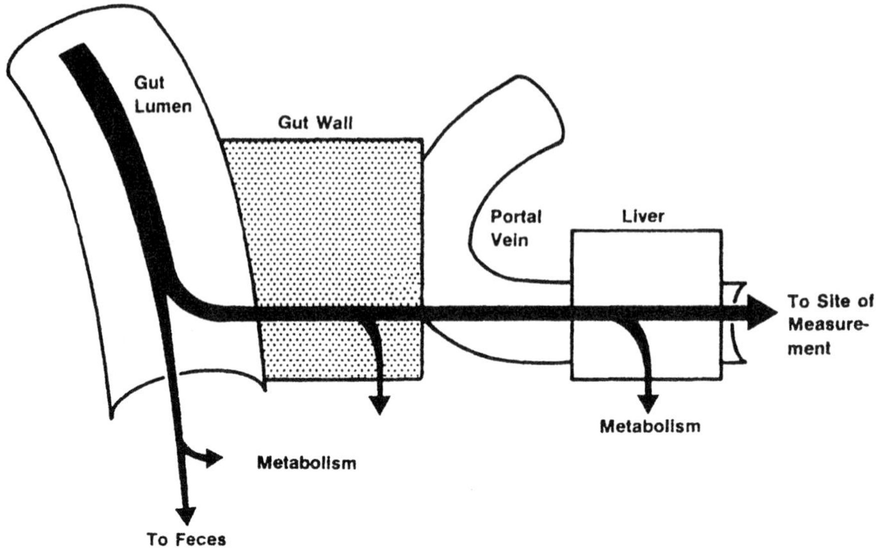

Figure 14 Schematic depiction of events occurring during absorption after oral administration of a drug. On dissolution, the drug, in addition to having to permeate the intestinal wall, must pass through the liver to reach the systemic circulation and subsequent sites within the body. Loss of the drug can occur at any of these sites, leading to a loss of oral bioavailability. *Source*: From Ref. 1.

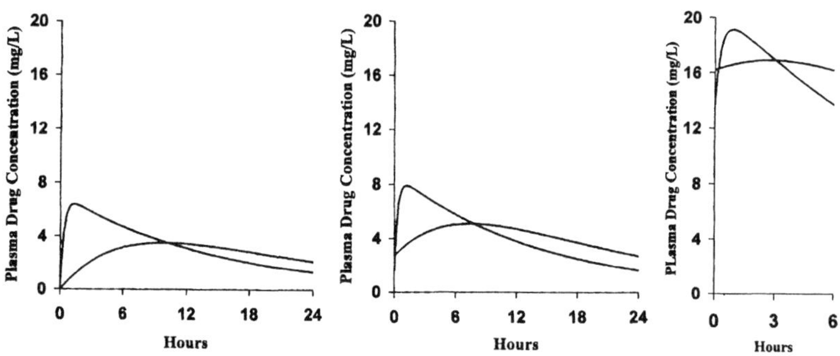

Figure 15 Impact of dosing frequency on the kinetics at plateau. Although clear differences are seen after a single dose (left panel), these will also be seen at plateau only if the drug is dosed relatively infrequently (once every 24 hours in this scenario), when little accumulation occurs (middle panel). With frequent dosing (once every 6 hours), accumulation is extensive, so changes in absorption kinetics now have only a minor effect at plateau (right panel).

drug is given once or as a multiple-dosing regimen. A slowing in absorption kinetics will always result in a lower and later peak concentration, which could be critical if the affected drug is intended for rapid onset of action, such as for the relief of a headache. However, whether this difference is sustained on multiple dosing depends heavily on the dosing frequency of the affected drug relative to its half-life. When it is given infrequently, there is little accumulation, so the events at plateau are similar to those seen following a single dose. However, when given relatively frequently, because of extensive accumulation the amount absorbed from any one dose is such a small fraction of that in the body at plateau that events at plateau are insensitive to changes in absorption kinetics. In contrast, changes in the extent of absorption seen during single-dose administration, whatever the cause, will still be seen on multiple dosing, irrespective of the frequency of drug administration.

There are many causes of low, particularly oral, bioavailability, F. Some of these occur in the gastrointestinal lumen, affecting the dissolution of a solid or its stability by changing, for example, pH so that only a fraction (F_A) of the administered dose reaches the epithelial absorption sites. However, only a fraction of this dose may permeate through the intestinal wall into the portal blood (F_G), and then only another fraction(F_H) escapes the liver and enters the systemic circulation. Accordingly, because these sites of loss are arranged in series, it follows that the overall systemic oral bioavailability F is

$$F = F_A \cdot F_G \cdot F_H \tag{27}$$

Notice that overall bioavailability is zero if drug is made total unavailable at any one of the three sites. Also, while measurement of F is important, which in turn requires the administration of an intravenous dose, it is almost impossible to rationally interpret a drug interaction affecting oral bioavailability without some estimate of the events occurring at at least one of the three sites of loss. It usually requires additional studies to be undertaken to untangle the various events, such as comparing the interaction with both a solution and the usual solid dosage form of the affected drug. Clearly, if no difference is seen, it provides strong evidence that the interaction is not the one affecting the dissolution of the drug from the solid. Furthermore, the lack of an interaction following intravenous dosing of the affected drug would then strongly point to the interaction occurring within the intestinal wall.

C. Displacement

With many drugs highly bound to plasma and tissue proteins, and with activity residing in the unbound drug, there has been much concern that displacement of drug from its binding sites could have severe therapeutic consequences. In practice, this concern is somewhat unfounded. We have seen why this is so following a single dose of a drug (sec. II.A.2). It is also the case following chronic dosing. Consider again a drug of low clearance, administered as a constant infusion. Then, at steady state, when the rate of elimination ($CL_{int} \cdot C_u$) matches the rate of infusion, it follows that

$$\text{Rate of infusion} = CL \cdot C_{ss} = CL_{int} \cdot C_{uss} \quad (28)$$

Now, displacement, by increasing f_u, will increase CL (since $CL = f_u \cdot CL_{int}$). But because the events within the cell are unaffected by displacement, it follows that CL_{int} will not change and therefore neither will C_{uss}, the therapeutically important unbound concentration at steady state. Consequently, no change in response is expected. Indeed, had no plasma measurements been made, one would have been totally unaware that an interaction had occurred. Furthermore, if plasma measurements are made, it is important to determine the fraction of the unbound drug and its free concentration; otherwise, there is clearly a danger of misinterpretation of the interaction.

VI. ADDITIONAL COMPLEXITIES

There are a whole variety of factors that further complicate both the interpretation and quantitative prediction of the pharmacokinetic aspects of drug interactions. Most are either beyond the scope of this introductory chapter or are covered elsewhere in this book. Several, however, are worth mentioning here. One is that sometimes drug interactions are multidimensional, with more than one process affected. For example, although no longer prescribed, the anti-inflammatory compound phenylbutazone interacts with many drugs, as is well documented. One in particular is noteworthy here, namely, the interaction with warfarin causing an augmentation of its anticoagulant effect. On investigation, it was found that phenylbutazone not only markedly inhibits many of the metabolic pathways responsible for warfarin elimination, but also displaces warfarin from its major binding protein, albumin, making interpretation of the pharmacokinetic events based on total plasma concentration problematic (13,14). In such situations, and indeed whenever possible, interpretation should be based on the more relevant unbound drug.

Another complexity is the presence of multiple sites for drug elimination. For example, increasing evidence points to the small intestine, in addition to the liver, having sufficient metabolic activity to cause appreciable loss in the oral bioavailability of some drugs. Then unambiguous quantitation of the degree of involvement of each organ in an interaction in vivo becomes difficult, unless one has a way of separating them physically, such as by sampling the hepatic portal vein, which drains the intestine, to assess the amount passing across the intestinal wall, as well as the systemic circulation to assess the loss of the drug on passage through the liver.

Still another is the metabolites themselves, which may possess pharmacological and toxicological activity in their own right. Each metabolite has its own kinetic profile, which is often altered during an interaction, through a change either in its formation or occasionally in its elimination and distribution. Despite these complexities, however, measurement of both a drug and its metabolites can often be very informative and provide more definitive insights into an interaction than gained from measurement of the drug alone (5).

The last complexity mentioned here is the pharmacokinetics of the interacting drug itself, be it an inhibitor, an inducer, or a displacer. Given that drug interactions are graded and recognizing that individuals vary widely in their degree of interaction for a given dosage regimen of each drug, it would seem sensible to measure both of them when characterizing an interaction. Unfortunately, this is rarely done. Even in vitro, all too often it is assumed that the concentration of the interactant is that added, without any regard to the possibility that it may bind extensively to components in the system or be metabolically degraded. In both cases, the unbound compound of the interacting drug is lower than assumed and if ignored may give a false sense of comfort, suggesting that higher (unbound) concentrations are needed to produce a given degree of interaction than is actually the case. When measured in vivo, it is usually the interacting drug in the circulating plasma rather than at the site of the interaction, such as the hepatocyte, that is inaccessible. In addition, the liver receives the drug primarily from the portal blood, where the concentration may be much higher than in plasma during the absorption phase of the interactant, making any attempt to generate a meaningful concentration-response relationship difficult. Finally, because many drug interactions involve competitive processes, the possibility always exists that the interaction is mutual, with both drugs affecting each other, the degree of effect exerted by each on the other depending on the relative concentrations of the two compounds.

Despite these complexities, all is not lost. Through careful planning and subsequent analysis of both in vitro and in vivo data, progress is being made in our understanding of the mechanisms and pharmacokinetic aspects of drug interactions.

VII. PHARMACODYNAMIC CONSIDERATIONS

Although when related to a dose the clinical outcome of a drug interaction may appear the same, it is useful to distinguish between pharmacokinetic and pharmacodynamic causes of the interaction. In the former case, the change in response is caused by a change in the concentration of the affected drug, together perhaps with one or more metabolites. In the latter, there may be no change in pharmacokinetics at all.

One feature commonly experienced in pharmacodynamics but much less in pharmacokinetics is saturability, giving rise to nonlinearity. Typically in pharmacodynamics, on raising the concentration of drug, the magnitude of response rises initially sharply and then more slowly on approach to the maximum effect,

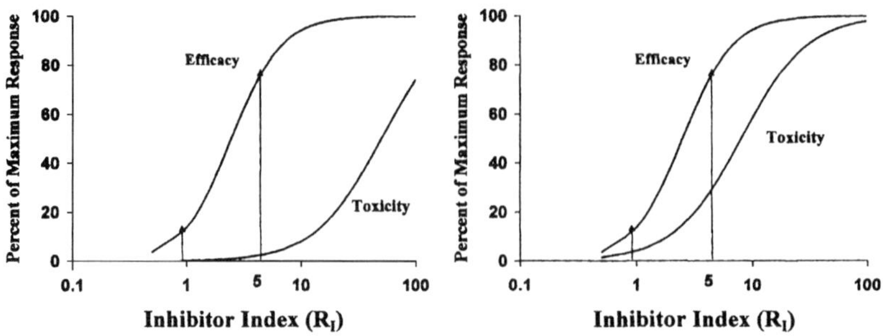

Figure 16 The wider the therapeutic index of a drug, the smaller the impact that a given degree of inhibition, expressed in terms of the inhibitor index R_I, has on the likelihood of an increase in the frequency and severity of side effects. In this example, whereas a fivefold increase in R_I [from 1 (drug alone) to 5] produces a substantial increase in efficacy, it causes only a marked increase in toxicity for the drug with a narrow therapeutic index (right panel). The increase in toxicity for a drug with a wide therapeutic window is minimal (left panel).

E_{max}. This relationship is characterized in its simplest form, and displayed graphically in Figure 16, by

$$\text{Effect}, E = \frac{E_{max} \cdot C}{\text{EC}_{50} + C} \tag{29}$$

where EC_{50} is the concentration of drug that causes 50% of the maximum response; it may be regarded as a measure of potency. This relationship is of the same hyperbolic form as that used to describe the Michaelis-Menten enzyme kinetics. The reason why saturability is almost the norm in pharmacodynamics and not in pharmacokinetics in vivo is that a drug's affinity for its receptor is often many orders of magnitude greater than that for metabolic enzymes, so EC_{50} values tend to be much lower than K_m values. Accordingly, the concentrations needed to produce the often-desired 50–80% of E_{max}, which are already in the saturable part of the concentration-response relationship, are well below the K_m of the metabolic enzyme systems. It also follows that quite large differences in the plasma concentration of drugs when operating in the 50–80% E_{max} range will produce relatively small changes in response. So why the concern for pharmacokinetic drug interactions? The answer is complex, but one reason is that as one pushes further toward the maximum possible response, E_{max}, the body sometimes goes into a hazardous state, putting the patient at risk. An example of this is seen with warfarin, which is used to lower the concentrations of the clotting factors, thereby decreasing the tendency to form clots, through inhibition of the production of these clotting factors. Normally, inhibition is modest. However, if it is too severe, the clotting factors fall to such low concentrations that internal hemorrhage may occur, with potential fatal

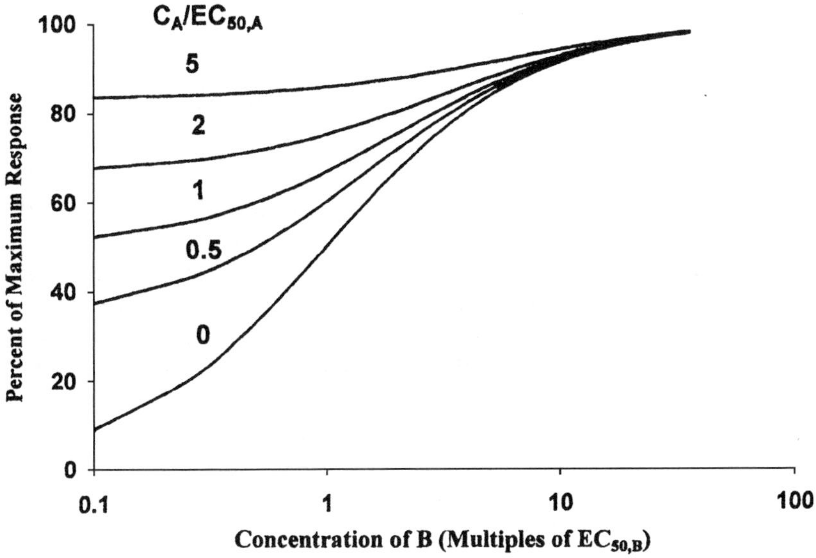

Figure 17 When two drugs, drug A and drug B, are full competitive agonists (or antagonists), the effect of drug B on drug A depends on the fraction of the maximum effect achieved by drug A in the absence of drug B. As can readily be seen, the closer to E_{max} achieved by drug A alone, the smaller the impact of drug B.

consequences. This condition is clearly an example of the adverse effect being the direct extension of the pharmacological properties of the drug.

In many other cases, the limiting toxicity is not an extension of its desired effect but rather arises from a different effect of the drug, such as excessive intestinal bleeding associated with some anti-inflammatory agents. And, as stated in the introduction and illustrated in Fig. 16, the likelihood of a clinically significant interaction occurring for a given change in plasma concentration of the drug depends on its therapeutic window. The wider the window, the bigger the increase in plasma concentration of a drug needed to produce a significant interaction.

Pharmacodynamic interactions occur when one drug modifies the pharmacodynamic response to the same concentration of another. In most cases the mechanism of the effect of each is known, so the outcome is predictable and the combination is either used in therapy to benefit or is contraindicated if it is anticipated to produce undesirable effects. The interaction can result in additivity, but also sometimes in synergism or antagonism, when the response is either greater or less than expected for additivity (16–19). Additivity occurs when the increase in response produced by the addition of the second drug is that expected from the concentration-response curve for each substance. A common example of additivity is seen with full agonists and antagonisms competing for the same receptor. Then the response to the mixture of compounds A and B for full agonists, for example, is

$$\text{Effect}, E = \frac{E_{\max}(C_{\text{A}}/EC_{50,\text{A}} + C_{\text{B}}/EC_{50,\text{B}})}{1 + C_{\text{A}}/EC_{50,\text{A}} + C_{\text{B}}/EC_{50,\text{B}}} \tag{30}$$

The important features of this type of interaction are that each drug alone produces the same maximum response, $E_{\max}$, and that each drug effectively increases the EC_{50} value of the other. Accordingly, in terms of drug interactions, as shown in Figure 17, however much drug B is added to drug A, one cannot exceed $E_{\max}$. The nearer the effect is to $E_{\max}$, with one drug alone, the lower the impact of the addition of the other. In summary, a sound understanding of pharmacokinetic and pharmacodynamic concepts not only enables one to place in vitro information into an in vivo framework, but also helps in both the design and the interpretation of in vitro and in vivo drug interaction studies.

REFERENCES

1. Rowland M, Tozer TN. Clinical Pharmacokinetics: Concepts and Applications. 3rd ed. Baltimore, MD: Williams & Wilkins, 1995.
2. Evans WE, Schentag JJ, Jusko WJ, eds. Applied Pharmacokinetics. 3rd ed. San Francisco, CA: Applied Therapeutics, 1992.
3. Wilkinson GR. Clearance approaches in pharmacology. Pharmacol Rev 1987; 39: 1–47.
4. Rowland M, Matin SB. Kinetics of drug-drug interactions. J Pharmacokinet Biopharm 1973; 1:553–567.
5. Shaw PN, Houston JB. Kinetics of drug metabolism inhibition: use of metabolite concentration-time profiles. J Pharmacokinet Biopharm 1987; 15:497–510.
6. O'Reilly RA. Interaction of sodium warfarin and rifampin. Ann Int Med 1974; 81:337–340.
7. Alvan G, Piafsky K, Lind M, et al. Effect of pentobarbital on the disposition of alprenolol. Clin Pharmacol Ther 1977; 22:316–321.
8. Tozer TN. Concepts basic to pharmacokinetics. Pharmacol Ther 1982; 12:109–131.
9. Rowland M, Benet LZ, Graham GG. Clearance concepts in pharmacokinetics. J Pharmacokinet Biopharm 1973; 1:123–136.
10. Wilkinson GR, Shand DG. A physiological approach to hepatic drug clearance. Clin Pharmacol Ther 1975; 18:377–390.
11. Pang KS, Rowland M. Hepatic clearance of drugs. I. Theoretical considerations of the well-stirred and parallel-tube model. Influence of hepatic blood flow, plasma and blood cell binding, and hepatocellular enzymatic activity on hepatic drug clearance. J Pharmacokinet Biopharm 1977; 5:625–653.
12. Roberts MS, Donaldson JD, Rowland M. Models of hepatic elimination: a comparison of stochastic models to describe residence time distributions and to predict the influence of drug distribution, enzyme heterogeneity, and systemic recycling on hepatic elimination. J Pharmacokinet Biopharm 1988; 16:41–83.
13. Tanaka C, Kawai R, Rowland M. Dose-dependent pharmacokinetics of cyclosporine A in rat: events in tissues. Drug Metab Dispos 2000; 28:582–589.

14. Banfield C, O'Reilly RE, Chan E, et al. Phenylbutazone-warfarin interaction in man: further stereochemical and metabolic considerations. Br J Clin Pharmacol 1983; 16:669–675.
15. Chan E, McLachlan AJ, O'Reilly R, et al. Stereochemical aspects of warfarin drug interactions: use of a combined pharmacokinetic-pharmacodynamic model. Clin Pharmacol Ther 1994; 56:286–294.
16. Holford NHG, Sheiner LB. Kinetics of pharmacological response. Pharmacol Ther 1982; 16:143–166.
17. Greco WR, Bravo G, Parsons JC. The search for strategy: a critical review from a response surface perspective. Pharmacol Rev 1995; 47:331–385.
18. Koizumi T, Kakemi M, Katayama K. Kinetics of combined drug effects. J Pharmacokinet Biopharm 1993; 21:593–607.
19. Berenbaum MC. The expected effect of a combination of agents. J Theor Biol 1985; 114:413–431.

2

In Vitro Enzyme Kinetics Applied to Drug-Metabolizing Enzymes

Kenneth R. Korzekwa
Preclinical Research and Development, AllChemie Inc., Wayne, Pennsylvania, U.S.A.

I. INTRODUCTION

Most new drugs enter clinical trials with varying amounts of information on the human enzymes that may be involved in their metabolism. Most of this information is obtained from (1) animal studies, (2) human tissue preparations in conjunction with chemical inhibitors or antibodies, and (3) expressed enzymes. This chapter will focus on the techniques used to characterize the in vitro metabolism of drugs. Although many enzymes may play some role in drug metabolism, this chapter will focus on the cytochrome P450 (P450) enzymes. The P450 superfamily of enzymes represents the most important enzymes in the metabolism of hydrophobic drugs and other foreign compounds, and many drug-drug interactions result from altering the activities of these enzymes (1). Although not studied as extensively as the P450 enzymes, other drug-metabolizing enzymes, transporters, and xenobiotic receptors share a characteristics that is relatively unique in biochemistry: broad substrate selectivity. This versatility has a profound influence on the enzymology and kinetics of these proteins. Therefore, many of the techniques described for the P450s may apply to other drug-metabolizing enzymes, transporters, and xenobiotic receptors as well.

In the area of drug metabolism, there is a substantial amount of effort toward predicting in vivo pharmacokinetic and pharmacodynamic characteristics from in vitro data (2–6). If valid, these in vitro–in vivo correlations could be used to predict the potential for drug interactions as well as the genotypic and phenotypic variations in the population. A very significant advancement in preclinical drug metabolism is the cloning and expression of the human P450 enzymes. This phenomenon allows the individual human enzymes involved in the metabolism of a particular drug or other xenobiotic to be identified directly and their kinetic properties (K_m and V_m) characterized. This information can be used to predict which enzymes may be involved at physiologically relevant concentrations, drug-drug interactions, and population variability due to variations in genotype and phenotype.

A simple approach to screen a new drug for metabolism or potential drug interactions is to determine the inhibition kinetics for a standard assay. The use of standard assays precludes the need to develop assays for the metabolites of new drug candidates and allows many compounds to be screened rapidly. With this approach, a standard assay is developed for each P450 enzyme. Metabolism is observed in the presence of varying concentrations of the new compound. Competitive inhibition kinetics suggests that the compound is bound to the P450 active site. If the inhibition constant (K_i) is within physiologically relevant concentrations, the compound is likely to be a substrate for that P450 and is likely to have interactions with other drugs metabolized by that P450. The kinetic constants (K_m and V_m) can then be determined for the enzymes that are likely to be important.

Most P450 oxidations and drug interactions can be predicted from inhibition studies, since most P450 inhibitors show competitive Michaelis-Menten kinetics. However, there are examples of unusual kinetics, and most of these are associated with CYP3A oxidations. In this chapter, both Michaelis-Menten kinetics and more complex kinetics will be discussed. General experimental protocols that can be used to obtain and analyze kinetic data will be presented, and the implications of the results when predicting drug interactions will be discussed.

II. MICHAELIS-MENTEN KINETICS

A drug that binds reversibly to a protein, as shown in Figure 1A, displays hyperbolic saturation kinetics. At equilibrium, the fraction bound is as described by Eq. (1), where $K_b = k_{21}/k_{12}$, ES is the enzyme-substrate complex and E_t is the total enzyme:

$$\frac{[\text{ES}]}{[\text{E}_t]} = \frac{[\text{S}]}{(K_b + [\text{S}])} \tag{1}$$

The binding affinity, and therefore the concentration dependence of the process, is described by the binding constant K_b. Likewise, when a drug binds

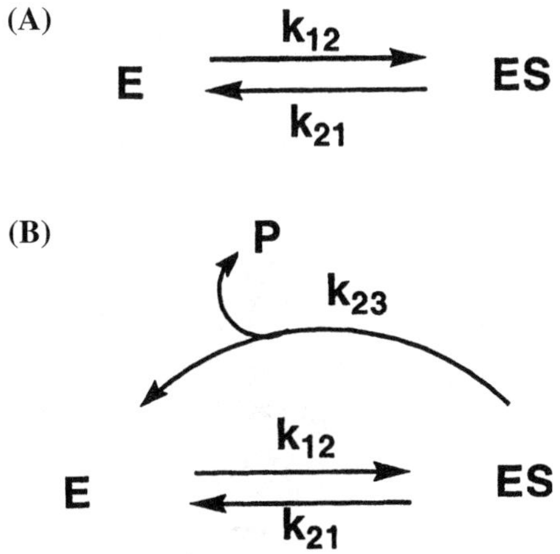

Figure 1 Simple schemes for (**A**) protein binding and (**B**) enzyme catalysis.

reversibly to an enzyme, the reaction velocity usually shows hyperbolic saturation kinetics. Under steady-state conditions, the velocity of the simple reaction shown in Figure 1B can be described by the Michaelis-Menten equation:

$$\frac{v}{\mathrm{E}_t} = \frac{V_\mathrm{m}[\mathrm{S}]}{K_\mathrm{m} + [\mathrm{S}]} \tag{2}$$

In this equation, a hyperbolic saturation curve is described by two constants, V_m and K_m. In the simple example in Figure 1B, v is velocity, V_m is simply $k_{23}[\mathrm{E}_t]$ and K_m is $(k_{21} + k_{23})/k_{12}$. V_{max} (or V_m) is the reaction velocity at saturating concentrations of substrate, and K_m is the concentration of the substrate that achieves half the maximum velocity. Although the constant K_m is the most useful descriptor of the affinity of the substrate for the enzyme, it is important to note the difference between K_m and K_b. Even for the simplest reaction scheme (Fig. 1B), the K_m term contains the rate constant for conversion of substrate to product (k_{23}). If the rate of equilibrium is fast relative to k_{23}, then K_m approaches K_b.

More complex enzymatic reactions usually display Michaelis-Menten kinetics and can be described by Eq. (2). However, the forms of constants K_m and V_m can be very complicated, consisting of many individual rate constants. King and Altman (7) have provided a method to readily derive the steady-state equations for enzymatic reactions, including the forms that describe K_m and V_m. The advent of symbolic mathematics programs makes the implementation of these methods routine, even for very complex reaction schemes. The P450 catalytic cycle (Fig. 2) is an example of a very complicated reaction scheme. However, most P450-mediated reactions display standard hyperbolic saturation kinetics. Therefore, although the rate constants that determine K_m and V_m are

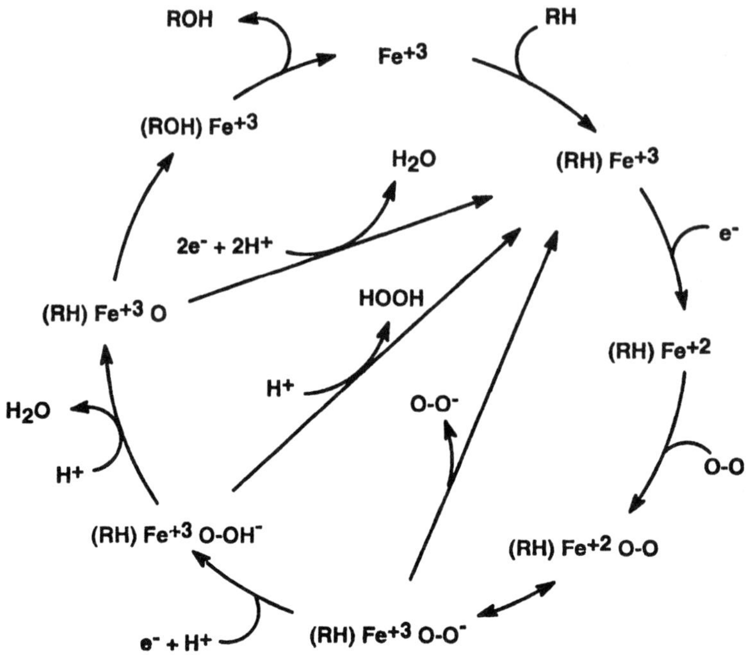

Figure 2 P450 catalytic cycle.

generally unknown for the P450 enzymes, the values of K_m and V_m can be experimentally determined. Another constant that has important implications in drug metabolism is the ratio of V_m to K_m, or V/K. This ratio is the slope of the hyperbolic saturation curve at low substrate concentrations. Since most P450-mediated reactions have relatively high K_m values, most drug metabolism occurs in the linear or V/K region of the saturation curve.

A. Experimental Determination of In Vitro Kinetic Parameters

1. P450 Enzyme Preparations

The P450 enzymes are found primarily in the outer membrane of the endoplasmic reticulum. Enzyme activity requires that the enzyme be integrated into a membrane that contains P450 reductase and, for some reactions, cytochrome b_5. Characterization of the saturation kinetics for the P450 enzymes can be determined using a variety of enzyme preparations, including tissue slices, whole cells, microsomes, and reconstituted, purified enzymes. The more intact the in vitro preparation, the more it is likely that the environment of the enzyme will represent the in vivo environment. However, intact cell preparations do not generally give kinetic parameters that are observed with microsomal preparations. This could be due to factors such as limiting diffusion into the cells, binding to intracellular proteins, or differences in membrane partitioning. Therefore, when whole-cell preparations are used, observed kinetic characteristics may not provide the true kinetic constants for the enzyme being studied.

Microsomal preparations generally provide reproducible kinetic analyses when only one enzyme is involved in the reaction. However, microsomal preparations (and other intact preparations) contain many different P450 enzymes. Although this characteristic is useful when trying to mimic the metabolic characteristics of an organ, it is a drawback when trying to characterize the kinetic constants of an individual P450 enzyme or when trying to determine which enzyme is involved in the metabolism of a particular drug. Because of the generally broad substrate selectivities of the P450 enzymes, most observed metabolic reactions can be catalyzed by more than one enzyme. Interindividual variability in the content of the different P450s makes it even more difficult to determine the different kinetic parameters when more than one enzyme is involved in a given reaction.

Preparations containing a single P450 isozyme are available as either expression systems or purified, reconstituted enzymes. The P450s have been expressed in bacterial, yeast, insect, and mammalian cells (8). Most of these enzymes can be used in the membranes in which they are expressed. However, in order to obtain adequate enzyme activity for most expression systems, it is necessary to supplement the membranes with reductase and in some cases cytochrome b_5. This is accomplished by either supplementing the membranes with purified coenzymes or by coexpression of the coenzymes. Alternatively, the P450 enzymes can be purified and reconstituted with coenzymes into artificial membranes.

Every enzyme preparation has its advantages and disadvantages. Microsomes may more closely represent the in vivo activity of a particular organ, but kinetic analyses are complicated by the presence of multiple enzymes. It is not possible to spectrally quantitate the content of any individual enzyme when a mixture of enzymes is present. Expression systems provide isozymically pure preparations, but they also have their disadvantages. The P450 enzymes are membrane bound, and for the nonmammalian expression systems the membranes may have different interactions with the P450 proteins. Although expression levels in most of the systems are adequate for spectral quantitation, coexpression of the coenzymes adds variability to different batches. Reconstituted enzymes allow for the exact control of enzyme and coenzyme content. However, the membranes are artificial and can have an influence on enzyme activity. For example, whereas most P450 enzymes can be reconstituted into dilaurylphosphatidylcholine (DLPC) vesicles, the CYP3A enzymes require the presence of both unsaturated lipid and a small amount of nonionic detergent (9). Finally, these differences are further complicated by unpredictable influences of ionic strength, pH, etc., of the incubation medium, as will be discussed next.

2. Incubation Conditions

Enzyme kinetics are normally determined under steady-state, initial-rate conditions, which place several constraints on the incubation conditions. First, the amount of substrate should greatly exceed the enzyme concentration, and the consumption of substrate should be held to a minimum. Generally, the amount of substrate consumed should be held to less than 10%. This constraint ensures that accurate substrate concentration data are available for the kinetic analyses and minimizes the probability that product inhibition of the reaction will occur. This constraint can be problematic when the K_m of the reaction is low, since the amount of product (10% of a low substrate concentration) may be below that needed for accurate product quantitation. One method to increase the substrate amount available is to use larger incubation volumes. For example, a 10-mL incubation has 10 times more substrate available than a 1-mL incubation. Another method is to increase the sensitivity of the assay, e.g., using mass spectral or radioisotope assays. When more than 10% of the substrate is consumed, the substrate concentration can be corrected via the integrated form of the rate equation (Dr. James Gillette, personal communication):

$$\frac{v}{E_t} = \frac{V_m[S]}{K_m + [S]} \tag{3}$$

$$S' = \frac{[S]_0 - [S]_f}{\ln[S]_0/[S]_f} \tag{4}$$

In Eq. (3) $[S]_0$ and $[S]_f$ are starting and ending substrate concentrations. S′ approaches [S] when substrate consumption is minimal, and S′ is substituted for [S] to correct for excess substrate consumption. In these analyses, however, substrate inhibition can be a problem if the product has a similar affinity to the substrate. Fortunately, most P450 oxidations produce products that are less hydrophobic than the substrates, resulting in lower affinities to the enzymes. There are exceptions, including desaturation reactions that produce alkenes from alkanes (10) and carbonyl compounds from alcohols. These products have hydrophobicities that are similar or increased relative to their substrates.

A second constraint is that the reaction remains linear with time. In the presence of reducing equivalents, the P450 enzymes will generally lose activity over time. Provided that the loss of enzyme is not dependent on substrate concentration, the V_m of the enzyme will change, but not the K_m. For P450 reactions, the presence of substrate in the active site can either protect the enzyme or increase its rate of deactivation. Substrate dependence on stability can generate inaccurate saturation curves. Enzyme stabilization can result in a sigmoidal saturation curve for an enzyme showing hyperbolic saturation kinetics, and enzyme destabilization can show substrate inhibition if the enzyme content varies over the incubation time. The reaction should also be linear with enzyme concentration to ensure that other processes, such as saturable, nonspecific binding, do not alter the enzyme saturation profile.

B. Analysis of Michaelis-Menten Kinetic Data

By far, the best method of determining kinetic parameters is to perform an appropriately weighted least-squares fit to the relevant rate equation (11). Although reciprocal plots are useful for determining initial parameters for the regression and for plotting the results, initial parameters for a single enzyme showing hyperbolic saturation kinetics can be obtained by inspection of the data. When more than one enzyme is present, e.g., in microsomes, the data can be fit to combined Michaelis-Menten equations:

$$\frac{v}{[E_t]} = \frac{V_{m1}[S]}{K_{m1} + [S]} + \frac{V_{m2}[S]}{K_{m2} + [S]} + \cdots + \frac{V_{mn}[S]}{K_{mn} + [S]} \tag{5}$$

If the highest substrate concentration shows a linear increase in velocity, the last component of the rate equation should be V/K, i.e., $v_n = (V/K)_n$. Inclusion of additional rate components should be justified by statistical methods, such as comparing F values for the regression analyses or the minimum Akaike information criterion estimation (MAICE) (12,13).

C. Reaction Conditions

In addition to the preceding complexities, the P450 enzymes have some unique characteristics that complicate the design of experimental protocols. Because of the broad substrate selectivities for these enzymes, the enzymes are not optimized for the metabolism of a particular substrate. Therefore, the reaction conditions (i.e., pH, ionic strength, temperature) that result in optimum velocities for a given reaction are dependent on both the enzyme and the substrate. To further complicate matters, the velocities for these enzymes tend to vary greatly with changes in these reaction conditions. This variation may well be due to the dependence of the reaction velocity on several pathways in the catalytic cycle.

It is generally accepted that the overall flux through the catalytic cycle (Fig. 2) is dependent on the rates of reduction by P450 reductase (14,15). However, the actual rates of substrate oxidation are probably dependent on three additional rates: the rate of substrate oxidation and the rates of the decoupling pathways (hydrogen peroxide formation and excess water formation). Thus, the efficiency of the reaction plays a major role in determining the velocity of a P450 oxidation (16,17). The sensitivity of the reaction velocities to incubation conditions may be due to changes in the reduction rate as well as to changes in the enzyme efficiency.

Although many P450 reactions show optimal activity in the pH range of 7 to 8, both chlorobenzene and octane metabolism show optimum activity at pH 8.2 in rat liver microsomes (18,19). This is also the pH at which P450 oxidoreductase optimally reduces cytochrome *c*. In addition, whereas essentially all in vitro metabolism studies are carried out at 37°C, both these reactions occur much faster at 25°C. For a given enzyme, the optimum ionic strength is a function of the substrate. For example, the rate of benzphetamine metabolism by reconstituted CYP2B1 increases with increasing ionic strength (20), whereas the optimum for testosterone metabolism by this enzyme is 20 mM potassium phosphate (KPi) buffer and decreases with increasing ionic strength (unpublished results).

Even the optimum ratio of reductase to P450 depends on the substrate and the enzyme. Whereas most reactions are saturated by a reductase/P450 ratio of 10:1, testosterone metabolism by CYP2A1 saturates at much higher reductase ratios. In contrast, essentially all reactions that have a cytochrome b_5 dependence are saturated at a b_5/P450 ratio of 1:1.

Thus, many P450 oxidations show a substantial and variable dependence on reaction conditions, which makes it impractical to optimize each reaction. In fact, the optimum reaction conditions may not represent the in vivo reaction environment. It would be difficult to justify a reaction temperature of 25°C in an experiment that will be used for in vitro–in vivo correlations. A more practical approach would be to use a consistent set of reaction conditions that provide adequate velocities. Common reaction conditions include 100 mM KPi, pH 7.4, 37°C, a reductase/P450 ratio of 2:1, and a cytochrome b_5/P450 ratio of 1:1.

III. INHIBITION: MICHAELIS-MENTEN KINETICS

For a detailed review of simple to complex enzyme kinetics, a book by Segel (21) is recommended. Most P450 oxidations show hyperbolic saturation kinetics and competitive inhibition between substrates. Therefore, both K_m values and drug interactions can be predicted from inhibition studies. Competitive inhibition suggests that the enzymes have a single binding site and only one substrate can bind at any one time. For the inhibition of substrate A by substrate B to be competitive, the following must be observed:

1. Substrate A has a hyperbolic saturation curve: Enzymes that bind to only one substrate molecule will show hyperbolic saturation kinetics. However, the observation of hyperbolic saturation kinetics does not necessarily mean that only one substrate molecule is interacting with the enzyme (see discussion of non-Michaelis-Menten kinetics in sec. IV).
2. The presence of substrate B changes the apparent K_m but not the V_m for substrate A: Saturating concentrations of A must be able to completely displace B from the active site.
3. Complete inhibition of metabolism is achieved with saturating concentrations of substrate B: Saturating concentrations of B must be able to completely displace A from the active site.

4. Substrate B does not change the regioselectivity of substrate A: The regioselectivity of the enzyme is determined by the interactions between the substrate and the active site. Since the substrate saturation curve is defined by the K_m of the enzyme, regioselectivity cannot be a function of substrate or inhibitor concentration [I].

One standard equation for competitive inhibition is given in Eq. (6). This equation shows that the presence of the inhibitor modifies the observed K_m but not the observed V_m. A double reciprocal plot gives an x intercept of $-1/K_m$ and a y intercept of $1/V_m$.

$$\frac{v}{[E_t]} = \frac{V_m[S]}{K_m[1 + I/K_i] + [S]} \tag{6}$$

Equation (7) gives the fraction activity remaining in the presence of an inhibitor relative to its absence (v_i/v_0):

$$\frac{v_i}{v_0} = \frac{K_m + [S]}{K_m(1 + I/K_i) + [S]} \tag{7}$$

Equation (8) describes the fraction of inhibition, or $1 - (v_i/v_0)$.

$$i = 1 - \left(\frac{v_i}{v_0}\right) = \frac{[I]}{[I] + K_i(1 + [S]/K_m)} \tag{8}$$

Finally, many reports provide IC_{50} values (concentration of inhibitor required to achieve 50% inhibition), which are dependent on both substrate concentration and K_m [Eq. (9)]. Equation (9) shows that when $[S] = K_m$, then $IC_{50} = 2K_i$:

$$IC_{50} = K_i\left(1 + \frac{[S]}{K_m}\right) \tag{9}$$

A. Experimental Design and Analysis of Inhibition Data

By far the best method for characterizing inhibition data is to vary both substrate and inhibitor concentration. The resulting rate data is fit to Eq. (6) by weighted least-squares regression. Initial estimates for the parameters can be obtained from the control (no inhibitor) data and by a double reciprocal plot. This analysis provides estimates of V_m, K_m, and K_i from a single experiment. If a minimum of effort is required, the K_m of the reaction is known, and competitive inhibition is assumed. Equations (6) to (9) can be used to determine the K_i by varying [I] at a single substrate concentration. However, neither the K_m nor the type of inhibition can be validated. Only an observation of partial inhibition or nonhyperbolic kinetics indicates that simple competitive inhibition is not involved. If both substrate and inhibitor concentration are varied, the data can also be fit to equations for other types of inhibition, e.g., noncompetitive and mixed type, and the fits can be compared. For the P450 enzymes, the second most prevalent type of inhibition is the partial mixed type of inhibition, which will be discussed later.

IV. NON-MICHAELIS-MENTEN KINETICS

Most P450 oxidations show standard saturation kinetics and competitive inhibition between substrates. However, some P450 reactions show unusual enzyme kinetics, and most of those identified so far are associated with CYP3A oxidations (22). The unusual kinetic characteristics of the CYP3A enzymes (and less frequently other enzymes) include five categories: activation, autoactivation, partial inhibition, biphasic saturation kinetics, and substrate inhibition. Activation is the ability to be activated by certain compounds, i.e., the rates of a reaction are increased in the presence of another compound. Autoactivation occurs when the activator is the substrate itself, resulting in sigmoidal saturation kinetics. For partial inhibition, saturation of the inhibitor does not completely inhibit substrate metabolism. Substrate inhibition occurs when increasing the substrate beyond a certain concentration results in a decrease in metabolism.

Although most of the observed kinetics are consistent with allosteric binding at two distinct sites (23), previous studies suggest that the activation of metabolism involves the simultaneous binding of both the activator and the substrate in the same active site (24,25). The possibility of binding two substrate molecules to a P450 active site could almost be expected, given the relatively nonspecific nature of the P450-substrate interactions. For example, CYP1A1 is a P450 that metabolizes polycyclic aromatic hydrocarbons (PAHs). The size of the PAHs can vary between naphthalene (two aromatic rings) to very large substrates, such as dibenzopyrenes (six rings). If an active site can accommodate very large substrates, it can be expected that more than one naphthalene molecule can be bound. Indeed, naphthalene metabolism by CYPlAl has a sigmoidal saturation curve (unpublished results). Finally, it has been shown by NMR studies that both pyridine and imidazole can coexist in the P450cam active site (26). Thus, even a P450 with rigid structural requirements can simultaneously bind two small substrates.

If enzyme activation and the other unusual kinetic characteristics result from multiple substrates in the active site, kinetic parameters will be difficult to characterize and drug interactions will be more difficult to predict, since they are a function of the enzyme and of both the substrates. In addition, there are some indications that non-Michaelis-Menten kinetics can be seen in vivo (27–29).

A. Non-Michaelis-Menten Kinetics for a Single Substrate

If non-Michaelis-Menten kinetics for all P450 enzymes are a result of multiple substrates binding to the enzyme, then the reaction kinetics for the binding of two substrates to an active site can be complicated. A number of analyses of

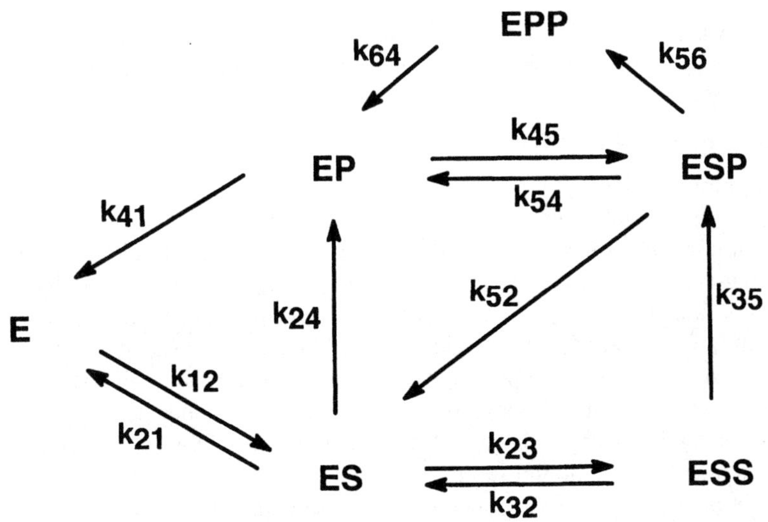

Figure 3 Proposed kinetic scheme for an enzyme with two binding sites within an active site and a single substrate. *Source*: From Ref. 17.

varying complexity have been published and a review of this topic is available. Differences in analyses are due to different numbers of distinct binding sites and distinct binding constants. For this section and the next, we make the assumption that two compounds can bind to the active site with different affinities, but the binding sites are not defined regions of the active site. These assumptions can describe all observed kinetic characteristics and are still simple enough to allow for the determination of kinetic constants.

The full kinetic scheme for the two-substrate model is given in Figure 3. If product release is fast relative to the oxidation rates, the velocity equation is simplified to Eq. (10):

$$\frac{v}{\mathrm{E}_t} = \frac{k_{25}[\mathrm{S}]/K_{\mathrm{m1}} + k_{35}[\mathrm{S}]^2/K_{\mathrm{m1}}K_{\mathrm{m2}}}{1 + [\mathrm{S}]/K_{\mathrm{m1}} + [\mathrm{S}]^2/K_{\mathrm{m1}}K_{\mathrm{m2}}} \tag{10}$$

In this equation, $K_{\mathrm{m1}} = (k_{21} + k_{23})/k_{12}$ and $K_{\mathrm{m2}} = (k_{23} + k_{35})/k_{32}$. K_{m1} would be the standard Michaelis constant for the binding of the first substrate, if [ESS] = 0. K_{m2} would be the standard Michaelis constant for the binding of the second substrate, if [E] = 0 (i.e., the first binding site is saturated). In the complete equation, these constants are not true K_{m} values, but their form (i.e., $K_{\mathrm{m1}} = (k_{21} + k_{25})/k_{12}$) and significance are analogous. Likewise, k_{25} and k_{35} are $V_{\mathrm{m1}}/\mathrm{E}_t$ and $V_{\mathrm{m2}}/\mathrm{E}_t$ terms when the enzyme is saturated with one and two substrate molecules, respectively. Equation (10) describes several non-Michaelis-Menten kinetic profiles. Autoactivation (sigmoidal saturation curve) occurs when $k_{35} > k_{24}$ or $K_{\mathrm{m2}} < K_{\mathrm{m1}}$, substrate inhibition occurs when $k_{24} > k_{35}$, and a biphasic saturation curve results when $k_{35} > k_{24}$ and $K_{m2} \gg K_{m1}$. This equation was used to fit experimental data for the metabolism of several other substrates, as described next.

1. Sigmoidal Saturation Kinetics

Although sigmoidal binding kinetics can be discussed in terms of binding cooperativity, this is not always the case for enzymes. Sigmoidal saturation kinetics of an enzyme can result when either the second substrate binds to the enzyme with greater affinity than the first or the ESS complex is metabolized at a faster rate than the ES complex. There have been several reports that describe sigmoidal saturation curves for P450 oxidations (23,30,31) and carbamazepine is a classic CYP3A substrate that shows sigmoidal saturation kinetics (Fig. 4). This figure also shows that quinine converts the sigmoidal curve into a hyperbolic curve. This conversion will be discussed in section V, on interactions between different substrates. For sigmoidal saturation curves, a unique solution for a fit to Eq. (10) is not possible (25). This fit becomes apparent when the influence of the second substrate is considered. For this discussion, K_{m1}, K_{m2}, V_{m1}, and V_{m2} are

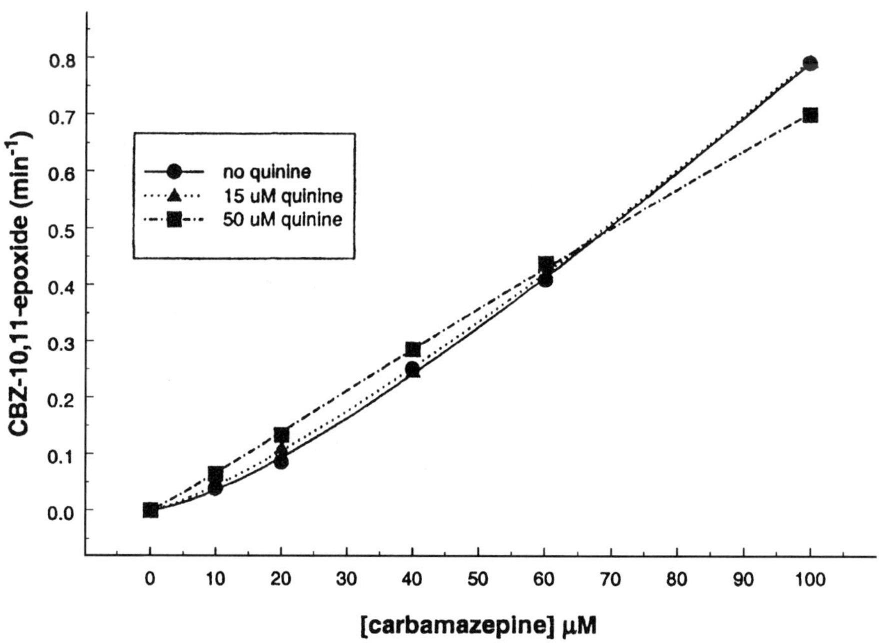

Figure 4 Effect of quinine on the carbamazepine saturation curve. Quinine makes the sigmoidal saturation curve more hyperbolic. *Source*: Courtesy of K. Nandigama and K. Korzekwa (unpublished results).

defined as described for Eq. (10). If the second substrate binds with a lower K_m than the first substrate and has the same rate of product formation, the slope will equal $(V/K)_1$ at low substrate concentrations, since only one substrate will be bound. As the substrate concentration increases into the range of the second K_m, much of the ES complex becomes ESS. Since the ratio of [E] to [ES] is determined by the first K_m, the ESS complex increases at the expense of E. Therefore, the enzyme becomes saturated faster, resulting in a concave-upward region in the saturation curve. Likewise, if the second substrate binds with a K_m identical to that of the first substrate but has a higher V_m, the linear portion of the curve will again have a slope of $(V/K)_1$. As the substrate concentration approaches K_{m2}, [ESS] increases. Since the rate of product formation is higher for ESS, a concave-upward region results. From a sigmoidal saturation curve, one can determine $(V/K)_1$ from the slope at low substrate concentrations, and V_{m2} at saturating substrate concentrations. However, V_{m1}, K_{m1}, and K_{m2} remain undetermined, since $(V/K)_1$ can have either a K_{m1} higher than K_{m2} or a V_{m1} lower than V_{m2}. Therefore, multiple solutions are possible when sigmoidal saturation data are fit to Eq. (10).

If a sigmoidal saturation curve is obtained, information relevant to in vitro–in vivo correlations can be obtained from appropriately designed experimental data. The values of $(V/K)_1$, V_{m2}, and the concave-upward region should be defined if they occur within the therapeutic concentration range. The $(V/K)_1$ region will define the rate of metabolism at low substrate concentrations. If the concave-upward region occurs in the therapeutic range, a dose-dependent increase in drug clearance can be expected. On the other hand, if enzyme saturation occurs, a dose-dependent decrease in clearance can be expected. If there is no linear range (i.e., the slope constantly increases at low substrate concentrations), then $(V/K)_1 = 0$. This is probably due to $V_{m1} = 0$, since an enzyme with a very high K_m will not be very active at moderate substrate concentrations.

2. Biphasic Saturation Kinetics

A second type of nonhyperbolic saturation kinetics became apparent during studies on the metabolism of naproxen to desmethylnaproxen (32). Studies with human liver microsomes showed that naproxen metabolism has biphasic kinetics and is activated by dapsone (T. Tracy, unpublished results). The unactivated data shows what appears to be a typical concentration profile for metabolism by at least two different enzymes. However, a similar biphasic profile was obtained with expressed enzyme (25). This biphasic kinetic profile is observed with the two-substrate model when $V_{m2} > V_{m1}$ and $K_{m2} \gg K_{m1}$. The appropriate equation for the two-site model when $[S] < K_{m2}$ is

$$\frac{v}{E_t} = \frac{V_{m1}[S] + V_{m2}/K_{m2}[S]^2}{K_{m1} + [S]} \tag{11}$$

This equation can be compared to that when two enzymes are present, one with a very high K_m:

$$\frac{v}{[E_t]} = \frac{V_{m1}[S]}{K_{m1} + [S]} + \frac{V_{m2}}{K_{m2}}[S] \tag{12}$$

Fits of experimental data to the two equations are almost indistinguishable. Therefore, saturation kinetic data alone cannot determine the appropriate model when multiple enzymes are present. In addition, higher concentrations of dapsone result in hyperbolic naproxen demethylation kinetics (T. Tracy, unpublished results), suggesting that dapsone is occupying one of the two naproxen-binding regions in the CYP2C9 active site. Again, this will be discussed in section V, on interactions between different substrates.

3. Substrate Inhibition

Another kinetic profile, substrate inhibition, occurs when the velocity from ESS is lower than that of ES (Fig. 5). In this case, the saturation curve will increase to a maximum and then decrease before leveling off at V_{m2}. For the P450 enzymes, V_{m2} is usually not zero when sub-millimolar concentrations of substrate are involved. This observation suggests that ESS still has some activity. If substrate inhibition occurs at very high substrate concentrations, non-active-site interactions should be suspected. Substrate inhibition profiles are easily identified, provided that the observed concentration range is appropriate and K_{m1} is not much smaller than K_{m2} (Fig. 5). However, determining the kinetic constants in Eq. (10) requires

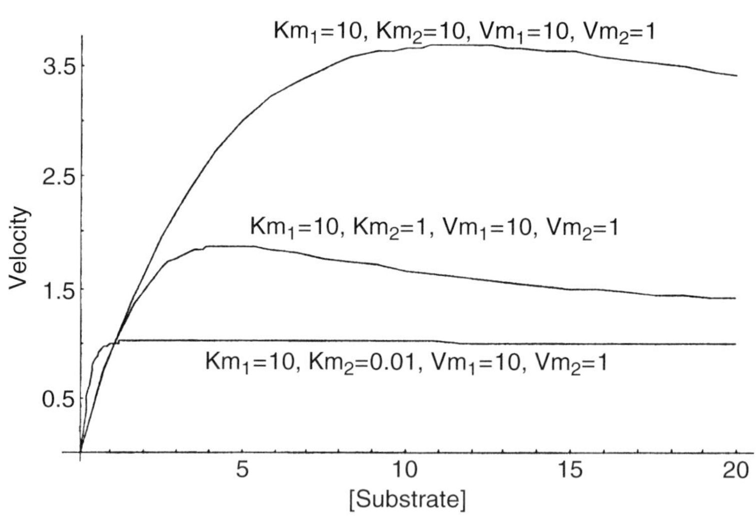

Figure 5 Substrate inhibition saturation curves.

adequate experimental data. The number and concentration of data points must be sufficient to define four regions in the saturation curve: the $(V/K)_1$ region, the concave-downward region, the concave-upward region, and V_{m2}.

V. SIMULTANEOUS BINDING OF DIFFERENT SUBSTRATES TO THE P450 ACTIVE SITES

If two different substrates bind simultaneously to the active site, then the standard Michaelis-Menten equations and competitive inhibition kinetics do not apply. Instead it is necessary to base the kinetic analyses on a more complex kinetic scheme. The scheme in Figure 6 is a simplified representation of a substrate and an effector binding to an enzyme, with the assumption that product release is fast. In Figure 6, S is the substrate and B is the effector molecule. Product can be formed from both the ES and ESB complexes. If the rates of product formation are slow relative to the binding equilibrium, we can consider each substrate independently (i.e., we do not include the formation of the effector metabolites from EB and ESB in the kinetic derivations). This results in the following relatively simple equation for the velocity:

$$\frac{v}{\mathrm{E}_t} = \frac{V_\mathrm{m}[\mathrm{S}]}{K_\mathrm{m}\dfrac{(1+[\mathrm{B}]/K_\mathrm{b})}{(1+\beta[\mathrm{B}]/\alpha K_\mathrm{b})} + [S]\dfrac{(1+[\mathrm{B}]/\alpha K_\mathrm{b})}{(1+\beta[\mathrm{B}]/\alpha K_\mathrm{b})}} \tag{13}$$

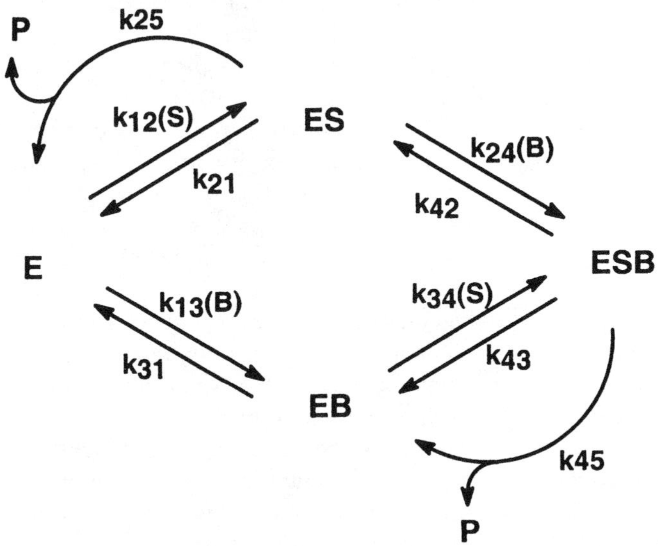

Figure 6 Simplified kinetic scheme for the interaction between a substrate and an effector molecule for an enzyme with two binding sites within the active site. *Source*: From Ref. 17.

In this equation, S is the substrate, B is the effector, $V_m = k_{25}E_t$, $K_m = (k_{21} + k_{25})/k_{12}$ (kinetic constants for substrate metabolism), $K_b = k_{31}/k_{13}$ (binding constant for effector), α is the change in K_m resulting from effector binding, and β is the change in V_m from effector binding. For inhibitors, $\beta < 1$; for activators, $\beta > 1$.

The scheme in Figure 6 provides a general description of the interaction of two molecules with an enzyme, including both inhibition and activation. Since we are considering only the metabolism of S, the effector molecule can be binding at any other site on the enzyme, e.g., an allosteric site. With respect to P450 activation, at least some P450 effectors are also substrates for the enzymes (24,25). Also, saturating concentrations of S will not completely inhibit the metabolism of B, and saturating concentrations of B cannot completely inhibit the metabolism of S. Since the P450 enzymes have only one active site, these data suggest that both molecules bind simultaneously to the active site (i.e., they have access to the reactive oxygen). The observation of partial inhibition by another P450 substrate is also consistent with this hypothesis.

To experimentally define these kinds of interactions, it is necessary to vary both substrate and effector concentrations. For Eq. (13), initial parameters can be obtained by first performing double reciprocal plots and then replotting 1/slope and 1/intercept versus 1/[I] (21). The intercept of the 1/intercept replot is $\beta V_m/(1 - \beta)$, which can be used to solve for β. The value for α can then be obtained from the 1/slope intercept $= [\beta V_m/K_m(\alpha - \beta)]$.

If the metabolism of both substrate and effector are measured, the validity of treating the two processes independently can be tested. For example, we reported that 7,8-benzoflavone dramatically increases the V_m of phenanthrene metabolism by CYP3A4 and that phenanthrene is a partial inhibitor of 7,8-benzoflavone metabolism (24,25). If the scheme in Figure 6 is valid, then the K_m when phenanthrene is analyzed as the substrate should equal K_b when 7,8-benzoflavone is analyzed as the substrate. In addition, since any thermodynamic state is path independent, the α values and $K_m\alpha K_b$ values should be similar between experiments. For this pair of substrates, these relationships were shown to be true.

The situation becomes even more complicated when one of the substrates can bind twice to the enzyme, as represented in Figure 7. In this case, inhibition or activation is combined with the nonhyperbolic saturation kinetics for a single substrate described earlier. Analysis of the equation derived for the scheme in Figure 7 suggests that some compounds would be activators at low substrate concentrations and inhibitors at high substrate concentrations. This situation can occur when the rate of product formation from the intermediates has the order ES < ESB < ESS. At low substrate concentrations, the reaction is activated by B by converting ES to ESB. At high substrate concentrations, the reaction is inhibited by B by converting ESS to ESB. This is precisely what has been observed in Figure 4. In this figure, quinine converts the sigmoidal carbamazepine saturation curve to a hyperbolic curve (linear double-reciprocal plot), by

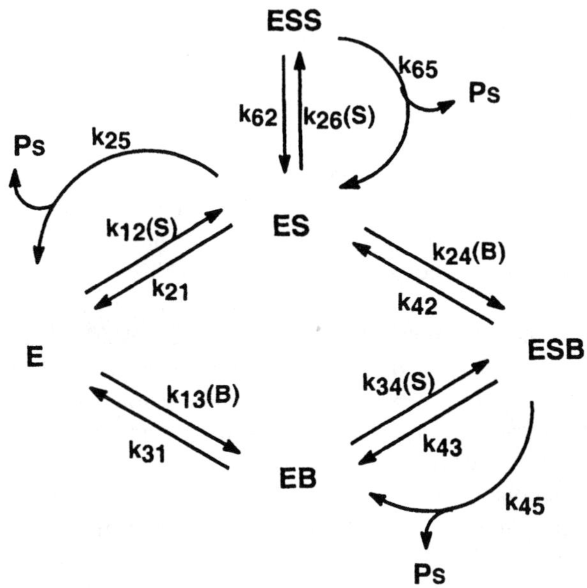

Figure 7 Kinetic scheme for an enzyme with two binding sites that can bind two substrate molecules and one effector molecule. *Source*: From Ref. 17.

apparently binding one of the substrate-binding sites. The presence of quinine results in significant activation at low substrate concentrations and inhibition at high substrate concentrations. This observation suggests that the reaction velocities from the various substrate complexes have the order ES < EB < ESS, where S is carbamazepine and B is quinine.

Two other examples of sigmoidal reactions that are made linear by an activator include a report by Johnson et al. (31), who showed that pregnenolone has a nonlinear double-reciprocal plot that was made linear by the presence of 5 μM 7,8-benzoflavone, and Ueng et al. (23), who showed that aflatoxin Bl has sigmoidal saturation curve that is made more hyperbolic by 7,8-benzoflavone. As with the effect of quinine on carbamazepine metabolism, 7,8-benzoflavone is an activator at low aflatoxin Bl concentrations and an inhibitor at high aflatoxin Bl concentrations.

Another example of reactions that can be described by Figure 7 is the effect of dapsone on naproxen metabolism by CYP2C9. In this case, dapsone makes the biphasic naproxen curve more hyperbolic. Finally, one can expect similar influences on reactions that show substrate inhibition. If ESB has a metabolic rate similar to ES, one would expect activation at high substrate concentrations. Conversely, if the rate is similar to ESS, inhibition would be expected at intermediate substrate concentrations, with little effect at V_m.

VI. INFLUENCE OF ATYPICAL KINETICS ON INHIBITION AND DRUG INTERACTION STUDIES

In vitro studies of drug metabolism with human enzymes are becoming an increasingly important part of preclinical drug development, since they can provide information on the expected genotypic and phenotypic variation within the population and can be used to predict drug interactions. It is common practice to use inhibition of standard assays to determine if a substrate will interact with a particular P450. This practice is based on the assumption that competitive inhibition occurs and that a given inhibitor will have a K_i value that is independent of the substrate being inhibited. Although this assumption is true for most P450 oxidations, there are an increasing number of examples where non-Michaelis-Menten kinetics are observed. The foregoing discussion suggests that an effector can either increase or decrease either V_m or K_m or both. It is also possible for an effector to bind to the active site and have no influence on a reaction. This can be seen by the effect of quinine on pyrene metabolism by CYP3A4 (Fig. 8). Although quinine is a known CYP3A4 substrate, it appears to have no effect on the reaction. However, if pyrene metabolism is first activated by testosterone or 7,8-benzoflavone, quinine displaces the activator, causing inhibition. This suggests that negative results for one drug cannot always be extrapolated to predict interactions with other drugs. In

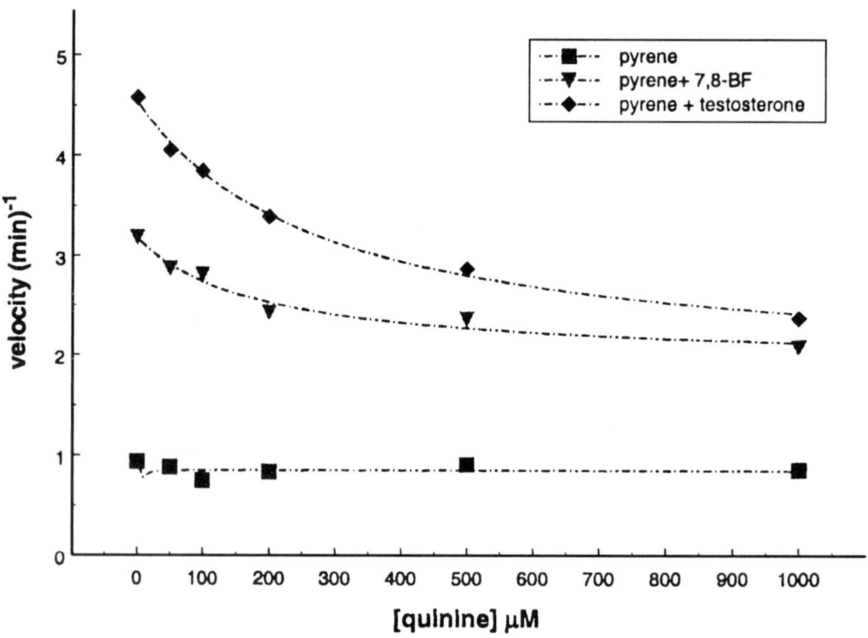

Figure 8 Effect of quinine on pyrene metabolism. *Source*: Courtesy of K. Nandigama and K. Korzekwa (unpublished results).

general, since both α and β are substrate-pair dependent, drug interactions cannot be extrapolated to other substrates for enzymes that show non-Michaelis-Menten kinetics. This does not mean that inhibition studies are not useful in predicting drug metabolism or drug interactions, but only that the limitations of the data should be understood. At an early stage of drug development, it is not practical to perform the extensive kinetic analyses that may be required to define all relevant kinetic parameters. It is still useful to conduct inhibition studies with standard assays to determine the enzymes involved and their approximate binding constants. However, a common result of complex kinetics is the observation of partial inhibition and, less frequently, activation. When inhibition occurs, an approximate binding constant for the inhibitor at the given substrate concentration can be obtained by fitting inhibition from the following equation, where β_{app} is the fraction of activity remaining at saturating [I]:

$$\frac{v}{v_0} = 1 - \frac{\left(1 - \beta_{\text{app}}\right)[\mathrm{I}]}{\mathrm{IC}_{50,\text{app}} + [\mathrm{I}]} \tag{14}$$

More complex kinetics that does not fit hyperbolic inhibition or activation are also possible. These cases usually involve combinations of activation or inhibition with a second component resulting from two-substrate kinetics, e.g., sigmoidal, biphasic, or substrate inhibition kinetics. An example is activation followed by inhibition. The inhibition component occurs when two substrates in the active site displaces the inhibitor.

It would be desirable to determine all binding constants from the simple experiments, but values for K_{i}, α, and β cannot be obtained without performing more complex experiments. More importantly, the observation of partial inhibition or activation indicates that multisubstrate kinetic mechanisms are likely to be involved, and care should be taken in the interpretation of the data and the design of future experiments.

VII. SUMMARY

Most P450-catalyzed reactions show hyperbolic saturation kinetics and competitive inhibition kinetics. Therefore, binding constants can be obtained by inhibition of standard assays. Some P450-catalyzed reactions show atypical kinetics, including activation, autoactivation, partial inhibition, biphasic saturation kinetics, and substrate inhibition. Although atypical kinetics are for metabolism with any P450 enzyme, these phenomena occur most frequently for the CYP3A enzymes. In general, an observation of non-Michaelis-Menten kinetics makes it difficult to interpret results and makes in vitro–in vivo correlations difficult. In particular, the interactions between two substrates and an enzyme are dependent on both substrates, which can result in both false negatives and false positives when predicting drug interactions with inhibition studies.

REFERENCES

1. Ortiz de Montellano PR. Cytochrome P450: Structure, Mechanism, and Biochemistry. New York: Springer, 2005.
2. Grime K, Riley RJ. The impact of in vitro binding on in vitro - in vivo extrapolations, projections of metabolic clearance and clinical drug-drug interactions. Curr Drug Metab 2006; 7(3):251–264.
3. Brown HS, Ito K, Galetin A, et al. Prediction of in vivo drug-drug interactions from in vitro data: impact of incorporating parallel pathways of drug elimination and inhibitor absorption rate constant. Br J Clin Pharmacol 2005; 60(5):508–518.
4. Wienkers LC, Heath TG. Predicting in vivo drug interactions from in vitro drug discovery data. Nat Rev Drug Discov 2005; 4(10):825–833.
5. Obach RS, Walsky RL, Venkatakrishnan K, et al. The utility of in vitro cytochrome p450 inhibition data in the prediction of drug-drug interactions. J Pharmacol Exp Ther 2006; 316(1):336–348.
6. Shou M. Prediction of pharmacokinetics and drug-drug interactions from in vitro metabolism data. Curr Opin Drug Discov Dev 2005; 8(1):66–77.
7. King EL, Altman C. A schematic method of deriving the rate laws for enzyme-catalyzed reactions. J Chem Phys 1956; 60:1375–1378.
8. Gonzalez FJ, Korzekwa KR. Cytochromes P450 expression systems. Annu Rev Pharmacol Toxicol 1995; 35:369–390.
9. Eberhart DC, Parkinson A. Cytochrome P450 IIIA1 (P450p) requires cytochrome b_5 and phospholipid with unsaturated fatty acids. Arch Biochem Biophys 1991; 291 (2):231–240.
10. Hanioka N, Korzekwa K, Gonzalez FJ. Sequence requirements for cytochromes P450IIA1 and P450IIA2 catalytic activity: evidence for both specific and non-specific substrate binding interactions through use of chimeric cDNAs and cDNA expression. Protein Eng 1990; 3(7):571–575.
11. Cleland WW. Statistical analysis of enzyme kinetic data. Methods Enzymol 1979; 63:103–138.
12. Akaike T. A new look at the stastistical model identification. IEEE Trans Automat Contr 1974; 19:716–723.
13. Yamaoka K, Nakagawa T, Uno T. Application of Akaike's information criterion (AIC) in the evaluation of linear pharmacokinetic equations. J Pharmacokinet Biopharm 1978; 6(2):165–175.
14. Peterson JA, Ebel RE, O'Keeffe DH, et al. Temperature dependence of cytochrome P-450 reduction. A model for NADPH-cytochrome P-450 reductase:cytochrome P-450 interaction. J Biol Chem 1976; 251(13):4010–4016.
15. Grogan J, Shou M, Zhou D, et al. Use of aromatase (CYP19) metabolite ratios to characterize electron transfer from NADPH-cytochrome P450 reductase. Biochemistry 1993; 32(45):12007–12012.
16. Hanioka N, Gonzalez FJ, Lindberg NA, et al. Site-directed mutagenesis of cytochrome P450s CYP2A1 and CYP2A2: influence of the distal helix on the kinetics of testosterone hydroxylation. Biochemistry 1992; 31(13):3364–3370.
17. Gorsky LD, Koop DR, Coon MJ. On the stoichiometry of the oxidase and monooxygenase reactions catalyzed by liver microsomal cytochrome P-450. Products of oxygen reduction. J Biol Chem 1984; 259(11):6812–6817.

18. Korzekwa KR, Swinney DC, Trager WF. Isotopically labeled chlorobenzenes as probes for the mechanism of cytochrome P-450 catalyzed aromatic hydroxylation. Biochemistry 1989; 28(23):9019–9027.
19. Jones JP, Korzekwa KR, Rettie AE, et al. Isotopically sensitive branching and its effect on the observed intramolecular isotope effects in cytochrome-p-450 catalyzed-reactions—a new method for the estimation of intrinsic isotope effects. J Am Chem Soc 1986; 108(22):7074–7078.
20. Voznesensky AI, Schenkman JB. The cytochrome P450 2B4-NADPH cytochrome P450 reductase electron transfer complex is not formed by charge-pairing. J Biol Chem 1992; 267(21):14669–14676.
21. Segel IH. Enzyme Kinetics. New York: John Wiley and Sons, 1975.
22. Korzekwa KR, Jones JP. Predicting the cytochrome P450 mediated metabolism of xenobiotics. Pharmacogenetics 1993; 3(1):1–18.
23. Ueng YF, Kuwabara T, Chun YJ, et al. Cooperativity in oxidations catalyzed by cytochrome P450 3A4. Biochemistry 1997; 36(2):370–381.
24. Shou M, Grogan J, Mancewicz JA, et al. Activation of CYP3A4: evidence for the simultaneous binding of two substrates in a cytochrome P450 active site. Biochemistry 1994; 33(21):6450–6455.
25. Korzekwa KR, Krishnamachary N, Shou M, et al. Evaluation of atypical cytochrome P450 kinetics with two-substrate models: evidence that multiple substrates can simultaneously bind to cytochrome P450 active sites. Biochemistry 1998; 37(12):4137–4147.
26. Banci L, Betini I, Marconi S, et al. Cytochrome P450 and aromatic bases: a 1H NMR study. J Am Chem Soc 1994; 116(11):4866–4873.
27. Tang W, Stearns RA, Kwei GY, et al. Interaction of diclofenac and quinidine in monkeys: stimulation of diclofenac metabolism. J Pharmacol Exp Ther 1999; 291 (3):1068–1074.
28. Hutzler JM, Frye RF, Korzekwa KR, et al. Minimal in vivo activation of CYP2C9-mediated flurbiprofen metabolism by dapsone. Eur J Pharm Sci 2001; 14(1):47–52.
29. Egnell AC, Houston B, Boyer S. In vivo CYP3A4 heteroactivation is a possible mechanism for the drug interaction between felbamate and carbamazepine. J Pharmacol Exp Ther 2003; 305(3):1251–1262.
30. Johnson EF, Schwab GE, Vickery LE. Positive effectors of the binding of an active site-directed amino steroid to rabbit cytochrome P-450 3c. J Biol Chem 1988; 263 (33):17672–17677.
31. Schwab GE, Raucy JL, Johnson EF. Modulation of rabbit and human hepatic cytochrome P-450-catalyzed steroid hydroxylations by alpha-naphthoflavone. Mol Pharmacol 1988; 33(5):493–499.
32. Tracy TS, Marra C, Wrighton SA, et al. Involvement of multiple cytochrome P450 isoforms in naproxen O-demethylation. Eur J Clin Pharmacol 1997; 52(4):293–298.

3

Human Cytochromes P450 and Their Role in Metabolism-Based Drug-Drug Interactions

Stephen E. Clarke
GlaxoSmithKline Pharmaceuticals, Ware, U.K.
Barry C. Jones
Pfizer Global Research & Development, Kent, U.K.

I. INTRODUCTION

Cytochrome P450 (P450) binding is now widely recognized as a major focus for drug-drug interactions in the pharmaceutical industry. P450 metabolism-based drug-drug interactions, in vitro and in vivo, are now routinely part of the product labeling and advertising copy, often in incomprehensible detail. Although this focus has led, on more than one occasion, to undue emphasis on clinically insignificant effects, there does exist in many circumstances a significant risk to patients arising from interactions with the P450 enzyme system. What is more, these interactions can be reasonably well predicted from in vitro data and extrapolated from drug to drug, thanks to the large body of literature information. From the authors' survey of the available data on the elimination pathways for 438 drugs marketed in the United States and Europe, the overall importance of P450-mediated clearance can be determined. The elimination of unchanged drug via urine (the most commonly defined), bile, expired air, or feces represented, on average, approximately 25% of the total elimination of dose for these compounds. P450-mediated metabolism represented 55%, with all other metabolic processes making up the remaining 20%. Thus, this focus (or perhaps obsessive compulsion) on studying P450 is justified.

II. CYTOCHROME P450 SUPERFAMILY

P450s are ubiquitous throughout nature: they are present in bacteria, plants, and mammals, and there are hundreds of known enzymes that can show tissue- and species-specific expression. This diversity of enzymes has necessitated a systematic nomenclature system (1). The root name given to all cytochrome P450 enzymes is CYP (or *CYP* for the gene). Enzymes showing greater than 40% amino acid sequence homology are placed in the same family, designated by an Arabic numeral. When two or more subfamilies are known to exist within the family, then enzymes with greater than 60% homology are placed in the same subfamily, designated with a letter. Finally this letter is followed by an Arabic number, representing the individual enzyme, which is assigned on an incremental basis, i.e., first come, first served. As of October 2006 there were 6422 P450 enzymes, organized into 708 families, which were identified in species from alfalfa to the zebra finch, although only 2279 in 99 families in animals (2). Only the 50 P450 enzymes described in man (Table 1) are likely to be of any clinical relevance, and even then only the P450s in families 1, 2, and 3 appear to

Table 1 Human Cytochrome P450 Superfamily

Family	Subfamilies	Number of enzymes	Best-described substrates
1	A, B	3	Xenobiotics
2	A, B, C, D, E, F, J, R, S	15	Xenobiotics
3	A	4	Xenobiotics
4	A, B, F, X, Z	9	Fatty acids/leukotrienes
5	A	1	Thromboxane
7	A, B	2	Cholesterol
8	A, B	2	Prostacyclin
11	A, B	3	Steroids
17	—	1	Steroids
19	—	1	Estrogen
21	A	1	Steroids
24	—	1	Vitamin D/steroids
26	A, B	2	Retinoic acid
27	A, B	2	Vitamin D/steroids
39	A	1	Cholesterol
46	—	1	Cholesterol
51	—	1	Steroids

be responsible for the metabolism of drugs and therefore are potential sites for drug interactions. The P450 enzymes from the other families are generally involved in endogenous processes, particularly hormone biosynthesis. An interaction with these enzymes could have significant toxicological effects, but a pharmacokinetic drug-drug interaction between two exogenous pharmacological agents is unlikely. Even of the 22 P450 enzymes in families 1, 2, and 3, perhaps only five or six are quantitatively relevant in the metabolism of pharmaceuticals.

III. TISSUE DISTRIBUTION AND ABUNDANCE

P450 enzymes can be found throughout the body, particularly at interfaces, such as the intestine, nasal epithelia, and skin. The liver and the intestinal epithelia are the predominant sites for P450-mediated drug elimination and are also the sites worth considering in most detail with respect to drug-drug interactions. Although P450 enzymes have been well characterized in many other tissues, it is unlikely that these play a significant role in the overall elimination of drugs. These tissues and their P450s may play a role, for example, in tissue-specific production of reactive species and thereby toxicity, but they are unlikely to represent a concern for pharmacokinetic drug interactions.

The complement of intestinal P450s appears to be more restricted than that in the liver. Despite this restriction, many different P450 enzymes have been detected (by activity or mRNA) in the intestine from various species, including man. The available data would suggest that there are measurable levels of at least CYP1Al, CYP2C9, CYP2D6, CYP2E1, and representatives of subfamilies CYP2J and CYP4B present in the intestinal epithelia (3–9); however, overwhelmingly, the most significant P450 enzymes in human intestine are from the CYP3A family (10–13). The other P450 enzymes are clearly present in low quantities and/or are not capable of contributing to the pharmacokinetic profile (e.g., limiting oral bioavailability) via intestinal metabolism. That CYP3A4, in particular, is the P450 enzyme of significant concern for drug-drug interactions in the intestine is supported by a number of pharmacokinetic studies.

Although it is not a trivial task to clearly demonstrate the role of a human P450 enzyme in intestinal presystemic elimination, this has been shown for several drugs metabolized by CYP3A4, e.g., cyclosporin (14,15), tacrolimus (16,17), sirolimus (18), midazolam (19), saquinavir (20), felodipine (21,22), and nefazadone (23). Interestingly, grapefruit juice has been shown to have a significant interaction with a number of these drugs (24). Grapefruit juice's effect is believed to be limited to the intestine and to be specifically CYP3A4 mediated (22,25,26). Psoralen derivatives and related compounds are thought to be involved as the active ingredients in grapefruit juice interactions (27–32). Interestingly, these components are very potent inhibitors (submicromolar inhibitory constants) of CYP1A2, CYP2C9, CYP2C19, and CYP2D6, in addition to any effects they have on CYP3A4 (H. Oldham, personal communication, 1998). Yet the reports of significant interactions in vivo appear to be limited to

CYP3A4 substrates. This supports the contention that the effect is solely on the intestine, not the liver, and that CYP3A4 is the only P450 that plays a significant role in the intestinal metabolism of drugs. Therefore, the intestine is an important site for P450 drug interactions, but only those mediated via CYP3A4.

In the human liver, the relative content of the major P450 enzymes has been determined in several studies, and a general consensus has emerged. On average, CYP3A4 is quantitatively the most important, with CYP2C8, CYP2C9, CYP2A6, CYP2E1, and CYP1A2 present in somewhat lower quantities; CYP2C19 and CYP2D6 are of relatively minor quantitative importance (Fig. 1A) (33). However, a very different picture emerges when evaluating the extent to which P450 enzymes are responsible for drug elimination processes (Fig. 1B). CYP3A4 is responsible for approximately 50% of the P450-mediated metabolism of marketed pharmaceuticals, and CYP2D6 has a disproportionate share ($\sim$25%) in comparison with the amount of enzyme present in the liver. CYP2C9 and CYP1A2 make up a progressively less significant proportion of the whole. All the other P450 enzymes make somewhat minor contributions.

It is notable that CYP3A4 appears to be more frequently cited for newly developed drugs than CYP2D6. This increase in the incidence of CYP3A4 substrates follows the increase in lipophilicity, probably a consequence of the paradigm shift in the pharmaceutical industry's drug discovery process, which is now driven by in vitro pharmacological screening. It is easy to understand why such a large number of CYP2D6 substrates have been identified. Because of the polymorphic nature of CYP2D6, substrates of this enzyme were among the first and easiest to be defined, even before the molecular basis of the polymorphism was known. Lately, because of the current impracticality of personalizing doses, CYP2D6 substrates are being engineered out or deselected during the drug discovery and optimization phase wherever this might provide a competitive advantage. For other P450 enzymes, such as CYP2C8, the tools to investigate and identify interactions at the enzyme (specific substrates and inhibitors suitable for in vitro and in vivo use) have been available only relatively recently, and the importance of these enzymes may be underestimated. These considerations and the data for those drugs whose mechanisms of elimination have yet to be fully elucidated might be expected to alter this overall distribution somewhat; however, it is unlikely that the current picture will change for at least the medium-term future. Thus, from the pharmaceutical industry's perspective, CYP1A2, CYP2C9, CYP2D6, and CYP3A4 address the overwhelming majority of the P450 issues and a little over 50% of the total target for pharmacokinetic drug-drug interaction studies.

IV. PHARMACOKINETIC CONSIDERATIONS

The pharmacokinetics of drug-drug interactions has been described in detail in another chapter (see chap. 1); however, a number of points are worth briefly reiterating in the context of P450. For an inhibition interaction, the affected drug

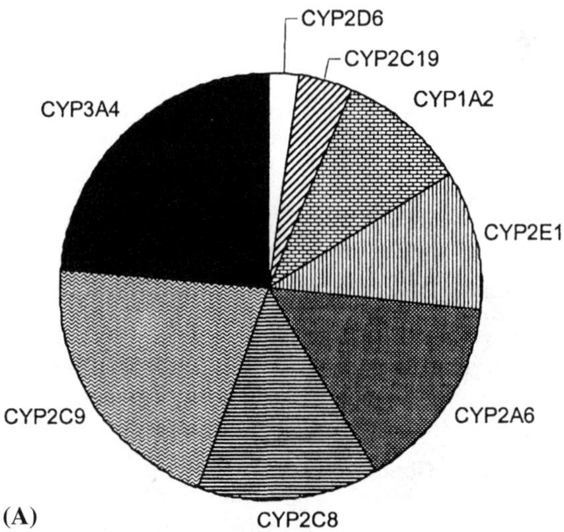

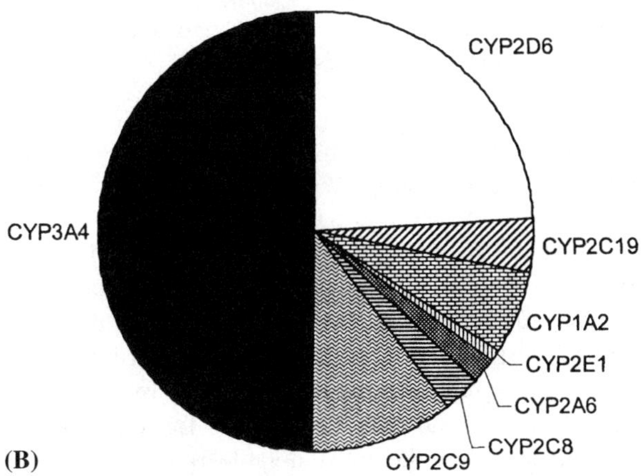

Figure 1 (**A**) Relative hepatic abundance of the major cytochromes P450 in man. (**B**) Relative significance of the major hepatic cytochromes P450 in the P450-mediated clearance of marketed drugs. This figure represents the author's survey of 438 drugs marketed in the United States and/or Europe. Rather than the number of drugs, the values represent the average proportion of drug clearance that each P450 enzyme is responsible for. *Source*: Part A adapted from Ref. 33.

clearly must have an appreciable proportion of its clearance (f_m, fraction metabolized by inhibited P450) via the P450 enzyme being inhibited, i.e., $f_m > 0.3$. For example, if the P450-mediated metabolism was only 20% of the total clearance of a compound, a fivefold reduction in its activity would have a limited

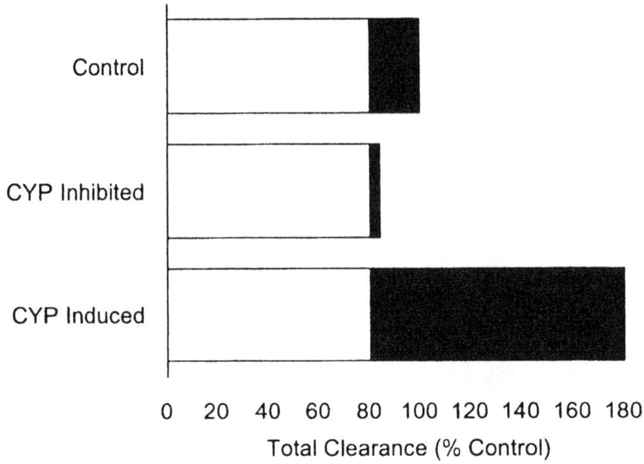

Figure 2 Influence of f_m (fraction metabolized by inhibited P450) on drug-drug interactions. The control represents a model drug for which cytochrome P450 (*dark bar*) is responsible for 20% of the clearance, with the remaining 80% being non-P450 mediated (*white bar*). "CYP inhibited" and "CYP induced" illustrate the effect on total clearance of a fivefold reduction or increase in the P450 activity, respectively.

effect overall (Fig. 2). Therefore, for inhibition interactions the relative importance of the individual P450 enzymes is simply described by Figure 1B. For induction interactions, the degree of effect is less sensitive to the f_m, and significant pharmacokinetic changes can be seen even if the induced P450 is normally a relatively minor contributor to overall clearance. Using the same example as for inhibition, a fivefold increase in the P450 activity has a significant effect on total clearance, despite the normally minor contribution to clearance (Fig. 2). In such cases the degree of sensitivity is defined by the extent of induction as well as the f_m. There is evidence of induction for a number of P450 enzymes in man, although some of the most notable inductive effects involve CYP3A4.

It is often thought that drugs with an appreciable f_m by CYP2D6, which have dangerous interaction potential, have been generally identified (because of the polymorphic nature of this enzyme) and withdrawn. This has been the case with perhexiline (34,35) and phenformin (36). But it has long been recognized that CYP2D6 poor metabolizers (PMs) and extensive metabolizers (EMs) coadministered with potent CYP2D6 inhibitors are at particular risk of adverse drug reactions (37). There are still a large number of CYP2D6 substrates marketed, and serious if not acutely fatal interactions are possible, despite the existence of a "canary" population that will exhibit very different pharmacokinetics to warn of potential consequences of drug interactions.

The clearance of the target drug can be the most significant arbiter of the severity of interaction for systemic interactions. Using the venous equilibrium

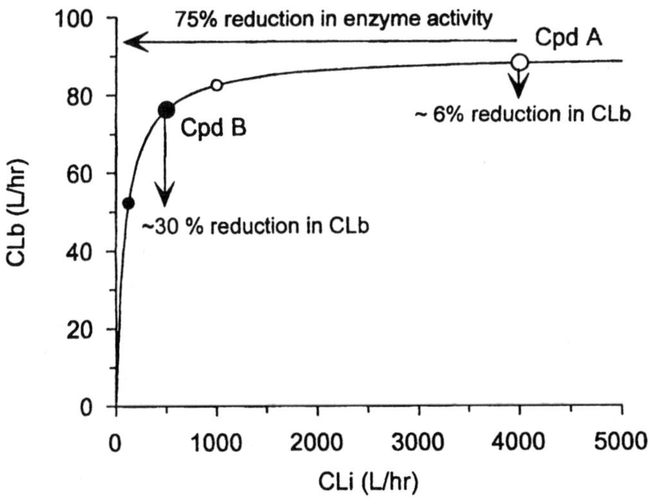

Figure 3 Influence of clearance on systemic drug-drug interactions. For model compound A (*open circles*) and compound B (*closed circles*), the effect on blood clearance of a 75% reduction in intrinsic enzyme activity (CL) is illustrated. The line represents the relationship between CL_i and CL_b that is described by the venous equilibrium, or "well-stirred," model of hepatic extraction.

model of hepatic elimination, a very highly intrinsically cleared compound (e.g., compound A in Fig. 3) would be relatively insensitive to inhibition interactions. In this case, a 75% reduction in enzyme activity would result in virtually no change (~6%) in blood clearance. For a significantly less readily metabolized substrate (e.g., compound B in Fig. 3), such a reduction in enzyme activity would have a significant effect (~30%) on blood clearance. For low-clearance drugs (assuming f_m is 1), the reduction in clearance exactly reflects the reduction in enzyme activity.

Although systemically low-clearance drugs would be expected to be the most sensitive to drug-drug interactions, such compounds frequently have high oral bioavailability. As such, a coadministered inhibitor will cause little alteration of the C_{max} on a single oral dose but would need to be able to maintain inhibitory levels throughout the dosing interval. At steady state, a large inhibitory effect could be mediated, but the maximum initial "jump" in blood levels of the target drug would be twofold, with each subsequent dose adding at most another unit until the steady state was reached. Such a relatively gentle rate of elevation of blood levels might enable, in some circumstances, known tolerated adverse effects to be identified before serious toxicity is encountered. Many CYP2C9 substrates are high-bioavailability, low-clearance drugs, e.g., glyburide, tolbutamide, phenytoin, and warfarin, as are some CYP1A2 substrates, e.g., caffeine and theophylline. There are also higher-clearance CYP1A2 substrates, e.g., ropinirole and tacrine, although most published interaction studies have involved caffeine or theophylline. CYP2D6 and particularly CYP3A4 substrates exhibit a wide range of pharmacokinetic properties, in the latter case involving some of the highest-clearance drugs.

Blood-flow-limited drugs are not only theoretically systemic drug-interaction resistant but also rarely make good drugs (because of a low oral bioavailability and a high likelihood of a short half-life), and there are few drugs marketed, except prodrugs. However, on oral dosing, a putative inhibitor of the metabolism of such drugs need only be effective during the first-pass phase to cause a very significant effect. High levels of inhibitory blockade can be achieved because of the concentrations that can be achieved in the gut and the liver during absorption. Since the target drug has a low bioavailability, changes in blood C_{max} can be quite sudden and of an order of magnitude or more. Currently, the greatest concern for low-bioavailability, high-clearance drugs is with certain CYP3A4 substrates. The best-known example is the interaction between potent CYP3A4 inhibitors and terfenadine, where plasma levels of terfenadine have become greatly elevated (38,39) and can result in fatal effects because of the cardiotoxicity of terfenadine.

V. INCIDENCE OF INHIBITION

P450 inhibitors can be readily identified by in vitro methods (see chaps. 2 and 7), and in the authors' laboratories approximately 400 marketed drugs have been identified. For comparison, the probit plots showing the incidence versus potency of these drugs and approximately 2000 typical pharmaceutical company compounds (ca. 1998) are given in Figure 4.

For the marketed drugs, only 5% had an IC_{50} of less than 10 μM against CYP1A2, and this incidence was increased to approximately 10% for CYP2C9, CYP2C19, and CYP3A4. Many more drugs had a significant inhibitory effect on CYP2D6, with 20% of marketed drugs having an in vitro IC_{50} of less than 10 μM. To some degree these results reflect the relative importance of the P450 enzymes in drug clearance (Fig. 1B); however, the results for CYP3A4 are somewhat at odds with this. Although there is much concern about CYP3A4-mediated drug interactions, not many marketed drugs are potent inhibitors of this enzyme. Certainly the majority of research in this area has generally focused on a limited set of HIV protease inhibitors, azole antifungals, and a few macrolide antibiotics. CYP3A4 often has the role of a high-capacity, low-affinity drug-metabolizing enzyme. Equally high-affinity compounds (and therefore potent inhibitors) may have poor pharmacokinetic properties (very high V_{max}/K_m, therefore high CL_i) that limit their application as pharmaceutical agents, and hence the relatively low incidence of CYP3A4 inhibitors in the marketed drugs.

A more interesting comparison is that of marketed drugs and pharmaceutical company compounds. There is a particularly dramatic difference in the incidence of CYP3A4 inhibition (Fig. 4E). Typical pharmaceutical company compounds are very much more inhibitory to CYP3A4 than are marketed drugs.

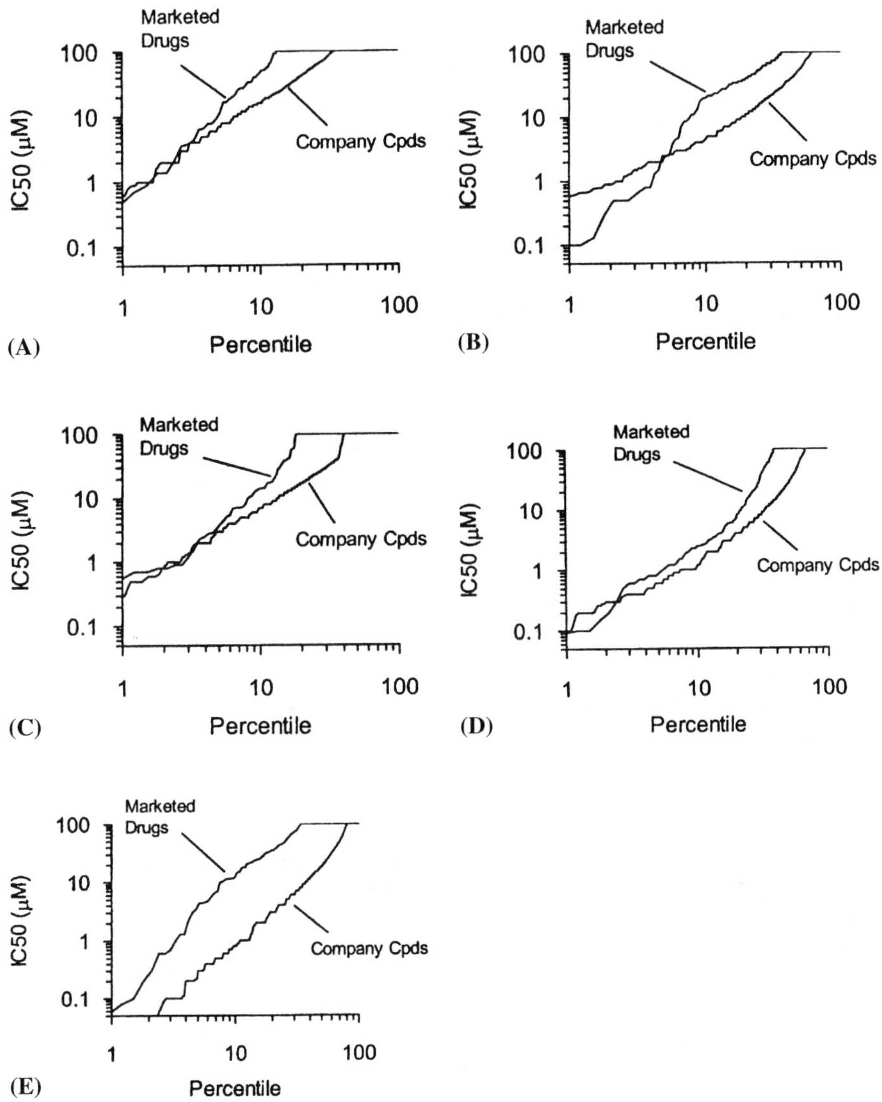

Figure 4 Incidence of P450 inhibition. Probit plots generated from in vitro P450-inhibition data in the authors' laboratories using heterologously expressed P450s in microsomal membranes. The plots represent data from approximately 400 marketed drugs and 2000 pharmaceutical company compounds synthesized in 1998. (**A**) CYP1A2, (**B**) CYP2C9, (**C**) CYP2C19, (**D**) CYP2D6, and (**E**) CYP3A4.

As in vitro high-throughput screening supports drug discovery activity more and more, DMSO solubility has become the only limitation to testing. Thus, with high lipophilicity no longer a barrier to testing and the trend to increasing molecular weight, as medicinal chemists ''build'' additional functionality and selectivity onto their molecular templates, a greater proportion of compounds fulfill the structural requirements for CYP3A4 substrates and inhibitors. This observation is similar to what has been described in the context of permeability and absorption and is part of the basis of the ''Lipinski rule of five'' (40).

The differences between marketed drugs and pharmaceutical company compounds are less marked for the other major P450 enzymes. For CYP1A2, there are few changes in the incidence of very potent inhibitors, as might be expected. Any increase in lipophilicity, which should improve the affinity of a compound for any P450 enzyme, would be countered by the increased molecular weight, which would make a compound less suitable for the CYP1A2 active site. In fact, CYP1A2, CYP2C9, CYP2C19, and CYP2D6 show broadly similar patterns to one another. There is no increase in the incidence of very potent inhibitors of these P450s in the contemporary company compounds compared with currently marketed drugs. Clearly the specific QSAR attributes that these P450 enzymes exhibit are being no more consistently met now than over the last 20 to 30 years. However, there are now many more ''midrange'' inhibitors and many less ''clean'' compounds than have been seen previously, primarily because of the general increase in lipophilicity. It is noteworthy that the more recently developed selective serotonin reuptake inhibitors (SSRIs) and HIV protease inhibitors are less like the majority of other marketed drugs and have a particularly high incidence of interactions with P450. Overall these data would suggest that, unchecked, CYP3A4 inhibition is likely to be a significant drug-drug interaction challenge facing the pharmaceutical industry in coming years.

Overall, in the authors' opinion, interactions with CYP3A4 are of the most concern, followed by CYP2C9, CYP2D6, and CYP1A2 in that order. However, the interaction profile of the next clinically or commercially important drug will always be of the most immediate significance, even if it concerns an otherwise relatively insignificant P450 enzyme.

VI. CYP1A2

A. Selectivity

Initial studies on the CYP1A family characterized the substrates as being lipophilic planar polyaromatic/heteroaromatic molecules, with a small depth and a large area/depth ratio. Later studies have suggested that caffeine interacts with the CYP1A2 via three hydrogen bonds, which orient the molecule so that it can undergo N-3-demethylation. Protein homology modeling suggests that the active sites of the CYP1A enzymes are composed of several aromatic residues, which form a rectangular slot and restrict the size and shape of the cavity, so only planar structures are able to occupy the binding site. This is in keeping with the initial observation and could explain the preference of CYP1A enzymes for hydrophobic, planar aromatic species that are able to partake in $\pi-\pi$ interactions with these aromatic residues. In addition to the aromatic residues, there are several residues able to form hydrogen bonds with substrate molecules. Such a model is able to rationalize that caffeine is N-demethylated at the 1, 3, and 7 positions by CYP1A2, of which the N-3-demethylation is the major pathway. Hence, it appears that binding to the active site of CYP1A2 requires certain molecular dimensions and hydrophobicity, together with defined hydrogen bonding and $\pi-\pi$ interactions.

The domination of the $\pi-\pi$ interactions is also evident in the inhibitor selectivity of the enzyme. The quinolone antibacterial enoxacin is an inhibitor that directly coordinates via the 4′-nitrogen atom on the piperazine function to the heme iron. In addition, there are aromatic regions and hydrogen bonding functions within the molecule that could be important in forming interactions with residues in the enzyme active site. Indeed, a comparison of a series of quinolone antibiotics has indicated that the keto group, the carboxylate group, and the core nitrogen at position 1 are able to form a similar pattern of hydrogen bonds with the active site, as has been suggested for caffeine.

Unlike some of the other P450s, CYP1A2 does not have a clear preference for acidic or basic molecules. It is able to metabolize basic compounds such as imipramine, but is inhibited by acidic compounds such as enoxacin. It is perhaps not surprising, then, that octanol/buffer partition coefficients or overall lipophilicity is not reflective per se of the interaction between CYPlA2 and its substrates or inhibitors.

B. Induction

Though CYP1A2 appears to be nonpolymorphic in man (41), it is inducible by environmental factors, such as cigarette smoking (42), which leads to an increased variability of this enzyme. In terms of induction by pharmaceutical agents, probably the most significant example is omeprazole. Omeprazole has been shown to be a CYP1A2 inducer in human hepatocytes (43). In vivo at higher omeprazole doses (40 and 120 mg for 7 days) there was a significant increase in caffeine metabolism, as shown by urinary metabolic ratios, the caffeine breath test, and caffeine clearance (44). However, at a low dose of omeprazole (20 mg/day for 7 days), there was no effect on caffeine metabolic ratios (45) or on phenacetin-mediated CYP1A2 metabolism (46), suggesting that omeprazole is a dose-dependent inducer of CYP1A2 in man.

C. Inhibition

Furafylline, a structural analogue of theophylline, was produced as a long-acting substitute for theophylline. Early clinical studies showed that the compound produced marked inhibition of caffeine metabolism. Further in vitro studies showed that furafylline is a selective mechanism-based inhibitor of CYP1A2 (47,48). Detailed mechanistic studies have indicated that metabolic processing of the C-8 methyl group is involved in the inactivation (48).

The interaction between the quinolone antibacterials and CYPlA2 has been studied in some depth for enoxacin and pefloxacin. Both compounds have been shown to inhibit CYPlA2-mediated metabolism of caffeine in vitro (49). This in vitro inhibition translated into a twofold decrease in caffeine clearance by pefloxacin and a sixfold decrease in clearance by enoxacin (50). Because pefloxacin undergoes N-demethylation to norfloxacin (51) and norfloxacin is much more potent as an inhibitor than pefloxacin (50), the observed in vivo interaction seen for pefloxacin may, in part, be due to norfloxacin. Many other quinolone antibacterial agents have been investigated for their interaction with theophylline, and ciprofloxacin has also been shown to have notable inhibitory effects (52).

There have been a number of investigations into the ability of the SSRIs to inhibit CYP1A2 (53–55). In general these studies agree that fluvoxamine is the most potent CYP1A2 inhibitor in this class, with $K_i \sim 0.2$ μM. Other members of the class, such as fluoxetine, paroxetine, and sertraline, have been shown to be at least tenfold less potent, with nefazodone and venlaflaxine showing low inhibitory potential against CYP1A2. The potent inhibition of caffeine metabolism by fluvoxamine results in an approximate fivefold decrease in caffeine clearance and sixfold increase in half-life (56).

D. Substrates

CYP1A2 metabolizes several drug substrates, including phenacetin, tacrine, ropinirole, riluzole, theophylline, and caffeine. Caffeine, although not used therapeutically, is, given the worldwide consumption of tea, coffee, and other caffeine-containing beverages, of significant interest.

The relative safety of caffeine has lead to its widespread use as an in vivo probe for CYPlA2 activity in man. The primary route of caffeine metabolism is via N-demethylation to paraxanthine, theophylline, and theobromine. The major route of caffeine clearance in man is to paraxanthine (57). The N-3-demethylation of caffeine to paraxanthine has been shown to be mediated by CYP1A2 (58). However, paraxanthine is further metabolized to a number of different products, and as a consequence urinary metabolic ratios are often used to describe an individual CYP1A2 phenotype.

Such approaches have been used successfully to demonstrate the induction of CYP1A2 by smoking (42). In addition, this study showed that oral contraceptives produce a small but significant inhibition of CYP1A2. Urinary metabolic ratios have also been used to show that oral AUC of clozapine was correlated with caffeine N-3-demethylation (59), a finding supported by some recent in vitro data, which has shown that clozapine N-demethylation is mediated by CYP1A2 (60).

VII. CYP2C9

A. Selectivity

CYP2C9 drug substrates include phenytoin, tolbutamide, various nonsteroidal anti-inflammatory drugs (NSAIDs), and (*S*)-warfarin. In terms of physicochemistry, the majority of the CYP2C9 substrates are acidic or contain areas of hydrogen bonding potential. Therefore, it has been proposed that these groups are important in binding to the active site of CYP2C9. There are a number of substrate template models for CYP2C9, which typically produces template models where the hydrogen bonding groups are positioned at a distance of approximately 8 Å and at an angle of 82° from the site of oxidation (61).

A homology model based on CYPl02 has suggested that there may be two serine residues within the active site that are key substrate residues. In addition, there is the suggestion that $\pi-\pi$ stacking interactions also occur between some of the substrates and the active site (62).

B. Polymorphism

There are three allelic variants of CYP2C9 that show significantly altered catalytic properties. These variants are termed CYP2C9*1 (wild type), CYP2C9*2 (Arg to Cys at position 144), and CYP2C9*3 (Ile to Leu at position 359). In general, CYP2C9*2 and CYP2C9*3 show reduced rates of metabolism toward substrates, relative to CYP2C9*1 (63,64).

Warfarin perhaps best exemplifies the impact of this reduced rate of metabolism. Warfarin is administered as a racemate, with different P450 enzymes being involved in the metabolism of the different enantiomers. (*R*)-Warfarin is metabolized by various P450s, including CYP1A2, CYP2C19, and CYP3A4 (65–67). (*S*)-Warfarin, however, is metabolized predominantly by CYP2C9 (68). Patients who are homozygous for CYP2C9*1 typically receive doses of between 4 and 8 mg of warfarin per day and have plasma (*S*)-warfarin/(*R*)-warfarin ratios of 0.5. Patients with the CYP2C9*3 allele are more sensitive to warfarin effects (69), and an individual who was homozygous for CYP2C9*3 could not receive more than 0.5 mg/day and even at this dose had a plasma (*S*)-warfarin/(*R*)-warfarin ratio of 4 (70).

C. Inhibition

Sulfaphenazole is perhaps the most potent and selective inhibitor of CYP2C9 (71). The mode of inhibition is via ligation to the heme iron of CYP2C9. Sulfaphenazole is a very commonly used in vitro diagnostic inhibitor for CYP2C9 activity, but it has been used in vivo for this purpose. The azole antifungal fluconazole also inhibits CYP2C9, and a series of studies has demonstrated the relationship between in vitro K_i values and the in vivo effect on warfarin clearance (72–74).

There are several other drug classes that have been shown to be inhibitors of CYP2C9. One example is the HMG-CoA reductase inhibitors, which inhibit CYP2C9 in vitro (75). These compounds are generally lipophilic carboxylic acids and hence might be expected to interact with the CYP2C9-active site. In fact, many of these compounds are relatively weak inhibitors of the enzyme, with the exception of fluvastatin. Racemic fluvastatin was a potent inhibitor of CYP2C9 activity ($K_i < 1$ μM), with the (+)-enantiomer being five times more potent than the (–)-enantiomer (75). This inhibition was also observed in vivo when diclofenac and fluvastatin were coadministered. In this case, there was an increase in diclofenac C_{max}, a reduction in oral clearance, and a decrease in the 4′-hydroxydiclofenac/diclofenac urinary ratio (76).

D. Substrates

There are a number of CYP2C9 substrates; however, the use of some of these agents is complicated by their narrow therapeutic margin, e.g., warfarin. This makes the enzyme an important target for drug-drug interactions, but also somewhat less straightforward to investigate clinically, at least if a significant interaction was to be pursued to steady state. Other than warfarin, there are a substantial number of studies using phenytoin and tolbutamide.

1. Phenytoin

Phenytoin is an anticonvulsant that has been shown to be preferentially hydroxylated in the pro-(*S*) ring by CYP2C9 (77), which accounts for approximately 80% of its clearance in man (78). The use of phenytoin is complicated by virtue of its nonlinear kinetics, long half-life, and narrow therapeutic margin. However, it has been used to confirm the in vitro finding that phenytoin and tolbutamide are metabolized by the same P450 enzyme (79).

2. Tolbutamide

Tolbutamide is metabolized by hydroxylation of the methyl tolyl group in man (80), forming hydroxytolbutamide. Hydroxytolbutamide is further metabolized to carboxytolbutamide (80,81). However, it is the initial hydroxylation that is rate limiting for elimination, accounting for approximately 85% of the clearance in man. This elimination pattern has enabled urinary ratios to be used to assess tolbutamide interactions, which gave a good correlation with total clearance on coadministration with sulfaphenazole (82).

VIII. CYP2C19

A. Selectivity

Substrates for this enzyme include (*R*)-mephobarbital, moclobemide, proguanil, diazepam, omeprazole, and imipramine, which do not show obvious structural or physicochemical similarities. Some inferences can be made when the differences between the CYP2C9 substrate phenytoin and the CYP2C19 substrate (*S*)-mephenytoin are considered. Phenytoin is para-hydroxylated on the *pro-(S)* phenyl ring by CYP2C9, and the (*S*)-enantiomer of mephenytoin is para-hydroxylated by CYP2C19. While (*S*)-mephenytoin is structurally similar to phenytoin, the *N*-methyl function in mephenytoin makes donation of a hydrogen bond impossible, which may be why mephenytoin is not a substrate for CYP2C9. CYP2C19 can bind compounds that are weakly basic like diazepam ($pK_a = 3.4$), strongly basic like imipramine ($pK_a = 9.5$), or acidic compounds such as (*R*)-warfarin ($pK_a = 5.0$). One possibility is that CYP2C19 binds substrates via hydrogen bonds, but in a combination of a hydrogen bond donor and acceptor mechanisms.

B. Polymorphism

The frequency of the CYP2C19 polymorphism shows marked interracial differences, with an occurrence of approximately 3% in Caucasians and between 18 and 23% in Orientals (83). CYP2C19 PMs lack any functional CYP2C19 activity (84). The mechanism of this polymorphism has been ascribed largely to two defects in the CYP2C19 gene: a G^{681}-to-A mutation in exon 5, resulting in an aberrant splice site, which accounts for between 75 and 85% of PMs in Caucasian and Japanese populations, and a G^{636}-to-A mutation in exon 4, which accounts for the remaining PMs in the Japanese population (85). Further alleles, particularly those accounting for Caucasian PMs, e.g., CYP2C19*6, and those requiring the subdivision of previously assigned alleles, e.g., CYP2C19*2a and CYP2C19*2b, have been identified (86,87).

C. Inhibition

There are relatively few clinically relevant inhibitors of CYP2C19, the most significant being the SSRIs. In an in vitro study citalopram appeared to be a weak inhibitor ($K_i > 50$ μM), with the remaining compounds all having K_i values of less than 10 μM (88). A corresponding study indicated that fluoxetine and fluvoxamine were able to inhibit CYP2C19 in vivo (89), although neither compound is selective, since they have marked effects on CYP2D6 and CYP1A2.

D. Substrates

The metabolic activity of CYP2C19 has most frequently been probed, both in vivo and in vitro, using (*S*)-mephenytoin hydroxylation or mephenytoin *S*/*R* ratios. However, other substrates for this enzyme, including diazepam and imipramine, have been identified that have the potential to be used as probes (90,91). However, the most widely used identified CYP2C19 substrate is omeprazole (92).

1. Mephenytoin

Racemic mephenytoin is stereoselectively metabolized in man, with the (*S*)-enantiomer being rapidly hydroxylated in the 4′-position by CYP2C19 and the (*R*)-enantiomer being slowly metabolized. The (*S*)-mephenytoin phenotype (genotypically conferred or by administration of an inhibitor) is determined following an oral dose by measuring the ratio of (*S*)-mephenytoin to (*R*)-mephenytoin in the 0- to 8-hour urine (93).

2. Imipramine

Imipramine is metabolized mainly by N-demethylation and 2-hydroxylation in man. The N-demethylation pathway has been shown, in vitro, to be mediated by CYP2C19 at low imipramine concentrations (91). In vivo the partial clearance of imipramine, via N-demethylation, was shown to be significantly reduced in PMs of (*S*)-mephenytoin (94). In addition, a much larger study showed that the *S*/*R* ratio for mephenytoin correlated with the N-demethylation of imipramine (95).

3. Omeprazole

Omeprazole has been shown, in vitro, to be metabolized to a number of products, one of which, the 5-hydroxy metabolite, appears to be formed at least in part by CYP2C19 (92). These in vitro metabolism studies correlate with in vivo studies that showed that the oral clearance of omeprazole and the formation of the 5-hydroxy metabolite in three ethnic groups were directly related to CYP2C19 phenotype status (96).

IX. CYP2D6

A. Selectivity

The overwhelming majority of CYP2D6 substrates contain a basic nitrogen atom ($pK_a > 8$), which is ionized at physiological pH. It is the ionic interaction between this protonated nitrogen atom and an aspartic acid residue that governs the binding. All the models of CYP2D6 show essentially the same characteristics, in which there is a 5 to 7 Å distance between this basic nitrogen atom and the site of metabolism. The relative strength of this ionic interaction means that the affinity for substrates can be high and that this P450 enzyme tends to have many examples of low K_m and low K_i interactions. Although most of the substrates for CYP2D6 are basic, there are still marked differences in binding affinity. Once the ionic interaction is formed, any difference in binding affinity could be attributed to other π–π or hydrophobic interactions. In addition, for very potent CYP2D6 inhibitors, such as ajmalicine, there is a hydrogen acceptor site, in addition to the ion pair and hydrophobic/lipophilic interaction, which increases the inhibitory potency.

B. Polymorphism

CYP2D6 was perhaps the first and best characterized of the polymorphic P450 enzymes. The PM phenotype is characterized clinically by a marked deficiency in the metabolism of certain compounds, which can result in drug toxicity or reduced efficacy. The prevalence of the PM phenotype shows marked ethnic differences, with a mean value of approximately 7% in Caucasian populations (97) but 1% or less in Orientals (98). There are many different CYP2D6 alleles identified, including some that result in an ultrarapid metaboliser phenotype (99), and the typically applied genotyping methodologies are 90% predictive of phenotype (100).

C. Inhibition

CYP2D6 is inhibited by very low concentrations of quinidine. Although not metabolized significantly by CYP2D6, quinidine conforms closely to the structural requirements of the enzyme (101), but based on template models, the quinoline nitrogen occupies the position most likely for oxidative attack. Although quinidine is one of the most potent inhibitors of CYP2D6, the most studied class of inhibitory drugs are the SSRIs.

Several studies have been carried out using different substrate probes to determine the inhibitory potency of various members of this class against CYP2D6 (102–105). The potential implications of CYP2D6 (and other P450 enzymes) inhibition by this class of drugs has been exhaustively reviewed (106–116) and is not considered further here.

Not all CYP2D6 inhibitors have a basic nitrogen atom. The HIV-1 protease inhibitor ritonavir has a weakly basic center but a relatively strong interaction with CYP2D6 (117). However, the molecule does have a number of hydrogen bonding groups, which, if there are complementary hydrogen bonding sites in the CYP2D6-active site, may explain the inhibitory potency.

D. Substrates

There is a wide choice of drugs that are substrates for CYP2D6, but sparteine, debrisoquine, desipramine, dextromethorphan, and metoprolol have been used most frequently, both in vitro and in vivo. One advantage for in vivo drug-drug interaction studies is that most of the substrates were identified in the clinic rather than by the use of a battery of in vitro methods.

1. Debrisoquine

It was the identification of a group of subjects unable to metabolize debrisoquine (118,119), resulting in a potentially life-threatening drop in blood pressure, which lead to the identification of the CYP2D6 polymorphism (120). Debrisoquine is metabolized specifically by CYP2D6 (121) to produce 4-hydroxydebrisoquine.

Following an oral dose, the metabolite is excreted in the urine along with unchanged drug, and it is this ratio that can determine the CYP2D6 phenotype or the extent of drug interaction. With compromised CYP2D6, debrisoquine is excreted largely unchanged, resulting in a high ratio.

2. Dextromethorphan

Dextromethorphan is well tolerated, with few clinically relevant side effects, and it is a readily accessible drug in a large number of countries, making it ideal for drug-drug interaction studies. The major route of metabolism, O-demethylation to dextrorphan, has been shown, both in vitro and in vivo, to be mediated by CYP2D6 (122). Dextromethorphan metabolic ratios have been used primarily to identify CYP2D6 PMs, where a metabolic ratio of greater than 0.3 would be indicative of the PM phenotype (123).

3. Metoprolol

Metoprolol is a β-blocker that has been proposed as a pharmacokinetic alternative to debrisoquine in countries where it is difficult to use debrisoquine. Metoprolol is metabolized to desmethylmetroprolol and α-hydroxymetoprolol by CYP2D6 (124). The α-hydroxymetoprolol metabolite has been shown to be bimodally distributed and to correlate with the debrisoquine oxidation phenotype (125). Again, metoprolol has been used primarily to distinguish between CYP2D6 EMs and PMs. However, in African populations, the metoprolol metabolic ratio failed to predict the PMs of debrisoquine (126). These studies would suggest that in some ethnic groups metoprolol may not be a suitable probe.

X. CYP3A4

A. Selectivity

CYP3A4 appears to metabolize lipophilic drugs in positions largely dictated by the ease of hydrogen abstraction in the case of carbon hydroxylation, or electron abstraction in the case of N-dealkylation reactions. There are many drugs that are predominantly eliminated by CYP3A4 and many others where CYP3A4 is a secondary mechanism. The binding of substrates to CYP3A4 seems to be due essentially to lipophilic forces. Generally such binding, if based solely on hydrophilic interactions, is relatively weak and without specific interactions, which allows motion of the substrate in the active site. Thus, a single substrate may be able to adopt more than one orientation in the active site, and there can be several products of the reaction. Moreover, there is considerable evidence for allosteric behavior, due possibly to the simultaneous binding of two or more substrate molecules to the CYP3A4 active site (127–131). Such binding can lead to atypical enzyme kinetics and inconsistent drug-drug interactions and is almost diagnostic of CYP3A4 involvement, although other P450 enzymes may, more rarely, be able to exhibit such properties (130,131). Alternatively, the CYP3A4-active site may undergo substrate-dependent conformational changes (132–134), or there may be an alteration in the pool of active enzyme (135). Whatever the case, it is not surprising that there is no useful template model for CYP3A4 substrates.

Protein homology models for CYP3A4 have been produced using the soluble bacterial enzymes CYP101 and CYP102. These models suggest the active site pocket to be large and open and made up predominantly of hydrophobic and some neutral residues, together with a small number of polar side chains. The large number of aromatic side residues allows for the possibility of $\pi-\pi$ interactions with aromatic substrates. In addition, the presence of polar residues suggests the possibility of hydrogen bonds between substrates and the active site.

B. Induction

CYP3A4 activity can vary considerably between individuals. CYP3A4 can be modulated by dietary factors and hormones as well as pharmaceutical agents, and significant genetic polymorphisms have been identified in the 5′ regulatory region (136), which may contribute to this variability. In addition to the upstream response elements, a human orphan nuclear receptor, termed the pregnane X receptor (PXR), has been shown to be involved in the inductive mechanism (137).

It is interesting that most of the pharmaceutical inducers of CYP3A4, in man, either accumulate significantly on multiple dosing, are given at doses of hundreds of milligrams, or both, e.g., phenobarbital, felbamate, rifampin, phenytoin, carbamezepine, and troglitazone. Therefore the total body burden or liver levels are likely to be high, suggesting that no marketed drugs are highly potent ligands for PXR. Since there are high-throughput screens (138) and a drive in the pharmaceutical industry for highly potent and selective compounds, if these deliver lower therapeutic doses for new drugs then new clinically relevant CYP3A4 inducers may become rare.

Meanwhile, the currently marketed CYP3A4 inducers can profoundly affect the pharmacokinetics of coadministered CYP3A4 substrates, e.g., rifampin on midazolam (139) or triazolam (140). Clearly, the most frequent outcome is a loss of efficacy, which is perhaps less serious than inhibition interactions, although the consequences of coadministering rifampin with the oral contraceptive pill can lead to contraceptive failure (141–143).

C. Inhibition

Ketoconazole is a potent, somewhat selective inhibitor of CYP3A4 and is often used in vitro and in vivo as a diagnostic inhibitor. The drug is basic, partially ionized at physiological pH, and highly lipophilic, and it is also a substrate for the enzyme, being metabolized in the imidazole ring, the site of its ligation to the heme (144). This high-energy interaction results in a high potency of enzyme inhibition, with K_i values typically substantially less than 1 μM. Not surprisingly, oral ketoconazole is contraindicated with many CYP3A4 substrates and can cause life-threatening drug-drug interactions (38). Other azole antifungals (e.g., itraconazole) also have CYP3A4 inhibitory effects through similar mechanisms, and the drug-drug interactions of these molecules have been extensively reviewed (145,146).

Mechanism-based inhibitors or suicide substrates seem to be particularly prevalent with CYP3A4. Such compounds are substrates for the enzyme, but metabolism is believed to form products that deactivate the enzyme. Several macrolide antibiotics, generally involving a tertiary amine function, are able to inhibit CYP3A4 in this manner (147,148). Erythromycin is one of the most widely used examples of this type of interaction, although there are other commonly prescribed agents that inactivate CYP3A4 (149–151), and a consideration of this phenomenon partially explains a number of interactions that are not readily explained by the conventional in vitro data (152).

Because of the large number of drug molecules metabolized by CYP3A4, potent inhibition, by whatever mechanism, can have a detrimental effect on a compound's marketability. This effect is exemplified by mibefradil, which was withdrawn from the market during its first year of sales because of its extensive CYP3A4 drug interactions (153–156).

D. Substrates

There is an enormous choice of CYP3A4 substrates with a wide variety of clinical indications and structural features. Some of these substrates are not ideal targets for investigations of drug-drug interactions, because of potential safety concerns upon inhibition, e.g., terfenadine, or efficacy issues upon induction, e.g., the oral contraceptive pill. Additionally, there are increasing concerns about the predictivity of one substrate to another because of the emerging understanding of the apparent allosteric behavior of CYP3A4. However, the major structural types of CYP3A4 substrates can perhaps be covered by large molecular weight molecules derived from natural products, e.g., the macrolides, the benzodiazepines, and the dihydropyridine calcium channel blockers.

1. Erythromycin

Although the rate of elimination of this CYP3A substrate can be determined from plasma pharmacokinetics, the erythromycin breath test (ERMBT) is less invasive (157). The ERMBT involves the intravenous administration of a trace amount of ^{14}C-*N*-methyl erythromycin. At specified time points, the subject breathes through a one-way valve, into a CO_2-trapping solution, and the ^{14}C-CO_2 is subsequently measured by liquid scintillation counting. This test shows fairly good correlations with trough cyclosporin concentrations (158) and clearly demonstrates the inductive effect of rifampin (157). However, there was a poor correlation between the ERMBT and the clearance of the CYP3A4 substrate alfentanil (159,160). The test is still somewhat invasive (intravenous administration) and does not assess presystemic effects; a further limitation is the need to administer radioactivity.

2. Midazolam

A dose of midazolam in man is eliminated renally (98%), with 1-hydroxymidazolam (the product of CYP3A metabolism) accounting for half of the urinary elimination (161). Midazolam clearance provides a good estimate of CYP3A activity, which has been found to correlate with the concentration of CYP3A immunoreactive protein in liver biopsies (162), cyclosporin clearance (163), and the ERMBT (161). Midazolam clearance has been increased in patients receiving phenytoin (164) and reduced in patients receiving erythromycin (165) or itraconazole (166), showing wide utility for drug-drug interaction studies. This suitability of midazolam has led to its being the most widely used CYP3A in vivo probe, and the large literature precedence enables it to be the benchmark interaction for assessing CYP3A inhibitors. It is suggested that inhibitors be classified as weak (<2-fold), moderate (2- to 5-fold), or potent (>5-fold) inhibitors on the basis of the change in midazolam oral AUC.

3. Nifedipine

Nifedipine was one of the first CYP3A4 substrates to be identified (167,168) and has been the subject of a large number of drug-drug interaction studies both in vitro and in vivo. Pharmacokinetic studies with nifedipine clearly identify inhibitors, such as itraconazole (169) and grapefruit juice (170), and inducers, such as the barbiturates (171) and rifampin (172).

XI. OTHER CYP ENZYMES

Several other P450 enzymes are involved in the metabolism of pharmaceuticals, although they are still regarded as minor enzymes.

CYP2B6 is proposed as a major contributor to bupropion clearance (173–175), efavirenz (176), and cyclophosphamide (177) metabolism and has been implicated in the partial metabolism of many other drugs. For example, CYP2B6 contributes to the 4-hydroxylation of propofol (178); however, other P450 enzymes can contribute (178,179), and the major pathway of propofol elimination is glucuronidation. It is unlikely that the fraction of propofol AUC defined by CYP2B6 activity would be greater than 0.2 and as a consequence would be the cause of a significant drug-drug interaction. Even bupropion, which is probably the best clinically described CYP2B6 probe, has limited drug-drug interactions. Ticlopidine is a submicromolar inhibitor of CYP2B6 (180,181) and does cause a measurable drug interaction in vivo (182); however, the scale of interaction in terms of bupropion AUC was small and relatively insignificant to those observed with some CYP3A4 substrate inhibitor pairs. Clearly, other metabolic pathways or mechanisms of clearance are also contributing to bupropion clearance in vivo. This pattern of in vitro identified substrates with limited in vivo consequences that can be ascribed to the enzyme is typical of reported CYP2B6 substrates. The same is true of many inhibitors, which if potent generally lack specificity for CYP2B6 (183) and limit their use as in vitro or in vivo tools for P450 drug-drug interactions.

Perhaps a better case for a previously neglected P450 enzyme being classified as a significant contributor to drug-drug interactions is CYP2C8. This enzyme has a growing list of structurally diverse substrates, including some major therapeutic agents such as the glitazones, repaglinide, paclitaxel, and cerivastatin, certainly enough to build substrate pharmacophores (184). Furthermore, the reported clinical drug-drug interactions result in more than a twofold increase in AUC, e.g., gemfibrozil on rosiglitazone (185), pioglitazone (186,187), and repaglinide (188). The interaction of gemfibrozil with cerivastatin (189) led to the withdrawal of this statin from the market. It has been clearly shown that both gemfibrozil and its glucuronide metabolite inhibit CYP2C8 (190); however, gemfibrozil has also been shown to inhibit the hepatic uptake transporter OATP1B1 as well as other transporters (191), perhaps more potently than CYP2C8. The largest CYP2C8 implicated drug-drug interactions (gemfibrozil on repaglinide and cerivastatin), probably involve a large contribution from the transporter-mediated effect, with the CYP2C8 element being somewhat more modest. The potential CYP2C8 drug-drug interaction liability may be better reflected by the two- to threefold effect of gemfibrozil on pioglitazone and rosiglitazone (186–188) or the less than twofold effects of trimethoprim on repaglinide (192) or rosiglitazone (193).

XII. CONCLUSIONS

There is clear evidence of the extensive involvement of the P450 enzyme system in the elimination of pharmaceutical agents and an enormous body of information demonstrating the modulation of activity, via inhibition or induction, with polypharmacy. This fully justifies the intensive research in this area and the pharmaceutical industry's focus on such drug-drug interactions. This focus is reinforced in this volume, in which P450 is either the major or the most significant subject of over half the chapters, and inhibition and induction, in vitro and in vivo, are further exemplified and discussed.

REFERENCES

1. Nelson DR, Kamataki T, Waxman DJ, et al. The P450 superfamily: update on new sequences, gene mapping, accession numbers, early trivial names of enzymes, and nomenclature. DNA Cell Biol 1993; 12:1–51.
2. David Nelson Cytochrome P450 Homepage. http://drnelson.utmem.edu/CytochromeP450.html

3. Yamamoto Y, Ishizuka M, Takada A, et al. Cloning, tissue distribution, and functional expression of two novel rabbit cytochrome P450 isozymes, CYP2D23 and CYP2D24. J Biochem 1998; 124:503–508.
4. Zhang QY, Raner G, Ding XX, et al. Characterization of the cytochrome P450 CYP2J4-expression in rat small intestine and role in retinoic acid biotransformation from retinal. Arch Biochem Biophys 1998; 353:257–264.
5. Hiroi T, Imaoka S, Chow T, et al. Tissue distributions of CYP2D1, 2D2, 2D3 and 2D4 mRNA in rats detected by RT-PCR. Biochim Biophys Acta 1998; 1380:305–312.
6. Zeldin DC, Foley J, Goldsworthy SM, et al. CYP2J subfamily cytochrome P450s in the gastrointestinal tract-expression, localization, and potential functional significance. Mol Pharmacol 1997; 51:931–943.
7. Zhang QY, Wikoff J, Dunbar D, et al. Regulation of cytochrome P4501A1 expression in rat small intestine. Drug Metab Dispos 1997; 25:21–26.
8. Prueksaritanont T, Gorham LM, Hochman JH, et al. Comparative studies of drug-metabolizing enzymes in dog, monkey, and human small intestines, and in Caco-2 cells. Drug Metab Dispos 1996; 24:634–642.
9. Kaminsky LS, Fasco MJ. Small intestinal cytochromes P450. Crit Rev Toxicol 1991; 21:407–422.
10. Watkins PB, Wrighton SA, Schuetz EG, et al. Identification of gluticoid-inducible cytochromes P-450 in intestinal mucosa of rats and man. J Clin Invest 1987; 90:1871–1878.
11. Kolars JC, Schmiedlin-Ren P, Schuetz JD, et al. Identification of rifampin-inducible P450IIIA4 (CYP3A4) in human small bowel enterocytes. J Clin Invest 1992; 90:1871–1878.
12. Kolars JC, Lown KS, Schmiedlinren P, et al. CYP3A gene-expression in human gut epithelium. Pharmacogenetics 1994; 4:247–259.
13. Lown KS, Kolars JC, Thummel KE, et al. Interpatient heterogeneity in expression of CYP3A4 and CYP3A5 in small bowel. Lack of prediction by the erythromycin breath test. Drug Metab Dispos 1994; 22:947–955.
14. Wu CY, Benet LZ, Hebert MF, et al. Differentiation of absorption, first-pass gut and hepatic metabolism in man: studies with cyclosporine. Clin Pharmacol Ther 1995; 58:492–497.
15. Benet LZ, Wu CY, Hebert MF, et al. Intestinal drug-metabolism and anti-transport processes-a potential paradigm shift in oral-drug delivery. J Cont Rel 1996; 39:139–143.
16. Hashimoto Y, Sasa H, Shimomura M, et al. Effects of intestinal and hepaticmetabolism on the bioavailability of tacrolimus in rats. Pharm Res 1998; 15:1609–1613.
17. Lampen A, Christians U, Guengerich FP, et al. Metabolism of the immunosuppressant tacrolimus in the small intestine: cytochrome P450, drug interactions, and interindividual variability. Drug Metab Dispos 1995; 23:1315–1324.
18. Lampen A, Zhang Y, Hackbarth I, et al. Metabolism and transport of the macrolide immunosuppressant sirolimus in the small intestine. J Pharmacol Exp Ther 1998; 285:1104–1112.
19. Paine MF, Shen DD, Kunze KL, et al. First-pass metabolism of midazolam by the human intestine. Clin Pharmacol Ther 1996; 60:14–24.

20. Wacher VJ, Silverman JA, Zhang Y, et al. Role of P-glycoprotein and cytochrome P450 3A in limiting oral absorption of peptides and peptidomimetics. J Pharm Sci 1998; 87:1322–1330.
21. Wang SX, Sutfm TA, Baarnhielm C, et al. Contribution of the intestine to the first-pass metabolism of felodipine in the rat. J Pharmacol Exp Ther 1989; 250:632–636.
22. Lown KS, Bailey DG, Fontana RJ, et al. Grapefruit juice increases felodipine oral availability in humans by decreasing intestinal CYP3A protein expression. J Clin Invest 1997; 99:2545–2553.
23. Marathe PH, Salazar DE, Greene DS, et al. Absorption and presystemic metabolism of nefazodone administered at different regions in the gastrointestinal tract of humans. Pharm Res 1995; 12:1716–1721.
24. Ameer B, Weintraub RA. Drug interactions with grapefruit juice. Clin Pharmacokinet 1997; 33:103–121.
25. Fuhr U. Drug interactions with grapefruit juice. Extent, probable mechanism and clinical relevance. Drug Saf 1998; 18:251–272.
26. Feldman EB. How grapefruit juice potentiates drug bioavailability. Nutr Rev 1997; 55:398–400.
27. Edwards DJ, Fitzsimmons ME, Schuetz EG, et al. 6′,7′-Dihydroxybergamottin in grapefruit juice and Seville orange juice: effects on cyclosporine disposition, enterocyte CYP3A4, and P-glycoprotein. Clin Pharmacol Ther 1999; 65:237–244.
28. Bailey DG, Kreeft JH, Munoz C, et al. Grapefruitjuice-felodipine interaction: effect of naringin and 6′,7′-dihydroxybergamottin in humans. Clin Pharmacol Ther 1998; 64:248–256.
29. He K, Iyer KR, Hayes RN, et al. Inactivation of cytochrome P450 3A4 by bergamottin, a component of grapefruit juice. Chem Res Toxicol 1998; 11:252–259.
30. Schmiedlin-Ren P, Edwards DJ, Fitzsimmons ME, et al. Mechanisms of enhanced oral availability of CYP3A4 substrates by grapefruit constituents. Decreased enterocyte CYP3A4 concentration and mechanism-based inactivation by furanocoumarins. Drug Metab Dispos 1997; 25:1228–1233.
31. Fukuda K, Ohta T, Oshima Y, et al. Specific CYP3A4 inhibitors in grapefruit juice: furocoumarin dimers as components of drug interaction. Pharmacogenetics 1997; 7:391–396.
32. Fukuda K, Ohta T, Yamazoe Y. Grapefruit component interacting with rat and human P450 CYP3A: possible involvement of non-flavonoid components in drug interaction. Biol Pharm Bull 1997; 20:560–564.
33. Rodrigues AD. Integrated P450 reaction phenotyping: attempting to bridge the gap between cDNA-expressed cytochromes P450 and native human liver microsomes. Biochem Pharmacol 1999; 57:465–480.
34. Morgan MY, Reshef R, Shah RR, et al. Impaired oxidation of debrisoquine in patients with perhexiline liver injury. Gut 1984; 25:1057–1064.
35. Shah RR, Oates NS, Idle JR, et al. Impaired oxidation of debrisoquine in patients with perhexiline neuropathy. Br Med J (Clin Res Ed) 1982; 284:295–2999.
36. Oates NS, Shah RR, Idle JR, et al. Phenformin-induced lacticacidosis associated with impaired debrisoquine hydroxylation. Lancet 1981; 1(8224):837–838.
37. Idle JR, Oates NS, Shah RR, et al. Protecting poor metabolizers, a group at high risk of adverse drug reactions. Lancet 1983; 1(8338):1388.

38. Honig PK, Wortham DC, Zamani K, et al. Terfenadine-ketoconazole interaction. Pharmacokinetic and electrocardiographic consequences. JAMA 1993; 269: 1513–1518.
39. Honig PK, Woosley RL, Zamani K, et al. Changes in the pharmacokinetics and electrocardiographic pharmacodynamics of terfenadine with concomitant administration of erythromycin. Clin Pharmacol Ther 1992; 52:231–238.
40. Lipinski CA, Lombardo F, Dominy BW, et al. Experimental and computational approaches to estimate solubility and permeability in drug discovery and development settings. Adv Drug Delivery Rev 1997; 23:3–25.
41. Catteau A, Bechet YC, Poisson N, et al. Population and family study of CYP1A2 using caffeine urinary metabolites. Eur J Clin Pharmacol 1995; 47:423–430.
42. Campbell ME, Speilberg SP, Kalow W. A urinary metabolite ratio that reflects systemic caffeine clearance. Clin Pharmacol Ther 1987; 42:157–165.
43. Curi-Pedrosa R, Daujat M, Pichard L, et al. Omeprazole and lansoprazole are mixed inducers of CYP1A and CYP3A in human hepatocytes in primary culture. J Pharmacol Exp Ther 1994; 269:384–392.
44. Rost KL, Roots I. Accelerated caffeine metabolism after omeprazole treatment as indicated by urinary metabolite ratios: coincidence with plasma clearance and breath test. Clin Pharmacol Ther 1994; 55:402–411.
45. Andersson T. Omeprazole drug interaction studies. Clin Pharmacokinet 1991; 21:195–212.
46. Xiaodong S, Gatti G, Bartoli A, et al. Omeprazole does not enhance the metabolism of phenacetin, a marker of CYP1A2 activity, in healthy volunteers. Ther Drug Monit 1994; 16:248–250.
47. Clarke SE, Ayrton AD, Chenery RJ. Characterisation of the inhibition of 1A2 by furafylline. Xenobiotica 1994; 24:517–526.
48. Kunze KL, Trager WF. Isoform selective mechanism based inhibition of human cytochrome P4501A2 by furafylline. Chem Res Toxicol 1993; 6:649–656.
49. Fuhr U, Anders EM, Mahr G, et al. Inhibitory potency of quinolone antibacterial agents against cytochrome P450IA2 activity in vivo and in vitro. Antimicrob Agents Chemother 1992; 36:942–948.
50. Kinzig-Schippers M, Fuhr U, Zaigler U, et al. Interaction of pefloxacin and enoxacin with the human cytochrome P450 enzyme CYP1A2. Clin Pharmacol Ther 1999; 65:262–274.
51. Nikolaidis P, Walker SE, Dombros N, et al. Single-dose pefloxacin pharmacokinetics and metabolism in patients undergoing continuous ambulatory peritoneal dialysis (CAPD). Perit Dial Int 1991; 11:59–63.
52. Robson RA, Begg EJ, Atkinson HC, et al. Comparative effects of ciprofloxacin and lomefloxacin on the oxidative metabolism of theophylline. Br J Clin Pharmacol 1990; 29:491–493.
53. Brosen K, Skjelbo E, Rasmussen BB, et al. Fluvoxamine is a potent inhibitor of cytochrome P4501A2. Biochem Pharmacol 1993; 45:1211–1214.
54. Von Moltke LL, Greenblatt DJ, Duan SX, et al. Phenacetin O-deethylation by human liver micro-somes in vitro: inhibition by chemical probes, SSRI antidepressants, nefazodone and venlafaxine. Psychopharmacology 1996; 128:398–407.
55. Rasmussen BB, Nielsen TL, Brosen K. Fluvoxamine is a potent inhibitor of the metabolism of caffeine in vitro. Pharmacol Toxicol 1998; 83:240–245.

56. Jeppesen U, Loft S, Poulsen HE, et al. A fluvoxamine-caffeine interaction study. Pharmacogenetics 1996; 6:213–222.
57. Lelo A, Miners JO, Robson RA, et al. Quantitative assessment of caffeine partial clearances in man. Br J Clin Pharmacol 1986; 22:183–186.
58. Butler MA, Iwasaki M, Guengerich FP, et al. Human cytochrome P450PA (P450IA2), the phenacetin O-deethylase, is primarily responsible for the hepatic 3-demethylation of caffeine and N-oxidation of carcinogenic arylamines. Proc Natl Acad Sci U S A 1989; 86:7696–7700.
59. Bertilsson L, Carrillo JA, Dahl ML, et al. Clozapine disposition covaries with CYP1A2 activity determined by a caffeine test. Br J Clin Pharmacol 1994; 38: 471–473.
60. Pirmohamed M, Williams D, Madden S, et al. Metabolism and bioactivation of clozapine by human liver in vitro. J Pharmacol Exp Ther 1995; 272:984–990.
61. Jones BC, Hawksworth G, Horne VA, et al. Putative active site model for CYP2C9 (tolbutamide hydroxylase). Drug Metab Dispos 1996; 24:1–7.
62. Haining RL, Jones JP, Henne KR, et al. Enzymatic determinants of the substrate specificity of CYP2C9: role of B′-C loop residues in providing the pi-stacking anchor site for warfarin binding. Biochem J 1999; 38:3285–3292.
63. Haining RL, Hunter AP, Veronese ME, et al. Allelic variants of human cytochrome P450 2C9-baculovirus-mediated expression, purification, structural characterization, substrate stereoselectivity, and prochiral selectivity of the wild-type and 13591 mutant forms. Arch Biochem Biophys 1996; 333:447–458.
64. Yamazaki H, Inoue K, Chiba K, et al. Comparative studies on the catalytic roles of cytochrome P450 2C9 and its Cys- and Leu-variants in the oxidation of warfarin, flurbiprofen, and diclofenac by human liver microsomes. Biochem Pharmacol 1998; 56:243–251.
65. Kaminsky LS. Warfarin as a probe of cytochromes P-450 function. Drug Metab Rev 1989; 20:479–487.
66. Zhang Z, Fasco MJ, Huang Z, et al. Human cytochromes P4501A1 and P4501A2—R-warfarin metabolism as a probe. Drug Metab Dispos 1995; 23:1339–1345.
67. Wienkers LC, Wurden CJ, Storch E, et al. Formation of (R)-8-hydroxywarfarin in human liver - microsomes - a new metabolic marker for the (S)-mephenytoin hydroxylase, P4502C19. Drug Metab Dispos 1996; 24:610–614.
68. Rettie AE, Korzekwa KR, Kunze KL, et al. Hydroxylation of warfarin by human cDNA-expressed cytochrome P-450: a role for P-4502C9 in the etiology of (S)-warfarin drug interactions. Chem Res Toxicol 1992; 5:54–59.
69. Haining RL, Steward DJ, Henne KR, et al. Correlation of the cytochrome-P450 2C9 Leu(359) defect with warfarin sensitivity. FASEB J 1997; 11:130–139.
70. Steward DJ, Haining RL, Henne KR, et al. Genetic association between sensitivity to warfarin and expression of CYP2C9*3. Pharmacogenetics 1997; 7:361–367.
71. Baldwin SJ, Bloomer JC, Smith GJ, et al. Ketoconazole and sulfaphenazole as the respective selective inhibitors of P4503A and 2C9. Xenobiotica 1995; 25:261–270.
72. Kunze KL, Wienkers LC, Thummel KE, et al. Warfarin-fluconazole. 1. Inhibition of the human cytochrome-P450-dependent metabolism of warfarin by fluconazole - in vitro studies. Drug Metab Dispos 1996; 24:414–421.
73. Black DJ, Kunze KL, Wienkers LC, et al. Warfarin-fluconazole. 2. A metabolically based drug-interaction-in vivo studies. Drug Metab Dispos 1996; 24:422–428.

74. Kunze KL, Trager WF. Warfarin-fluconazole. 3. A rational approach to management of a metabolically based drug-interaction. Drug Metab Dispos 1996; 24:429–435.
75. Transon C, Leemann T, Dayer P. In vitro comparative inhibition profiles of major human drug-metabolizing cytochrome-P450 isozymes (CYP2C9, CYP2D6 and CYP3A4) by HMG-CoA reductase inhibitors. Eur J Clin Pharmacol 1996; 50:209–215.
76. Transon C, Leemann T, Vogt N, et al. In vivo inhibition profile of cytochrome P450 (tb) (CYP2C9) by (+/-)-fluvastatin. Clin Pharmacol Ther 1995; 58:412–417.
77. Doecke CJ, Veronese ME, Pond SM, et al. Relationship between phenytoin and tolbutamide hydroxylations in human liver microsomes. Br J Clin Pharmacol 1991; 31:124–130.
78. Dickerson RG, Hooper WD, Patterson M, et al. Extent of urinary excretion of p-hydroxyphenytoin in healthy subjects given phenytoin. Ther Drug Monit 1985; 7:283–289.
79. Tassaneeyakul W, Veronese ME, Birkett DJ, et al. Coregulation of phenytoin and tolbutamide metabolism in humans. Br J Clin Pharmacol 1992; 34:494–498.
80. Thomas RC, Ikeda GJ. The metabolic fate of tolbutamide in man and rat. J Med Chem 1966; 9:507–510.
81. Nelson E, O'Reilly I. Kinetics of carboxytolbutamide excretion following tolbutamide and carboxytolbutamide administration. J Pharmacol Exp Ther 1961; 132:103–109.
82. Veronese ME, Miners JO, Randies D, et al. Validation of the tolbutamide metabolic ratio for population screening with the use of sulfaphenazole to produce model phenotypic poor metabolizers. Clin Pharmacol Ther 1990; 47:403–411.
83. Bertilsson L. Geographical/interracial differences in polymorphic drug oxidation. Clin Pharmacokinet 1995; 29:192–209.
84. Goldstein JA, Faletto MB, Romkes-Sparks M, et al. Evidence that CYP2C19 is the major (S)-mephenytoin 4′-hydroxylase in humans. Biochemistry 1994; 33:1743–1752.
85. Goldstein JA, De Morais SMF. Biochemistry and molecular biology of the human CYP2C subfamily. Pharmacogenetics 1994; 4:285–299.
86. Ibeanu GC, Blaisdell J, Ghanayem BI, et al. An additional defective allele, CYP2C19*5, contributes to the S-mephenytoin poor metabolizer phenotype in Caucasians. Pharmacogenetics 1998; 8:129–135.
87. Ibeanu GC, Goldstein JA, Meyer U, et al. Identification of new human CYP2C19 alleles (CYP2C19*6 and CYP2C19*2b) in a Caucasian poor metabolizer of mephenytoin. J Pharmacol Exp Ther 1998; 286:1490–1495.
88. Kobayashi K, Yamamoto T, Chiba K, et al. The effects of selective serotonin reuptake inhibitors and their metabolites on S-mephenytoin 4′-hydroxylase activity in human liver-microsomes. Br J Clin Pharmacol 1995; 40:481–485.
89. Jeppesen U, Gram LF, Vistisen K, et al. Dose-dependent inhibition of CYP1A2, CYP2C19 and CYP2D6 by citalopram, fluoxetine, fluvoxamine and paroxetine. Eur J Clin Pharmacol 1996; 51:73–78.
90. Andersson T, Miners JO, Veronese ME, et al. Diazepam metabolism by human liver microsomes is mediated by both (5)-mephenytoin hydroxylase and CYP3A isoforms. Br J Clin Pharmacol 1994; 38:131–137.
91. Chiba K, Saitoh A, Koyama E, et al. The role of (S)-mephenytoin 4′-hydroxylase in imipramine metabolism by human liver microsomes: a two-enzyme analysis of N-demethylation and 2-hydroxylation. Br J Clin Pharmacol 1994; 37:237–242.

92. Andersson T, Miners JO, Veronese ME, et al. Identification of human liver cytochrome-P450 isoforms mediating omeprazole metabolism. Br J Clin Pharmacol 1993; 36:521–530.
93. Tybring G, Bertilsson L. A methodological investigation on the estimation of the (S)-mephenytoin hydroxylation phenotype using the urinary S/R ratio. Pharmacogenetics 1992; 2:241–243.
94. Skjelbo E, Brosen K, Hallas J, et al. The mephenytoin oxidation polymorphism is partially responsible for the N-demethylation of imipramine. Clin Pharmacol Ther 1991; 49:18–23.
95. Skjelbo E, Gram LF, Brosen K. The N-demethylation of imipramine correlates with the oxidation of (5)-mephenytoin (S/R ratio). A population study. Br J Clin Pharmacol 1993; 35:331–334.
96. Balian JD, Sukhova N, Harris JW, et al. The hydroxylation of omeprazole correlates with S-mephenytoin metabolism: a population study. Clin Pharmacol Ther 1995; 57:662–669.
97. Alvan G, Bechtel B, Iselius L, et al. Hydroxylation polymorphism of debrisoquine and mephenytoin in European populations. Eur J Clin Pharmacol 1990; 39:533–537.
98. Sohn DR, Shin SG, Park CW, et al. Metoprolol oxidation polymorphism in a Korean population: comparison with native Japanese and Chinese populations. Br J Clin Pharmacol 1991; 32:504–507.
99. Dahl ML, Johansson I, Bertilsson L, et al. Ultrarapid hydroxylation of debrisoquine in a Swedish population. Analysis of the molecular genetic basis. J Pharmacol Exp Ther 1995; 274:516–520.
100. Leathart JS, London SJ, Steward A, et al. CYP2D6 phenotype-genotype relationships in African-Americans and Caucasians in Los Angeles. Pharmacogenetics 1998; 8:529–541.
101. Smith DA, Jones BC. Speculations on the substrate structure-activity relationship (SSAR) of cytochrome P450 enzymes. Biochem Pharmacol 1992; 44:2089–2098.
102. Crewe HK, Lennard MS, Tucker GT, et al. The effect of selective serotonin reuptake inhibitors on cytochrome P4502D6 (CYP2D6) activity in human liver microsomes. Br J Clin Pharmacol 1992; 34:262–265.
103. Skjelbo E, Brosen K. Inhibitors of imipramine metabolism by human liver microsomes. Br J Clin Pharmacol 1992; 34:256–261.
104. Von Moltke LL, Greenblatt DJ, Cotreaubibbo MM, et al. Inhibition of desipramine hydroxylation in vitro by serotonin-reuptake inhibitor antidepressants, and by quinidine and ketoconazole—a model system to predict drug-interactions in vivo. J Pharmacol Exp Ther 1994; 268:1278–1283.
105. Belpaire FM, Wijnant P, Temmerman A, et al. The oxidative metabolism of metoprolol in human liver microsomes-inhibition by the selective serotonin reuptake inhibitors. Eur J Clin Pharmacol 1998; 54:261–264.
106. Zapotoczky HG, Simhandl CA. Interactions of antidepressants. Wien Klin Wochenschr 1995; 107:293–300.
107. Brosen K. Are pharmacokinetic drug-interactions with the SSRIs an issue. Int Clin Psychopharmacol 1996; 11:23–27.
108. Catterson ML, Preskorn SH. Pharmacokinetics of selective serotonin reuptake inhibitors-clinical relevance. Pharmacol Toxicol 1996; 78:203–208.
109. Ereshefsky L, Riesenman C, Lam YWF. Serotonin selective reuptake inhibitor drug-interactions and the cytochrome-P450 system. J Clin Psychiatry 1996; 57:17–25.

110. Lane RM. Pharmacokinetic drug-interaction potential of selective serotonin reuptake inhibitors. Int Clin Psychopharmacol 1996; 11:31–61.
111. Lane R, Baldwin D. Selective serotonin reuptake inhibitor-induced serotonin syndrome. J Clin Psychopharmacol 1997; 17:208–221.
112. Lopezmunoz F, Alamo C, Cuenca E, et al. Effect of antidepressant drugs on cytochrome-P-450 isoenzymes and its clinical relevance-differential profile. Actas Luso Esp Neurol Psiquiatr Cienc Afines 1997; 25:397–409.
113. Mitchell PB. Drug interactions of clinical significance with selective serotonin reuptake inhibitors. Drug Saf 1997; 17:390–406.
114. Sproule BA, Naranjo CA, Bremner KE, et al. Selective serotonin reuptake inhibitors and CNS drug interactions-a critical review of the evidence. Clin Pharmacokinet 1997; 33:454–471.
115. Baker GB, Fang J, Sinha S, et al. Metabolic drug interactions with selective serotonin reuptake inhibitor (SSRI) antidepressants. Neurosci Biobehav Rev 1998; 22:325–333.
116. Caccia S. Metabolism of the newer antidepressants-an overview of the pharmacological and pharmacokinetic implications. Clin Pharmacokinet 1998; 34:281–302.
117. Von Moltke LL, Greenblatt DJ, Grassi JM, et al. Protease inhibitors as inhibitors of human cytochromes P450-high-risk associated with ritonavir. J Clin Pharmacol 1998; 38:106–111.
118. Mahgoub A, Idle JR, Dring LG, et al. Polymorphic hydroxylation of debrisoquine in man. Lancet 1977; 2:584–586.
119. Angelo M, Dring LG, Lancaster R, et al. A correlation between the response to debrisoquine and the amount of unchanged drug excreted in the urine. Br J Pharmacol 1975; 55:264.
120. Smith RL. Introduction. Xenobiotica 1986; 16:361–365.
121. Gut J, Catin T, Dayer P, et al. Debrisoquine/sparteine type polymorphism of drug oxidation. J Biol Chem 1986; 261:11734–11743.
122. Dayer P, Leemann T, Striberni R. Dextromethorphan O-demethylation in liver microsomes as a prototype reaction to monitor cytochrome P-450 dbl activity. Clin Pharmacol Ther 1989; 45:34–40.
123. Schmid B, Bircher J, Preisig R, et al. Polymorphic dextromethorphan metabolism: co-segregation of oxidative O-demethylation with debrisoquine hydroxylation. Clin Pharmacol Ther 1985; 38:618–624.
124. Otton SV, Crewe HK, Lennard MS, et al. Use of quinidine inhibition to define the role of the sparteine/debrisoquine cytochrome P450 in metoprolol oxidation by human liver microsomes. J Pharmacol Exp Ther 1988; 247:242–247.
125. McGourty JC, Silas JH, Lennard MS, et al. Metoprolol metabolism and debrisoquine oxidation polymorphism-population and family studies. Br J Clin Pharmacol 1985; 20:555–566.
126. Lennard MS, Iyun AO, Jackson PR, et al. Evidence for a dissociation in the control of sparteine, debrisoquine and metoprolol metabolism in Nigerians. Pharmacogenetics 1992; 2:89–92.
127. Schwab GE, Raucy JL, Johnson EF. Modulation of rabbit and human hepatic cytochrome P-450-catalyzed steroid hydroxylations by alpha-naphthoflavone. Mol Pharmacol 1988; 33:493–499.

128. Shou M, Grogan J, Mancewicz JA, et al. Activation of CYP3A4-evidence for the simultaneous binding of 2 substrates in a cytochrome-P450 active site. Biochemistry 1994; 33:6450–6455.
129. Ueng YF, Kuwabara T, Chun YJ, et al. Cooperativity in oxidations catalyzed by cytochrome-P450 3A4. Biochemistry 1997; 36:370–381.
130. Korzekwa KR, Krishnamachary N, Shou M, et al. Evaluation of atypical cytochrome-P450 kinetics with 2-substrate models-evidence that multiple substrates can simultaneously bind to cytochrome-P450 active sites. Biochemistry 1998; 37:4137–4147.
131. Ekins S, Ring BJ, Binkley SN, et al. Autoactivation and activation of the cytochrome P450s. Int J Clin Pharmacol Ther 1998; 36:642–651.
132. Koley AP, Buters JTM, Robinson RC, et al. CO Binding kinetics of human cytochrome-P450 3A4-specific interaction of substrates with kinetically distinguishable conformers. J Biol Chem 1995; 270:5014–5018.
133. Koley AP, Robinson RC, Friedman FK. Cytochrome-P450 conformation and substrate interactions as probed by CO binding-kinetics. Biochimie 1996; 78:706–713.
134. Koley AP, Robinson RC, Markowitz A, et al. Drug-drug interactions effect of quinidine on nifedipine binding to human cytochrome-P450 3A4. Biochem Pharmacol 1997; 53:455–460.
135. Koley AP, Buters JTM, Robinson RC, et al. Differential mechanisms of cytochrome-P450 inhibition and activation by alpha-naphthoflavone. J Biol Chem 1997; 272:3149–3152.
136. Rebbeck TR, Jaffe JM, Walker AH, et al. Modification of clinical presentation of prostate tumors by a novel genetic variant in CYP3A4. J Natl Cancer Inst 1998; 90:1225–1229.
137. Lehmann JM, McKee DD, Watson MA, et al. The human orphan nuclear receptor PXR is activated by compounds that regulate CYP3A4 gene expression and cause drug interactions. J Clin Invest 1998; 102:1016–1023.
138. Ogg MS, Gray TJ, Gibson GG. Development of an in vitro reporter gene assay to assess xenobiotic induction of the human CYP3A4 gene. Eur J Drug Metab Pharmacokinet 1997; 22:311–313.
139. Backman JT, Olkkola KT, Neuvonen PJ. Rifampin drastically reduces plasma concentrations and effects of oral midazolam. Clin Pharmacol Ther 1996; 59:7–13.
140. Villikka K, Kivisto KT, Backman JT, et al. Triazolam is ineffective in patients taking rifampin. Clin Pharmacol Ther 1997; 61:8–14.
141. Dommisse J. Oral contraceptive failure due to drug interaction. S Afr Med J 1976; 50:796.
142. Back DJ, Breckenridge AM, Crawford FE, et al. The effect of rifampicin on the pharmacokinetics of ethynylestradiol in women. Contraception 1980; 21:135–143.
143. Fazio A. Oral contraceptive drug interactions: important considerations. South Med J 1991; 84:997–1002.
144. Daneshmend TK, Warnock DW. Clinical pharmacokinetics of ketoconazole. Clin Pharmacokinet 1988; 14:13–34.
145. Lomaestro BM, Piatek MA. Update on drug interactions with azole antifungal agents. Ann Pharmacother 1998; 32:915–928.
146. Albengres E, Le Louet H, Tillement JP. Systemic antifungal agents. Drug interactions of clinical significance. Drug Saf 1998; 18:83–97.
147. Ludden TM. Pharmacokinetic interactions of the macrolide antibiotics. Clin Pharmacokinet 1985; 10:63–79.

148. Nahata M. Drug interactions with azithromycin and the macrolides: an overview. J Antimicrob Chemother 1996; 37:133–142.
149. He K, Woolf TF, Hollenberg PF. Mechanism-based inactivation of cytochrome P-450-3A4 by mifepristone (RU486). J Pharmacol Exp Ther 1999; 288:791–797.
150. Voorman RL, Maio SM, Payne NA, et al. Microsomal metabolism of delavirdine: evidence for mechanism-based inactivation of human cytochrome P450 3A. J Pharmacol Exp Ther 1998; 287:381–388.
151. Koudriakova T, Iatsimirskaia E, Utkin I, et al. Metabolism of the human immunodeficiency virus protease inhibitors indinavir and ritonavir by human intestinal microsomes and expressed cytochrome P4503A4/3A5: mechanism-based inactivation of cytochrome P4503A by ritonavir. Drug Metab Dispos 1998; 26:552–561.
152. Galetin A, Burt H, Gibbons L, et al. Prediction of time-dependent CYP3A4 drug-drug interactions: impact of enzyme degradation, parallel elimination pathways, and intestinal inhibition. Drug Metab Dispos 2006; 34:166–175.
153. Spoendlin M, Peters J, Welker H, et al. Pharmacokinetic interaction between oral cyclosporin and mibefradil in stabilized post-renal-transplant patients. Nephrol Dial Transplant 1998; 13:1787–1791.
154. Schmassmann-Suhijar D, Bullingham R, Gasser R, et al. Rhabdomyolysis due to interaction of simvastatin with mibefradil. Lancet 1998; 351:1929–1930.
155. Mullins ME, Horowitz BZ, Linden DH, et al. Life-threatening interaction of mibefradil and beta-blockers with dihydropyridine calcium channel blockers. JAMA 1998; 280:157–158.
156. Roche, FDA announce new drug-interaction warnings for mibefradil. Am J Health Syst Pharm 1998; 55:210.
157. Watkins PB, Murray SA, Winkelman LG, et al. Erythromycin breath test as an assay of glucurocorticoid-inducible liver cytochromes P450. J Clin Invest 1989; 83:688–697.
158. Watkins PB, Hamilton TA, Annesley TM, et al. The erythromycin breath test as a predictor of cyclosporin A blood levels. Clin Pharmacol Ther 1990; 48:120–129.
159. Yun CH, Wood M, Wood AJJ, et al. Identification of the pharmacogenetic determinants of alfentanil metabolism: cytochrome P-4503A4. An explanation of the variable elimination clearance. Anesthesiology 1992; 77:467–474.
160. Krivoruk Y, Kinirons MT, Wood AJJ, et al. Metabolism of cytochrome P4503A substrates in vivo administered by the same route: lack of correlation between alfentanil clearance and erythromycin breath test. Clin Pharmacol Ther 1994; 56:608–614.
161. Lown KS, Thummel KE, Benedict PE, et al. The erythromycin breath test predicts the clearance of midazolam. Clin Pharmacol Ther 1995; 57:16–24.
162. Thummel KE, Shen DD, Carithers RL, et al. Prediction of in vivo midazolam clearance from hepatic CYP3A content and midazolam 1-hydroxylation activity in liver transplant patients. ISSX Proc 1993; 4:235.
163. Thummel KE, Shen DD, Podoll TD, et al. Use of midazolam as a human cytochrome P450 3A probe: 1. in vitro—in vivo correlations inliver transplant patients. J Pharmacol Exp Ther 1994; 271:549–556.
164. Thummel KE, Shen DD, Bacchi CE, et al. Induction of human hepatic P4503A3/4 by phenytoin. Hepatology 1992; 16:160A.
165. Olkkola KT, Aranko K, Luurila H, et al. A potentially hazardous interaction between erythromycin and midazolam. Clin Pharmacol Ther 1993; 53:298–305.

166. Ahonen J, Olkkola KT, Neuvonen PJ. Effects of itraconazole and terbinafine on the pharmacokinetics and pharmacodynamics of midazolam in healthy volunteers. Br J Clin Pharmacol 1995; 40:270–272.
167. Guengerich FP, Martin MV, Beaune PH, et al. Characterization of rat and human liver microsomal cytochrome P-450 forms involved in nifedipine oxidation, a prototype for genetic polymorphism in oxidative drug metabolism. J Biol Chem 1986; 261:5051–5060.
168. Bork RW, Muto T, Beaune PH, et al. Characterization of mRNA species related to human liver cytochrome P-450 nifedipine oxidase and the regulation of catalytic activity. J Biol Chem 1989; 264:910–919.
169. Tailor SA, Gupta AK, Walker SE, et al. Peripheral edema due to nifedipine itraconazole interaction: a case report. Arch Dermatol 1996; 132:350.
170. Bailey DG, Spence JD, Munoz C, et al. Interaction of citrus juices with felodipine and nifedipine. Lancet 1991; 337:268–269.
171. Schellens JH, van der Wart JH, Brugman M, et al. Influence of enzyme induction and inhibition on the oxidation of nifedipine, sparteine, mephenytoin and antipyrine in humans as assessed by a ''cocktail'' study design. J Pharmacol Exp Ther 1989; 249:638–645.
172. Holtbecker N, Fromm MF, Kroemer HK, et al. The nifedipine-rifampin interaction. Evidence for induction of gut wall metabolism. Drug Metab Dispos 1996; 24: 1121–1123.
173. Ekins S, VandenBranden M, Ring BJ, et al. Examination of purported probes of human CYP2B6. Pharmacogenetics 1997; 7:165–179.
174. Hesse LM, Venkatakrishnan K, Court MH, et al. CYP2B6 mediates the in vitro hydroxylation of bupropion: potential drug interactions with other antidepressants. Drug Metab Dispos 2000; 28:1176–1183.
175. Faucette SR, Hawke RL, Lecluyse EL, et al. Validation of bupropion hydroxylation as a selective marker of human cytochrome P450 2B6 catalytic activity. Drug Metab Dispos 2000; 28:1222–1230.
176. Ward BA, Gorski JC, Jones DR, et al. The cytochrome P450 2B6 (CYP2B6) is the main catalyst of efavirenz primary and secondary metabolism: implication for HIV/ AIDS therapy and utility of efavirenz as a substrate marker of CYP2B6 catalytic activity. J Pharmacol Exp Ther 2003; 306:287–300.
177. Huang Z, Roy P, Waxman DJ. Role of human liver microsomal CYP3A4 and CYP2B6 in catalyzing N-dechloroethylation of cyclophosphamide and ifosfamide. Biochem Pharmacol 2000; 59:961–972.
178. Court MH, Duan SX, Hesse LM, et al. Cytochrome P-450 2B6 is responsible for interindividual variability of propofol hydroxylation by human liver microsomes. Anesthesiology 2001; 94:110–119.
179. Oda Y, Hamaoka N, Hiroi T, et al. Involvement of human liver cytochrome P4502B6 in the metabolism of propofol. Br J Clin Pharmacol 2001; 51:281–285.
180. Turpeinen M, Nieminen R, Juntunen T, et al. Selective inhibition of CYP2B6-catalyzed bupropion hydroxylation in human liver microsomes in vitro. Drug Metab Dispos 2004; 32:626–631.
181. Richter T, Murdter TE, Heinkele G, et al. Potent mechanism-based inhibition of human CYP2B6 by clopidogrel and ticlopidine. J Pharmacol Exp Ther 2004; 308:189–197.

182. Turpeinen M, Tolonen A, Uusitalo J, et al. Effect of clopidogrel and ticlopidine on cytochrome P450 2B6 activity as measured by bupropion hydroxylation. Clin Pharmacol Ther 2005; 77:553–559.
183. Turpeinen M, Nieminen R, Juntunen T, et al. Selective inhibition of CYP2B6-catalyzed bupropion hydroxylation in human liver microsomes in vitro. Drug Metab Dispos 2004; 32:626–631.
184. Melet A, Marques-Soares C, Schoch GA, et al. Analysis of human cytochrome P450 2C8 substrate specificity using a substrate pharmacophore and site-directed mutants. Biochemistry 2004; 43:15379–15392.
185. Niemi M, Backman JT, Granfors M, et al. Gemfibrozil considerably increases the plasma concentrations of rosiglitazone. Diabetologia 2003; 46:1319–1323.
186. Jaakkola T, Backman JT, Neuvonen M, et al. Effects of gemfibrozil, itraconazole, and their combination on the pharmacokinetics of pioglitazone. Clin Pharmacol Ther 2005; 77:404–414.
187. Deng LJ, Wang F, Li HD. Effect of gemfibrozil on the pharmacokinetics of pioglitazone. Eur J Clin Pharmacol 2005; 61:831–836.
188. Niemi M, Backman JT, Neuvonen M, et al. Effects of gemfibrozil, itraconazole, and their combination on the pharmacokinetics and pharmacodynamics of repaglinide: potentially hazardous interaction between gemfibrozil and repaglinide. Diabetologia 2003; 46:347–351.
189. Backman JT, Kyrklund C, Neuvonen M, et al. Gemfibrozil greatly increases plasma concentrations of cerivastatin. Clin Pharmacol Ther 2002; 72:685–691.
190. Ogilvie BW, Zhang D, Li W, et al. Glucuronidation converts gemfibrozil to a potent, metabolism-dependent inhibitor of CYP2C8: implications for drug-drug interactions. Drug Metab Dispos 2006; 34:191–197.
191. Yamazaki M, Li B, Louie SW, et al. Effects of fibrates on human organic anion-transporting polypeptide 1B1-, multidrug resistance protein 2- and P-glycoprotein-mediated transport. Xenobiotica 2005; 35:737–753.
192. Niemi M, Kajosaari LI, Neuvonen M, et al. The CYP2C8 inhibitor trimethoprim increases the plasma concentrations of repaglinide in healthy subjects. Br J Clin Pharmacol 2004; 57:441–447.
193. Niemi M, Backman JT, Neuvonen PJ. Effects of trimethoprim and rifampin on the pharmacokinetics of the cytochrome P450 2C8 substrate rosiglitazone. Clin Pharmacol Ther 2004; 76:239–249.

4

UDP-Glucuronosyltransferases

Rory P. Remmel and Jin Zhou
Department of Medicinal Chemistry, College of Pharmacy, University of Minnesota, Minneapolis, Minnesota, U.S.A.

Upendra A. Argikar
Novartis Pharmaceuticals, Cambridge, Massachusetts, U.S.A.

I. INTRODUCTION

The uridine diphosphate (UDP)-glycosyltransferases (EC2.4.21.17) are a group of enzymes that catalyze the transfer of sugars (glucuronic acid, glucose, and xylose) to a variety of acceptor molecules (aglycones). The sugars may be attached at aromatic and aliphatic alcohols, carboxylic acids, thiols, primary, secondary, tertiary, and aromatic amino groups, and acidic carbon atoms. In vivo, the most common reaction occurs by transfer of glucuronic acid moiety from UDP glucuronic acid (UDPGA) to an acceptor molecule. This process is termed either *glucuronidation* or *glucuronosylation.* When the enzymes catalyze this reaction, they are also referred to as UDP-glucuronosyltransferases (UGTs). The structure and function of the enzymes have been the subject of several reviews (1–4). This chapter reviews the role of these enzymes in drug-drug interactions that occur in humans.

Glucuronidation is an important step in the elimination of many important endogenous substances from the body, including bilirubin, bile acids, steroid hormones, thyroid hormones, retinoic acids, and biogenic amines such as serotonin. Many of these compounds are also substrates for sulfonyltransferases (SULTs) (2).

The interplay between glucuronidation and sulfonylation (sulfation) of steroid and thyroid hormones and the corresponding hydrolytic enzymes, β-glucuronidase and sulfatase, may play an important role in development and regulation. The UGTs are expressed in many tissues, including liver, kidney, intestine, colon, adrenals, spleen, lung, skin, testes, ovaries, olfactory glands, and brain. Interactions between drugs at the enzymatic level are most likely to occur during the absorption phase in the intestine and liver or systemically in the liver, kidney, or intestine.

Given the broad array of substrates and the variety of molecular diversity, it is not surprising that there are multiple UGTs. The UGTs have been divided into two families (UGT1 and UGT2) on the basis of their sequence homology. All members of a family have at least 50% sequence identity to one another (3). The UGT1A family is encoded by a gene complex located on chromosome 2. The large UGT1A gene complex contains 13 variable region exons that are spliced onto four constant region exons that encode for amino acids on the C-terminus of the enzyme. Consequently, all enzymes in the UGT1 family have an identical C-terminus (encoding for the UDPGA binding site), but the N-terminus is highly variable, with a sequence homology of only 24–49% (3). The UGT1A enzymes are generally named in order of their proximity to the four constant region exons, i.e., UGT1A1 through UGT1A13. The arrangement (Fig. 1) appears to be conserved across all mammalian species studied to date. In humans, all of the gene products are functions except for pseudogenes UGT1A2, UGT1A11, UGT1A12, and UGT1A13. Pseudogenes encoding for inactive proteins vary from species to species. For example, UGT1A6 is a pseudogene in cats (6), whereas UGT1A3 and UGT1A4 are pseudogenes in rats and mice. The *UGT1A* gene complex is located on human chromosome 2 at 2q.37. Nomenclature for these enzymes in other species can be found on the UGT Web site at http://som.flinders.edu.au/FUSA/ClinPharm/UGT/.

The UGT2A subfamily represents olfactory UGTs and will not be discussed further in this review. Human UGT2A was originally cloned by Burchell and coworkers (7). The UGT2B subfamily is encoded in a series of complete UGT genes located at 4q12 on chromosome 4. Like the UGT1A enzymes, the C-terminus is highly conserved among all members of the *UGT2B* genes, with greater variation in the N-terminal half of the protein. Several human UGT2B enzymes have been cloned, expressed, and characterized for a variety of substrates. The nomenclature for

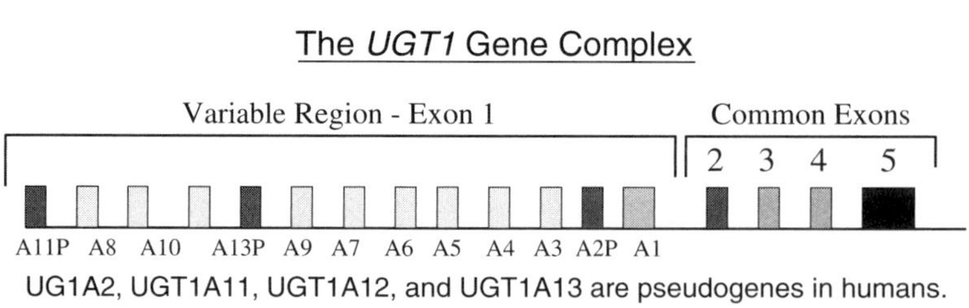

Figure 1 The *UGT1* gene complex.

the UGT2B genes has been assigned on the basis of the order of their discovery and submission to the nomenclature committee similar to that for CYP2 and CYP3 family enzymes. The human UGT2B enzymes are UGT2B4, UGT2B7, UGT2B10, UGT2B11, UGT2B15, UGT2B17, and UGT2B43.

Inhibitory interactions involving glucuronidation have been described in a number of clinical and in vitro studies and have been recently reviewed (8). Apparent decreases in the amount of glucuronide excreted in urine or bile or apparent increases in the AUC (decreased clearance) have been demonstrated in clinical studies. These apparent effects on glucuronidation could occur via several different mechanisms as follows:

1. Direct inhibition of the enzyme by competition with substrate or with UDPGA
2. Induction of the individual UGT enzymes resulting in increased clearance
3. Depletion of the UDPGA cofactor
4. Inhibition of the transport of UDPGA into the endoplasmic reticulum (ER)
5. Inhibition of the renal excretion of the glucuronide, with subsequent reconversion to the parent aglycone by β-glucuronidases (futile cycling)
6. Alteration of ER transport, sinusoidal membrane transport, or bile canalicular membrane transport of the glucuronides
7. Inhibition of the intestinal microflora, resulting in interruption of enterohepatic recycling and increased fecal excretion of the glucuronide metabolite.

Major interactions involving individual UGT enzymes will be discussed in detail along with a brief discussion of the function of each enzyme. A table of substrates, inducers, and inhibitors for the UGT enzymes is provided in the appendix to this chapter.

II. UGT1A1

UGT1A1 is an important enzyme that is primarily responsible for the glucuronidation of bilirubin in the liver. Cloned, expressed UGT1A1 is a glycosyltransferase that is also capable of catalyzing the formation of bilirubin xylosides and glycosides in the presence of UDP-xylose and UDP-glucose, respectively (9). In vivo, glucuronidation predominates, but bilirubin xylosides and glucosides have been identified in human bile. Polymorphisms in the UGT1A1 gene have been extensively studied because of a rare inborn error of bilirubin metabolism resulting in Crigler-Najjar syndrome. Type I Crigler-Najjar patients typically require liver transplantation, whereas Type II patients can be treated with UGT1A1 inducers such as phenobarbital. Gilbert syndrome is an asymptomatic unconjugated hyperbilirubinemia that is most often caused by a genetic polymorphism in the promoter region of the UGT1A1 gene in Caucasians and Africans. Decreased expression of UGT1A1 in Gilbert's patients is a result of the presence of a $(TA)_7TAA$ allele (*UGT1A1*28*) in place of the more prevalent $(TA)_6TAA$ allele (10,11). Persons who are homozygous for the $(TA)_7TAA$ express approximately 70% less UGT1A1 enzyme in the liver. A second mutation at –3279 C>T in a phenobarbital response enhancer module (PBREM) also is linked with Gilbert syndrome and is often in linkage disequilibrium with UGT1A1*28 in Caucasians and Japanese (12–14). Larger screening studies have demonstrated that this regulatory defect occurs in approximately 2–19% of various populations (11). In Asian patients, other mutations in the *UGT1A1* gene besides the $(TA)_7TAA$ genotype contribute significantly to hyperbilirubinemia, including *UGT1A1*6* (211 G>A, G71R) (15,16). Drugs that are substrates for or inhibit UGT1A1 may cause a further increase of unconjugated bilirubin concentrations, especially in patients with Gilbert syndrome. For example, the HIV protease inhibitors atazanavir and indinavir are known to increase bilirubin levels (17). Lankisch et al. recently found that atazanavir treatment increased median bilirubin concentrations from 10 to 41 μM ($p = 0.001$) (18). Bilirubin levels exceeding 43 μM were observed in 37% of the 106 patients. Hyperbilirubinemia >43 μM was significantly associated with three non-1A1 mutations UGT1A3-66C, UGT1A7-57G, and *UGT1A7*2* along with *UGT1A1*28*, although these variants are not typically in linkage disequilibrium in other populations. Six patients expressing all four mutations had bilirubin levels >87 μM, a level that may require discontinuation or dosage adjustment. UGT1A3 is a weak catalyst of bilirubin glucuronidation, whereas UGT1A7 would not be expected to contribute given its extrahepatic tissue distribution.

Older studies in persons with mild hyperbilirubinemia (meeting the criteria for Gilbert syndrome, but not genetically determined) demonstrated a decreased clearance rate for drugs that are glucuronidated. Clearance of acetaminophen (APAP; also catalyzed by other UGT enzymes, especially UGT1A5) was decreased by 30% in six subjects with Gilbert syndrome (19). In contrast, a small study by Ullrich et al. demonstrated no difference in the APAP-glucuronide/acetaminophen ratio in urine of 11 persons with Gilbert syndrome (20). A more recent study in genotyped patients also found no difference in the glucuronide/acetaminophen urinary ratio (21). Racemic (*S*/*R*) lorazepam clearance (catalyzed by UGT2B7 and UGT2B15) was 30–40% lower in persons with Gilbert syndrome (22). A modest decrease (32%) in lamotrigine oral clearance was observed in persons with Gilbert's syndrome (23). However, lamotrigine is glucuronidated by cloned, expressed UGT1A3 and UGT1A4, but not by UGT1A1 (24,25). In general, these studies were conducted in a small number of Gilbert syndrome subjects. A distinct heterogeneity may be present in persons exhibiting mild hyperbilirubinemia that could include patients with Crigler-Najjar Type II syndrome who have mutations in the *UGT1A1*-coding region, persons who are homozygous for *UGT1A1*28*, or in patients with a higher than normal breakdown of heme.

The role of *UGT1A1*28* polymorphism and irinotecan toxicity has been extensively investigated in Japan by Ando et al. (26) and in the United States by

Innocenti et al. (27). Irinotecan is a prodrug that is rapidly converted by esterases to active phenolic compound, SN-38. SN-38 glucuronidation is catalyzed primarily by UGT1A1 in studies with cloned, expressed enzymes. Iyer et al. compared the liver microsomal glucuronidation rate of SN-38 and bilirubin in 44 patients genotyped for the $(TA)_7TAA$ allele (*UGT1A1*28*) and found a high correlation ($r = 0.9$) (28). Patients with the *UGT1A1*28* allele who take irinotecan have a significantly higher risk for neutropenia, and the FDA has recently recommended that patients should be genotyped prior to use of irinotecan.

Evidence for drug-drug or herb-drug interactions involving UGT1A1 and irinotecan are limited (29). Case reports have suggested that inducers (e.g., phenytoin, carbamazepine, or rifampin) acting via the constitutive androstane receptor (CAR) or pregnenolone-16α-nitrile-X-receptor (pregnane-X-receptor; PXR) reduce exposure to SN-38; however, this could be due to enhanced CYP3A4-mediated metabolism of irinotecan to 7-ethyl-10-[4-N-[(5-aminopentanoic acid)-1-piperidino]-carbonyloxy-camptothecin (APC) (30) or by glucuronidation (31,32). Similar findings by Mathijssen et al. have implicated induction of SN-38 metabolism by St. John's wort (contains hyperforin, a potent PXR ligand) (33); however, evidence of increased glucuronidation in humans is lacking even though UGT1A1 is inducible by both PXR and CAR activation. Milk thistle (sylibinin) had no effect on SN-38 or SN-38 glucuronide levels (34). Sylibinin is metabolized by UGT1A1, but bioavailability is low and circulating levels are probably not high enough to affect glucuronidation. Gefitinib enhances irinotecan (SN-38) bioavailability in mice apparently via inhibition of the ABCG2 transporter (BCRP) (35). In a small study of etoposide and irinotecan, Ohtsu reported that all three patients receiving the combination had grade 3 or 4 toxicities (one neutropenia, one hepatotoxicity, and one hyperbilirubinemia) (36). Etoposide was recently shown to be a UGT1A1 substrate (37,38), so this combination should be avoided. In a single patient case report, an interaction between lopinavir/ritonavir and irinotecan was reported resulting in increased SN-38 AUC, most likely because of inhibition of CYP3A4 to APC (29,39). No reports of interactions between atazanavir or indinavir (known inhibitors of UGT1A1) and irinotecan have surfaced.

III. UGT1A3 AND UGT1A4

UGT1A3 and UGT1A4 appear to be important enzymes involved in the catalysis of many tertiary amine or aromatic heterocycles to form quaternary ammonium glucuronides (24,25). UGT1A3, UGT1A4, and UGT1A5 share a high nucleic acid sequence homology of 93–94% in the first variable-region exon and probably have arisen by gene duplication. The first exon of this group of enzymes appears to have diverged considerably from UGT1A1 (58% homology to 1A4), UGT1A5, and UGT1A7-10. UGT1A4 is expressed in human liver, intestine, and colon, although the level of expression of UGT1A4 mRNA is lower than that of UGT1A1 mRNA. UGT1A3 is expressed in liver, biliary epithelium, colon, and gastric tissue. UGT1A4 has low activity for bilirubin compared with UGT1A1 and has sometimes been designated as a minor bilirubin form. Although the N-glucuronidation of UGT1A3 and UGT1A4 for a variety of tertiary amines such as imipramine, cyproheptadine, amitriptyline, tripelennamine, and diphenhydramine overlaps ($K_m = 0.2$–2 mM), some differences have been observed. UGT1A3 catalyzes the glucuronidation of buprenorphine, norbuprenorphine (low K_m values), morphine (3-position only), and naltrexone. Only UGT1A3 is capable of forming carboxyl-linked glucuronides of bile acids and nonsteroidal anti-inflammatory drugs (NSAIDs) (25). Fulvestrant appears to be a highly selective substrate for this enzyme (40). In contrast, N-glucuronidation of trifluoperazine and tamoxifen are selectively catalyzed by UGT1A4 and the steroidal sapogenins, hecogenin, and tigogenin are low K_m substrates (7–20 μM) for 1A4, but not 1A3. UGT1A4 has good activity for progestins, especially 5α-pregnane-3α,20α-diol and androgens such as 5α-androstane-3α,17β-diol.

Assuming that UGT1A3 and UGT1A4 are primarily responsible for the glucuronidation of tertiary amine antihistamines and antidepressants, significant drug interactions involving glucuronidation with these substrates have not been reported. This is not unexpected because $<25\%$ of the dose is excreted as a direct quaternary ammonium glucuronide in urine. The formation of quaternary ammonium glucuronides appears to be highly species specific, with the highest activity in humans and monkeys. Rats and mice are generally incapable of forming quaternary ammonium glucuronides (UGT1A3 and UGT1A4 are pseudogenes in these rodents). Lamotrigine, a novel triazine anticonvulsant, is extensively glucuronidated at the 2-position of the triazine ring in humans ($>80\%$ of the dose is excreted in human urine) (41). It is not significantly glucuronidated in rats or dogs, but 60% of the dose is excreted in guinea pig urine as the 2-*N*-glucuronide (42). Several significant interactions have been reported for lamotrigine in humans. Lamotrigine glucuronidation is induced in patients taking phenobarbital, phenytoin, or carbamazepine (CAR inducers), resulting in a twofold decrease in apparent half-life from 25 hours to approximately 12 hours (43). In contrast, valproic acid inhibits lamotrigine glucuronidation resulting in a two- to threefold increase in half-life (44). Valproic acid is a weak substrate for UGT1A4 and UGT1A3 (U Argikar, PhD thesis, University of Minnesota, 2006), but has higher affinity for UGT2B7. Lamotrigine had a small, but significant effect (25% increase) on the apparent oral clearance of valproic acid (44). This increase could be due to induction of the UGTs responsible for valproic acid glucuronidation, since chronic treatment with lamotrigine results in autoinduction. The interaction between APAP and lamotrigine has also been studied. Surprisingly, APAP decreased the lamotrigine AUC by approximately 20% after multiple oral doses in human volunteers. Lamotrigine clearance was 32% lower in seven patients with Gilbert syndrome compared with persons with normal bilirubin levels, but it does not appear to be a substrate for UGT1A1 (23).

Polymorphisms have been identified in both UGT1A3 and UGT1A4. Iwai et al. identified four nonsynonymous single-nucleotide polymorphisms (SNPs) in the *UGT1A3* sequence of a Japanese population ($n = 100$) at Q6R, W11R, R45W, and V47A (45). Five allele combinations with frequencies of 0.055 to 0.13 were identified. The intrinsic clearances of estrone glucuronidation for the cloned, expressed variants were determined and the only significant difference was in the W11R-V47A variant (*UGT1A3**2) that showed an increase of 369% due to a fivefold lower K_m value (allele frequency = 0.125). In contrast, Ehmer et al. reported that the W11R and V47A variants were much more common in German Caucasians (allele frequency = 0.65 and 0.58, respectively) (46). Chen et al. extended this work and examined activities of the variants with several other flavonoid substrates, including quercetin, luteolin, and kaempferol, and also found increased activity (47). They found that the R45W variant had 3.5 to 4.7 times higher intrinsic clearance toward the flavonoids, whereas for estrone, activity was reduced to 70% of control (47). Regioselectivity in the glucuronidation of quercetin was also altered between variants. Two common variants in the UGT1A4 gene have also been identified, but the effect on activity appears to vary depending on the substrate. Ehmer et al. found two major variants at P24T and L48V (allele frequencies of 0.07–0.1 in Caucasians). The L48V mutant completely lost dihydrotestosterone glucuronidation activity (46), but was more efficient for 4-(methylnitrosamino)-1-(3-pyridyl)-1-butanol (NNAL) (48) and clozapine compared with wild-type UGT1A4 (49). Catalytic efficiencies for substrates such as *trans*-androsterone, imipramine, cyproheptadine, and tigogenin also changed (49).

Regulation of UGT1A4 and UGT1A3 has been recently investigated in a transgenic human UGT1A knock-in mouse model (50). UGT1A3 bile acid glucuronidation was highly upregulated by peroxisome proliferator activated receptor (PPAR)-α agonists (51). UGT1A4 activity and mRNA expression was inducible by PXR and CAR agonists. Consequently, induction interactions are likely to occur and have been demonstrated in humans as demonstrated by lamotrigine interactions with inducing anticonvulsants.

IV. UGT1A6

UGT1A6 is the most important enzyme for the conjugation of planar phenols and amines. It displays high activity for a variety of aromatic alcohols, including 1-naphthol, 4-nitrophenol, 4-methylumbelliferone, and APAP. However, these planar phenols are substrates for several other UGT enzymes. Immunoinhibition studies with an antibody raised against the 120 amino acid N-terminal region UGT1A6 peptide fused to *Staphylococcus aureus* protein A revealed that approximately 50% of the 1-naphthol glucuronidation activity in human liver microsomes (HLMs) could be inhibited (52). Cats are highly susceptible to APAP liver toxicity because UGT1A6 is a pseudogene in this species (6). Serotonin appears to be a highly selective endogenous substrate for this enzyme (53). The first exon sequence of UGT1A6 is divergent from other UGT1A sequences, being most similar to UGT1A9 with only a 54% homology. In rats,

UGT1A6 is inducible by polycyclic aromatic hydrocarbons (PAH). UGT1A6 was also induced in human hepatocytes by β-naphthoflavone and in some, but not all, hepatocyte preparations by rifampin. APAP glucuronidation appears to be increased in smokers, perhaps due to PAH-mediated induction of UGT1A6. Serotonin glucuronidation was doubled in microsomes from persons with moderate-to-heavy alcohol use (54).

Krishnaswamy discovered several variants in the UGT1A6 gene (55). The UGT1A6*2 variant (S7A/T181A/R184S) showed a twofold higher activity (lower K_m) for several substrates (serotonin 4-nitrophenol, APAP, valproic acid) when cloned and expressed in HEK-293 cells compared with wild-type enzyme; however, the K_m was higher than wild type in (*2/*2) HLMs (54). Allele frequencies in Caucasians for the S7A, T181A, and R184S variants were 0.32 to 0.37. In Japanese, the frequency of these mutations is somewhat lower (0.22) (56). Response elements for HNF1-α, Nrf-2, AhR, PXR/CAR have been identified in the regulatory region of this gene (55). In a small study of 15 β-thalassemia/hemoglobin E patients, those subjects with a *UGT1A6*2* variant without *UGT1A1*28* showed a significant, lower AUC of APAP, APAP-glucuronide, and APAP-sulfate than those of the patients with wild-type UGT1A1 and UGT1A6 (58).

Interactions involving APAP and its glucuronidation are listed in Table 1. Approximately 50% of a typical dose of APAP is glucuronidated (59). UGT1A1, UGT1A6, and UGT1A9 are the principal UGTs involved in glucuronidation.

Table 1 Interactions Affecting APAP Glucuronidation

Precipitant drug	Object drug	Effect	Comments	Reference
Propranolol	APAP		Fractional clearance to the glucuronide reduced by 27%. Overall CL decreased by 14%	57
Oral contraceptives	APAP		Oral metabolic clearance increased 22–61% due to increased glucuronidation	143
Phenytoin	APAP		CL increased by 46%, half-life decreased by 28% glucuronide/ APAP ratio in urine increased by 41%	144,145
Probenecid	APAP		Renal elimination of glucronide decreased from 260 to 84 mg/day	146
Rifampin	APAP		Glucuronide/APAP ratio increased by 37%	144,147

UGT1A6 is a high-affinity ($K_m = 2.2$ mM), low-capacity enzyme. UGT1A1 has intermediate affinity (9 mM) with high capacity, and UGT1A9 is a low-affinity, high-capacity enzyme (21 mM) (59). With a kinetic model, Court et al. estimated that at typical therapeutic concentrations (0.05–5 mM), UGT1A9 was the most important enzyme (>55% of total activity). Consequently, the mechanism of induction by oral contraceptives, phenytoin, and rifampin is unclear and may involve multiple enzymes.

V. UGT1A7, UGT1A8, UGT1A9, AND UGT1A10

There is a 93–94% sequence homology in the first exon of UGT1A7 to UGT1A10; however, these enzymes show great variation in the level of tissue expression. This group of UGT1A enzymes is highly divergent from UGT1A3 to UGT1A5 with approximately 50% identity in the first exon compared with UGT1A9. UGT1A9 is expressed in human hepatic and kidney tissues, whereas UGT1A7, UGT1A8, and UGT1A10 are expressed extrahepatically. Liver expression appears to be controlled by the presence of an HNF4-α response element at –372 to –360, that is present only in UGT1A9 and a distal response element to HNF-1 (60,61). UGT1A8 and UGT1A10 are intestinal forms (and UGT1A7 is expressed in esophagus and gastric epithelium). In both rat and rabbit, UGT1A7 is expressed in liver. The rabbit (legomorph) enzyme (UGT1A7l) displays high activity for a variety of small phenolic compounds such as 4-methylumbelliferone, p-nitrophenol, vanillin, 4-*tert*-butylphenol, and octylgallate. In addition, the rabbit enzyme is capable of catalyzing the N-glucronidation of imipramine to a quaternary ammonium glucuronide, similar to UGT1A4 (62). Rat UGT1A7 catalyzes the glucuronidation of benzo(a)pyrene phenols and is inducible by both 3-methylcholanthrene (3-MC) and oltipraz. Ciotti demonstrated that human UGT1A7 has very high activity for the glucuronidation of 7-ethyl-10-hydroxycamptothecin (SN-38), the active metabolite of irinotecan, and therefore may play a role in the gastrointestinal first-pass metabolism of this drug along with UGT1A8 and UGT1A10 (63).

UGT1A8 mRNA is expressed in human jejunum, ileum, and colon, but not in the liver or kidney. Intestinal expression of both UGT1A8 and UGT1A10 appears to be due to caudal-related homeodomain protein (Cdx2) consensus site in the respective promoters (64). UGT1A8 catalyzes the glucuronidation of a variety of planar and bulky phenols, coumarins, flavonoids, anthroquinones, and primary aromatic amines (65). It also catalyzes the glucuronidation of several endogenous compounds, including dihydrotestosterone, 2-OH and 4-OH-estrone, estradiol, hypocholic acid, *trans*-retinoic acid, and 4-OH-retinoic acid. Several drugs are also substrates, including opioids (e.g., buprenorphine, morphine, naloxone, and naltrexone), ciprofibrate, diflunisal, furosemide, mycophenolic acid (MPA), phenolphthalein, propofol, raloxifene, 4-OH-tamoxifen, and tolcapone (65). Cloned, expressed UGT1A8 has high intrinsic clearance for the conjugation of flavonoids such as apigenin and narigenin; thus, drug-food interactions are possible, particularly if the drugs display extensive first-pass metabolism in the intestine (65).

UGT1A9 is expressed in human liver, kidney, and colon. UGT1A9 is expressed in greater amounts in kidney than in liver and is the most prevalent UGT in renal tissue. UGT1A9 is largely responsible for the glucuronidation of a variety of bulky phenols, e.g., *tert*-butylphenol and the anaesthetic agent, propofol (2,6-diisopropylphenol, commonly used as a marker substrate). Propofol is a selective substrate for UGT1A8 and UGT1A9, but extrahepatic metabolism of propofol appears to be important because propofol glucuronide is formed in substantial amounts in patients during the anhepatic phase of liver transplantation (66,67). Propofol clearance is greater than liver blood flow, also suggesting that extrahepatic metabolism is important for this compound. It is glucuronidated in vitro by human kidney and small intestinal microsomes. The V_{max} was 3 to 3.5 times higher in human kidney microsomes compared with liver or small intestine microsomes on a milligram per microsomal protein basis. A number of pharmacodynamic interactions have been reported between propofol and benzodiazepines or opoids such as fentanyl and alfentanil (68–70). Pharmacokinetic interaction studies in humans with fentanyl or alfentanil revealed a modest decrease in propofol clearance (20–50%).

UGT1A9 also catalyzes the glucuronidation of clofibric acid *S*-oxazepam, propranolol, raloxifene, valproic acid, *cis*-4-OH-tamoxifen, and several NSAIDs. These acidic drugs appear to be glucuronidated at a much faster rate by cloned, expressed UGT1A9 than by UGT2B7 on a milligram protein basis (assuming equivalent levels of expression). Formation of the phenolic ether glucuronide of MPA is catalyzed by UGT1A8 and UGT1A9, whereas the acyl glucuronide formation of MPA (a minor metabolite in HLMs) is attributable to UGT2B7. The 7-*O*-glucuronide is the predominant conjugate formed in vivo and is the major excretory metabolite of mycophenolate (90% of the dose in human urine). Tacrolimus and cyclosporine (agents commonly used with mycophenolate in transplant patients) have been shown to inhibit mycophenolate glucuronidation in vitro (71) and were later shown to be substrates for intestinal UGT2B7 (72). In renal transplant patients, cyclosporine increased MPA AUC by 1.8-fold, and sirolimus increased the AUC by 1.5-fold (73). Several investigators suggested that the effect of cyclosporine was due to inhibition of biliary excretion of the glucuronide metabolites by inhibition of organic anion transporters such as MRP2 (73–75). Relatively few clinical drug interactions with NSAIDs have been reported, although probenecid may inhibit glucuronidation directly and cause modest increases in NSAIDs concentrations (see sec. IX on probenecid).

UGT1A9 is an inducible enzyme. In a case study report, rifampin decreased MPA AUC by greater than twofold and increased the AUC of both the phenolic and acyl glucuronides (76), suggesting that there is a PXR response element in the human *UGT1A9* gene. Klaassen et al. had previously shown that mouse Ugt1a9 is upregulated by PXR agonists (77). In rat, phenobarbital is a good general inducer of the glucuronidation of bulky phenols catalyzed by

Table 2 Polymorphisms in the UGT1A8 gene

Allele	Sequence change	Amino acid substitution	Frequency
*UGT1A8*1*			0.551
*UGT1A8*1a*	765A>G	T255T	0.282
*UGT1A8*2*	518C>G	A173G	0.145
*UGT1A8*3*	830G>A	C277Y	0.022

Table 3 Polymorphisms in the UGT1A9 gene

Allele	Sequence change	Amino acid substitution	Frequency ($n = 288$) (%)
*UGT1A9*1*			97.8 (Caucasians)
*UGT1A9*2*	8G>A	C3Y	2.5 (Africans) 0 (Caucasians)
*UGT1A9*3*	98T>C	M33T	2.2–3.6 (Caucasians)
*UGT1A9*4*	726T>G	(Truncated protein)	
*UGT1A9*5*	766G>A	D256N	1.7 (Asians)

UGT1A9. UGT1A9 along with UGT1A6 were inducible by 10 μM tetrachlorodibenzodioxin (TCDD) in Caco-2 cells, a human-derived colon carcinoma cell line (78).

Four genotypes of UGT1A8 have been identified but one mutation is silent (T^{255} A>G, *UGT1A8*1a*), while the other mutations lead to base pair changes: $A^{173}C^{277}$ (UGT1A8*1), $G^{173}C^{277}$ (*UGT1A8**2), and $A^{173}Y^{277}$ (UGT1A8*3) (80). Allele frequencies are: *1 = 0.551, *1a = 0.282, *2 = 0.145, *3 = 0.022 (Table 2). UGT1A8*1 and 1A8*2 appear to exhibit similar activities toward a variety of substrates (e.g., estrone, 4-methylumbelliferone, 17α-ethinylestradiol, hydroxybenzo(a)pyrene (all positions), benzo(a)pyrene *cis*- and *trans*-diols, and hydroxyacetylaminofluorenes). However, little activity toward any substrate was noted with the *3 variant (79). Thibaudeau et al. also found substantially lower activity with 4-OH-estradiol (2- to 3-fold lower intrinsic clearance) and 4-OH-estrone (8- to 13-fold lower intrinsic clearance) in vitro with this variant enzyme (80).

Several polymorphisms in the UGT1A9 gene have been identified (see http://galien.pha.ulaval.ca/alleles/UGT1A/UGT1A9.htm). Coding region mutants and relative frequencies are shown in Table 3. Allele frequencies for the coding region mutations are relatively uncommon (<5%). Functionally, expressed UGT1A9.3 had a drastically reduced intrinsic clearance for SN-38 glucuronidation (Table 4) (81).

Table 4 Kinetic Constants for SN-38 Glucuronidation of Wild-type UGT1A9 Vs. 256N Variant

UGT1A9 variant	K_m (μM)	$V_{max}{}^{a}$ (pmol/ min/mg protein)	Normalized V_{max}/K_m (nL/min/mg protein)
Wild-type 256D	19.3	2.94	153
256N variant	44.4	0.24	7.1

[a] V_{max} and V_{max}/K_m ratios normalized for expression differences.

A small sequencing study of Japanese cancer patients ($n = 61$) was carried out to examine the role of potential UGT1A9 polymorphisms on irinotecan metabolism. Jinno et al. reported that one patient carried a genetic variant 766 G>A, resulting in a nonsynonymous mutation of D256N (82). This variant protein was expressed in COS cells and was characterized with regard to SN-38 glucuronidation. Expression of the protein was slightly lower in COS-1 cells relative to wild type. Kinetic characterization showed large differences in SN-38 glucuronidation.

In vitro studies have indicated that two regulatory region mutations at –275 T>A and –2152 C>T may result in increased expression of UGT1A9 (81). The role of these mutations has been studied in addition to a more rare (<5% allele frequency) coding region mutation, *UGT1A9*3*, (T98C) on MPA kinetics in kidney transplant patients. The two regulatory region mutations are more common appearing in >15% of Caucasians and may result in increased protein expression. In a population of 95 kidney transplant recipients, (83) 16/95 carried only the –275 T>A mutation, 12/95 had only the –2152 C>T mutation, and 11/95 carried both mutations, although Innocenti et al. reported far lower frequencies, 0.0.4 and 0.03, respectively, in 132 Caucasians (84). The kinetics of MPA were not significantly altered at a 1-g dose, but in a smaller number of patients at the 2-g dose, the CL/*F* (apparent oral clearance) was increased (decreased AUC) suggesting that these regulatory region mutations increased enzyme or mRNA expression. In three heterozygote patients with a *UGT1A9*3* allele, MPA AUC increased in accordance with the low activity observed in vitro (83). Innocenti reported a linkage disequilibrium between the two regulatory mutations of UGT1A9 and the –53 $(TA)_7$ mutation of UGT1A1 (84). Glucuronidation of 4-OH-catechol estrogens was not affected in the UGT1A9.2 enzyme, but the Thr33Met mutation resulted in a 9- to 12-fold decrease in intrinsic clearance for 4-OH-estrone glucuronidation and a four- to sixfold decrease in intrinsic clearance in 4-OH-estradiol glucuronidation due to a dramatic decrease in V_{max} (81).

Like UGT1A1, there is also a common TATA box polymorphism in the UGT1A9 gene. The *UGT1A9*22* mutation contains a AT(10)AT [$-118(T)_{9>10}$] repeat instead of the more common AT_9AT repeat (85). Allele frequencies were 60% in Japanese ($n = 87$), 39% in Caucasians ($n = 50$), and 44% in African Americans ($n = 50$). Innocenti found similar frequencies [53% in Asians ($n = 200$) and 39% in Caucasians ($n = 254$)] (84). When transfected into HepG2 cells, the expression level of UGT1A9.22 by Western blotting was 2.6-fold higher. Further studies will be needed to determine if this is true in vivo.

VI. UGT1A10

UGT1A10 is expressed in intestine and kidney and is closely related to UGT1A7 to UGT1A9. Mojarrabi and Mackenzie cloned the cDNA from human colon, and it was 90% homologous to UGT1A9 (86). It is an important enzyme in the extrahepatic metabolism of estrogens (estrone and estradiol) as well as the catechol estrogens with much higher activity than other UGT1 enzymes (87). The binding motif of F90-M91-V92-F93 in UGT1A10 is essential for enzyme activity toward estrogens. When tranfected into COS-7 cells, the enzyme was very active in the conjugation of MPA, the major active metabolite of the prodrug, mycophenolate mofetil, an immunosuppressant agent used for the treatment of allograft rejection and bone marrow transplants. In vitro, the enzyme was shown to catalyze conjugation at both the phenolic hydroxyl at the 7-position and the carboxylic acid moiety to form an acyl glucuronide. Zucker et al. studied the interaction between tacrolimus and MPA in vitro and demonstrated that MPA glucuronidation was 100-fold higher in human kidney microsomes compared with HLMs (71). With a partially purified preparation of the kidney UGT, tacrolimus was found to be a potent inhibitor of MPA glucuronidation ($K_i = 27.3$ ng/mL compared with 2158 ng/mL for cyclosporin A). Both UGT1A9 and UGT1A10 are expressed in human kidney. Tacrolimus would also be expected to affect first-pass metabolism of MPA in the intestine and liver, resulting in an increased C_{max} and AUC. Intestinal first-pass metabolism may be more attributable to UGT1A8 than UGT1A10 because Cheng et al. reported that the formation of MPA-glucuronide was 1900 pmol/min/mg protein for UGT1A8 versus 93 pmol/min/mg protein for UGT1A10 (88). UGT1A10 appears to be less active than UGT1A8 for flavonoids such as alizarin and scopoletin, but further studies will be needed to determine the relative expression levels of the enzymes in the gut. UGT1A10 has not been as extensively examined for other metabolic activities, but it may be an important enzyme in the extrahepatic metabolism of other drugs such as propofol and dobutamine. Compared with UGT1A9, a surprising opposite stereoselectivity for propranolol enantiomers was observed. UGT1A9 prefers *S*-propranolol as a substrate, whereas UGT1A10 prefers *R*-propranolol with relatively equal affinity between the two enzymes. Consequently, HLMs glucuronidate *S*-propranolol selectively, and human intestinal microsomes selectively glucuronidate the R isomer (89). Raloxifene 4-O-glucuronidation is the predominant metabolite formed by both UGT1A10 and human intestinal microsomes (90). In contrast, the 6-*O*-glucuronide of raloxifene was the major metabolite formed in Caco-2 cell lysate and no UGT1A10 mRNA was found in Caco-2 cells. These data suggest that the Caco-2 cell system may not be the optimal model to predict small intestinal glucuronidation. The very low bioavailability of raloxifene in humans (2%) is therefore attributable to UGT1A10 as well as UGT1A9 in the liver (90). Structure-activity relationships for the regioselectivity of UGT1A10 for bioflavonoids were recently studied by Lewinsky et al. (91). Thirty-four out of 42 bioflavonoids tested were UGT1A10 substrates and the 6- and 7-OH groups on the A ring were the preferred sites for glucuronidation. Thus, food-drug interactions may be problematic with substrates of this enzyme.

Variants in *UGT1A10* gene have been recently identified. Lazarus et al. have shown that the Glu139Lys mutant (*UGT1A10*2*) had significantly lower activity for *p*-nitrophenol and phenols of PAH (92). The allele frequency of this variant is rare in Caucasians (0.01%) and more prevalent but also rare in African-Americans (0.05%). Two other coding region SNPs (T202I and M59I) with a frequency of 2.1% were identified in a Japanese population (93). The V_{max} values for the M59I variant were about half of wild type for 17β-estradiol glucuronidation with a similar K_m value (93,94).

VII. UGT2B7

UGT2B7 is an important enzyme involved in the glucuronidation of several drug substrates, including NSAIDs, morphine, 3-OH-benzodiazepines, and zidovudine (ZDV). UGT2B7 has 82% sequence homology to UGT2B4, but has <50% homology to the UGT1A family enzymes. UGT2B4 has limited activity for drug substrates such as 3-O-glucuronidation of morphine and ZDV-5′-O-glucuronidation, but is the primary catalyst for hyodeoxycholic acid glucuronidation. Ritter et al. initially cloned and expressed UGT2B7(H), a protein with a His at amino acid 268 (95). This enzyme had activity toward several steroidal substrates, including estriol and androsterone, with low activity for the bile acid and hyodeoxycholic acid. Jin et al. cloned and expressed a polymorphic variant from the same cDNA library, UGT2B7(Y), with a His268Tyr substitution. UGT2B7(Y) was expressed in COS-7 cells and was more extensively characterized for activity against a variety of drug substrates (96). The enzyme catalyzed the conjugation of several NSAIDs (naproxen, ketoprofen, ibuprofen, fenoprofen, zomipirac, diflunisal, and indomethacin) and 3-OH-benzodiazepines (temazepam, lorazepam, and oxazepam). Tephly et al. demonstrated that UGT2B7 catalyzed both the 3-O- and 6-O-glucuronidation of morphine, 6-O-glucuronidation of codeine, and the conjugation of several other opiods (97). This group also compared the activities of UGT2B7(Y) and UGT2B7(H) that were stably expressed in HEK293 cells. Both isoforms displayed similar activity for a range of compounds. Endogenous substrates for UGT2B7(H) include 4-OH estrone, hyodeoxycholic acid, estriol, androsterone, and epitestosterone. Testosterone is a poor substrate. Other xenobiotic substrates for UGT2B7 are listed below:

Phenols and aliphatic alcohols: abacavir, APAP, almokalant, carvedilol, chloramphenicol, epirubicin, 1′-OH-estragole, 5-OH-rofecoxib, lorazepam, menthol, 4-methylumbelliferone, 1-naphthol (low), 4-nitrophenol, octylgallate, *R*-oxazepam, propranolol, temazepam, ZDV

Carboxylic acid–containing drugs: a variety of NSAIDs, chloramphenicol, ciprofibrate, clofibric acid, dimethylxanthenone-4-acetic acid (DMXAA), MPA (acyl glucuronide), pitavastatin, simvastatin acid, tiaprofenic acid, and valproic acid

Other drugs: almokalant, carvedilol, carbamazepine (N-glucuronidation), clonixin, cyclosporin A, epirubicin, ezetimibe, Maxipost, tacrolimus

On the basis of the substrate activity for a variety of important drugs, one might expect that several interactions could result from competition for UGT2B7. Morphine glucuronidation has been well studied; however, relatively few clinical drug-drug interaction with morphine have been reported. In HLMs, the 3-O-glucuronidation of morphine is biphasic with a high K_m of 2 to 7 μM and a low K_m of 700 to 1600 μM. UGT2B7 is the only human UGT expressed in liver that has been shown to glucuronidate morphine to its pharmacologically active metabolite, morphine-6-glucuronide. Morphine-6-glucuronide is much more potent in binding to the μ receptor in the CNS than morphine (30- to 50-fold more potent). However, morphine-6-glucuronide has poor ability to cross the blood-brain barrier, with a permeability coefficient in rats that 1/57 that of morphine. Morphine-6-glucuronide has potency similar to the analgesic effects of morphine when administered to rats on a mg/kg basis. Since rats are unable to make morphine-6-glucuronide, this reflects a balance of poor permeability and higher CNS potency. In humans, both morphine-3-glucuronide (lacking analgesic activity) and morphine-6-glucuronide are present in higher concentrations in plasma than morphine at steady state. Competitive inhibition with other UGT2B7 substrate may not result in a significant effect on analgesic efficiency of morphine, since morphine levels would rise while morphine-6-glucuronide levels would fall. Morphine glucuronidation is inhibited by various benzodiazepines in vitro in rats, and oxazepam (20 mg/kg PO) was shown to lower the morphine-3-glucuronide/morpine ratio in urine. In vitro, the 6-O-glcuronidation of codeine in HLMs is inhibited by morphine, amitriptyline, diazepam, probenecid, and chloramphenicol with K_i values of 3.5, 0.13, 0.18, 1.7, and 0.27 mM, respectively.

Benzodiazepines containing a hydroxyl group at the 3-position, such as lorazepam, oxazepam, and temazepam, are glucuronidated by UGT2B7. (*S*)-oxazepam is a better substrate for glucuronidation in HLMs than the R isomer with a V_{max}/K_m ratio of 1.125 mL/(min·mg) protein versus 0.25 mL/(min·mg) protein. Inhibition studies with racemic ketoprofen in HLMs demonstrated competitive inhibition for (*S*)-oxazepam, with weaker inhibition of (*R*)-oxazepam glucuronidation. The data did not fit to a simple hyperbolic fit expected of a competitive inhibitor of single enzyme. (*S*)-oxazepam glucuronidation was inhibited (in order of potency) by hyodeoxycholic acid, estriol, (*S*)-naproxen, ketoprofen, ibuprofen, fenoprofen, and clofibric acid. Since these initial findings, Court et al. demonstrated that UGT2B15 is the primary catalyst for (*S*)-oxazepam glucuronidation (98). Drug interaction studies with lorazepam and clofibric acid in humans have been reported and are summarized in Table 5.

Table 5 Interactions Involving UGT2B7 Substrates

Precipitant drug	Object drug	Effect	Comments	Reference
Valproate	Lorazepam	⇑	20% increase in lorazepam AUC, 31% decrease in formation CL of lorazepam glucuronide; 40% decrease in lorazepam CL	148,149
Probenecid	Lorazepam	⇑	Lorazepam CL decreased twofold, half-life increased from 14 to 33 hr	146
Neomycin + Cholestyramine	Lorazepam	⇓	Half-life decreased 19–26%, 34% increase in free oral CL/F, effect attributed to decreased enterohepatic circulation	150
Probenecid	Clofibric acid	⇑	Nonrenal CL decreased by 72%, free clofibric acid C_{ss} increased 3.6-fold	113
Probenecid	Diflunisal	⇑	Formation CL of phenol glucuronide and acyl glucuronide decreased 45% and 54%, respectively	114
Probenecid	Zomepirac	⇑	Zomepirac CL declined by 64%, zomepirac glucuronide CL formation decreased by 71%, urinary excretion of zomepirac glucuronide decreased from 72% to 58%	121
Probenecid	Naproxen	⇑	Decreased naproxen CL	120
Oral contraceptives	Clofibric acid	⇓	Clofibric acid CL increased 48% in women receiving oral contraceptives	156

Abbreviations: CL, clearance. C_{ss}, concentration at steady-state. AUC, area-under the concentration-time curve.

Table 6 Allele Frequency of UGT2B Variants in Caucasians and Asians

UGT2B variant	Frequency in Caucasians ($n = 202$)	Frequency in Asians ($n = 32$)	Percent homozygous for variant protein
UGT2B4 (D458)	0.75	1.00	Caucasian = 8.4 Asian = 0
UGT2B7 (H268)	0.46	0.73	Caucasian = 29.2 Asians = 9.4
UGTB15 (D85)	0.45	0.64	Caucasian = 32.2 Asians = 18.7

Source: From Ref. 111.

A number of clinical drug-drug interactions with ZDV, another selective UGT2B7 substrate, have also been reported and are discussed in section X. Kiang et al. have recently reviewed the literature concerning drug-drug interactions for several UGT2B7 substrates and readers are referred to this extensive review as an additional source of information (8).

A highly prevalent polymorphism has been observed in UGT2B7. The variant of UGT2B7 with a tyrosine at position 268 instead of a histidine (UGT2B7.2) appears to affect the activity of the enzyme toward some substrates, but not all, and is highly prevalent in Caucasians and Asians. Polymorphisms for three UGT2B enzymes UGT2B4 (D458E), UGT2B7 (H268Y), and UGT2B15 (D85Y) have been identified and are shown in Table 6.

Miners et al. reported an ethnic difference in the His268Tyr (802 C>T) variant (99). In 91 Caucasians, the allele frequency for *UGT2B7*2* (802 C>T) was 0.482 versus 0.268 for 84 Japanese subjects. Patel et al. reported a potential polymorphism in the ratio of (*R*)- and (*S*)-oxazepam glucuronides in urine (100). While (*R*)-oxazepam is a substrate for UGT2B7, the turnover is very low and there was no difference between the UGT2B7 variants in terms of stereoselectivity (99). More recent data indicates that (*S*)-oxazepam is a UGT2B17 substrate.

The Tyr268 variant, UGT2B7(Y), glucuronidates menthol and androsterone, compounds not glucuronidated by UGT2B7(H), (UGT2B7.1), UGT2B7(Y), and UGT2B7(H) have similar activities toward opioid and catechol estrogen substrates, except for normorphine, buprenorphine, and norbuprenorphine (101). The location of this amino acid change is near the junction of the variable and constant regions (99). Court et al. found no difference in enzyme kinetics for ZDV, morphine, or codeine between UGT2B7.1 and UGT2B7.2 (Table 7) (102). However, UGT2B7.1 had an 11-fold higher intrinsic clearance (V_{max}/K_m) for aldosterone glucuronidation compared with UGT2B7.2 (157).

Holthe et al. screened 239 Norwegian cancer patient for sequence variation in the coding and regulatory region of UGT2B7 (103). The impact of genetic

Table 7 Kinetics of Buprenorphine and Morphine-3-O-glucuronidation in UGT2B7 Variants

	Buprenorphine		Morphine-3-O-glucuronidation	
UGT2B7 variant	K_m (μM)	V_{max} (pmol/min/mg protein)	K_m (μM)	V_{max} (pmol/min/mg protein)
UGT2B7(H) (UGT2B7.1)	22 ± 6	400 ± 40	633, 331	4779, 3054
UGT2B7(Y) (UGT2B7.2)	3, 1	580, 900	458, 490	5050, 5900

variant of morphine glucuronidation was studied in 175 patients receiving oral morphine. They found 12 SNPs (only one of which was in the coding region—H268Y). There was no functional polymorphism observed for seven common genotypes and the three main haplotypes with regard to the morphine-6-glucuronide/morphine ratio. The authors concluded that factors other than UGT2B7 polymorphisms are responsible for the variability in morphine glucuronidation (104). A similar study on the effect of polymorphisms on morphine kinetics was done in the United States by Sawyer et al. (105). They found that the 802 C>T variant (*UGT2B7*2*) was in complete linkage disequilibrium with a –161 C>T mutation in the regulatory region of UGT2B7. In this study, morphine-6-glucuronide and morphine-3-glucuronide concentrations were significantly lower in C/C patients (105).

VIII. UGT2B15 AND UGTB17

UGT2B15 and UGT2B17 (96% homologous) were initially identified by screening for UGT androgen glucuronidation activity in prostate cells by Belanger et al. (106). *UGT2B17* cDNA was first cloned in 1996, and mRNA was also detected in liver and kidney (107). UGT2B15 specifically catalyzes the conjugation at the 17-OH position of 5α-androgens (dihydrotestosterone, androstane-3α-17β-diol), but can also catalyze the glucuronidation of hydroxy-androgens with high to moderate K_m values. Also, 2- and 4-OH-catechol estrogens are substrates, but with low efficiency. UGT2B17 glucuronidates at both the 3- and the 17-OH positions of androgens as well as (*S*)-oxazepam (98).

UGT2B15 and UGT2B17 are major UGTs in human prostate. UGT2B15 is expressed in adipose tissue, and clearance of racemic oxazepam is faster in obese patients (108) and in women compared to lean men. UGT2B17 is also expressed in liver, kidney, skin, brain, mammary gland ovaries, and uterus. The UGT2B gene cluster is located on chromosome 4q13. Androgens, epidermal growth factor, and interleukin-1 downregulate UGT2B15 and UGT2B17 expression in LnCAP cells (prostate cancer cell line) (109).

Table 8 Frequency of UGT2B15 Variants in Caucasians and Asians

UGT2B15 variant	Frequency in Caucasians ($n = 48$)	Frequency in Asians ($n = 32$)	Alleles (%)
*UGT2B15*2* (D85Y)	0.55	0.72	Caucasian = 27
*UGT2B15*3* (L86S)	–	–	Japanese < 1
*UGTB15*4* (K523T)	0.35	0.64	Caucasian = 11
*UGT2B15*5* (D85Y/K523T)	–	–	Caucasians = 14
*UGT2B15*6* (T352I)	0.02	0.73	Caucasian = 2

*UGT2B15*1* represented 17% of alleles.
Source: Adapted from Ref. 110.

Table 9 Kinetics of *S*-oxazepam in UGT2B15 Variants

UGT2B7 variant	*S*-oxazepam mean velocity (pmol/min/mg protein)
UGT2B15.1 (85D/D)	131
UGT2B15.2 (85Y/Y)	49
UGT2B15.1 (352T/T)	64
UGT2B15.1/6 (352T/I)	135 and 210
UGT2B15.4 (523 K/K)	77
UGT2B15.1 (523 T/T)	65

Source: Adapted from Ref. 110.

A polymorphism has been observed in UGT2B15 (Table 8). The common allele, *UGT2B15*2* results in approximately 50% lower activity in genotyped microsomes with the substrate *S*-oxazepam (see Table 9), but shows increased activity with androgens (110). In contrast, the rare variant, *UGT2B15*6*, may result in a more active or efficient enzyme. No significant difference in velocity was observed in the *UGT2B15*4* variant enzyme (see Table 9) (110). The UGT2B15*2 variant (D85Y) is more prevalent in Asians than in Caucasians (111). Court et al. also identified a gender difference in human liver microsomal samples. Median rates of glucuronidation were 65 pmol/min/mg protein in male samples (25–75% range of 49–112, $n = 38$) versus 39 in females (25–75% range of 30–72, $n = 16$), $p = 0.042$ (110).

Wilson et al. have determined that in some DNA samples, no UGT2B17 DNA could be identified. Further investigation found that a 170 kB stretch of

DNA encompassing the entire UGT2B17 locus was deleted in some individuals (*UGT2B17*2*) (112).

IX. INTERACTIONS WITH PROBENECID

Probenecid is a uricosuric agent that is used in the treatment of gout. Probenecid inhibits the active tubular secretion of a number of organic anions, including uric acid and glucuronides of several different drugs. Detailed studies of clinical interactions between prebenecid and several drugs, including clofibric acid, ZDV, and NSAIDs, have demonstrated that the rate of excretion of glucuronides into the urine is decreased, which coincides with the known effects of probenecid upon organic anion transport. Clinical interactions between probenecid and clofibric acid (113), diflunisal, (114), ketoprofen (115), indomethacin (116), carprofen (117,118), isofezolac (119), naproxen (120), zomepirac (121), and ZDV (122) have been described. In addition to the expected effect of a decreased rate of glucuronide excretion, these studies have also revealed that the clearance of the parent aglycone is also decreased. In several cases, it has been demonstrated that probenecid affects both the nonrenal and renal clearance of the parent aglycones, suggesting that there are multiple mechanisms for the probenecid effect. The apparent decrease in clearance of the parent drugs has been attributed to three basic mechanisms: (1) inhibition of the renal clearance of the parent drug, (2) direct inhibition of the UGT enzyme responsible for the glucuronidation of the parent drugs, and (3) inhibition of the active secretion of the glucuronide and subsequent hydrolysis of the glucuronide back to the aglycone, resulting in reversible metabolism. Several interactions between NSAIDs and probenecid have been reported (referenced above). Inhibition of direct renal excretion may occur but probably does not significantly contribute, since the urinary excretion of unchanged NSAIDs is negligible (115). Consequently, alternate mechanisms have been proposed. Probenecid has been shown to inhibit the formation clearance of zomepirac glucuronide by 78% in humans, suggesting a direct effect on the UGT enzyme responsible for glucuronidation. Similarly, both the phenolic and acyl glucuronide formation clearance of diflunisal was reduced by approximately 50% (114). Glucuronidation of NSAIDs is catalyzed by several UGT enzymes, including UGT1A9 and UGT2B7, although UGT1A9 may be the most important enzyme for these drugs. An alternate mechanism involving hydrolysis of the glucuronide back to the parent aglycone has also been proposed. The reversible metabolism (futile cycle) hypothesis has been well studied with clofibric acid in a uranyl nitrate–induced renal failure model in rabbits (123).

The interaction between ZDV and probenecid has been extensively studied in vitro and in several species. The interaction is complex. Probenecid inhibits the renal tubular secretion of both ZDV and ZDV glucuronide. Probenecid also directly affects the glucuronidation step, thus decreasing the nonrenal clearance of ZDV. For example, the nonrenal clearance of ZDV was significantly decreased from 10.5 ± 2.1 mL/min/kg to 7.8 ± 3.3 mL/min/kg by probenecid in a rabbit model. Probenecid has been demonstrated to be a direct inhibitor of the glucuronidation of ZDV in HLMs. In freshly isolated rat hepatocytes, probenecid decreased ZDV glucuronide by 10-fold. Probenecid also appears to inhibit the efflux of ZDV from the brain, presumably at the choroid plexus.

X. INTERACTIONS WITH ZIDOVUDINE

Zidovudine (3-azido-deoxythymidine, AZT or ZDV) is an important nucleoside used in the treatment of AIDS. It was the first drug approved for the treatment of AIDS, and as such there is a number of in vitro and in vivo drug interaction studies conducted with this compound. Zidovudine (ZDV) is eliminated in humans primarily by glucuronidation; approximately 75% of the dose is excreted as the glucuronide, with the rest excreted unchanged in urine. A small portion of the drug is reduced to 3′-amino-3′-deoxythymidine, a reaction catalyzed by CYP3A4. The enzyme responsible for ZDV glucuronidation is UGT2B7 with a small contribution of UGT2B4 (102,124). HLMs from Crigler-Najar Type I patients and Gunn rat liver microsomes did not show diminished ZDV glucuronidation rates, suggesting that the enzyme responsible was not a member of the UGT1A family of enzymes. In rats, ZDV glucuronidation was inducible by phenobarbital, but not by 3-MC or clofibrate and the activity was inhibited by morphine. The enzyme responsible for ZDV glucuronidation in human is UGT2B7 with a small contribution of UGT2B4 and the activity was inhibited by morphine and probenecid in human liver microsomes (158).

Several in vitro drug interaction studies have been conducted in HLMs. In HLMs, the K_m for ZDV glucuronidation is approximately 2 to 3 mM, a concentration well above the typical therapeutic concentration of 0.5 to 2 μM (159). Turnover of the substrate is also quite slow, which belies the relatively high clearance observed in vivo. On the basis of the determination of K_i in *N*-octyl-β-D-glucoside solubilized HLMs and comparison to therapeutic concentrations in plasma, Resetar et al. predicted potential interactions of more than 10% with probenecid, chloramphenicol, and (+)-naproxen out of 17 drugs tested (159). Rajaonarison et al. examined the inhibitory potential of 55 different drugs on ZDV glucuronidation (125). By comparison of the relevant therapeutic concentrations, interactions were predicted for cefoperazone, penicillin G, amoxicillin, piperacillin, chloramphenicol, vancomycin, miconazole, rifampicin, phenobarbital, carbamezepine, phenytoin, valproic acid, quinidine, phenylbutazone, ketoprofen, probenecid, and propofol. Interactions with β-lactam antibiotics and vancomycin are not likely to be significant because these compounds do not penetrate into cells well and are excreted primarily by direct renal elimination, except for cefoperazone. A similar study was conducted by Sim et al. (126). Indomethacin, naproxen, chloramphenicol, probenecid, and ethinylestradiol decreased the glucuronidation of ZDV (2.5 mM) by over 90% at supratherapeutic concentrations of 10 mM.

Table 10 Clinical Interactions Affecting ZDV Glucuronidation

Precipitant drug	Object drug	Effect	Comments	Reference
Atovaquone	ZDV	⇑	ZDV CL/F decreased by 25%, $AUC_{(m)}/AUC_p$ ratio declined from 4.48 ± 1.94 to 3.12 ±1.1 with atovaquone	151
Fluconazole (400 mg)	ZDV	⇑	Decreased CL/F by 46%, decreased ZDV-G CL_f by 48%, $A_{e(m)}/A_e$ decreased by 34%	152
Methadone	ZDV	⇑	Oral AUC increased by 41%, IV AUC by 19%, Chronic methadone decreased CL by 26%, ZDV-G CL_f decreased by 17%	153
ZDV	Methadone	N. S.	No significant change in methadone levels	153
Naproxen	ZDV	N. S.	No alteration in ZDV pharmacokinetics, ZDV-G AUC significantly decreased by 21%	154
Probenecid	ZDV	⇑	ZDV AUC significantly increased more than twofold	122
Rifampicin	ZDV	⇓	Decreased AUC of ZDV by 2- to 4-fold ($n = 4$), AUC ratio of ZDV-G/ZDV increased in three patients, ratio returned to baseline in one patient discontinuing rifampin	139
Valproate	ZDV	⇑	ZDV AUC increased twofold, $A_{e(m)}/A_e$ in urine decreased by >50%	155

Abbreviations: ZDV, zidovudine; AUC, area under concentration-time curve; CL, clearance; CL_f, formation clearance; $AUC_{(m)}$, AUC of the metabolite; AUC_p, AUC of parent; A_e, amount excreted unchanged in urine; $A_{e(m)}$, amount of metabolite excreted in urine; ZDV-G, zidovudine glucuronide.

Other compounds producing some inhibition of ZDV conjugation were oxazepam, salicylic acid, and acetylsalicyclic acid. More recently, Trapnell et al. examined the inhibition of ZDV at a more relevant concentration of 20 μM in bovine serum albumin (BSA)-activated microsomes by atovaquone, methadone, fluconazole, and valproic acid at therapeutically relevant concentrations (127). Both fluconazole and valproic acid inhibited ZDV glucuronidation by more than 50% at therapeutic concentrations. Clinical interaction studies have been conducted with methadone, fluconazole, naproxen, probenecid, rifampicin, and valproic acid (see Table 10).

XI. IN VITRO APPROACHES TO PREDICTION OF DRUG-DRUG INTERACTIONS

UGTs are membrane-bound enzymes located intracellularly in the endoplasmic reticulum (ER). Unlike cytochrome P450, the active site is located in the lumen of the ER, and there is good evidence for the existence of an ER transporter for UDPGA, the polar, charged cofactor that is produced in the cytosol. Similarly, the polar glucuronides that are formed in the lumen may require specific transporters for drug efflux from the ER. Microsomes maintain this membrane integrity, and thus both UDPGA and substrate access may be limited in incubations. Consequently, a variety of techniques have been used to "active enzyme" or to "remove enzyme latency" in vitro. The previously cited in vitro studies with ZDV can be used to illustrate these approaches.

ZDV glucuronidation has been stimulated by the addition of detergents such as asoleoyl lysophosphatidylcholine (0.8 mg/mg protein optimal), Brij 58 (0.5 mg/mg protein), and *N*-octyl-β-D-glucoside (0.05%) (128). Trapnell et al. reported a 15-fold increase in ZDV glucuronidation rate with 2.25% BSA (127). In our laboratory, we have used a pore-forming antibiotic, alamethacin, to stimulate the glucuronidation of ZDV in HLMs. The advantage of alamethacin is that isozyme-dependent inhibition by detergents can be avoided, but it is still important to determine the optimal concentration for activation for an individual substrate. In our hands, alamethacin stimulated ZDV glucuronidation activity three- to fourfold, to a slightly higher extent than Fraction V BSA (Remmel RP and Streich JA, unpublished data). Addition of BSA to alamethacin did not substantially increase activation. When low-endotoxin, fatty acid–free BSA was used, almost no activation was observed, suggesting that endotoxin or fatty acids may be involved in a detergent-like effect. Recently, Rowland et al. reported that long-chain free fatty acids acted as inhibitors of ZDV or 4-methylumbelliferone glucuronidation resulting in higher K_m/S50 values (128). Alamethacin is now used routinely by many investigators in the field to overcome latency and allow access of UDPGA into the interior of microsomal vescicles (129,130).

Unlike the situation with cytochrome P450, specific and selective inhibitors of individual UGT enzymes may not be available. Furthermore, inhibitory antibodies have not been developed because of the high similarity in amino acid content (identical in all UGT1 enzymes) in the constant region containing the UDPGA binding site. Consequently, at this time the only method available to identify isozyme selectivity is to conduct studies with cloned, expressed enzymes. Fortunately, many of these enzymes have recently been commercially available as microsomes prepared from lymphocytes, mammalian cells, insect cells, or bacteria. Procedures for "activation" of UGT activity in cloned, expressed cell systems also vary, but sonication of whole-cell lysates has been commonly used as a convenient method for screening.

XII. INTERACTIONS INVOLVING DEPLETION OF UDPGA

An alternate mechanism of drug-drug interactions involving glucuronidation may involve depletion of the required cofactor, UDPGA. Several drugs and chemicals have been shown to deplete UDPGA in the rat, including D-galactosamine, diethylether, ethanol, and APAP. In the mouse, Howell et al. demonstrated that valproic acid, chloramphenicol, and salicylamide depleted hepatic UDPGA by >90% at doses of 1 to 2 mmol/kg. Maximal decreases were noted at 7 to 15 minutes after injection, but rebounded toward control levels by two to four hours after injection (131). Once depleted, UDPGA levels will be replaced by the breakdown of glycogen stores in the liver. For drugs that are glucuronidated but are given at relatively low doses, UDPGA depletion is not likely to be of major importance. Extrahepatic glucuronidation may be more susceptible to depletion of UDPGA, since UDPGA concentrations in liver (279 μmol/kg) were reportedly 15 times higher than intestine, kidney, or lung (160). However, in patients receiving high doses of certain drugs, such as the NSAIDs, ethanol, APAP, and valproate, depletion of UDPGA stores may influence the rate of glucuronidation, especially if glycogen stores are low. For example, lamotrigine clearance is decreased two- to threefold in patients also taking valproic acid (44). Lamotrigine has shown to be glucuronidated by UGT1A4 and may also be a substrate for UGT1A3, which also catalyzes the glucuronidation of many tertiary amine drugs. Valproic acid is a slow substrate for UGT1A3 and is weak inhibitor of lamotrigine glucuronidation in microsomes containing excess UDPGA (161). The maximum recommended dose of valproic acid is 60 mg/kg/day (4200 mg/day), which is equivalent to a dose of 0.14 mmol/kg. Thus, it is conceivable that UDPGA depletion may play a role in interactions involving valproic acid. A similar case could be made for patients taking high dose of APAP, although in the case of lamotrigine, coadministration of APAP resulted in an unexpected 20% decrease in lamotrigine AUC. Evidence for UDPGA depletion by any drug in humans is lacking, and thus the clinical relevance of this mechanism is unclear.

XIII. INTERACTIONS INVOLVING INDUCTION OF UGT ENZYMES

Regulation of the UGT enzymes has been well studied in animals, especially in the rat. It is clear that many of the enzymes involved in metabolism of xenobiotics share common regulatory sequences (response elements) in the 5′ promoter region that respond to classic inducers such as 3-MC, phenobarbital, clofibrate, dexamethasone, and rifampin. Treatment of rats with PAH, such as β-naphthoflavone (β-NF), or 3-MC has been shown to increase the transcription of *UGT1A6*, an enzyme that conjugates a variety of planar phenols, such as 1-naphthol. UGT1A6, the PAH-inducible cytochrome P450 enzymes, CYP1A1 and CYP1A2, glutathione transferase Ya (GSTA1-1), NAD (P)H-menadione oxidoreductase, and class 3 aldehyde reductase (ALDH3) are members of an Ah-receptor gene battery because all of the genes encoding these enzyme contain a xenobiotic response element (XRE) in their 5′ promoter regions. In humans, omeprazole and cigarette smoking have been shown to induce CYP1A1/2. Cigarette smoking modestly induces the glucuronidation of APAP, codeine, mexiletine, and propranolol. In smokers or patients receiving omeprazole treatment, the in vitro glucuronidation of 4-methylumbelliferone (a general substrate for UGT activity) was not significantly induced in duodenal mucosal biopsies. 1-Naphthol glucuronidation (a marker substrate for UGT1A6) was induced fourfold by β-NF in Caco-2 cells, a human colon carcinoma cell line. In contrast, CYP1A1 activity (ethoxyresorufin-deethylation) was induced by more than 100-fold in the same cell line. 1-Naphthol glucuronidation was not affected by the addition of rifampin or clofibrate. Induction of UGT1A6 mRNA and 1-naphthol glucuronidation by β-NF was observed in MZ-Hep-1 cells, another human hepatocarcinoma line. Rifampin (100 μM) significantly increased this activity in MZ-Hep-1 cells, but not in KYN-2 cells. A variable response to induction by rifampin and β-NF was also observed in cultured hepatocyes isolated from five different donors. Fabre et al. also reported that inducibility of glucuronidation of 1-naphthol by β-NF in human hepatocytes was variable (132).

Induction of glucuronidation by anticonvulsant drugs such as phenobarbital, phenytoin, and carbamazepine has been demonstrated for a number of different drugs, including APAP, chloramphenicol, irinotecan, lamotrigine, valproic acid, and ZDV. HLMs obtained from patients treated with phenytoin or phenobarbital displayed two or three times higher activity for the glucuronidation of bilirubin, 4-methylumbelliferone, and 1-naphthol compared with control HLMs. Less is known about the response to induction of the mRNA concentrations of individual genes, but Sutherland et al. (133) reported that the UGT1A1 mRNA was elevated in livers from individuals treated with phenytoin and phenobarbital. Bilirubin conjugation is also elevated in microsomes prepared from patients taking phenobarbital and phenytoin, and rat bilirubin UGT activity was inducible by phenobarbital and clofibrate in H4IIE rat hepatoma cells. However, when a proximal 611 bp UGT1A1 promoter/luciferase reporter gene construct was transfected into H4IIE cells, no induction was observed upon treatment with phenobarbital. Retinoic acid and a combination of retinoic acid and WY 14643 (a potent PPAR-α ligand) both increased luciferase activity. Patients with Crigler-Najjar Type II syndrome (a genetic deficiency in UGT1A1) have been treated with phenobarbital or clofibrate in order to increase bilirubin glucuronidation. The beneficial effect could arise either by increasing the transcription of a poorly expressed UGT1A1 or by inducing UGT1A4 (the minor builirubin enzyme). Lamotrigine, a triazine anticonvulsant that metabolizes to a quaternary ammonium is increased approximately twofold in patients taking other inducing anticonvulsants, suggesting that UGT1A4 is inducible by CAR activators such as phenobarbital, phenytoin, and carbamazepine.

Induction of the glucuronidation of several drugs, including lamotrigine by oral contraceptive steroids (OCSs), has been observed (134). The formation clearance to the acyl glucuronide of diflunisal increased from 3.01 mL/min in control women compared with 4.81 mL/min in OCS users (135). The urinary recovery of phenprocoumon glucuronide was 14% of the dose in age-matched controls compared with 21% of the dose in OCS users. Ethinylestradiol doubled the fraction of propranolol metabolized to the glucuronide without affecting total body clearance (136). Oral contraceptives have also been shown to induce the metabolism of APAP, clofibric acid, and temazepam.

Rifampin is a potent inducer of several cytochrome P450 enzymes via PXR activation and also appears to be an inducer of several UGTs such as UGT1A1, UGT1A4, UGT1A9, and UGT2B7. Several case reports have documented an induction of methadone withdrawal symptoms upon introduction of anti-tuberculosis therapy that included rifampin. Fromm et al. studied the effect of rifampin (600 mg/day for 18 days) on morphine analgesia and pharmacokinetics in healthy volunteers (137). Morphine CL/F was increased from 3.58 ± 0.97 L/min initially to 5.49 ± 2.97 L/min during rifampin treatment. The AUC of both morphine-6-glucuronide (an active metabolite) and morphine-3-glucuronide were significantly reduced, although the ratio of the morphine AUC/AUCs of the glucuronide was not significantly increased. Since the metabolite/parent ratios in blood were not affected, the authors suggested that rifampin may have affected the absorption of morphine, perhaps by induction of MDR1 (P-glycoprotein) or an alternate pathway of metabolism or excretion was enhanced, since the urinary recovery of both the glucuronide was decreased. The area under the pain threshold–time curve (cold pressor test) was also significantly reduced by rifampin treatment. Both methadone and morphine are reported substrates for UGT2B7. Rifampicin has also been shown to double the oral clearance of lamotrigine, a UGT1A4 substrate (138). Rifampin appears to significantly increase the glucuronidation of zidovudine (ZDV) in humans (139). Burger et al. reported a higher CL/F and significantly increased ratio of ZDV-glucuronide/ZDV in plasma in four AIDS patients on rifampin compared with untreated controls (140). In one patient, who had stopped rifampin, the metabolite/parent AUC ratio also decreased. Rifabutin, a new rifamycin analog, has been reported to decrease ZDV C_{max} and AUC by 48% and 37%, respectively. However, Gallicano et al. reported that 300 mg of rifabutin/day for 7 or 14 days had no significant effect on ZDV pharmacokinetics, except for a statistically significant decrease in half-life from 1.5 to 1.1 hours (139). Culture of human hepatocytes with 15-μM rifabutin for 48 hours modestly increased the rate of ZDV glucuronidation (28% increase) in one of two donors, but no significant induction was observed with either rifampin or rifapentine, which were more potent inducers of CYP3A4 and CYP2C8/9 in vitro.

XIV. METABOLIC SWITCHING AND INHIBITION OF GLUCURONIDATION

Glucuronidation is normally a primary detoxification pathway. In cases where glucuronidation becomes saturated or inhibited, metabolic switching to form reactive metabolites (typically catalyzed by cytochrome P450 enzymes) can occur.

APAP is the classic example of a drug that at high doses is hepatotoxic because saturation of phase II pathways (glucuronidation and sulfation) due to metabolic switching to a CYP2E1-mediated pathway to form *N*-acetylbenzoquinoneimine. Our laboratory has recently shown that inhibition of naltrexone metabolism by NSAIDs can lead to hepatotoxicity. In vitro experiments have revealed that naltrexone is metabolized by CYP3A4 to form a catechol metabolite that is rapidly oxidized to a quinone and quinonemethide as evidenced by the formation of two glutathione conjugates in a microsomal incubation (Kalyanaraman, Kim, and Remmel, unpublished). Naltrexone glucuronidation was inhibited by NSAIDs, especially fenamates, and the reduction to β-naltrexol (the primary metabolic pathway) is also inhibited by NSAIDs (162). Glucuronides can also be substrates for cytochrome P450 enzymes. Gemfibrozil glucuronide was shown to be a potent inhibitor of CYP2C8, and inhibition of CYP2C8 and competition of the UGT-catalyzed lactonization of statins is the mechanism for the interaction between cerivastatin and gemfibrozil (142). This interaction was an important factor in the removal of cerivastatin (Baycol®) from the market.

XV. CONCLUSIONS

It is clear from the examples just discussed that interactions involving glucuronidation are possible, especially for drugs that extensively excreted as glucuronides. Because of the overlapping substrate specificity among different UGTs, most interactions (particularly with phenolic substrates) are likely to be relatively modest. Prediction of interactions is possible in HLMs, but it is important to conduct these studies at relevant therapeutic concentrations. With the availability of cloned, expressed enzymes, detailed kinetic studies of inhibitory interactions may be carried out. Induction potential may be accomplished in human hepatocytes or perhaps by utilization of a reporter gene assay similar to studies conducted with cytochrome P450 enzymes. While outside the scope of this review, interactions involving glucuronide transport may be important as well.

REFERENCES

1. Burchell B, Brierley CH, Rance D. Specificity of human UDP-glucuronosyltransferases and xenobiotic glucuronidation. Life Sci 1995; 57(20):1819–1831.
2. Burchell B, Coughtrie MW. Genetic and environmental factors associated with variation of human xenobiotic glucuronidation and sulfation. Environ Health Perspect 1997; 105(suppl 4):739–747.
3. Mackenzie PI, Owens IS, Burchell B, et al. The UDP glycosyltransferase gene superfamily: recommended nomenclature update based on evolutionary divergence. Pharmacogenetics 1997; 7(4):255–269.
4. Tukey RH, Strassburg CP. Human UDP-glucuronosyltransferases: metabolism, expression, and disease. Annu Rev Pharmacol Toxicol 2000; 40:581–616.

5. Burchell B, Brierley CH, Monaghan G, et al. The structure and function of the UDP-glucuronosyltransferase gene family. Adv Pharmacol 1998; 42:335–338.
6. Court MH, Greenblatt DJ. Molecular genetic basis for deficient acetaminophen glucuronidation by cats: UGT1A6 is a pseudogene, and evidence for reduced diversity of expressed hepatic UGT1A isoforms. Pharmacogenetics 2000; 10(4):355–369.
7. Jedlitschky G, Cassidy AJ, Sales M, et al. Cloning and characterization of a novel human olfactory UDP-glucuronosyltransferase. Biochem J 1999; 340(pt 3): 837–843.
8. Kiang TK, Ensom MH, Chang TK. UDP-glucuronosyltransferases and clinical drug-drug interactions. Pharmacol Ther 2005; 106(1):97–132.
9. Senafi SB, Clarke DJ, Burchell B. Investigation of the substrate specificity of a cloned expressed human bilirubin UDP-glucuronosyltransferase: UDP-sugar specificity and involvement in steroid and xenobiotic glucuronidation. Biochem J 1994; 303(pt 1):233–240.
10. Monaghan G, Ryan M, Seddon R, et al. Genetic variation in bilirubin UPD-glucuronosyltransferase gene promoter and Gilbert's syndrome [see comments]. Lancet 1996; 347(9001):578–581.
11. Guillemette C. Pharmacogenomics of human UDP-glucuronosyltransferase enzymes. Pharmacogenomics J 2003; 3(3):136–158.
12. Kitagawa C, Ando M, Ando Y, et al. Genetic polymorphism in the phenobarbital-responsive enhancer module of the UDP-glucuronosyltransferase 1A1 gene and irinotecan toxicity. Pharmacogenet Genomics 2005; 15(1):35–41.
13. Ferraris A, D'Amato G, Nobili V, et al. Combined test for UGT1A1 -3279T->G and A(TA)nTAA polymorphisms best predicts Gilbert's syndrome in Italian pediatric patients. Genet Test 2006; 10(2):121–125.
14. Costa E. Hematologically important mutations: bilirubin UDP-glucuronosyltransferase gene mutations in Gilbert and Crigler-Najjar syndromes. Blood Cells Mol Dis 2006; 36(1):77–80.
15. Urawa N, Kobayashi Y, Araki J, et al. Linkage disequilibrium of UGT1A1 *6 and UGT1A1 *28 in relation to UGT1A6 and UGT1A7 polymorphisms. Oncol Rep 2006; 16(4):801–806.
16. Akaba K, Kimura T, Sasaki A, et al. Neonatal hyperbilirubinemia and mutation of the bilirubin uridine diphosphate-glucuronosyltransferase gene: a common missense mutation among Japanese, Koreans and Chinese. Biochem Mol Biol Int 1998; 46 (1):21–26.
17. Rotger M, Taffe P, Bleiber G, et al. Gilbert syndrome and the development of antiretroviral therapy-associated hyperbilirubinemia. J Infect Dis 2005; 192(8): 1381–1386.
18. Lankisch TO, Moebius U, Wehmeier M, et al. Gilbert's disease and atazanavir: from phenotype to UDP-glucuronosyltransferase haplotype. Hepatology 2006; 44(5):1324–1332.
19. de Morais SM, Uetrecht JP, Wells PG. Decreased glucuronidation and increased bioactivation of acetaminophen in Gilbert's syndrome. Gastroenterology 1992; 102(2): 577–586.
20. Ullrich D, Sieg A, Blume R, et al. Normal pathways for glucuronidation, sulphation and oxidation of paracetamol in Gilbert's syndrome. Eur J Clin Invest 1987; 17(3): 237–240.

21. Rauchschwalbe SK, Zuhlsdorf MT, Wensing G, et al. Glucuronidation of acetaminophen is independent of UGT1A1 promotor genotype. Int J Clin Pharmacol Ther 2004; 42(2):73–77.
22. Herman RJ, Chaudhary A, Szakacs CB. Disposition of lorazepam in Gilbert's syndrome: effects of fasting, feeding, and enterohepatic circulation. J Clin Pharmacol 1994; 34(10):978–984.
23. Posner J, Cohen AF, Land G, et al. The pharmacokinetics of lamotrigine (BW430C) in healthy subjects with unconjugated hyperbilirubinaemia (Gilbert's syndrome). Br J Clin Pharmacol 1989; 28(1):117–120.
24. Green MD, Tephly TR. Glucuronidation of amines and hydroxylated xenobiotics and endobiotics catalyzed by expressed human UGT1.4 protein. Drug Metab Dispos 1996; 24(3):356–363.
25. Green MD, King CD, Mojarrabi B, et al. Glucuronidation of amines and other xenobiotics catalyzed by expressed human UDP-glucuronosyltransferase 1A3. Drug Metab Dispos 1998; 26:507–512.
26. Ando Y, Saka H, Asia G, et al. UGT1A1 genotypes and glucuronidation of SN-38, the active metabolite of irinotecan. Ann Oncol 1998; 9(8):845–847.
27. Innocenti F, Ratain MJ. Pharmacogenetics of irinotecan: clinical perspectives on the utility of genotyping. Pharmacogenomics 2006; 7(8):1211–1221.
28. Iyer L, King CD, Whitington PF, et al. Genetic predisposition to the metabolism of irinotecan (CPT-11). Role of uridine diphosphate glucuronosyltransferase isoform 1A1 in the glucuronidation of its active metabolite (SN-38) in human liver microsomes. J Clin Invest 1998; 101(4):847–854.
29. Toffoli G, Cecchin E, Corona G, et al. The role of UGT1A1*28 polymorphism in the pharmacodynamics and pharmacokinetics of irinotecan in patients with metastatic colorectal cancer. J Clin Oncol 2006; 24(19):3061–3068.
30. Murry DJ, Cherrick I, Salama V, et al. Influence of phenytoin on the disposition of irinotecan: a case report. J Pediatr Hematol Oncol 2002; 24(2):130–133.
31. Mathijssen RH, Verweij J, Loos WJ, et al. Irinotecan pharmacokinetics-pharmacodynamics: the clinical relevance of prolonged exposure to SN-38. Br J Cancer 2002; 87(2):144–150.
32. Kuhn JG. Influence of anticonvulsants on the metabolism and elimination of irinotecan. A North American Brain Tumor Consortium preliminary report. Oncology (Williston Park) 2002; 16(8 suppl 7):33–40.
33. Mathijssen RH, Verweij J, de Bruijn P, et al. Effects of St. John's wort on irinotecan metabolism. J Natl Cancer Inst 2002; 94(16):1247–1249.
34. van Erp NP, Baker SD, Zhao M, et al. Effect of milk thistle (Silybum marianum) on the pharmacokinetics of irinotecan. Clin Cancer Res 2005; 11(21):7800–7806.
35. Stewart CF, Leggas M, Schuetz JD, et al. Gefitinib enhances the antitumor activity and oral bioavailability of irinotecan in mice. Cancer Res 2004; 64(20):7491–7499.
36. Ohtsu T, Sasaki Y, Igarashi T, et al. Unexpected hepatotoxicities in patients with non-Hodgkin's lymphoma treated with irinotecan (CPT-11) and etoposide. Jpn J Clin Oncol 1998; 28(8):502–506.
37. Wen Z, Tallman MN, Ali SY, et al. UDP-glucuronosyltransferase 1A1 is the principal enzyme responsible for etoposide glucuronidation in human liver and intestinal microsomes: structural characterization of phenolic and alcoholic glucuronides of etoposide and estimation of enzyme kinetics. Drug Metab Dispos 2007; 35(3):371–380.

38. Watanabe Y, Nakajima M, Ohashi N, et al. Glucuronidation of etoposide in human liver microsomes is specifically catalyzed by UDP-glucuronosyltransferase 1A1. Drug Metab Dispos 2003; 31(5):589–595.
39. Corona G, Vaccher E, Cattarossi G, et al. Potential hazard of pharmacokinetic interactions between lopinavir-ritonavir protease inhibitors and irinotecan. Aids 2005; 19(17):2043–2044.
40. Chouinard S, Tessier M, Vernouillet G, et al. Inactivation of the pure antiestrogen fulvestrant and other synthetic estrogen molecules by UDP-glucuronosyltransferase 1A enzymes expressed in breast tissue. Mol Pharmacol 2006; 69(3):908–920.
41. Sinz MW. Animal model systems for the study of selected antiepileptic drug interactions. In: Ph.D thesis. Minneapolis: University of Minnesota, 1991.
42. Remmel RP, Sinz MW. A quaternary ammonium glucuronide is the major metabolite of lamotrigine in guinea pigs. In vitro and in vivo studies. Drug Metab Disp 1991; 19(3):630–636.
43. Jawad S, Yuen WC, Peck AW, et al. Lamotrigine: single-dose pharmacokinetics and initial 1 week experience in refractory epilepsy. Epilepsy Res 1987; 1(3):194–201.
44. Anderson GC, Yau MK, Gidal BE, et al. Bidirectional interaction of valproate and lamotrigine in healthy subjects. Clin Pharmacol Ther 1996; 60:145–156.
45. Iwai M, Maruo Y, Ito M, et al. Six novel UDP-glucuronosyltransferase (UGT1A3) polymorphisms with varying activity. J Hum Genet 2004; 49(3):123–128.
46. Ehmer U, Vogel A, Schütte JK, et al. Variation of hepatic glucuronidation: Novel functional polymorphisms of the UDP-glucuronosyltransferase UGT1A4. Hepatology 2004; 39(4):970–977.
47. Chen Y, Chen S, Li X, et al. Genetic variants of human UGT1A3: functional characterization and frequency distribution in a Chinese Han population. Drug Metab Dispos 2006; 34(9):1462–1467.
48. Wiener D, Doerge DR, Fang JL, et al. Characterization of N-glucuronidation of the lung carcinogen 4-(methylnitrosamino)-1-(3-pyridyl)-1-butanol (NNAL) in human liver: importance of UDP-glucuronosyltransferase 1A4. Drug Metab Dispos 2004; 32(1): 72–79.
49. Mori A, Maruo Y, Iwai M, et al. UDP-glucuronosyltransferase 1A4 polymorphisms in a Japanese population and kinetics of clozapine glucuronidation. Drug Metab Dispos 2005; 33(5):672–675.
50. Chen S, Beaton D, Nguyen N, et al. Tissue-specific, inducible, and hormonal control of the human UDP-glucuronosyltransferase-1 (UGT1) locus. J Biol Chem 2005; 280(45):37547–37557.
51. Senekeo-Effenberger K, Chen S, Magdalou J, et al. Expression of the human UGT1 locus in transgenic mice by 4-chloro-6-(2,3-xylidino)-2-pyrimidinylthioacetic acid (WY-14643) and implications on drug metabolism through peroxisome proliferator-activated receptor alpha activation. Drug Metab Dispos 2007; 35(3): 419–427.
52. Ouzzine M, Pillot T, Fournel-Gigleux S, et al. Expression and role of the human liver UDP-glucuronosyltransferase UGT1*6 analyzed by specific antibodies raised against a hybrid protein produced in Escherichia coli. Arch Biochem Biophys 1994; 310(1):196–204.
53. Krishnaswamy S, Duan SX, von Moltke LL, et al. Validation of serotonin (5-hydroxtryptamine) as an in vitro substrate probe for human UDP-glucuronosyltransferase (UGT) 1A6. Drug Metab Dispos 2003; 31(1):133–139.
54. Krishnaswamy S, Hao Q, Al-Rohaimi A, et al. UDP glucuronosyltransferase (UGT) 1A6 pharmacogenetics: I. Identification of polymorphisms in the 5′-regulatory and

exon 1 regions, and association with human liver UGT1A6 gene expression and glucuronidation. J Pharmacol Exp Ther 2005; 313(3):1331–1339.

55. Krishnaswamy S, Hao Q, Al-Rohaimi A, et al. UDP glucuronosyltransferase (UGT) 1A6 pharmacogenetics: II. Functional impact of the three most common nonsynonymous UGT1A6 polymorphisms (S7A, T181A, and R184S). J Pharmacol Exp Ther 2005; 313(3):1340–1346.
56. Saeki M, Saito Y, Jinno H, et al. Genetic polymorphisms of UGT1A6 in a Japanese population. Drug Metab Pharmacokinet 2005; 20(1):85–90.
57. Baraka OZ, Truman CA, Ford JM, et al. The effect of propranolol on paracetamol metabolism in man. Br J Clin Pharmacol 1990; 29(2):261–264.
58. Tankanitlert J, Morales NP, Howard TA, et al. Effects of combined UDP-Glucuronosyltransferase (UGT) 1A1*28 and 1A6*2 on paracetamol pharmacokinetics in beta-thalassemia/HbE. Pharmacology 2007; 79(2):97–103.
59. Court MH, Duan SX, von Moltke LL, et al. Interindividual variability in acetaminophen glucuronidation by human liver microsomes: identification of relevant acetaminophen UDP-glucuronosyltransferase isoforms. J Pharmacol Exp Ther 2001; 299(3):998–1006.
60. Barbier O, Girard H, Inoue Y, et al. Hepatic expression of the UGT1A9 gene is governed by hepatocyte nuclear factor 4alpha. Mol Pharmacol 2005; 67(1):241–249.
61. Gardner-Stephen DA, Mackenzie PI. Hepatocyte nuclear factor1 transcription factors are essential for the UDP-glucuronosyltransferase 1A9 promoter response to hepatocyte nuclear factor 4alpha. Pharmacogenet Genomics 2007; 17(1):25–36.
62. Bruck M, Li Q, Lamb JG, et al. Characterization of rabbit UDP-glucuronosyltransferase UGT1A7: tertiary amine glucuronidation is catalyzed by UGT1A7 and UGT1A4. Arch Biochem Biophys 1997; 348(2):357–364.
63. Ciotti M, Basu N, Brangi M, et al. Glucuronidation of 7-ethyl-10-hydroxycamptothecin (SN-38) by the human UDP-glucuronosyltransferases encoded at the UGT1 locus. Biochem Biophys Res Commun 1999; 260(1):199–202.
64. Gregory PA, Lewinsky RH, Gardner-Stephen DA, et al. Coordinate regulation of the human UDP-glucuronosyltransferase 1A8, 1A9, and 1A10 genes by hepatocyte nuclear factor 1alpha and the caudal-related homeodomain protein 2. Mol Pharmacol 2004; 65(4):953–963.
65. Cheng Z, Radominska-Pandya A, Tephly TR. Cloning and expression of human UDP-glucuronosyltransferase (UGT) 1A8. Arch Biochem Biophys 1998; 356(2):301–305.
66. Veroli P, O'Kelly B, Bertrand F, et al. Extrahepatic metabolism of propofol in man during the anhepatic phase of orthotopic liver transplantation. Br J Anaesth 1992; 68(2):183–186.
67. Takizawa D, Sato E, Hiraoka H, et al. Changes in apparent systemic clearance of propofol during transplantation of living related donor liver. Br J Anaesth 2005; 95 (5):643–647.
68. Pavlin DJ, Coda B, Shen DD, et al. Effects of combining propofol and alfentanil on ventilation, analgesia, sedation, and emesis in human volunteers. Anesthesiology 1996; 84(1):23–37.
69. Cockshott ID, Briggs LP, Douglas EJ, et al. Pharmacokinetics of propofol in female patients. Studies using single bolus injections. Br J Anaesth 1987; 59(9):1103–1110.
70. Gepts E, Camu F, Cockshott ID, et al. Disposition of propofol administered as constant rate intravenous infusions in humans. Anesth Analg 1987; 66(12):1256–1263.

71. Zucker K, Tsaroucha A, Olson L, et al. Evidence that tacrolimus augments the bioavailability of mycophenolate mofetil through the inhibition of mycophenolic acid glucuronidation. Ther Drug Monit 1999; 21(1):35–43.
72. Strassburg CP, Barut A, Obermayer-Straub P, et al. Identification of cyclosporine A and tacrolimus glucuronidation in human liver and the gastrointestinal tract by a differentially expressed UDP-glucuronosyltransferase: UGT2B7. J Hepatol 2001; 34(6):865–872.
73. Picard N, Premaud A, Rousseau A, et al. A comparison of the effect of ciclosporin and sirolimus on the pharmokinetics of mycophenolate in renal transplant patients. Br J Clin Pharmacol 2006; 62(4):477–484.
74. Westley IS, Brogan LR, Morris RG, et al. Role of Mrp2 in the hepatic disposition of mycophenolic acid and its glucuronide metabolites: effect of cyclosporine. Drug Metab Dispos 2006; 34(2):261–266.
75. Hesselink DA, van Hest RM, Mathot RA, et al. Cyclosporine interacts with mycophenolic acid by inhibiting the multidrug resistance-associated protein 2. Am J Transplant 2005; 5(5):987–994.
76. Naesens M, Kuypers DR, Streit F, et al. Rifampin induces alterations in mycophenolic acid glucuronidation and elimination: implications for drug exposure in renal allograft recipients. Clin Pharmacol Ther 2006; 80(5):509–521.
77. Kuypers DR, Naesens M, Vermeire S, et al. The impact of uridine diphosphate-glucuronosyltransferase 1A9 (UGT1A9) gene promoter region single-nucleotide polymorphisms T-275A and C-2152T on early mycophenolic acid dose-interval exposure in de novo renal allograft recipients. Clin Pharmacol Ther 2005; 78(4):351–361.
78. Chen C, Staudinger JL, Klaassen CD. Nuclear receptor, pregname X receptor, is required for induction of UDP-glucuronosyltranferases in mouse liver by pregnenolone-16 alpha-carbonitrile. Drug Metab Dispos 2003; 31(7):908–915.
79. Munzel PA, Schmohl S, Heel H, et al. Induction of human UDP glucuronosyltransferases (UGT1A6, UGT1A9, and UGT2B7) by t-butylhydroquinone and 2,3,7, 8-tetrachlorodibenzo-p-dioxin in Caco-2 cells. Drug Metab Dispos 1999; 27(5): 569–573.
80. Huang YH, Galijatovic A, Nguyen N, et al. Identification and functional characterization of UDP-glucuronosyltransferases UGT1A8*1, UGT1A8*2 and UGT1A8*3. Pharmacogenetics 2002; 12(4):287–297.
81. Thibaudeau J, Lepine J, Tojcic J, et al. Characterization of common UGT1A8, UGT1A9, and UGT2B7 variants with different capacities to inactivate mutagenic 4-hydroxylated metabolites of estradiol and estrone. Cancer Res 2006; 66(1):125–133.
82. Girard H, Court MH, Bernard O, et al. Identification of common polymorphisms in the promoter of the UGT1A9 gene: evidence that UGT1A9 protein and activity levels are strongly genetically controlled in the liver. Pharmacogenetics 2004; 14 (8):501–515.
83. Jinno H, Saeki M, Saito Y, et al. Functional characterization of human UDP-glucuronosyltransferase 1A9 variant, D256N, found in Japanese cancer patients. J Pharmacol Exp Ther 2003; 306(2):688–693.
84. Innocenti F, Liu W, Chen P, et al. Haplotypes of variants in the UDP-glucurnosyltransferase1A9 and 1A1 genes. Pharmacogenet Genomics 2005; 15(5): 295–301.

85. Yamanaka H, Nakajima M, Katoh M, et al. A novel polymorphism in the promoter region of human UGT1A9 gene (UGT1A9*22) and its effects on the transcriptional activity. Pharmacogenetics 2004; 14(5):329–332.
86. Mojarrabi B, Mackenzie PI. The human UDP glucuronosyltransferase, UGT1A10, glucuronidates mycophenolic acid. Biochem Biophys Res Commun 1997; 238 (3):775–778.
87. Starlard-Davenport A, Xiong Y, Bratton S, et al. Phenylalanine(90) and phenylalanine(93) are crucial amino acids within the estrogen binding site of the human UDP-glucuronosyltransferase 1A10. Steroids 2007; 72(1):85–94.
88. Cheng Z, Radominska-Pandya A, Tephly TR. Studies on the substrate specificity of human intestinal UDP- lucuronosyltransferases 1A8 and 1A10. Drug Metab Dispos 1999; 27(10):1165–1170.
89. Sten T, Qvisen S, Uutela P, et al. Prominent but reverse stereoselectivity in propranolol glucuronidation by human UDP-glucuronosyltransferases 1A9 and 1A10. Drug Metab Dispos 2006; 34(9):1488–1494.
90. Jeong EJ, Liu Y, Lin H, et al. Species- and disposition model-dependent metabolism of raloxifene in gut and liver: role of UGT1A10. Drug Metab Dispos 2005; 33 (6):785–794.
91. Lewinsky RH, Smith PA, Mackenzie PI. Glucuronidation of bioflavonoids by human UGT1A10: structure-function relationships. Xenobiotica 2005; 35(2):117–129.
92. Elahi A, Bendaly J, Zheng Z, et al. Detection of UGT1A10 polymorphisms and their association with orolaryngeal carcinoma risk. Cancer 2003; 98(4):872–880.
93. Saeki M, Ozawa S, Saito Y, et al. Three novel single nucleotide polymorphisms in UGT1A10. Drug Metab Pharmacokinet 2002; 17(5):488–490.
94. Jinno H, Saeki M, Tanaka-Kagawa T, et al. Functional characterization of wild-type and variant (T202I and M59I) human UDP-glucuronosyltransferase 1A10. Drug Metab Dispos 2003; 31(5):528–532.
95. Ritter JK, Chen F, Sheen YY, et al. Two human liver cDNAs encode UDP-glucuronosyltransferases with 2 log differences in activity toward parallel substrates including hyodeoxycholic acid and certain estrogen derivatives. Biochemistry 1992; 31(13):3409–3414.
96. Jin C, Miners JO, Lillywhite KG, et al. Complementary deoxyribonucleic acid cloning and expression of a human liver uridine diphosphate-glucuronosyltransferase glucuronidating carboxylic acid-containing drugs. J Pharmacol Exp Ther 1993; 264 (1):475–479.
97. Coffman BL, Rios GR, King CD, et al. Human UGT2B7 catalyzes morphine glucuronidation. Drug Metab Dispos 1997; 25(1):1–4.
98. Court MH, Duan SX, Guillemette C, et al. Stereoselective conjugation of oxazepam by human UDP-glucuronosyltransferases (UGTs): S-oxazepam is glucuronidated by UGT2B15, while R-oxazepam is glucuronidated by UGT2B7 and UGT1A9. Drug Metab Dispos 2002; 30(11):1257–1265.
99. Bhasker CR, McKinnon W, Stone A, et al. Genetic polymorphism of UDP-glucuronosyltransferase 2B7 (UGT2B7) at amino acid 268: ethnic diversity of alleles and potential clinical significance. Pharmacogenetics 2000; 10(8):679–685.
100. Patel M, Tang BK, Kalow W. (S)oxazepam glucuronidation is inhibited by ketoprofen and other substrates of UGT2B7. Pharmacogenetics 1995; 5(1):43–49.

101. Coffman BL, King CD, Rios GR, et al. The glucuronidation of opioids, other xenobiotics, and androgens by human UGT2B7Y(268) and UGT2B7H(268). Drug Metab Dispos 1998; 26(1):73–77.
102. Court MH, Krishnaswamy S, Hao Q, et al. Evaluation of 3′-azido-3′-deoxythymidine, morphine, and codeine as probe substrates for UDP-glucuronosyltransferase 2B7 (UGT2B7) in human liver microsomes: specificity and influence of the UGT2B7*2 polymorphism. Drug Metab Dispos 2003; 31(9):1125–1133.
103. Holthe M, Rakvag TN, Klepstad P, et al. Sequence variations in the UDP-glucuronosyltransferase 2B7 (UGT2B7) gene: identification of 10 novel single nucleotide polymorphisms (SNPs) and analysis of their relevance to morphine glucuronidation in cancer patients. Pharmacogenomics J 2003; 3(1):17–26.
104. Holthe M, Klepstad P, Zahlsen K, et al. Morphine glucuronide-to-morphine plasma ratios are unaffected by the UGT2B7 H268Y and UGT1A1*28 polymorphisms in cancer patients on chronic morphine therapy. Eur J Clin Pharmacol 2002; 58(5): 353–356.
105. Sawyer MB, Innocenti F, Das S, et al. A pharmacogenetic study of uridine diphosphate-glucuronosyltransferase 2B7 in patients receiving morphine. Clin Pharmacol Ther 2003; 73(6):566–574.
106. Belanger A, Pelletier G, Labrie F, et al. Inactivation of androgens by UDP-glucuronosyltransferase enzymes in humans. Trends Endocrinol Metab 2003; 14 (10):473–479.
107. Beaulieu M, Levesque E, Hum DW, et al. Isolation and characterization of a novel cDNA encoding a human UDP-glucuronosyltransferase active on C19 steroids. J Biol Chem 1996; 271(37):22855–22862.
108. Abernethy DR, Greenblatt DJ, Divoll M, et al. Enhanced glucuronide conjugation of drugs in obesity: studies of lorazepam, oxazepam, and acetaminophen. J Lab Clin Med 1983; 101(6):873–880.
109. Chouinard S, Pelletier G, Bélanger A, et al. Isoform-specific regulation of uridine diphosphate-glucuronosyltransferase 2B enzymes in the human prostate: differential consequences for androgen and bioactive lipid inactivation. Endocrinology 2006; 147(11):5431–5442.
110. Court MH, Hao Q, Krishnaswamy S, et al. UDP-glucuronosyltransferase (UGT) 2B15 pharmacogenetics: UGT2B15 D85Y genotype and gender are major determinants of oxazepam glucuronidation by human liver. J Pharmacol Exp Ther 2004; 310(2):656–665.
111. Lampe JW, Bigler J, Bush AC, et al. Prevalence of polymorphisms in the human UDP-glucuronosyltransferase 2B family: UGT2B4(D458E), UGT2B7(H268Y), and UGT2B15(D85Y). Cancer Epidemiol Biomarkers Prev 2000; 9(3):329–333.
112. Wilson W III, Pardo-Manuel de Villena F, Lyn-Cook BD, et al. Characterization of a common deletion polymorphism of the UGT2B17 gene linked to UGT2B15. Genomics 2004; 84(4):707–714.
113. Veenendaal JR, Brooks PM, Meffin PJ. Probenecid-clofibrate interaction. Clin Pharmacol Ther 1981; 29(3):351–358.
114. Macdonald JI, Wallace SM, Herman RJ, et al. Effect of probenecid on the formation and elimination kinetics of the sulphate and glucuronide conjugates of diflunisal. Eur J Clin Pharmacol 1995; 47(6):519–523.
115. Upton RA, Buskin JN, Williams RL, et al. Negligible excretion of unchanged ketoprofen, naproxen, and probenecid in urine. J Pharm Sci 1980; 69(11):1254–1257.

116. Baber N, Halliday L, Sibeon R, et al. The interaction between indomethacin and probenecid. A clinical and pharmacokinetic study. Clin Pharmacol Ther 1978; 24 (3):298–307.
117. Yu TF, Perel J. Pharmacokinetic and clinical studies of carprofen in gout. J Clin Pharmacol 1980; 20(5–6 pt 1):347–351.
118. Spahn H, Spahn I, Benet LZ. Probenecid-induced changes in the clearance of carprofen enantiomers: a preliminary study. Clin Pharmacol Ther 1989; 45(5):500–505.
119. Bannier A, Comet F, Soubeyrand J, et al. Effect of probenecid on isofezolac kinetics. Eur J Clin Pharmacol 1985; 28(4):433–437.
120. Runkel R, Mroszczak E, Chaplin M, et al. Naproxen-probenecid interaction. Clin Pharmacol Ther 1978; 24(6):706–713.
121. Smith PC, Langendijk PN, Boss JA, et al. Effect of probenecid on the formation and elimination of acyl glucuronides: studies with zomepirac. Clin Pharmacol Ther 1985; 38(2):121–127.
122. de Miranda P, Good SS, Yarchoan R, et al. Alteration of ZDV pharmacokinetics by probenecid in patients with AIDS or AIDS-related complex. Clin Pharmacol Ther 1989; 46(5):494–500.
123. Meffin PJ, Zilm DM, Veenendaal JR, et al. A renal mechanism for the clofibric acid-probenecid interaction. J Pharmacol Exp Ther 1983; 227(3):739–742.
124. Barbier O, Turgeon D, Girard C, et al. 3′-azido-3′-deoxythimidine (ZDV) is glucuronidated by human UDP-glucuronosyltransferase 2B7 (UGT2B7). Drug Metab Dispos 2000; 28(5):497–502.
125. Rajaonarison JF, Lacarelle B, De Sousa G, et al. In vitro glucuronidation of 3′-azido-3′-deoxythymidine by human liver. Role of UDP-glucuronosyltransferase 2 form. Drug Metab Dispos 1991; 19(4):809–815.
126. Sim SM, Back DJ, Breckenridge AM. The effect of various drugs on the glucuronidation of ZDV (azidothymidine; ZDV) by human liver microsomes. Br J Clin Pharmacol 1991; 32(1):17–21.
127. Trapnell CB, Klecker RW, Jamis-Dow C, et al. Glucuronidation of 3′-azido-3′-deoxythymidine (ZDV) by human liver microsomes: relevance to clinical pharmacokinetic interactions with atovaquone, fluconazole, methadone, and valproic acid. Antimicrob Agents Chemother 1998; 42(7):1592–1596.
128. Rowland A, Gaganis P, Elliot J, et al. Binding of inhibitory fatty acids is responsible for the enhancement of UDP-glucuronosyltransferase 2B7 activity by albumin: implications for in vitro-in vivo extrapolation. J Pharmacol Exp Ther 2007; 321 (1):137–147.
129. Boase S, Miners JO. In vitro-in vivo correlations for drugs eliminated by glucuronidation: investigations with the model substrate ZDV. Br J Clin Pharmacol 2002; 54(5):493–503.
130. Soars MG, Ring BJ, Wrighton SA. The effect of incubation conditions on the enzyme kinetics of udp-glucuronosyltransferases. Drug Metab Dispos 2003; 31(6): 762–767.
131. Howell SR, Hazelton GA, Klaassen CD. Depletion of hepatic UDP-glucuronic acid by drugs that are glucuronidated. J Pharmacol Exp Ther 1986; 236(3):610–614.
132. Fabre G, Combalbert J, Berger Y, et al. Human hepatocytes as a key in vitro model to improve preclinical drug development. Eur J Drug Metab Pharmacokinet 1990; 15(2):165–171.

133. Sutherland L, Ebner T, Burchell B. The expression of UDP-glucuronosyltransferases of the UGT1 family in human liver and kidney and in response to drugs. Biochem Pharmacol 1993; 45(2):295–301.
134. Sabers A, Ohman I, Christensen J, et al. Oral contraceptives reduce lamotrigine plasma levels. Neurology 2003; 61(4):570–571.
135. Macdonald JI, Herman RJ, Verbeeck RK. Sex-difference and the effects of smoking and oral contraceptive steroids on the kinetics of diflunisal. Eur J Clin Pharmacol 1990; 38(2):175–179.
136. Walle T, Fagan TC, Walle UK, et al. Stimulatory as well as inhibitory effects of ethinyloestradiol on the metabolic clearances of propranolol in young women. Br J Clin Pharmacol 1996; 41(4):305–309.
137. Fromm MF, Eckhardt K, Li S, et al. Loss of analgesic effect of morphine due to coadministration of rifampin. Pain 1997; 72(1–2):261–267.
138. Ebert U, Thong NQ, Oertel R, et al. Effects of rifampicin and cimetidine on pharmacokinetics and pharmacodynamics of lamotrigine in healthy subjects. Eur J Clin Pharmacol 2000; 56(4):299–304.
139. Gallicano KD, Sahai J, Shukla VK, et al. Induction of ZDV glucuronidation and amination pathways by rifampicin in HIV-infected patients. Br J Clin Pharmacol 1999; 48(2):168–179.
140. Burger DM, Meenhorst PL, Koks CH, et al. Pharmacokinetic interaction between rifampin and ZDV. Antimicrob Agents Chemother 1993; 37(7):1426–1431.
141. Ogilvie BW, Zhang D, Li W, et al. Glucuronidation converts gemfibrozil to a potent, metabolism-dependent inhibitor of CYP2C8: implications for drug-drug interactions. Drug Metab Dispos 2006; 34(1):191–197.
142. Prueksaritanont T, Subramanian R, Fang X, et al. Glucuronidation of statins in animals and humans: a novel mechanism of statin lactonization. Drug Metab Dispos 2002; 30(5):505–512.
143. Miners JO, Attwood J, Birkett DJ. Influence of sex and oral contraceptive steroids on paracetamol metabolism. Br J Clin Pharmacol 1983; 16(5):503–509.
144. Miners JO, Attwood J, Birkett DJ. Determinants of acetaminophen metabolism: effect of inducers and inhibitors of drug metabolism on acetaminophen's metabolic pathways. Clin Pharmacol Ther 1984; 35(4):480–486.
145. Bock KW, Bock-Hennig BS. Differential induction of human liver UDP-glucuronosyltransferase activities by phenobarbital-type inducers. Biochem Pharmacol 1987; 36(23):4137–4143.
146. Abernethy DR, Greenblatt DJ, Ameer B, et al. Probenecid impairment of acetaminophen and lorazepam clearance: direct inhibition of ether glucuronide formation. J Pharmacol Exp Ther 1985; 234(2):345–349.
147. Bock KW, Wiltfang J, Blume R, et al. Paracetamol as a test drug to determine glucuronide formation in man. Effects of inducers and of smoking. Eur J Clin Pharmacol 1987; 31(6):677–683.
148. Samara EE, Granneman RG, Witt GF, et al. Effect of valproate on the pharmacokinetics and pharmacodynamics of lorazepam. J Clin Pharmacol 1997; 37(5):442–450.
149. Anderson GD, Gidal BE, Kantor ED, et al. Lorazepam-valproate interaction: studies in normal subjects and isolated perfused rat liver. Epilepsia 1994; 35(1):221–225.
150. Herman RJ, Van Pham JD, Szakacs CB. Disposition of lorazepam in human beings: enterohepatic recirculation and first-pass effect. Clin Pharmacol Ther 1989; 46(1):18–25.

151. Lee BL, Tauber MG, Sadler B, et al. Atovaquone inhibits the glucuronidation and increases the plasma concentrations of ZDV. Clin Pharmacol Ther 1996; 59(1): 14–21.
152. Sahai J, Gallicano K, Pakuts A, et al. Effect of fluconazole on ZDV pharmacokinetics in patients infected with human immunodeficiency virus. J Infect Dis 1994; 169 (5):1103–1107.
153. McCance-Katz EF, Rainey PM, Jatlow P, et al. Methadone effects on ZDV disposition (AIDS Clinical Trials Group 262). J Acquir Immune Defic Syndr Hum Retrovirol 1998; 18(5):435–443.
154. Barry M, Howe J, Back D, et al. The effects of indomethacin and naproxen on ZDV pharmacokinetics. Br J Clin Pharmacol 1993; 36(1):82–85.
155. Lertora JJ, Rege AB, Greenspan DL, et al. Pharmacokinetic interaction between zidovudine and valproic acid in patients infected with human immunodeficiency virus. Clin Pharmacol Ther 1994; 56(3):272–278.
156. Miners JO, Robson RA, Birkett DJ. Gender and oral contraceptive steroids as determinants of drug glucuronidation: effects on clofibric acid elimination. Br J Clin Pharmacol 1984; 18(2):240–243.
157. Girard C, Barbier O, Veilleux G, et al. Human uridine diphosphate-glucuronosyltransferase UGT2B7 conjugates mineralocorticoid and glucocorticoid metabolites. Endocrinology 2003; 144(6):2659–2668.
158. Haumont M, Magdalou J, Lafaurie C, et al. Phenobarbital inducible UDP-glucuronosyltransferase is responsible for glucuronidation of 3′-azido-3′-deoxythymidine: characterization of the enzyme in human and rat liver microsomes. Arch Biochem Biophys 1990; 281(2):264–270.
159. Resetar A, Spector T. Glucuronidation of 3′-azido-3′-deoxythymidine: human and rat enzyme specificity. Biochem Pharmacol 1989; 38(9):1389–93.
160. Cappiello M, Giuliani L, Pacifici GM. Distribution of UDP-glucuronosyltransferase and its endogenous substrate uridine 5′-diphosphoglucuronic acid in human tissues. Eur J Clin Pharmacol 1991; 41(4):345–350.
161. Argikar U. Effects of age, induction, regulation and polymorphisms on the metabolism of antiepileptic drugs. In: PhD. thesis. Minneapolis: University of Minnesota, 2006.
162. Kalyanaraman N. Inhibition of naltrexone metabolism by NSAIDs leading to metabolic switching and reactive metabolite formation. In: MS. thesis. Minneapolis: University of Minnesota, 2004.

Appendix

Isoenzyme	Trivial names	Tissue expression	Endogenous substrates	Drug or xenobiotic substrates	Inducers	Inhibitors
UGT1A1	HP3 $HUG_{Br\text{-}1}$ UGTBr1	Liver, small intestine, mammary glands	Bilirubin, estradiol (3-OH), 2-OH-estrone, 2-OH-estradiol *trans*-retinoic acid, Catechol estrogens (2-OH & 4-OH)15-OH-eicosa-tetraenoic acid, 20-OH-eicosa-tetraenoic acid, arachidonic acid, prostaglandin B1	Ethinyl estradiol, buprenorphine ferulic acid, genistein naltrexone (low), naloxone (low), SN-38 (active metabolite of irinotecan) alizarin, quinalizarin, retigabine	Bilirubin, chloropheno-xypropionic acid, chrysin, clofibrate, 3-MC, oltipraz, phenylpropionic acid, phenobarbital, clotrimazole, rifampin, and St. John's wort. WY-14643 Response elements for AhR, CAR, GR, PPAR-α, PXR, Nrf2 (antioxidant response element) have been identified	Atazanavir, indinavir, ketoconazole ($K_i = 3$ μM)
UGT1A2		Pseudogene in humans	–	–	–	–
UGT1A3		Liver (lower), small intestine, kidney, prostate, testes	*Bile acids*: (carboxyl functional group), e.g, lithocholic acid, chenodeoxycholic acid Bilirubin (low) *Catechol estrogens*:(2-OH > 4-OH), 2-OH-estrone, 2-OH-estradiol, decanoic acid, dodecanoic acid, 15-OH-eicosatetraenoic acid, arachidonic acid	Afloqualone, alizarin, buprenorphine, norbuprenorphine, bropirimine, cyproheptadine, diphenylamine, diprenorphine, emodin, esculetin, eugenol, ezetimibe, fulvestrant, fisetin, genistein, 5,6,7,3′,4′,5′-hexamethoxyflavone, 3-hydroxydesloratadine, 7-hydroxyflavone, hydromorphone, 4-methylumbelliferone, morphine, nalorphine, naloxone, naltrexone, naringenin, quercetin (3->4-> 3′- >7-glucuronide), scopoletin, thymol, umbelliferone	β-Naphthoflavone, rifampin, WY-14643 Response elements for AhR, PPAR-α, PXR, have been identified. Fibrates are potent inducers	Bile acids

				Substrates with carboxyl groups: clofibrate, ciprofibrate, etodolac, fenoprofen, ibuprofen, ketoprofen, naproxen (racemic > S), valproic acid and formation of simvastatin and atorvastatin lactones via an intermediate acyl glucuronide, (fourfold lower turnover by UGT2B17)		
UGT1A4	$HUG_{Br\text{-}2}$	Liver, small intestine	*Estrogens:* 2-OH-estrone, 2-OH estradiol, 4-OH catechol estrogens (low), estriol *Progestins*: 5α-pregnan-3α,20α-diol, 16α-hydroxy-pregnenolone, 19-hydroxy- and 21-hydroxy-pregnenolone, pregnenolone, androsterone, epiandrosterone, etiocholanone *Androgens:* dehydroepiandrosterone, dihydrotestosterone, epitestosterone, testostereone, 5α-androstan-3α,17β-diol, 5β-androstan-3α,11α,17β-triol; bilirubin (very low), F_6-1α,23S,25 $(OH)_3D_3$—a hexafluorinated Vit D_3 analog, 15-OH-eicosatetraenoic acid, 20-OH-eicosatetraenoic acid, arachidonic acid	*Tertiary amines:* Afloqualone, amitriptyline, chlorpheniramine, chlorpromazine, clozapine, cyproheptadine, diphenylamine, doxepin, imipramine, ketotifen, loxapine, olanzapine, promethazine, tamoxifen, tripelennamine, trifluoperazine *Aromatic heterocyclic amines:* croconazole, lamotrigine, nicotine (30X> velocity than UGT1A3), 1-phenylimidazole, posaconazole, retigabine *Primary and secondary amines*: 2- and 4-aminobiphenyl, diphenylamine, desmethylclozapine *Alcoholic and phenolic substrates:* borneol, carveol, carvacrol diosgenin, hecogenin, isomenthol, menthol, neomenthol, 1- and 2-naphthol (low), *p*-nitrophenol (low), nopol, tigogenin	Phenobarbital, phenytoin, and carbamazepine, TCDD, PCN (transgenic mice), WY-14643 (weak) In vivo induction experiments indicate response elements are present for CAR, PXR, and PPAR-α ligands	Hecogenin Trifluoroperazine

Continued

Appendix *Continued*

Isoenzyme	Trivial names	Tissue expression	Endogenous substrates	Drug or xenobiotic substrates	Inducers	Inhibitors
UGT1A5				4-Methyl-umbelliferone (low), scopoletin (low), 1-hydroxypyrene	Rifampin, 3-methylchloranthrene	
UGT1A6	HP1 HlugP1 UGT1-6	Liver, kidney, intestine, brain, ovaries, testes, skin, spleen	Serotonin, 3-hydroxy-methyl DOPA	*Phenols:* APAP, 2-amino-5-nitro-4-trifluoromethylphenol (flutamide metabolite), BHA, BHT, 7-hydroxy-coumarin, 4-hydroxy-coumarin (low), dobutamine, 4-ethylphenol, 3-ethylphenol, 4-fluorocatechol, 2-OH-biphenyl, 4-iodophenol, 4-isopropylphenol (low), 4-methylcatechol, 4-methylphenol, methylsalicylate, 4-methylumbelliferone, 4-nitrophenol, 4-nitrocatechol, octylgallate, phenol, 4-propylphenol (low), *cis*-resveratrol, salicylate, 4-*tert*-butylphenol (low), tetrachlorocatechol, vanillin *Amines:* 4-aminobiphenyl, 1-naphthylamine >2-naphthylamine, *N*-OH-2-naphthylamine *Drugs:* APAP, β-blocking adrenergic agents (low activity) such as atenolol, labetolol, metoprolol, pindolol, propranolol, naproxen (R≫S for rat 1A6), salicylate, valproic acid	TCDD, β-naphthoflavone, 3-MC Response elements for AhR have been identified	α-napthol, 4-*tert*-butyl phenol, 4-methyl-umbelliferone, 7-hydroxy–coumarin

				Flavonoids: chrysin, 7-hydroxyflavone, naringenin	
UGT1A7		Gastric epithelium, oesophagus	Estriol, 2-OH-estradiol, 4-OH-estrone	Benzo(a)pyrene phenols (7-OH≫9-OH>3-OH), benzo (a)pyrene-*tert*-7,8-dihydrodiol (7R-glucuronide, low affinity), 2-OH-biphenyl, 4-methylumbelliferone, 1- and 2-naphthol, 4-nitrophenol, octylgallate, vanillin	TCDD
UGT1A8	HP4	Liver, kidney, ovaries, testes, skin, spleen, oesophagus	2-OH-estrone, 4-OH-estrone, 2-OH-estradiol, 4-OH-estradiol, estrone, dihydrotestosterone, *trans*-retinoic acid, 4-hydroxy-retinoic acid, hyocholic acid, hyodeoxycholic, testosterone, LTB4	Alizarin, anthraflavic acid, apigenin, benzo(a)pyrene-*tert*-7,8-dihydrodiol (7*R*- and 8*S*-glucuronides), emodin, fisetin, flavoperidol, genistein, naringenin, quercetin, quinalizarin, 4-methylumbelliferone, scopoletin, carvacrol, eugenol, 1-naphthol, *p*-nitrophenol, 4-aminobiphenyl, 2-OH-, 3-OH-, and 4-OH-biphenyl, buprenorphine (low), morphine (low), naloxone, naltrexone, ciprofibrate, diflunisal, diphenylamine, furosemide, MPA (high), phenolphthalein, propofol, valproic acid, nandrolone, 1-methyl-5α-androst-1-en-17β-ol-3-one (metabolite of metenolone), 5α-androstane-3α,17β-diol (metabolite of testosterone), (-)-epigallocatechin gallate (tea phenol), SN-38 (low) (metabolite of irinotecan),	3-Methyl-cholanthrene

Continued

Appendix *Continued*

Isoenzyme	Trivial names	Tissue expression	Endogenous substrates	Drug or xenobiotic substrates	Inducers	Inhibitors
				troglitazone (moderate), raloxifene (both 6β- and 4′-β-glucuronides), quercetin, luteolin		
UGT1A9		Intestine, oesophagus	Retinoic acid, thyroxine (T4), tri-iodothyronine (T3; minor), 4-OH-estrone, 4-OH-estradiol (major), 15-OH-eicosatetraenoic acid, arachidonic acid, prostaglandin B1	*Planar Phenols:* Phenol, APAP,2-OH-biphenyl, 4-iodophenol,4-propylphenol, 4-isopropylphenol (low), 4-ethylphenol, 3-ethylphenol, 4-methylphenol, 4-nitrophenol, methylsalicylate, salicylate, mono(ethylhexyl) phthalate, BHA, BHT, vanillin,7-hydroxy-coumarin,4-hydroxy-coumarin (low), 4-methyl-umbelliferone *Bulky Phenols:* Phenol red, phenolphthalein, fluorescein 4-*tert*-butylphenol (low), propofol (2,6-di-isopropylphenol) *Simple catechols:* Octyl gallate, propyl gallate *Primary amines:* 4-Aminobiphenyl *Xenobiotics:* APAP, bumetanide, carbidopa, clofibric acid, ciprofibric acid, dobutamine, dopamine, entacapone, ethinyl estradiol–(minor), fenofibric acid, furosemide, gemfibrozil, levodopa, MPA, propofol, atenolol, labetolol, metoprolol, pindolol, propranolol, *R*-oxazepam, *p*-HPPH (phenytoin metabolite), raloxifene,	TCDD, tetrabutyl hydroquinone, clofibric acid	High concentrations of propofol, flurbiprofen

			retigabine, SN-38 (active metabolite of irinotecan), troglitazone, zidovudine *NSAIDs*: (Low activity against all NSAIDs) diflunisal, fenoprofen, flurbiprofen, ibuprofen, ketoprofen, mefenamic acid, naproxen *Flavonoids:* Emodin, chrysin, 7-hydroxyflavone, galangin, naringenin, quercetin carveol, nopol, citronellol, 6-hydroxychrysene, retigabine, quercetin (3->7->3′- >4′-glucuronide)		
UGT1A10	Intestine	2-OH-estrone (low), 4-OH estrone (low), dihydrotestosterone, testosterone, 15-OH-eicosatetraenoic acid, arachidonic acid, prostaglandin B1	Alizarin, anthraflavic acid, apigenin, benzo(a)pyrene-*tert*-7,8-dihydrodiol (7*R*- and 8*S*-glucuronides, high affinity), emodin, fisetin, genistein, naringenin, quercetin, quinalizarin, 4-methylumbelliferone, scopoletin, carvacrol, eugenol, MPA, 17β-methyl-5β-androst-4-ene-3α,17α-diol (metabolite of metadienone), nandrolone, 1-methyl-5α-androst-1-en-17β-ol-3-one (metabolite of metenolone), 5α-androstane-3a,17β-diol (metabolite of testosterone), SN-38 (minor), raloxifene (4′-β-glucuronide only)		
UGT1A11	pseudogene		–	–	–
UGT1A12	pseudogene		–	–	–

Continued

Appendix *Continued*

Isoenzyme	Trivial Names	Species	Human Counter-part	Tissue Expression	Endogenous Substrates	Drug or Xenobiotic Substrates	Inducers	Inhibitors
UGT2B4		Human, variant of UGT2B11, 82% homologous with UGT2B7		Liver, small intestine, aerodigestive tract (tongue, mouth)	6α-hydroxy bile acids, 3α-hydroxy pregnanes, 3α-, 16α-, 17β-androgens, metabolites of PUFA, arachidonic and linoleic acids, estriol, 2-OH-estriol, 4-OH-estrone	*Phenols:* Eugenol, 4-nitrophenol, 2-aminophenol, 4-methylumbelliferone, morphine, zidovudine (low)	Fenofibric acid, chenodeoxycholic acid activated Response elements for FXR and PPAR-α have been identified	
UGT2B7		Human		Liver, kidney, intestine, brain (cerebellum), esophagus	*Arachidonic acid metabolites*: Arachidonic acid, LTB4, 5-HETE, 12-HETE, 15-HETE, 20-HETE, and 13-HODE) *Bile acids*: hyodexycholic acid *Estrogens*: Estriol, estradiol (17β-hydroxy),4-OH-estrone (high) 2-OH-estrone, 2-OH-estriol *Pregnanes:* 3α-hydroxy pregnanes *Androgens*: 3α -, 16α -, 17β-androgens *Others*: 5α- and 5β-dihydroaldosterone,	*R*-oxazepam, naproxen, menthol, abacavir, APAP, almokalant, AZT, carvedilol, chloramphenicol, epirubicin, 1′-OH-estragole, gemcabene, 5-OH-rofecoxib, lorazepam, menthol, 4-methylumbelliferone, Maxipost, 1-naphthol (low), 4-nitrophenol, octylgallate, propranolol, temazepam *Carboxylic acid-containing drugs*: benoxaprofen, ciprofibrate, clofibric acid, diflunisal, DMXAA, fenoprofen, ibuprofen,	Rifampin, phenobarbital, HNF1α	*R*-oxazepam and zidovudine (competitive), flunitrazepam relatively potent (K_i ~50–90 μM), but also inhibits UGT1A3 (K_i = 20–30 μM for 2-OH-estrogens) and UGT1A1 (K_i > 200 μM), diclofenac, etonitazenyl

				trans-retinoic acid, prostaglandin B1	indomethacin, ketoprofen, naproxen, pitavastatin, simvastatin acid, tiaprofenic acid, valproic acid, zaltoprofen, zomepirac, *S*-flurbiprofen *Opioids*: morphine 3-OH>6-OH, buprenorphine, nalorphine, naltrexone, codeine (low), and naloxone		
UGT2B10	Humans			*Arachidonic acid metabolites*: Arachidonic acid, LTB4, 5-HETE, 12-HETE, 15-HETE, 20-HETE, and HODE			
UGT2B11	Humans 91% identical to UGT2B10, 76% identical to UGT2B15 and UGT2B17		Liver, kidney, prostrate, lungs, mammary gland, skin, adipose tissue, adrenal glands	*Arachidonic acid metabolites*: Arachidonic acid, LTB4, HETE, 12-HETE, 15-HETE, 20-HETE, and HODE			
UGT2B15	Human		Liver, prostate, testes, adipose tissue	*Steroids:* UGT2B17 glucuronidates preferentially at the 17-OH positions of androgens (>3OH) Aldosterone, 5α- and 5β-dihydroaldosterone,	*S*-oxazepam, temazepam, S-OH-rofecoxib, E-4-OH-tamoxifen, eugenol, nansroline, phenolphthalein		Androgens, epidermal growth factor, and IL-1α downregulate UGT2B15 and UGT2B17 expression in LnCAP cells (prostate cancer cell line)

Continued

Appendix *Continued*

Isoenzyme	Trivial Names	Species	Human Counter-part	Tissue Expression	Endogenous Substrates	Drug or Xenobiotic Substrates	Inducers	Inhibitors
					androsterone (low), dihydrotestosterone (3 and 17-O-glucuronidation), androstane-3α-17β-diol (17-O-glucuronidation), 3α-, 16α-, 17β–OH-androgens, 3α-hydroxypregnanes; estriol, 4-OH-estrone (high) 2-OH-estrone, 2-OH estriol *Bile acids*: hyodeoxycholic acid (hydroxyl glucuronide), lithocholic acid (hydroxyl>carboxyl) Retinoic acid			
UGT2B17		Human		Liver, kidney, prostrate, testes, placenta, uterus, mammary glands, skin, adrenal glands	Dihydrotestosterone, Testosterone, androsterone UGT2B17 glucuronidates at both the 3- and the 17-OH positions of androgens	Ibuprofen *Anthraquinones, coumarins, flavonoids, and terpenoids*: alizarin, anthraflavic acid, borneol, chrysin, emodin, eugenol, 4-ethylphenol, galangin, 7-hydroxyflavone, menthol (low), 4-methylumbelliferone (low), 1-naphthol, naringen (low),		Androgens, epidermal growth factor, and IL-1α downregulate UGT2B15 and UGT2B17 expression in LnCAP cells (prostate cancer cell line)

4-nitrophenol, phenol red, 4-propylphenol, scopoletin, and umbelliferone (low). Alizarin, borneol, galangin, and scopoletin are higher turnover compounds

---

AhR activators in humans—TCDD, β-naphthoflavone, 3-methylcloranthrene.
PXR activators in rodents—PCN, dexamethasone.
PXR activators in humans—clotrimazole, rifampin, and St. John's wort.
CAR activators in humans—3-MC, phenylpropionic acid, phenobarbital, phenytoin, carbamazepine.
PPARα activator in humans—clofibric acid, fenofibric acid, pirinixic acid (WY-14643).
PPARα activator in humans—rosiglitazone.
FXR activators in humans—chenodeoxycholic acid.
Underlined substrates denote the most commonly used probes for enzymatic activity.
Abbreviations: AhR, aromatic hydrocarbon receptor; TCDD, tetrachlorodibenzodioxin; PXR, pregnenolone-16α-nitrile-X-receptor; PCN, pregnenolone-16α-nitrile; CAR, constitutive androstane receptor; MC, methylcholanthrene; PPARα, peroxisome proliferated-activated receptor-α; FXR, farnesoid-X-receptor; LXR, liver-X-receptor; RXR, retinod-X-receptor; 5-HETE, 5-OH-eicosatetraenoic acid; LTB4, leukotriene B4; 13-HODE, hydroxyoctadecadienoic acid; DMXAA, dimethylxanthenone-4-acetic acid.

5

Drug-Drug Interactions Involving the Membrane Transport Process

Hiroyuki Kusuhara and Yuichi Sugiyama
Department of Molecular Pharmacokinetics, Graduate School of Pharmaceutical Sciences, The University of Tokyo, Tokyo, Japan

I. INTRODUCTION

Transporters are membrane proteins regulating the influx and efflux of organic solute across the plasma membrane. Transporters, particularly involved in the drug disposition, are characterized by broad substrate specificities and accept structurally unrelated compounds. Cumulative studies have elucidated the importance of the transporters as one of the determinant factors for the pharmacokinetic properties of drugs in the body, e.g., site of absorption (small intestine), clearance organs (liver and kidney) and the peripheral tissues (1). During two decades, a number of transporters have been cloned and subjected to functional analysis (summarized in Table 1). They are classified into the solute carrier (SLC) family and ATP-binding cassette (ABC) transporter family; the SLC family includes facilitated and secondary active transporter (a special issue has been published in Pflugers Arch, 2004, and online at http://www.bioparadigms.org/slc/menu.asp), while ABC transporter family includes primary active transporters with evolutionally preserved cytosolic catalytic domain (ABC) (a special issue has been published in Pflugers Arch, 2006, and online at http://www.ncbi.nlm.nih.gov/books/bv.fcgi?rid=mono_001. chapter.137). Cloning of transporters together with functional analyses have made a great contribution to elucidate the molecular characteristics of the transporter involved in the hepatobiliary transport and tubular secretion in the kidney, and barrier functions in the blood-tissue barriers such as blood-brain, cerebrospinal, and placenta barriers.

Some drugs have been found that modulate the function of transporters at clinical dosage. Concomitant use of such drugs will affect the drug disposition of substrate drugs in which the transporters are deeply involved. The possible sites for drug-drug interactions involving transports are summarized in Table 2. Drug-drug interactions in the liver, kidney, and small intestine affect the drug exposure in the circulating blood, while those in the peripheral organs affect the tissue concentrations only in the peripheral organs, leading to enhancement/attenuation of pharmacological effect and/or incidence of adverse effect. In most cases, the drug-drug interactions in peripheral tissues hardly affect the drug exposure in the circulating blood because of small contribution of transporters in peripheral tissues to the clearance mechanism and distribution volume. The impact of the drug-drug interaction depends on the pharmacokinetic properties of the substrate drug and the contribution of the transporter to the net membrane transport process in addition to the concentration of the inhibitors.

This chapter describes recent advances in the prediction of transporter-mediated drug-drug interactions and methods for their evaluation.

II. PREDICTION OF DRUG-DRUG INTERACTIONS FROM IN VITRO EXPERIMENTS

This section describes the theoretical part of the prediction of drug-drug interaction (Fig. 1). Unlike channels, transporters form intermediate complex with its substrate, and thus, the membrane transport involving transporters is characterized by saturation, reaching the maximum transport velocity by increasing the substrate concentrations. The intrinsic clearance of the membrane transport involving transporters (PS_{int}) follows Michaelis-Menten equation (Eq. 1).

$$PS_{int} = \frac{V_{max}}{K_m + C_u} \tag{1}$$

where K_m represents the Michaelis constant, C_u represents the unbound concentration of substrate drug, and V_{max} represents maximal transport velocity. There are two types of inhibition, competitive and noncompetitive. Competitive inhibition occurs when substrates and inhibitors share a common binding site on the transporter, resulting in an increase in the apparent K_m value in the presence of inhibitor (Eq. 2). Noncompetitive inhibition assumes that the inhibitor has an allosteric effect on the transporter, does not inhibit the formation of an

(text continues on page 146)

Table 1 Transporters Responsible for the Drug Disposition

Family		Species	Gene name	Alias	Gene ID	OMIM	Tissue distribution	Localization	Driving force
SLC family (facilitated transport or secondary active transporters)									
SLC15	Pept1	Mouse	*Slc15a1*	–	56643	–	Small intestine	Small intestine: BBM	
		Rat			117261	–	Small intestine, kidney	ND	H^+ symport
	PEPT1	Human	*SLC15A1*	–	6564	600544	Ileum, kidney, liver	Small intestine: BBM	
	Pept2	Mouse	*Slc15a2*	–	57738	–	Kidney, brain, CPx, lung, spleen	Kidney, CPx: BBM	
		Rat			60577	–	Kidney, brain, CPx, lung	Kidney, CPx: BBM	H^+ symport
	PEPT2	Human	*SLC15A2*	–	6565	602339	Kidney	ND	
SLC17	NaPi-1	Mouse	*Slc17a1*	NPT1	20504	–	–	–	Cl^- antiport(?)/ facilitated transport
		Rat			171080	–	Kidney, liver	Liver: SM	
		Human	*SLC17A1*		6568	182308	Kidney	Kidney: BBM	
SLCO	Oatp1a1	Mouse	*Slco1a1*	Oatp1	28248	–	–	–	
		Rat			50572	–	Liver, kidney (male)	Liver: SM Kidney: BBM	ND
	Oatp1a3	Rat	*Slco1a3*	Oat-k1	80899	–	Kidney (proximal tubule)	BBM	
		Rat		Oat-k2	–	–	Kidney (proximal tubule)	ND	

Continued

Table 1 *Continued*

Family	Species	Gene name	Alias	Gene ID	OMIM	Tissue distribution	Localization	Driving force
Oatp1a4	Mouse	*Slco1a4*	Oatp2	28250	–	Liver, kidney, brain	–	
	Rat			170698	–	Liver, brain, retina	Liver, CPx: basal BBB: LM, ALM	
Oatp1a5	Mouse	*Slco1a5*	Oatp3	108096	–	Very low (?)	–	
	Rat			80900	–	Retina, brain(CPx and female cortex), small intestine	Small intestine, CPx: BBM	
Oatp1a6	Mouse	*Slco1a6*	Oatp5	28254	–	Kidney	ND	
	Rat			84608	–	Kidney		
OATP1A2	Human	*SLCO1A2*	OATP OATP-A	6579	602883	Brain, liver (very low)	BBB: LM	
Oatp1b2	Mouse	*Slco1b2*	Lst-1	28253	–	Liver	–	
	Rat		Oatp 4	58978	–	Liver	SM	
OATP1B1	Human	*SLCO1B2*	LST-1 OATP-C OATP 2	10599	604843	Liver	SM	
OATP1B3	Human	*SLCO1B3*	OATP 8	25234	605495	Liver	SM	
Oatp1c1	Mouse	*Slco1c1*		58807	–	Brain (BBB, CPx)	CPx: BLM	
	Rat		Oatp14	84511	–	Brain (BBB, CPx)	BBB: LM/ALM	
OATP1C1	Human	*SLCO1C1*	OATP-F	53919	–	Brain	ND	

	Oatp2b1	–	*Slco2b1*	Oatp9	101488	–	Ubiquitous	–	
		–			140860	–	Ubiquitous	–	
	OATP2B1	Human	*SLCO2B1*	OATP-B	11309	604988	Ubiquitous	Liver: SM Small intestine: BBM BBB: LM	
	Oatp4c1	Mouse	*Slco4c1*	–	227394	–	Kidney, lung	–	
		Rat		–	432363	–	Kidney, lung	BLM	
	OATP4C1	Human	*SLCO4C1*	–	353189	609013	Kidney	ND	
SLC22	Oct1	Mouse	*Slc22a1*	–	20517	–	Kidney (proximal tubule), liver, intestine	Liver: SM Kidney: BLM	
		Rat		–	24904	–	Kidney (proximal tubule), liver, colon	–	
	OCT1	Human	*SLC22A1*	–	6580	602607	Liver	ND	
	Oct2	Mouse	*Slc22a2*	–	20518	–	Kidney	BLM	Facilitated transport
		Rat		–	29503	–	Kidney, brain	BLM	
	OCT2	Human	*SLC22A2*	–	6582	602608	Kidney	BLM	
	Oct3	Mouse	*Slc22a3*	–	20519	–	Ubiquitous (very low)	ND	
		Rat		–	29504	–	Ubiquitous (very low)	ND	
	OCT3	Human	*SLC22A3*	–	6581	604842	Placenta, heart, brain, small intestine	ND	

Continued

Table 1 *Continued*

Family	Species	Gene name	Alias	Gene ID	OMIM	Tissue distribution	Localization	Driving force
Octn1	Mouse	*Slc22a4*	–	30805	–	Ubiquitous	ND	
	Rat		–	64037	–	Ubiquitous	ND	H^+ antiport
OCTN1	Human	*SLC22A4*	–	6583	604190	Ubiquitous (except adult liver), fetal liver (fetal)	ND	
Octn2	Mouse	*Slc22a5*	–	20520	–	Kidney, small intestine	Kidney, small intestine: BBM	Na^+(carnitine)/ H^+ antiport (organic cation)
	Rat		CT1	29726	–	Kidney, small intestine	ND	
OCTN2	Human	*SLC22A5*	–	6584	603377	Kidney, skeletal muscle	Kidney: BBM(?)	
Octn3	Mouse	*Slc22a21*	–	56517	–	Testis	–	–
Oat1	Mouse	*Slc22a6*	NKT	18399	–	Kidney	Kidney: BLM	
	Rat		–	29509	–	Kidney	Kidney: BLM	Dicarboxylate antiport
OAT1	Human	*SLC22A6*	–	9356	607582	Kidney	Kidney: BLM	
Oat2	Mouse	*Slc22a7*	–	108114	–	Kidney, liver (female > male?)	ND	
	Rat		NLT	89776	–	Liver, kidney (female > male)	Liver: SM Kidney: BBM	ND
OAT2	Human	*SLC22A7*	–	10864	604995	Liver, kidney	Liver: SM Kidney: BBM	
Oat3	Mouse	*Slc22a8*	Roct	19879	–	Kidney, brain, eye	Kidney: BLM	
	Rat		–	83500	–	Liver(male), kidney, brain, eye	Kidney, BBB: basal CPx:BBM	Dicarboxylate antiport

	OAT3	Human	*SLC22A8*	–	9376	607581	Kidney	BLM	
	Oat5	Mouse	*Slc22a19*	–	207151	–	–	–	ND
		Rat		–	286961	–	Kidney	BBM	Succinate antiport
	OAT4	Human	*SLC22A11*	–	55867	607097	Kidney, placenta	BBM	Dicarboxylate antiport
SLC47	MATE1	Mouse	–	–	–	–	Liver, kidney, heart	Apical (CM, BBM)	H^+ antiport
		Rat	–	–	360539	–	Kidney	ND	H^+ antiport
		Human	–	–	55244	609832	Liver, kidney, skeletal muscle	Apical (CM, BBM)	H^+ antiport
	MATE2	Mouse	–	–	–	–	Testis	ND	H^+ antiport
		Human	–	–	146802	609833	Kidney	BBM	H^+ antiport
ABC transporters (primary active transport)									
ABCB	Mdr1a	Mouse	*Abcb1a*	–	18671	–	Small intestine, heart, brain, liver, kidney, lung, testis	Apical (CM, BBM, LM)	Mg^{2+}/ATP
		Rat		–	17913	–	Large intestine > ileum > jejunum > duodenum, brain, kidney	–	
	Mdr1b	Mouse	*Abcb1b*	–	18669	–	Placenta (during pregnancy), adrenal gland, kidney, heart	Apical (CM, BBM)	
		Rat		–	24646	–	Large intestine > ileum > jejunum > duodenum, liver, kidney, brain	–	
	MDR1	Human	*ABCB1*	–	5243	171050	Brain, liver, kidney, intestine	Apical (CM, BBM, LM)	

Continued

Table 1 *Continued*

Family		Species	Gene name	Alias	Gene ID	OMIM	Tissue distribution	Localization	Driving force
ABCC	Mrp1	Mouse	*Abcc1*	–	17250	–	Muscle, lung, testis, heart, kidney, spleen, brain	Kidney, CPx:BLM	
		Rat		–	24565	–	–	–	
	MRP1	Human	*ABCC1*	–	4363	158343	Lung, spleen, thyroid gland, testis, bladder, adrenal gland	–	
	Mrp2	Mouse	*Abcc2*	cMOAT cMRP	12780	–	Liver, kidney, duodenum > jejunum > ileum	–	
		Rat			25303	–	Liver, kidney, duodenum > jejunum	Liver: CM kidney, small intestine: BBM	
	MRP2	Human	*ABCC2*		1244	601107	Liver, kidney, duodenum	Liver: CM kidney: BBM	
	Mrp3	Mouse	*Abcc3*	–	76408	–	Colon > duodenum > jejunum > ileum, liver, stomach	BLM	
		Rat		–	140668	–	Ileum > jejunum, liver (EHBR, TR)	SM/BLM	

	MRP3	Human	*ABCC3*	–	8714	604323	Liver, intestine	Liver: SM
	Mrp4	Mouse	*Abcc4*	–	239273	–	Kidney, liver (very low), BBB, CPx	Kidney, BBB: apical liver, CPx:basal
		Rat		–	170924	–	Kidney, liver	Kidney, BBB: apical liver, CPx: basal
	MRP4	Human	*ABCC4*	–	10257	605250	Kidney, liver	Kidney: BBM liver, CPx: basal
ABCG	Bcrp	Mouse	*Abcg2*	MXR ABCP	26357	–	Small and large intestine, kidney, liver, brain	Liver: CM, kidney, small intestine: BBM BBB: LM
		Rat			312382	–	Small and large intestine, kidney, liver, brain	ND
	BCRP	Human	*ABCG2*		9429	603756	Ubiquitous	Liver: CM, small intestine: BBM BBB: LM

Abbreviations: (?), there is a discrepancy; ND, not determined; SLC, solute carrier; SM, sinusoidal membrane; CM, canalicular membrane; BLM, basolateral membrane; BBM, brush border membrane; LM, luminal membrane; ALM, abluminal membrane; BBB, brain capillary endothelial cells; CPx, choroid plexus.

Table 2 Possible Sites for Drug-Drug Interaction and the In Vitro Transport Models

	Transport direction			
Tissue	Process	From	To	In vitro transport experiment
Liver	Uptake	Blood	Parenchymal cells	Isolated, cultured cryopreserved hepatocytes, sinusoidal membrane vesicles, transporter expressions system
	Efflux	Parenchymal cells	Blood	–
	Excretion	Parenchymal cells	Bile	Canalicular membrane vesicles, transporter expression system
Kidney	Uptake	Blood	Epithelial cells	Kidney slices, isolated and cultured renal epithelial cells, basolateral membrane vesicles, transporter expressions system
	Efflux	Epithelial cells	Blood	–
	Excretion	Epithelial cells	Urine	Brush border membrane vesicles, transporter expression system
	Reabsorption	Urine	Epithelial cells	Brush border membrane vesicles, transporter expression system
Small intestine	Uptake	Digestive tract	Epithelial cells	Everted sac, Ussing-chamber experiments using intestinal epithelium, brush border membrane vesicles, Caco-2 cells monolayer, transporter expression system
	Efflux	Epithelial cells	Digestive tract	–
	Absorption	Epithelial cells	Blood	Everted sac, Ussing-chamber experiments using intestinal epithelium, basolateral membrane vesicles, Caco-2 cells monolayer

	Excretion	Blood	Epithelial cells	Everted sac, Ussing-chamber experiments using intestinal epithelium, basolateral membrane vesicles, Caco-2 cells monolayer
BBB	Uptake	Blood	Endothelial cells	Primary cultured cerebral endothelial cells, immortalized cell line
	Uptake	Endothelial cells	Brain parenchyma	–
	Efflux	Brain parenchyma	Endothelial cells	–
	Efflux	Endothelial cells	Blood	Primary cultured cerebral endothelial cells, immortalized cell line
BCSFB	Uptake	Blood	Epithelial cells	Primary cultured choroid epithelial cells, immortalized cell line
	Uptake	Epithelial cells	Brain parenchyma	–
	Efflux	Brain parenchyma	Epithelial cells	–
	Efflux	Epithelial cells	Blood	Primary cultured choroid epithelial cells, immortalized cell line
Tumor	Uptake	Blood	Tumor	Cell line, membrane vesicles
	Efflux	Tumor	Blood	Cell line, membrane vesicles

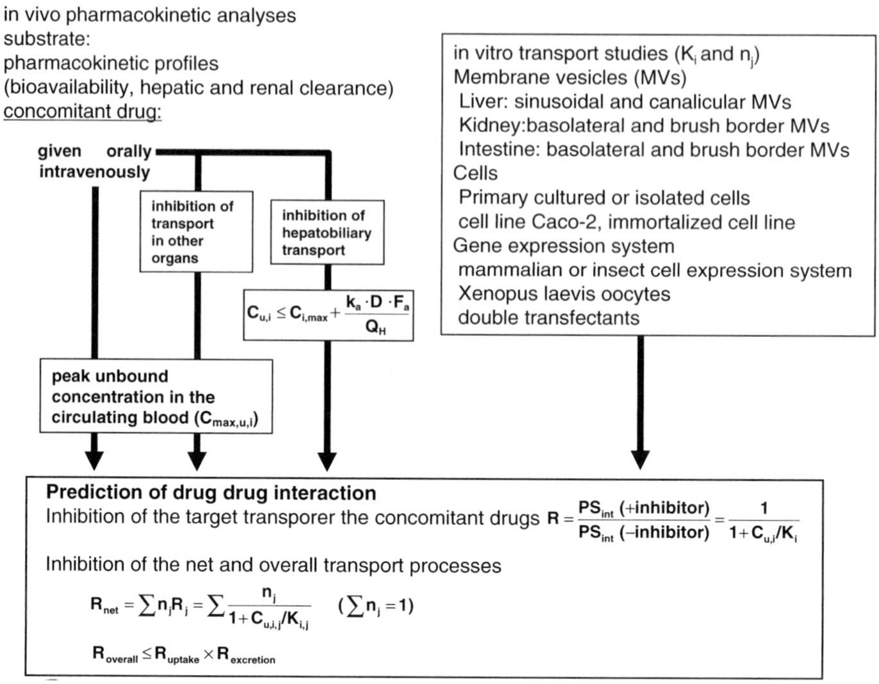

Figure 1 The schematic diagram for the prediction of drug-drug interactions involving membrane transport from in vitro transport experiments.

intermediate complex of substrate and transporter, but does inhibit the subsequent translocation process (Eq. 2).

$$PS_{int} = \frac{V_{max}/(1 + C_{u,i}/K_i)}{K_m + C_u} \quad \text{competitive}$$

$$PS_{int} = \frac{V_{max}}{K_m(1 + C_{u,i}/K_i) + C_u} \quad \text{noncompetitive} \tag{2}$$

where $C_{u,i}$ and K_i represent the unbound concentration of an inhibitor around a transporter and its inhibition constant, respectively. When the substrate concentration is much lower than the K_m value (so-called linear condition, this assumption holds true for many drugs at their clinical dosages), the intrinsic membrane transport clearance can be expressed by the following equation, independently of the type of inhibition.

$$PS_{int} = \frac{V_{max}}{K_m(1 + C_{u,i}/K_i)} \tag{3}$$

The degree of inhibition (R) is defined as follows

$$R = \frac{\mathrm{PS_{int}}\ (+\text{inhibitor})}{\mathrm{PS_{int}}\ (-\text{inhibitor})} = \frac{1}{1 + C_{\mathrm{u,i}}/K_{\mathrm{i}}} \tag{4}$$

where $\mathrm{PS_{int}}$(+inhibitor) and $\mathrm{PS_{int}}$(– inhibitor) represent the intrinsic membrane transport clearance in the presence and absence of inhibitor, respectively. Finally, the unbound concentration of inhibitors at clinical dosage and inhibition constant (K_i) for the target transporter are necessary to predict the interaction in vivo. The inhibition constant can be determined by kinetic analysis of the data from an in vitro transport study using isolated or cultured cells, membrane vesicles, and gene expression systems, etc. (Table 2). It is recommended to use human-based experimental systems to obtain kinetic parameters. Although animal-based experimental systems are readily available, species differences in the kinetic parameters and the relative contribution of the transporters cannot be ruled out.

When multiple transporters participate in the membrane transport of a drug, not only the degree of inhibition of the target transporter but the contribution of the transporter to the net transport process, is taken into consideration for the prediction (Eq. 5) (2).

$$R_{\mathrm{net}} = \sum n_j R_j = \sum \frac{n_j}{1 + C_{\mathrm{u,i},j}/K_{\mathrm{i},j}} \quad \left(\sum n_j = 1\right) \tag{5}$$

where R_j represents the degree of inhibition for each transporter and n_j represents the contribution of the transporter to the net membrane transport. In the case of hepatobiliary and tubular secretion where transporters are involved both in the uptake and efflux processes, the overall degree of inhibition can be approximated by multiplying the degrees of inhibition at the uptake and efflux processes (Eq. 6) (2).

$$R_{\mathrm{overall}} \le R_{\mathrm{uptake}} \times R_{\mathrm{excretion}} \tag{6}$$

Strictly speaking, the calculation of $R_{\mathrm{excretion}}$ requires the unbound concentration in the tissue, which is not available in most of the case. It is recommended to perform sensitivity analysis of $R_{\mathrm{excretion}}$ by changing the tissue concentration from the plasma unbound concentration to the 10-fold greater values. When even 10-fold greater concentration does not affect $R_{\mathrm{excretion}}$ significantly, inhibition will not occur.

We have previously proposed a simple method for predicting in vivo drug-drug interactions involving cytochrome P450 (CYP)-mediated metabolism based on in vitro experiments, using Eq. (4), for prescreening of the drug-drug interaction (3,4). For the drug-drug interactions in the renal transport and the efflux transport at the blood-brain barriers (BBB), the peak unbound concentration in the blood has been used, which gives maximum inhibition of the transporter at the dosage. For hepatic transport, when inhibitors are given intravenously, the peak unbound concentration in the blood will also provide the degree of inhibition of hepatic transport. However, when inhibitors are given orally, the concentration in the inlet to the liver is often higher than the peak concentration in the circulating blood, and thus, maximum inhibition should be predicted using the inlet concentration. To avoid false negative predictions, maximum unbound concentration of inhibitors in the inlet to the liver ($C_{u,i}$) can be approximated by the following equation (3,4).

$$C_{u,i} \leq C_{i,max} + \frac{k_a \cdot D \cdot F_a}{Q_H} \tag{7}$$

where k_a, F_a, and Q_H represent the absorption rate constant, the fraction absorbed from the gastrointestinal tract into the portal vein, and the hepatic blood flow rate, respectively. It should be noted that this approximation overestimates the $C_{u,i}$, and thereby, the degree of inhibition. When the predicted R value is close to unity, the possibility of a drug-drug interaction can be excluded. In other cases, more detailed analysis using physiologically based pharmacokinetic model is required for more precise prediction.

III. METHODS TO EVALUATE TRANSPORTER-MEDIATED DRUG INTERACTIONS

Table 2 shows the in vitro methods for evaluating drug-drug interactions. Details of the experimental conditions are readily available in the references cited in this section.

A. In Vitro Transport Systems Using Tissues, Cells, and Membrane Vesicles

1. Isolated/Cultured Hepatocytes

Hepatocytes freshly prepared are subjected to the transport study using a centrifugal filtration technique. After incubating the hepatocytes with test compounds, the reaction was terminated by separating the cells from the medium by passing through the layer of a mixture of silicone and mineral oil (density: 1.015) by centrifugation. The hepatic uptake of peptidic endothelin antagonists by freshly isolated rat hepatocytes was extrapolated to give the in vivo uptake clearance based on the assumption of a well-stirred model; they were very close to those obtained by in vivo integration plot analysis (Fig. 2) (5). Thus, isolated hepatocytes are a good model for evaluating hepatic uptake clearance. Because of progress in cryopreservation techniques, cryopreserved human hepatocytes are now available from several commercial sources for transport studies. Shitara et al. demonstrated that cryopreserved human hepatocytes retained saturable uptake of typical organic anions, such as estradiol-17β-glucuronide (E217βG) and taurocholate (TCA), and sodium-dependence of TCA uptake (6). Cryopreserved hepatocytes are now frequently used for the characterization of hepatic uptake of drugs in human. Since there is a large interbatch difference, it is recommended to prescreen the cryopreserved human hepatocytes with high

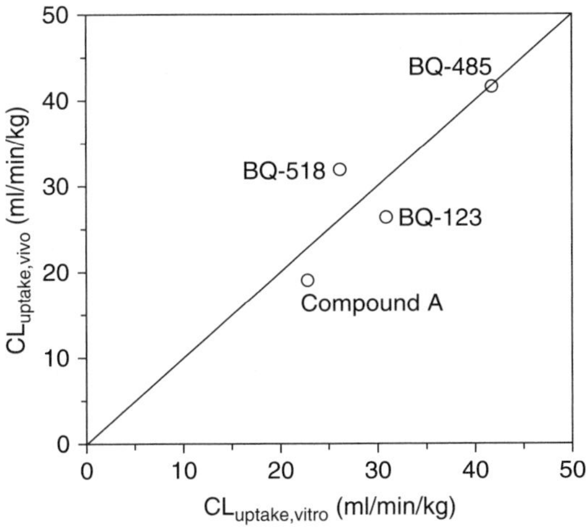

Figure 2 Comparison between the uptake clearance obtained in vivo and that extrapolated from the in vitro transport study of endothelin antagonists. In vivo uptake clearance of endothelin antagonists (BQ-123, BQ-518, BQ-485, compound A) was evaluated by integration plot analysis using the plasma concentration–time profile after intravenous administration (500 nmol/kg) and the amount of drug in the liver and that excreted in the bile. In vitro hepatic uptake clearance was measured using isolated rat hepatocytes and was extrapolated to the in vivo uptake clearance assuming the well-stirred model. *Source*: From Ref. 5.

uptake activities of E217βG and TCA, typical substrates for OATP1B1 and NTCP respectively, and determine the uptake of test compounds, at least, three batches of hepatocytes (7).

Cultured hepatocytes can be applied to measure the hepatic uptake of compounds. Since they attach to the cell culture dish, it can be washed several times to remove extracellular compounds. The disadvantage of this system is that the expression levels of transporters decrease during culture: a saturable component for the uptake of pravastatin into cultured rat hepatocytes is reduced to 70% by a 6-hour culture, and to 33% by a 24-hour culture, although the non-saturable component remained constant during culture (8). The time of culture should be no more than four to six hours, the minimum time for cell attachment. Cultured hepatocytes on the collagen-coated dish do not form bile canaliculi. LeCluyse et al. demonstrated that a collagen-sandwich configuration made hepatocytes form bile canaliculi (9). The transport activity was retained to some extent even in 96-hour cultured rat hepatocytes (10). The cell accumulation of methotrexate (MTX), [D-pen^{2,5}]enkephalin, and TCA was 1/5∼1/2 that in a three-hour culture of hepatocytes, while the uptake of salicylate was comparable (10). Incubating the hepatocytes in the absence of Ca^{2+} for 10 min disrupts the bile canaliculi (11). The cumulative biliary excretion of drug in this system is obtained by comparing the cumulative accumulation of drugs with or without preincubation of Ca^{2+} free butter. Liu et al. compared in vitro biliary excretion clearance with in vivo intrinsic clearance obtained from biliary excretion clearance based on the well-stirred model and found a good correlation for the five compounds examined, inulin, salicylate, MTX, [D-pen^{2,5}]enkephalin, and TCA, using this system (10). In sandwich-cultured rat hepatocytes, the P-glycoprotein (P-gp) expression was increased during six days of incubation, while their uptake transporters (Oatp1a1 and Oatp1a4) were similar or rather decreased during incubation (12). Human hepatocytes also form canalicular network following a four-day incubation in sandwich culture (12). The expression of uptake transporters (OATP1B1 and OATP1B3) and canalicular ABC transporters, such as P-gp and multidrug-resistance-associated protein 2 (MRP2), increased for 6 days in comparison with that in day 1.

2. Membrane Vesicles

The methods for preparing brush border membrane vesicles from intestine, kidney, and choroid plexus, basolateral membrane vesicles from kidney, sinusoidal and canalicular membrane vesicles from liver and luminal and abluminal membrane of the brain capillary endothelial cells are readily available in the literature (13–21). The advantages of using membrane vesicles for transport studies are (1) its suitability for examining the driving force of transport by changing the ion composition or ATP concentration, (2) its suitability for measuring the transport across the basolateral or brush border membranes separately, and (3) negligible intracellular binding and metabolism. It is important to characterize the preparation of membrane vesicles in terms of purity and orientation. Purity can be estimated by the enrichment of the relative activity of marker enzymes for the target plasma membrane (13–21). Orientation is particularly important for measuring primary active transport. There are two orientations in the membrane vesicles, i.e., physiological (right-side out) and inverted (inside-out) orientation (13–21). Since ATP binding sites are located in the intracellular domain, the domain is exposed to the transport butter only in the membrane vesicles with inside orientation, allowing access of ATP, and accumulation of substrate drugs into the membrane vesicles. Indeed, Kamimoto et al. demonstrated that inside-out-oriented, but not right-side-out-oriented, canalicular membrane vesicles exhibit ATP-dependent uptake of daunomycin (22). Therefore, a low fraction of inside-out membrane vesicles makes it difficult to detect the ATP-dependent uptake of drugs. Generally speaking, as far as secondary or tertiary active transporters are concerned, orientation is not important, because the transport mediated by these transporters is bidirectional.

3. Kidney Slices

Kidney slices have been widely used to characterize renal uptake. The extracellular marker compounds, such as methoxyinulin and sucrose, were below the limit of detection in the luminal space of the proximal tubules, while they could be detected in the extracellular space (23). Therefore, the kidney slices allow only a limited access of drugs from the luminal space in the kidney slices, but free access from the basolateral side. In vitro studies using kidney slices have proved its usefulness for examining uptake mechanisms of drugs. Hasegawa et al. demonstrated that the uptake of *p*-aminohippurate (PAH) and pravastatin by rat kidney slices is mediated by different transports by mutual inhibition study and an inhibition study using benzylpenicillin (PCG) (24). Fleck et al. prepared kidney slices from human kidney and demonstrated the active accumulation of PAH and MTX, suggesting that human kidney slices also retain the activities of organic anion transporters (25–27). Nozaki et al. determined mRNA expression of OAT1 and OAT3 and the uptake of OAT1 and OAT3 substrates in human kidney slices (28). Although there was large interbatch difference, OAT1 and OAT3 mRNA levels correlated well, and there was a good correlation between the uptakes of PAH and benzylpenicillin by kidney slices. Thus, human kidney slices retain the contribution of OAT1 and OAT3, and can be used to investigate the renal uptake mechanism of drugs. However, the possible impact of disease state and patient drug treatments on OAT function in the available source tissues is unknown, and caution must be used when extrapolating such data to quantitative evaluation of the normal human response.

4. Everted Sac

This method is used to measure drug absorption from the mucosal to serosal side (29). A segment of intestine is everted and, thus, the mucosal side is turned to the outside. Drug absorption is evaluated by measuring the amount of drug that appears inside the sac when the everted sac is incubated in the presence of test compound. Since a segment of intestine is used for the assay, not only transport but also metabolism should be taken into consideration. Barr et al. improved this method so that they could measure the drug concentration–time profile in one everted intestine (30).

5. Ussing Chamber Method

A segment of small intestine is opened along the mesenteric border to expose the epithelial cells and is mounted on the diffusion cell chamber after the longitudinal muscle fibers have been carefully stripped from the serosal side. The transcellular transport of test compound from the mucosal to serosal side, and vice versa, is measured to evaluate the drug absorption. There are two routes, i.e., the transcellular and paracellular routes, connecting the mucosal and serosal sides. Ussing chamber method allows the determination of electrophysical parameters such as membrane electroresistance, membrane potential and short circuit current, and the transport via the transcellular and paracellular routes can be evaluated separately (31,32). The transport of ionized drug via the paracellular route is sensitive to the potential difference, while that via the transcellular route is not, because of the high electrical resistance of plasma membrane. By measuring the transport rate at different potential difference (the voltage clamp method), the contribution of transport via the paracellular route can be evaluated. Also, in this system, metabolism should be taken into account.

6. Caco-2 Cells

Caco-2 cells, which are derived from human colorectal tumor, are used as an in vitro system for the intestine (33–35). Caco-2 cells retain the specific features of intestinal epithelial cells and differentiate to form tight junction and microvilli, but without a mutin layer. When Caco-2 cells are cultivated on a porous filter, they differentiate and form tight junctions and microvilli (36), and the membrane electroresistance and the permeability of mannitol (a marker for paracellular leakage) reach a plateau 15 days after seeding (36). Thus, at least, a 15-day culture period is needed for such transport studies. Absorption can be evaluated by measuring transcellular transport across a monolayer of Caco-2 cells cultured on a porous filter. Gres et al. examined the correlation between the fraction absorbed and the permeability from the apical-to-basal side of Caco-2 cells using 20 different compounds and showed that the compounds with high permeability were highly absorbed (Fig. 3) (37). The expression of dipeptide transporter (PEPT1) (38), amino acid transporter (39), monocarboxylic acid transporter (40), P-gp (41), MRP2 (42), and breast cancer resistance protein (BCRP) (43) has

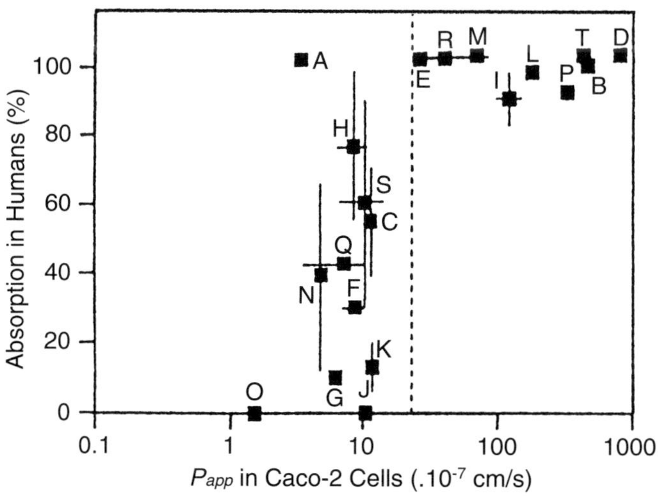

Figure 3 Correlation between the fraction absorbed and the membrane permeability in Caco-2 cells. P_{app} represents the membrane permeability of following 20 compounds, and was obtained by measuring the transcellular transport from the apical-to-basal side in Caco-2 cells. The fraction absorbed was obtained from literature. A: amoxicillin, B: antipyrine, C: atenolol, D: caffein, E: cephalexin, F: cyclosporin A, G: enalaprilate, H: L-glutamine, I: hydrocortisone, J: inulin, K: D-mannitol, L: metoprolol, M: L-phenylalanine, N: PEG-400, O: PEG-4000, P: propranolol, Q: sucrose, R: taurocholate, S: terbutaline, T: testosterone. *Source*: From Ref. 37.

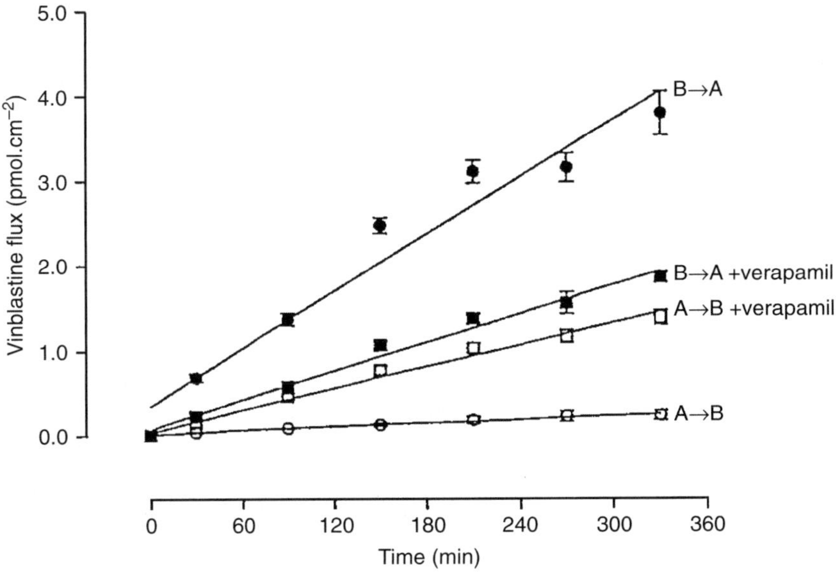

Figure 4 Time profiles of the transcellular transport of vinblastine in Caco-2 cells and the effect of verapamil on this transport. The transcellular transport of vinblastine in the presence (+verapamil) and absence of verapamil (100 μM) was measured across a monolayer of Caco-2 cells cultured on a porous filter for 14 to 15 days. B→A corresponds to the transport from the basal-to-apical and A→B is in the opposite direction. *Source*: From Ref. 44.

been confirmed on the apical membrane of Caco-2 cells. For instance, the permeability of P-gp substrates from the apical-to-basal side is lower than that in the opposite direction due to active efflux on the apical side (44), which was diminished in the presence of P-gp inhibitors (verapamil in Fig. 4) (44). Therefore, the Caco-2 cell is a useful model for evaluating drug-drug interactions where these transporters are involved.

7. Brain Capillary Endothelial Cells

Primary cultured porcine or bovine brain capillary endothelial cells have been used as an in vitro model for the BBB. Recently, an immortalized cell line has been established from mouse, rat, and human brain capillary endothelial cells by infection with Simian virus 40 or transfection of SV40 large T antigen (45–47). Tatsuta et al. established an immortalized mouse brain capillary endothelial cell line (MBEC4). The activity of γ-glutamyl transpeptidase and alkaline phosphatase, specific marker enzymes for brain capillary endothelial cells, was half that in the brain capillary (45). Also, P-gp was expressed on the apical membrane of MBEC4 cells, which corresponds to the abluminal membrane of the brain capillary (45). These indicate that MBEC4 cells retain some of the characteristics of brain capillary endothelial cells. It should be noted that Mdr1b, but not Mdr1a, is expressed in MBEC4 cells, although mdr1a is a predominant subclass in mouse brain capillary endothelial cells (45). The expression level of Mdr1b increases in primary cultured rat brain capillary endothelial cells, while that of mdr1a decreases (47). In addition, immortalization and culture increase the expression of multidrug resistance associated protein 1 (Mrp1) (48,49).

B. Gene Expression Systems

The advantage of using a gene expression system is that the kinetic parameters for the target transporter can be obtained. Once the responsible transporters are identified, the possibility of drug-drug interactions can be examined using gene expression systems comprehensively. This will save time and materials, otherwise the uptake or excretion needs to be examined in vivo with many possible combinations of drugs. According to our prediction method, the maximum unbound concentration and K_i are needed to determine the degree of inhibition for each transporter under clinical conditions. They can be obtained from the pharmacokinetic data in clinical trials and from in vitro transport studies, respectively. As mentioned previously, when a drug is transported by several transporters, the contribution of each needs to be estimated to predict the degree of overall drug-drug interaction. To determine the contribution, gene deficient/knockout animals are helpful compared with normal/wild-type animals, according to the pharmacokinetic profile of both. Animals such as Mdr1a(–/–), Mdr1a/1b(–/–), Mrp1(–/–), Mrp2(–/–), Mrp3(–/–), Mrp4(–/–), Bcrp(–/–), Oct1(–/–), Oct2(–/–), Oat1(–/–), Oat3(–/–) and Pept2(–/–) mice, Octn2-deficient mutant mice (*jvs*), and Mrp2-deficient mutant rats [TR^- and Eisai hyperbilirubinemic rats (EHBR)] have been established.

1. RAF Method

To evaluate the contribution of uptake process, relative activity factor (RAF) method has been used in hepatocytes and kidney slices. The scheme for this method is shown in Figure 5. Assuming that the transport activities of test compounds relative to that of reference compound for specific transporters is preserved between hepatocytes/kidney slices and cDNA-transfectants, multiplying the transport activities of test compounds in the cDNA transfectants by the ratio of the transport activities of reference compounds in the cDNA transfectants and hepatocytes/kidney slices gives the transport activities of test compounds mediated by the specific transporters in the hepatocytes/kidney slices. Thus, comparing the predicted transport activity among candidate transporters will allow the rough estimation of the contribution of each transporter. Kouzuki et al. applied this concept to evaluate the contribution of Oatp1a1 and sodium taurocholate transporting polypeptide (Ntcp) to the net uptake of organic anion and bile acids in primary cultured rat hepatocytes and cDNA-transfected COS-7 cells (50,51). They used E217βG and TCA as reference compounds for

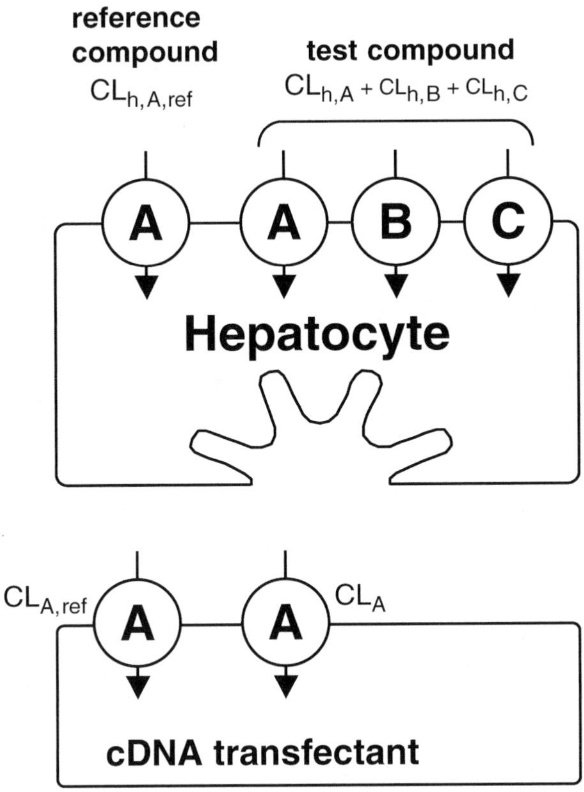

$$\text{contribution} = \frac{\dfrac{CL_{h,A,ref}}{CL_{A,ref}} \bullet CL_A}{CL_{h,A} + CL_{h,B} + CL_{h,C}}$$

Figure 5 The schematic diagram to evaluate the contribution of the specific transporter to the hepatic and renal uptake of drugs using RAF method.

Oatp1a1 and Ntcp, respectively, and found that they account for the part of the hepatic uptake. Hirano applied this method to evaluate the contribution of OATP1B1 and OATP1B3 to the uptake of pitavastatin by cryopreserved human hepatocytes using estrone sulfate and cholecystokinin (CCK-8) as reference compounds for OATP1B1 and OATP1B3, respectively (7). The sum of the predicted values was comparable with the observed values in the cryopreserved human hepatocytes, and, by comparing the predicted transport activities by OATP1B1 and OATP1B3, they concluded that the hepatic uptake of pitavastatin is mainly mediated by OATP1B1. This estimation was supported by inhibition by E217βG (OATP1B1/OATP2B1 inhibitor) and estrone sulfate (OATP1B1/OATP1B3 inhibitor) (52). Hasegawa et al. also applied this RAF method for evaluating the contribution of OAT1 and OAT3 to the net uptake of drugs in rat kidney slices using PAH and pravastatin as reference compounds for OAT1 and OAT3, respectively (53).

2. Double Transfectants

Hepatobiliary and tubular secretions in the kidney are characterized by vectorial transport across the epithelial cells from blood side to the luminal side. Except lipophilic compounds, uptake and efflux transporters coordinately form this vectorial transport (Fig. 6). Coexpression of uptake and efflux transporters in the polarized cell line (LLC-PK1, MDCK, and MDCK II cells) allows evaluation of the hepatobiliary and tubular secretion by measuring the transcellular transport across the double transfectants cultured on a porous membrane. Such double transfectants have been established in the following combinations; OATP1B1/MRP2 (54–55), OATP1B1/P-gp (56), OATP1B1/BCRP (56), OATP1B3/MRP2 (55,57), OATP2B1/MRP2 (55), OATP2B1/BCRP (58), OATP1B1/1B3/2B1/MRP2 (55), Oatp1b2/Mrp2 (59), Ntcp/Bsep (60), NTCP/BSEP (61) for hepatobiliary transport of organic anions and bile acids, Oat3/RST for tubular secretion of organic anions in the kidney (62), and ASBT/OSTα/β for intestinal transport of bile acids (63). Considering the scaling factor, the clearance values for in vitro transcellular transport across the monolayers of Oatp1b2/Mrp2 cells correlated well with those for in vivo biliary clearance (Fig. 6) (59).

IV. DRUG TRANSPORTERS

A. Secondary or Tertiary Active Transporters (SLC Family)

1. Organic Anion Transporting Polypeptide (OATP/SLCO) Family

OATP/SLCO superfamily is classified into six families in mammalians (64). This chapter described the characteristics of three families (OATP1, OATP2, and OATP4), which have been suggested to be involved in the drug disposition, i.e., hepatic uptake process, basolateral uptake and reabsorption in the kidney, intestinal uptake, and efflux transport in the barriers of central nervous system.

Oatp1/OATP1 family is comprised of three subfamilies. There is great interspecies difference in the number of genes forming the subfamily "a" and "b" between human and rodents. Rodent Oatp1 subfamily a consists of five isoforms (Oatp1a1, Oatp1a3, Oatp1a4, Oatp1a5, and Oatp1a6), which exhibit high amino acid, identical to each other (>70%), while only OATP1A2 is the human isoform.

Oatp1a1 was isolated from rat liver as a candidate for sodium-independent uptake of organic anions (65). Oatp1a1 is localized to the sinusoidal membrane in the rat liver and the brush border membrane in the male kidney (66). Cumulative studies have elucidated its broad substrate specificity, including

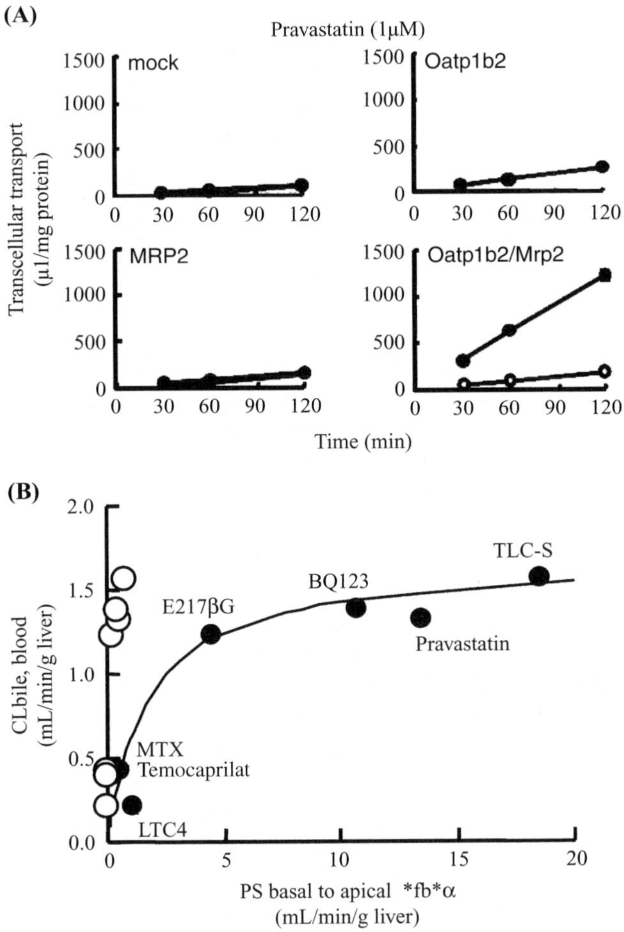

Figure 6 Directional transport of pravastatin in Oatp1b2/Mrp2 double transfectants in the apical direction (**A**), and comparison of in vivo biliary excretion clearance and in vitro transcellular transport clearance across the double transfectant (**B**). (**A**) Transcellular transport across the monolayers of MDCK II cells was determined in the basal-to-apical and the opposite direction. (**B**) The x axis represents CL_{int} determined in vitro multiplied by f_B and the scaling factor, and the y axis represents the in vivo biliary clearance defined for the blood ligand concentrations. The symbol (•) represents data whose x axis values were corrected for the scaling factor (α = 17.9). The solid line represents the theoretical curve, and the symbol (○), the observed data. *Source*: From Ref. 59.

amphipathic organic anions, such as bile acids and steroid conjugated with sulfate or glucuronide, and type II organic cations, such as *N*-(4,4′-azo-*n*-pentyl)-21-deoxyajmalinium (APDA), *N*-methyl-quinine, and rocuronium (65,67). Oatp1a1 mediates active transport; however, the driving force has not been identified yet. An outward concentration gradient of glutathione has been suggested as driving force since uptake of TCA and leukotriene C_4 by Oatp1a1 was influenced by the intracellular concentration of reduced glutathione in *Xenopus laevis* oocytes (68).

Oatp1a3 consists of two variants (Oat-k1 and Oat-k2) in the kidney (69,70). Oat-k2 lacks 172 amino acids at the amino terminal (70). The localization of Oat-k1 has been suggested to be brush border membrane of the renal tubules since polyclonal antibody detected Oat-k1 only in the brush border membrane–enriched fraction from the kidney (71). In contrast to other Oatps, Oat-k1 mediates facilitated transport since the uptake by Oat-k1 was insensitive to an ATP depleter (sodium azide) (69). Oat-k1 accepts only folate derivatives such as MTX and folate, while the substrates of Oat-k2 include TCA and prostaglandin E_2 in addition to these folate derivatives (69,70).

Oatp1a4 is expressed in the liver and brain (72–74), sinusoidal membrane of the hepatocytes around the central vein (75), the luminal and abluminal membrane of the brain capillaries (76), and the basolateral membrane of choroid plexus epithelial cells (76). Substrate specificity of Oatp1a4 is similar to Oatp1a1 (67,73,75,77) except for digoxin, which is a high-affinity substrate of Oatp1a4 (72). In the brain, Oat1a4 has been suggested to be involved in the efflux transport of amphipathic organic anions across the BBB (78–81) and in the uptake of [D-pen^{2,5}]enkephalin from the blood circulation (82). The brain uptake of [D-pen^{2,5}]enkephalin is limited by P-gp under normal condition: Knockout of Mdr1a increased the brain uptake of [D-pen^{2,5}]enkephalin, which was inhibited by Oatp substrates including digoxin.

Oatp1a5 is expressed in the rat female cerebral cortex (83), choroid plexus (84), and small intestine (83,85), but is very low in mouse tissues (74). Reverse transcriptase polymerase chain reaction (RT-PCR) analyses have shown that Oatp1a5 is expressed in the brain capillary and that immunofluorescence by Oatp1a5 antibody detected protein expression in the brain capillaries, although the exact membrane localization has not been determined (86). In the choroid plexus and small intestine, it is expressed on the brush border membrane (84,85). Functional expression studies of Oatp1a5 have confirmed its broad substrate specificity for amphipathic organic anions, such as bile acids and steroid conjugates, and thyroid hormones (73,84,85,87). It has been suggested to play a major role in the uptake of amphipathic organic anions by the choroid plexus from the cerebrospinal fluid (84) and also to mediate the intestinal uptake of fexofenadine (88), while the role of Oatp1a5 in the brain capillaries remains to be elucidated.

Human OATP1A2 was originally isolated from the liver (89). However, its expression in the liver is low in comparison with OATP1B1 (90,91); rather it is abundantly expressed in the brain capillary endothelial cells (89,92). As in the case of rodent homologs, OATP1A2 accepts organic anions such as bile acids, a neutral compound such as ouabain, and type II organic cations such as APDA, *N*-methyl-quinidine, *N*-methyl-quinine, and rocuronium as substrates (89,93).

Oatp1/OATP1 subfamily b consists of three members: one in rodents (Oatp1b2) and two in humans (OATP1B1 and OATP1B3). OATP1B1 and OATP1B3 are expressed predominantly in the liver, where it is localized to the sinusoidal membrane (90,94–98), and Oatp1b2 is also predominantly expressed in the liver (74,99,100). The members of Oatp1/OATP1 subfamily b exhibit broad substrate specificities, amphipathic organic anions, such as bile acids, steroid conjugated with sulfate and glucuronide, statins, and sartans (87,90,94–98,101,102). Estrone-3-sulfate is selectively transported by OATP1B1 (103), while CCK-8 is selectively transported by OATP1B3, but not by OATP1A1, OATP1B1, and OATP2B1 (104). Therefore, they were used for the reference compound for probing OATP1B1 and OATP1B3 activities in human hepatocytes (7). Oatp1b2 accepts both estrone-3-sulfate and CCK-8 as substrate (87,104).

Oatp1c/OATP1C1, of Oatp1/OATP1 subfamily "c", is the brain specific isoform and has been considered to be involved in the thyroid hormone (thyroxine) transport (105–107).

OATP2B1 is ubiquitously expressed in the normal tissues (74,96). In comparison with human OATP1 family, OATP2B1 exhibited narrow substrate specificity (108). It is expressed in the sinusoidal membrane on the hepatocytes (108) and brush border membrane in the small intestine (109). OATP2B1 has been considered to mediate the intestinal absorption of fexofenadine and estrone sulfate (109,110).

OATP4C1 is the isoform predominantly expressed in the human kidney, and its rat isoform is expressed in the lung and kidney (111). In the kidney, it is mainly expressed in the basolateral membrane of the proximal tubules, and functional analysis elucidated that it accepts digoxin and T3 as substrates (111).

2. Organic Cation Transporter (OCT/SLC22)

The Oct/OCT family consists of three members: Oct1/OCT1 (*SLC22A1*), Oct2/OCT2 (*SLC22A2*), and Oct3/OCT3 (*SLC22A3*).

Oct1 (*Slc22a1*) is expressed in the liver and kidney (112,113), while OCT1 is expressed predominately in the kidney (114). Oct1 is localized to the sinusoidal membrane of the hepatocytes surrounding the central vein and basolateral membrane in the kidney (115,116). Although the membrane localization has not been determined, Oct1 is likely expressed in the basolateral membrane of the small intestine since the distribution of metformin and intestinal excretion of tetraethylammonium (TEA) following intravenous injection was decreased in Oct1(–/–) mice (117,118). The reduction in the distribution of metformin in Oct1(–/–) was the most prominent in the duodenum followed by jejunum and ileum, but unchanged in the colon, which was consistent with the mRNA of Oct1 distribution from duodenum to ileum (119). Oct1 mediates the uptake of TEA, which was sensitive to the membrane potential, and thus, it is classified as facilitated transporter (112). Following drugs have been shown to be Oct1/OCT1 substrate: biguanides (metformin, buformin, and phenformin) (117), H2 receptor antagonists (cimetidine, ranitidine, and famotidine) (120,121), acyclic guanosine derivatives (acyclovir and ganciclovir) (122). In Oct1(–/–) mice, the distribution of TEA, MPP^+, metaiodobenzylguanidine, and metformin in the liver was significantly decreased, while that of cimetidine and choline was unchanged (117,118).

Oct2/OCT2 is predominantly expressed in the kidney (113,119), where it is localized in the basolateral membrane of the proximal tubules (123,124). Oct2/OCT2 exhibited overlapped substrate specificity with Oct1/OCT1. Comparison of substrate recognition between Oct1 and Oct2 was performed using a gene expression system. Although the inhibition constants of MPP^+, cimetidine, quinidine, nicotine, NMN, guanidine on Oct1- or Oct2-mediated TEA transport were very similar (115), the relative transport activity is different in gene transfected HEK-293 cells: the transport activity of choline relative to MPP^+ was higher in Oct1 than in Oct2, and vice versa for cimetidine, creatinine, and guanidine (121). Oct2/OCT2 more efficiently transports metformin because of greater V_{max} value (125). There was no difference in the transport activities of cimetidine, ranitidine, and famotidine by Oct1, while Oct2 efficiently transports cimetidine rather than ranitidine and famotidine (120). Both Oct1 and Oct2 are expressed in the rodent kidney and are involved in the net uptake of TEA. At steady state, the kidney concentration of TEA was decreased in both Oct1(–/–) and Oct2(–/–) mice with equal degree, and further decreased in Oct1/2(–/–) mice, and renal clearance of TEA was decreased to glomerular filtration rate in Oct1/2(–/–) mice (126).

OCT3 was isolated from the placenta, and when expressed in *X. laevis* oocytes, it mediates the uptake of TEA and guanidine in a membrane voltage-dependent manner (127). Oct3 is ubiquitously expressed in normal tissue with low level, and, among them, gonads (testes and ovaries), placenta, and uterus exhibited relatively high expression (119). In Oct3(–/–) mice, only heart and fetus exhibited reduced accumulation of MPP^+ in comparison with wild-type mice (128). The role of Oct3 in drug disposition remains unclear.

3. Organic Anion Transporter (OAT/SLC22)

Oat/OAT family consists of five members: Oat1/OAT1 (*SLC22A6*), Oat2/OAT2 (*SLC22A7*), Oat3/OAT3 (*SLC22A8*), OAT4 (*SLC22A11*), and Oat5 (*Slc22a19*).

Oat1 was cloned by expression cloning using *X. laevis* oocytes by coexpression of sodium-dicarboxylate transporter, which forms outward concentration gradient of dicarboxylate to drive Oat1-mediated uptake (129). Oat1 is predominantly expressed in the kidney, where it is localized in basolateral membrane of the proximal tubules (130). Oat1 is a multispecific transporter, and it accepts PAH, a typical substrate for this transporter, and relatively hydrophilic small organic anions, including nonsteroidal anti-inflammatory drugs and cephalosporins, and acyclic nucleotides analogs (129,131–134). Oat1-mediated transport is characterized by *trans*-stimulation. Preincubation of oocytes expressing Oat1 in the presence of α-ketoglutarate stimulated the initial uptake velocity of PAH, which is a typical character of the basolateral organic anion transporter in the kidney (129). The slices from the kidney of Oat1(–/–) mice exhibited marked reduction in the uptake of PAH and slight reduction in the uptake of fluorescein (135). The renal clearance of PAH was decreased to the glomerular filtration rate, and the renal excretion of furosemide was also decreased, resulting in attenuation of its diuretic effect (135).

Tissue distribution and membrane localization of Oat2/OAT2 exhibit gender and interspecies difference. The kidney expression is markedly higher in female rats than in male rats with similar hepatic expression (136,137), whereas the hepatic expression exhibits gender difference in mice, high in female and almost absent in male (138), although a controversial result was also obtained (139). Oat2 is localized on the sinusoidal membrane of the rat hepatocytes (140). Functional analyses of Oat2 elucidated that it exhibits substrate specificity similar to Oat1 (141), and accepts nonsteroidal anti-inflammatory drugs, such as salicylate, ketoprofen, and indomethacin as substrate (141–143). Oat2 has been suggested to be involved in the uptake of indomethacin and ketoprofen by rat hepatocytes (142,143).

Rat Oat3 is expressed in the kidney, liver, eye, and brain (144), while its human counterpart is detected predominantly in the kidney (145,146). Oat3/OAT3 is expressed in the basolateral membrane of the proximal tubule in the rat (24) and human kidneys (123,146). In rat brain, Oat3 is expressed in brain capillaries and choroid plexus, where it is localized on the abluminal and brush border membranes of the brain capillaries and choroid plexus epithelial cells, respectively, and accounts for the uptake of hydrophilic organic anions (81,147–150). In comparison with Oat1/OAT1, the substrate specificities of Oat3/OAT3 is more broad and accepts hydrophilic organic anions such as PAH, cephalosporins, 2,4-dichlorophenoxyacetate and hippurate, and amphipathic organic anions, pravastatin, pitavastatin, E217βG, estrone sulfate, dehydroepiandrosterone sulfate (DHEAS), and ochratoxin A (24,53,79,133,144,146,148,151,152). In addition, it accepts some cationic compounds, cimetidine, ranitidine, and famotidine (144,150), which have been known as bisubstrate and recognized by both organic anion and cation transporters (153). Using rat kidney slices, it has been suggested that Oat3 is responsible for the uptake of amphipathic organic anions, such as pravastatin, and steroid conjugated with sulfate (24,53,151). In Oat3(–/–) mice, the renal uptake of amphipathic anions, such as estrone sulfate and TCA was markedly decreased and also that of PAH decreased slightly (154).

OAT4 is expressed in the kidney and placenta (155), and in the kidney, unlike other human isoforms, it is expressed on the brush border membrane of the proximal tubules (156). It accepts sulfate conjugates, ochratoxin A, and PAH, although the transport activity of PAH is quite low (155). As in the case of

OAT4, Oat5 is localized on the brush border membrane of the proximal tubules and accepts sulfate conjugates (157).

4. OCTN1/OCTN2 (SLC22A4/5)

OCTN1 (*SLC22A4*) is strongly expressed in kidney, trachea, bone marrow, and fetal liver, but not in adult liver (158). When OCTN1 cDNA was transfected to HEK-293 cells, the uptake of TEA was increased in a pH-sensitive manner (158). An inward proton concentration gradient stimulated the efflux of TEA in OCTN1 expressed oocytes indicating that OCTN1-mediated transport couples with proton antiport (159). The membrane localization of OCTN1 in the kidney has not yet been described. Since the transport characteristics seem to be consistent with the previous observation using brush border membrane vesicles, it has been considered to be expressed on the brush border membrane of the kidney. The substrates include quinidine and adriamycin as well as TEA (159). OCTN2 (*SLC22A5*) was isolated from human placenta (160). Although OCTN2 can accept TEA, the transport activity is not as high as that of OCTN1. Carnitine, a cofactor essential for β-oxidation of fatty acids, has been shown to be an endogenous substrate of OCTN2 (161). Striking difference was observed in ion requirement for the transport of carnitine and organic cation via OCTN2; the transport of carnitine via OCTN2 is coupled with synport of Na^+, while that of cationic compounds is coupled with antiport of H^+ (161,162). In addition to TEA and carnitine, cephaloridine and other cationic compounds, such as verapamil, quinidine, and phyrilamine, are substrates of OCTN2 (163,164). Functional impairment of OCTN2 is associated with systemic carnitine deficiency (OMIM 212140) due to impairment of the reabsorption of carnitine from the urine (165). Octn2 is hereditarily deficient in a mouse strain, *jvs* mice, which exhibits the similar symptoms of systemic carnitine deficiency (165). The renal clearance of TEA was significantly decreased in *jvs* mice in comparison with normal mice, while that of cefazolin was unchanged. Therefore, Octn2/OCTN2 has been considered to mediate luminal efflux of organic cations in the kidney in addition to reabsorption of carnitine from the urine (166).

5. MATE

In human, two isoform have been identified, multidrug and toxin extrusion 1 and 2 (MATE1 and MATE2) (167,168). MATE1 is expressed in the liver and kidney, where it is localized on the apical membranes (canalicular membrane in the liver and brush border membrane of the proximal and distal convoluted tubules in the kidney), while MATE2 is predominantly expressed in the kidney (the brush border membranes of the proximal tubules). In rodents, MATE1 is expressed in the liver and kidney, and MATE2 only in the testis (167). Both MATE1 and MATE2 mediate antiport of organic cations with H^+, and thus, they have been considered to serve the efflux transport of hydrophilic organic cations (167,168).

6. Peptide Transporter

PEPT1 (*SLC15A1*) is expressed in the intestine (duodenum, jejunum, and ileum), kidney, and liver (169,170) and is localized to the brush border membrane (170–172). The driving force of PEPT1 is an inward H^+ concentration gradient (169). PEPT1 accepts not only di- and tripeptides but also several peptide-mimetic β-lactam antibiotics (173). PEPT1 has attracted attention as a target for drug delivery systems (DDS). Valinyl esterification of the antiviral agent acyclovir showed a three- to fivefold increase in bioavailability (174–176). Since valacyclovir is a substrate of PEPT1 (177,178), this increase has been ascribed to PEPT1-mediated transport. In addition, this approach has succeeded in the improvement of intestinal absorption of 2,3-dideoxyazidothymidine (AZT) and L-dopa modified with L-valine and L-phenylalanine, respectively (177,179).

Unlike PEPT1, PEPT2 is not expressed in the small intestine, but in the kidney and brain (180,181). In the kidney, PEPT1 is expressed in the early part of the proximal tubule (pars convoluta), while PEPT2 is expressed further along the proximal tubule (pars recta) and localized to the brush border membrane (171,182), and in the brain it is expressed in the glial cells and choroid plexus (183,184). The transport via PEPT2 is also coupled with the synport of H^+ (180,181,185). PEPT2 generally has a higher affinity for peptides and β-lactam antibiotics except cefdinir, ceftibuten, and cefixime, whose affinities were similar for PEPT1 and PEPT2 (186,187). There are high and low affinity sites responsible for the reabsorption of glycylsarcosine in the brush border membrane of the proximal tubule, and these may correspond to PEPT2 and PEPT1, respectively (188).

7. Sodium Phosphate Cotransporter (SLC17A1)

NaPi-1 (*SLC17A1*), alternatively referred to as NPT1, was originally cloned as a transporter involved in the reabsorption of phosphate in the body. Expression of NaPi-1 in *X. laevis* oocytes induced saturable uptake of benzylpenicillin (189). This uptake does not depend on Na^+ and H^+, but on Cl^- (190), and increasing extracellular concentration of chloride reduced the uptake of benzylpenicillin (190). The substrates include faropenem, foscarnet, and mevalonate, as well as benzylpenicillin (190). In contrast to the kidney, the expression is localized to the sinusoidal membrane of the liver (190). When the direction of the concentration gradient of Cl^- is taken into consideration, the transport direction mediated by NaPi-1 is efflux from inside the cells to the blood and urine in the liver and kidney, respectively.

B. Primary Active Transporters (ABC Transporters)

1. P-gp

P-gp was originally found as overexpressed protein on the plasma membrane of multidrug-resistant tumor cells, and confers multidrug resistance by actively extruding anticancer drugs to the outside (191,192). In normal tissue, P-gp is expressed in the clearance organs (liver and kidney), the site of absorption (small and large intestine), and tissue barriers (brain capillary endothelial cells), where it is localized to the luminal side, i.e., the brush border membrane in the kidney and intestine and the canalicular membrane in the liver and luminal membrane of the brain capillaries (193–199). The rodent P-gp consists of two isoforms, i.e., Mdr1a (*Abcb1a*) and Mdr1b (*Abcb1b*) (200). In the small intestine and brain capillaries, Mdr1a is the predominant isoform, while both isoforms are expressed in the liver and kidney (200). P-gp expression exhibits regional difference; it increases from the duodenum to the colon, both in rodent (201–203) and human (204–206). This expression pattern is associated with functional activity, namely, lowest activity in the duodenum and highest in the ileum (202) and colon (203).

The substrate specificity of P-gp is quite broad, and a number of compounds have been identified as P-gp substrates, generally overall positive charge or neutral compounds (193–199,207). The tissue distribution and membrane localization suggest that P-gp limits oral absorption and penetration into the brain and mediates biliary and urinary excretion of drugs. This has been supported by an in vivo finding using Mdr1a(–/–) and Mdr1a/1b(–/–) mice. The biliary excretion clearance and intestinal excretion clearance of tri-*n*-butylmethylammonium, azidoprocainamide methoiodide and vecuronium was decreased in Mdr1a(–/–) mice, and the renal clearance of tri-*n*-butylmethylammonium and azidoprocainamide methoiodide was also decreased in Mdr1a(–/–)(208). For digoxin, the amount excreted into the intestine fell markedly, while that into the bile and urine was unchanged in Mdr1a(–/–) mice (209), but fell to half the normal value in the Mdr1a/1b(–/–) (210). Following oral administration, the plasma concentration of ivermectin (200), paclitaxel (211), and fexofenadine (212) was greater in Mdr1a(–/–) mice. In situ intestinal perfusion study elucidated that the outflow concentrations of quinidine, ritonavir, cyclosporin A, daunomycin, loperamide, and verapamil (for some time points) was decreased in Mdr1a/1b(–/–) mice, indicating that the intestinal absorption of these drugs is limited by P-gp. In addition, the brain uptake of many P-gp substrates increased by inhibiting P-gp activity or in Mdr1a(–/–) and Mdr1a/1b(–/–), but not Mdr1b(–/–), (195,198–200). Since the integrity of the BBB is maintained in the Mdr1a(–/–) mouse (214), this was attributed to dysfunction of P-gp in the BBB.

Clinical studies also suggest the role of P-gp in normal human tissues. C3435T is a well-known polymorphism of MDR1 gene, which is associated with P-gp expression (TT < CC) (215). The oral absorption of digoxin is greater in healthy volunteers with the TT allele than those with CC allele, and vice versa for the renal clearance (215,216). Respiratory depression, an opioid central nervous system effect, produced by loperamide was induced by the simultaneous administration of quinidine to healthy volunteers (217). Cyclosporin A significantly increased the brain concentration of ^{11}C-verapamil (218). These have been suggested to involve inhibition of P-gp at the human BBB.

2. MRP1

MRP1 was isolated from non-P-gp multidrug resistance tumor cells, HL60AR (219). Northern blot analysis and RNase protection assay indicated that MRP1 is expressed in the lung, spleen, thymus, testis, bladder, and adrenal gland (220) and mMrp1 is abundantly expressed in muscle (221). Overexpression of MRP1 confers resistance to doxorubicin, daunorubicin, epirubicin, vincristine, vinblastine, and etoposide (221,222). In addition to anticancer drugs, MPR1 accepts amphipathic glucuronide and glutathione conjugates (223). Involvement of Mrp1 in the efflux transport in the BBB and blood-cerebrospinal fluid barrier has been suggested. The concentration of etoposide in the cerebrospinal fluid in the Mdr1a/1b/Mrp1(–/–) mice was 10-fold greater than that in Mdr1a/1b(–/–) mice, while there was no significant difference in the plasma concentration (224). The efflux transport of E217βG from the brain was significantly delayed in Mrp1(–/–) mice (225), while there was no significant change in the elimination of E217βG from the cerebrospinal fluid (226).

3. MRP2

The mutant rats, such as TR^- rats and EHBR, exhibit hyperbilirubinemia because of a deficiency in biliary excretion of bilirubin glucuronide (227–229). These mutant rats are animal model of Dubin-Johnson syndrome (OMIM 237500). Canalicular multispecific organic anion transporter (cMOAT) had been characterized by comparison of in vivo biliary excretion clearance, and ATP-dependent uptake by the canalicular membrane vesicles between normal and mutant rats. It turned out that the biliary excretion of amphipathic organic anions, such as glutathione conjugates, glucuronides, and relatively lipophilic nonconjugated organic anions, is mediated by primary active transport, and deficient in the mutant strains (228,230–232). The cDNA encoding cMOAT was isolated using homology cloning assuming a similarity with MRP1 on the basis of a similar substrate specificities (233–235). Comparison of amino acid sequence elucidated that cMOAT is a homolog of MRP1, and thus, cMOAT is renamed as MRP2. MRP2 is also expressed in the canalicular membrane of the hepatocytes, and a mutation in MRP2 gene was found in the patient suffering from Dubin-Johnson syndrome (236). The transport activity of MRP2 was compared with that of the rat counterpart using canalicular membrane vesicles. The ATP-dependent uptake clearance of glutathione conjugates was 10- to 40-fold lower in humans than that in rats, because of greater K_m values while that of glucuronide conjugates was more comparable with that in rats (2- to 4-fold lower) (237).

In addition to the liver, MRP2 is expressed in small intestine and kidney. In the small intestine, the Mrp2 expression is higher in the duodenum than that in the jejunum in rodent (234,238) and higher or similar to that in the ileum in human (204,206). Mrp2 is localized on the brush border membrane (239). Functional analysis was performed in vitro using Ussing chamber and everted sac (240). DNP-SG (2,4-dinitrophenyl-*S*-glutathione) showed 1.5-fold greater serosal-to-mucosal flux than the opposite direction in normal rats, whereas a similar flux was observed in both directions in EHBR. In everted sac studies, intestinal secretion clearance, defined as the efflux rate of DNP-SG into the mucosal side divided by the area under the curve on the serosal side, was significantly lower in the jejunum of EHBR than that in normal rats.

Schaub et al. demonstrated that Mrp2/MRP2 is expressed in the proximal tubules in the kidney (241,242). In vivo study and clinical study supports that Mrp2/MRP2 is involved in the tubular secretion of organic anions. The urinary excretion rates of calcein and fluo-3 were three to four times lower in perfused kidneys from TR^- rats compared with normal rats, and the renal excretion of lucifer yellow was delayed in TR^- rats (243). Hulot et al. identified a heterozygous mutation, which results in a loss of function of MRP2, in the patient who showed delay of renal MTX elimination (244).

4. MRP3

MRP3 is expressed in the small and large intestine in all species (238,245–248), while the hepatic expression exhibits interspecies difference. MRP3 is constitutively expressed in normal liver in mouse and human (245,247,248), while it was undetectable in rat normal liver, but high in the liver of Mrp2-deficient mutant strain, EHBR (246). Furthermore, hepatic expression of Mrp3 was subjected to induction by bile duct ligation and the treatments of α-naphthylisothiocyanate, phenobarbital, or bilirubin in rats (249), while that of Mrp3 was unchanged by bile duct ligation in mice (245). MRP3 was identified on the sinusoidal membrane of the hepatocytes in two patients with Dubin-Johnson syndrome (250) and on the basolateral membrane of rat's small and large intestine (239). Unlike MRP1 and MRP2, the transport activity of Mrp3 for glutathione conjugates was quite low, while glucuronides are good substrates of Mrp3 (251). In addition, the substrates of Mrp3/MRP3 include bile acids, taurolithocholate sulfate, and MTX (252–254). Akita et al. demonstrated the positive correlation between the protein expression of Mrp3 and sinusoidal efflux clearance of TCA (255). Using Mrp3(–/–) mice, it was shown that Mrp3 is involved in the sinusoidal efflux of glucuronide conjugates of morphine, acetoaminophen, and 4-methylumbelliferone in the liver (256–258). Unlike the liver, the role of Mrp3/MRP3 in the gastrointestine remains unclear. ATP-dependent uptake of E217βG was observed in the basolateral membrane vesicles from rat ileum, which has been considered to involve Mrp3 (259), but *trans*-ileal transport of TCA and fecal bile acid excretion was unchanged in Mrp3(–/–) mice (245).

5. MRP4

MRP4 is abundantly expressed in the kidney followed by the liver (238,260). The membrane localization of Mrp4 is tissue dependent: sinusoidal membrane in the hepatocytes (261), brush border membrane of the renal tubules (262,263), luminal membrane of the brain capillaries (262), and basolateral membrane of the choroid epithelial cells (262).

MRP4 substrates include organic anions, such as E217βG, DHEAS and PAH, and prostaglandins, cyclic nucleotide (cAMP and cGMP), diuretics (furosemide and hydrochlorothiazide), and acyclic nucleotide analogs (adefovir and tenofovir) as substrates, (263–268). In particular, TCA uptake by MRP4 expressing membrane vesicles requires reduced glutathione or its analog, *S*-methyl-glutathione in addition to ATP (261). Leggas et al. found that the elimination rate of topotecan from the brain was delayed in Mrp4(–/–) mice, although the brain concentration at early sampling points exhibited no difference (262). In addition, the concentration of topotecan in the cerebrospinal fluid was markedly increased in Mrp4(–/–) mice (262). In the kidney, the renal clearance of furosemide with regard to the plasma concentration was decreased, and the kidney concentrations of hydrochlorothiazide, adefovir, and tenofovir were significantly increased in Mrp4(–/–) mice (267,268).

6. Breast Cancer Resistance Protein (BCRP/ABCG2)

BCRP is classified in ABCG subfamily; other members of this subfamily are involved in sterol transport (269). Unlike P-gp and MRPs, BCRP consists of a single ABC cassette in the amino terminal followed by six putative transmembrane domains; however, it forms a homodimer linked by a disulfide bond in the plasma membrane (270,271). Initially, ABCG2 was identified as an mRNA expressed in placenta (272) and as a non-MDR1- and non-MRP-type resistance factor from cell lines selected in the presence of anthracyclines and mitoxantrone (273). BCRP is expressed widely in the normal tissues (274) and localized on the canalicular membrane of the hepatocytes and apical membranes of epithelial cells (274,275) and brain capillary endothelial cells (276,277).

BCRP exhibits broad substrate specificity for various anticancer drugs, such as mitoxantrone and topotecan (278), drugs such as pitavastatin, sulfasalazine, cimetidine and AZT, fluoroquinolones (279–281,282, and glucuronide- and sulfate conjugates (397), and dietary carcinogens (283,284). Cumulative in vivo studies, particularly using Bcrp(–/–) mice, have shown the importance of BCRP in drug disposition. BCRP limits the oral absorption of topotecan (275), sulfasalzine (280), and ciprofloxacin (281). Bcrp has been shown to account for the efflux of intracellularly formed glucuronide and sulfate conjugates (E3040 glucuronide, E3040 sulfate, and 4-methyumbelliferone sulfate) (285), and the active form of the ester-type prodrug of ME3277 (286) in the small intestine, and the biliary excretion of drugs, such as nitrofurantoin (287), MTX (288), pitavastatin (282), and sulfasalazine (280). BCRP limits the brain penetration of imatinib, but not other BCRP substrates, such as mitoxantrone and dehydroepiandro sterone sulfate (277) and pitavastatin (282,289). Unlike human, Bcrp is expressed in the brush border membrane of renal tubules (275), and it is involved in the tubular secretion of E3040 sulfate (290) and MTX (288).

V. EXAMPLES OF DRUG-DRUG INTERACTIONS INVOLVING MEMBRANE TRANSPORT

A. Direct Inhibition

1. Digoxin-Quinidine and Digoxin-Quinine

Digoxin undergoes both biliary and urinary excretion in human (291). The drug-drug interactions between digoxin and quinidine or quinine (a stereoisomer of quinidine) are very well known (291). The degree of inhibition by quinidine and quinine of the biliary and urinary excretion of digoxin are different; quinine reduced the biliary excretion clearance of digoxin to 65% of the control value, while quinidine reduced both the biliary and renal clearance to 42% and 60%, respectively (Fig. 7) (291). In proportion to the reduction in total body clearance, coadministration of quinine and quinidine increases the plasma concentration of digoxin by 1.1-fold and 1.5-fold, respectively (291). In addition to these agents, verapamil also has an inhibitory effect, but specifically on the biliary excretion (292), has only a slight inhibition of renal excretion (293).

No inhibitory effect of quinine and quinidine was obtained in isolated human hepatocytes at a concentration of 50 μM (294), whereas stereoselective inhibition of quinine and quinidine has been observed in isolated rat hepatocytes (295). Quinine inhibits uptake into isolated hepatocytes at the concentration of 50 μM, while the effect of quinidine was minimal (at most a 20% reduction)

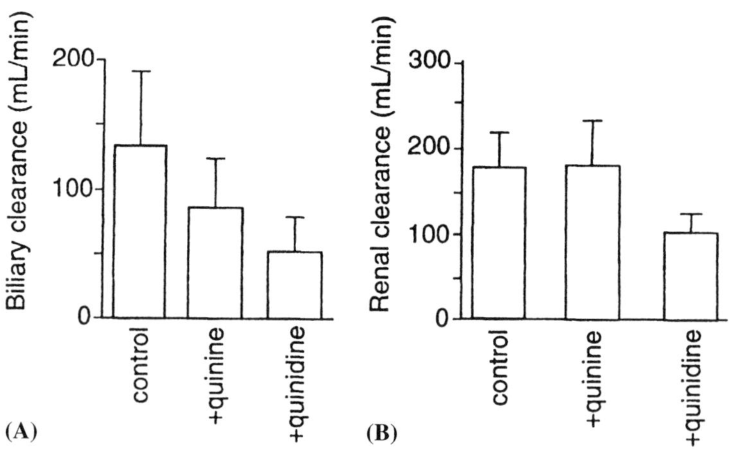

Figure 7 Change in the biliary and renal clearance of digoxin caused by quinidine or quinine treatment. After a steady state concentration of quinine or quinidine was achieved by multiple oral administrations, the plasma concentration and biliary and urinary excretion of digoxin after oral administration were measured in healthy volunteers. The steady state concentrations of quinine and quinidine were 7.0 ± 2.5 and 4.5 ± 0.5 μM, respectively. *Source*: From Ref. 291.

(295). Substrates of P-gp, such as vinblastine, daunorubicin, and reserpine, as well as quinine, quinidine, and verapamil, also inhibit the renal excretion of digoxin in rats, although typical substrates for organic cation and anion transporter on the basolateral membrane (TEA and PAH) do not (296). On the basis of the animal (209,210) and clinical (216) observations, P-gp has been suggested to be the candidate transporter for the biliary and urinary excretion of digoxin. The role of P-gp in this drug-drug interaction has been examined using the Mdr1a(–/–) mice (297). Coadministration of quinidine caused a 73% increase in the plasma concentration of digoxin in normal mice, whereas it had little effect (20% increase) in the Mdr1a(–/–) mice at the same plasma concentration of quinidine (Fig. 9) (297).

The drug-drug interaction between digoxin and quinidine has been also suggested in the intestinal absorption of digoxin in rats (298). The appearance rate of digoxin on the basolateral side of an everted sac of the jejunum and ileum increased in the presence of quinidine or an unhydrolyzed ATP analogue, AMPPNP, and intestinal secretion of digoxin was also inhibited by quinidine. These results indicate that digoxin undergoes active efflux in the small intestine (298). Indeed, the intestinal secretion of digoxin was significantly reduced in Mdr1a(–/–) and Mdr1a/1b(–/–) mice (209,210). The area under the curve of the plasma concentration of digoxin following oral administration is associated with genetic polymorphism of MDR1 gene (C3435T, AUCpo TT > CC) (215). Therefore, the interaction of quinidine and digoxin involving intestinal absorption may be due to the inhibition of P-gp function.

2. Fexofenadine-Itraconazole/Verapamil/Ritonavir

Fexofenadine is mainly excreted into the bile and urine without metabolism. Many transporters are involved in the pharmacokinetics of fexofenadine. OATP1A2 (212), OATP2B1 (299), OATP1B3 (300), OAT3 (301), and P-gp (212) have been suggested to accept fexofenadine as substrate. On the basis of in vivo study using Mdr1a and Mdr1a/1b(–/–) mice, it has been shown that P-gp limits intestinal absorption and brain penetration of fexofenadine, but makes only a limited contribution to the biliary and urinary excretion (212,302). Furthermore, inhibition of P-gp in the intestine allowed detection of saturable uptake of fexofenadine and inhibition by Oatp inhibitor in rats (88).

Drug-drug interactions involving fexofenadine have been reported which includes not only interactions with concomitant drugs, but also those with fruit juices. Concomitant use of itraconazole (303), verapamil (304), and ritonavir (305) increased the area under the curve of the plasma concentration (AUC) and peak plasma concentration ($C_{\max}$) of fexofenadine following oral administration, but did not affect the elimination half-life. Itraconazole and verapamil did not affect the renal clearance of fexofenadine, while the effect of ritonavir on the renal clearance was not examined. Considering the absence of the effect on the renal clearance, these interactions will include the inhibition of intestinal efflux and/or hepatobiliary transport. Since itraconazole [K_i ~ 2 μM (306,307)], verapamil [K_i ~ 8 μM (308)], and ritonavir [K_i ~ 4 and 12 μM (306,309)] are inhibitors of P-gp, it is possible that these drug-drug interactions involve inhibition of P-gp-mediated efflux in the small intestine. Fruit juice made from grapefruit, orange, or apple decreased the AUC and C_{max} of fexofenadine, without affecting the renal clearance (310–312). This has been suggested to include OATP-mediated uptake in the intestine (311).

3. HMG-CoA Reductase Inhibitor, Cerivastatin-Cyclosporin A/Gemfibrozil

In the kidney transport recipients treated with cyclosporin A, the AUC of cerivastatin was 3.8-fold larger than that in healthy volunteers (313). Initially, inhibition of CYP3A4 and CYP2C8, major metabolic enzymes for cerivastatin, has been considered as underlying mechanism. Finally, an inhibition of the hepatic uptake process of cerivastatin mediated by OATP1B1 has been suggested as underlying mechanism. OATP1B1 accepts cerivastatin as substrate (314). The K_i values of cyclosporin A for the uptake of cerivastatin in two lots of cryopreserved human hepatocytes were comparable with that for OATP1B1 (0.28 and 0.68 μM vs. 0.25 μM), while cyclosporin A did not affect the metabolic rate of cerivastatin by pooled human microsomes by 3 μM (314). OATP1B1 plays a significant role in the hepatic uptake of other open acid form of statins, such as pravastatin (315), pitavastatin (316), and simvastatin (317), but not fluvastatin (318), and thus, cyclosporin A increased the plasma concentration of pravastatin by 5- to 8-fold and pitavastatin by 4.5-fold (319). In addition to statins, OATP1B1 is also involved in the hepatic uptake of valsartan (320) and repaglinide (321). Cyclosporin A also increased the total area under the plasma concentration–time curve of repaglinide by 2.4-fold, but this may include an inhibition of metabolism as well as inhibition of hepatic uptake process (322). In addition to cyclosporin A, rifampicin and rifamycin SV will have potent inhibitory effect of OATP1B1 by their clinical concentrations (52). Rifampicin is a well-known drug causing induction of drug metabolizing enzymes and transporters by repeated administration, but it may also inhibit hepatic uptake process by a concomitant usage.

Gemfibrozil increased the plasma concentrations of cerivastatin. The effect of gemfibrozil on the plasma concentration–time profile of cerivastatin following oral administration is different from that of cyclosporin A (319). Cyclosporin A increased C_{max} without affecting the elimination half-life, while gemfibrozil prolonged the elimination half-life. The interaction between gemfibrozil and cerivastatin may include the inhibition of hepatic uptake, but this effect is considered to be weak considering their clinical concentrations and IC_{50} values for the hepatic uptake. Rather, inhibition of CYP2C8 (mechanism based inhibition) by gemfibrozil glucuronide has been suggested as an underlying mechanism for this drug-drug interaction, considering that volunteers were given gemfibrozil for 4 days (twice a day) before cerivastatin administration (323,324).

4. Interaction with Probenecid

Probenecid has been reported to inhibit renal elimination of many drugs: acyclovir (325,326), allopurinol (327), bumetanide (328), cephalosporins (329–334), cidofovir (335), ciprofloxacin (336), famotidine (337), fexofenadine (338), furosemide (339), and oseltamivir (Ro 64–0802) (340). Recent studies have elucidated that probenecid is a potent inhibitor of renal organic anion transporters (OAT1 and OAT3) with the K_i values lower than the unbound plasma concentration of probenecid, indicating the interaction with probenecid includes inhibition of the basolateral uptake process mediated by OAT1 and/or OAT3.

5. Furosemide/Cidofovir/Oseltamivir-Probenecid

Furosemide undergoes both of renal excretion and glucuronidation. Probenecid reduced the renal clearance of furosemide to 34% of the normal value, resulted in a 2.7-fold increase in the AUC of plasma furosemide, following oral administration to healthy volunteers (339). Since furosemide is actively secreted from blood to the lumen by organic anion transport systems and exhibit diuretic effects by inhibiting the reabsorption of ions mediated by Na^+-K^+-$2Cl^-$ cotransporter in the loop of Henle (341), this drug-drug interaction also inhibits the diuretic action in humans (339,342). Oat1 has been suggested to be responsible for renal uptake of furosemide since the renal excretion of furosemide was markedly reduced in the Oat1(–/–) mice (135). The fact that probenecid is a potent inhibitor of Oat1/OAT1 with K_i value of 4 μM (343) and 13 μM (344) suggests that this drug-drug interaction will include an inhibition of uptake process mediated by OAT1. In addition to furosemide, drug-drug interactions of cidofovir and oseltamivir with probenecid has been suggested to involve inhibition of OAT1, since they are substrates of OAT1 (340,345).

6. H2 Receptor Antagonists (Famotidine/Ranitidine)/ Fexofenadine-Probenecid

H2 receptor antagonists are weak base or cationic compounds at physiological pH. They have been known as bisubstrates, which are substrates of both renal organic anion and cation transporters. Indeed, they are substrates of Oat3/OAT3 (120,144) and Oct1/OCT1 and Oct2/OCT2 (120). The renal elimination of H2 receptor antagonists is the major elimination pathway and both glomerular filtration and tubular secretion are involved (337,346). Probenecid exhibited different inhibition potency to the renal elimination of cimetidine and famotidine; probenecid significantly decreased the renal clearance of famotidine and the tubular secretion clearance was decreased to 10% of the control value (Fig. 8), while it did not affect the renal clearance of cimetidine (337,346). Considering that probenecid is a potent inhibitor of OAT3, but not OCTs, and that the unbound probenecid concentration of probenecid ranged from 30 to 90 μM is sufficient to inhibit OAT3 (347), this is likely ascribed to the difference in the contribution of OAT3 and

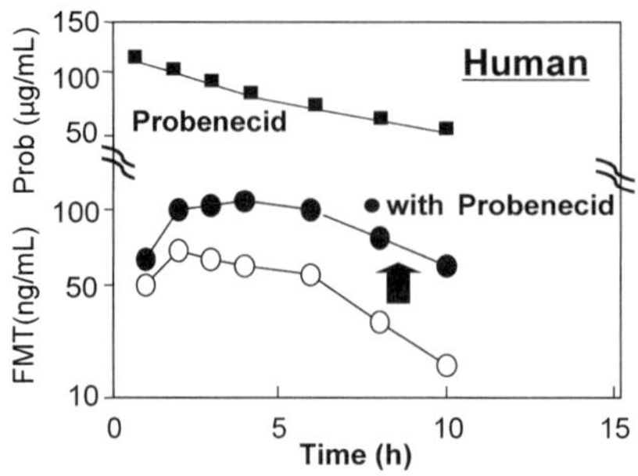

Figure 8 Effect of probenecid on the plasma concentration of famotidine in healthy volunteers. Plasma concentration of famotidine was determined in healthy subjects treated with or without probenecid. The renal and tubular secretion clearances were decreased by the probenecid treatment (CL_{renal} 279 vs. 107 mL/min and CL_{sec} 196 vs. 22 mL/min). *Source*: (A) from Ref. 348 and (B) from Ref. 337.

OCT2 to the tubular secretion of cimetidine and famotidine. Furthermore, a great interspecies difference was found in the effect of probenecid, which had no effect on the tubular secretion of famotidine and cimetidine in rats (Fig. 8) (348). Two factors have been proposed for this interspecies difference (1) expression of Oct1 only in rodent kidney and (2) greater transport activity of famotidine by OAT3 than by Oat3 (120). In monkey, as in the case in human, probenecid had significant effect on the renal elimination of famotidine, but not for cimetidine (152).

The renal clearance of ranitidine accounts for 53% of the total body clearance in the beagle dog. Although ranitidine is a cationic compound, probenecid treatment reduced the total body clearance and renal clearance to 60% and 52% of the control value, respectively (349). According to analysis using a physiological pharmacokinetic model, the drug-drug interaction between ranitidine and probenecid is due to inhibition of transport across the basolateral membrane. Presumably, this drug-drug interaction also involves OAT3 as suggested for famotidine.

In addition to H_2 receptor antagonists, drug-drug interaction between fexofenadine and probenecid has been suggested to involve an inhibition of OAT3 based on an in vitro observation that fexofenadine is a substrate of OAT3, but not OAT1 and OCT2 (301). Cimetidine has been reported to inhibit the renal clearance of fexofenadine by 39% on average in healthy subjects (338). Cimetidine is a substrate of OAT3; however, the clinical plasma concentration of unbound cimetidine at a dose of 400 mg was reported to be, at most, 5.2 μM (398), far below its K_m and IC_{50} values for OAT3 (113 μM (120)). It is unlikely that the interaction involves OAT3, and presumably, cimetidine inhibits efflux process across the brush border membrane of the proximal tubules.

7. Benzylpenicillin-Probenecid

Benzylpenicillin disappears from the blood very rapidly (the elimination half-life is 30 minute in the adult), and 60–90% of dose is excreted in the urine (350). The renal clearance is approximately equal to the blood flow rate, indicating a high secretion clearance (350). Probenecid and phenylbutazone reduced its renal clearance to 60%, while sulfinpyrazone reduced it to 40% of the control value (351). In rat kidney, Oat3 has been suggested to be responsible for the uptake of benzylpenicillin (53). As discussed above, inhibition of uptake process mediated by OAT3 is likely mechanics underlying this interaction.

8. Ciprofloxacin-Probenecid

Renal clearance accounts for 61% of the total body clearance of ciprofloxacin in humans (350). Coadministration of probenecid reduces the total body and renal clearance to 59% and 36% of the control value, respectively, but has no effect on the nonrenal clearance (336). The transporters involved in the renal elimination of ciprofloxacin remains unknown.

9. MTX-Organic Anions

Urinary excretion is the major elimination pathway of MTX in humans (350). The renal clearance of MTX was three times greater than the glomerular filtration clearance in the monkey, indicating secretion is involved in the renal excretion (352). Since the renal excretion of MTX is saturable, transporters are responsible for the renal secretion of MTX (352). Coadministration of probenecid 700 mg/m^2 reduced the renal clearance to the glomerular filtration clearance (352). The site, where MTX undergoes secretion, was examined using the stop-flow method (353). A peak appeared at the site corresponding to the proximal tubule in the monkey, indicating that excretion of MTX occurs at the proximal tubule, and benzylpenicillin reduced the peak value to 33% of the control value (353). The interaction between MTX and benzylpenicillin was also examined using kidney slices (353). The uptake of MTX into kidney slices was inhibited by benzylpenicillin in a concentration-dependent manner, and the saturable component was completely inhibited by benzylpenicillin (353). Takeda et al. suggested that salicylate, phenylbutazone, and indomethacin inhibited OAT3-mediated MTX uptake at the concentrations comparable with therapeutically relevant unbound plasma concentrations (354), suggesting that the interactions, at least, involving these drugs includes inhibition of basolateral uptake process. Nozaki et al. also reported that, in addition to Oat3, reduced folate carrier is also involved in the uptake of MTX in rat kidney, which is hardly inhibited by nonsteroidal anti-inflammatory drugs, and thus, the inhibition of the net uptake is not so potent in the kidney as expected (355). However, some nonsteroidal anti-inflammatory drugs are more potent inhibitors in human kidney slices, and expected to have significant effect on the MTX uptake in the kidney at clinical dose (396). Whole interactions involving MTX cannot be explained by inhibition of uptake process. Recently, MRP4 and MRP2, candidate transporters for the luminal efflux of MTX, have been also suggested to be involved in the interaction of MTX with nonsteroidal anti-inflammatory drugs (NSAIDs) based on a inhibition potency for the ATP-dependent uptake of MTX by MRP2 and MRP4, although the clinical relevance remains unknown (356,396).

10. Cefadroxil-Cephalexin

Both the dose normalized AUC of the plasma concentration for two hours after administration and the maximum plasma concentration exhibited nonlinearity, when cefadroxil, a β-lactam antibiotic, was administered at different oral doses from 5 to 30 mg/kg orally (357). Coadministration of cephalexin (15 mg/kg) reduced both the AUC and C_{max} of cephadroxil (357). Since cefadroxil and cephalexin are substrates of PEPT1 (358), this interaction may be accounted for by an interaction at the binding site of PEPT1(357).

Both cefadroxil (5 mg/kg) and cephalexin (45 mg/kg) were administered as a 200-mL suspension (357). Assuming the suspension not to undergo any dilution during transit into the small intestine, the free substrate and inhibitor concentrations were estimated to be 3.9 and 28.3 mM, respectively. The K_m value of cefadroxil was found to be 5.9 mM using the rat in situ perfusion method (359), the substrate concentration is not low enough. The K_m value of cephalexin (7.5 mM), determined using Caco-2 cells, is used as the K_i value in this prediction (33), which is lower than the estimated luminal concentration, indicating significant inhibition of PEPT1 in the small intestine.

11. Loperamide-Quinidine and Verapamil-Cyclosporin A

Respiratory depression, an opioid central nervous system effect, produced by loperamide was induced by the simultaneous administration of quinidine to healthy volunteers (600 mg/kg) (217). Since the time profile of the plasma concentration of loperamide was similar irrespective of quinidine administration when respiratory depression was induced, inhibition of P-gp at the BBB by quinidine has been suggested as underlying mechanism. The IC_{50} values of quinidine for P-gp vary depending on the substrates, ranging from 0.4 to 20 μM (summarized in TP-Search, http://tp-search.jp). The broad range of IC_{50} values of quinidine may be due to the multiple substrate recognition sites, one is high affinity for the Hoechst compound, but low affinity for rhodamine 123 (H site), and vice versa for the other (R site) (360). The unbound concentration of quinidine is estimated to be at most 1 μM in the clinical study, which will be sufficient to inhibit P-gp if loperamide is recognized by P-gp at the site exhibiting lower IC_{50} values against quinidine.

Positron emission tomography (PET) using ^{11}C labeled P-gp substrates allows noninvasive and sequential determination of brain concentrations in nonhuman primates and humans (361). Using this technical advance, Sasongko et al. demonstrated that cyclosporin A (given by intravenous constant infusion

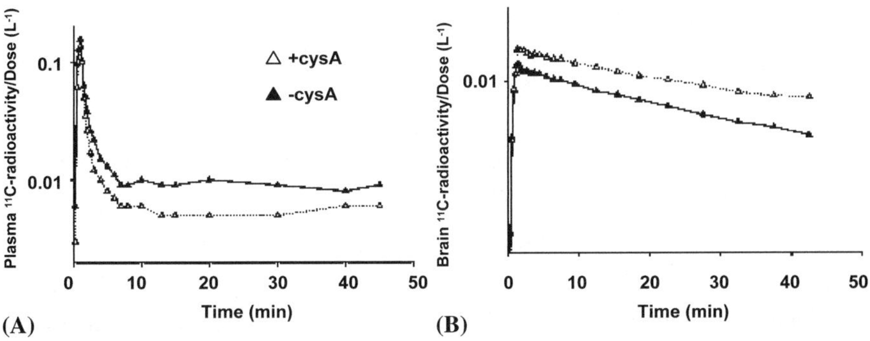

(A) (B)

Figure 9 Effect of cyclosporin A on the brain (**A**) and plasma (**B**) concentration of ^{11}C verapamil in healthy volunteers. (**A**) ^{11}C-Verapamil ($\sim$0.2 mCi/kg) was administered to healthy volunteers intravenously, approximately one minute before and after one-hour infusion of cyclosporin A (2.5 mg/kg/h). (**B**) PET images of a normal human brain after ^{11}C-verapamil administration in the absence or presence of cyclosporin A. Images shown are in SUV summed over a period of 5 to 25 minutes, which is an index of regional radioactivity uptake normalized to the administered dose and weight of the subject. *Abbreviations*: PET, positron emission tomography; SUV, standardized uptake value. *Source*: From Ref. 218.

at 2.5 mg/h/kg for 1 h) significantly increased the brain concentration of ^{11}C-verapamil and the ratio of the area under the curve of the brain and plasma concentrations, representing the brain-to-plasma partition coefficient, resulting in an 87% increase on average by cyclosporin A (Fig. 9) (218). Compared with the in vivo results using Mdr1a(–/–) mice, the size of the increase was less in humans. This is probably because of incomplete inhibition since the unbound concentration of cyclosporin A was approximately 0.2 μM, similar to or lower than the previously reported IC_{50} values of cyclosporin A for P-gp ranging from 0.4 to 4 μM (summarized in TP-Search, http://tp-search.jp).

12. Transport Via the Large Neutral Amino Acid Transporter Is Affected by Diet

The pharmacological effect of L-dopa is affected by diet (362). The ''off'' period in Parkinsonian patients treated with L-dopa is a clinical problem, since the efficacy of the drug suddenly fails. Because of the inverse relationship between the plasma levels of large neutral amino acid (LNAA) and the clinical performance of Parkinsonian patients (362) and the fact that the transcellular transport of L-leucine is inhibited by L-dopa (363) across primary cultured bovine brain capillary endothelial cells, the ''off'' period may be attributed to the membrane transport of L-dopa via LNAAT at the BBB. In addition to L-dopa, baclofen and melphalan are suggested to be taken up into the brain via amino acid transporter (363,364), and thereby, their brain transport might be also affected by the plasma concentration of large neutral amino acids.

B. Indirect Interaction

In addition to the direct interaction with drug transporters, administration of some kinds of drugs (so-called inducers) modulates the expression of drug transporters, and thereby, affects the pharmacokinetics of other drugs; one is modulating membrane trafficking of transporter protein and the other is induction/downregulation of transporter mRNA.

1. Modulation of Membrane Trafficking-Genipin/Mrp2

Genipin is an intestinal bacterial metabolite of geniposide, a major ingredient of a herbal medicine, Inchin-ko-to, which have potent choleretic effects, and it rapidly stimulates redistribution of Mrp2 to the canalicular membrane in rats (365). Infusion of genipin for 30 minutes significantly increased the biliary excretion of glutathione in normal rats. The effect of genipin is associated with Mrp2 function since genipin had no effect in Mrp2-deficient mutant rats (EHBR). Genipin did not affect the mRNA expression of Mrp2, whereas it significantly increased Mrp2 protein in the canalicular membrane, resulting in a significant increase in and ATP-dependent uptake of Mrp2 substrates by canalicular membrane vesicles. Accordingly, genipin treatment increases an insertion of Mrp2 to the canalicular membrane and/or decreases internalization by known mechanism.

2. Induction

Recent studies have revealed the importance of orphan receptors, which form heterodimer with the 9-*cis* retinoic acid receptor (RXR) in regulating drug metabolism enzymes and transporters. Such orphan receptors include pregnane X receptor (PXR/NR1I2), constitutive androstane receptor (CAR/NR1I3), farnesoid X receptor (FXR/NR1H4), and peroxisome proliferator-activated receptor α (PPARα/NR1C1) (366–370) (Table 3). Except CAR, they act as ligand-activated nuclear receptor and bind a specific element in the enhancer of the target genes as a heterodimer with RXR, while CAR shows a constitutive transcriptional activity and undergo translocation from cytosol into nucleus upon activation (366–370). Kato et al. investigated this for the quantitative prediction of CYP enzymes in the liver on the basis of in vitro study taking in vivo exposure of inducers, which was in good agreement with in vivo observation (371). The same strategy will be also effective in predicting induction of drug transporters.

a. PXR (NR1I2). PXR is expressed abundantly in the liver and to a lesser extent in the small intestine and colon. PXR is activated by various compounds, including drugs such as rifampicin and food such as St. John's wort, and its major antidepressant constituent, hyperforin (367–369). Repeated administration of rifampicin for nine days increased MDR1 P-gp expression in the duodenum both in the mRNA and protein levels (372,373), and thereby, caused a decrease in oral bioavailability of digoxin (Fig. 10) (372) and fexofenadine (374) in healthy volunteers and a decrease in the AUC of another substrate, talinolol, both

Table 3 Nuclear Receptors Involved in the Induction of Xenobiotic Transporters

Nuclear receptor	Gene name	OMIM	Gene ID	Major organ[a]	Typical agonist
PXR	NR1I2	603065	8856	Liver, small intestine, and colon	Rifampicin, St. John's wort
CAR	NR1I3	603881	65035	Liver, kidney	TCPOBOP inverse agonist: androstenol and androstanol
FXR	NR1H4	603826	9971	Liver, kidney, small intestine	Bile acids, GW4064
PPARα	NR1C1	170998	5465	Ubiquitous	Fibrates

[a]Adapted from Ref. 395.
Abbreviations: PXR, pregnane X receptor; CAR; constitutive androstane receptor; FXR, farnesoid X receptor.

after intravenous and oral administration (373). The induction of P-gp by rifampicin occurs via activation of promoter activity. Promoter assay revealed that the induction occurred via binding of a heterodimer complex of PXR and RXR to a *cis*-element in the enhancer of MDR1 P-gp (375). PXR is also involved in the induction of other ABC transporters MRP2, MRP3, and BCRP, and uptake transporter OATP1B1. The mRNA expression of MRP2 was increased in duodenum in healthy volunteers treated with rifampicin (376), and, upon the treatment of PXR agonists (rifampicin or hyperforin), mRNA level of MRP2 was also increased in the primary cultured human hepatocytes (377–379). Rifampicin treatment has also induced the mRNA expression of OATP1B1 (2.4-fold), BCRP (2.7-fold), and MRP3 (1.7-fold), but had no effect on OATP1B3, OCT1, and OAT2 in human hepatocytes (379). A PXR agonist, pregnenolone 16α-carbonitrile (PCN)-treatment induced mRNA expression of Oatp1a4 (380–383) accompanied with an increase in hepatic uptake of digoxin in rats (382), while it did not affect mRNA expression of Oatp1b2 (380,383). PCN treatment also induced mRNA of Mrp3 in mouse liver (380), but not in rat liver (384). Unlike MRP2, rodent Mrp2 was unchanged by PCN-treatment in vivo (380,384), but induced in primary cultured rat hepatocytes (377).

b. CAR/NR1I3. CAR is involved in the induction by phenobarbital and antagonized by endogenous ligands such as androstenol and androstanol (so-called inverse agonists) (367–370). Phenobarbital treatment enhanced mRNA expression of MDR1, MRP2, and BCRP in human hepatocytes (379). In rodent liver, phenobarbital as well as other CAR activator/ligand, such as TCPOBOP (a synthetic CAR agonist) induced mRNA expression of Mrp3 (380,381,384,385), Mrp4 (385,386), and Oatp1a4 (381,387), but not for Oatp1b2 (383,387),

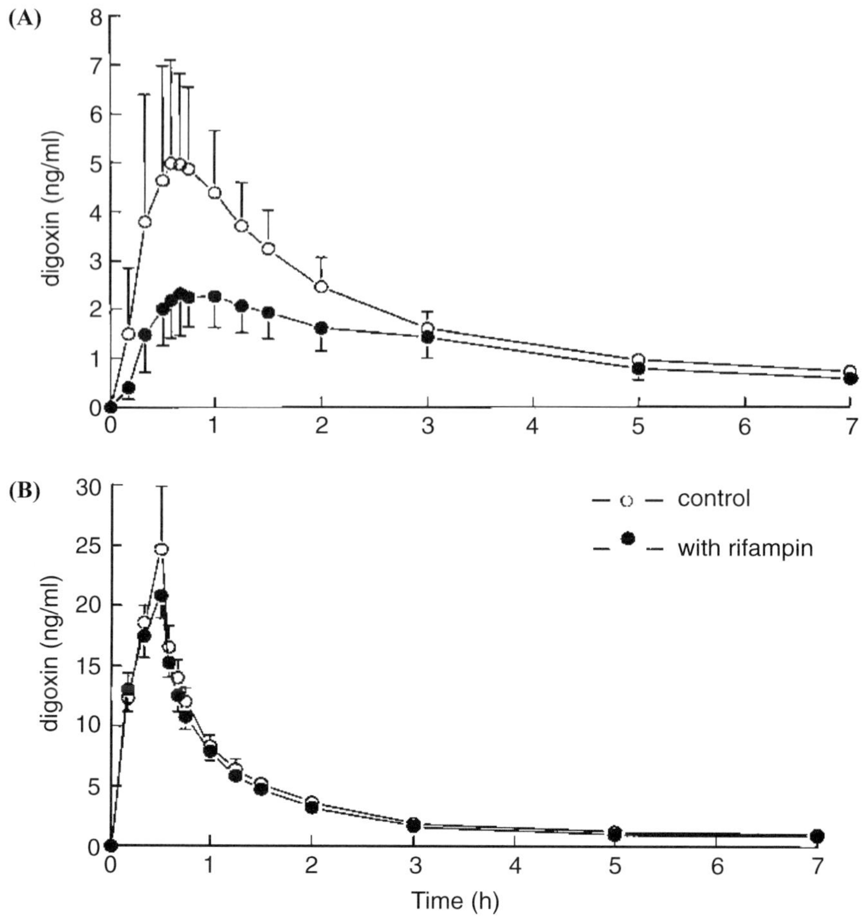

Figure 10 Effect of repeated administration of rifampicin on the time profile of the plasma concentration of digoxin following intravenous and oral administration. Eight healthy male volunteers [age 29 ± 5 years, body weight 84 ± 9 kg (mean ± SD)] were included in the study. Plasma concentration (mean ± SD) time curves of digoxin given orally (1 mg) (**A**) and intravenous infusion over 30 minutes (1 mg) (**B**) before (*open circles*) and during (*filled circles*, day 11) coadministration of rifampicin (600 mg, once daily orally for 16 days). *Source*: From Ref. 372.

although induction of Mrp3 by phenobarbital includes CAR-independent mechanism (388,389).

c. FXR/NR1H4. FXR is activated by bile acids, such as chenodeoxycholate, and a synthetic agonist, GW4064, and plays an important role in regulating the bile acid homeostasis (366,367). FXR enhanced OATP1B3 promoter through the binding to the FXR response element, and chenodeoxycholate induces an expression of OATP1B3 in human hepatoma cells (390,391). On the other hand, chenodeoxycholate suppresses the expression of OATP1B1 through the suppression of HNF1α (392), which is critically involved in the expression of OATP1B1 (393).

d. PPARα/NR1C1. PPARα is the target molecule of hypolipidemic drugs (fibrates), which have been used to reduce triglycerides and cholesterol in patients with hyperlipidemia. In mice, PPARα is involved in the induction of mRNA of hepatic ABC transporters, such as Mdr1a, Bcrp, Mrp3, and Mrp4, by clofibrate in the liver, resulting in a significant increase in protein expression of P-gp, Mrp3, and Mrp4 (394).

VI. SUMMARY

Transporters are membrane proteins regulating the influx and efflux of organic solute across the plasma membrane, and thereby, act as one of the determinant factors for the drug disposition. They play important roles in the hepatobiliary transport and tubular secretions in the kidney, absorption in the small intestine, and efflux transport in the tissue barriers, such as BBB. Most transporters involved in the drug disposition are characterized by broad substrate specificities and accept structurally unrelated compounds. Molecular cloning has elucidated the molecular characteristics of such drug transporters, which include members of SLC family, such as OATP (*SLCO*), OCT/OAT/OCTN (*SLC22*), PEPT (*SLC15*) and MATE (*SLC47*), and ABC transporters such as P-gp (*ABCB1*), MRPs (*ABCC*), and BCRP (*ABCG2*). Using gene knockout/deficient animals and selective inhibitors, scientists have investigated the roles of transporters in drug disposition. Drug-drug interactions involving transporters include direct inhibition or indirect modulation, and thereby, affect the pharmacokinetics of the substrate drugs. For direct inhibition, using unbound concentration of inhibitors and inhibition constant of the target transporter, one can quantitatively evaluate the degree of inhibition of the target transporter. This rough estimation will be helpful for prescreening of drug-drug interaction and evaluation of in vivo relevance of such inhibition in the drug-drug interactions. As indirect modulation, the role of nuclear receptors, such as PXR, CAR, FXR, and PPARα, forming heterodimer with RXR, has been suggested to transactivate the promoter of the drug transporters. This chapter focused on the molecular characteristics of drug transporters and drug-drug interaction involving these drug transporters.

ACKNOWLEDGMENTS

We would like to thank Hiroshi Suzuki, Yukio Kato, Kiyomi Ito, Kosei Ito, Yoshihisa Shitara, Daisuke Sugiyama, and Yoko Ootsubo-Mano for sharing meaningful discussions with us, giving us useful suggestions, and helping collect information to prepare this manuscript and to Atsushi Ose, Takami Saji, Sayaka Ichihara, and Etsuro Watanabe for their kind help.

REFERENCES

1. Giacomini KM, Sugiyama, Y. Membrane transporters and drug response. In: Brunton LL, Lazo JS, Parker KL, eds. Goodman & Gilman's The Pharmacological Basis of Therapeutics. 11th ed. New York: McGraw-Hill, 2006:41–70.
2. Ueda K, Kato Y, Komatsu K, et al. Inhibition of biliary excretion of methotrexate by probenecid in rats: quantitative prediction of interaction from in vitro data. J Pharmacol Exp Ther 2001; 297:1036–1043.
3. Ito K, Iwatsubo T, Ueda K, et al. Quantative prediction of in vivo drug clearance and drug interactions from in vitro data on metabolism together with binding and transport. Ann Rev Pharmacol Toxicol 1998; 38:461–499.
4. Ito K, Iwatsubo T, Kanamitsu S, et al. Prediction of pharmacokinetic alterations caused by drug-drug interactions: metabolic interaction in the liver. Pharmacol Rev 1998; 50:387–412.
5. Kato Y, Akhteruzzaman S, Hisaka A, et al. Hepatobiliary transport governs overall elimination of peptidic endothelin antagonists in rats. J Pharmacol Exp Ther 1999; 288:568–574.
6. Shitara Y, Li AP, Kato Y, et al. Function of uptake transporters for taurocholate and estradiol 17b - D-glucuronide in cryopreserved human hepatocytes. Drug Metab Pharmacokinet 2003; 18:33–41.
7. Hirano M, Maeda K, Shitara Y, et al. Contribution of OATP2 (OATP1B1) and OATP8 (OATP1B3) to the hepatic uptake of pitavastatin in humans. J Pharmacol Exp Ther 2004; 311:139–146.
8. Ishigami M, Tokui T, Komai T, et al. Evaluation of the uptake of pravastatin by perfused rat liver and primary cultured rat hepatocytes. Pharm Res 1995; 12: 1741–1745.
9. LeCluyse EL, Audus KL, Hochman JH. Formation of extensive canalicular networks by rat hepatocytes cultured in collagen-sandwich configuration. Am J Physiol 1994; 266:C1764–C1774.
10. Liu X, Chism JP, LeCluyse EL, et al. Correlation of biliary excretion in sandwich-cultured rat hepatocytes and in vivo in rats. Drug Metab Dispos 1999; 27:637–644.
11. Liu X, LeCluyse EL, Brouwer KR, et al. Use of Ca^{2+} modulation to evaluate biliary excretion in sandwich- cultured rat hepatocytes. J Pharmacol Exp Ther 1999; 289:1592–1599.
12. Hoffmaster KA, Turncliff RZ, LeCluyse EL, et al. P-glycoprotein expression, localization, and function in sandwich-cultured primary rat and human hepatocytes: relevance to the hepatobiliary disposition of a model opioid peptide. Pharm Res 2004; 21:1294–1302.
13. Kannan R, Mittur A, Bao Y, et al. GSH transport in immortalized mouse brain endothelial cells: evidence for apical localization of a sodium-dependent GSH transporter. J Neurochem 1999; 73:390–399.
14. Kusuhara H, Suzuki H, Naito M, et al. Characterization of efflux transport of organic anions in a mouse brain capillary endothelial cell line. J Pharmacol Exp Ther 1998; 285:1260–1265.
15. Pritchard JB, Miller DS. Mechanisms mediating renal secretion of organic anions and cations. Physiol Rev 1993; 73:765–796.

16. Sanchez del Pino MM, Hawkins RA, Peterson DR. Neutral amino acid transport by the blood-brain barrier. Membrane vesicle studies. J Biol Chem 1992; 267: 25951–25957.
17. Murer H, Gmaj P, Steiger B, et al. Transport studies with renal proximal tubular and small intestinal brush border and basolateral membrane vesicles: vesicle heterogeneity, coexistence of transport system. Methods Enzymol 1989; 172:346–364.
18. Boyer JL, Meier PJ. Characterizing mechanisms of hepatic bile acid transport utilizing isolated membrane vesicles. Methods Enzymol 1990; 192:517–533.
19. Kinne-Saffran E, Kinne RK. Isolation of lumenal and contralumenal plasma membrane vesicles from kidney. Methods Enzymol 1990; 191:450–469.
20. Meier PJ, Boyer JL. Preparation of basolateral (sinusoidal) and canalicular plasma membrane vesicles for the study of hepatic transport processes. Methods Enzymol 1990; 192:534–545.
21. Meier PJ, St. Meier-Abt A, Barrett C, et al. Mechanisms of taurocholate transport in canalicular and basolateral rat liver plasma membrane vesicles. Evidence for an electrogenic canalicular organic anion carrier. J Biol Chem 1984; 259:10614–10622.
22. Kamimoto Y, Gatmaitan Z, Hsu J, et al. The function of Gp170, the multidrug resistance gene product, in rat liver canalicular membrane vesicles. J Biol Chem 1989; 264:11693–11698.
23. Wedeen RP, Weiner B. The distribution of p-aminohippuric acid in rat kidney slices. I. Tubular localization. Kidney Int 1973; 3:205–213.
24. Hasegawa M, Kusuhara H, Sugiyama D, et al. Functional involvement of rat organic anion transporter 3 (rOat3; Slc22a8) in the renal uptake of organic anions. J Pharmacol Exp Ther 2002; 300:746–753.
25. Fleck C, Bachner B, Gockeritz S, et al. Ex vivo stimulation of renal tubular PAH transport by dexamethasone and triiodothyronine in human renal cell carcinoma. Urol Res 2000; 28:383–390.
26. Fleck C, Gockeritz S, Schubert J. Tubular PAH transport capacity in human kidney tissue and in renal cell carcinoma: correlation with various clinical and morphological parameters of the tumor. Urol Res 1997; 25:167–171.
27. Fleck C, Hilger R, Jurkutat S, et al. Ex vivo stimulation of renal transport of the cytostatic drugs methotrexate, cisplatin, topotecan (Hycamtin) and raltitrexed (Tomudex) by dexamethasone, T3 and EGF in intact human and rat kidney tissue and in human renal cell carcinoma. Urol Res 2002; 30:256–262.
28. Nozaki Y, Kusuhara H, Kondo T, et al. Characterization of the uptake of OAT1 and OAT3 substrates by human kidney slices. J Pharmacol Exp Ther 2007; 321:362–369.
29. Wilson T, Wiselman G. The use of sacs of everted small intestine for the study of the transference of substances from the mucosal to serosal surface. J Physiol 1954; 123:116–125.
30. Barr WH, Riegelman S. Intestinal drug absorption and metabolism. II. Kinetic aspects of intestinal glucuronide conjugation. J Pharm Sci 1970; 59:164–168.
31. Frizzell RA, Koch MJ, Schultz SG. Ion transport by rabbit colon. I. Active and passive components. J Membr Biol 1976; 27:297–316.
32. Frizzell RA, Schultz SG. Ionic conductances of extracellular shunt pathway in rabbit ileum. Influence of shunt on transmural sodium transport and electrical potential differences. J Gen Physiol 1972; 59:318–346.

33. Dantzig AH, Bergin L. Uptake of the cephalosporin, cephalexin, by a dipeptide transport carrier in the human intestinal cell line, Caco-2. Biochim Biophys Acta 1990; 1027:211–217.
34. Wilson G. Cell culture techniques for the study of drug transport. Eur J Drug Metab Pharmacokinet 1990; 15:159–163.
35. Barthe L, Woodley J, Houin G. Gastrointestinal absorption of drugs: methods and studies. Fundam Clin Pharmacol 1999; 13:154–168.
36. Meunier V, Bourrie M, Berger Y, et al. The human intestinal epithelial cell line Caco-2; pharmacological and pharmacokinetic applications. Cell Biol Toxicol 1995; 11:187–194.
37. Gres MC, Julian B, Bourrie M, et al. Correlation between oral drug absorption in humans, and apparent drug permeability in TC-7 cells, a human epithelial intestinal cell line: comparison with the parental Caco-2 cell line. Pharm Res 1998; 15:726–733.
38. Basu SK, Shen J, Elbert KJ, et al. Development and utility of anti-PepT1 anti-peptide polyclonal antibodies. Pharm Res 1998; 15:338–342.
39. Kekuda R, Torres-Zamorano V, Fei YJ, et al. Molecular and functional characterization of intestinal Na(+)-dependent neutral amino acid transporter B0. Am J Physiol 1997; 272:G1463–G1472.
40. Tamai I, Takanaga H, Maeda H, et al. Proton-cotransport of pravastatin across intestinal brush-border membrane. Pharm Res 1995; 12:1727–1732.
41. Hunter J, Jepson MA, Tsuruo T, et al. Functional expression of P-glycoprotein in apical membranes of human intestinal Caco-2 cells. Kinetics of vinblastine secretion and interaction with modulators. J Biol Chem 1993; 268:14991–14997.
42. Hirohashi T, Suzuki H, Chu XY, et al. Function and expression of multidrug resistance-associated protein family in human colon adenocarcinoma cells (Caco-2). J Pharmacol Exp Ther 2000; 292:265–270.
43. Xia CQ, Liu N, Yang D, et al. Expression, localization, and functional characteristics of breast cancer resistance protein in Caco-2 cells. Drug Metab Dispos 2005; 33:637–643.
44. Hunter J, Hirst BH, Simmons NL. Drug absorption limited by P-glycoprotein-mediated secretory drug transport in human intestinal epithelial Caco-2 cell layer. Pharm Res 1993; 10:743–749.
45. Tatsuta T, Naito M, Oh-hara T, et al. Functional involvement of P-glycoprotein in blood-brain barrier. J Biol Chem 1992; 267:20383–20391.
46. Hosoya KI, Takashima T, Tetsuka K, et al. mRNA expression and transport characterization of conditionally immortalized rat brain capillary endothelial cell lines; a new in vitro BBB model for drug targeting. J Drug Target 2000; 8:357–370.
47. Barrand MA, Robertson KJ, von Weikersthal SF. Comparisons of P-glycoprotein expression in isolated rat brain microvessels and in primary cultures of endothelial cells derived from microvasculature of rat brain, epididymal fat pad and from aorta. FEBS Lett 1995; 374:179–183.
48. Seetharaman S, Barrand MA, Maskell L, et al. Multidrug resistance-related transport proteins in isolated human brain microvessels and in cells cultured from these isolates. J Neurochem 1998; 70:1151–1159.
49. Regina A, Koman A, Piciotti M, et al. Mrp1 multidrug resistance-associated protein and P-glycoprotein expression in rat brain microvessel endothelial cells. J Neurochem 1998; 71:705–715.

50. Kouzuki H, Suzuki H, Ito K, et al. Contribution of sodium taurocholate co-transporting polypeptide to the uptake of its possible substrates into rat hepatocytes. J Pharmacol Exp Ther 1998; 286:1043–1050.
51. Kouzuki H, Suzuki H, Ito K, et al. Contribution of organic anion transporting polypeptide to uptake of its possible substrates into rat hepatocytes. J Pharmacol Exp Ther 1999; 288:627–634.
52. Hirano M, Maeda K, Shitara Y, et al. Drug-drug interaction between pitavastatin and various drugs via OATP1B1. Drug Metab Dispos 2006; 34:1229–1236.
53. Hasegawa M, Kusuhara H, Endou H, et al. Contribution of organic anion transporters to the renal uptake of anionic compounds and nucleoside derivatives in rat. J Pharmacol Exp Ther 2003; 305:1087–1097.
54. Sasaki M, Suzuki H, Ito K, et al. Transcellular transport of organic anions across a double-transfected Madin-Darby canine kidney II cell monolayer expressing both human organic anion-transporting polypeptide (OATP2/SLC21A6) and multidrug resistance-associated protein 2 (MRP2/ABCC2). J Biol Chem 2002; 277:6497–6503.
55. Kopplow K, Letschert K, Konig J, et al. Human hepatobiliary transport of organic anions analyzed by quadruple-transfected cells. Mol Pharmacol 2005; 68:1031–1038.
56. Matsushima S, Maeda K, Kondo C, et al. Identification of the hepatic efflux transporters of organic anions using double-transfected Madin-Darby canine kidney II cells expressing human organic anion-transporting polypeptide 1B1 (OATP1B1)/multidrug resistance-associated protein 2, OATP1B1/multidrug resistance 1, and OATP1B1/ breast cancer resistance protein. J Pharmacol Exp Ther 2005; 314:1059–1067.
57. Cui Y, Konig J, Keppler D. Vectorial transport by double-transfected cells expressing the human uptake transporter SLC21A8 and the apical export pump ABCC2. Mol Pharmacol 2001; 60:934–943.
58. Grube M, Reuther S, Meyer Zu, et al. Organic anion transporting polypeptide 2B1 and breast cancer resistance protein interact in the transepithelial transport of steroid sulfates in human placenta. Drug Metab Dispos 2007; 35:30–35.
59. Sasaki M, Suzuki H, Aoki J, et al. Prediction of in vivo biliary clearance from the in vitro transcellular transport of organic anions across a double-transfected Madin-Darby canine kidney II monolayer expressing both rat organic anion transporting polypeptide 4 and multidrug resistance associated protein 2. Mol Pharmacol 2004; 66:450–459.
60. Mita S, Suzuki H, Akita H, et al. Vectorial transport of bile salts across MDCK cells expressing both rat Na^+-taurocholate cotransporting polypeptide and rat bile salt export pump. Am J Physiol Gastrointest Liver Physiol 2005; 288: G159–G167.
61. Mita S, Suzuki H, Akita H, et al. Vectorial transport of unconjugated and conjugated bile salts by monolayers of LLC-PK1 cells doubly transfected with human NTCP and BSEP or with rat Ntcp and Bsep. Am J Physiol Gastrointest Liver Physiol 2006; 290:G550–G556.
62. Imaoka T, Kusuhara H, Adachi-Akahane S, et al. The renal-specific transporter mediates facilitative transport of organic anions at the brush border membrane of mouse renal tubules. J Am Soc Nephrol 2004; 15:2012–2022.
63. Dawson PA, Hubbert M, Haywood J, et al. The heteromeric organic solute transporter alpha-beta, Ostalpha-Ostbeta, is an ileal basolateral bile acid transporter. J Biol Chem 2005; 280:6960–6968.

64. Hagenbuch B, Meier PJ. The superfamily of organic anion transporting polypeptides. Biochim Biophys Acta 2003; 1609:1–18.
65. Jacquemin E, Hagenbuch B, Stieger B, et al. Expression cloning of a rat liver Na(+)-independent organic anion transporter. Proc Natl Acad Sci U S A 1994; 91:133–137.
66. Bergwerk AJ, Shi X, Ford AC, et al. Immunologic distribution of an organic anion transport protein in rat liver and kidney. Am J Physiol 1996; 271:G231–G238.
67. van Montfoort JE, Hagenbuch B, Fattinger KE, et al. Polyspecific organic anion transporting polypeptides mediate hepatic uptake of amphipathic type II organic cations. J Pharmacol Exp Ther 1999; 291:147–152.
68. Li L, Lee TK, Meier PJ, et al. Identification of glutathione as a driving force and leukotriene C4 as a substrate for oatp1, the hepatic sinusoidal organic solute transporter. J Biol Chem 1998; 273:16184–16191.
69. Saito H, Masuda S, Inui K. Cloning and functional characterization of a novel rat organic anion transporter mediating basolateral uptake of methotrexate in the kidney. J Biol Chem 1996; 271:20719–20725.
70. Masuda S, Ibaramoto K, Takeuchi A, et al. Cloning and functional characterization of a new multispecific organic anion transporter, OAT-K2, in rat kidney. Mol Pharmacol 1999; 55:743–752.
71. Masuda S, Saito H, Nonoguchi H, et al. mRNA distribution and membrane localization of the OAT-K1 organic anion transporter in rat renal tubules. FEBS Lett 1997; 407:127–131.
72. Noe B, Hagenbuch B, Stieger B, et al. Isolation of a multispecific organic anion and cardiac glycoside transporter from rat brain. Proc Natl Acad Sci U S A 1997; 94:10346–10350.
73. Abe T, Kakyo M, Sakagami H, et al. Molecular characterization and tissue distribution of a new organic anion transporter subtype (oatp3) that transports thyroid hormones and taurocholate and comparison with oatp2. J Biol Chem 1998; 273:22395–22401.
74. Cheng X, Maher J, Chen C, et al. Tissue distribution and ontogeny of mouse organic anion transporting polypeptides (Oatps). Drug Metab Dispos 2005; 33:1062–1073.
75. Reichel C, Gao B, Van Montfoort J, et al. Localization and function of the organic anion-transporting polypeptide Oatp2 in rat liver. Gastroenterology 1999; 117: 688–695.
76. Gao B, Stieger B, Noe B, et al. Localization of the organic anion transporting polypeptide 2 (Oatp2) in capillary endothelium and choroid plexus epithelium of rat brain. J Histochem Cytochem 1999; 47:1255–1264.
77. Gao B, Hagenbuch B, Kullak-Ublick GA, et al. Organic anion-transporting polypeptides mediate transport of opioid peptides across blood-brain barrier. J Pharmacol Exp Ther 2000; 294:73–79.
78. Kitazawa T, Terasaki T, Suzuki H, et al. Efflux of taurocholic acid across the blood-brain barrier: interaction with cyclic peptides. J Pharmacol Exp Ther 1998; 286:890–895.
79. Sugiyama D, Kusuhara H, Shitara Y, et al. Characterization of the efflux transport of 17beta-estradiol-D-17beta- glucuronide from the brain across the blood-brain barrier. J Pharmacol Exp Ther 2001; 298:316–322.

80. Asaba H, Hosoya K, Takanaga H, et al. Blood-brain barrier is involved in the efflux transport of a neuroactive steroid, dehydroepiandrosterone sulfate, via organic anion transporting polypeptide 2. J Neurochem 2000; 75:1907–1916.
81. Kikuchi R, Kusuhara H, Abe T, et al. Involvement of multiple transporters in the efflux of 3-hydroxy-3-methylglutaryl-CoA reductase inhibitors across the blood-brain barrier. J Pharmacol Exp Ther 2004; 311:1147–1153.
82. Dagenais C, Zong J, Ducharme J, et al. Effect of mdr1a P-glycoprotein gene disruption, gender, and substrate concentration on brain uptake of selected compounds. Pharm Res 2001; 18:957–963.
83. Li N, Hartley DP, Cherrington NJ, et al. Tissue expression, ontogeny, and inducibility of rat organic anion transporting polypeptide 4. J Pharmacol Exp Ther 2002; 301:551–560.
84. Kusuhara H, He Z, Nagata Y, et al. Expression and functional involvement of organic anion transporting polypeptide subtype 3 (Slc21a7) in rat choroid plexus. Pharm Res 2003; 20:720–727.
85. Walters HC, Craddock AL, Fusegawa H, et al. Expression, transport properties, and chromosomal location of organic anion transporter subtype 3. Am J Physiol Gastrointest Liver Physiol 2000; 279:G1188–G1200.
86. Ohtsuki S, Takizawa T, Takanaga H, et al. Localization of organic anion transporting polypeptide 3 (oatp3) in mouse brain parenchymal and capillary endothelial cells. J Neurochem 2004; 90:743–749.
87. Cattori V, van Montfoort JE, Stieger B, et al. Localization of organic anion transporting polypeptide 4 (Oatp4) in rat liver and comparison of its substrate specificity with Oatp1, Oatp2 and Oatp3. Pflugers Arch 2001; 443:188–195.
88. Kikuchi A, Nozawa T, Wakasawa T, et al. Transporter-mediated intestinal absorption of fexofenadine in rats. Drug Metab Pharmacokinet 2006; 21:308–314.
89. Kullak-Ublick GA, Hagenbuch B, Stieger B, et al. Molecular and functional characterization of an organic anion transporting polypeptide cloned from human liver. Gastroenterology 1995; 109:1274–1282.
90. Abe T, Kakyo M, Tokui T, et al. Identification of a novel gene family encoding human liver-specific organic anion transporter LST-1. J Biol Chem 1999; 274:17159–17163.
91. Alcorn J, Lu X, Moscow JA, et al. Transporter gene expression in lactating and nonlactating human mammary epithelial cells using real-time reverse transcription-polymerase chain reaction. J Pharmacol Exp Ther 2002; 303:487–496.
92. Bronger H, Konig J, Kopplow K, et al. ABCC drug efflux pumps and organic anion uptake transporters in human gliomas and the blood-tumor barrier. Cancer Res 2005; 65:11419–11428.
93. Bossuyt X, Muller M, Meier PJ. Multispecific amphipathic substrate transport by an organic anion transporter of human liver. J Hepatol 1996; 25:733–738.
94. Konig J, Cui Y, Nies AT, et al. A novel human organic anion transporting polypeptide localized to the basolateral hepatocyte membrane. Am J Physiol Gastrointest Liver Physiol 2000; 278:G156–G164.
95. Hsiang B, Zhu Y, Wang Z, et al. A novel human hepatic organic anion transporting polypeptide (OATP2). Identification of a liver-specific human organic anion transporting polypeptide and identification of rat and human hydroxymethylglutaryl-CoA reductase inhibitor transporters. J Biol Chem 1999; 274:37161–37168.

96. Tamai I, Nezu J, Uchino H, et al. Molecular identification and characterization of novel members of the human organic anion transporter (OATP) family. Biochem Biophys Res Commun 2000; 273:251–260.
97. Konig J, Cui Y, Nies AT, et al. Localization and genomic organization of a new hepatocellular organic anion transporting polypeptide. J Biol Chem 2000; 275:23161–23168.
98. Abe T, Unno M, Onogawa T, et al. LST-2, a human liver-specific organic anion transporter, determines methotrexate sensitivity in gastrointestinal cancers. Gastroenterology 2001; 120:1689–1699.
99. Cattori V, Hagenbuch B, Hagenbuch N, et al. Identification of organic anion transporting polypeptide 4 (Oatp4) as a major full-length isoform of the liver-specific transporter-1 (rlst-1) in rat liver. FEBS Lett 2000; 474:242–245.
100. Kakyo M, Unno M, Tokui T, et al. Molecular characterization and functional regulation of a novel rat liver-specific organic anion transporter rlst-1. Gastroenterology 1999; 117:770–775.
101. Ishiguro N, Maeda K, Kishimoto W, et al. Predominant contribution of OATP1B3 to the hepatic uptake of telmisartan, an angiotensin II receptor antagonist, in humans. Drug Metab Dispos 2006; 34:1109–1115.
102. Yamashiro W, Maeda K, Hirouchi M, et al. Involvement of transporters in the hepatic uptake and biliary excretion of valsartan, a selective antagonist of the angiotensin II AT1-receptor, in humans. Drug Metab Dispos 2006; 34:1247–1254.
103. Cui Y, Konig J, Leier I, et al. Hepatic uptake of bilirubin and its conjugates by the human organic anion transporter SLC21A6. J Biol Chem 2001; 276:9626–9630.
104. Ismair MG, Stieger B, Cattori V, et al. Hepatic uptake of cholecystokinin octapeptide by organic anion-transporting polypeptides OATP4 and OATP8 of rat and human liver. Gastroenterology 2001; 121:1185–1190.
105. Sugiyama D, Kusuhara H, Taniguchi H, et al. Functional characterization of rat brain-specific organic anion transporter (Oatp14) at the blood-brain barrier: high affinity transporter for thyroxine. J Biol Chem 2003; 278:43489–43495.
106. Pizzagalli F, Hagenbuch B, Stieger B, et al. Identification of a novel human organic anion transporting polypeptide as a high affinity thyroxine transporter. Mol Endocrinol 2002; 16:2283–2296.
107. Tohyama K, Kusuhara H, Sugiyama Y. Involvement of multispecific organic anion transporter, Oatp14 (Slc21a14), in the transport of thyroxine across the blood-brain barrier. Endocrinology 2004; 145:4384–4391.
108. Kullak-Ublick GA, Ismair MG, Stieger B, et al. Organic anion-transporting polypeptide B (OATP-B) and its functional comparison with three other OATPs of human liver. Gastroenterology 2001; 120:525–533.
109. Kobayashi D, Nozawa T, Imai K, et al. Involvement of human organic anion transporting polypeptide OATP-B (SLC21A9) in pH-dependent transport across intestinal apical membrane. J Pharmacol Exp Ther 2003; 306:703–708.
110. Sai Y, Kaneko Y, Ito S, et al. Predominant contribution of organic anion transporting polypeptide OATP-B (OATP2B1) to apical uptake of estrone-3-sulfate by human intestinal Caco-2 cells. Drug Metab Dispos 2006; 34:1423–1431.
111. Mikkaichi T, Suzuki T, Onogawa T, et al. Isolation and characterization of a digoxin transporter and its rat homologue expressed in the kidney. Proc Natl Acad Sci U S A 2004; 101:3569–3574.

112. Grundemann D, Gorboulev V, Gambaryan S, et al. Drug excretion mediated by a new prototype of polyspecific transporter. Nature 1994; 372:549–552.
113. Slitt AL, Cherrington NJ, Hartley DP, et al. Tissue distribution and renal developmental changes in rat organic cation transporter mRNA levels. Drug Metab Dispos 2002; 30:212–219.
114. Gorboulev V, Ulzheimer JC, Akhoundova A, et al. Cloning and characterization of two human polyspecific organic cation transporters. DNA Cell Biol 1997; 16:871–881.
115. Urakami Y, Okuda M, Masuda S, et al. Functional characteristics and membrane localization of rat multispecific organic cation transporters, OCT1 and OCT2, mediating tubular secretion of cationic drugs. J Pharmacol Exp Ther 1998; 287:800–805.
116. Meyer-Wentrup F, Karbach U, Gorboulev V, et al. Membrane localization of the electrogenic cation transporter rOCT1 in rat liver. Biochem Biophys Res Commun 1998; 248:673–678.
117. Wang DS, Jonker JW, Kato Y, et al. Involvement of organic cation transporter 1 in hepatic and intestinal distribution of metformin. J Pharmacol Exp Ther 2002; 302:510–515.
118. Jonker JW, Wagenaar E, Mol CA, et al. Reduced hepatic uptake and intestinal excretion of organic cations in mice with a targeted disruption of the organic cation transporter 1 (Oct1 [Slc22a1]) gene. Mol Cell Biol 2001; 21:5471–5477.
119. Alnouti Y, Petrick JS, Klaassen CD. Tissue distribution and ontogeny of organic cation transporters in mice. Drug Metab Dispos 2006; 34:477–482.
120. Tahara H, Kusuhara H, Endou H, et al. A species difference in the transport activities of H2 receptor antagonists by rat and human renal organic anion and cation transporters. J Pharmacol Exp Ther 2005; 315:337–345.
121. Grundemann D, Liebich G, Kiefer N, et al. Selective substrates for non-neuronal monoamine transporters. Mol Pharmacol 1999; 56:1–10.
122. Takeda M, Khamdang S, Narikawa S, et al. Human organic anion transporters and human organic cation transporters mediate renal antiviral transport. J Pharmacol Exp Ther 2002; 300:918–924.
123. Motohashi H, Sakurai Y, Saito H, et al. Gene expression levels and immunolocalization of organic ion transporters in the human kidney. J Am Soc Nephrol 2002; 13:866–874.
124. Sweet DH, Miller DS, Pritchard JB. Basolateral localization of organic cation transporter 2 in intact renal proximal tubules. Am J Physiol Renal Physiol 2000; 279:F826–F834.
125. Kimura N, Masuda S, Tanihara Y, et al. Metformin is a superior substrate for renal organic cation transporter OCT2 rather than hepatic OCT1. Drug Metab Pharmacokinet 2005; 20:379–386.
126. Jonker JW, Wagenaar E, Van Eijl S, et al. Deficiency in the organic cation transporters 1 and 2 (Oct1/Oct2 [Slc22a1/Slc22a2]) in mice abolishes renal secretion of organic cations. Mol Cell Biol 2003; 23:7902–7908.
127. Kekuda R, Prasad PD, Wu X, et al. Cloning and functional characterization of a potential-sensitive, polyspecific organic cation transporter (OCT3) most abundantly expressed in placenta. J Biol Chem 1998; 273:15971–15979.
128. Zwart R, Verhaagh S, Buitelaar M, et al. Impaired activity of the extraneuronal monoamine transporter system known as uptake-2 in Orct3/Slc22a3-deficient mice. Mol Cell Biol 2001; 21:4188–4196.

129. Sekine T, Watanabe N, Hosoyamada M, et al. Expression cloning and characterization of a novel multispecific organic anion transporter. J Biol Chem 1997; 272:18526–18529.
130. Tojo A, Sekine T, Nakajima N, et al. Immunohistochemical localization of multispecific renal organic anion transporter 1 in rat kidney. J Am Soc Nephrol 1999; 10:464–471.
131. Apiwattanakul N, Sekine T, Chairoungdua A, et al. Transport properties of nonsteroidal anti-inflammatory drugs by organic anion transporter 1 expressed in Xenopus laevis oocytes. Mol Pharmacol 1999; 55:847–854.
132. Jariyawat S, Sekine T, Takeda M, et al. The interaction and transport of beta-lactam antibiotics with the cloned rat renal organic anion transporter 1. J Pharmacol Exp Ther 1999; 290:672–677.
133. Ueo H, Motohashi H, Katsura T, et al. Human organic anion transporter hOAT3 is a potent transporter of cephalosporin antibiotics, in comparison with hOAT1. Biochem Pharmacol 2005; 70:1104–1113.
134. Cihlar T, Lin DC, Pritchard JB, et al. The antiviral nucleotide analogs cidofovir and adefovir are novel substrates for human and rat renal organic anion transporter 1. Mol Pharmacol 1999; 56:570–580.
135. Eraly SA, Vallon V, Vaughn DA, et al. Decreased renal organic anion secretion and plasma accumulation of endogenous organic anions in OAT1 knock-out mice. J Biol Chem 2006; 281:5072–5083.
136. Buist SC, Cherrington NJ, Choudhuri S, et al. Gender-specific and developmental influences on the expression of rat organic anion transporters. J Pharmacol Exp Ther 2002; 301:145–151.
137. Kato Y, Kuge K, Kusuhara H, et al. Gender difference in the urinary excretion of organic anions in rats. J Pharmacol Exp Ther 2002; 302:483–489.
138. Kobayashi Y, Ohshiro N, Shibusawa A, et al. Isolation, characterization and differential gene expression of multispecific organic anion transporter 2 in mice. Mol Pharmacol 2002; 62:7–14.
139. Buist SC, Klaassen CD. Rat and mouse differences in gender-predominant expression of organic anion transporter (Oat1-3; Slc22a6-8) mRNA levels. Drug Metab Dispos 2004; 32:620–625.
140. Simon N, Dailly E, Combes S, et al. Role of lipoprotein in the plasma binding of SDZ PSC 833, a novel multidrug resistance-reversing cyclosporin. Br J Clin Pharmacol 1998; 45:173–175.
141. Sekine T, Cha SH, Tsuda M, et al. Identification of multispecific organic anion transporter 2 expressed predominantly in the liver. FEBS Lett 1998; 429:179–182.
142. Morita N, Kusuhara H, Nozaki Y, et al. Functional involvement of rat organic anion transporter 2 (Slc22a7) in the hepatic uptake of the nonsteroidal anti-inflammatory drug ketoprofen. Drug Metab Dispos 2005; 33:1151–1157.
143. Morita N, Kusuhara H, Sekine T, et al. Functional characterization of rat organic anion transporter 2 in LLC-PK1 cells. J Pharmacol Exp Ther 2001; 298:1179–1184.
144. Kusuhara H, Sekine T, Utsunomiya-Tate N, et al. Molecular cloning and characterization of a new multispecific organic anion transporter from rat brain. J Biol Chem 1999; 274:13675–13680.
145. Race JE, Grassl SM, Williams WJ, et al. Molecular cloning and characterization of two novel human renal organic anion transporters (hOAT1 and hOAT3). Biochem Biophys Res Commun 1999; 255:508–514.

146. Cha SH, Sekine T, Fukushima JI, et al. Identification and characterization of human organic anion transporter 3 expressing predominantly in the kidney. Mol Pharmacol 2001; 59:1277–1286.
147. Mori S, Takanaga H, Ohtsuki S, et al. Rat organic anion transporter 3 (rOAT3) is responsible for brain-to-blood efflux of homovanillic acid at the abluminal membrane of brain capillary endothelial cells. J Cereb Blood Flow Metab 2003; 23:432–440.
148. Kikuchi R, Kusuhara H, Sugiyama D, et al. Contribution of organic anion transporter 3 (Slc22a8) to the elimination of p-aminohippuric acid and benzylpenicillin across the blood-brain barrier. J Pharmacol Exp Ther 2003; 306:51–58.
149. Nagata Y, Kusuhara H, Endou H, et al. Expression and functional characterization of rat organic anion transporter 3 (rOat3) in the choroid plexus. Mol Pharmacol 2002; 61:982–988.
150. Nagata Y, Kusuhara H, Hirono S, et al. Carrier-mediated uptake of H2-receptor antagonists by the rat choroid plexus: involvement of rat organic anion transporter 3. Drug Metab Dispos 2004; 32:1040–1047.
151. Deguchi T, Kusuhara H, Takadate A, et al. Characterization of uremic toxin transport by organic anion transporters in the kidney. Kidney Int 2004; 65:162–174.
152. Tahara H, Kusuhara H, Chida M, et al. Is the monkey an appropriate animal model to examine drug-drug interactions involving renal clearance? Effect of probenecid on the renal elimination of H2 receptor antagonists. J Pharmacol Exp Ther 2006; 316:1187–1194.
153. Ullrich KJ, Rumrich G, David C, et al. Bisubstrates: substances that interact with renal contraluminal organic anion and organic cation transport systems. I. Amines, piperidines, piperazines, azepines, pyridines, quinolines, imidazoles, thiazoles, guanidines and hydrazines. Pflugers Arch 1993; 425:280–299.
154. Sweet DH, Miller DS, Pritchard JB, et al. Impaired organic anion transport in kidney and choroid plexus of organic anion transporter 3 (Oat3 (Slc22a8)) knockout mice. J Biol Chem 2002; 277:26934–26943.
155. Cha SH, Sekine T, Kusuhara H, et al. Molecular cloning and characterization of multispecific organic anion transporter 4 expressed in the placenta. J Biol Chem 2000; 275:4507–4512.
156. Babu E, Takeda M, Narikawa S, et al. Role of human organic anion transporter 4 in the transport of ochratoxin A. Biochim Biophys Acta 2002; 1590:64–75.
157. Anzai N, Jutabha P, Enomoto A, et al. Functional characterization of rat organic anion transporter 5 (Slc22a19) at the apical membrane of renal proximal tubules. J Pharmacol Exp Ther 2005; 315:534–544.
158. Tamai I, Yabuuchi H, Nezu J, et al. Cloning and characterization of a novel human pH-dependent organic cation transporter, OCTN1. FEBS Lett 1997; 419:107–111.
159. Yabuuchi H, Tamai I, Nezu J, et al. Novel membrane transporter OCTN1 mediates multispecific, bidirectional, and pH-dependent transport of organic cations. J Pharmacol Exp Ther 1999; 289:768–773.
160. Wu X, Prasad PD, Leibach FH, et al. cDNA sequence, transport function, and genomic organization of human OCTN2, a new member of the organic cation transporter family. Biochem Biophys Res Commun 1998; 246:589–595.
161. Tamai I, Ohashi R, Nezu J, et al. Molecular and functional identification of sodium ion-dependent, high affinity human carnitine transporter OCTN2. J Biol Chem 1998; 273:20378–20382.

162. Wu X, Huang W, Prasad PD, et al. Functional characteristics and tissue distribution pattern of organic cation transporter 2 (OCTN2), an organic cation/carnitine transporter. J Pharmacol Exp Ther 1999; 290:1482–1492.
163. Ohashi R, Tamai I, Yabuuchi H, et al. Na(+)-dependent carnitine transport by organic cation transporter (OCTN2): its pharmacological and toxicological relevance. J Pharmacol Exp Ther 1999; 291:778–784.
164. Ganapathy ME, Huang W, Rajan DP, et al. Beta-lactam antibiotics as substrates for OCTN2, an organic cation/carnitine transporter. J Biol Chem 2000; 275:1699–1707.
165. Nezu J, Tamai I, Oku A, et al. Primary systemic carnitine deficiency is caused by mutations in a gene encoding sodium ion-dependent carnitine transporter. Nat Genet 1999; 21:91–94.
166. Ohashi R, Tamai I, Nezu Ji J, et al. Molecular and physiological evidence for multifunctionality of carnitine/organic cation transporter OCTN2. Mol Pharmacol 2001; 59:358–366.
167. Otsuka M, Matsumoto T, Morimoto R, et al. A human transporter protein that mediates the final excretion step for toxic organic cations. Proc Natl Acad Sci U S A 2005; 102:17923–17928.
168. Masuda S, Terada T, Yonezawa A, et al. Identification and functional characterization of a new human kidney-specific H+/organic cation antiporter, kidney-specific multidrug and toxin extrusion 2. J Am Soc Nephrol 2006; 17:2127–2135.
169. Fei YJ, Kanai Y, Nussberger S, et al. Expression cloning of a mammalian proton-coupled oligopeptide transporter. Nature 1994; 368:563–566.
170. Ogihara H, Saito H, Shin BC, et al. Immuno-localization of H+/peptide cotransporter in rat digestive tract. Biochem Biophys Res Commun 1996; 220:848–852.
171. Shen H, Smith DE, Yang T, et al. Localization of PEPT1 and PEPT2 proton-coupled oligopeptide transporter mRNA and protein in rat kidney. Am J Physiol 1999; 276:F658–F665.
172. Groneberg DA, Doring F, Eynott PR, et al. Intestinal peptide transport: ex vivo uptake studies and localization of peptide carrier PEPT1. Am J Physiol Gastrointest Liver Physiol 2001; 281:G697–G704.
173. Boll M, Markovich D, Weber WM, et al. Expression cloning of a cDNA from rabbit small intestine related to proton-coupled transport of peptides, beta-lactam antibiotics and ACE- inhibitors. Pflugers Arch 1994; 429:146–149.
174. Sinko PJ, Balimane PV. Carrier-mediated intestinal absorption of valacyclovir, the L-valyl ester prodrug of acyclovir: 1. Interactions with peptides, organic anions and organic cations in rats. Biopharm Drug Dispos 1998; 19:209–217.
175. Soul-Lawton J, Seaber E, On N, et al. Absolute bioavailability and metabolic disposition of valaciclovir, the L-valyl ester of acyclovir, following oral administration to humans. Antimicrob Agents Chemother 1995; 39:2759–2764.
176. Weller S, Blum MR, Doucette M, et al. Pharmacokinetics of the acyclovir pro-drug valaciclovir after escalating single- and multiple-dose administration to normal volunteers. Clin Pharmacol Ther 1993; 54:595–605.
177. Han H, de Vrueh RL, Rhie JK, et al. 5′-Amino acid esters of antiviral nucleosides, acyclovir, and AZT are absorbed by the intestinal PEPT1 peptide transporter. Pharm Res 1998; 15:1154–1159.
178. Ganapathy ME, Huang W, Wang H, et al. Valacyclovir: a substrate for the intestinal and renal peptide transporters PEPT1 and PEPT2. Biochem Biophys Res Commun 1998; 246:470–475.

179. Tamai I, Nakanishi T, Nakahara H, et al. Improvement of L-dopa absorption by dipeptidyl derivation, utilizing peptide transporter PepT1. J Pharm Sci 1998; 87:1542–1546.
180. Saito H, Terada T, Okuda M, et al. Molecular cloning and tissue distribution of rat peptide transporter PEPT2. Biochim Biophys Acta 1996; 1280:173–177.
181. Boll M, Herget M, Wagener M, et al. Expression cloning and functional characterization of the kidney cortex high-affinity proton-coupled peptide transporter. Proc Natl Acad Sci U S A 1996; 93:284–289.
182. Smith DE, Pavlova A, Berger UV, et al. Tubular localization and tissue distribution of peptide transporters in rat kidney. Pharm Res 1998; 15:1244–1249.
183. Dieck ST, Heuer H, Ehrchen J, et al. The peptide transporter PepT2 is expressed in rat brain and mediates the accumulation of the florescent dipeptide derivative. Glia 1998; 25:10–20.
184. Novotny A, Xiang J, Stummer W, et al. Mechanisms of 5-aminolevulinic acid uptake at the choroid plexus. J Neurochem 2000; 75:321–328.
185. Chen XZ, Zhu T, Smith DE, et al. Stoichiometry and kinetics of the high-affinity H+-coupled peptide transporter PepT2. J Biol Chem 1999; 274:2773–2779.
186. Terada T, Saito H, Mukai M, et al. Recognition of beta-lactam antibiotics by rat peptide transporters, PEPT1 and PEPT2, in LLC-PK1 cells. Am J Physiol 1997; 273:F706–F711.
187. Ramamoorthy S, Liu W, Ma YY, et al. Proton/peptide cotransporter (PEPT 2) from human kidney: functional characterization and chromosomal localization. Biochim Biophys Acta 1995; 1240:1–4.
188. Takahashi K, Nakamura N, Terada T, et al. Interaction of beta-lactam antibiotics with H+/peptide cotransporters in rat renal brush-border membranes. J Pharmacol Exp Ther 1998; 286:1037–1042.
189. Busch AE, Schuster A, Waldegger S, et al. Expression of a renal type I sodium/ phosphate transporter (NaPi-1) induces a conductance in Xenopus oocytes permeable for organic and inorganic anions. Proc Natl Acad Sci U S A 1996; 93:5347–5351.
190. Yabuuchi H, Tamai I, Morita K, et al. Hepatic sinusoidal membrane transport of anionic drugs mediated by anion transporter Npt1. J Pharmacol Exp Ther 1998; 286:1391–1396.
191. Dano K. Active outward transport of daunomycin in resistant Ehrlich ascites tumor cells. Biochim Biophys Acta 1973; 323:466–483.
192. Skovsgaard T. Mechanisms of resistance to daunorubicin in Ehrlich ascites tumor cells. Cancer Res 1978; 38:1785–1791.
193. Kusuhara H, Suzuki H, Sugiyama Y. The role of P-glycoprotein and canalicular multispecific organic anion transporter in the hepatobiliary excretion of drugs. J Pharm Sci 1998; 87:1025–1040.
194. Oude Elferink RP, Meijer DK, Kuipers F, et al. Hepatobiliary secretion of organic compounds; molecular mechanisms of membrane transport. Biochim Biophys Acta 1995; 1241:215–268.
195. Tsuji A, Tamai I. Blood-brain barrier function of P-glycoprotein. Adv Drug Deliv Rev 1997; 25:285–298.
196. Simons NL, Hunter J, Jepson MA. Renal secretion of xenobiotics mediated by P-glycoprotein: importance to renal function in health and exploitation for targeted

drug delivery to epithelial cysts in polycystic kidney disease. Adv Drug Deliv Rev 1997; 25:243–256.

197. Hunter J, Hirst BH. Intestinal secretion of drugs. The role of P-glycoprotein and related drug efflux systems in limiting oral drug absorption. Adv Drug Deliv Rev 1997; 25:129–157.
198. Mizuno N, Niwa T, Yotsumoto Y, et al. Impact of drug transporter studies on drug discovery and development. Pharmacol Rev 2003; 55:425–461.
199. Schinkel AH. The roles of P-glycoprotein and MRP1 in the blood-brain and blood-cerebrospinal fluid barriers. Adv Exp Med Biol 2001; 500:365–372.
200. Schinkel AH, Smit JJ, van Tellingen O, et al. Disruption of the mouse mdr1a P-glycoprotein gene leads to a deficiency in the blood-brain barrier and to increased sensitivity to drugs. Cell 1994; 77:491–502.
201. Cao X, Yu LX, Barbaciru C, et al. Permeability dominates in vivo intestinal absorption of P-gp substrate with high solubility and high permeability. Mol Pharmacol 2005; 2:329–340.
202. Liu S, Tam D, Chen X, et al. P-glycoprotein and an unstirred water layer barring digoxin absorption in the vascularly perfused rat small intestine preparation: induction studies with pregnenolone-16alpha-carbonitrile. Drug Metab Dispos 2006; 34:1468–1479.
203. Ohashi R, Kamikozawa Y, Sugiura M, et al. Effect of P-glycoprotein on intestinal absorption and brain penetration of antiallergic agent bepotastine besilate. Drug Metab Dispos 2006; 34:793–799.
204. Zimmermann C, Gutmann H, Hruz P, et al. Mapping of multidrug resistance gene 1 and multidrug resistance-associated protein isoform 1 to 5 mRNA expression along the human intestinal tract. Drug Metab Dispos 2005; 33:219–224.
205. Thorn M, Finnstrom N, Lundgren S, et al. Cytochromes P450 and MDR1 mRNA expression along the human gastrointestinal tract. Br J Clin Pharmacol 2005; 60:54–60.
206. Englund G, Rorsman F, Ronnblom A, et al. Regional levels of drug transporters along the human intestinal tract: co-expression of ABC and SLC transporters and comparison with Caco-2 cells. Eur J Pharm Sci 2006; 29:269–277.
207. Kim RB. Drugs as P-glycoprotein substrates, inhibitors, and inducers. Drug Metab Rev 2002; 34:47–54.
208. Smit JW, Schinkel AH, Muller M, et al. Contribution of the murine mdr1a p-glycoprotein to hepatobiliary and intestinal elimination of cationic drugs as measured in mice with an mdr1a gene disruption. Hepatology 1998; 27:1056–1063.
209. Mayer U, Wagenaar E, Beijnen JH, et al. Substantial excretion of digoxin via the intestinal mucosa and prevention of long-term digoxin accumulation in the brain by the mdr 1a P-glycoprotein. Br J Pharmacol 1996; 119:1038–1044.
210. Schinkel AH, Mayer U, Wagenaar E, et al. Normal viability and altered pharmacokinetics in mice lacking mdr1-type (drug-transporting) P-glycoproteins. Proc Natl Acad Sci U S A 1997; 94:4028–4033.
211. Sparreboom A, van Asperen J, Mayer U, et al. Limited oral bioavailability and active epithelial excretion of paclitaxel (Taxol) caused by P-glycoprotein in the intestine. Proc Natl Acad Sci U S A 1997; 94:2031–2035.
212. Cvetkovic M, Leake B, Fromm MF, et al. OATP and P-glycoprotein transporters mediate the cellular uptake and excretion of fexofenadine. Drug Metab Dispos 1999; 27:866–871.

213. Adachi Y, Suzuki H, Sugiyama Y. Quantitative evaluation of the function of small intestinal P-glycoprotein: comparative studies between in situ and in vitro. Pharm Res 2003; 20:1163–1169.
214. de Lange EC, de Bock G, Schinkel AH, et al. BBB transport and P-glycoprotein functionality using MDR1A (-/-) and wild-type mice. Total brain versus microdialysis concentration profiles of rhodamine-123. Pharm Res 1998; 15:1657–1665.
215. Hoffmeyer S, Burk O, von Richter O, et al. Functional polymorphisms of the human multidrug-resistance gene: multiple sequence variations and correlation of one allele with P-glycoprotein expression and activity in vivo. Proc Natl Acad Sci U S A 2000; 97:3473–3478.
216. Kurata Y, Ieiri I, Kimura M, et al. Role of human MDR1 gene polymorphism in bioavailability and interaction of digoxin, a substrate of P-glycoprotein. Clin Pharmacol Ther 2002; 72:209–219.
217. Sadeque AJ, Wandel C, He H, et al. Increased drug delivery to the brain by P-glycoprotein inhibition. Clin Pharmacol Ther 2000; 68:231–237.
218. Sasongko L, Link JM, Muzi M, et al. Imaging P-glycoprotein transport activity at the human blood-brain barrier with positron emission tomography. Clin Pharmacol Ther 2005; 77:503–514.
219. Cole SP, Bhardwaj G, Gerlach JH, et al. Overexpression of a transporter gene in a multidrug-resistant human lung cancer cell line [see comments]. Science 1992; 258:1650–1654.
220. Zaman GJ, Versantvoort CH, Smit JJ, et al. Analysis of the expression of MRP, the gene for a new putative transmembrane drug transporter, in human multidrug resistant lung cancer cell lines. Cancer Res 1993; 53:1747–1750.
221. Stride BD, Grant CE, Loe DW, et al. Pharmacological characterization of the murine and human orthologs of multidrug-resistance protein in transfected human embryonic kidney cells. Mol Pharmacol 1997; 52:344–353.
222. Cole SP, Sparks KE, Fraser K, et al. Pharmacological characterization of multidrug resistant MRP-transfected human tumor cells. Cancer Res 1994; 54:5902–5910.
223. Deeley RG, Cole SP. Substrate recognition and transport by multidrug resistance protein 1 (ABCC1). FEBS Lett 2006; 580:1103–1111.
224. Wijnholds J, de Lange EC, Scheffer GL, et al. Multidrug resistance protein 1 protects the choroid plexus epithelium and contributes to the blood-cerebrospinal fluid barrier. J Clin Invest 2000; 105:279–285.
225. Sugiyama D, Kusuhara H, Lee YJ, et al. Involvement of multidrug resistance associated protein 1 (Mrp1) in the efflux transport of 17beta estradiol-D-17beta-glucuronide (E217betaG) across the blood-brain barrier. Pharm Res 2003; 20: 1394–1400.
226. Lee YJ, Kusuhara H, Sugiyama Y. Do multidrug resistance-associated protein-1 and -2 play any role in the elimination of estradiol-17 beta-glucuronide and 2,4-dinitrophenyl-S-glutathione across the blood-cerebrospinal fluid barrier? J Pharm Sci 2004; 93:99–107.
227. Mikami T, Nozaki T, Tagaya O, et al. The characters of a new mutant in rats with hyperbilirubinemic syndrome. Cong Anom 1986; 26:250–251.
228. Jansen PL, Peters WH, Lamers WH. Hereditary chronic conjugated hyperbilirubinemia in mutant rats caused by defective hepatic anion transport. Hepatology 1985; 5:573–579.

229. Hosokawa S, Tagaya O, Mikami T, et al. A new rat mutant with chronic conjugated hyperbilirubinemia and renal glomerular lesions. Lab Anim Sci 1992; 42:27–34.
230. Elferink RP, Ottenhoff R, Liefting W, et al. Hepatobiliary transport of glutathione and glutathione conjugate in rats with hereditary hyperbilirubinemia. J Clin Invest 1989; 84:476–483.
231. Sathirakul K, Suzuki H, Yasuda K, et al. Kinetic analysis of hepatobiliary transport of organic anions in Eisai hyperbilirubinemic mutant rats. J Pharmacol Exp Ther 1993; 265:1301–1312.
232. Suzuki H, Sugiyama Y. Excretion of GSSG and glutathione conjugates mediated by MRP1 and cMOAT/MRP2. Semin Liver Dis 1998; 18:359–376.
233. Paulusma CC, Bosma PJ, Zaman GJ, et al. Congenital jaundice in rats with a mutation in a multidrug resistance-associated protein gene. Science 1996; 271: 1126–1128.
234. Ito K, Suzuki H, Hirohashi T, et al. Molecular cloning of canalicular multispecific organic anion transporter defective in EHBR. Am J Physiol 1997; 272:G16–G22.
235. Buchler M, Konig J, Brom M, et al. cDNA cloning of the hepatocyte canalicular isoform of the multidrug resistance protein, cMrp, reveals a novel conjugate export pump deficient in hyperbilirubinemic mutant rats. J Biol Chem 1996; 271:15091–15098.
236. Paulusma CC, Kool M, Bosma PJ, et al. A mutation in the human canalicular multispecific organic anion transporter gene causes the Dubin-Johnson syndrome. Hepatology 1997; 25:1539–1542.
237. Niinuma K, Kato Y, Suzuki H, et al. Sugiyama. Primary active transport of organic anions on bile canalicular membrane in humans. Am J Physiol 1999; 276:G1153–G1164.
238. Maher JM, Slitt AL, Cherrington NJ, et al. Tissue distribution and hepatic and renal ontogeny of the multidrug resistance-associated protein (Mrp) family in mice. Drug Metab Dispos 2005; 33:947–955.
239. Rost D, Mahner S, Sugiyama Y, et al. Expression and localization of the multidrug resistance-associated protein 3 in rat small and large intestine. Am J Physiol Gastrointest Liver Physiol 2002; 282:G720–G726.
240. Gotoh Y, Suzuki H, Kinoshita S, et al. Involvement of an organic anion transporter (canalicular multispecific organic anion transporter/multidrug resistance-associated protein 2) in gastrointestinal secretion of glutathione conjugates in rats. J Pharmacol Exp Ther 2000; 292:433–439.
241. Schaub TP, Kartenbeck J, Konig J, et al. Expression of the MRP2 gene-encoded conjugate export pump in human kidney proximal tubules and in renal cell carcinoma. J Am Soc Nephrol 1999; 10:1159–1169.
242. Schaub TP, Kartenbeck J, Konig J, et al. Expression of the conjugate export pump encoded by the mrp2 gene in the apical membrane of kidney proximal tubules. J Am Soc Nephrol 1997; 8:1213–1221.
243. Masereeuw R, Notenboom S, Smeets PH, et al. Impaired renal secretion of substrates for the multidrug resistance protein 2 in mutant transport-deficient (TR-) rats. J Am Soc Nephrol 2003; 14:2741–2749.
244. Hulot JS, Villard E, Maguy A, et al. A mutation in the drug transporter gene ABCC2 associated with impaired methotrexate elimination. Pharmacogenet Genom 2005; 15:277–285.

245. Zelcer N, van de Wetering K, de Waart R, et al. Mice lacking Mrp3 (Abcc3) have normal bile salt transport, but altered hepatic transport of endogenous glucuronides. J Hepatol 2006; 44:768–775.
246. Hirohashi T, Suzuki H, Ito K, et al. Hepatic expression of multidrug resistance-associated protein-like proteins maintained in eisai hyperbilirubinemic rats. Mol Pharmacol 1998; 53:1068–1075.
247. Kiuchi Y, Suzuki H, Hirohashi T, et al. cDNA cloning and inducible expression of human multidrug resistance associated protein 3 (MRP3). FEBS Lett 1998; 433:149–152.
248. Uchiumi T, Hinoshita E, Haga S, et al. Isolation of a novel human canalicular multispecific organic anion transporter, cMOAT2/MRP3, and its expression in cisplatin-resistant cancer cells with decreased ATP-dependent drug transport. Biochem Biophys Res Commun 1998; 252:103–110.
249. Ogawa K, Suzuki H, Hirohashi T, et al. Characterization of inducible nature of MRP3 in rat liver. Am J Physiol Gastrointest Liver Physiol 2000; 278:G438–G446.
250. Konig J, Rost D, Cui Y, et al. Characterization of the human multidrug resistance protein isoform MRP3 localized to the basolateral hepatocyte membrane. Hepatology 1999; 29:1156–1163.
251. Hirohashi T, Suzuki H, Sugiyama Y. Characterization of the transport properties of cloned rat multidrug resistance-associated protein 3 (MRP3). J Biol Chem 1999; 274:15181–15185.
252. Hirohashi T, Suzuki H, Takikawa H, et al. ATP-dependent transport of bile salts by rat multidrug resistance-associated protein 3 (Mrp3). J Biol Chem 2000; 275: 2905–2910.
253. Zeng H, Chen ZS, Belinsky MG, et al. Transport of methotrexate (MTX) and folates by multidrug resistance protein (MRP) 3 and MRP1: effect of polyglutamylation on MTX transport. Cancer Res 2001; 61:7225–7232.
254. Akita H, Suzuki H, Hirohashi T, et al. Transport activity of human MRP3 expressed in Sf9 cells: comparative studies with rat MRP3. Pharm Res 2002; 19:34–41.
255. Akita H, Suzuki H, Sugiyama Y. Sinusoidal efflux of taurocholate is enhanced in Mrp2-deficient rat liver. Pharm Res 2001; 18:1119–1125.
256. Zelcer N, van de Wetering K, Hillebrand M, et al. Mice lacking multidrug resistance protein 3 show altered morphine pharmacokinetics and morphine-6-glucuronide antinociception. Proc Natl Acad Sci U S A 2005; 102:7274–7279.
257. Zamek-Gliszczynski MJ, Nezasa K, Tian X, et al. Evaluation of the role of multidrug resistance-associated protein (Mrp) 3 and Mrp4 in hepatic basolateral excretion of sulfate and glucuronide metabolites of acetaminophen, 4-methylumbelliferone, and harmol in Abcc3-/- and Abcc4-/- mice. J Pharmacol Exp Ther 2006; 319:1485–1491.
258. Manautou JE, de Waart DR, Kunne C, et al. Altered disposition of acetaminophen in mice with a disruption of the Mrp3 gene. Hepatology 2005; 42:1091–1098.
259. Shoji T, Suzuki H, Kusuhara H, et al. ATP-dependent transport of organic anions into isolated basolateral membrane vesicles from rat intestine. Am J Physiol Gastrointest Liver Physiol 2004; 287:G749–G756.
260. Kool M, de Haas M, Scheffer GL, et al. Analysis of expression of cMOAT (MRP2), MRP3, MRP4, and MRP5, homologues of the multidrug resistance-associated protein gene (MRP1), in human cancer cell lines. Cancer Res 1997; 57:3537–3547.

261. Rius M, Nies AT, Hummel-Eisenbeiss J, et al. Cotransport of reduced glutathione with bile salts by MRP4 (ABCC4) localized to the basolateral hepatocyte membrane. Hepatology 2003; 38:374–384.
262. Leggas M, Adachi M, Scheffer GL, et al. Mrp4 confers resistance to topotecan and protects the brain from chemotherapy. Mol Cell Biol 2004; 24:7612–7621.
263. van Aubel RA, Smeets PH, Peters JH, et al. The MRP4/ABCC4 gene encodes a novel apical organic anion transporter in human kidney proximal tubules: putative efflux pump for urinary cAMP and cGMP. J Am Soc Nephrol 2002; 13:595–603.
264. Reid G, Wielinga P, Zelcer N, et al. Characterization of the transport of nucleoside analog drugs by the human multidrug resistance proteins MRP4 and MRP5. Mol Pharmacol 2003; 63:1094–1103.
265. Chen ZS, Lee K, Walther S, et al. Analysis of methotrexate and folate transport by multidrug resistance protein 4 (ABCC4): MRP4 is a component of the methotrexate efflux system. Cancer Res 2002; 62:3144–3150.
266. Smeets PH, van Aubel RA, Wouterse AC, et al. Contribution of multidrug resistance protein 2 (MRP2/ABCC2) to the renal excretion of p-aminohippurate (PAH) and identification of MRP4 (ABCC4) as a novel PAH transporter. J Am Soc Nephrol 2004; 15:2828–2835.
267. Hasegawa M, Kusuhara H, Adachi M, et al. Multidrug resistance-associated protein 4 is involved in the urinary excretion of hydrochlorothiazide and furosemide. J Am Soc Nephrol 2007; 18:37–45.
268. Imaoka T, Kusuhara H, Adachi M, et al. Functional involvement of multidrug resistance associated protein 4 (MRP4/ABCC4) in the renal elimination of the antiviral drugs, adefovir and tenofovir. Mol Pharmacol 2007; 71:619–627.
269. Kusuhara H. Sugiyama Y. ATP-binding cassette, subfamily G (ABCG family). Pflugers Arch 2007; 453:735–744.
270. Mitomo H, Kato R, Ito A, et al. A functional study on polymorphism of the ATP-binding cassette transporter ABCG2: critical role of arginine-482 in methotrexate transport. Biochem J 2003; 373:767–774.
271. Kage K, Tsukahara S, Sugiyama T, et al. Dominant-negative inhibition of breast cancer resistance protein as drug efflux pump through the inhibition of S-S dependent homodimerization. Int J Cancer 2002; 97:626–630.
272. Allikmets R, Schriml LM, Hutchinson A, et al. A human placenta-specific ATP-binding cassette gene (ABCP) on chromosome 4q22 that is involved in multidrug resistance. Cancer Res 1998; 58:5337–5339.
273. Doyle LA, Yang W, Abruzzo LV, et al. A multidrug resistance transporter from human MCF-7 breast cancer cells. Proc Natl Acad Sci U S A 1998; 95:15665–15670.
274. Maliepaard M, Scheffer GL, Faneyte IF, et al. Subcellular localization and distribution of the breast cancer resistance protein transporter in normal human tissues. Cancer Res 2001; 61:3458–3464.
275. Jonker JW, Buitelaar M, Wagenaar E, et al. The breast cancer resistance protein protects against a major chlorophyll-derived dietary phototoxin and protoporphyria. Proc Natl Acad Sci U S A 2002; 99:15649–15654.
276. Hori S, Ohtsuki S, Tachikawa M, et al. Functional expression of rat ABCG2 on the luminal side of brain capillaries and its enhancement by astrocyte-derived soluble factor(s). J Neurochem 2004; 90:526–536.
277. Lee YJ, Kusuhara H, Jonker JW, et al. Investigation of efflux transport of dehydroepiandrosterone sulfate and mitoxantrone at the mouse blood-brain barrier:

a minor role of breast cancer resistance protein. J Pharmacol Exp Ther 2005; 312: 44–52.

278. Doyle LA, Ross DD. Multidrug resistance mediated by the breast cancer resistance protein BCRP (ABCG2). Oncogene 2003; 22:7340–7358.
279. Jonker JW, Merino G, Musters S, et al. The breast cancer resistance protein BCRP (ABCG2) concentrates drugs and carcinogenic xenotoxins into milk. Nat Med 2005; 11:127–129.
280. Zaher H, Khan AA, Palandra J, et al. Breast cancer resistance protein (Bcrp/abcg2) is a major determinant of sulfasalazine absorption and elimination in the mouse. Mol Pharmacol 2006; 3:55–61.
281. Merino G, Alvarez AI, Pulido MM, et al. Breast cancer resistance protein (BCRP/ABCG2) transports fluoroquinolone antibiotics and affects their oral availability, pharmacokinetics, and milk secretion. Drug Metab Dispos 2006; 34:690–695.
282. Hirano M, Maeda K, Matsushima S, et al. Involvement of BCRP (ABCG2) in the biliary excretion of pitavastatin. Mol Pharmacol 2005; 68:800–807.
283. van Herwaarden AE, Wagenaar E, Karnekamp B, et al. Breast cancer resistance protein (Bcrp1/Abcg2) reduces systemic exposure of the dietary carcinogens aflatoxin B1, IQ and Trp-P-1 but also mediates their secretion into breast milk. Carcinogenesis 2006; 27:123–130.
284. van Herwaarden AE, Jonker JW, Wagenaar E, et al. The breast cancer resistance protein (Bcrp1/Abcg2) restricts exposure to the dietary carcinogen 2-amino-1-methyl-6-phenylimidazo[4,5-b]pyridine. Cancer Res 2003; 63:6447–6452.
285. Adachi Y, Suzuki H, Schinkel AH, et al. Role of breast cancer resistance protein (Bcrp1/Abcg2) in the extrusion of glucuronide and sulfate conjugates from enterocytes to intestinal lumen. Mol Pharmacol 2005; 67:923–928.
286. Kondo C, Onuki R, Kusuhara H, et al. Lack of improvement of oral absorption of ME3277 by prodrug formation is ascribed to the intestinal efflux mediated by breast cancer resistant protein (BCRP/ABCG2). Pharm Res 2005; 22:613–618.
287. Merino G, Jonker JW, Wagenaar E, et al. The breast cancer resistance protein (BCRP/ABCG2) affects pharmacokinetics, hepatobiliary excretion, and milk secretion of the antibiotic nitrofurantoin. Mol Pharmacol 2005; 67:1758–1764.
288. Breedveld P, Zelcer N, Pluim D, et al. Mechanism of the pharmacokinetic interaction between methotrexate and benzimidazoles: potential role for breast cancer resistance protein in clinical drug-drug interactions. Cancer Res 2004; 64:5804–5811.
289. Breedveld P, Pluim D, Cipriani G, et al. The effect of Bcrp1 (Abcg2) on the in vivo pharmacokinetics and brain penetration of imatinib mesylate (Gleevec): implications for the use of breast cancer resistance protein and P-glycoprotein inhibitors to enable the brain penetration of imatinib in patients. Cancer Res 2005; 65:2577–2582.
290. Mizuno N, Suzuki M, Kusuhara H, et al. Impaired renal excretion of 6-hydroxy-5,7-dimethyl-2-methylamino-4-(3-pyridylmethyl) benzothiazole (e3040) sulfate in breast cancer resistance protein (bcrp1/abcg2) knockout mice. Drug Metab Dispos 2004; 32:898–901.
291. Hedman A, Angelin B, Arvidsson A, et al. Interactions in the renal and biliary elimination of digoxin: stereoselective difference between quinine and quinidine. Clin Pharmacol Ther 1990; 47:20–26.

292. Hedman A, Angelin B, Arvidsson A, et al. Digoxin-verapamil interaction: reduction of biliary but not renal digoxin clearance in humans. Clin Pharmacol Ther 1991; 49:256–262.
293. Kuhlmann J. Effects of verapamil, diltiazem, and nifedipine on plasma levels and renal excretion of digitoxin. Clin Pharmacol Ther 1985; 38:667–673.
294. Olinga P, Merema M, Hof IH, et al. Characterization of the uptake of rocuronium and digoxin in human hepatocytes: carrier specificity and comparison with in vivo data. J Pharmacol Exp Ther 1998; 285:506–510.
295. Hedman A, Meijer DK. Stereoselective inhibition by the diastereomers quinidine and quinine of uptake of cardiac glycosides into isolated rat hepatocytes. J Pharm Sci 1998; 87:457–461.
296. Hori R, Okamura N, Aiba T, et al. Role of P-glycoprotein in renal tubular secretion of digoxin in the isolated perfused rat kidney. J Pharmacol Exp Ther 1993; 266:1620–1625.
297. Fromm MF, Kim RB, Stein CM, et al. Inhibition of P-glycoprotein-mediated drug transport: a unifying mechanism to explain the interaction between digoxin and quinidine [see comments]. Circulation 1999; 99:552–557.
298. Emi Y, Tsunashima D, Ogawara K, et al. Role of P-glycoprotein as a secretory mechanism in quinidine absorption from rat small intestine. J Pharm Sci 1998; 87:295–299.
299. Nozawa T, Imai K, Nezu J, et al. Functional characterization of pH-sensitive organic anion transporting polypeptide OATP-B in human. J Pharmacol Exp Ther 2004; 308:438–445.
300. Shimizu M, Fuse K, Okudaira K, et al. Contribution of oatp (organic anion-transporting polypeptide) family transporters to the hepatic uptake of fexofenadine in humans. Drug Metab Dispos 2005; 33:1477–1481.
301. Tahara H, Kusuhara H, Maeda K, et al. Inhibition of oat3-mediated renal uptake as a mechanism for drug-drug interaction between fexofenadine and probenecid. Drug Metab Dispos 2006; 34:743–747.
302. Tahara H, Kusuhara H, Fuse E, et al. P-glycoprotein plays a major role in the efflux of fexofenadine in the small intestine and blood-brain barrier, but only a limited role in its biliary excretion. Drug Metab Dispos 2005; 33:963–968.
303. Uno T, Shimizu M, Sugawara K, et al. Lack of dose-dependent effects of itraconazole on the pharmacokinetic interaction with fexofenadine. Drug Metab Dispos 2006; 34:1875–1879.
304. Lemma GL, Wang Z, Hamman MA, et al. The effect of short- and long-term administration of verapamil on the disposition of cytochrome P450 3A and P-glycoprotein substrates. Clin Pharmacol Ther 2006; 79:218–230.
305. van Heeswijk RP, Bourbeau M, Campbell P, et al. Time-dependent interaction between lopinavir/ritonavir and fexofenadine. J Clin Pharmacol 2006; 46:758–767.
306. Schwab D, Fischer H, Tabatabaei A, et al. Comparison of in vitro P-glycoprotein screening assays: recommendations for their use in drug discovery. J Med Chem 2003; 46:1716–1725.
307. Wang EJ, Lew K, Casciano CN, et al. Interaction of common azole antifungals with P glycoprotein. Antimicrob Agents Chemother 2002; 46:160–165.
308. Petri N, Tannergren C, Rungstad D, et al. Transport characteristics of fexofenadine in the Caco-2 cell model. Pharm Res 2004; 21:1398–1404.

309. Choo EF, Leake B, Wandel C, et al. Pharmacological inhibition of P-glycoprotein transport enhances the distribution of HIV-1 protease inhibitors into brain and testes. Drug Metab Dispos 2000; 28:655–660.
310. Dresser GK, Kim RB, Bailey DG. Effect of grapefruit juice volume on the reduction of fexofenadine bioavailability: possible role of organic anion transporting polypeptides. Clin Pharmacol Ther 2005; 77:170–177.
311. Dresser GK, Wacher V, Wong S, et al. Evaluation of peppermint oil and ascorbyl palmitate as inhibitors of cytochrome P4503A4 activity in vitro and in vivo. Clin Pharmacol Ther 2002; 72:247–255.
312. Banfield C, Gupta S, Marino M, et al. Grapefruit juice reduces the oral bioavailability of fexofenadine but not desloratadine. Clin Pharmacokinet 2002; 41:311–318.
313. Muck W, Mai I, Fritsche L, et al. Increase in cerivastatin systemic exposure after single and multiple dosing in cyclosporine-treated kidney transplant recipients. Clin Pharmacol Ther 1999; 65:251–261.
314. Shitara Y, Itoh T, Sato H, et al. Inhibition of transporter-mediated hepatic uptake as a mechanism for drug-drug interaction between cerivastatin and cyclosporin A. J Pharmacol Exp Ther 2003; 304:610–616.
315. Nishizato Y, Ieiri I, Suzuki H, et al. Polymorphisms of OATP-C (SLC21A6) and OAT3 (SLC22A8) genes: consequences for pravastatin pharmacokinetics. Clin Pharmacol Ther 2003; 73:554–565.
316. Chung JY, Cho JY, Yu KS, et al. Effect of OATP1B1 (SLCO1B1) variant alleles on the pharmacokinetics of pitavastatin in healthy volunteers. Clin Pharmacol Ther 2005; 78:342–350.
317. Pasanen MK, Neuvonen M, Neuvonen PJ, et al. SLCO1B1 polymorphism markedly affects the pharmacokinetics of simvastatin acid. Pharmacogenet Genom 2006; 16:873–879.
318. Niemi M, Pasanen MK, Neuvonen PJ. SLCO1B1 polymorphism and sex affect the pharmacokinetics of pravastatin but not fluvastatin. Clin Pharmacol Ther 2006; 80:356–366.
319. Shitara Y, Sugiyama Y. Pharmacokinetic and pharmacodynamic alterations of 3-hydroxy-3-methylglutaryl coenzyme A (HMG-CoA) reductase inhibitors: drug-drug interactions and interindividual differences in transporter and metabolic enzyme functions. Pharmacol Ther 2006; 112:71–105.
320. Maeda K, Ieiri I, Yasuda K, et al. Effects of organic anion transporting polypeptide 1B1 haplotype on pharmacokinetics of pravastatin, valsartan, and temocapril. Clin Pharmacol Ther 2006; 79:427–439.
321. Niemi M, Backman JT, Kajosaari LI, et al. Polymorphic organic anion transporting polypeptide 1B1 is a major determinant of repaglinide pharmacokinetics. Clin Pharmacol Ther 2005; 77:468–478.
322. Kajosaari LI, Niemi M, Neuvonen M, et al. Cyclosporine markedly raises the plasma concentrations of repaglinide. Clin Pharmacol Ther 2005; 78:388–399.
323. Shitara Y, Hirano M, Sato H, et al. Gemfibrozil and its glucuronide inhibit the organic anion transporting polypeptide 2 (OATP2/OATP1B1:SLC21A6)-mediated hepatic uptake and CYP2C8-mediated metabolism of cerivastatin: analysis of the mechanism of the clinically relevant drug-drug interaction between cerivastatin and gemfibrozil. J Pharmacol Exp Ther 2004; 311:228–236.

324. Ogilvie BW, Zhang D, Li W, et al. Glucuronidation converts gemfibrozil to a potent, metabolism-dependent inhibitor of CYP2C8: implications for drug-drug interactions. Drug Metab Dispos 2006; 34:191–197.
325. De Bony F, Tod M, Bidault R, et al. Multiple interactions of cimetidine and probenecid with valaciclovir and its metabolite acyclovir. Antimicrob Agents Chemother 2002; 46:458–463.
326. Laskin OL, de Miranda P, King DH, et al. Effects of probenecid on the pharmacokinetics and elimination of acyclovir in humans. Antimicrob Agents Chemother 1982; 21:804–807.
327. Were JB, Shapiro TA. Effects of probenecid on the pharmacokinetics of allopurinol riboside. Antimicrob Agents Chemother 1993; 37:1193–1196.
328. Odlind B, Beermann B, Lindstrom B. Coupling between renal tubular secretion and effect of bumetanide. Clin Pharmacol Ther 1983; 34:805–809.
329. Sennello LT, Quinn D, Rollins DE, et al. Effect of probenecid on the pharmacokinetics of cefmenoxime. Antimicrob Agents Chemother 1983; 23:803–807.
330. Ko H, Cathcart KS, Griffith DL, et al. Pharmacokinetics of intravenously administered cefmetazole and cefoxitin and effects of probenecid on cefmetazole elimination. Antimicrob Agents Chemother 1989; 33:356–361.
331. Vlasses PH, Holbrook AM, Schrogie JJ, et al. Effect of orally administered probenecid on the pharmacokinetics of cefoxitin. Antimicrob Agents Chemother 1980; 17:847–855.
332. Shukla UA, Pittman KA, Barbhaiya RH. Pharmacokinetic interactions of cefprozil with food, propantheline, metoclopramide, and probenecid in healthy volunteers. J Clin Pharmacol 1992; 32:725–731.
333. LeBel M, Paone RP, Lewis GP. Effect of probenecid on the pharmacokinetics of ceftizoxime. J Antimicrob Chemother 1983; 12:147–155.
334. Garton AM, Rennie RP, Gilpin J, et al. Comparison of dose doubling with probenecid for sustaining serum cefuroxime levels. J Antimicrob Chemother 1997; 40: 903–906.
335. Cundy KC, Petty BG, Flaherty J, et al. Clinical pharmacokinetics of cidofovir in human immunodeficiency virus-infected patients. Antimicrob Agents Chemother 1995; 39:1247–1252.
336. Jaehde U, Sorgel F, Reiter A, et al. Effect of probenecid on the distribution and elimination of ciprofloxacin in humans. Clin Phamacol Ther 1995; 58:532–541.
337. Inotsume N, Nishimura M, Nakano M, et al. The inhibitory effect of probenecid on renal excretion of famotidine in young, healthy volunteers. J Clin Pharmacol 1990; 30:50–56.
338. Yasui-Furukori N, Uno T, Sugawara K, et al. Different effects of three transporting inhibitors, verapamil, cimetidine, and probenecid, on fexofenadine pharmacokinetics. Clin Pharmacol Ther 2005; 77:17–23.
339. Vree TB, van den Biggelaar-Martea M, Verwey-van Wissen CP. Probenecid inhibits the renal clearance of frusemide and its acyl glucuronide. Br J Clin Pharmacol 1995; 39:692–695.
340. Hill G, Cihlar T, Oo C, et al. The anti-influenza drug oseltamivir exhibits low potential to induce pharmacokinetic drug interactions via renal secretion-correlation of in vivo and in vitro studies. Drug Metab Dispos 2002; 30:13–19.

341. Jackson E Diuretics, In: Hardman JG, Limberd LE, Gilman AG, eds. Goodman & Gilaman's The Pharmacological of Basis of Therapeutics. 10th ed. New York: McGraw-Hill, 2001:757–787.
342. Honari J, Blair AD, Cutler RE. Effects of probenecid on furosemide kinetics and natriuresis in man. Clin Pharmacol Ther 1977; 22:395–401.
343. Cihlar T, Ho ES. Fluorescence-based assay for the interaction of small molecules with the human renal organic anion transporter 1. Anal Biochem 2000; 283:49–55.
344. Ichida K, Hosoyamada M, Kimura H, et al. Urate transport via human PAH transporter hOAT1 and its gene structure. Kidney Int 2003; 63:143–155.
345. Ho ES, Lin DC, Mendel DB, et al. Cytotoxicity of antiviral nucleotides adefovir and cidofovir is induced by the expression of human renal organic anion transporter 1. J Am Soc Nephrol 2000; 11:383–393.
346. Gisclon LG, Boyd RA, Williams RL, et al. The effect of probenecid on the renal elimination of cimetidine. Clin Pharmacol Ther 1989; 45:444–452.
347. Emanuelsson BM, Beermann B, Paalzow LK. Non-linear elimination and protein binding of probenecid. Eur J Clin Pharmacol 1987; 32:395–401.
348. Lin JH, Los LE, Ulm EH, et al. Kinetic studies on the competition between famotidine and cimetidine in rats. Evidence of multiple renal secretory systems for organic cations. Drug Metab Dispos 1988; 16:52–56.
349. Boom SP, Meyer I, Wouterse AC, et al. A physiologically based kidney model for the renal clearance of ranitidine and the interaction with cimetidine and probenecid in the dog. Biopharm Drug Dispos 1998; 19:199–208.
350. Goodman and Gilman's the Pharmacological Basis of Therapeutics 8th ed. edited by Gilman AG, Rall TW, Nies AS, Taylor P, New York: McGraw-Hill, 1990.
351. Nierenerg DW. Drug inhibition of penicillin tubular secretion: concordance between in vitro and clinical findings. J Pharmacol Exp Ther 1987; 240:712–716.
352. Bourke RS, Chheda G, Bremer A, et al. Inhibition of renal tubular transport of methotrexate by probenecid. Cancer Res 1975; 35:110–116.
353. Williams WM, Chen TS, Huang KC. Effect of penicillin on the renal tubular secretion of methotrexate in the monkey. Cancer Res 1984; 44:1913–1917.
354. Takeda M, Khamdang S, Narikawa S, et al. Characterization of methotrexate transport and its drug interactions with human organic anion transporters. J Pharmacol Exp Ther 2002; 302:666–671.
355. Nozaki Y, Kusuhara H, Endou H, et al. Quantitative evaluation of the drug-drug interactions between methotrexate and nonsteroidal anti-inflammatory drugs in the renal uptake process based on the contribution of organic anion transporters and reduced folate carrier. J Pharmacol Exp Ther 2004; 309:226–234.
356. El-Sheikh AA, van den Heuvel JJ, Koenderink JB, et al. Interaction of nonsteroidal anti-inflammatory drugs with multidrug resistance protein (MRP) 2/ABCC2- and MRP4/ABCC4-mediated methotrexate transport. J Pharmacol Exp Ther 2007; 320:229–235.
357. Garrigues TM, Martin U, Peris-Ribera JE, et al. Dose-dependent absorption and elimination of cefadroxil in man. Eur J Clin Pharmacol 1991; 41:179–183.
358. Bai JPF, Amidon GL. Structural specificity of mucosal-cell transport and metabolism of peptide drugs: implication for oral peptide drug delivery. Pharm Res 1992; 9:969–978.

359. Sinko PJ, Amidon GL. Characterization of the oral absorption of beta-lactam antibiotics. I. Cephalosporins: determination of intrinsic membrane absorption parameters in the rat intestine in situ. Pharm Res 1988; 5:645–650.
360. Shapiro AB, Ling V. Positively cooperative sites for drug transport by P-glycoprotein with distinct drug specificities. Eur J Biochem 1997; 250:130–137.
361. Lee YJ, Maeda J, Kusuhara H, et al. In vivo evaluation of P-glycoprotein function at the blood-brain barrier in nonhuman primates using [11C]verapamil. J Pharmacol Exp Ther 2006; 316:647–653.
362. Pincus JH, Barry K. Protein distribution diet restores motor function in patients with dopa-resistant ''off'' period. Neurology 1988; 38:481–483.
363. Audus KL, Borchardt RT. Characterization of the large neutral amino acid transport system of bovine brain microvessel endothelial cell monolayers. J Neurochem 1986; 47:484–488.
364. Greig NH, Momma S, Sweeney DJ, et al. Facilitated transport of melphalan at the rat blood-brain barrier by the large neutral amino acid carrier system. Cancer Res 1987; 47:1571–1576.
365. Shoda J, Miura T, Utsunomiya H, et al. Genipin enhances Mrp2 (Abcc2)-mediated bile formation and organic anion transport in rat liver. Hepatology 2004; 39: 167–178.
366. Zollner G, Marschall HU, Wagner M, et al. Role of nuclear receptors in the adaptive response to bile acids and cholestasis: pathogenetic and therapeutic considerations. Mol Pharmacol 2006; 3:231–251.
367. Eloranta JJ, Meier PJ, Kullak-Ublick GA. Coordinate transcriptional regulation of transport and metabolism. Methods Enzymol 2005; 400:511–530.
368. Willson TM, Kliewer SA. PXR, CAR and drug metabolism. Nat Rev Drug Discov 2002; 1:259–266.
369. Stanley LA, Horsburgh BC, Ross J, et al. PXR and CAR: nuclear receptors which play a pivotal role in drug disposition and chemical toxicity. Drug Metab Rev 2006; 38:515–597.
370. Kodama S, Negishi M. Phenobarbital confers its diverse effects by activating the orphan nuclear receptor car. Drug Metab Rev 2006; 38:75–87.
371. Kato M, Chiba K, Horikawa M, et al. The quantitative prediction of in vivo enzyme-induction caused by drug exposure from in vitro information on human hepatocytes. Drug Metab Pharmacokinet 2005; 20:236–243.
372. Greiner B, Eichelbaum M, Fritz P, et al. The role of intestinal P-glycoprotein in the interaction of digoxin and rifampin. J Clin Invest 1999; 104:147–153.
373. Westphal K, Weinbrenner A, Zschiesche M, et al. Induction of P-glycoprotein by rifampin increases intestinal secretion of talinolol in human beings: a new type of drug/drug interaction. Clin Pharmacol Ther 2000; 68:345–355.
374. Hamman MA, Bruce MA, Haehner-Daniels BD, et al. The effect of rifampin administration on the disposition of fexofenadine. Clin Pharmacol Ther 2001; 69:114–121.
375. Geick A, Eichelbaum M, Burk O. Nuclear receptor response elements mediate induction of intestinal MDR1 by rifampin. J Biol Chem 2001; 276:14581–14587.
376. Fromm MF, Kauffmann HM, Fritz P, et al. The effect of rifampin treatment on intestinal expression of human MRP transporters. Am J Pathol 2000; 157: 1575–1580.

377. Kast HR, Goodwin B, Tarr PT, et al. Regulation of multidrug resistance-associated protein 2 (ABCC2) by the nuclear receptors pregnane X receptor, farnesoid X-activated receptor, and constitutive androstane receptor. J Biol Chem 2002; 277: 2908–2915.
378. Dussault I, Lin M, Hollister K, et al. Peptide mimetic HIV protease inhibitors are ligands for the orphan receptor SXR. J Biol Chem 2001; 276:33309–33312.
379. Jigorel E, Le Vee M, Boursier-Neyret C, et al. Differential regulation of sinusoidal and canalicular hepatic drug transporter expression by xenobiotics activating drug-sensing receptors in primary human hepatocytes. Drug Metab Dispos 2006; 34:1756–1763.
380. Wagner M, Halilbasic E, Marschall HU, et al. CAR and PXR agonists stimulate hepatic bile acid and bilirubin detoxification and elimination pathways in mice. Hepatology 2005; 42:420–430.
381. Staudinger JL, Madan A, Carol KM, et al. Regulation of drug transporter gene expression by nuclear receptors. Drug Metab Dispos 2003; 31:523–527.
382. Rausch-Derra LC, Hartley DP, Meier PJ, et al. Differential effects of microsomal enzyme-inducing chemicals on the hepatic expression of rat organic anion transporters, OATP1 and OATP2. Hepatology 2001; 33:1469–1478.
383. Cheng X, Maher J, Dieter MZ, et al. Regulation of mouse organic anion-transporting polypeptides (Oatps) in liver by prototypical microsomal enzyme inducers that activate distinct transcription factor pathways. Drug Metab Dispos 2005; 33:1276–1282.
384. Cherrington NJ, Hartley DP, Li N, et al. Organ distribution of multidrug resistance proteins 1, 2, and 3 (Mrp1, 2, and 3) mRNA and hepatic induction of Mrp3 by constitutive androstane receptor activators in rats. J Pharmacol Exp Ther 2002; 300:97–104.
385. Maher JM, Cheng X, Slitt AL, et al. Induction of the multidrug resistance-associated protein family of transporters by chemical activators of receptor-mediated pathways in mouse liver. Drug Metab Dispos 2005; 33:956–962.
386. Assem M, Schuetz EG, Leggas M, et al. Interactions between hepatic Mrp4 and Sult2a as revealed by the constitutive androstane receptor and Mrp4 knockout mice. J Biol Chem 2004; 279:22250–22257.
387. Hagenbuch N, Reichel C, Stieger B, et al. Effect of phenobarbital on the expression of bile salt and organic anion transporters of rat liver. J Hepatol 2001; 34:881–887.
388. Cherrington NJ, Slitt AL, Maher JM, et al. Induction of multidrug resistance protein 3 (mrp3) in vivo is independent of constitutive androstane receptor. Drug Metab Dispos 2003; 31:1315–1319.
389. Xiong H, Yoshinari K, Brouwer KL, et al. Role of constitutive androstane receptor in the in vivo induction of Mrp3 and CYP2B1/2 by phenobarbital. Drug Metab Dispos 2002; 30:918–923.
390. Jung D, Podvinec M, Meyer UA, et al. Human organic anion transporting polypeptide 8 promoter is transactivated by the farnesoid X receptor/bile acid receptor. Gastroenterology 2002; 122:1954–1966.
391. Ohtsuka H, Abe T, Onogawa T, et al. Farnesoid X receptor, hepatocyte nuclear factors 1alpha and 3beta are essential for transcriptional activation of the liver-specific organic anion transporter-2 gene. J Gastroenterol 2006; 41:369–377.

392. Jung D, Kullak-Ublick GA. Hepatocyte nuclear factor 1 alpha: a key mediator of the effect of bile acids on gene expression. Hepatology 2003; 37:622–631.
393. Jung D, Hagenbuch B, Gresh L, et al. Characterization of the human OATP-C (SLC21A6) gene promoter and regulation of liver-specific OATP genes by hepatocyte nuclear factor 1 alpha. J Biol Chem 2001; 276:37206–37214.
394. Moffit JS, Aleksunes LM, Maher JM, et al. Induction of hepatic transporters multidrug resistance-associated proteins (Mrp) 3 and 4 by clofibrate is regulated by peroxisome proliferator-activated receptor alpha. J Pharmacol Exp Ther 2006; 317:537–545.
395. Nishimura M, Naito S, Yokoi T. Tissue-specific mRNA expression profiles of human nuclear receptor subfamilies. Drug Metab Pharmacokinet 2004; 19:135–149.
396. Nozaki Y, Kusuhara H, Kondo T, et al. Species difference in the inhibitory effect of non-steroidal anti-inflammatory drugs on the uptake of methotrexate by human kidney slices. J. Pharmacol Exp Ther 2007; 322:1162–70.
397. Suzuki M, Suzuki H, Sugimoto Y, et al. ABCG2 transports sulfated conjugates of steroids and xenobiotics. J Biol Chem 2003; 278:22644–9.
398. Crugten J, Bochner F, Keal J, et al. Selectivity of the cimetidine-induced alterations in the renal handling of organic substrates in humans. Studies with anionic, cationic and zwitterionic drugs. J Pharmacol Exp Ther. 1986; 236:481–7.

6

In Vitro Models for Studying Induction of Cytochrome P450 Enzymes

Jose M. Silva
Johnson and Johnson, Raritan, New Jersey, U.S.A.

Deborah A. Nicoll-Griffith
Merck Research Laboratories, Rahway, New Jersey, U.S.A.

I. INTRODUCTION

Cytochrome P450 (CYP) enzymes form a gene superfamily that are involved in the metabolism of a variety of chemically diverse substances ranging from endogenous compounds to xenobiotics, including drugs, carcinogens, and environmental pollutants. Although CYP regulation is only now beginning to be understood, it is well known that several of the CYP genes are induced by many drugs. This may cause variability in enzymatic activity, with different groups of patients producing unexpected pharmacological outcomes of some drugs as a result of drug-drug interactions (1,2). For example, induction of CYP3A has been shown to result in a significant loss of efficacy for the contraceptive steroids (3,4). Thus, regulatory agencies now request that new drugs be tested for their potential to induce CYP enzymes. Until recently, this involved treating laboratory animals with the test compound, followed by analysis of liver CYP enzymes ex vivo. This raises four major issues. First, there is the requirement of large number of animals; reduction in animal usage should be encouraged where possible. Second, large amount of test compounds have to be synthesized; this imposes a heavy burden on the synthetic chemistry efforts and is not compatible with combinatorial chemistry strategies. Third, in vivo studies are not high throughput, this in a time where advancements in combinatorial chemical synthesis have greatly increased the number of drug candidates being produced at the drug discovery stage. And finally, it's well known that species differences in CYP induction exist (5), making the extrapolation from animals to humans unreliable. Therefore, it is desirable to have in vitro models, in particular of human origin, to address CYP induction of drug candidates before costly clinical trials are conducted.

Unfortunately, there are no hepatoma cell lines able to express most of the major forms of adult CYP enzymes. However, various in vitro models for assessing enzyme induction have been described and include precision-cut liver slices, primary hepatocytes, and reporter gene constructs. The last model involves transfecting recombinant DNA encoding a reporter enzyme, such as chloramphenicol acetyl transferase, under the control of the regulatory element of the specific CYP of interest. In this chapter all three models are discussed, with focus mostly on the primary hepatocyte model, which, in our opinion, is the gold standard for predicting CYP induction in both laboratory animals and human. In addition, a case study involving a drug candidate (DFP) is discussed along with strategies for managing CYP induction in drug candidates.

II. MODELS

A. Primary Hepatocytes

Isolation of viable hepatocytes was first demonstrated by Howard et al. and rapidly increased in popularity with the further development of a high-yield preparative technique by Berry and Friend (6,7). Compared with liver slices, isolated hepatocytes are easier to manipulate and show a superior range of activities (8). For a detailed description of rat and human hepatocyte isolation techniques, the reader is referred to other reviews (8,9).

While primary hepatocytes maintained under conventional culture conditions tend to undergo rapid loss of liver-specific functions, great progress has been made in the last decade to slow this process. In our opinion, the three most important factors in retaining CYP responsiveness in primary hepatocyte are: media formulation, matrix composition, and cell-cell contacts (10–13).

There are several commercially available media that have been reported to support CYP-inducible hepatocytes in culture, including Dubecco's modified Eagle's medium, Liebovitz L-15 medium, Waymouth 752/1 medium, and modified Williams' E medium, to name a few (11). In summary, these are all enriched media containing a high amino acid content. High concentrations of certain amino acids have been reported to decrease protein degradation while stabilizing some levels of mRNA (14). Serum has routinely been used as a media supplement with many immortalized cell lines and is thought to improve cell attachment, survival, and morphology. However, with primary hepatocytes, serum is generally thought to promote growth and therefore has a dedifferentiation effect on hepatocytes, resulting in a loss of CYP expression (15). As a result, serum is used in the initial cell attachment stage (<24 hours) but is usually not included for the duration of the culture. Other supplements usually include dexamethasone and insulin. A low concentration of dexamethasone (<100 nM) has been reported to improve the viability of hepatocytes in culture (8) as well as to improve the responsiveness to CYP inducers (12,16,17). Insulin is also considered to be beneficial for the long-term survival of cultured hepatocytes (11).

Another culture condition known to be important in the maintenance of differentiated hepatocytes is the extracellular matrix. These comprise simple matrices, such as rat tail collagen, as well as more complex matrices, including fibronectin (18), extracts from rat liver (19), and, more recently, Matrigel, a biomatrix preparation derived from the Engelbreth-Holm-Swarm sarcoma (10,12). In our laboratory, we initially compared both rat and human primary hepatocytes cultured on collagen compared to Matrigel and found that while CYP3A responsiveness was not affected, basal CYP3A levels were better maintained in hepatocytes on Matrigel. In contrast, responsiveness to CYP2B1 in rat hepatocytes was markedly affected by the substratum used. As shown in Figure 1, Western blots of rat hepatocytes treated with phenobarbital show marked induction in CYP2B1 protein, while a poor response was observed in cells cultured on collagen (12).

Another substratum model developed to preserve liver function in hepatocytes in culture is the collagen-sandwich model. It was first demonstrated by Dunn et al. (20) that overlaying cultured rat hepatocytes with a top layer of collagen preserved the liver-specific phenotype, including CYP inducibility (21). In addition, cells cultured under these conditions reestablished cell polarity and developed a structurally and functionally normal bile canalicular network (22). More recently, LeCluyse et al. (23) reported that the matrix conditions considered to be optimal for maintaining cellular integrity, protein yields, and CYP enzyme induction in primary human hepatocytes are a collagen-sandwich model in combination with modified Chee's medium containing insulin and dexamethasone.

Figure 2 illustrates the standard induction protocol that we follow in our laboratory. Freshly isolated hepatocytes are cultured on 60-mm dishes or multiwell plates precoated with Matrigel for a minimum of two days. This allows the cells to recover from the damage endured during isolation and allows the cells to adapt to the culture environment. It's been reported that during this initial culture period a rapid loss in constitutive CYP expression is observed in the first 24 hours, followed by a recovery period after which the cells are capable to respond to CYP inducers (12,24). The cells are then challenged with the test compounds and allowed to incubate for a period of 24 to 48 hours. Response to inducers is rapid, as shown by the Northern blots of rat hepatocytes treated with dexamethasone and phenobarbital in Figure 3. CYP3A and CYP2B1 mRNA levels increased within two hours, reaching a maximum at 24 hours. Corresponding CYP protein induction requires at least eight hours before a significant rise is observed.

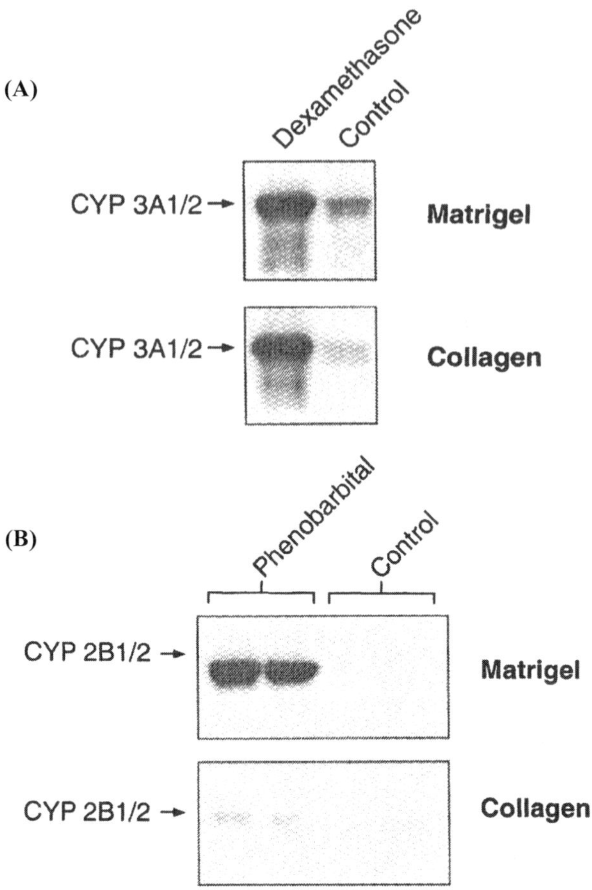

Figure 1 Immunoblot analysis of CYP3A (**A**) and CYP2B (**B**) protein in rat hepatocytes cultured on Matrigel and collagen. Cells were incubated in the presence of 10-μM dexamethasone or 50-μM phenobarbital for 48 hours. *Source*: From Ref. 12.

1. Interpretation of Induction Data

Induction of CYP expression by xenobiotics has been reported in mainly three ways: (1) induction potential (fold induction over control), (2) EC_{50} (effective concentration for 50% maximal induction), and (3) ''potency index'' (the ratio of induction response of the test compound compared to that of a gold standard). In our laboratory, we have defined CYP induction as a potency index or a percentage of a classic inducer rather than as fold increase over a control (induction potential). The reason for this is twofold. First, the basal levels of some CYPs may be low and therefore difficult to accurately quantitate. Second, we, and others, have found that basal CYP levels in culture may be highly

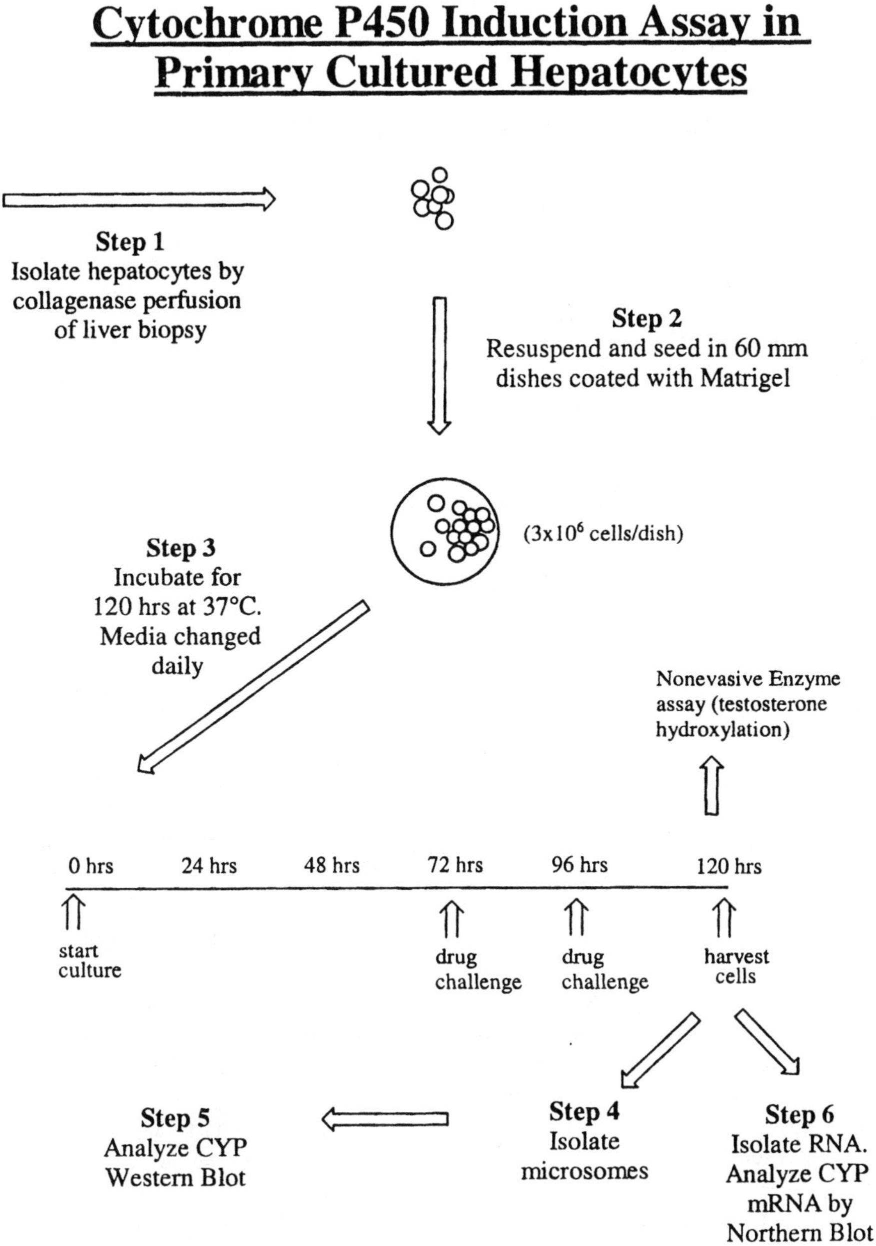

Figure 2 Hepatocyte induction protocol.

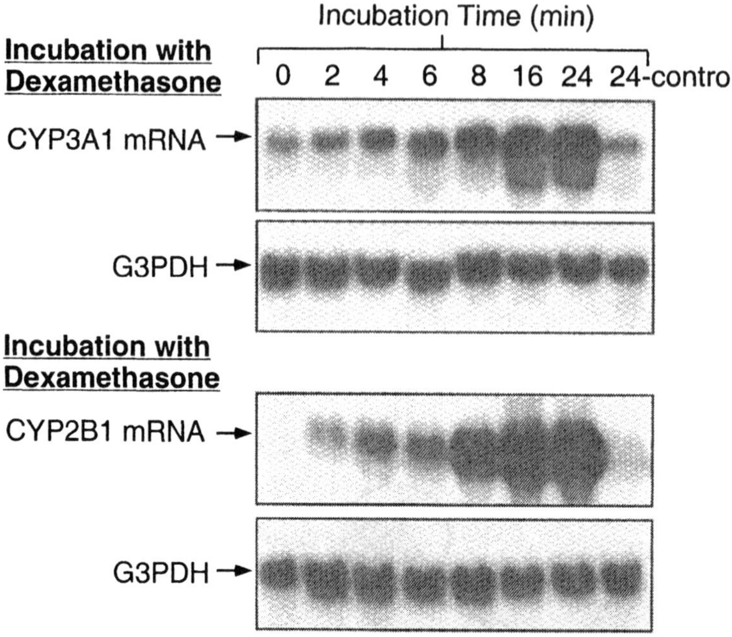

Figure 3 Northern blot analysis of CYP3A and CYP2B mRNAs in rat hepatocytes cultured on Matrigel. Cells were incubated in the presence of 10-μM dexamethasone or 50-μM phenobarbital for 48 hours. *Source*: From Ref. 12.

variable between different hepatocyte preparations, while the maximum induction levels are more consistent. For example, we have found that induction of CYP3A protein in dexamethasone-treated rat hepatocyte cultures from different preparations varies from an approximately 7- to 20-fold increase (Fig. 4) (12). In human primary cultures, induction of CYP3A4 by a drug candidate, calculated as a fold increase, also varied from a two- to eightfold increase in hepatocytes from four different donors (12). In contrast, when the results were expressed as a percentage of a classic inducer (rifampicin), the range was from 16% to 34% (12). Interestingly, Kostrubsky et al. reported that variation in the basal level of CYP3A4 expression in human primary hepatocytes was up to fivefold between different donors (25). However, the maximal CYP3A activity detected after treatment with rifampicin was similar in six separate human hepatocyte cultures. Another study, by Chang et al., reported that induction of oxazaphosphorine 4-hydroxylation activity by rifampicin in human hepatocyte cultures was inversely related to the basal activity (26). These results suggest that CYP activity after maximal induction is similar between separate cultures and that differences in fold induction result from variation in basal expression. It is therefore prudent to include a positive control to address the variability between different hepatocyte preparations. This is particularly important when comparing

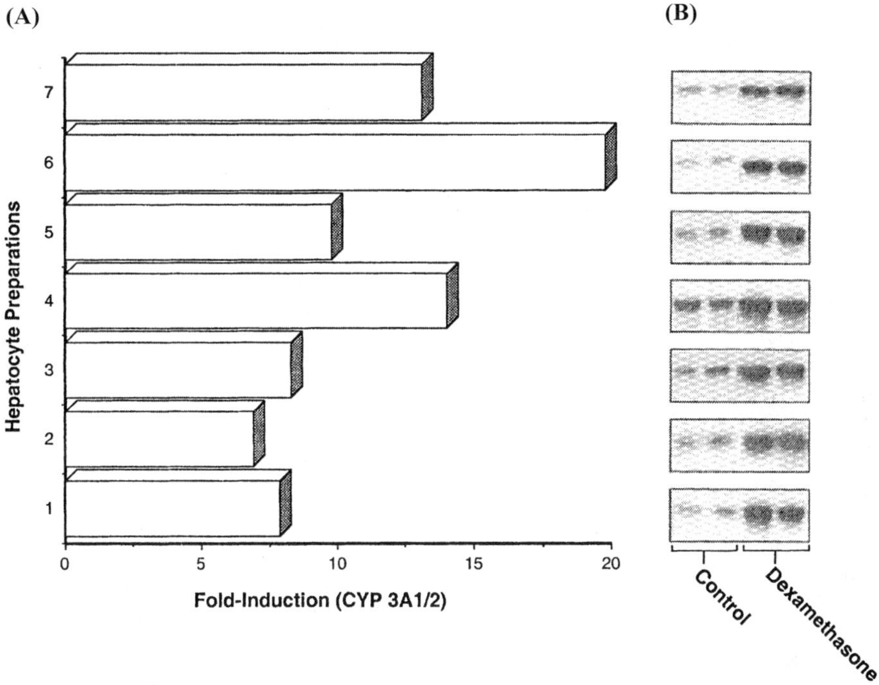

Figure 4 Variability in CYP3A induction in rat hepatocytes from seven different preparations. (**A**) Represents fold increase in CYP3A protein from dexamethasone-treated cells over nontreated cells. *Source*: From Ref. 12.

the potency indices of different drug candidates that may not have been incubated with the cells from the same donor.

In order to further validate this approach, we recently compared induction potency indices for a series of compounds in vivo, in rats, with those obtained in the rat hepatocyte model (12). As shown in Figure 5, results demonstrated an excellent correlation for CYP3A and CYP2B expression.

2. In Vitro Induction Screening

One of the drawbacks of using protein level measurements in hepatocytes for induction screening is the relatively large amount of time and labor required for cell harvest and preparation of samples for CYP analysis. In addition, large numbers of cells are required per dish. This is particularly undesirable when using human hepatocytes, an increasingly limited resource. The immediate challenge, therefore, is to modify the model to accommodate a higher-throughput format. In our laboratory this has been achieved by developing a 96-well format and using analytical methodology that allows for the measurement of CYP expression in fewer cells.

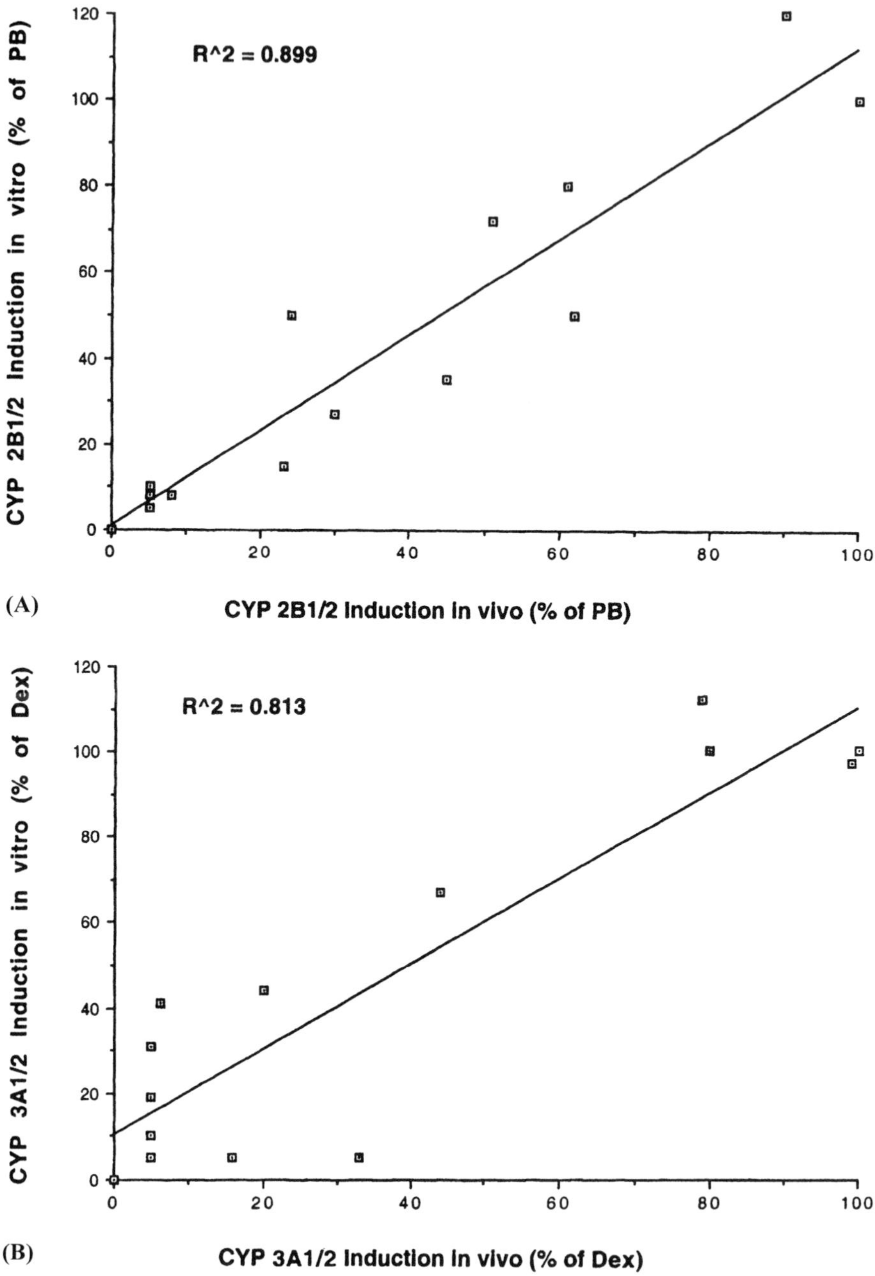

Figure 5 In vitro versus in vivo induction of CYP2B (**A**) and CYP3A (**B**). Cultured rat hepatocytes and Sprague Dawley rats were treated with 13 drug candidates at a dose of 50 μM and 400 mg/kg, respectively. Potency indexes for all the compounds in vitro were compared to ones found in vivo. *Source*: From Ref. 12.

In regards to culturing hepatocytes in a 96-well plate format, we have adopted the same conditions that we used when culturing cells in 60-mm dishes and 24-well plates (12) and simply scaled them down to a 96-well plate format. The 96-well plates are precoated with Matrigel and are commercially available (Collaborative Biomedical Products, Boston, Massachusetts, U.S.) or, alternatively, normal plates can be coated with diluted Matrigel and dried overnight (27). Hepatocytes cultured on collagen-coated 96-well plates have also been reported to be suitable for CYP induction (28).

With respect to higher-throughput analytical methodologies, we have taken two approaches. The first involves the addition of CYP-selective substrates to cell culture and measuring the formation of the relevant metabolites in the media (CYP activity assays). The second approach is to measure CYP mRNA levels using newly developed technologies compatible with 96-well culture formats.

a. CYP activity assays. The first example of using activity probes for determining CYP expression in intact hepatocytes cultured on 96-well plates was described by Donato et al. (28). In this study, the authors used two derivatives of phenoxazone, namely, 7-ethoxyresorufin (EROD) and 7-pentoxyresorufin (PROD), to determine activity of CYP1A1 and CYP2B1, respectively, in rat and human hepatocytes. These two compounds are specifically O-dealkylated to a highly fluorescent metabolite, resorufin. Therefore, in this assay the substrates are added directly to the cells in the presence of dicumarol, to prevent further reduction of the quinone moiety by DT-diaphorase, and incubated for a period of time (~30 minutes). Aliquots of the media are then transferred to microplates to fluorometrically determine amount of product (resorufin) formed. Because resorufin is also known to be further conjugated by glucuronic acid and sulfate in the intact cell, a mixture of β-glucuronidase and arylsulfatase is added to the microplate to hydrolyze either conjugate back to resorufin. Validation of this method was examined by comparing the results obtained in intact cultured hepatocytes with the activity determined in the microsomal fraction. An excellent correlation between the two assays was found for EROD ($r = 0.95$) and PROD ($r = 0.94$) activities (28).

The classical CYP3A probe is testosterone, which is known to undergo CYP3A4-dependent 6β-hydroxylation (29). This probe has been well characterized and is widely used to determine CYP3A activity in human liver microsomes. Testosterone has also been used to determine CYP3A4 activity in human primary hepatocytes (as carefully described in Ref. 25). In our laboratory we have used both Western blot analysis and testosterone 6β-hydroxylation activity assays to determine CYP3A4 induction in human hepatocytes and have found good agreement (Fig. 6). However, HPLC or LC/MS analysis is required for the quantification of the 6β-hydroxytestosterone metabolite, resulting in a tedious and time-consuming assay. Two other probes, benzyloxyquinoline and benzyloxytrifluorocoumarin (BFC), have also been identified as potential CYP3A fluorescent probes (30). A recent study by Price et al. demonstrated that BFC is metabolized in microsomes from cells expressing recombinant human CYP1A,

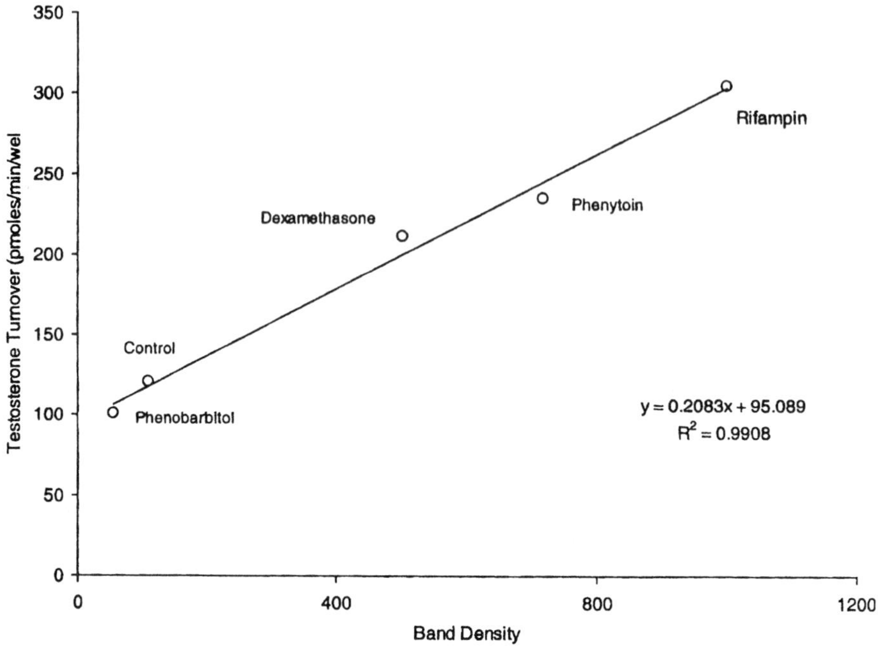

Figure 6 Correlation between testosterone 6β-hydroxylation and CYP3A protein levels, as determined by Western blot, in human hepatocytes incubated with several prototypical inducers.

CYP2B, CYP2C, and CYP3A isoforms. In primary rat hepatocytes, however, BFC was shown to be a good substrate for assessing the induction of CYP1A, CYP1A2, and CYP2B1 isoforms but not CYP3A (31).

A recent paper by Chauret et al. described the discovery of a novel fluorescent probe that is selectively metabolized by CYP3A in human liver microsomes (32). This probe, DFB [3-[(3,4-difluorobenzyl)oxy]-5,5-dimethyl-4-[4-(methylsulfonyl) phenyl]furan-2(5*H*)-one], is metabolized to DFH [3-hydroxy-5,5-dimethyl-4-[4-(methylsulfonyl)phenyl]furan-2(5*H*)-one], which has fluorescent characteristics (Fig. 7). In vitro CYP reaction phenotyping studies (cDNA-expressed CYP proteins and immunoinhibition experiments with highly selective anti-CYP3A4 antibodies) demonstrated that DFB was metabolized primarily by CYP3A4 (Fig. 8). Furthermore, metabolism studies performed with human liver microsomes obtained from different donors indicated that DFB dealkylation and testosterone 6β-hydroxylation correlated well (Fig. 9).

In our laboratory we have further characterized the use of this probe for assessing CYP3A4 induction in cultured human hepatocytes (33,34). In this assay, hepatocytes cultured in 96-well plates are incubated with DFB for 15 minutes. An aliquot of the media is then transferred to a microplate and DFH

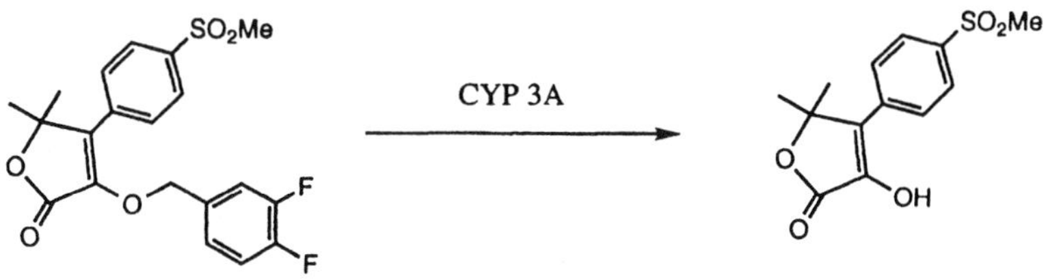

Figure 7 Metabolic pathway for DFB. *Abbreviation*: DFB, [3-[(3,4-difluorobenzyl)oxy]-5,5-dimethyl-4-[4-(methylsulfonyl)phenyl]furan-2(5*H*)-one].

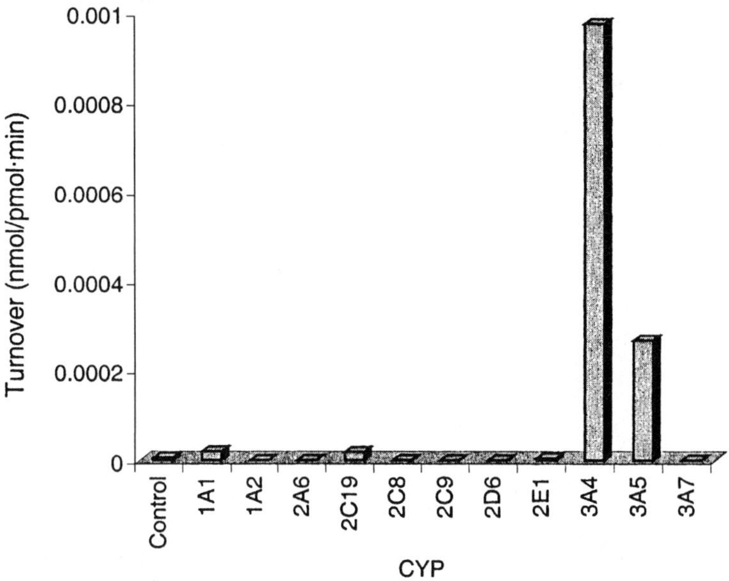

Figure 8 Turnover of DFB to DFH in microsomes prepared from cell lines expressing a single CYP. *Abbreviations*: DFB, [3-[(3,4-difluorobenzyl)oxy]-5,5-dimethyl-4-[4-(methylsulfonyl)phenyl]furan-2(5*H*)-one]; DFH, [3-hydroxy-5,5-dimethyl-4-[4-(methylsulfonyl) phenyl]furan-2(5*H*)-one]; CYP, cytochrome P450. *Source*: From Ref. 32.

quantified using a fluorescent plate reader. During the course of the reaction, the fluorescent metabolite DFH is not metabolized and there is no need for further manipulation of the sample. Figure 10 shows the correlation of CYP3A4 activity obtained with DFB and testosterone in human hepatocytes treated with several inducers. The DFB assays afford a quick and simple readout of CYP3A4 activity. Furthermore, because the cells are not adversely affected, multiple assays can be performed at different times. Indeed, it may even be possible to use the same cells to test more than one compound after an adequate washout period. Ferrini et al. have described culture conditions to maintain human hepatocytes

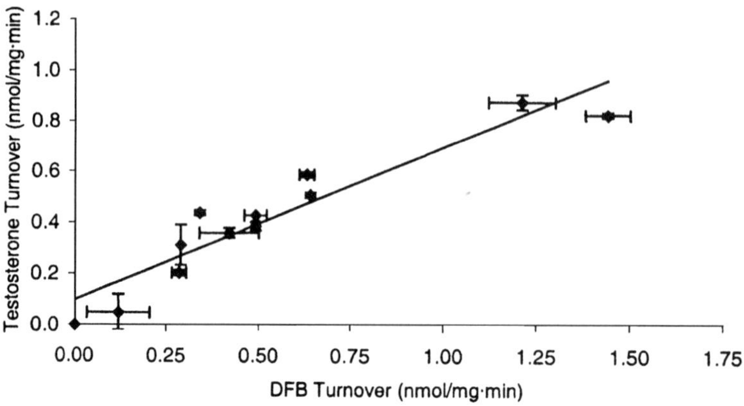

Figure 9 Correlation between testosterone 6β-hydroxylation and DFB debenzylation in various human liver microsomes. *Abbreviation*: DFB, [3-[(3,4-difluorobenzyl)oxy]-5,5-dimethyl-4-[4-(methylsulfonyl)phenyl]furan-2(5*H*)-one]. *Source*: From Ref. 32.

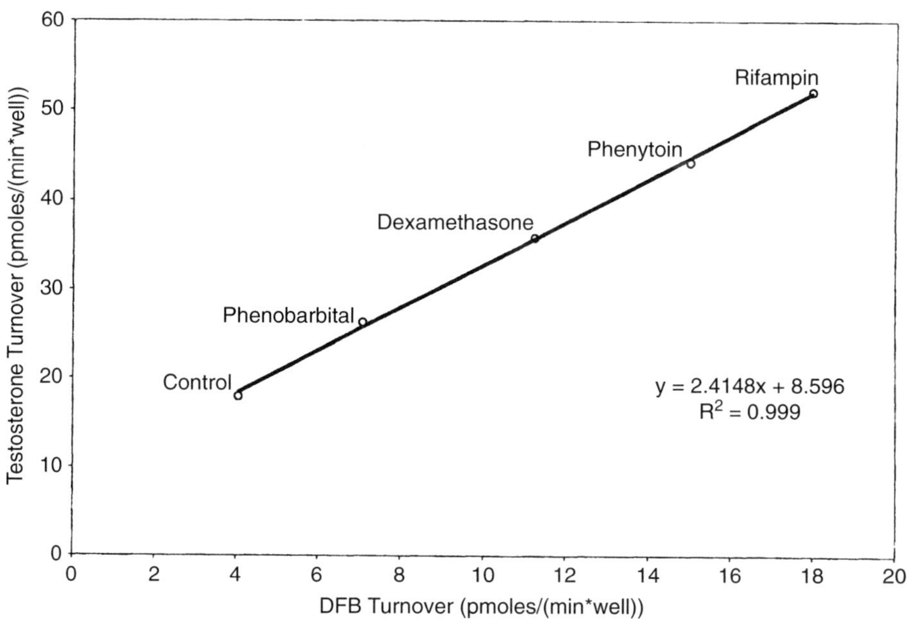

Figure 10 Correlation between testosterone 6β-hydroxylation and DFB debenzylation in human hepatocytes treated with several prototypical inducers. *Abbreviation*: DFB, [3-[(3,4-difluorobenzyl)oxy]-5,5-dimethyl-4-[4-(methylsulfonyl)phenyl]furan-2(5*H*)-one].

for several weeks while retaining CYP inducibility (35). In this latter study, the authors demonstrated that CYP3A4 expression would return to basal levels after removal of the inducer, at which time another round of testing by other compounds could be attempted.

Some compounds may have inducing properties as well as being mechanism-based inhibitors of the same CYP enzyme. It is therefore prudent, when analyzing CYP expression with activity probes, to verify that the compound being tested does not inhibit the CYP enzyme activity. In our laboratory we routinely test for this by incubating cells for approximately two hours with the test compound prior to measurement of CYP enzyme activity. If, after thoroughly washing the cells, the activity in induced cells is reduced, it is likely that inhibition of CYP3A4 has occurred. This will indicate whether the test compound is a time-dependent inhibitor.

b. mRNA analysis. Quantitative real-time reverse transcriptase-polymerase chain reaction (RT-PCR): In addition to using immunodetection of apoprotein and substrate metabolism, it is possible to screen for induction by analyzing expression of CYP mRNA. However, such methodology does not detect CYP induction resulting from posttranslational stabilization of proteins. An example of this latter case is with induction of CYP2E1 by isoniazid where an increase in mRNA is not observed (36).

Precise quantification of mRNA expression is difficult using conventional methods such as Northern blotting. By comparison, RT-PCR is more quantitative; however, the methodology is not suitable for a reasonable throughput (37,38) and may lead to semiquantitative data (39). Recently, a more efficient technology for precise analysis of mRNA has been developed in the form of real-time RT-PCR (40,41). The assay is based on the use of a 5′-nuclease assay and the detection of fluorescent PCR products (42). The method uses the 5′-nuclease activity of Taq polymerase to cleave a nonextendable oligonucleotide probe that hybridizes to the target cDNA and is labeled with a fluorescent reporter dye [6-FAM (6-carboxyfluorescein)] on the 5′ end and a quencher dye on the 3′ end [TAMRA (6-carboxy-tetramethyl-rhodamine)] (Fig. 11). When the probe is intact, the fluorescent signal is quenched due to the close proximity of the fluorescent and quencher dyes. However, during PCR, the nuclease degradation of the hybridization probe releases the quenching of the reporter dye, resulting in an increase in fluorescent emission. Real-time analysis of fluorescent products after each PCR cycle is determined using the ABI Prism 7700 Sequence Detection System (PE Biosystems, Foster City, California, U.S.). The PCR amplification is done in a 96-well plate format, and accumulation of PCR products is determined in real time by fluorescence detection. The mRNA copy number of the targeted gene is obtained by determining the PCR threshold cycle number generated when the fluorescent signal reaches a threshold value (42). Commonly, induction of gene expression is obtained by comparing fold increase of targeted mRNA from treated cells over mRNA from untreated cells.

The application of this technology for determining CYP induction in primary hepatocytes was first described by Strong et al. (40). To further demonstrate the potential of this technology, a study was conducted in our laboratory using primary human primary hepatocytes cultured on a 96-well plate precoated

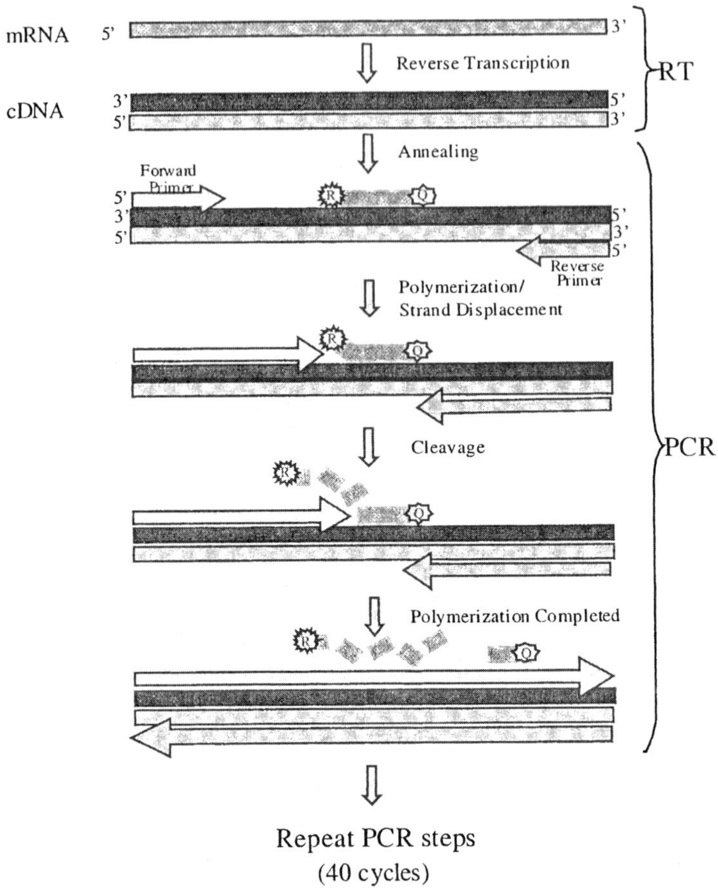

Figure 11 TaqMan RT-PCR assay. *Abbreviation*: RT-PCR, real-time reverse transcriptase-polymerase chain reaction.

with Matrigel (34). Cells were treated with increasing doses of rifampicin (0.08–50 μM) and at various intervals (3, 6, 12, and 24 hours). CYP3A4 activity was assessed with DFB prior to RNA isolation. CYP3A4, CYP3A5, and CYP3A7 mRNA analysis using real-time TaqMan PCR was then conducted. As shown in Figure 12, induction of CYP3A4 activity was clearly demonstrated after 24 hours in a dose-dependent manner. However, CYP3A4 mRNA was markedly elevated three hours after rifampicin dosing and continued to increase over 24 hours (Fig. 13A). These results demonstrate not only that rifampicin causes induction in CYP3A4 mRNA leading to a concomitant increase in CYP3A4 activity but also that the increase in mRNA is a much earlier event compared to alteration in enzyme activity. In contrast to CYP3A4 mRNA, CYP3A5 and CYP3A7 mRNA were not significantly elevated by rifampicin (Fig. 13B, C). This clearly

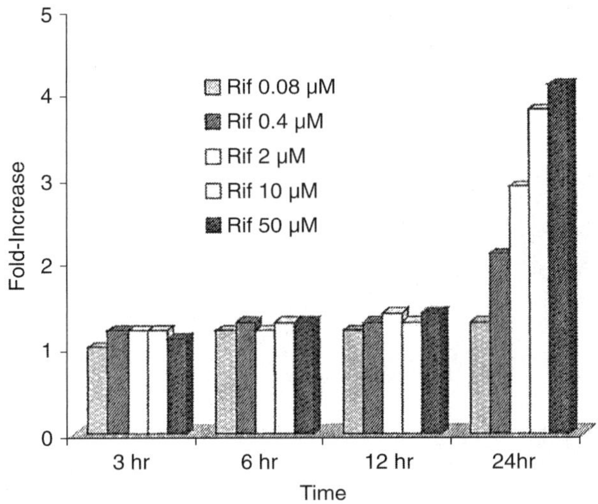

Figure 12 Dose- and time-dependent induction of CYP3A activity in human cultured hepatocytes incubated with rifampicin. Cells cultured on Matrigel-coated 96-well plates were incubated with increasing doses of rifampicin and CYP3A was determined by probing the cells with DFB prior to RNA isolation. *Abbreviation*: DFB, [3-[(3,4-difluorobenzyl)oxy]-5,5-dimethyl-4-[4-(methylsulfonyl)phenyl]furan-2(5*H*)-one].

demonstrates the advantage of this technology in its ability to differentiate between closely related CYPs. A recent study by Bowen et al. has also demonstrated the use of quantitative real-time PCR to measure the expression of CYP1A1 and CYP3A4 in human hepatocytes (41). The more conventional analytical methodology exemplified by Western blot and substrate probes lack the sensitivity and selectivity to profile all CYPs. Another advantage of this method is the ability to store the isolated mRNAs in order to perform further analysis of other genes at a later date. The cells, however, are terminated at the end of the experiment and, therefore, cannot be recycled for further studies.

c. Ribonuclease protection assays. Surry et al. reported a new assay to quantify mRNA levels for CYP isoforms 1A1, 1A2, 3A, and 4A1 in rat hepatocytes (43). This assay uses a set of oligonucleotide probes end labeled with [^{35}S]dATP to hybridize to mRNA in rat hepatocytes cultured on Cytostar-T 96-well scintillating microplates precoated with Matrigel. After treating the cells with potential inducers, hepatocytes were fixed with formaldehyde followed by in situ hybridization with specific [^{35}S] ATP-labeled oligonucleotide probes developed to hybridize to specific sites on CYP mRNA. While the probes for CYP1A1, CYP1A2, and CYP4A1 were selective, the set for CYP3A did not discriminate between CYP3A1, CYP3A2, CYP3A18, and CYP3A23. In this

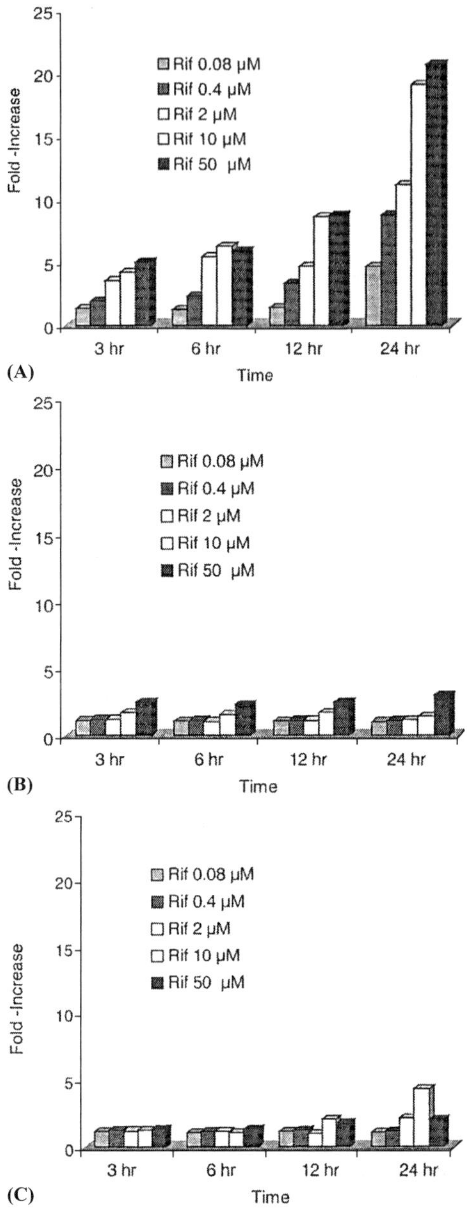

Figure 13 Dose- and time-dependent induction of CYP3A4 (**A**), CYP3A5 (**B**), and CYP3A7 (**C**) mRNAs in human cultured hepatocytes incubated with rifampicin. Cells cultured on Matrigel-coated 96-well plates were incubated with increasing doses of rifampicin, and RNA was harvested at times indicated. Specific CYP mRNAs were determined by TaqMan RT-PCR. *Abbreviations*: CYP, cytochrome P450; RT-PCR, real-time reverse transcriptase-polymerase chain reaction.

study, the authors demonstrated that the CYP3A mRNA levels obtained in rat hepatocytes treated with various compounds correlated well with testosterone 6β-hydroxylase activities in hepatic microsomes from in vivo studies (43).

The advantage of such a technique is that mRNA does not need to be isolated. The procedure, from culture to hybridization to detection, takes place within a single 96-well plate. A limitation to this assay is that only one CYP may be analyzed per well and samples cannot be stored for analysis of other genes at a later date.

d. bDNA technology. Another recently dxeveloped technology to measure CYP mRNA levels is the branched DNA (bDNA) signal amplification assay (44,45). This technology involves a nonpolymerase chain reaction and nonradioactive detection method resembling the enzyme-linked immunosorbent assay (ELISA). One of the advantages of this assay is the capability to use total RNA or cell extract for the analysis. The assay comprises a multiple of oligonucleotides to capture the mRNA of interest (see Ref. 45 for details). Three types of hybrid target probes are used and include capture probes, label extender, and blocker probes. Capture probes are designed so that a portion hybridizes to an oligonucleotide that is fixed to the well surface of a 96-well plate and another portion hybridizes to the target mRNA. Label extender oligonucleotide probes are designed so that a portion hybridizes to the mRNA target and the other portion hybridizes to the bDNA molecule that is essential for the amplification of the hybridization signal. Blocker oligonucleotide robes fill in the gaps in the mRNA between the capture and the label extender probes (which minimizes RNase-mediated mRNA degradation). Detection of target CYP mRNA is then accomplished by adding an enzyme (alkaline phosphatase) conjugated to an oligonucleotide, which hybridizes to the branches of the bDNA molecule. On addition of a substrate, dioxetane, a chemiluminescent signal is produced and measured. This technology has recently been used to analyze the expression of CYP1A1, CYP1A2, CYP2B1/2, CYP2E1, CYP3A1/23, and CYP4A2/3 in rats treated with classical enzyme-inducing compounds (45).

B. Slices

Tissue slices have been used for several decades to study basic pathways of intermediary metabolism as well as hepatotoxicity (46–48). However, procurement of the slices was performed by handheld instruments, and therefore the quality of the slices tended to vary between different preparations as well as between different laboratories (49). It was not until 1985 that the first paper described the development of a mechanical tissue slicer, where it was possible to obtain reproducible slices of specified thickness (50). More recently, this model has gained popularity and acceptance. There are now two commercially available instruments to produce slices of reproducible thickness, the Krumdieck slicer (Alabama Research and Development Corp., Munford, Alabama, U.S.) and the Brendel-Vitron slicer (Vitron Inc., Tucson, Arizona, U.S.) (see Ref. 49 for a review). Although liver slices have been widely used for drug metabolism and toxicity studies, their use for CYP induction studies have been limited. Several groups have shown that it is possible to culture slices for several days while retaining CYP inducibility (51–54). To overcome the problems associated with long-term culture of tissue slices, a roller culture system has been developed that allows the upper and lower surfaces of the cultured slice to be exposed to the gas phase during the course of incubation (49). Precision-cut rat and human liver slices cultured in this way are reported to survive for up to 72 hours while still retaining CYP inducibility (55). The same authors demonstrated induction of CYP2B1/2 and CYP3A in rat liver slices when treated with phenobarbital, CYP1A2 when treated with β-naphthoflavone, and CYP1A2, CYP2B1/2, and CYP3A when treated with Aroclor 1254 (55). In cultured human liver slices, rifampicin has also been shown to induce CYP 3A4 (56).

This model offers the advantage of maintaining tissue architecture and cell-to-cell communication. Moreover, slices may be prepared from a range of tissues, including liver, heart, kidney, lung, and spleen, from laboratory animals and humans. The main disadvantage of this model is in the handling of the slices. Because a complex culture system is required, the number of samples that can be handled at any one time is limited. The model is also not amenable to automation, unlike a 96-well cell-culture format. In addition, slices have a limited lifespan in culture (~7 days), and several investigators have expressed concerns about the ability of a test compound to penetrate through a layer of damaged cells to reach viable cells (57,58).

C. Reporter Gene Constructs

Kliewer et al. first identified a new member of the steroid/thyroid receptor family termed PXR (pregnane X receptor) to be responsible for mediating the activation of CYP3A gene expression (59). Their conclusions were based on three lines of evidence. First, both dexamethasone and pregnenolone 16α-carbonitrile (PCN) were potent activators of PXR; second, PXR binds as a heterodimer with RXR (retinoic acid receptor) to the conserved DR-3 motifs in the CYP3A23 and CYP3A2 gene promoters; and finally, PXR was found to be tissue selective, expressed mainly in liver, intestine, and kidney. These tissues are the ones reported to express inducible CYP3A genes in response to both dexamethasone and PCN (60). This was immediately followed by another report by the same group (61) with the identification of a human PXR that bound to the rifampicin response element in the CYP3A4 promoter as a heterodimer with RXR. Comparison of the human PXR with the recently cloned mouse PXR revealed significant differences in their activation by several drugs. This further supports the hypothesis that the molecular reasoning for the observed species differences in CYP3A expression was due to species specificity in PXR. Furthermore, with the cloning of PXRs from other species, including rabbit and rat, observed species-specific xenobiotic activation of CYP3A in vivo and in primary hepatocytes has so far correlated with the activation of PXR in vitro (62). This prompted the suggestion to use PXR binding and activation assays to assess the potential for drug candidates to induce CYP3A. A recent study on St. John's wort, a herbal remedy used for the treatment of depression, demonstrated that hyperforin, a constituent of St. John's wort, induced CYP3A4 in human primary hepatocytes and was a potent PXR ligand (63). Moreover, CV-1 cells cotransfected with an expression vector for human PXR and a reporter gene resulted in the activation of PXR comparable to that achieved with rifampicin. These data further support the hypothesis that PXR is a key regulator of CYP3A expression in different species. However, it's important to note that activation of PXR may not be the only possible mechanism resulting in the induction of CYP3A.

These type of assays have great potential as screening tools for CYP induction without having to use valuable human hepatocytes. The human hepatocyte model could be reserved to confirm results of a lead compound after exhaustive screening with such reporter gene construct models.

III. CASE STUDY

An example of an in vitro–in vivo correlation was recently demonstrated in a study involving autoinduction (64). The major oxidative pathway of a cyclooxygenase-2 inhibitor, DFP [(5,5-dimethyl-3-(2-propoxy)-4-(4-methanesulfonylphenyl)-2(5*H*)-furanone)], gives rise to DFH, a dealkylated product (Fig. 14). This process is mediated by CYP3A enzymes in the rat, as demonstrated by incubations of DFP with hepatic microsomes from rats treated with dexamethasone (CYP3A23) and with recombinant rat CYP3A1 and CYP3A2. DFP is also a potent inducer of CYP3A in the rat hepatocyte induction model as measured by Western blot or enzyme activity, using both testosterone and DFP as probe substrates (Fig. 15). Thus, the CYP3A-mediated pathway of DFP was induced in hepatocytes that had been treated for 48 hours with 2-, 10-, and 50-μM DFP in a dose-dependent manner. In vivo rat pharmacokinetic studies at oral doses of 10, 30, and 100 mg/kg gave C_{max} concentrations of circa 20, 40, and 80 μM, respectively (Fig. 16A), indicating that the in vitro concentrations approximated in vivo concentrations. On the basis of these data, it was predicted that autoinduction should occur in vivo, giving rise to altered pharmacokinetic parameters such as lowered C_{max} and area under the plasma concentration–time curve (AUC) values.

Induction of rat CYP3A was confirmed in vivo by dosing rats with DFP at 100 mg/kg for four days. Microsomes prepared from the excised livers showed that DFP gave 55% (±7% S.D., $n = 4$) of the induction observed with dexamethasone, as determined by Western blot analysis. In vivo treatment of rats with DFP (10–100 mg/(kg·day) for 13 weeks) indicated that DFP induced its own metabolism. The C_{max} and plasma drug AUC values during the 13th week were significantly lower than that on the first day, and the effect was dose dependent (Fig. 16). Thus, at the lowest dose, the changes in C_{max} and AUC were modest or insignificant. However, at the top dose, reductions in both parameters were

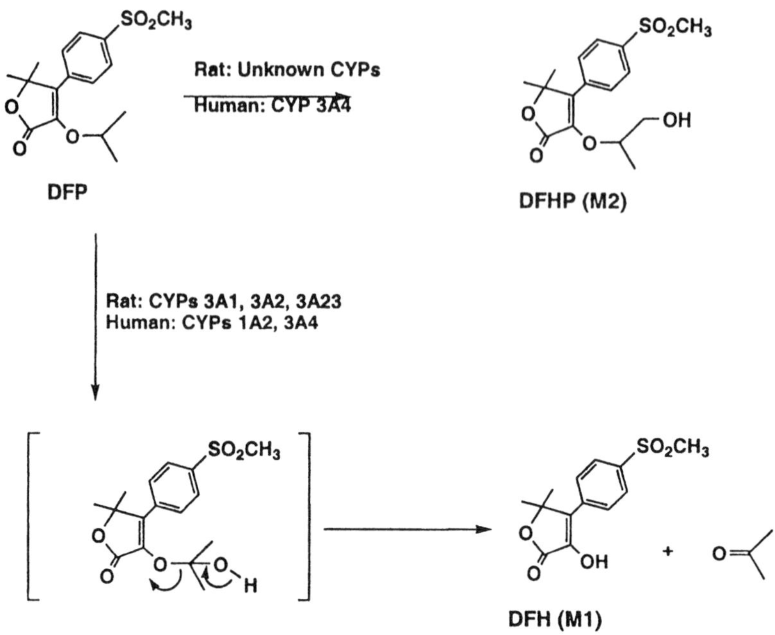

Figure 14 Proposed metabolic pathway for DFP. *Abbreviation*: DFP, [(5,5-dimethyl-3-(2-propoxy)-4-(4-methanesulfonylphenyl)-2(5*H*)-furanone)]. *Source*: From Ref. 64.

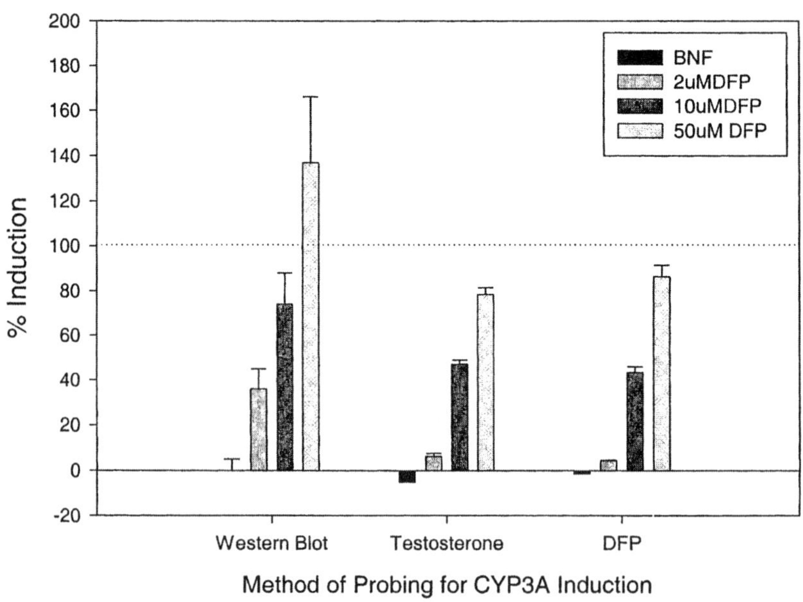

Figure 15 CYP3A potency indices of DFP in cultured rat hepatocytes as determined by Western blots, testosterone 6β-hydroxylation, and DFP turnover. *Abbreviation*: DFP, [(5,5-dimethyl-3-(2-propoxy)-4-(4-methanesulfonylphenyl)-2(5*H*)-furanone)]. *Source*: From Ref. 64.

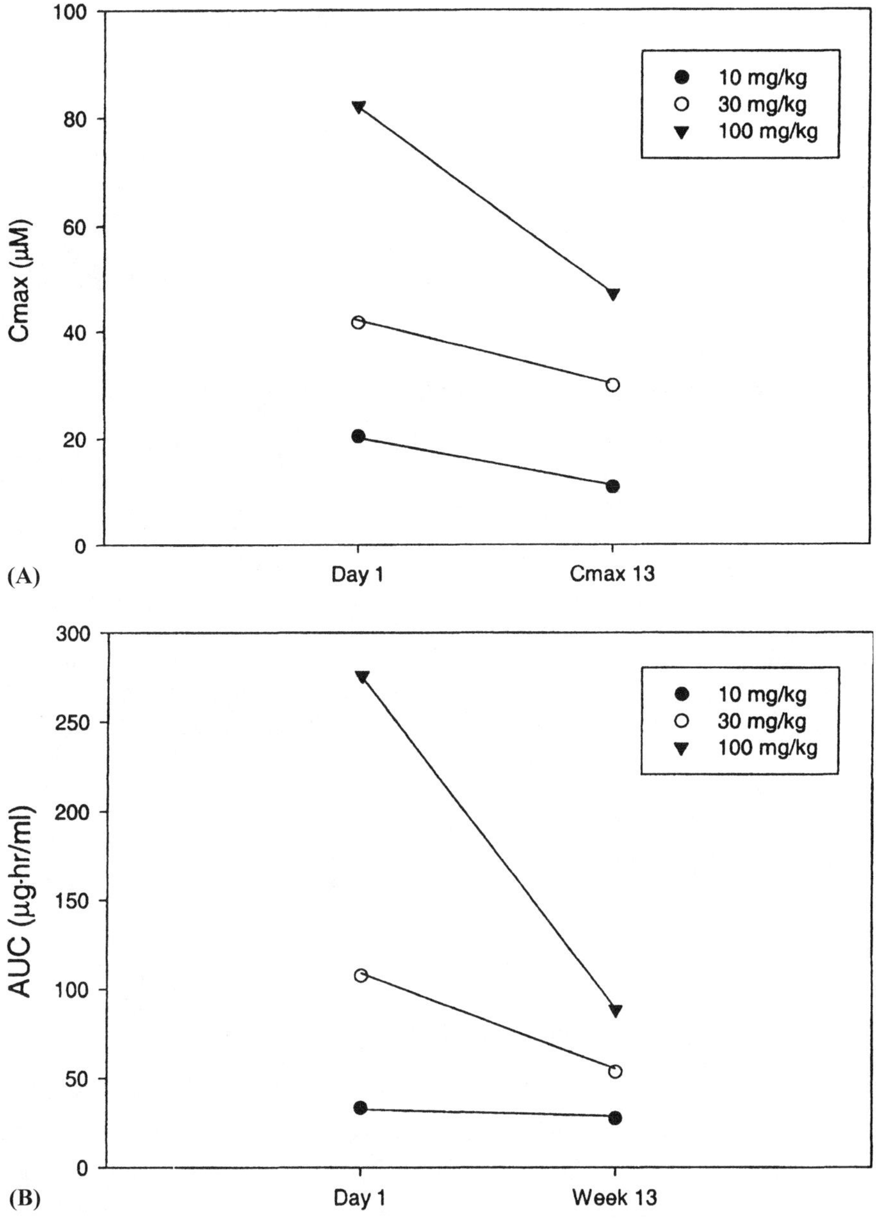

Figure 16 Changes to pharmacokinetic parameters over 13-week dosing regimen. (**A**) Maximum plasma concentration ($C_{\max}$) of DFP determined after single doses of 10, 30, or 100 mg/kg of DFP compared to the $C_{\max}$ after 13 weeks of dosing. (**B**) AUC after a single dose of 10, 30, or 100 mg/kg of DFP compared with the $C_{\max}$ after 13 weeks of dosing. *Abbreviations*: DFP, [(5,5-dimethyl-3-(2-propoxy)-4-(4-methanesulfonylphenyl)-2(5*H*)-furanone)]; AUC, area under the plasma concentration–time curve. *Source*: From Ref. 64.

marked. The C_{max} was reduced by about 50% and the AUC was reduced by about 80%. Thus, autoinduction was realized in vitro and in vivo. In both cases, the observed metabolic autoinduction is consistent with the hypothesis that it is caused by CYP3A induction.

Subsequent studies with DFP indicated that the human oxidative pathways were catalyzed by CYP3A and CYP1A, as demonstrated by turnover with recombinant CYP enzymes. Induction studies in the human hepatocyte model demonstrated that human CYP3A was not significantly induced by DFP. Therefore, despite the autoinduction observed in rats, this compound was carried forward into clinical trials.

IV. STRATEGIES FOR MANAGING POTENTIAL CYP INDUCERS

Figure 17 summarizes strategies for screening of and dealing with possible CYP inducers at the drug discovery and development stages. The first step involves incubating a lead compound in the human hepatocyte model to determine CYP potency indices over a wide range of concentrations. If human hepatocytes are not readily available, hepatocytes from a relevant species may be used. For example, rabbit hepatocytes appear to be a better predictor for CYP3A4 induction compared to rat hepatocytes (5,12). If results show that the compound is a potent inducer of a particular CYP, then other closely related analogs may be tested to determine the feasibility of reducing the induction potential within a chemical series. In cases where it may not be possible to identify a noninducing

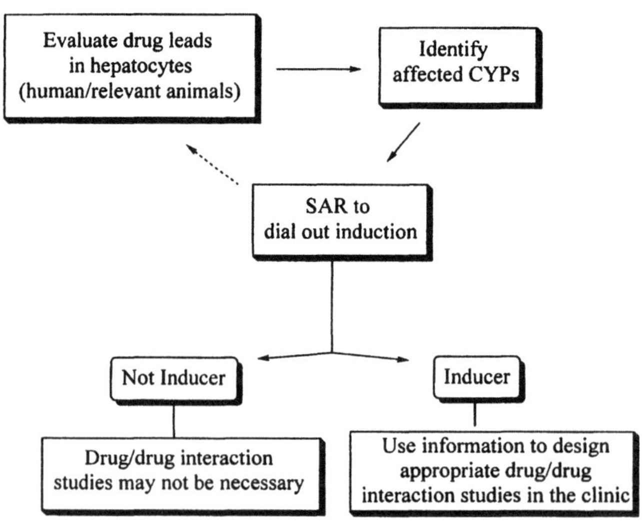

Figure 17 Strategies for dealing with CYP induction in drug discovery. *Abbreviation*: CYP, cytochrome P450.

analog, this information can be used as a guide to plan relevant drug-drug interaction studies in the clinic.

ACKNOWLEDGMENTS

The authors would like to thank Dr. Thomas Rushmore, Dr. Karen Richards, and Ms. Kristie Strong-Basalyga for their contribution in the development of the quantitative real-time reverse transcriptase-polymerase chain reaction assay for CYP analysis in primary hepatocytes.

REFERENCES

1. Wadhwa NK, Schroeder TJ, Pesce AJ, et al. Cyclosporin drug interactions: a review. Ther Drug Monit 1987; 9:399–406.
2. Whitlock JP, Denison MS. Induction of cytochrome P450 enzymes that metabolize xenobiotics. In: Ortiz de Montellano PR, ed. Cytochrome P450: Structure, Mechanism and Biochemistry. 2nd ed. New York: Plenum Press, 1995:367–390.
3. Breckenridge AM, Back DJ, Crawford FE, et al. Drug interactions with oral contraceptives: clinical and experimental studies. Int Congress Symposia Series—Royal Soc Med 1980; 31:1–11.
4. Guengerich FP. Oxidation of 17 alpha ethenylestradiol by human liver cytochrome P450. Mol Pharmacol 1988; 33:500–508.
5. Kocarek TA, Schuetz EG, Strom SC, et al. Comparative analysis of cytochrome CYP3A induction in primary cultures of rat, rabbit, and human hepatocytes. Drug Metab Dispos 1995; 23:415–421.
6. Howard RB, Christensen AK, Gibbs FA, et al. The enzymatic preparation of isolated intact parenchymal cells from rat liver. J Cell Biol 1967; 35:675–684.
7. Berry MN, Friend DS. High-yield preparation of isolated rat liver parenchymal cells: a biochemical and fine structural study. J Cell Biol 1969; 43:506–520.
8. Berry MN, Edwards AM, Barrit GK, eds. Isolated Hepatocytes: Preparation, Properties and Applications. New York: Elsevier, 1991.
9. Li AP, Roque MA, Beck DJ, et al. Isolation and culturing of hepatocytes from human livers. J Tissue Cult Methods 1992; 14:139–146.
10. Schuetz EG, Li D, Omiecinski CJ, et al. Regulation of gene expression in adult rat hepatocytes cultured on basement membrane matrix. J Cell Physiol 1988; 143: 309–323.
11. LeCluyse EL, Bullock PL, Parkinson A, et al. Cultured rat hepatocytes. Pharm Biotechnol 1996; 8:121–159.
12. Silva JM, Morin PE, Day SH, et al. Refinement of an in vitro cell model for cytochrome P450 induction. Drug Metab Dispos 1998; 26:490–496.
13. Maurel P. The use of adult human hepatocytes in primary culture and other in vitro systems to investigate drug metabolism in man. Adv Drug Dev Rev 1996; 22:105–132.
14. Seglen PO, Gordon PB, Poli A. Amino acid inhibition of autophagic/lysosomal pathway of protein degradation in isolated rat hepatocytes. Biochim Biophys Acta 1980; 630:103–118.

15. Enat R, Jefferson DM, Ruiz-Opaza N, et al. Hepatocyte proliferation in vitro: its dependence on the use of serum-free hormonally defined medium and substrata of extracellular matrix. Proc Natl Acad Sci U S A 1984; 81:1411–1415.
16. Waxman DJ, Morrissey JJ, Naik S, et al. Phenobarbital induction of cytochrome P450: high-level long-term responsiveness of primary rat hepatocyte cultures to drug induction, and glucocorticoid dependence of the phenobarbital response. Biochem J 1990; 271:113–119.
17. Sidhu JS, Omiecinski CJ. Modulation of xenobiotic-inducible cytochrome CYP gene expression by dexamethasone in primary rat hepatocytes. Pharmacogenetics 1995; 5:24–36.
18. Johansson S, Hook M. Substrate adhesion of rat hepatocytes: on the mechanism of attachment of fibronectin. J Cell Biol 1984; 98:810–817.
19. Reid LM, Gaitmaitan Z, Arias I, et al. Long-term cultures of normal rat hepatocytes on liver biomarkers. Ann N Y Acad Sci 1980; 349:70–76.
20. Dunn JC, Yarmush ML, Kowbe HG, et al. Hepatocyte function and extracellular matrix geometry: long-term culture in a sandwich configuration. FASEB J 1989; 3:174–177.
21. Sidhu JS, Farm FM, Omiecinski CJ. Influence of extracellular matrix overlay on phenobarbital-mediated induction of CYP2B1, 2B2, and 3Al genes in primary adult rat hepatocyte culture. Arch Biochem Biophys 1993; 30:103–113.
22. LeCluyse EL, Audus KL, JK Hochman. Formation of extensive canalicular networks by rat hepatocytes cultured on collagen-sandwich configuration. Am J Physiol 1994; 266:C1764–C1774.
23. LeCluyse EL, Madan A, Hamilton G, et al. Expression and regulation of cytochrome P450 enzymes in primary cultures of human hepatocytes. J Biochem Mol Toxicol 2000; 14:177–188.
24. Kocarek TA, Schuetz EG, Guzelian PS. Expression of multiple forms of cytochrome P450 mRNAs in primary cultures of rat hepatocytes maintained on Matrigel. Mol Pharmacol 1992; 43:328–334.
25. Kostrubsky VE, Ramachandran V, Venkataramanan R, et al. The use of human hepatocyte cultures to study the induction of cytochrome P-450. Drug Metab Dispos 1999; 27:887–894.
26. Chang TK, Yu L, Maurel P, et al. Enhance cyclosphosphamide and ifosfamide activation in primary human hepatocyte cultures: response of cytochrome P-450 inducers and autoinduction by oxazaphosphorines. Cancer Res 1997; 57: 1946–1954.
27. Surry DD, McAllister G, Meneses-Lorente G, et al. High-throughput ribonuclease protection assay for the determination of CYP3A mRNA induction in cultured rat hepatocytes. Xenobiotica 1999; 29:827–838.
28. Donato MT, Gomez-Lechon MJ, Castell JV. A microassay for measuring cytochrome P4501A1 and P4502B1 activities in intact human and rat hepatocytes cultured on 96-well plates. Anal Biochem 1993; 213:29–33.
29. Waxman DJ, Attisano C, Guengerich FP, et al. Human liver microsomal steroid metabolism: identification of the major microsomal steroid hormone 6 betahydroxylase cytochrome P-450 enzyme. Arch Biochem Biophys 1988; 263:424–436.
30. Crespi CL, Miller VP, Ackermann JM, et al. Novel high-throughput fluorescent P450 assays. Toxicologist 1999; 48:323 (abstr).

31. Price RJ, Surry D, Renwick RB, et al. CYP isoform induction screening in 96-well plates: use of 7-benzyloxy-4-trifluoromethylcoumarin as a substrate for studies with rat hepatocytes. Xenobiotica 2000; 30:781–795.
32. Chauret N, Tremblay N, Lackman R, et al. Description of a 96-well plate assay to measure cytochrome P4503A inhibition in human liver microsomes using a selective fluorescent probe. Anal Biochem 1999; 276:215–226.
33. Silva JM, Day S, Chauret N, et al. Fluorescent probe for CYP3A activity in human hepatocytes cultured on a 96-well plate. ISSX Proc 1999; 15:204.
34. Silva JM. HTS for assessing cytochrome P450 induction in primary hepatocytes. Drug Metab Rev 2000; 32:168.
35. Ferrini JB, Pilchard L, Domergue J, et al. Long-term primary cultures of adult human hepatocytes. Chem Biol Interact 1997; 107:31–45.
36. Park KS, Sohn DH, Veech RL, et al. Translational activation of ethanol-inducible cytochrome P450 (CYP2E1) by isoniazid. Eur J Pharmacol 1993; 248:7–14.
37. Greuet J, Pichard L, Ourlin JC, et al. Effect of cell density and epidermal growth factor on the inducible expression of CYP3A and CYP1A genes in human hepatocytes in primary culture. Hepatology 1997; 25:1166–1175.
38. Donato MT, Gomez-Lechon MJ, Jover R, et al. Human hepatocyte growth factor down-regulates the expression of cytochrome P450 isozymes in human hepatocytes in primary culture. J Pharmacol Exp Ther 1998; 284:760–767.
39. Murphy LD, Herzog CE, Rudick JB, et al. Use of the polymerase chain reaction in the quantitation of mdr-1 gene expression. Biochemistry 1990; 29:10351–10356.
40. Strong KL, Rushmore TH, Richards KM. Quantitation of human CYP3A4, 3A5 and 3A7 RNA levels in cultured primary hepatocytes using real-time RT-PCR. ISSX Proc 1999; 15:69.
41. Bowen WP, Carey J, Miah A, et al. Measurement of cytochrome P450 gene induction in human hepatocytes using quantitative real-time reverse transcriptase-polymerase chain reaction. Drug Metab Dispos 2000; 28:781–788.
42. Heid CA, Stevens J, Livak KJ, et al. Real time quantitative PCR. Genome Res 1996; 6:986–994.
43. Surry DD, Meneses-Lorente G, Heavans R, et al. Rapid determination of rat hepatocyte mRNA induction potential using oligonucleotide probes for CYP1A1, 1A2, 3A and 4A1. Xenobiotica 2000; 30:441–456.
44. Burris TP, Pelton PD, Zhou L, et al. A novel method for analysis of nuclear receptor function at natural promoters: peroxisome proliferator-activated receptor y agonist actions on a P2 gene expression detected using branched DNA messenger RNA quantitation. Mol Endocrinol 1999; 13:410–417.
45. Hartley DP, Klaassen CD. Detection of chemical-induced differential expression of rat hepatic cytochrome P450 mRNA transcripts using branched DNA signal amplification technology. Drug Metab Dispos 2000; 28:608–616.
46. Warburg O. Experiments on surviving carcinoma tissue. Methods Biochem Z 1923; 142:317–333.
47. Krebs HA. Body size and tissue respiration. Biochim Biophys Acta 1950; 4:249–269.
48. Fraga CG, Leibovitz BE, Tappel AL. Halogenated compounds as inducers of lipid peroxidation in tissue slices. Free Radic Biol Med 1987; 3:119–123.
49. Parrish AR, Gandolfi AJ, Brendel K. Minireview precision-cut tissue slices: applications in pharmacology and toxicology. Life Sci 1995; 57:1887–1905.

50. Smith PF, Gandolfi AJ, Krumdieck CL, et al. Dynamic organ culture of precision liver slices for in vitro toxicology. Life Sci 1985; 13:1367–1375.
51. Lake BG, Beamand JA, Japenga AC, et al. Induction of cytochrome P-450-dependent enzyme activities in cultured rat liver slices. Food Chem Toxicol 1993; 31:377–386.
52. Gokhale MS, Bunton TE, Zurlo J, et al. Cytochrome P-450 1 Al/1 A2 induction, albumin secretion, and histological changes in cultured liver slices. In Vitro Toxicol 1995; 8:357–368.
53. Drahushuk AT, McGarrigle BP, Tai HL, et al. Validation of precision-cut liver slices in a dynamic organ culture as an in vitro model for studying CYP1Al and CYP1A2 induction. Toxicol Appl Pharmacol 1996; 140:393–403.
54. Muller D, Glockner R, Rost M. Monooxygenation, cytochrome P4501Al and P4501Al-mRNA in rat liver slices exposed to beta-naphthoflavone and dexamethasone in vitro. Exp Toxicol Pathol 1996; 48:433–438.
55. Lake BG, Charzat C, Tredger JM, et al. Induction of cytochrome P450 isoenzymes in cultured precision-cut rat and human liver slices. Xenobiotica 1996; 26:297–306.
56. Lake BG, Ball SE, Renwick AB, et al. Induction of CYP3A isoforms in cultured precision-cut human liver slices. Xenobiotica 1997; 27:1165–1173.
57. Ekins S, Murray GI, Williams JA, et al. Quantitative differences in phase I and II enzyme activities between rat precision-cut liver slices and isolated hepatocytes. Drug Metab Dispos 1995; 23:1274–1279.
58. Ekins S, Williams JA, Murray GI, et al. Xenobiotic metabolism in rat, dog and human precision-cut liver slices, freshly isolated hepatocytes, and vitrified precision-cut liver slices. Drug Metab Dispos 1996; 24:990–995.
59. Kliewer SA, Moore JT, Wade L, et al. An orphan nuclear receptor activated by pregnanes defines a novel steroid signaling pathway. Cell 1998; 92:73–82.
60. Debri K, Boobis AR, Davis DS, et al. Distribution and induction of CYP3Al and CYP3A2 in rat liver and extrahepatic tissues. Biochem Pharmacol 1995; 50: 2047–2056.
61. Lehmann JM, McKee DD, Watson MA, et al. The human orphan nuclear receptor PXR is activated by compounds that regulate CYP3A4 gene expression and cause drug interactions. J Clin Invest 1998; 102:1016–1023.
62. Jones SA, Moore LB, Shenk JL, et al. The pregnane X receptor: a promiscuous xenobiotic receptor that has diverged during evolution. Mol Endocrinol 2000; 14:27–39.
63. Moore LB, Goodwin B, Jones SA, et al. St. John's wort induces hepatic drug metabolism through activation of the pregnane X receptor. Proc Natl Acad Sci U S A 2000; 97:7500–7502.
64. Nicoll-Griffith DA, Silva JM, Chauret N, et al. Application of rat hepatocyte culture to predict in vivo metabolic auto-induction: studies with DEP, a cyclooxygenase-2 inhibitor. Drug Metab Dispos 2001; 29:159–165.

7

In Vitro Approaches for Studying the Inhibition of Drug-Metabolizing Enzymes and Identifying the Drug-Metabolizing Enzymes Responsible for the Metabolism of Drugs (Reaction Phenotyping) with Emphasis on Cytochrome P450

Brian W. Ogilvie, Etsuko Usuki, Phyllis Yerino, and Andrew Parkinson
XenoTech LLC, Lenexa, Kansas, U.S.A.

I. INTRODUCTION

From a drug interaction and pharmacogenetic perspective, drugs can be evaluated for their victim and perpetrator potential. *Victims* are those drugs whose clearance is predominantly determined by a single route of elimination, such as metabolism by a single cytochrome P450 (CYP) enzyme. Such drugs have a high victim potential because a diminution or loss of that elimination pathway, either due to a genetic deficiency in the relevant CYP enzyme or due to its inhibition by another concomitantly administered drug, will result in a large decrease in clearance and a correspondingly large increase in exposure to the victim drug (e.g., area under the plasma concentration–time curve or AUC). *Perpetrators* are those drugs (or other environmental factors) that inhibit or induce the enzyme that is otherwise responsible for clearing a victim drug. Genetic polymorphisms that result in the partial or complete loss of enzyme activity (i.e., the intermediate and poor metabolizer genotypes) can also be viewed as perpetrators because they have the same effect as an enzyme inhibitor, i.e., they cause a decrease in the clearance of—and an increase in exposure to—victim drugs. Likewise, genetic polymorphisms that result in the overexpression of enzyme activity [i.e., the ultrarapid metabolizer (UM) genotype] can be viewed as perpetrators because they have the same effect as an enzyme inducer: they cause an increase in the clearance of—and a decrease in exposure to—victim drugs. Several drugs, whose elimination is largely determined by their metabolism by CYP2C9, CYP2C19, or CYP2D6, three genetically polymorphic enzymes, are victim drugs because their clearance is diminished in poor metabolizers (PMs), i.e., individuals who are genetically deficient in one of these enzymes. Drugs whose disposition is dependent on uptake or efflux by a transporter or on metabolism by a drug-metabolizing enzyme other than CYP can also be considered from the victim/perpetrator perspective. From a drug interaction perspective, victim drugs are also known as *objects*, whereas perpetrators are also known as *precipitants*.

This chapter focuses on in vitro reaction phenotyping (also known as enzyme mapping) and CYP inhibition. Reaction phenotyping is the process of identifying the enzyme or enzymes that are largely responsible for metabolizing a drug candidate. When biotransformation is known or suspected to play a significant role in the clearance of a drug candidate (which applies to most drug candidates with a log $D_{7.4} \geq 1.0$), then reaction phenotyping is required prior to approval by the FDA and other regulatory agencies (1–6). Reaction phenotyping allows an assessment of the victim potential of a drug candidate. The FDA also requires drug candidates be evaluated for their potential to inhibit the major CYP enzymes involved in drug metabolism. This allows an assessment of the perpetrator potential of the drug candidate. Drugs can also cause pharmacokinetic drug interactions by inducing CYP and other drug-metabolizing enzymes and/or drug transporters. The assessment of drugs as enzyme inducers is reviewed in Chapter 6.

Terfenadine, cisapride, astemizole, and cerivastatin are all victim drugs, so much so that they have all been withdrawn from the market or, in the case of cisapride, made available only with severe restrictions. The first three are all victim drugs because they are extensively metabolized by CYP3A4. Inhibition of CYP3A4 by various antimycotic drugs, such as ketoconazole, and antibiotic drugs, such as erythromycin, decrease the clearance of terfenadine, cisapride, and astemizole, and increase their plasma concentrations to levels that, in some individuals, cause ventricular arrhythmias (QT prolongation and torsade de pointes), which can result in fatal heart attacks. Cerivastatin is extensively metabolized by CYP2C8. Its hepatic uptake by OATP1B1 and CYP2C8-mediated metabolism is inhibited by gemfibrozil (actually by gemfibrozil glucuronide), and the combination of cerivastatin (Baycol®) and gemfibrozil (Lopid®) was associated with a high incidence of fatal, cerivastatin-induced rhabdomyolysis, which prompted the worldwide withdrawal of cerivastatin (7–11).

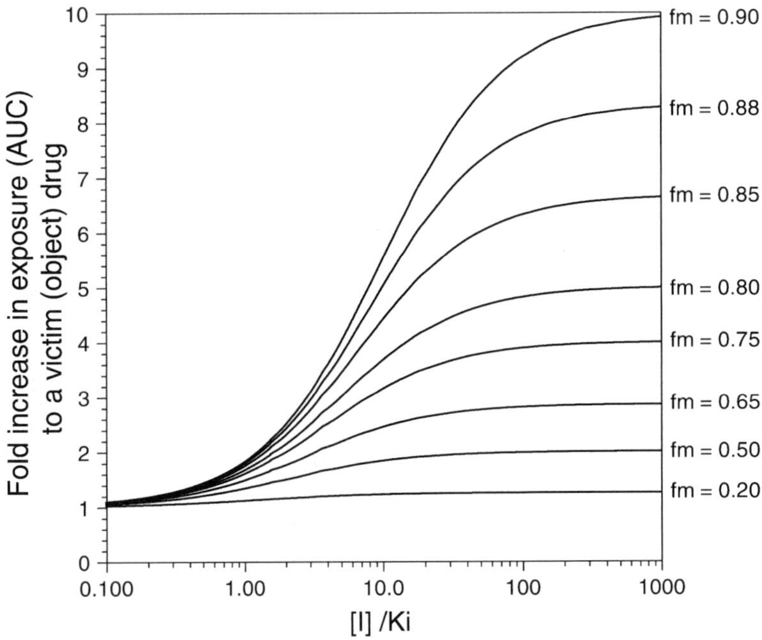

Figure 1 The effect of fractional metabolism by CYP, $f_{m(CYP)}$, on the theoretical increase in exposure to a victim drug with increasing inhibition by a perpetrator drug based on $[I]/K_i$ values. *Abbreviation*: CYP, cytochrome P450.

Posicor® (mibefradil) is the only drug withdrawn from the U.S. market largely because of its perpetrator potential. This calcium channel blocker not only caused *extensive* inhibition of CYP3A4, it also caused *prolonged* inhibition of the enzyme by virtue of being a metabolism-dependent inhibitor of CYP3A4. By inactivating CYP3A4 in an irreversible manner, so that restoration of normal CYP3A4 activity required the synthesis of new enzyme, mibefradil inhibited CYP3A4 long after treatment with the drug was discontinued.

Victim potential can be quantified on the basis of fractional metabolism according to the following equation:

$$\text{Fold increase in exposure} = \frac{\text{AUC}_\text{i} \text{ or AUC}_\text{PM}}{\text{AUC}_\text{ui} \text{ or AUC}_\text{EM}} = \frac{1}{1 - f_\text{m}} \qquad (1)$$

where AUC_i (AUC inhibited) and AUC_{ui} (AUC uninhibited) are the plasma AUC values in the presence and absence of inhibitor, respectively, and where AUC_{PM} and AUC_{EM} are plasma AUC values in genetically determined PMs and extensive metabolizers (EMs), respectively. The relationship between fractional metabolism by a single enzyme and the fold increase in drug exposure that results from the loss of that enzyme is shown in Figure 1, which shows that the relationship is not a linear one. Loss of an enzyme that accounts for 50% of a drug's clearance ($f_m = 0.5$) results in a twofold increase in AUC, whereas it results in a 10-fold increase in AUC when the affected enzyme accounts for 90% of a drug's clearance ($f_m = 0.9$). In the case of oral drugs that undergo significant presystemic clearance (i.e., first-pass metabolism in the intestine and/or liver), the impact of enzyme inhibition (or the PM genotype) can be twofold: it can increase AUC of the victim drug by (1) decreasing presystemic clearance (which increases oral bioavailability) and (2) decreasing systemic clearance. In the case of drugs administered intravenously, enzyme inhibition increases AUC only by decreasing systemic clearance. Consequently, the magnitude of the increase in AUC for certain drugs depends on their route of administration, as illustrated by the interaction between ketoconazole (perpetrator) and midazolam (victim): ketoconazole, a CYP3A4 inhibitor, increases the AUC of midazolam three- to fivefold when midazolam is administered intravenously, but it causes a 10- to 16-fold increase when midazolam is administered orally (12). The same is true when the perpetrator is an enzyme inducer. For example, the CYP3A4 inducer rifampin decreases the AUC of midazolam by a factor of 9.7 when midazolam is administered orally, but it decreases the AUC by a factor of only 2.2 when midazolam is administered intravenously (13). The difference caused by route of administration is more dramatic for the inductive effect of rifampin on the AUC of nifedipine (12-fold for oral vs. 1.4-fold for intravenous) (14) and even more dramatic for *S*-verapamil (30-fold for oral vs. 1.3-fold for intravenous) (15).

When an enzyme accounts for only 20% of a drug's clearance ($f_m = 0.2$), complete loss of the enzyme activity causes only a 25% increase in AUC, which is normally considered to be bioequivalent (the so-called bioequivalence goalposts range from 80% to 125%, meaning that AUC values within this range can be considered equivalent and, therefore, acceptable). Therefore, an "unacceptable" increase in AUC requires an f_m of greater than 0.2. Actually, the FDA urges the characterization of all elimination pathways that account for 25% or more of a drug's clearance (i.e., $f_m \geq 0.25$).

Fractional metabolism by an enzyme determines the magnitude of the increase in drug exposure in individuals lacking the enzyme, but it does not determine its pharmacological or toxicological consequences. These are a function of therapeutic index, which is a measure of the difference between the levels of drug associated with the desired therapeutic effect and the levels of drug associated with adverse events. For drugs with a large therapeutic index, a high degree of clearance by a polymorphically expressed enzyme is not necessarily an obstacle to regulatory approval. For example, dextromethorphan is extensively metabolized by CYP2D6. Its fractional metabolism is estimated to be 0.93 to 0.98 such that the AUC of dextromethorphan in CYP2D6 PMs or in EMs administered quinidine is about 27 to 48 times greater than that in EMs (16,17). Dextromethorphan has a large therapeutic index, hence, despite this large increase in exposure in CYP2D6 PMs, dextromethorphan is an ingredient in a large number of over-the-counter (OTC) medications. Strattera® (atomoxetine) is another drug whose clearance is largely determined by CYP2D6

($f_{m(CYP2D6)} \approx 0.9$). Its AUC in CYP2D6 PMs is about 10 times that of EMs, but Strattera has a sufficiently large therapeutic index and was approved by the FDA only in the last few years (i.e., 2002).

Genetic polymorphisms give rise to four basic phenotypes on the basis of the combination of allelic variants that encode a fully functional enzyme (the wild-type or *1 allele designated "+"), a partially active enzyme (designated "*"), or an inactive enzyme (designated "−"). These four basic phenotypes are: (1) EMs, individuals who have at least one functional allele (+/+, +/*, or +/−), (2) PMs, individuals who have no functional alleles (−/−), (3) intermediate metabolizers (IMs), individuals who have two partially functional alleles or one partially functional and one nonfunctional allele (*/* or */−), and (4) UMs, individuals who, through gene duplication, have multiple copies of the functional gene [(+/+)*n*]. This traditional classification scheme has been revised recently on the basis of an activity score, which assigns to each allelic variant a functional activity value from one (for the wild-type or *1 allele) to zero (for any completely nonfunctional allele), as reviewed by Zineh et al. (18). The basis of the activity score, as it applies to CYP2D6, is illustrated in Table 1.

A wide range of activity is also observed for many of the CYP enzymes that have a very low incidence of genetic polymorphisms. This variation arises because CYP inhibitors can produce the equivalent of the PM phenotype, whereas CYP inducers can produce the equivalent of the UM phenotype. For example, CYP3A4 shows a low incidence of functionally significant genetic polymorphisms, but PMs can be produced pharmacologically with inhibitors, such as ketoconazole, erythromycin, and mibefradil, whereas UMs can be produced pharmacologically with inducers such as rifampin, St. John's wort, and the enzyme-inducing antiepileptic drugs (EIAEDs), such as phenobarbital, phenytoin, carbamazepine, and felbamate.

The variation in drug exposure that results from genetic polymorphisms or CYP inhibition/induction and the importance of therapeutic index to drug safety is illustrated in Figure 2. When a drug has a large therapeutic index, it is possible that no dosage adjustment is required either to achieve therapeutic efficacy in UMs or to prevent adverse effects in PMs (assuming that the therapeutic and adverse effects are both mediated by the parent drug, which is often but not always the case). When a drug has a narrow therapeutic index, the standard dosage may need to be increased in UMs (to achieve a therapeutic effect) or decreased in PMs (to prevent an adverse effect). The FDA defines a narrow therapeutic range as either "less than a 2-fold difference in median lethal dose (LD_{50}) and median effective dose (ED_{50}) values" or "less than a 2-fold difference in the minimum toxic concentrations and minimum effective concentrations in the blood, and safe and effective use of the drug products require careful titration and patient monitoring."

Drugs with a narrow therapeutic index are candidates for therapeutic drug monitoring, as in the case of the anticoagulant warfarin. The adverse effect of warfarin, namely excessive anticoagulation that can result in fatal hemorrhaging, is an extension of its pharmacological effect (inhibition of the synthesis of

Table 1 The Relationship Between Genotype and Phenotype for a Polymorphically Expressed Enzyme with Active (wt), Partially Active (*x), and Inactive (*y) Alleles

Genotype	Alleles	Conventional phenotype[a]	Activity score[b]	Activity score phenotype[c]
Duplication of active alleles ($n = 2$ or more)	(wt/wt)n	UM	$2 \times n$	UM
Two fully active wild-type (wt) alleles	wt/wt	EM	$1 + 1 = 2$	High EM
One fully active + one partially active allele	wt/*x	EM	$1 + 0.5 = 1.5$	Medium EM
One fully active + one inactive allele	wt/*y	EM	$1 + 0 = 1$	Low EM
Two partially active alleles	*x/*x	EM or IM	$0.5 + 0.5 = 1$	Low EM
One partially active + one inactive allele	*x/*y	IM	$0.5 + 0 = 0.5$	IM
Two inactive alleles	*y/*y	PM	$0 + 0 = 0$	PM

[a] The phenotypes are ultrarapid metabolizer (UM), extensive metabolizer (EM), intermediate metabolizer (IM), and poor metabolizer (PM), based on the particular combination of alleles that are fully active (wt), partially active (*x), or inactive (*y).

[b] In the case of CYP2D6 [Zineh et al. (18)], various activity scores have been determined experimentally as follows:
Activity score = 1.0 for each *1 (wt), *2, *35, and *41 [2988G]
Activity score = 0.75 for each *9, *29, *45, and *46
Activity score = 0.5 for each *10, *17, *41 [2988A]
Activity score = 0 for each *3, *4, *5, *6, *7, *8, *11, *12, *15, *36, *40, *42

[c] Activity scores are classified as follows [from Zineh et al. (18)]:
UM activity score = > 2.0 (e.g., [*1/*1]n where n is 2 or more gene duplications)
High EM activity score = 1.75 to 2.0 (e.g., *1/*1)
Medium EM activity score = 1.5 (e.g., *1/*17 or *9/*9)
Low EM activity score = 1.0 to 1.25 (e.g., *1/*4, *17/*17, or *9/*17)
IM activity score = 0.5 to 0.75 (e.g., *4/*9 or *4/*17)
PM activity score = 0 (e.g., *4/*4)

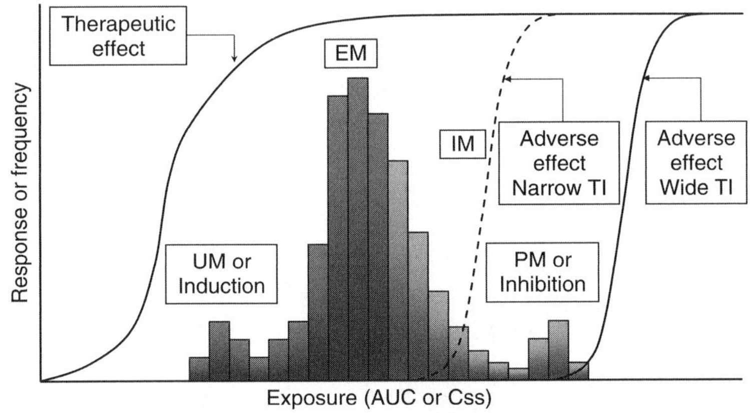

Figure 2 The impact of genetic polymorphisms (the UM, EM, IM, and PM genotypes) and the corresponding impact of enzyme induction or inhibition on exposure to a drug. *Abbreviations*: UM, ultrarapid metabolizer; EM, extensive metabolizer; IM, intermediate metabolizer; PM, poor metabolizer.

vitamin K-dependent clotting factors, which results in a prolongation of partial thromboplastin time or PTT). Warfarin is a racemic drug (a mixture of *R*- and *S*-enantiomers). The *S*-enantiomer, whose disposition is largely determined by CYP2C9, is four times more pharmacologically active than the *R*-enantiomer, whose disposition is largely determined by enzymes other than CYP2C9. The anticoagulant effect of warfarin [measured as an increase in prothrombin time on the basis of the international normalized ratio (INR)] is monitored during initial treatment with a low dose of warfarin, during dose escalation and then periodically during maintenance dosing to select the appropriate dose of anticoagulant on an individual-by-individual basis. Genotyping analysis provides useful information on dose selection. Individuals who are homozygous or heterozygous for certain allelic variants (the *3 and, to a lesser extent, the *2 allele) are CYP2C9 PMs or IMs and, as such, require less warfarin compared with an EM [an individual who is homozygous for the *1 (or wild-type) allele] (19). CYP2C9 genotype is not the only factor that influences dosing with warfarin. Genetic polymorphisms in the therapeutic target, vitamin K epoxide reductase (VKOR, gene symbol: VKORC1), can also impact warfarin dosing because different levels of warfarin are required to inhibit the variants of VKOR.

To achieve the same degree of anticoagulation, the dose of warfarin must be decreased during concomitant therapy with a CYP2C9 inhibitor and, conversely, it must be increased during concomitant therapy with a CYP2C9 inducer. This is why drug candidates that are identified in vitro as CYP2C9 inhibitors or inducers may need to be examined for their ability to cause significant interactions with warfarin in the clinic. Warfarin is a victim drug; its disposition is heavily reliant on CYP2C9 activity. It has been identified by the FDA as a *CYP2C9 substrate with a narrow therapeutic range*. Table 2 provides a

Table 2 Examples of Clinically Relevant Substrates, Inhibitors, and Inducers of the Major Human Liver Microsomal P450 Enzymes Involved in Xenobiotic Biotransformation

	CYP1A2	CYP2A6	CYP2B6	CYP2C8	CYP2C9	CYP2C19	CYP2E1	CYP2D6	CYP3A4			
Substrates	Alosetron[g] Caffeine[b,c] Duloxetine[g] 7-Ethoxyresorufin[b] Phenacetin[a] Tacrine[b] Tizanidine[h] Theophylline[b,c,h]	Coumarin[a] Nicotine[a]	Bupropion[a] Efavirenz[a,c] Propofol[b] *S*-Mephenytoin[b] Cyclophosphamide Ketamine Meperidine Nevirapine	Amodiaquine[b] Cerivastatin Paclitaxel[a,h] Rosiglitazone[b,c] Repaglinide[c,g]	Diclofenac[a] Fluoxetine[b] Flurbiprofen[b] Phenytoin[h] Tolbutamide[a,c] *S*-Warfarin[a,c,h]	Fluoxetine[b] *S*-Mephenytoin[a,h] Lansoprazole[c] Moclobemide Omeprazole[b,c,g] Pantoprazole[c]	Aniline[b] Chlorzoxazone[a] Lauric acid[b] 4-Nitrophenol[b]	Atomoxetine[c] Amitriptyline Aripiprazole Brofaromine (±)-Bufuralol[a] (*S*)-Chlorpheniramine Chlorpromazine Clomipramine Codeine Debrisoquine[b] Desipramine[c,g] Dextromethorphan[a,c] Dolasetron Duloxetine Fentanyl Haloperidol (reduced) Imipramine Loperamide (*R*)-Metoprolol Methylphenidate Mexiletine Morphine Nortriptyline Ondansetron Paroxetine Perhexilene Pimozide Propafenone (+)-Propranolol Sparteine Tamoxifen Thioridazine[h] Timolol Tramadol (*R*)-Venlafaxine	Alfentanil[h] Alfuzosin Alprazolam Amlodipine Amprenavir Aprepitant Artemether Astemiziole[h] Atazanavir Atorvastatin Azithromycin Barnidipine Bexarotene Bortezomib Brotizolam Budesonide[g] Buspirone[c,g] Capravirine Carbamazepine Cibenzoline Cilastazol Cisapride[h] Clarithromycin Clindamycin Clopidogrel Cyclosporine[h] Depsipeptide Dexamethasone Dextromethorphan[b] Diergotamine[h]	α-Dihydroergocriptine Disopyramide Docetaxel Domperidone Dutasteride Ebastine Eletriptan[c,g] Eplerenone[g] Ergotamine[h] Erlotinib Erythromycin[b] Eplerenone Ethosuximide Etoperidone Everolimus Ethinyl estradiol Etoricoxib Felodipine[c,g] Fentanyl[h] Fluticasone[g] Gallopamil Gefitinib Gepirone Granisetron Gestodene Halofantrine Laquinimod Imatinib Indinavir Isradipine	Itraconazole Karenitecin Ketamine Levomethadyl Lonafarnib Lopinavir Loperamide Lumefantrine Lovastatin[c,g] Medroxyprogesterone Methylprednisolone Mexazolam Midazolam[a,c,g] Mifepristone Mosapride Nicardipine Nifedipine[b] Nimoldipine Nisoldipine Nitrendipine Norethindrone Oxatomide Oxybutynin Perospirone Pimozide[h] Pranidipine Praziquantel Quetiapine Quinidine[h] Quinine	Reboxetine Rifabutin Ritonavir Rosuvastatin Ruboxistaurin Salmetrol Saquinavir[g] Sildenafil[c,g] Sibutramine Simvastatin[c,g] Sirolimus[h] Sunitinib Tacrolimus[h] Tadalafil Telithromycin Terfenadine[b,h] Testosterone[a] Tiagabine Tipranavir Tirilazad Tofisopam Triazolam[b,c,g] Trimetrexate Vardenafil[g] Vinblastine Vincristine Vinorelbine Ziprasidone Zonisamide

Inhibitors	Acyclovir[f] Cimetidine[f] Ciprofloxacin[e] Famotidine[f] Fluvoxamine[c,d] Furafylline[a] Mexilitene[e] α-Naphthoflavone[b] Norfloxacin[f] Propafenone[e] Verapamil[f] Zileuton[e]	Methoxsalen[a] Pilocarpine[b] Tranylcypromine[a] Tryptamine[b]	Clopidogrel[b] 3-Isopropenyl-3-methyl diamantane[b] 2-Isopropenyl-2-methyl-adamantane[b] Phencyclidine[b] Sertraline[b] Thio-TEPA[b] Ticlopidine[b] Phenylethyl-piperidine	Gemfibrozil[b,c,d] Montelukast[a] Pioglitazone[b] Quercetin[a] Rosiglitazone[b] Rosuvastatin Trimethoprim[b,f]	Amiodarone[c,e] Capecitabine Fluconazole[b,c,e] Fluoxetine[b] Fluvoxamine[b] Oxandrolone[e] Sulfaphenazole[a] Sulfinpyrazone[f] Tienilic acid	Fluvoxamine[c] Moclobemide[c] Nootkatone[b] Omeprazole[c,d] Ticlopidine[b]	Clomethiazole[b] Diallyldisulfide[b] Diethyldithiocarbamate[b] Disulfiram[c]	Amiodarone[f] Buproprion Chlorpheniramine Cimetidine Clomipramine Duloxetine[e] Haloperidol Fluoxetine[c,d] Methadone Mibefradil Paroxetine[c,d] Quinidine[a,c,] Sertraline[f] Terbinafine[e]	Amiodarone Amprenavir[e] Aprepitant[e] Atazanavir[c,d] Azamulin[b] Bosentan Cimetidine[f] Clarithromycin[c,d]	Diltiazem[e] Erythromycin[e] Felbamate Fluconazole[e] Fluvoxamine Fosamprenavir[e] Gestodene	Grapefruit Juice Ketoconazole[a,c,d] Indinavir[c,d] Itraconazole[a,c,d] Mibefradil Nefazodone[c,d] Nelfinavir[c,d]	Ritonavir[c,d] Roxithromycin[b] Saquinavir[c,d] St. John's wort Telithromycin[c,d] Troleandomycin Verapamil[b,e]
Inducers	3-Methylcholanthrene[a] β-Naphthoflavone[a] Omeprazole[a] Lansoprazole[b] TCDD	Dexamethasone[a] Pyrazole[b]	Phenobarbital[a] Phenytoin[b] Rifampin	Phenobarbital[b] Rifampin[a,c]	Phenobarbital Rifampin	Phenobarbital Rifampin[a,c]	Ethanol[c] Isoniazid	NA	Amprenavir Avasimibe Bosentan Carbamazepine[c] Clotrimazole Cyproterone acetate Dexamethasone[b] Efavirenz Etoposide	Guggulsterone Hyperforin Lovastatin Mifepristone Nelfinavir Nifedipine Omeprazole Paclitaxel[b]	PCBs Phenobarbital[b] Phenytoin[b] Rifabutin Rifampin[a,c] Rifapentine[b] Ritonavir Simvastatin	Spironolactone Sulfinpyrazole Topotecan Troglitazone[b] Troleandomycin Vitamin E Vitamin K2 Yin zhi wuang

Note: All FDA classifications are based on information available at the following URL: http://www.fda.gov/cder/drug/drugInteractions/tableSubstrates.htm#classInhibit

[a]FDA-preferred in vitro substrate, inhibitor, or inducer

[b]FDA-acceptable in vitro substrate, inhibitor, or inducer

[c]FDA-provided examples of in vivo substrates, inhibitors, or inducers for oral administration (Substrates in this category have plasma AUCs that are increased by at least twofold (5-fold for CYP3A4 substrates) when coadministered with inhibitors of the enzyme. Inhibitors in this category increase the AUC of substrates for that enzyme by at least twofold (5-fold for CYP3A4). Inducers in this category decrease the plasma AUC of substrates for that enzyme by at least 30%.

[d]Classified by the FDA as a "strong inhibitor" (i.e., caused a $\geq$ 5-fold increase in plasma AUC or $\geq$ 80% decrease in the clearance of CYP substrates in clinical evaluations)

[e]Classified by the FDA as a "moderate inhibitor" (i.e., caused a $\geq$ 2-fold but $<$ 5-fold increase in plasma AUC or 50–80% decrease in the clearance of *sensitive* CYP substrates when the inhibitor was given at the highest approved dose and the shortest dosing interval in clinical evaluations)

[f]Classified by the FDA as a "weak inhibitor" (i.e., caused a $\geq$ 1.25-fold but $<$ 2-fold increase in plasma AUC or 20–50% decrease in the clearance of *sensitive* CYP substrates when the inhibitor was given at the highest approved dose and the shortest dosing interval in clinical evaluations)

[g]Classified by the FDA as a "sensitive substrate" (i.e., drugs whose plasma AUC values have been shown to increase by $\geq$ 5-fold when co-administered with a known CYP inhibitor)

[h]Classified by the FDA as a "substrate with narrow therapeutic range" (i.e., drugs whose exposure response indicates that increases in their exposure levels by concomitant use of CYP inhibitors may lead to serious safety concerns such as torsades de pointes)

Abbreviations: CYP, cytochrome P450; AUC, area under the plasma concentration-time curve.

summary of selected CYP substrates, inhibitors, and inducers, many of which are recognized by the FDA, from an in vivo perspective, as sensitive substrates, potent inhibitors, or efficacious inducers, respectively.

When a potential victim drug is identified by reaction phenotyping, or when a potential perpetrator is identified by assessing CYP inhibition or induction, the information summarized in Table 2 can point to the type of clinical drug interaction study that would test in vivo the veracity and clinical significance of the in vitro data. For example, in the case of a drug candidate that is identified as a CYP2C9 substrate in vitro (i.e., a possible CYP2C9 victim drug), clinical studies might be carried out to assess whether the drug's disposition is affected by the same CYP2C9 genetic polymorphisms and the same CYP2C9 inhibitors/inducers that are known to influence the disposition of warfarin. Conversely, in the case of a drug candidate that is identified as a potent inhibitor (or efficacious inducer) of CYP2C9 (i.e., a CYP2C9 perpetrator), clinical studies might be carried out to assess whether the drug alters the disposition of warfarin (and other drugs whose clearance is dependent on CYP2C9).

The preceding discussion focuses only on drug-drug interactions that are pharmacokinetic in nature. Drug-drug interactions can also be pharmacodynamic in nature. For example, drugs that have antiplatelet activity and drugs that impede vitamin K absorption potentiate the anticoagulant effect of warfarin without necessarily impacting its disposition. This chapter focuses only on drug-drug interactions that are pharmacokinetic in nature and it further focuses mainly on drugs that are substrates for or inhibitors of CYP. Although the focus of this chapter is on the pathways that are determined by CYP-dependent metabolism, the same principles apply to other pathways of clearance, including metabolism by other enzymes, transport, biliary excretion, and urinary excretion.

II. EVALUATION OF DRUGS AS INHIBITORS OF CYP ENZYMES

In 1997 and 1999, the FDA issued two guidance documents entitled *Drug Metabolism/Drug Interaction Studies in the Drug Development Process: Studies In Vitro* and *In Vivo Drug Metabolism/Drug Interaction Studies—Study Design, Data Analysis, and Recommendations for Dosing and Labeling*. These documents reflected the FDA's thinking on these topics in the wake of the sometimes fatal interactions between terfenadine and drugs such as erythromycin and ketoconazole, and the announcement in January of 1997 that the FDA would withdraw approval of terfenadine. As many pharmaceutical companies and contract research organizations began adopting the general principles set forth in these guidance documents with a variety of experimental designs, it became increasingly clear that more direction was needed in order to optimize study designs. This observation was especially true with regard to time-dependent inhibition of CYP enzymes and examination of enzymes other than CYP1A2, CYP2C9, CYP2C19, CYP2D6, and CYP3A4. Workshops and conferences were held on the topic in 1997, 1999, and 2000, the latter of which specifically sought to achieve consensus on the conduct of in vitro and in vivo studies of metabolic and transport interactions and formed the basis of the 2001 "consensus paper" (4). Since 2001, other papers have reviewed many of the approaches commonly used to examine the potential for drug-drug interactions and these provided regulatory and industry perspectives as well as refinements to the original paper (5,20–22). In the absence of a revised formal guidance document on either in vitro or in vivo drug-drug interactions by 2004, the FDA's Center for Drug Evaluation and Research (CDER) issued a preliminary concept paper entitled *Drug Interaction Studies—Study Design, Data Analysis, and Implications for Dosing and Labeling* (1). This document formed the basis of the draft guidance document of the same title that replaced the earlier documents (2). Further refinements to this draft guidance will be posted online at the http://www.fda.gov/cder/drug/drugInteractions/default.htm. The classification of substrates, inhibitors, and inducers to be used in drug interactions studies as presented by the FDA appears in Table 2.

A. Guidelines for In Vitro CYP Inhibition Studies

The primary purpose of evaluating drugs as inhibitors of CYP enzymes in vitro is to determine their perpetrator or precipitant potential before advancing a candidate drug to a late stage of development. However, identifying a drug as an in vitro inhibitor of a given CYP enzyme does not imply that the drug will necessarily cause clinically relevant drug interactions. The clinical relevance of the inhibition must be considered in the following context:

1. The pharmacokinetics of the perpetrator (inhibitory) drug.
2. The potential of administering the perpetrator with the victim drug.
3. The extent to which the clearance of the victim drug is dependent on the inhibited CYP enzyme (i.e., $f_{\mathrm{m(CYP)}}$).
4. The potential for saturating the enzyme that metabolizes the victim drug.
5. The clinical consequences of altering the pharmacokinetics of the victim drug (which may or may not be a cause for concern depending on the drug's therapeutic index).
6. The therapeutic indication of the perpetrator and victim drug. Drugs used to treat life-threatening diseases (e.g., cancer, AIDS) are permitted more regulatory leeway than lifestyle-enhancing drugs (such as drugs to treat erectile dysfunction) or drugs that are not first-in-class [as in the case of Posicor (mibefradil), whose withdrawal was no doubt facilitated by its being one of many calcium channel blockers on the market].

The experimental studies described in the consensus papers and in this chapter provide tools for predicting the potential for inhibitory drug interactions.

Needless to say, a well-designed in vitro study can be a powerful predictor of clinical outcome. Unfortunately, it is all too easy to design an in vitro experiment that is analytically sound (it may even conform to the Bioanalytical Method Validation guidance document), but it is so seriously flawed that it provides meaningless data. For example, if coumarin (a high-turnover marker substrate for CYP2A6) is incubated under the same conditions that are sometimes used for *S*-mephenytoin (a low turnover marker substrate for CYP2C19), then 100% of the coumarin is converted to 7-hydroxycoumarin. Under these conditions, a drug candidate that partially inhibits CYP2A6 may go undetected because only marked inhibition of CYP2A6 will prevent complete metabolism of coumarin and decrease the amount of 7-hydroxycoumarin formed. This section will highlight the recommendations that pertain to selecting appropriate experimental conditions to assess the perpetrator potential of drug candidates on the basis of their ability to function as direct-acting and metabolism-dependent inhibitors of CYP enzymes.

1. Regulatory Perspective

The regulatory perspective will be covered in greater detail in chapter 16. This section will highlight the FDA's latest recommendations regarding in vitro CYP inhibition studies (1–3). It is expected that the final version of the guidance document entitled *Drug Interaction Studies—Study Design, Data Analysis, and Implications for Dosing and Labeling* may incorporate these recommendations as well as comments from industry and other refinements. This guidance document includes the FDA "preferred" or "acceptable" in vitro CYP substrates and inhibitors (also listed in Table 2). The FDA notes that the list is not exhaustive and that the choice of probe substrates for CYP enzymes should be based on selectivity (i.e., it is predominantly metabolized by a single enzyme in "pooled human liver microsomes or recombinant P450s") and those that have a simple metabolic scheme (i.e., no sequential metabolism). Practical considerations may guide the choice of substrate, such as the commercial availability of substrate and metabolite standards, and adequate turnover of the substrate to allow reasonable incubation times (1,2). The FDA defines preferred substrates as those meeting most of these criteria, with acceptable substrates meeting some and being judged acceptable by the scientific community. With regard to the in vitro examination of a drug candidate's ability to inhibit CYP3A4, the FDA recommends the "use of two structurally unrelated CYP3A4/5 substrates for evaluation of in vitro CYP3A inhibition. If the drug inhibits at least one CYP3A substrate in vitro, then in vivo evaluation is warranted" (1,2).

The additional recommendations contained in the draft guidance and on the FDA's website as they relate specifically to the design of in vitro CYP inhibition studies can be summarized as follows:

1. There is an increasing concern with inhibition of CYP2B6, and inhibition of this enzyme should be considered when warranted in addition to CYP1A2, CYP2C8, CYP2C9, CYP2C19, CYP2D6, and CYP3A4.

2. The FDA notes that CYP enzymes such as CYP2A6 and CYP2E1 are not as likely to be implicated in clinically relevant drug-drug interactions, but acknowledges that they should be examined when "appropriate."
3. Time-dependent inhibition should be examined. A 30-minute preincubation (i.e., with nicotinamide adenine dinucleotide phosphate (NADPH) enzyme, and drug candidate prior to addition of the probe substrate) is recommended.
4. The concentration of substrate and inhibitor should cover a range that brackets the K_m and K_i, respectively.
5. Microsomal protein concentrations should be <1 mg/mL.
6. Standardized assay conditions across all CYP enzymes are recommended because buffer strength, composition, and pH can affect enzyme kinetics.
7. Substrate or inhibitor depletion should be not more than 10–30%. The FDA acknowledges that this is difficult for low-K_m substrates.
8. For the chosen incubation conditions, metabolite formation should be linear with respect to incubation time and enzyme concentration.
9. Solvents should be kept to $\leq$1% (v/v) and preferably <0.1% (v/v) because some solvents inhibit one or more CYP enzymes. The experiment may include a no-solvent control as well as a solvent control to determine the effects of the solvent.
10. The use of positive controls is optional, but the FDA has nevertheless developed a list of preferred and acceptable inhibitors for use in reaction phenotyping studies that can be applied to CYP inhibition studies.

2. PhRMA Perspective

The guidance documents issued by the FDA in 1997, 1999, and 2006 do not provide extensive specific details of study design for in vitro or in vivo drug-drug interaction studies. As mentioned above, meetings were held in 1997, 1999, and 2000 in an attempt to address this need. The first consensus paper made the first published attempt to define study designs (4). A 2001 roundtable meeting between the Pharmaceutical Research and Manufacturers of America (PhRMA), Drug Metabolism/Clinical Pharmacology Technical Working Groups, and the FDA's CDER took place to further discuss these issues. During this meeting PhRMA was invited by the FDA to write a white paper detailing the industry perspective. In 2003, representatives of PhRMA published an industry perspective detailing the basic best practices for in vitro and in vivo pharmacokinetic-related drug-drug interaction studies to be conducted during drug development (as opposed to the earlier phase of drug discovery) and to define the data that are expected to be submitted to regulatory agencies (5,6). The PhRMA representatives did not want to limit innovation, however, in defining standards. The previous consensus paper formed the basis of the PhRMA perspective (4), so it will not be covered separately. This section will highlight key recommendations of the PhRMA perspective as they relate to the design of in vitro CYP inhibition studies and point out differences from recommendations published by the FDA to date.

Several minor differences in the list of preferred and acceptable in vitro substrates and inhibitors are apparent between the PhRMA perspective and the tables on the FDA's website. Most of these differences occur with the inclusion of a substrate or inhibitor on the preferred versus acceptable list. The substrates included on either of the FDA's lists but not the PhRMA lists include tacrine for CYP1A2, nicotine for CYP2A6, efavirenz and propofol for CYP2B6, amodiaquine and rosiglitazone for CYP2C8, flurbiprofen and phenytoin for CYP2C9, fluoxetine for CYP2C19, aniline for CYP2E1, and triazolam and dextromethorphan for CYP3A4/5. The PhRMA lists do not include the following inhibitors listed on the FDA's website: tranylcypromine, pilocarpine, and tryptamine for CYP2A6; 3-isopropenyl-3-methyl diamantane, 2-isopropenyl-2-methyl adamantane, thiotepa, phencyclidine, ticlopidine, and clopidogrel for CYP2B6; montelukast, trimethoprim, quercetin, and gemfibrozil[1] for CYP2C8; diethyldithiocarbamate and diallyldisulfide for CYP2E1; and itraconazole, azamulin, and verapamil for CYP3A4/5.

Recommendations contained in the PhRMA paper (which differ from the FDA documents), as they relate specifically to the design of in vitro CYP inhibition studies, can be summarized as follows:

1. Fluorogenic probes for in vitro studies are not recommended for regulatory submission.
2. Strict good laboratory practices (GLP) compliance was not recommended, but standard operating procedures (SOPs) and other GLP-like practices were recommended.
3. Bioanalytical method validation criteria should be applied to analytical methods whenever possible. It was also decided that long-term storage stability was not necessary, although short-term stability should be demonstrated.
4. Authentic metabolite standards should be used for calibration curves.
5. Human liver microsomes and recombinant human CYP enzymes are the preferred test systems because kinetic measurements are not confounded by cellular uptake and other metabolic processes present in hepatocytes and liver slices.
6. Protein concentrations of <0.5 mg/mL were recommended.
7. The rank order of the in vitro inhibition of various CYP enzymes can be assessed by either the determination of IC_{50} or K_i values.
8. The CYP marker substrate concentration should be equal to or less than the K_m for the reaction.

[1]It should be noted that gemfibrozil is a more potent direct-acting inhibitor of CYP2C9 than of CYP2C8 in vitro, whereas gemfibrozil glucuronide is a potent mechanism-based inhibitor of CYP2C8 (11).

9. The system chosen to conduct CYP inhibition studies should be well characterized. This procedure requires initial time-course experiments and determination of linearity of metabolite formation with the chosen incubation time and enzyme concentration. After these experiments, the kinetic parameters (i.e., K_m and V_{max}) for each substrate used with six or more concentrations spanning from 1/3 to $3K_m$ and inhibition potencies (i.e., IC_{50} or K_i) of typical inhibitors should be determined. This characterization does not need to be repeated for each batch or lot of test system.
10. For substrates exhibiting non-Michaelis-Menten kinetics (e.g., several CYP3A4 substrates), a wider range of substrate concentrations may be required to accurately determine reaction kinetics. CYP3A4 should be measured with at least two substrates, one exhibiting positive cooperativity (e.g., testosterone) and one exhibiting autoinhibition (e.g., midazolam).
11. No more than 20% of the CYP marker substrate should be consumed under the incubation conditions to be routinely employed.
12. The final concentration of organic solvents should be kept to $<0.5\%$ (v/v).
13. The concentration of the drug candidate will depend on solubility but should at least include the anticipated plasma concentration.
14. Control rates of reaction in each experiment should be compared with historical data.
15. Preliminary IC_{50} determinations may guide the design of K_i determinations and the inhibitor concentration should bracket the IC_{50} value.
16. Nonlinear regression should be used for curve-fitting. For K_i experiments, the inhibition equation that best fits the data determined by statistical criteria reflects the type of inhibition and K_i value.
17. It is recommended that time-dependent inhibition be examined when "deemed appropriate." Time-dependent inhibition should be examined with and without NADPH over an inhibitor concentration range of 1- to 10-fold the clinically relevant plasma concentrations. Various preincubation time points, such as 0, 15, 30, 45, and 60 minutes, should be utilized along with at least a 10-fold dilution step prior to the substrate incubation.

3. Additional Industry Perspectives on the Conduct of In Vitro CYP Inhibition Studies

In the last few years, several publications have described the use of automated and/or validated assays for CYP inhibition studies (20,22–26). Of these, the 2004 publication by Walsky and Obach (22) provides a comprehensive description of validated methods for assessing direct inhibition of CYP enzymes that meet or exceed the guidelines in the previous consensus papers. Walsky and Obach describe detailed methods for 12 assays for the 10 CYP enzymes most commonly involved in drug-drug interactions. The use of High performance liquid chromatography/tandem mass spectrometry (HPLC/MS/MS) methods (validated according to the FDA's 2001 Bioanalytical Method Validation criteria) were employed, which provide low limits of quantification (LOQ). The methods are sufficiently sensitive to permit the use of very low microsomal protein concentrations (i.e., 0.01–0.2 mg/mL) to minimize nonspecific binding, which is a frequent cause of erroneously high estimates of IC_{50} and K_i values and a major source of disagreement among values reported in the literature (22). Because Walsky and Obach used very low microsomal protein concentrations, their substrate incubation times were, in some cases, as long as 40 minutes. An additional advantage of HPLC/MS/MS is that deuterated metabolites can be used as internal standards, which virtually eliminates the possibility of analytical interference by the drug candidate under investigation. On the question of performing CYP inhibition assays in accordance with GLP regulations, the authors note that "although useful, good quality information can still be gathered in the absence of adherence to GLP, application of such practices can provide the highest possible assurance of the integrity of the data and a readily verifiable data audit trail," echoing the perspective provided by Bajpai and Esmay (20,22). The authors applied GLP-type criteria to experimental procedures and analytical methods. These criteria included: (1) determining initial rate conditions (i.e., $<15\%$ substrate metabolism); (2) determining enzyme kinetic constants (K_m and V_{max}) with ≥ 6 substrate concentrations spanning the K_m value by $\geq$ 3-fold; (3) processing data and performing nonlinear regression with validated computer software programs and using the Aikaike Information Criterion to assign the appropriate nonlinear model; (4) establishing control activities and interday variability with at least five initial assays; and (5) preparing SOPs once assays are validated.

As mentioned above, the Walsky and Obach paper strongly recommends the use of very low microsomal protein concentrations for CYP inhibition studies (e.g., as low as 0.01 mg/mL). The authors state that such low concentrations should "obviate the need to measure free fraction of inhibitor," which is in contrast to the recommendations of the Tucker consensus paper (4,22). However, the Tucker paper deemed protein concentrations up to 0.5 mg/mL to be acceptable, and at this concentration it can be important to determine the free drug fraction for basic lipophilic drugs. Nonspecific binding to microsomal protein and lipids is clearly a source of interlaboratory variation in IC_{50} or K_i values, as illustrated by the example of the highly lipophilic drug, montelukast. In the first published report of CYP inhibition by montelukast, a microsomal protein concentration of 1.0 mg/mL was utilized, providing K_i values for the inhibition of CYP2C9 and CYP3A4 of 15 and 200 μM, respectively, and >500 μM for CYP1A2, CYP2A6, CYP2C19, and CYP2D6 (CYP2C8 was not examined) (27). In contrast, Walsky et al. determined low IC_{50} values of 1.2 μM for *both* CYP2C9 and CYP3A4, a low value of 0.002 μM for CYP2C8, a high value of 180 μM for CYP2E1, and values from 7.9 to 32 μM for the other enzymes when examining montelukast as an inhibitor with very low microsomal protein concentrations (28). The K_i value for inhibition of recombinant CYP2C8 by montelukast was 0.0092 μM. Walsky et al. demonstrated that the IC_{50} value for inhibition of CYP2C8 increased linearly with

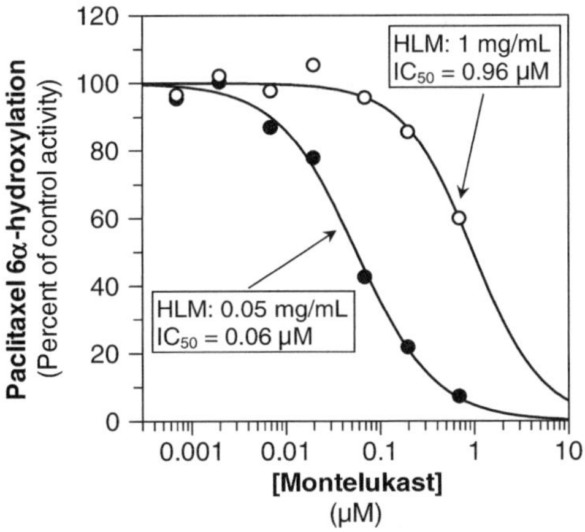

Figure 3 Effect of protein concentration on the inhibition of CYP2C8 (paclitaxel 6α-hydroxylation) by montelukast in human liver microsomes. The inhibitory effect of montelukast (CYP2C8 inhibitor) on the conversion of paclitaxel to 6α-hydroxypaclitaxel declined almost 20-fold when the microsomal protein concentration increased 20-fold due to nonspecific protein binding.

microsomal protein by 100-fold (i.e., 0.02–2.0 μM) as the protein concentration increased 80-fold from 0.025 to 2.0 mg/mL (28). As shown in Figure 3, we have confirmed that the IC_{50} value for CYP2C8 inhibition by montelukast increases almost 20-fold (i.e., 0.06–0.96 μM) with a 20-fold increase in microsomal protein concentration (i.e., 0.05–1.0 mg/mL). In such a case, it would seem only prudent to correct the in vitro K_i value by determining the free fraction of drug in the microsomal incubation. However, some highly lipophilic drugs are not amenable to a determination of free fraction in microsomal incubations because of binding to the equilibrium dialysis membrane or apparatus, which was the case with montelukast (28). Using the "uncorrected" K_i value for the inhibition of CYP2C8 by montelukast, Walsky et al. showed that predictions of increased exposure to a concomitantly administered drug with $f_{m(CYP2C8)} = 1.0$ ranged from 2.1- to 119-fold depending on the in vivo value used for inhibitor concentration (i.e., 0.01–1.09 μM for free vs. total, systemic, or portal C_{max} of montelukast). However, in vivo studies show that, when montelukast is coadministered to healthy volunteers at doses that produce plasma C_{max} values of approximately 0.9 μM, there is a negligible increase in the AUC for the CYP2C8 substrates pioglitazone and repaglinide, meaning that the extrapolation overpredicts the actual interaction (29,30). Therefore, in the case of montelukast at least, if correction of the in vitro K_i value for nonspecific binding to microsomal protein had been possible, the predicted interactions would have been even higher, since the corrected K_i would have been lower than the uncorrected K_i value. This scenario supports the idea that routine correction of in vitro K_i values for nonspecific binding to microsomal protein may not increase the predictive ability of in vitro CYP inhibition studies as long as the protein concentration used is low ($\leq$0.1 mg/mL). Nonspecific binding of candidate drugs to microsomal protein and lipids can also be predicted reasonably well on the basis of the compound's log P or log $D_{7.4}$ value, according to the following equation (31):

$$f_{u,inc} = \frac{1}{1 + C \cdot 10^{0.072 \cdot \log P/D + 0.067 \cdot \log P/D - 1.126}} \quad (2)$$

where C is the microsomal protein concentration in mg/mL, log P/D is the log P of the molecule if it is a base (basic pKa $>$ 7.4), and log $D_{7.4}$ of the molecule if it is neutral or an acid (acidic pKa $<$ 7.4).

B. Theoretical Concepts

Two major types of CYP inhibition are possible: direct inhibition and time-dependent inhibition. Direct inhibition occurs when a drug inhibits a CYP enzyme without significant delay (i.e., as soon as it binds to the CYP enzyme, which usually occurs in a matter of seconds) and without requiring biotransformation. Examples of direct inhibition include inhibition of CYP2D6 and CYP3A4 by quinidine and ketoconazole, respectively. Direct inhibition can occur with normal, Michaelis-Menten, or atypical kinetics, including partial inhibition and two-site binding with heterotrophic cooperation. Time-dependent inhibition occurs when the inhibitory potency of the drug candidate increases with incubation time, which may reflect a slow on-rate or more commonly the need for biotransformation. Time-dependent inhibition includes the quasi-irreversible and irreversible metabolism-dependent inhibition caused by drugs such as troleandomycin, mibefradil, diltiazem, tienilic acid, halothane, and furafylline.

1. Direct Inhibition

Direct inhibition can be subdivided into two types. The first involves competition between two drugs that are metabolized by the same CYP enzyme. For example, omeprazole and diazepam are both metabolized by CYP2C19. When the two drugs are administered simultaneously, omeprazole decreases the plasma clearance of diazepam and prolongs its plasma half-life. The inhibition of diazepam metabolism by omeprazole involves competition for metabolism by CYP2C19 and no such inhibition occurs in CYP2C19 PMs (individuals who lack CYP2C19) (32,33). The second type of direct inhibition is when the inhibitor is not a substrate for the affected CYP enzyme. The inhibition of dextromethorphan biotransformation by quinidine is a good example of this type of drug interaction. Dextromethorphan is O-demethylated by CYP2D6, and the clearance of dextromethorphan is impaired in individuals lacking this polymorphically expressed enzyme. The clearance of dextromethorphan in EMs is similarly impaired when this antitussive agent is taken with quinidine, a potent inhibitor of CYP2D6; quinidine causes up to a 48-fold increase in the AUC of dextromethorphan (16).

However, quinidine is not biotransformed by CYP2D6, even though it binds to this enzyme with high affinity [unbound $K_i < 1$ nM (34)]. Quinidine is actually biotransformed by CYP3A4 (35), and is a competitive inhibitor of this enzyme [K_i as low as 5.4 μM (36)], although its effects are highly dependent on the CYP3A4 substrate employed.

Direct inhibition, as defined above, can occur by at least four mechanisms: competitive, noncompetitive, mixed, and uncompetitive. Competitive inhibition occurs when the inhibitor and substrate compete for binding to the active site of the enzyme and is characterized by an increase in K_m with no change in V_{max}. Noncompetitive inhibition occurs when the inhibitor binds to a site on the enzyme that is different from the active site to which the substrate binds and is characterized by a decrease in V_{max} with no change in K_m. In the case of uncompetitive inhibition, which is rarely observed with CYP enzymes, the inhibitor binds to the enzyme when the substrate is bound to it and stabilizes the enzyme-substrate complex; the inhibitor binding site may be the same as or different from the active site (substrate binding site). Finally, mixed (competitive-noncompetitive) inhibition occurs when the inhibitor binds to the active site as well as to another site on the enzyme, or the inhibitor binds to the active site but does not block the binding of the substrate and is characterized by a decrease in V_{max} and an increase in K_m. The kinetics and the affinity with which an inhibitor binds to an enzyme are best described by the dissociation constant for the enzyme-inhibitor complex. This dissociation constant is referred to as the inhibition constant, K_i. In the past, linear transformations of the Michaelis-Menten equation (such as a Dixon plot or Lineweaver-Burk double-reciprocal plot) were used to calculate K_i values and assess the type of direct enzyme inhibition, but this has been supplanted by computer software that allows the use of nonlinear regression analysis to calculate kinetic constants. However, linear transformations, and in particular the Eadie-Hofstee plot, are still useful for visualizing the mechanism of inhibition (Fig. 4). More complex inhibition kinetics, as are occasionally found with CYP3A4 inhibitors, can be modeled by various multisite variations of the Michaelis-Menten equation. These models are beyond the scope of this chapter and are reviewed in detail by Galetin et al. (36–39).

The affinity with which an inhibitor binds to an enzyme is defined by its K_i value, whereas the affinity with which the substrate binds is generally defined by its K_m value. Both definitions are somewhat simplistic as they are based on three assumptions:

1. The dissociation of the enzyme-inhibitor or enzyme-substrate complex (as opposed to complex formation) is the rate-limiting step.
2. The concentration of the enzyme is negligible compared with the concentration of the substrate and inhibitor (so that binding of the substrate or inhibitor to the enzyme has a negligible effect on the free concentration of substrate or inhibitor).

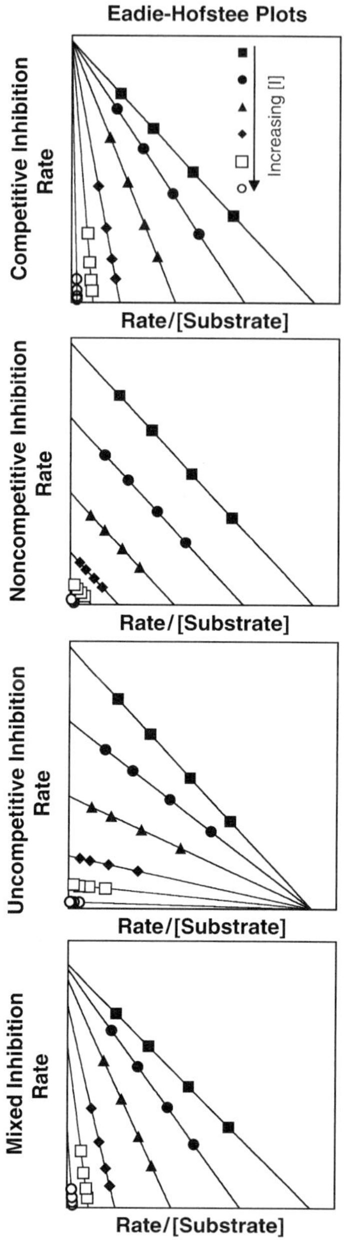

Figure 4 Graphical representation of enzyme inhibition: Eadie-Hofstee plots of theoretical K_i data. Eadie-Hofstee plots are useful in differentiating the various types of direct inhibition.

3. The free (unbound) concentration of the substrate/inhibitor is known or well approximated by the total concentration of substrate/inhibitor.

All three assumptions can be violated in the case of CYP enzymes, depending on the design of the in vitro CYP inhibition study. The first assumption can be potentially violated if the drug being tested is a time-dependent inhibitor (e.g., one with a slow on rate; see below). The potency of some inhibitors (e.g., the CYP3A inhibitors ketoconazole and clotrimazole) is such that the free concentration of the inhibitor tends to approach the concentration of the enzyme (40), a violation of the second assumption. In the case of such tight-binding inhibition, an apparent K_i value ($K_{i,app}$) can be estimated, as follows:

$$\frac{[I]_t}{1 - v_i/v_0} = K_{i,app} \times \frac{v_i}{v_0} + E_t \qquad (3)$$

where $[I]_t$ is the total inhibitor concentration, $1 - (v_i/v_0)$ is the fractional inhibition, and E_t is the total enzyme concentration. As noted in section II.A.3, a significant fraction of the substrate/inhibitor may also bind to the lipid membrane and/or protein if low concentrations of microsomal protein are not used, thereby violating assumption 3.

The above discussion puts an emphasis on the K_i value for inhibition rather than the IC_{50} value. The K_i value is an inhibition constant that defines the affinity of the inhibitor for the enzyme, whereas the IC_{50} is the concentration of inhibitor required to cause 50% inhibition under a given set of experimental conditions. Strictly speaking, then, it would be preferable to determine the K_i rather than an IC_{50} value for the following reasons:

1. K_i values are intrinsic constants, whereas IC_{50} values are extrinsic constants. Theoretically, IC_{50} values, in contrast to K_i values, are dependent on the type of substrate, the concentration of substrate, and incubation conditions (protein concentration or incubation times, etc).
2. Because they are intrinsic constants, K_i values can *theoretically* be reproduced from one laboratory to another.
3. The method of predicting the potential for drug interactions by a drug from K_i values and some measure of the in vivo concentrations of the drug is widely accepted (e.g., $AUC_i/AUC_{ui} = 1 + [I]/K_i$).

Despite the theoretical advantages of K_i determinations over IC_{50} determinations, it is generally more time and cost effective to determine IC_{50} values for the inhibition of several CYP enzymes by a drug candidate under conditions that permit a reasonably reliable estimate of K_i values. Once it is known which CYP enzymes are most potently inhibited, additional experiments can be conducted to determine K_i values for selected enzymes, with the IC_{50} value guiding the selection of drug candidate concentrations. In addition, if IC_{50} experiments are designed appropriately and the substrate concentration is equal to K_m for the marker reaction, the K_i value will be equal to one-half the IC_{50} value, if the inhibition is competitive, and equal to the K_i value, if inhibition is noncompetitive. This simple relationship provides more reason to begin an evaluation of CYP inhibition with IC_{50} rather than K_i determinations because a conservative estimate of the K_i value can be used to estimate the potential clinical significance of such in vitro inhibition.

2. Time-dependent Inhibition

An in vitro examination of time-dependent inhibition of the major drug-metabolizing CYP enzymes should be considered essential for drug candidates. Time-dependent inhibition occurs when the inhibitory potential of a drug candidate increases as the enzyme is exposed to the inhibitor over time. This type of inhibition may occur by several potential mechanisms, including the following:

1. Slow-binding (i.e., slow on rate) inhibition (e.g., inhibition of the steroidogenic enzyme CYP19A1 by 19-azido-androstenedione)
2. Nonenzymatic conversion of the drug candidate to an inhibitory product [e.g., rabeprazole, a sulfoxide that undergoes nonenzymatic reduction to the sulfide, which inhibits CYP2C9, CYP2C19, CYP2D6, and CYP3A4 with greater than 16-fold greater potency than the parent (41)]
3. Metabolism-dependent conversion of the drug candidate to a product that is a more potent direct-acting inhibitor than the parent (e.g., the conversion of fluoxetine to norfluoxetine, see below)
4. Metabolism-dependent conversion of the drug candidate to a metabolite that quasi-irreversibly coordinates with the heme iron (e.g., troleandomycin) or irreversibly (covalently) binds to amino acid residues or the heme moiety of a CYP enzyme in such a way as to completely prevent or significantly diminish catalytic activity (e.g., tienilic acid).

Slow-binding inhibition is a reversible process in which initial inhibition becomes more potent over time without any metabolism. In the case of CYP enzymes, slow-binding inhibition can often be followed spectrophotometrically by monitoring the development of type I, type II, or reverse type I spectra, which reflect substrate-induced changes in the spin state of the heme iron (low spin → high spin gives a type I spectrum, high spin → low spin gives a reverse type I spectrum) or which reflect coordination of the substrate (usually nitrogenous compounds) with the heme iron. This type of inhibition has rarely been reported with CYP enzymes, although 19-thiomethyl- and 19-azido-androstenedione were reported to be potent (i.e., K_i values <5 nM), slow-binding inhibitors of CYP19A1 (aromatase) for which the maximum spectrally apparent type II complex required up to six minutes to form (42). Nonenzymatic degradation to inhibitory or reactive products can occur with some unstable compounds, such as rabeprazole, or some acyl glucuronides, which can rapidly rearrange to form reactive aldehydes that form Schiff's bases (covalent adducts) with lysine residues on proteins (43). Inhibition that is only time dependent, such as slow-binding inhibition and the nonenzymatic formation of inhibitory products, are encountered less frequently than metabolism-dependent inhibition and will not be covered in detail in this chapter.

Throughout the remainder of this chapter, the phrase "metabolism-dependent inhibition" will be used to denote time-dependent inhibition that also requires one or more metabolic conversions (usually NADPH-dependent). Many researchers use the phrase "mechanism-based inhibition" to refer to any irreversible or quasi-irreversible metabolism-dependent inhibition of CYP enzymes. However, by definition, the phrase "mechanism-based inhibition" excludes the formation of metabolites that are simply more potent direct-acting inhibitors than the parent, whereas the term "metabolism-dependent inhibition" includes this type of time-dependent inhibition. Simply put, mechanism-based inactivators are substrates for a CYP enzyme that, during catalysis by the enzyme, are converted to one or more products, which immediately and irreversibly inactivate the enzyme and do not leave the active site (44). Strictly speaking, irreversible inhibitors that are affinity labeling agents, transition state analogs, and slow, tight-binding inhibitors (discussed above) are not mechanism-based inhibitors because they do not require a metabolic event to exert their effect (44). For a metabolism-dependent inhibitor of a CYP enzyme to be categorized as a mechanism-based inactivator, it must meet certain criteria that can be determined experimentally, according to Silverman (44), and as put in the context of CYP enzymes by a thorough review by Fontana et al. (45):

1. The CYP inhibition must be concentration- NADPH- and time-dependent.
2. Inactivation must occur prior to the release of the inhibitory metabolite. Any metabolite that is released from the active site cannot be the metabolite that inactivates the enzyme. (This criterion distinguishes mechanism-based inactivators from metabolism-dependent inhibitors that generate and release electrophilic metabolites. In such a case, inactivation may occur by binding to a site other than the active site, or by rearrangement of the metabolite prior to its return to the active site.) Furthermore, the addition of glutathione (GSH), radical scavengers, or other exogenous nucleophiles cannot prevent inactivation in the case of true mechanism-based inhibition, but they often abrogate the inhibition observed with other types of metabolism-dependent inhibition. Note that, in general, only a portion of the inactivator is converted to the species that covalently binds to the apoprotein or heme, and the rest is released from the active site. The amount converted to the inactivating species relative to other metabolites is known as the partition ratio and is generally >10 (i.e., one molecule of the inactivating species is produced for every 10 molecules that are metabolized). For instance, mibefradil, the potent inactivator of CYP3A4, has a partition ratio of 1.7, while suprofen, a mechanism-based inhibitor of CYP2C9, has a partition ratio of 101 (45).

3. Mechanism-based inhibition should be irreversible. Dialysis, ultrafiltration, or "washing" the protein (e.g., by isolating microsomes by centrifugation and resuspending them in drug-free buffer) will not restore enzyme activity, and the inhibition is highly resistant to sample dilution.
4. Mechanism-based inhibition should be saturable. The rate of inactivation is proportional to the concentration of the inactivator until all enzyme molecules are saturated, in accordance with Michaelis-Menten kinetics. Additionally, the decrease in enzymatic activity over time should follow pseudo-first-order kinetics.
5. Substrates should protect against mechanism-based inhibition. The addition of an alternative substrate or competitive inhibitor with good affinity for the enzyme will prevent or at least decrease the rate of inactivation.
6. There should be stoichiometric (ideally one-to-one) binding of inactivator to enzyme.
7. CYP content (as measured spectrophotometrically at 450 nm in the presence of sodium dithionite and carbon monoxide) is usually reduced.
8. Enzyme inactivation should be preceded by a catalytic event that converts the mechanism-based inhibitor to the inactivating metabolite.

For most cases encountered in drug development, it is not necessary to definitively prove all of the above criteria. Ideally, the experimental design of a definitive in vitro CYP inhibition study should allow the following questions to be answered in a single initial experiment:

1. Is the drug a direct-acting inhibitor of the CYP enzyme?
2. What is the potency of inhibition (e.g., IC_{50} or K_i value)?
3. Is the drug a metabolism-dependent inhibitor?

If necessary, subsequent experiments can then determine irreversibility and potency (e.g., K_I and k_{inact}) for metabolism-dependent inhibitors, as well as the potential for covalent binding. This general experimental strategy, including follow-up experiments, is illustrated in Figure 5. The design of experiments to determine K_I and k_{inact} will be covered in more detail in section II.C.7.c.

Metabolites formed in the active site of a CYP enzyme can cause irreversible or quasi-irreversible inhibition by three main mechanisms:

1. Tight coordination with the ferrous heme iron to form a metabolic-intermediate (MI) complex.
2. Covalent reaction with the porphyrin ring nitrogen atoms to form heme adducts.
3. Covalent reaction with nucleophilic amino acid residues in the active site (45).

MI complex formation most commonly occurs with alkylamines, heterocyclic amines, hydrazines, methylenedioxybenzenes (benzodioxoles), and macrolide

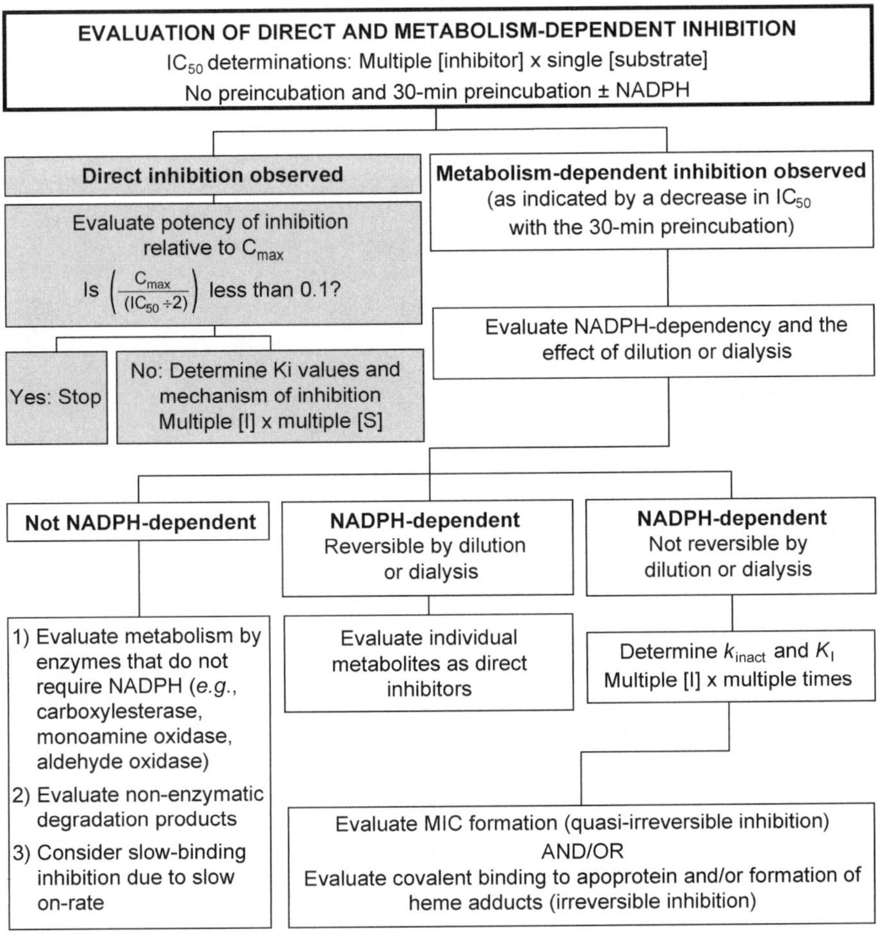

Figure 5 Flowchart for initial and possible follow-up CYP inhibition studies. The first box represents the IC_{50} experiment with and without a 30-minute preincubation with NADPH (the highest concentration of test article is also preincubated for 30 minutes without NADPH). Remaining boxes depict possible outcomes and follow-up experiments. *Abbreviations*: CYP, cytochrome P450; IC_{50}, concentration of inhibition causing 50% inhibition.

antibiotics, which can be metabolized by CYP enzymes to form stable complexes with the heme iron, thus inactivating the CYP enzyme in a quasi-irreversible (noncovalent) manner (Table 3) (45,46). The formation of MI complexes can usually be followed spectrophotometrically as an increase in absorbance maximum around 455 nm because the ferrous-metabolite complex resembles the ferrous-carbon monoxide complex that absorbs maximally around 450 nm. Figure 6 shows the spectrum of the MI complex that forms with CYP3A4 by incubating troleandomycin with NADPH-fortified human liver microsomes. Isolated furans and

Table 3 Structures Associated with Metabolism-Dependent Inhibition of CYP Enzymes

Chemical groups (examples)	Structures
Terminal (ω) and ω-1 acetylenes (Gestodene)	
Furans and thiophenes (Furafylline and tienilic acid)	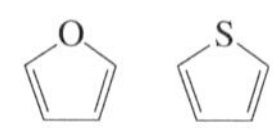
Epoxides (*R*-Bergamottin-6′,7′-epoxide)	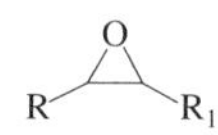
Dichloro- and trichloroethylenes (1,2,-Dichloroethylene and trichloroethylene)	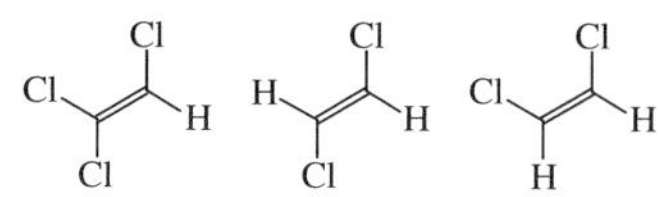
Secondary amines (Nortriptyline)	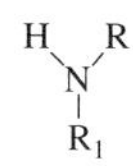
Benzodioxoles (Paroxetine)	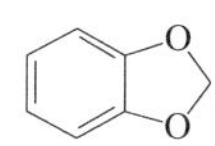
Isothiocyanates (Phenethyl isothiocyanate)	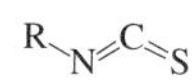
Thioamides (Methimazole)	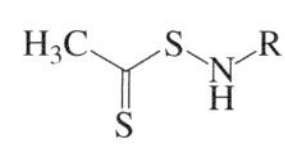
Dithiocarbamates (Disulfiram)	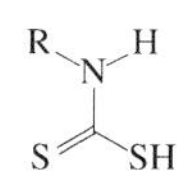
Conjugated structures (Rhapontigenin)	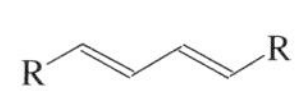
Terminal alkenes (Tiamulin)	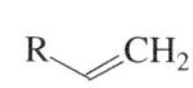

Abbreviation: CYP, cytochrome P450.
Source: Adapted from Ref. 45.

methylfurans, as well as terminal alkenes and alkynes (Table 3), which are small enough to directly access the heme of some CYP enzymes, can be oxidized to radical intermediates that alkylate heme, thus inhibiting the enzyme in an irreversible manner (46). Covalent binding to nucleophilic amino acids in the active site of CYP enzymes occurs most frequently with acetylenes, thiophenes, furans,

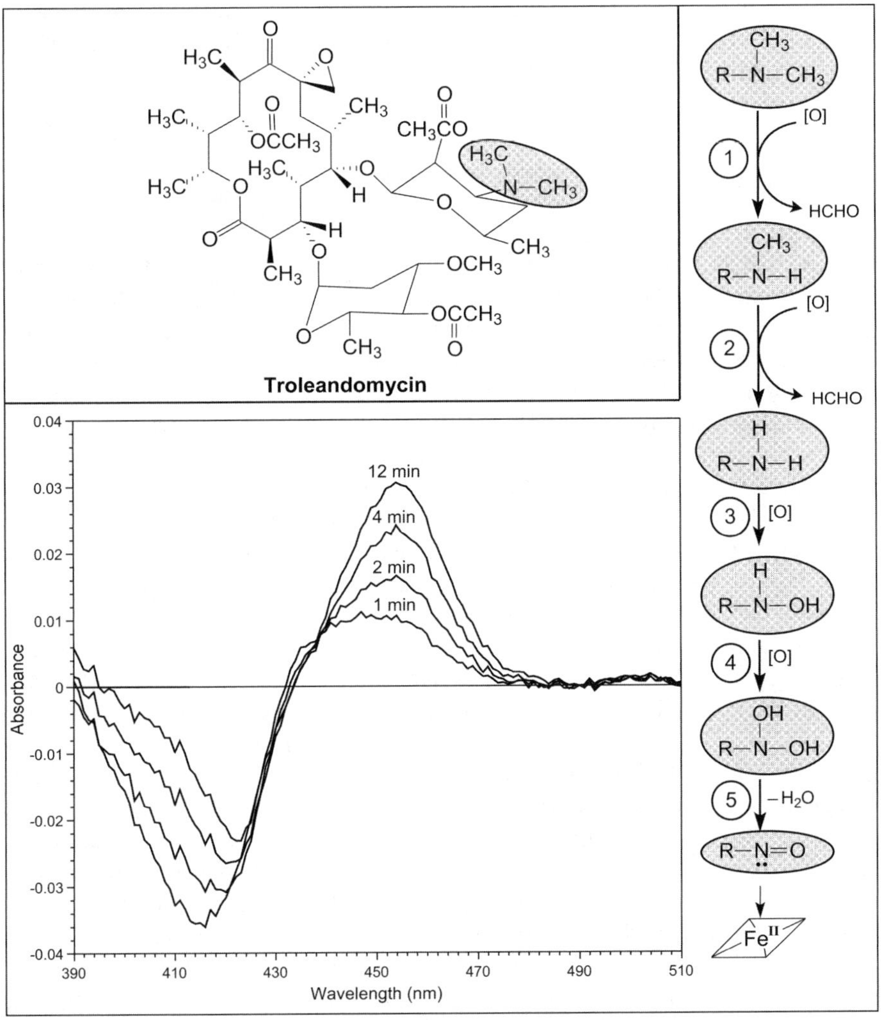

Figure 6 The structure of the potent quasi-irreversible CYP3A4 inhibitor troleandomycin (*top panel*), the metabolic steps required to convert troleandomycin into a nitroso metabolite that coordinates with the ferrous-heme of CYP3A4 (side panel), and the spectral changes associated with MI complex formation (*bottom panel*). Troleandomycin (50 μM) was incubated at 37°C with a human liver microsomal sample with high CYP3A4/5 activity (1 mg/mL) and NADPH (1 mM) for the times indicated. The reference cuvette contained the same components minus troleandomycin. Scans from 380 to 520 nm were recorded on a Varian Cary 100 BIO UV/Vis dual beam spectrophotometer. *Abbreviation*: MI, metabolic intermediate.

terminal alkenes, conjugated structures, and dichloro- and trichloroethylenes (Table 3). For example, tienilic acid is converted to a thiophene sulfoxide by CYP2C9, which is an electrophilic reactive intermediate that can covalently bind to a nucleophilic serine residue (Ser 365) in the active site of CYP2C9 (47,48). The mechanism of CYP inactivation caused by compounds containing a thiono moiety, such as methimazole, has not been clearly established, but may include heme or protein alkylation or cross-linking between the modified heme and amino acids in the active site (45).

Drugs such as tienilic acid not only pose a risk of prolonged inhibition of CYP enzymes, but they can also have wider implications. The diuretic tienilic acid (Selacryn®) was approved by the FDA in May 1979 and voluntarily withdrawn from the U.S. market within a year (January 1980) because of 500 cases of liver toxicity and 25 fatalities (49). This example of drug-induced liver injury (DILI) is caused in part by covalent binding of tienilic acid to CYP2C9, which induces a subsequent autoimmune-response involving formation of type 2 anti-liver kidney microsome (anti-KLM_2) antibodies in some individuals (an example of idiosyncratic DILI) (49). Idiosyncratic drug reactions (IDRs) are rare adverse events ($<0.1\%$) that do not involve an exaggerated pharmacological response, do not occur in most patients at any dose, and typically do not occur immediately after exposure but do so after weeks or months of repeated administration (50). IDRs are also known as type B reactions, and they can be divided into allergic and nonallergic IDRs. The former tend to develop relatively quickly (in days or weeks) and, after the drug is discontinued, patients respond robustly when rechallenged with the same or a closely related drug, whereas the latter tend to develop relatively slowly (with symptoms sometimes appearing after six months or more of drug treatment) and patients may or may not respond to rechallenge with the drug. Evidence for an immune component to allergic IDRs is often circumstantial or lacking, as is the evidence for the lack of an immune component to nonallergic IDRs. Uetrecht has proposed the general rule that drugs administered at a daily dose of 10 mg or less do not cause idiosyncratic toxicity (51). Hepatotoxicity is a prevalent IDR. Because of its potential to cause immune hepatitis, metabolism-dependent inhibition that involves covalent binding of a drug to a CYP enzyme (i.e., hapten formation) presents a greater obstacle to regulatory approval than metabolism-dependent inhibition that involves MI complex formation (which is noncovalent and quasi-irreversible). Therefore, when metabolism-dependent inhibition is observed, it is prudent to ascertain whether the mechanism involves irreversible inhibition with covalent modification of the CYP enzyme.

Another type of inactivation, namely *reversible* inactivation, deserves mention because of its unusual nature and its potential to impact the interpretation of mechanism-based inactivation of CYP enzymes. In 1995, alkylbenzene and 1-hexyne were reported to *N*-alkylate, the heme moiety of chloroperoxidase in a P450-like reaction (52). These compounds inactivated this heme-containing enzyme in a manner that met the criteria for a mechanism-based inhibitor as defined by Silverman (44). However, the inactivated enzyme spontaneously lost the heme adducts over several hours with a return of enzymatic activity and native heme, thus qualifying as "reversible inactivation." This is an unusual case of an enzyme being inactivated by covalent modification, but where the covalent modification is reversible. Two similar compounds, *tert*-butyl acetylene (*tert*-BA) and *tert*-butyl-1-methyl-2-propynyl ether were reported to inactivate CYP2E1 and an engineered variant of CYP2E1 (T303A, which lacks the conserved threonine directly over the heme) by covalently alkylating the heme moiety (by a combination of heme alkylation and protein adduction), or, in the case of *tert*-BA, by inactivating the CYP2E1 variant by *reversibly* alkylating the heme moiety (53). Furthermore, *tert*-BA also formed heme adducts with wild-type CYP2B4 that were partially reversible (20–30%) on dialysis. The heme adducts decomposed (became dealkylated) over time except under acidic conditions (53). In the case of the *tert*-BA-dependent reversible inactivation of the CYP2E1 variant, an absorbance maximum at 483 nm was observed, and this was also the case with 4-methyl-1-pentyne, which acted as a reversible inactivator (53). These examples illustrate how CYP enzymes, like chloroperoxidase, can be inactivated by a process involving reversible covalent modification of the heme moiety.

The formation of MI complexes is also known to be reversible with the addition of strong oxidants such as potassium ferricyanide (which oxidizes the ferrous iron to the ferric state and thereby alters the interaction between iron and the metabolite/ligand). In the case of benzotriazoles, MI complex formation is readily reversed once a source of reducing equivalents (e.g., NADPH) is exhausted. For this reason, CYP inactivation involving MI complex formation is considered quasi-irreversible (54). However, Silverman noted that no period of time is defined for how long an enzyme-inactivator complex must persist for the inactivator to be classified as irreversible (44). Despite being quasi-irreversible (noncovalent), MI complex formation is associated with several cases of prolonged and clinically significant inhibition. For example, the tertiary amine diltiazem increases the AUC of nifedipine threefold (55). The methylenedioxy-containing compound, paroxetine, forms a spectrally apparent MI complex with CYP2D6 (56) and causes an eightfold increase in the AUC of *R*-metoprolol (57) and a sevenfold increase in atomoxetine AUC (58). In such cases, the interactions can persist for some time after cessation of the perpetrator drug. For instance, it has been reported that CYP2D6 activity takes up to 20 days or more to return to baseline after cessation of paroxetine treatment (59).

The kinetics of metabolism-dependent irreversible or quasi-irreversible inhibition can be complex (46) and are covered in chapter 11 of this book. The kinetics of metabolism-dependent inhibition caused by metabolites that function as direct-acting inhibitors (e.g., norfluoxetine) are dependent on the pharmacokinetic properties and inhibitory potency of the metabolite. Therefore, when further examination of such inhibition is warranted, the inhibitory metabolite itself should be investigated.

C. Optimizing the Design of Automated and Validated In Vitro CYP Inhibition Studies

There are many published descriptions of rapid CYP inhibition studies based on a variety of test systems and methodologies (60–64). All of these designs offer certain advantages, but all have certain drawbacks as well, and most are intended to be used during drug discovery or early drug development. For instance, automated microtiter based assays with fluorogenic probes and recombinant human CYP enzymes offer the advantage of high throughput because of rapid analysis on a plate reader, but fluorescence interference can be problematic and, as mentioned in previous sections, these methods are not recommended for definitive studies. A high-throughput reactive oxygen species-based CYP3A4 assay has also been described (65), as well as CYP inhibition assays based on luminogenic probe substrates. These may offer some advantages over fluorogenic substrates, but they have not been widely utilized (66,67) and appear more suited for screening during early discovery. Although some of the available fluorogenic or luminogenic probe substrates are somewhat selective for certain CYP enzymes, the use of these substrates generally requires the use of recombinant human CYP enzymes to ensure that only a single CYP enzyme is examined.

Given the critical information that definitive CYP inhibition studies can provide when designed properly, and the potential loss of therapeutic benefit and revenue for each additional day a drug spends in development, there is great demand to perform such studies rapidly. For late-stage development compounds, such studies must not only be conducted rapidly, but with sufficient quality and documentation for inclusion in New Drug Application (NDA) submissions. The hallmark of methods that meet or exceed the latest recommendations of the consensus papers is the use of the CYP-selective conventional marker substrates shown in Table 2 (with the notable exception of 7-ethoxyresorufin, which is fluorogenic). Most of these substrates are not amenable to rapid analytical methods that make use of plate readers. Because of this, most definitive CYP inhibition studies must employ some type of separation technique such as HPLC, GC, capillary electrophoresis, etc. These methods can be coupled with flow-through (or occasionally stop-flow) detection such as UV, fluorescence, radiometric, mass spectrometry, etc. With the exception of LC/MS/MS methods, most of these methods require lengthy analytical run times, which limits throughput regardless of any automation applied to the incubation step. Because of this limitation of chromatographic-based separations, several groups have developed radiometric assays based on radiolabeled conventional substrates, often with an extraction step, followed by scintillation counting (26,68–70). Test article interference is seldom a problem with radiometric methods, but their sensitivity varies with the specific activity of the substrate, which may necessitate the use of undesirably high protein concentrations and/or lengthy incubation times. In addition, radiometric methods are undesirable from a waste management perspective.

Given the analytical sensitivity afforded by LC/MS/MS or LC/ESI/TOF-MS, with analytical run times on the order of just a few minutes, many groups have developed "cocktail" or "N-in-one" methods to accelerate the evaluation of CYP inhibition. Two types of cocktail approach are used: one involves preincubation pooling of multiple marker substrates; the other involves postincubation pooling of multiple samples that were incubated with individual marker substrates. In either case, the selectivity of LC/MS/MS permits simultaneous detection of multiple marker metabolites with relatively little chromatographic separation. A disadvantage of the preincubation pooling method (the substrate cocktail method), in which multiple substrates are present simultaneously in the incubation medium, is that certain marker substrates can inhibit each other's metabolism even though each one is converted to a metabolite by a single CYP enzyme. For instance, nifedipine is a selective substrate for CYP3A4 ($K_m = \sim 10\ \mu M$) and yet it is a potent inhibitor of CYP1A2 ($K_i = 0.47\ \mu M$) (71). Nifedipine would therefore not be a good choice to include in a substrate cocktail with a CYP1A2 substrate. Positive control inhibitors can also have effects on more than one enzyme. For example, the potent CYP2D6 inhibitor quinidine can markedly *increase* the activity of CYP2C9 and CYP3A4 toward certain substrates (72,73). These problems are solved by incubating the marker substrates individually and pooling the samples after the incubations are terminated. However, a disadvantage of such postincubation pooling is that the samples become significantly diluted depending on the number of samples pooled. Unless more sensitive LC/MS/MS methods are used to compensate, this dilution effect may necessitate forming more metabolite by incubating the marker substrates with undesirably high protein concentrations or lengthy incubation times.

In the absence of analytical equipment that can detect extremely low (subnanomolar) concentrations of all typical CYP marker metabolites (e.g., $\ll 0.5$ ng/mL), certain compromises in the design of CYP inhibition studies are required. The optimal design of a definitive CYP inhibition study will therefore be based on a balance of microsomal protein concentrations, incubation times, marker substrates, positive control inhibitors, buffer components, automated liquid-handling systems, and analytical techniques all chosen specifically to minimize the limitations of each component. The following sections outline an optimized approach to a definitive examination of drug candidates and other test articles as direct and time-dependent inhibitors of the major drug-metabolizing CYP enzymes in human liver microsomes. The starting point for these studies is a single experiment to determine two IC_{50} values from the same seven concentrations of drug candidate: one for direct inhibition (zero-minute preincubation with NADPH) and one for time-dependent inhibition (30-minute preincubation with NADPH). Figure 7 shows the results from this type of experiment with the CYP3A4 metabolism-dependent inhibitor, mibefradil, for which there is an 87-fold shift in IC_{50} value following a 30-minute preincubation with NADPH. Later sections will describe the design of follow-up studies, but most of the basic principles outlined below for the IC_{50} experiment will apply to those studies as well.

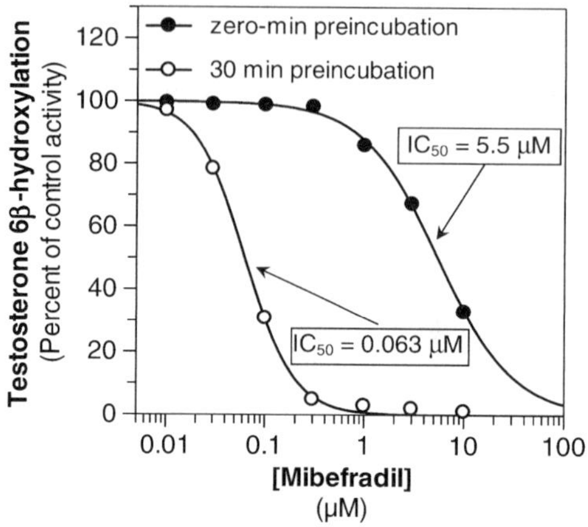

Figure 7 Metabolism-dependent inhibition of CYP3A4/5 by mibefradil. Mibefradil (0.01 to 10 μM) was examined as a direct-acting and metabolism-dependent inhibitor in human liver microsomes as outlined in the text. The IC_{50} value shifted approximately 87-fold after a 30-minute preincubation in the presence of NADPH. *Abbreviation*: IC_{50}, concentration of inhibition causing 50% inhibition.

1. Choice of Test System

The systems that have been used for CYP inhibition studies include purified-reconstituted CYP enzymes, microsomes from cell lines transfected with the cDNA encoding a single human CYP enzyme (recombinant human CYP enzymes), human liver microsomes, isolated/cultured hepatocytes, and human liver slices. The two systems that are used most often are human liver microsomes and recombinant human CYP enzymes. The choice of the in vitro system used for the evaluation of drugs as inhibitors of CYP enzymes can be a controversial subject. This controversy is partly because the principles of Michaelis-Menten enzyme kinetics (pure thoughts) are often applied to these (impure) systems. Systematic comparisons of all of the available systems have not been published for CYP inhibition studies. The advantages and disadvantages of each system are highlighted in the following sections.

a. Pooled human liver microsomes. The use of pooled human liver microsomes for CYP inhibition studies is well documented (2,22,23). Human liver microsomes contain all of the drug-metabolizing CYP enzymes expressed in human liver, although their levels can vary from one sample to the next. To circumvent the problem of variability, several individual samples of human liver microsomes are pooled, and this pool serves as the in vitro test system for evaluating drugs as inhibitors of human CYP enzymes. Because human liver microsomes are pooled from several individuals, they are more likely to contain the "average" levels of all CYP enzymes expressed in human livers. (Such pooled human liver microsomes are commercially available from several sources.) In addition, the ratio of NADPH-CYP reductase to CYP in human liver microsomes and the amount of cytochrome *b*5 and type of lipids are the same as those in the intact liver. Another advantage is that the same sample of pooled human liver microsomes (and often the same experimental conditions, i.e., protein concentration and incubation time) can be used to study all CYP enzymes of interest. Pooled human liver microsomes are also the system of choice for evaluating drugs as metabolism-dependent inhibitors of CYP enzymes, as they contain the complete enzymatic machinery to metabolize drugs that can inhibit CYP enzymes. This is an important consideration because the enzyme that converts a drug to an inhibitory metabolite may not be the one that is inhibited (covered in detail below).

A potential disadvantage of using pooled human liver microsomes is that these microsomes contain a large amount of lipid and protein that can decrease the free concentration of drug in the medium. However, to various degrees, this is a disadvantage of all available in vitro systems. This disadvantage can be largely overcome by using highly sensitive analytical methods (e.g., HPLC/MS/MS) to determine the amount of marker metabolite produced, so that low protein concentrations can be used (i.e., $\leq$0.1 mg/mL), as detailed in section II.A.3. Another potential disadvantage is that human liver microsomes are an exhaustible resource; therefore, each batch of microsomes is slightly different, although the variability can be minimized by pooling samples from a large number of individuals and by preparing large batches with careful selection of individual samples. Indeed, when these measures are taken, pooled human liver microsomes may be one of the most consistent in vitro systems, with a well-designed pool lasting for four years or more (sufficient for 200 definitive studies, one per week or more). Finally, with human liver microsomes, enzyme-selective substrates must be used. This is not a problem today because enzyme-selective substrates, metabolites, and their deuterated analogs (internal standards) are commercially available for all major CYP enzymes (2,3,11,22), but it is an issue with other enzyme systems such as UDP-glucuronosyltransferase enzymes (UGT enzymes). The use of conventional probe substrates with validated analytical methods is the expected choice for definitive CYP inhibition studies, so pooled human liver microsomes, conventional probe substrates, and validated LC/MS/MS methods represent the preferred test system for these studies.

b. Recombinant and purified human CYP enzymes. Membranes containing recombinant human CYP enzymes are commercially available from several sources. The major advantage of this system is its simplicity because such membranes contain only one human CYP enzyme. The same is true of purified human CYP enzymes (which must be reconstituted with NADPH-CYP reductase and lipid to be catalytically active). CYP enzymes that are expressed in extrahepatic tissues (e.g., CYP1A1 and CYP1B1) and allelic variants of polymorphically expressed enzymes, such as CYP2C9 and CYP2D6, are commercially available as recombinant CYP enzymes, as are drug-metabolizing CYP enzymes that are expressed in human liver but for which CYP-selective substrates have not been identified or well characterized (e.g., CYP2C18, CYP2J2, CYP4F2, and CYP4F3b). Another advantage of recombinant and purified CYP enzymes is that the selection of the substrate need not be limited to enzyme-specific substrates, as in the case of human liver microsomes (26,74,75). In fact, a substrate that is metabolized by all CYP enzymes would be particularly valuable for use with recombinant or purified enzymes.

One of the problems with recombinant and purified, reconstituted CYP enzymes is the variable levels of cytochrome b_5 and/or NADPH-CYP reductase, which can greatly affect turnover number (V_{max}) and, in some cases, K_m (76,77). Differences in kinetic constants have been observed between human liver microsomes and recombinant human CYP enzymes, and even between different expression systems for recombinant human CYP enzymes and purified enzymes. While the differences are sometimes artifacts of incubation conditions (especially those that are likely to violate the assumptions of the Michaelis-Menten equation), some differences appear to reflect genuine differences in the kinetics of reactions catalyzed by recombinant enzymes, purified enzymes, and human liver microsomes (Table 4). Kumar et al. compared the kinetics of the metabolism of four CYP2C9 substrates (diclofenac, *S*-warfarin, tolbutamide, and *S*-flurbiprofen) by various preparations of CYP2C9, namely human liver microsomes, recombinant CYP2C9 expressed in insect cells (Supersomes™), recombinant CYP2C9 expressed in *Escherichia coli* (Baculosomes®), and purified, reconstituted CYP2C9 (RECO®) (78). The authors found that the purified, reconstituted enzyme metabolized the substrates with K_m values that were 2- to 20-times higher than those obtained with human liver microsomes, whereas the two recombinant CYP2C9 preparations had K_m values that were within two- to threefold of the values obtained with human liver

Table 4 Comparison of the Apparent Affinity (K_m) with Which HLM and Recombinant CYP Enzymes (Bactosomes) Catalyze CYP-Selective Reactions

Enzyme	Marker reaction	HLM	Bactosomes
		K_m (μM)[a]	
CYP1A2	7-ethoxyresorufin *O*-dealkylation	0.347 ± 0.019	0.216 ± 0.013
CYP2A6	Coumarin 7-hydroxylation	0.879 ± 0.046	0.810 ± 0.071
CYP2B6	Bupropion hydroxylation	1290 ± 170	1570 ± 190
CYP2C8	Paclitaxel 6α-hydroxylation	9.71 ± 2.04	4.70 ± 1.61
CYP2C9	Diclofenac 4′-hydroxylation	4.97 ± 0.36	0.872 ± 0.040
CYP2C19	*S*-Mephenytoin 4′-hydroxylation	36.5 ± 1.6	13.0 ± 1.0
CYP2D6	Dextromethorphan *O*-demethylation	6.15 ± 0.62	1.04 ± 0.05
CYP2E1	Chlorzoxazone 6-hydroxylation	63.0 ± 1.9	409 ± 30
CYP3A4	Testosterone 6β-hydroxylation	105 ± 6	69.3 ± 8.1

[a]The K_m values were determined at XenoTech (unpublished data). Constants are shown ± standard error (rounded to 2 significant figures, with standard error values rounded to the same degree of accuracy as the constant), and were calculated using GraFit software, which utilized rates of product formation (triplicate data) at 13 substrate concentrations.

Abbreviation: HLM, human liver microsomes.

microsomes. In the case of diclofenac, the substrate inhibition characteristic of the reaction catalyzed by human liver microsomes was evident with RECO and Baculosomes, but not with Supersomes. The cause of this difference remains unclear, but it was postulated that it reflects differences in the access of diclofenac to an effector-binding site or differences in active site conformation. Baculosomes were found to have consistently higher V_{max} values across substrates, probably because they contained three times more NADPH-CYP reductase than did Supersomes. McGinnity et al. also reported differences in K_m values between human liver microsomes and recombinant CYP2C9 expressed in *E. coli*, as well as for CYP2D6 and CYP2E1 (78,79), but otherwise the K_m values were comparable between the two systems. Caution must be exercised when comparing the kinetics of a reaction catalyzed by a recombinant CYP enzyme and human liver microsomes when the latter reaction is catalyzed by two or more enzymes. For example, testosterone 6β-hydroxylation is catalyzed by both CYP3A4 and CYP3A5 in human liver microsomes, and recombinant CYP3A5 has nearly a 10-fold higher K_m than does CYP3A4 (80). For this reason, recombinant enzymes are more appropriate than human liver microsomes when it is necessary to differentiate the inhibitory potency of a drug toward two functionally similar enzymes, such as CYP3A4 and CYP3A5, and a substrate such as testosterone is used, which is not selective for either enzyme.

Differences in inhibitory potency were also found between in vitro systems, with human liver microsomes generally providing lower IC_{50} values across all enzymes (Table 5). Kumar et al. also determined the K_i values for inhibition of CYP2C9 by 12 competitive inhibitors in Supersomes, RECO CYP2C9, and human liver microsomes (78). K_i values in RECO CYP2C9 were approximately 3- to 14-fold higher than those determined with human liver microsomes, with the exception of fluvoxamine, ketoconazole, phenytoin, piroxicam, and tolbutamide (the latter two were 30–40% lower) (78). The K_i values in Supersomes were found to be within a factor of 3 of the values for human liver microsomes with the exception of fluvoxamine, ketoconazole, and piroxicam (in which case the K_i values were 9-, 5.5-, and 21-fold higher with human liver microsomes) and gemfibrozil and indomethacin (in which case the K_i values were 12.7- and 4.2-fold higher with recombinant CYP2C9) (78).

An important limitation of recombinant human CYP enzymes for CYP inhibition studies is that this test system fails to detect cases in which metabolites generated by one CYP enzyme inhibit another CYP enzyme (unless the two appropriate recombinant CYP enzymes are coincubated). CYP enzymes do in fact form metabolites that inhibit other CYP enzymes, and in some cases this occurs to a clinically significant extent. For example, fluoxetine is converted to norfluoxetine by CYP2C9, CYP2C19, and CYP2D6, the latter of which reaches plasma concentrations of ~1.6 μM with a half-life of ~6 days, whereas fluoxetine reaches plasma concentrations of ~1.7 μM with a half-life of ~53 hours (81). Although fluoxetine and norfluoxetine both cause clinically significant inhibition of CYP2D6, norfluoxetine also causes clinically significant inhibition of CYP3A4 and CYP2C19 (82–85).

Table 5 Direct Inhibition of CYP Enzymes in Human Liver Microsomes and Two Different Preparations of Recombinant Human CYP Enzymes

CYP enzyme	Marker reaction	Inhibitor	IC_{50} (μM)		
			HLM[a]	rCYPs[b]	rCYPs[c]
CYP1A2	7-Ethoxyresorufin O-dealkylation	α-Naphthoflavone	0.0149 ± 0.0006	0.0032 ± 0.0006	0.0081 ± 0.0002
CYP2A6	Coumarin 7-hydroxylation	Nicotine	44.2 ± 6.3	84.0 ± 2.5	144 ± 20
CYP2B6	Bupropion hydroxylation	Orphenadrine	165 ± 7	348 ± 12	236 ± 3
CYP2C8	CFD[d] O-dealkylation	Quercetin	2.69 ± 0.38	7.05 ± 0.84	7.22 ± 1.16
CYP2C9	Diclofenac 4′-hydroxylation	Sulfaphenazole	0.194 ± 0.014	1.20 ± 0.11	1.23 ± 0.07
CYP2C19	*S*-Mephenytoin 4′-hydroxylation	Modafinil	66.7 ± 6.6	85.1 ± 10.2	71.2 ± 7.5
CYP2D6	Dextromethorphan O-demethylation	Quinidine	0.071 ± 0.797	0.048 ± 1.07	0.032 ± 1.07
CYP2E1	Chlorzoxazone 6-hydroxylation	4-Methylpyrazole	0.549 ± 0.229	4.11 ± 0.47	1.40 ± 0.14
CYP3A4/5	Testosterone 6β-hydroxylation	Ketoconazole	0.041 ± 0.004	0.065 ± 0.002	0.165 ± 0.013

[a]Pooled human liver microsomes

[b]Recombinant human CYP enzymes expressed in bacteria (Bactosomes from Cypex)

[c]Recombinant human CYP enzymes expressed in insect cells (Supersomes™ from BD-Gentest)

[d]Chloromethyl fluorescein diethylether (216).

Amiodarone is another drug that is converted to metabolites that inhibit CYP enzymes other than those that form the inhibitory metabolites. Amiodarone is metabolized by CYP3A4 and CYP2C8 to desethylamiodarone (86). In recombinant human CYP enzymes, amiodarone inhibits CYP2D6 (K_i = 45 μM) followed by CYP2C9 (K_i = 95 μM) and CYP3A4 (K_i = 272 μM) (87). In human liver microsomes, however, amiodarone most potently inhibits CYP3A4 (IC_{50} = 15 μM) followed by CYP2C9 (IC_{50} = 25 μM) and CYP1A2 (IC_{50} = 86 μM) (88–90). Amiodarone is an irreversible metabolism-dependent inhibitor of recombinant human CYP2C8 (K_I = 1.5 μM, k_{inact} = 0.079 min^{-1}) and recombinant human CYP3A4 (K_I = 13.4 μM, k_{inact} = 0.06 min^{-1}) (87,91). Amiodarone also inactivates CYP2C8 (K_I = 51.2 μM, k_{inact} = 0.029 min^{-1}) and CYP3A4 (K_I = 10.2 μM, k_{inact} = 0.032 min^{-1}) in pooled human liver microsomes (91). Ohyama et al. reported that, compared with amiodarone, desethylamiodarone is a more potent direct-acting inhibitor of recombinant CYP1A2, CYP2A6, CYP2B6, CYP2C9, CYP2C19, CYP2D6, and CYP3A4, with K_i values ranging from 2.3 μM for CYP2C9 to 18.8 μM for CYP1A2 (87). Moreover, whereas amiodarone inactivates only recombinant human CYP3A4, desethylamiodarone also inactivates recombinant CYP1A2, CYP2B6, and CYP2D6 (87). Thus, desethylamiodarone is not only a more potent direct-acting inhibitor of more CYP enzymes than the parent drug, but it is also a metabolism-dependent inhibitor of more CYP enzymes than the parent drug. These effects were observed in human liver microsomes with the parent drug, but they were not observed in recombinant human CYP enzymes without directly examining the inhibitory effects of the metabolite. The lower IC_{50} or K_i values observed in human liver microsomes as compared with recombinant CYP enzymes for CYP1A2, CYP2C9, and CYP3A4 reflects the conversion of amiodarone to desethylamiodarone by CYP2C8 and CYP3A4 in human liver microsomes.

c. Hepatocytes. Cultured or suspended (fresh or cryopreserved) human hepatocytes contain a more complete and intact hepatic drug-metabolizing system than do human liver microsomes or recombinant human CYP enzymes, for which reason it is sometimes argued that hepatocytes are superior to pooled human liver microsomes or recombinant human CYP enzymes for the conduct of in vitro CYP inhibition studies, and therefore provide better predictive value than do the other test systems. Intuitively, it would appear that hepatocytes would yield superior data (i.e., better IC_{50} and K_i values) with which to make in vivo predictions, but surprisingly there is no compelling experimental evidence to support this viewpoint. Whereas hepatocytes have been shown on occasions to provide better predictions of clearance than human liver microsomes (92), this has not yet been convincingly demonstrated for CYP inhibition. In the clinical situation, the degree of CYP inhibition is typically determined by measuring the plasma AUC of a victim drug (usually administered orally) in the presence and absence of the inhibitor (also administered orally in most cases), which can be compared with predictions based on in vitro K_i values. K_i values can also be estimated in vivo, but the study design is more involved than that typically used to evaluate the inhibitory potential of a drug (93,94). In vivo K_i values can be determined in animals because it is possible to infuse a victim drug directly into the hepatic portal vein to accurately determine clearance, and subsequently to administer a wide range of bolus intravenous doses of a perpetrator drug in order to achieve a range of steady-state plasma concentrations. This approach was used in rats to determine that omeprazole inhibits the metabolism of diazepam with an in vivo K_i value of 21 μM (95), which is comparable to the in vitro K_i value determined in both rat liver microsomes and rat hepatocytes by a variety of experimental approaches (e.g., substrate depletion, individual metabolite formation, and weighted K_i determination) (96). Human hepatocytes have also been found to produce K_i and k_{inact} values for mechanism-based inhibition that are similar to those determined in human liver microsomes or recombinant human CYP enzymes (97). Inasmuch as the use of hepatocytes for in vitro CYP inhibition studies has not been shown to improve the prediction of in vivo inhibitory interactions, there is little to recommend this more complex and expensive system for in vitro studies. Furthermore, human hepatocytes do not offer many of the advantages afforded by human liver microsomes. In contrast to human liver microsomes, human hepatocytes are difficult to pool in sufficiently large quantities to permit a detailed analysis of the kinetics of each marker substrate. Consequently, the measurement of CYP activity in hepatocytes is often measured under non-Michaelis-Menten conditions or, if they are, with substrate concentrations that greatly exceed K_m (contrary to recommendations that the concentration of CYP marker substrate be $\leq K_m$). In hepatocytes, a portion of the metabolite formed from various maker substrates may be conjugated, which further complicates the analysis of enzyme kinetics. It is not practical to prepare a pool of human hepatocytes that might support inhibition studies for a year or more, which can easily be accomplished with pooled human liver microsomes. Finally, in contrast to the situation with microsomes, cell viability is an issue with hepatocytes.

d. Liver slices. Liver slices have been used for studies of drug metabolism (98–102), enzyme induction (98,103), and CYP inhibition (99). In addition to being plagued with the same problems as noted for isolated hepatocytes, liver slices cannot be pooled, and even precision-cut liver slices (~20 cells thick) present a barrier to drug, metabolite, nutrient, and oxygen diffusion. It is possible, therefore, that an inhibitor may not reach the same cells as those reached by the marker substrate, which will lead to an underestimation of inhibitory potential (102). Given these problems and the paucity of published information on CYP inhibition studies with liver slices, this system is not recommended over pooled human liver microsomes or recombinant human CYP enzymes for in vitro CYP inhibition studies.

2. Selection of Incubation Conditions

CYP enzymes have different pH optima and optimal ionic strengths. However, optimizing conditions for each individual CYP enzymes would be impractical, and in many cases, nonphysiological. It is more convenient and practical to choose a single set of incubation conditions that will support high but not necessarily maximal activity for all CYP enzymes. If the drug candidate is metabolized by more than one CYP enzyme, standard incubation conditions will help prevent differences in its metabolism from assay to assay. A buffer consisting of potassium phosphate (50 mM, pH 7.4), $MgCl_2$ (3 mM), and EDTA (1 mM) is a good compromise of the different optimal conditions ($MgCl_2$ is added to support the binding of NADPH to NADPH-CYP reductase, and ethylenediaminetetraacetic acid (EDTA) is added to chelate iron that, in the presence of microsomes and NADPH, can support lipid peroxidation). To start the reactions, an NADPH-generating system can be used, such as one consisting of NADP (1 mM), glucose-6-phosphate (5 mM), and glucose-6- phosphate dehydrogenase (1 U/mL). Reactions can be terminated with an appropriate volume (usually an equal volume) of an organic solvent that is compatible with the analytical method to be used. In an automated system, it is most convenient to include the internal standard (preferably deuterated forms of the marker metabolite) at an appropriate concentration in the stop reagent.

As mentioned in previous sections, microsomal protein concentrations and incubation times must be chosen in such a way that initial rate conditions are achieved and nonspecific binding to microsomal protein and lipids is minimized. With the use of highly sensitive LC/MS/MS methods it is possible to maintain protein concentrations at $\leq$0.1 mg/mL and substrate incubation times $\leq$5 minutes at the K_m for each marker reaction. The LC/MS/MS LOQ should be chosen with the goal of detecting the amount of metabolite that represents $\geq$90% inhibition at one-half the K_m for the reaction under such conditions. Table 6 summarizes the LOQ, incubation time, and microsomal protein concentrations used in our laboratory. The use of nearly uniform incubation conditions minimizes interassay differences in drug candidate metabolic stability and nonspecific binding. The use of highly sensitive analytical methods also allows for a short incubation time with marker substrate (e.g., $\leq$5 minutes) relative to the long preincubation times with drug candidates that are recommended to assess time-dependent inhibition (i.e., $\geq$30 minutes).

When less sensitive HPLC/UV or other methods are used, longer incubation times and higher microsomal protein concentrations must be used. The *S*-mephenytoin assay for CYP2C19 activity emerges as a prime example of a potential source of artificially high IC_{50} or K_i values because as much 1 mg/mL of microsomal protein and a 30-minute incubation with marker substrate must be used to detect the marker metabolite when HPLC/UV methods are used. As discussed in section II.A.3 and shown in Figure 3, the IC_{50} or K_i value of montelukast as an inhibitor of CYP2C8 can change dramatically with protein concentration. In the case of a compound such as montelukast, which is highly lipophilic, the apparent IC_{50} or K_i value is much higher than the actual value when conventional HPLC/UV methods are used because these typically require the use of a high protein concentration. A similar effect is observed with long substrate incubation times when the drug candidate is rapidly converted to less

Table 6 Typical In Vitro Incubation Conditions for CYP Inhibition Studies in Human Liver Microsomes and Selected Analytical and Kinetic Parameters

CYP enzyme	Marker reaction	Protein (mg/mL)[a]	K_m (μM)[b]	V_{max} (pmol/min/mg)[b]	LC/MS/MS ionization mode[c]	LC/MS/MS Limit of quantitation (ng/mL)
CYP1A2	Phenacetin *O*-deethylation	0.1	63 ± 3	1200 ± 0	ESI+	1.0
CYP2A6	Coumarin 7-hydroxylation	0.0125	0.77 ± 0.02	2100 ± 0	ESI-	0.30
CYP2B6	Bupropion hydroxylation	0.1	50 ± 2	2000 ± 0	ESI+	2.0
CYP2C8	Amodiaquine *N*-dealkylation	0.1	5.8 ± 0.2	5100 ± 100	ESI+	2.0
CYP2C8	Paclitaxel 6α-hydroxylation	0.1	11 ± 2	570 ± 40	ESI+	0.5
CYP2C9	Diclofenac 4′-hydroxylation	0.1	7.5 ± 0.4	2300 ± 100	ESI-	3.0
CYP2C19	*S*-Mephenytoin 4′-hydroxylation	0.1	42 ± 2	180 ± 10	ESI-	0.5
CYP2D6	Dextromethorphan *O*-demethylation	0.1	7.2 ± 0.6	210 ± 10	ESI+	0.5
CYP2D6	Bufuralol 1′-hydroxylation	0.1	12 ± 2	94 ± 9	ESI+	0.35
CYP2E1	Chlorzoxazone 6-hydroxylation	0.1	27 ± 3	1700 ± 100	ESI-	2.0
CYP3A4/5	Testosterone 6β-hydroxylation	0.1	130 ± 10	9500 ± 200	ESI+/−	10
CYP3A4/5	Midazolam 1′-hydroxylation	0.05	5.4 ± 0.3	2100 ± 0	ESI+	1.0
CYP3A4/5	Nifedipine oxidation	0.1	7.8 ± 0.2	.3100 ± 0	ESI+	5.0
CYP3A4/5	Atorvastatin *ortho*-hydroxylation	0.1	43 ± 4	800 ± 50	ESI+/−	2.0
CYP4A11	Lauric acid 12-hydroxylation	0.1	1.4 ± 0.2	1000 ± 100	ESI-	4.0

[a]Protein concentrations are less than 0.1 mg/mL in two cases to avoid over-metabolism of the substrate. The buffer consists of potassium phosphate (50 mM, pH 7.4), $MgCl_2$ (3 mM), EDTA (1 mM), and the NADPH-generating system [NADP (1 mM), glucose-6-phosphate (5 mM), and glucose-6-phosphate dehydrogenase (1 U/mL)]. The incubation time is five minutes for all assays.

[b]The kinetic constants were determined at XenoTech (unpublished data) with a pool of 16 human liver microsomal samples. Constants are shown ± standard error (rounded to two significant figures, with standard error values rounded to the same degree of accuracy as the constant), and were calculated using GraFit software, which utilized rates of product formation (triplicate data) at 13 substrate concentrations. The K_m values are rounded up or down for a convenient design of IC_{50} and K_i determinations.

[c]Designates either positive or negative ionization by ESI (electrospray ionization).

Abbreviation: CYP, cytochrome P450.

inhibitory metabolites (inhibitor depletion). Alternatively, if a drug candidate is a metabolism-dependent inhibitor, longer substrate-incubation times will artificially decrease its zero-minute preincubated IC_{50} or K_i value, and may result in a failure to detect time-dependent inhibition (discussed further in sec. II.C.7.c).

3. Importance of Multiple CYP3A4/5 Marker Substrates

In the case of CYP3A4, the inhibitory potency of a drug can depend on the choice of marker substrate (104–107). For example, the rank order of potency of CYP3A4 inhibition by 18 flavonoids differed depending on whether triazolam or testosterone was chosen as the marker substrate (108). In our laboratory, we have found up to a 19-fold difference in IC_{50} values for the inhibition of CYP3A4 by the same compound when four different CYP3A4 substrates (testosterone, midazolam, nifedipine, and atorvastatin) were used. Additionally, CYP3A4 is susceptible to homotropic activation or autoinhibition by certain substrates, which is demonstrated by an S-shaped or bell-shaped curve of rate versus substrate (109). The binding of one ligand (substrate, inhibitor, or activator) to CYP3A4 can cause conformational changes that almost double the size of the active site, which allows additional ligands to bind, possibly in a stacked or side-by-side configuration (110). This approach has been proposed to be the mechanism for homotropic (substrate-mediated) activation of CYP3A4, and for the heterotropic activation of CYP3A4 by agents such as flavonoids (e.g., α-naphthoflavone). This of course complicates the interpretation of data obtained with a single CYP3A4 marker substrate. Consequently, the inhibition of CYP3A4 should be evaluated with at least two structurally unrelated marker substrates (e.g., midazolam and testosterone, in accordance with the FDA's recommendations). In our laboratory, we routinely evaluate CYP3A4 inhibition with four marker substrates, namely midazolam, testosterone, nifedipine, and atorvastatin.

4. Selection of Inhibitor and Substrate Concentration, and Solvent Effects

For IC_{50} determinations, the substrate concentration should be close to the K_m for the marker reaction. As discussed previously, this choice of substrate concentration allows an estimate of the K_i value because $IC_{50} = 2K_i$ for competitive inhibition and $IC_{50} = K_i$ for noncompetitive inhibition. For K_i determinations, a common substrate concentration scheme is $K_m/3$, K_m, $3K_m$, $6K_m$, and $10K_m$. Assuming that the K_m for the reaction has been accurately determined, this range of substrate concentrations will provide an adequate spread of data on an Eadie-Hofstee plot to readily observe the mechanism of direct inhibition. For some substrates, solubility can become limiting at concentrations $>2K_m$. In such cases, it becomes necessary to choose alternate concentrations so that no fewer than five concentrations are used in a K_i determination. The choice of substrate concentration for determination of the metabolism-dependent inhibition parameters, k_{inact} and K_I, is discussed in greater detail in section II.C.7.c.

The choice of inhibitor concentration should ideally be based on known or anticipated plasma or hepatic concentrations of the drug candidate. The highest concentration examined in vitro should be at least 10-times higher than the maximum in vivo plasma concentration, and it is not uncommon to use a maximum in vitro concentration that is 100-fold higher. When such in vivo concentrations are not known, it is typical to use in vitro concentrations ranging from 0.1 to 100 μM (solubility permitting). The important consideration for IC_{50} determinations is that inhibitor concentrations should span at least two orders of magnitude, and preferably three, because this range of inhibitor concentrations is required to increase the degree of inhibition from 0% to nearly 100%. If IC_{50} values are already known, K_i values can be determined with concentrations ranging from 1/4 to 10 times the IC_{50} value, provided that the analytical methods are sufficiently sensitive to detect metabolite when the inhibitor concentration is 10 times the IC_{50} and the substrate concentration is one-third K_m.

Many drug candidates (and even some marker substrates) tend to have poor aqueous solubility at physiological pH. This limits the highest concentration of drug that can be achieved in vitro. To potentially circumvent this problem, the drug can be dissolved in an organic solvent [such as, methanol, acetonitrile, ethanol, dimethyl sulfoxide (DMSO), etc.] or weakly acidic or alkaline solutions and delivering the drug to the incubation mixtures. Several studies have demonstrated that organic solvents can potently and selectively inhibit CYP enzymes (111–114). This observation is not surprising because several organic solvents are substrates for CYP enzymes. Although organic solvents tend to be relatively weak CYP inhibitors, it is important to realize that a 1% (v/v) concentration of some organic solvents in the final incubation mixture translates to a molar concentration of >100 mM. The most susceptible enzyme is CYP2E1, which is almost completely inhibited by some organic solvents at a final concentration of 1% (v/v). Some solvents are generally more potent inhibitors of CYP enzymes than others. For example, 0.1% (v/v) DMSO markedly inhibits CYP2E1 and significantly inhibits CYP2C9, CYP2C19, and CYP3A4/5 (unpublished observations). In contrast, methanol does not potently inhibit CYP2C19 or CYP3A4/5, but it does markedly inhibit CYP2E1 and to a lesser extent CYP2C9. Acetonitrile inhibits, activates, or has no effect on CYP2C9 activity, depending on which substrate is used (115). The take-home point is that no single organic solvent is ideal for all CYP enzymes and that the final concentration of the organic solvent should be minimized as much as possible [<1.0% and preferably <0.1% (v/v)]. However, the use of an arbitrary cut-off in terms of percent solvent may not be appropriate for all CYP enzymes. Methanol concentrations as high as 2% (v/v) can be used for several CYP enzymes without significant inhibition, whereas 0.2% (v/v) DMSO can significantly inhibit several CYP enzymes. For instance, DMSO competitively inhibits CYP2E1 with a K_i value of 0.03% (v/v) (116). However, even when

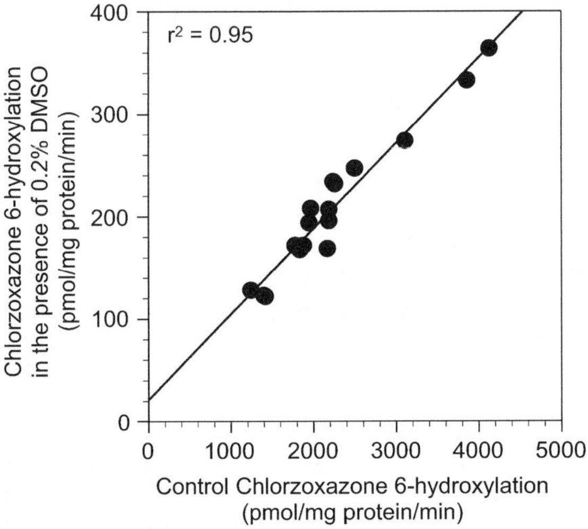

Figure 8 Correlation between chlorzoxazone 6-hydroxylase activity in the presence and absence of the CYP2E1 inhibitor DMSO [0.2% (v/v)]. Chlorzoxazone at the K_m for CYP2E1 (30 μM) was incubated with 16 individual samples of human liver microsomes in the presence of 0.2% DMSO. The strong correlation between residual and total chlorzoxazone 6-hydroxylase activity indicates that the reaction is still reasonably specific for CYP2E1 even in the presence of DMSO. *Abbreviation*: DMSO, dimethyl sulfoxide.

significant inhibition occurs, it may not necessarily affect the interpretation of additional inhibition by the drug candidate, as illustrated below.

With the use of highly sensitive LC/MS/MS methods, even substantial inhibition by an organic solvent may be acceptable provided the solvent does not *completely* inhibit the high affinity CYP enzyme responsible for metabolizing the marker substrate and provided the presence of solvent does not substantially alter the specificity of the marker reaction. CYP1A2 and CYP3A4 have been implicated as the low-affinity enzymes responsible for the 6-hydroxylation of chlorzoxazone by human liver microsomes, the typical marker reaction for CYP2E1 (117,118). In our laboratory, we have found that at a concentration routinely used to solubilize drug candidates for CYP inhibition studies [i.e., 0.2% (v/v)], DMSO causes approximately 75% inhibition of chlorzoxazone 6-hydroxylation at 30-μM chlorzoxazone (the K_{m} for CYP2E1), raising the possibility that CYP1A2 or CYP3A4 can now contribute significantly to the 6-hydroxylation of chlorzoxazone under such "low-CYP2E1" conditions. However, under such conditions, in human liver microsomes from 16 individuals, the residual chlorzoxazone 6-hydroxylase activity at a substrate concentration equal to K_{m} is still predominantly catalyzed by CYP2E1 as evidenced by the observation that the activity in the presence of 0.2% DMSO still correlates highly with chlorzoxazone 6-hydroxylase activity in the absence of DMSO ($r^2 = 0.95$), as shown in Figure 8, and it does not correlate well with CYP1A2, CYP3A4, or any other CYP marker activities ($r^2 < 0.38$). Because of the sensitivity afforded by the LC/MS/MS method used for this assay, further inhibition of the remaining activity can be easily detected, and most importantly, such inhibition represents inhibition of CYP2E1 when the assay is performed at a substrate concentration equal to K_m. Regardless of the assay, each experiment should include a no-vehicle control (no-solvent control) and a vehicle (solvent) control to assess the effect of the solvent under the conditions of a given experiment. The effect of the drug candidate is compared against the appropriate vehicle (solvent) control. It should be noted that although the presence of an organic solvent does not alter the specificity of the marker reaction, IC_{50} or Ki values for the inhibition of CYP2E1 may be artificially increased in the presence of high concentrations of organic solvents.

5. Intra-assay Controls

It is important to incorporate within each assay certain controls that prove that the test system is performing as expected. To verify that each assay is performed under initial rate conditions, incubations should be performed in the absence of the drug candidate at approximately half and twice the normal protein concentration and for approximately half and twice the normal incubation period.

Positive control inhibitors for each of the major CYP enzymes should also be included to further demonstrate that the test system is performing as expected. The direct-acting inhibitors used in our laboratory are summarized in Table 7, along with the IC_{50} values determined during assay validation and a comparison with literature values. It is worth noting that the positive control inhibitors used for CYP inhibition studies need not necessarily be CYP-selective inhibitors, in contrast to those used for reaction phenotyping, which should be CYP-selective inhibitors.

6. Automation, Analysis, and 96-Well-Plate Layout

A definitive, GLP-compliant CYP inhibition assay that examines both direct and time-dependent inhibition can be accommodated in a single 96-well plate. In our laboratory, a Tecan liquid-handling system (Tecan Inc., Research Triangle Park, North Carolina, U.S.) is used to pipette the substrate, NADPH, stop reagent (containing internal standard), and various premade buffer mixtures containing the microsomes and test article. Two different assays can be conducted simultaneously (i.e., two plates at a time) and this cycle can be easily repeated three or four times per day on a single liquid-handling system. Validated LC/MS/MS methods are used to accommodate the large number of samples generated by this automated procedure. High-throughput Shimadzu autosamplers (Shimadzu Scientific Instruments, Inc., Columbia, Maryland, U.S.) can accommodate four plates at a time, and with a sample analysis time of three to six minutes, all four plates can be analyzed in as little as 24 hours. The plate layout is optimized for LC/MS/MS analysis so that standard curve and quality control (QC) samples are present at the beginning, middle, and end of the analytical run (Fig. 9). Intra-assay controls as well as analytical controls are also included on the same plate.

Table 7 Inhibition of CYP Enzymes in Pooled Human Liver Microsomes by Direct Inhibitors (Positive Controls)

CYP enzyme	Marker reaction	Inhibitor	Average IC_{50}[a] (μM)	IC_{50} or K_i values reported in the literature (μM)
CYP1A2	Phenacetin *O*-deethylation	α-Naphthoflavone	0.0056 (7.1%)	0.01[b]
CYP2A6	Coumarin 7-hydroxylation	Nicotine	65 (5.2%)	24–50[c]
CYP2B6	Bupropion hydroxylation	Orphenadrine	180 (2.4%)	180–750[c]
CYP2C8	Amodiaquine *N*-dealkylation	Montelukast	0.036 (22%)	0.0092–0.15
CYP2C8	Paclitaxel 6α-hydroxylation	Quercetin	6.0 (9.2%)	1.1[b]
CYP2C9	Diclofenac 4′-hydroxylation	Sulfaphenazole	0.21 (17%)	0.3[b]
CYP2C19	*S*-Mephenytoin 4′-hydroxylation	Modafinil	72 (6.8%)	39[c]
CYP2D6	Dextromethorphan *O*-demethylation	Quinidine	0.079 (16%)	0.027–0.4[b]
CYP2D6	Bufuralol 1′-hydroxylation	Quinidine	0.13 (6.4%)	0.027–0.4[b]
CYP2E1	Chlorzoxazone 6-hydroxylation	4-Methylpyrazole	1.0 (14%)	1–50[c]
CYP3A4/5	Testosterone 6β-hydroxylation	Ketoconazole	0.046 (11%)	0.0037–0.18[b]
CYP3A4/5	Midazolam 1′-hydroxylation	Ketoconazole	0.015 (16%)	0.0037–0.18[b]
CYP3A4/5	Nifedipine oxidation	Ketoconazole	0.023 (13%)	0.0037–0.18[b]
CYP3A4/5	Atorvastatin *ortho*-hydroxylation	Ketoconazole	0.034 (14%)	0.0037–0.18[b]
CYP4A11	Lauric acid 12-hydroxylation	10-Imidazolyl decanoic acid	0.0093 (15%)	0.003[d]

[a]IC_{50} determinations were performed as described in the text. Values are rounded to two significant figures, and are the average of three or more determinations. Values in parentheses represent the relative standard deviation.
[b]These values were obtained from the FDA's Drug Interaction website: http://www.fda.gov/cder/drug/drugInteractions/tableSubstrates.htm#classInhibit.
[c]These values were obtained from the University of Washington's Metabolism and Transport Drug Interaction Database: http://www.druginteractioninfo.org
[d]Ogilvie et al. (217).
Abbreviation: CYP, cytochrome P450.

	1	2	3	4	5	6	7	8	9	10	11	12
A	CS6	0T	0P	QC1	TIC	SC	L2	L4	L6	SPD	SPI	CS6
B	ZS	0T	0P	QC2	TIC	SC	L2	L4	L6	SPD	SPI	ZS
C	CS5	0.5T	0.5P	QC3	**TIC**	**SC**	**L2**	**L4**	**L6**	**SPD**	QC2	CS5
D	CS4	0.5T	0.5P	QC1	**TIC**	**SC**	**L2**	**L4**	**L6**	**SPD**	QC3	CS4
E	CS3	1T	1P	TSC	NS	L1	L3	L5	L7	PD	PI	CS3
F	CS2	1T	1P	TSC	NS	L1	L3	L5	L7	PD	PI	CS2
G	CS1	2T	2P	**TSC**	**NS**	**L1**	**L3**	**L5**	**L7**	**PD**	**ZN**	CS1
H	BL	2T	2P	**TSC**	**NS**	**L1**	**L3**	**L5**	**L7**	**PD**	**ZN**	BL

Legend:

CS(*n*): Calibration standards	TIC: Test article interference check
ZS: Zero standard	TSC: Test article suppression check
BL: Blank	L(*n*): Test article concentration
(*x*)T: Time control samples	SPD: Solvent for MDI positive control
(*x*)P: Protein control samples	PD: MDI positive control
QC(*n*): QC samples	SPI: Solvent for direct positive control
NS: No Solvent Control	PI: Direct positive control
SC: Solvent Control	ZN: Zero-NADPH preincubation of L7

Shaded samples represent test-article samples and controls
Bolded samples are preincubated for 30 minutes

Figure 9 The 96-well microtiter-plate layout for a typical IC_{50} CYP inhibition experiment. *Abbreviations*: IC_{50}, concentration of inhibition causing 50% inhibition; CYP, cytochrome P450.

The analytical controls are intended to determine if the drug candidate causes ion suppression or chromatographic interference. Because the autosampler injects samples proceeding down the microtiter plate columns from left to right, the 0- and 30-minute preincubated samples are arranged so that they alternate, rather than placing all 30-minute preincubated samples at the end of the analytical run. This method minimizes bias that might result from slight changes in analytical response during the course of the analytical run. It is also for this reason that one set of standard curve samples is placed at the beginning and the other at the end of the analytical run. However, if deuterated forms of the metabolite standard are used as internal standard, changes in analytical response should affect the metabolite and internal standard to the same extent and therefore be corrected.

7. Follow-up Studies

Once an assessment of direct and time-dependent inhibition has been made with the initial IC_{50} determination, decisions regarding follow-up studies can be made as outlined in Figure 5. For direct inhibition, K_i determinations can be conducted for the most potently inhibited enzymes. Because K_i determinations provide information on the mechanism of inhibition (competitive, noncompetitive, etc.), the K_i values will allow a more precise rank-ordering of inhibitory potential and can better guide decisions regarding the clinical drug-drug interaction studies that may need to be performed when more than one CYP enzyme is inhibited with IC_{50} values that are within twofold of one another. If there is an indication of significant time-dependent inhibition, it may be necessary to perform additional experiments to further characterize this type of inhibition. The following sections will outline the rationale for choosing to perform follow-up studies and their experimental design.

a. K_i determinations. It is preferable that the decision to perform K_i determinations be based not only on IC_{50} values but also on the clinical concentration of the drug. An arbitrary cut-off for all drugs (e.g., K_i determination will be performed if $IC_{50} \leq 10$ μM) is not advisable. Instead, it is recommended that the cut-off point take the plasma concentration of the drug candidate into account. For example, the H_2-receptor antagonist, cimetidine, inhibits CYP1A2, CYP2C9, CYP2D6, and CYP3A4 with IC_{50} values as low as 300, 300, 130, and 230 μM, respectively. However, cimetidine can be administered in doses of up to 2400 mg/day and can reach plasma C_{max} values approaching 10 μM. Cimetidine has been found to cause clinically significant interactions with the CYP1A2 substrate, theophylline, the CYP2C9 substrate, tolbutamide, the CYP2D6 substrate desipramine, and the CYP3A4 substrate triazolam (119,120).

K_i determinations are conducted in essentially the same manner as IC_{50} determinations with respect to buffer components, analytical methods, incubation time, and microsomal protein concentration. The differences include the use of five substrate concentrations (i.e., $K_m/3$, K_m, $3K_m$, $6K_m$, and $10K_m$, solubility permitting), and the choice of inhibitor concentrations (i.e., one-fourth to 10 times the IC_{50} value, solubility permitting). As shown in Figure 10, K_i determinations require two 96-well microtiter plates, and the second plate can include a second set of K_m samples that are preincubated for 30 minutes with NADPH. This option allows for a definitive CYP inhibition study to be conducted for a drug candidate that has already been evaluated as a direct-acting inhibitor, but not as a metabolism-dependent inhibitor. Because K_i determinations are conducted at substrate concentrations from $K_m/3$ to $10K_m$, it is important to target an appropriate analytical range during development and validation of the analytical method. Ideally, the lower limit of quantitation should represent >90% inhibition at $K_m/3$ and the upper limit should normally represent the rate at $10K_m$ in the absence of inhibitor. A wide analytical range allows for a thorough characterization of inhibition to provide a more accurate K_i determination. The use of a well-characterized pool of several individual human liver microsomal samples (as discussed earlier) can obviate the need to change analytical ranges from one batch or lot to the next.

b. Time-dependent inhibition: NADPH-dependence and irreversibility. The initial IC_{50} experiment described previously permits a preliminary evaluation of whether any time-dependent inhibition occurs over seven concentrations of drug candidate and whether it is dependent on the presence of NADPH (for the

Plate 1

	1	2	3	4	5	6	7	8	9	10	11	12
A	CS6	0P	1T	1T	QC1	QC2	QC3	QC1	QC3	QC2	QC1	
B	ZS	0P	SC$_1$	SC$_1$	SC$_2$	SC$_2$	SC$_3$	SC$_3$	SC$_4$	SC$_4$	SC$_5$	SC$_5$
C	CS5	0.5P	L1$_1$	L1$_1$	L1$_2$	L1$_2$	L1$_3$	L1$_3$	L1$_4$	L1$_4$	L1$_5$	L1$_5$
D	CS4	0.5P	L2$_1$	L2$_1$	L2$_2$	L2$_2$	L2$_3$	L2$_3$	L2$_4$	L2$_4$	L2$_5$	L2$_5$
E	CS3	1P	L3$_1$	L3$_1$	L3$_2$	L3$_2$	L3$_3$	L3$_3$	L3$_4$	L3$_4$	L3$_5$	L3$_5$
F	CS2	1P	L4$_1$	L4$_1$	L4$_2$	L4$_2$	L4$_3$	L4$_3$	L4$_4$	L4$_4$	L4$_5$	L4$_5$
G	CS1	2P	L5$_1$	L5$_1$	L5$_2$	L5$_2$	L5$_3$	L5$_3$	L5$_4$	L5$_4$	L5$_5$	L5$_5$
H	BL	2P	L6$_1$	L6$_1$	L6$_2$	L6$_2$	L6$_3$	L6$_3$	L6$_4$	L6$_4$	L6$_5$	L6$_5$

Plate 2

	1	2	3	4	5	6	7	8	9	10	11	12
A	QC2	L7$_1$	NS$_1$	L7$_5$	TIC	TIC	**SC**	**SC**	**NS**	CS3		
B	QC3	L7$_1$	NS$_1$	L7$_5$	**TIC**	**TIC**	**L1**	**L1**	**NS**	CS2		
C	0T	L7$_2$	NS$_2$	NS$_5$	TSC	TSC	**L2**	**L2**	**ZN**	CS1		
D	0T	L7$_2$	NS$_2$	NS$_5$	**TSC**	**TSC**	**L3**	**L3**	**ZN**	BL		
E	0.5T	L7$_3$	NS$_3$	SPI	**SPD**	**SPD**	**L4**	**L4**	CS6	SS		
F	0.5T	L7$_3$	NS$_3$	SPI	**SPD**	**SPD**	**L5**	**L5**	ZS			
G	2T	L7$_4$	NS$_4$	PI	**PD**	**PD**	**L6**	**L6**	CS5			
H	2T	L7$_4$	NS$_4$	PI	**PD**	**PD**	**L7**	**L7**	CS4			

Legend:

CS(*n*): Calibration standards	TIC: Test article interference check
ZS: Zero standard	TSC: Test article suppression check
BL: Blank	L(n_s): Test article concentration, at $[S]_n$
(*x*)T: Time control samples	SPD: Solvent for MDI positive control
(*x*)P: Protein control samples	PD: MDI positive control
QC(*n*): QC samples	SPI: Solvent for direct positive control
NS: No Solvent Control	PI: Direct positive control
SC: Solvent Control	ZN: Zero-NADPH preincubation of L7

Shaded samples represent test-article samples and controls
Bolded samples are preincubated for 30 minutes
Black samples are blank

Figure 10 The 96-well microtiter-plate layout for a typical K_i CYP inhibition experiment. *Abbreviation*: CYP, cytochrome P450.

highest concentration of inhibitor). An additional experiment should be conducted to confirm NADPH dependence at multiple inhibitor concentrations and/ or preincubation times, and also to establish irreversibility of this inhibition prior to performing a full k_{inact} experiment (discussed in the next section) to avoid spending time on a larger experiment that may be unnecessary. A typical design for the first part of this experiment includes evaluating the drug candidate at the same concentration that provided the maximal change in percent inhibition from 0- to 30-minute preincubation in the initial experiment. Preincubations in the presence and absence of NADPH for 0, 15, and 30 minutes should be conducted under the conditions used in the initial IC_{50} experiment. In the second part of the experiment, the same concentration of drug candidate used in the first part is preincubated for the same of time, but the sample is diluted prior to measuring CYP activity. It is recommended that at least a 10-fold, and preferably a 25- or even a 50-fold dilution be used, which necessitates preincubating the drug candidate with a 10- to 50-fold higher protein concentration than used in the initial incubation. At the end of the preincubation period with this higher concentration of microsomal protein, an aliquot is removed and added to a normal incubation mixture, including the marker substrate, so that the 10- to 50-fold dilution produces the "normal" concentration of microsomal protein (typically 0.1 mg/mL), and the incubation is then continued for five minutes to allow for the formation of the marker metabolite. An increase in inhibition that is time- and NADPH-dependent and also resistant to dilution provides evidence that the drug candidate is an irreversible, metabolism-dependent inhibitor. In such cases, K_I and k_{inact} should be determined, as discussed in the following section.

The dilution method outlined above has some important limitations when examining mechanism-based inhibitors that are potent and highly protein-bound because in order to perform a dilution experiment, the microsomal protein concentration must first be increased during the preincubation. For instance, 8-methoxypsoralen is a very potent mechanism-based inhibitor of CYP2A6 ($k_{inact} = 2.1\ min^{-1}$, $K_I = 1.9\ \mu M$). The inhibition of CYP2A6 by 8-methoxypsoralen is irreversible; even overnight dialysis after a three-minute incubation with NADPH-fortified human liver microsomes and 8-methoxypsoralen (2.5 μM) does not restore CYP2A6 activity (121). For routine use as a positive control in the IC_{50} experiments (with a protein concentration of 0.0125 mg/mL), we have empirically determined that a low concentration of 8-methoxypsoralen (0.05 μM) causes extensive metabolism-dependent inhibition of CYP2A6 after a 30-minute preincubation with NADPH, and yet causes minimal direct inhibition in the zero-minute preincubation samples (Fig. 11A). However, after a 30-minute preincubation with 8-methoxypsoralen (0.05 μM) and a 25-fold higher protein concentration (0.3125 mg/mL) followed by a 25-fold dilution, no mechanism-based inhibition of CYP2A6 is apparent (Fig. 11B). In this case, increasing the concentration of 8-methoxypsoralen by 25-fold, to 1.25 μM, produces the expected result after a 25-fold dilution (Fig. 11D). However, in the absence of dilution, this concentration of 8-methoxypsoralen is too high because it causes complete inhibition of CYP2A6 after a 30-minute preincubation (Fig. 11C) and causes significant inhibition in the zero-minute samples (Fig. 11C, D). In any conventional approach to examine metabolism-dependent inhibition, the marker substrate (coumarin in the case of CYP2A6) must be added after the preincubation with inhibitor in order to measure the residual enzyme activity. Consequently, enzyme inactivation by the inhibitor still occurs during the substrate incubation period (in fact, it's virtually unavoidable), and it is especially pronounced for potent metabolism-dependent inhibitors like

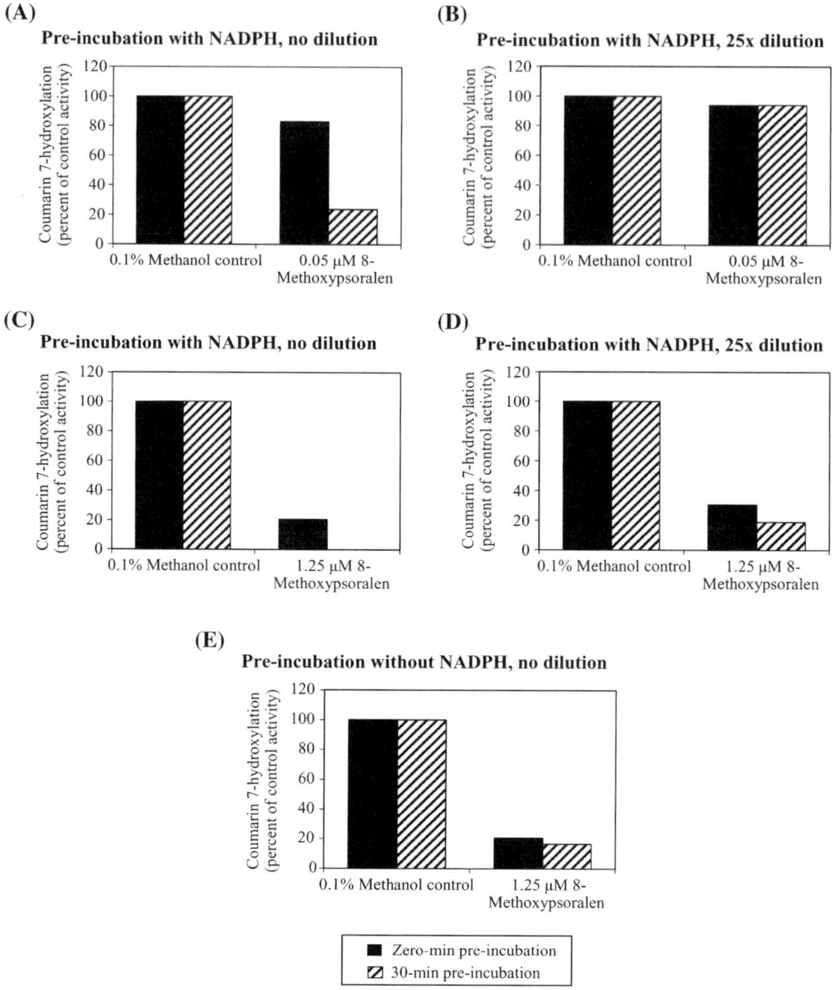

Figure 11 Irreversible inhibition of CYP2A6 by 8-methoxypsoralen at two concentrations with and without a dilution step. The overall design of the experiment is discussed in section II.C.7.b. Panels A and B show the effects of preincubating 8-methoxypsoralen (0.05 and 1.25 μM) for 30 minutes with NADPH-fortified human liver microsomes (0.0125 mg/mL) without a dilution prior to the incubation with substrate (coumarin). Panels B and D show the effects of preincubating 8-methoxypsoralen (0.05 and 1.25 μM) for 30 minutes with NADPH-fortified human liver microsomes (0.3125 mg/mL) with a 25-fold dilution prior to the incubation with substrate (coumarin). Panel E shows the effects of preincubating 8-methoxypsoralen (1.25 μM) for 30 minutes with pooled human liver microsomes (0.0125 mg/mL) in the absence of NADPH without a dilution step prior to the incubation with substrate (coumarin). Inhibition in the latter case is caused by inactivation of CYP2A6 during the substrate incubation step (5 minutes) because it occurs to the same extent in both the 0- and 30-minute preincubation samples.

8-methoxypsoralen. The final concentration of 8-methoxypsoralen after the 25-fold dilution is only 0.002 μM (when the initial concentration was 0.05 μM), which is apparently too low to cause significant inactivation of CYP2A6 during the five-minute substrate incubation (Fig. 11B). As shown in Figures 11C and D, 8-methoxypsoralen at the high concentration of 1.25 μM causes significant inactivation of CYP2A6 even in the zero-minute preincubation samples. The rapid inactivation of CYP2A6 during the substrate incubation becomes readily apparent when one compares the effect of preincubating human liver microsomes with 1.25-μM 8-methoxypsoralen in the presence and absence of NADPH: nearly the same degree of inhibition occurs in both the 0- and 30-minute preincubation samples because of the inactivation of CYP2A6 that occurs during the subsequent five-minute incubation with substrate (coumarin) and NADPH (Fig. 11E).

The lack of effect of 0.05-μM 8-methoxypsoralen after dilution (with the 25-fold higher protein concentration) is presumably due to nonspecific binding of 8-methoxypsoralen to microsomal protein and lipids and/or depletion of this small amount of inhibitor so that the number of CYP2A6 molecules inactivated is insignificant relative to the total amount of CYP2A6 present at the higher protein concentration. The effects of microsomal protein concentration on the inactivation of CYP1A2 by furafylline and of CYP2A6 by 8-methoxypsoralen are contrasted in Figure 12A and B, respectively. These experiments were conducted with microsomal protein concentrations of 0.1 and 1 mg/mL with a five-minute substrate incubation period. With this design, the marker reaction for CYP1A2 (phenacetin *O*-dealkylation) remains under initial rate conditions. The marker substrate for CYP2A6, on the other hand, is rapidly turned over (with a K_m of approximately 0.25 μM) and would be over-metabolized under such conditions. To avoid over-metabolism of coumarin, the substrate concentration was increased to 50 μM. These data show that inactivation of CYP1A2 by furafylline is virtually unaffected by microsomal protein concentration (over the range examined), whereas the apparent IC_{50} for inhibition of CYP2A6 after preincubation with 8-methoxypsoralen increased approximately 8-fold, with a 10-fold increase in microsomal protein concentration. For studies with potent inactivators that are also highly bound to protein, dialysis, rather than dilution, may be the preferred approach to investigate the irreversibility of metabolism-dependent inhibition.

c. k_{inact} determinations. Mechanism-based inhibition is characterized by the kinetic constants k_{inact} and K_I. The k_{inact} value is analogous to the Michaelis-Menten V_{max} and simply represents the maximal rate of enzyme inactivation at saturating concentrations of inhibitor. Likewise, K_I is analogous to K_m and represents the concentration of inactivator that supports half the maximal rate of inactivation. (Note that K_I for mechanism-based inhibition is neither the same symbol nor the same definition as K_i for direct inhibition). The results from the initial IC_{50} determination and the experiment to establish NADPH dependence and irreversibility can

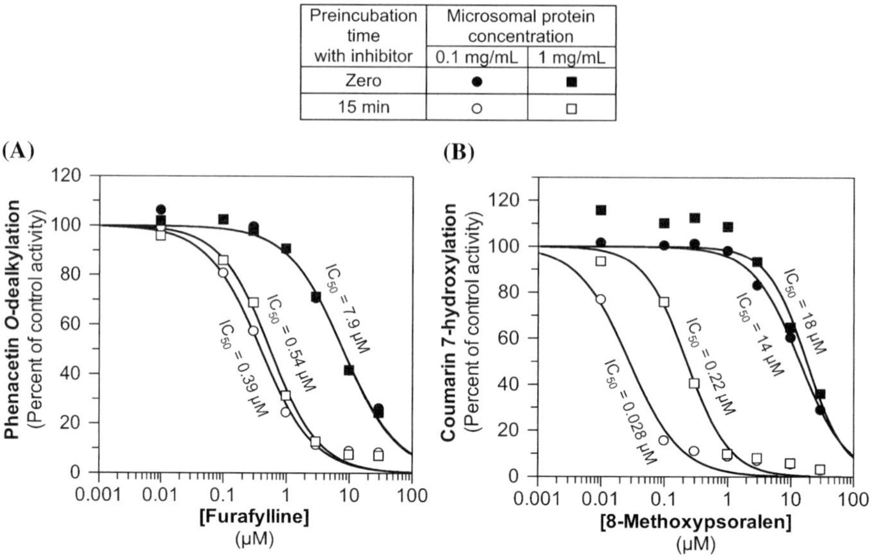

Figure 12 Effect of microsomal protein concentration on the mechanism-based inhibition of CYP1A2 by furafylline (**A**) and CYP2A6 by 8-methoxypsoralen (**B**). The typical IC_{50} experiment was conducted, except that microsomal protein concentrations of 0.1 and 1 mg/mL were utilized, a 15-minute preincubation period was used, and, to prevent over metabolism, coumarin (CYP2A6 substrate) was incubated at 50 μM. Phenacetin (60 μM) was used to measure CYP1A2 activity.

direct the choice of preincubation times and drug candidate concentrations to determine k_{inact} and K_I. It should be noted that because the choice of preincubation times is dependent on the particular mechanism-based inhibitor under investigation, the experimental design to determine k_{inact} and K_I is more difficult to automate than K_i and IC_{50} determinations for direct-acting inhibitors.

At a basic level, the design consists of choosing concentrations of drug candidate and preincubation times so that the percentage inhibition will range from 10% to 90% after preincubation, when possible. Beyond these basic experimental design characteristics, one must consider the assumptions inherent in a conventional experimental approach to the determination of k_{inact} and K_I values (i.e., an experimental approach based on determination of residual rates of enzyme activity with a marker substrate following various inactivation periods). These assumptions are: (1) there is negligible metabolism of the inhibitor during the preincubation stage, and (2) there is insignificant enzyme inactivation or direct inhibition during the substrate incubation stage. In fact, however, unless the inhibitor (drug candidate) is removed by dialysis prior to the substrate incubation, there is invariably some metabolism of the inhibitor during the substrate incubation period, and direct inhibition of the enzyme inevitably occurs to some extent because a mechanism-based inhibitor of an enzyme is, by definition, a substrate for that enzyme. The factors that should be optimized in order to provide the most accurate determination of k_{inact} and K_I values include the following:

1. The ratio of the preincubation time with inhibitor to the incubation time with maker substrate
2. The dilution factor
3. The marker substrate concentration relative to K_m
4. Normalization of data for the spontaneous time-dependent loss in enzyme activity in the absence of inhibitor
5. The method of data transformation (i.e., use of natural log transformation, rather than a base-10 log transformation)
6. Data analysis (e.g., nonlinear vs. linear regression)

Each of these factors is discussed below.

The substrate incubation time should be short relative to the preincubation time to minimize further inactivation of the enzyme after the preincubation stage. Therefore to maximize the ratio of substrate incubation time to preincubation time, the substrate incubation time should be as short as possible (e.g., 5 minutes or less), but the microsomal protein concentration should still be kept as low as possible. This means that analytical sensitivity is very important. In the case of metabolism-dependent inhibitors, the use of a long substrate incubation time can lead to an artifactual overestimation of direct inhibition and a corresponding underestimation of mechanism-based inhibition potential because there will be appreciable enzyme inactivation even in the zero-minute preincubation samples. This point is illustrated in Figure 13. In this case, mibefradil appears to have nearly fourfold greater potency as a direct inhibitor when a long substrate incubation period is used (i.e., 0-minute IC_{50} value of 1.6 μM with a 30-minute substrate incubation vs. 6.0 μM with a 5-minute substrate incubation). As a result, the shift in IC_{50} value is only 23-fold with the longer substrate incubation time compared with an 86-fold shift with a shorter substrate incubation time. For some metabolism-dependent inhibitors, the blunting effect of long substrate incubation times could be even more pronounced, possibly leading to the erroneous conclusion that no metabolism-dependent inhibition occurs.

After the preincubation stage, the samples should be diluted at least 10-fold (and preferably 25- to 50-fold) prior to the substrate incubation stage to reduce the concentration of inhibitor and thereby minimize its direct inhibitory effects. The general design of the experiment is shown in Figure 14. If the substrate incubation is carried out with 0.1 mg/mL protein for two to five minutes, then the preincubation must be carried out with 1 mg/mL protein for samples destined to be diluted 10-fold, and with 5 mg/mL protein for samples destined to be diluted 50-fold. These very high protein concentrations can dramatically decrease the free (unbound) concentration of drug candidate. Consequently, a correction for protein binding during the preincubation stage is warranted, especially for basic lipophilic

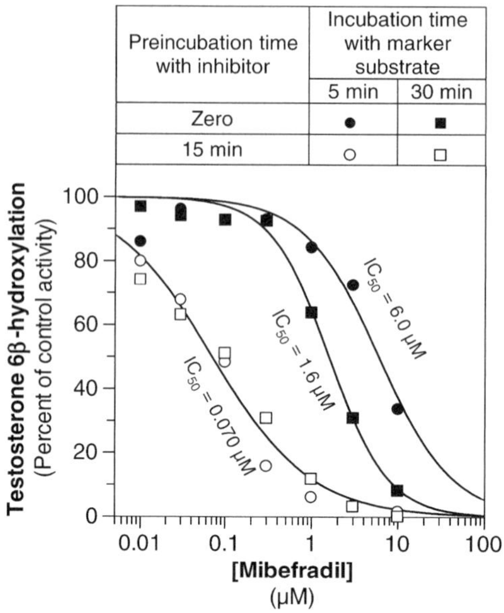

Figure 13 Effect of substrate incubation time on the metabolism-dependent inhibition of CYP3A4/5 by mibefradil. Mibefradil (0.01 to 10 μM) was examined as a direct-acting and metabolism-dependent inhibitor with either a 0- or 15-minute preincubation with NADPH followed by either a 5- or 30-minute incubation with testosterone (100 μM). In both cases, the IC_{50} value after a 15-minute preincubation with NADPH was approximately 0.07 μM. However, the IC_{50} value for direct inhibition (0-minute preincubation) varied nearly fourfold depending on the length of the substrate incubation, with a longer substrate incubation period diminishing the apparent impact of metabolism-dependent inhibition due to increased inactivation during the substrate incubation.

compounds (which are likely to bind to the negatively charged phosphate groups on the surface of the microsomal membrane). Van et al. investigated the effect of the dilution factor on the apparent inactivation kinetics of 3,4-methylenedioxyamphetamine and found that the k_{inact} value varied nearly threefold from 0.2 min^{-1} to 0.58 min^{-1} as the dilution factor increased 40-fold from 1.25- to 50-fold (122). The effect on K_I was also significant, ranging from 3.32 to 7.26 μM over this range of dilution factors.

To measure residual CYP activity, the substrate should be incubated at a high (saturating) concentration (e.g., 10–20K_m) to minimize the direct inhibitory effect of the mechanism-based inhibitor. However, there are two caveats to this rule: (1) the substrate must be soluble at this concentration and (2) the substrate must remain selective for the enzyme in question. If either solubility or selectivity is problematic at a high substrate concentration, then a lower substrate concentration must be used. The use of high substrate concentrations achieves two objectives: (1) it helps to diminish any competitive inhibition that might be caused by the remaining drug candidate and (2) it helps to decrease any further inactivation of the enzyme by diminishing the metabolism of the remaining drug candidate. Van et al. investigated the effect of the marker substrate concentration on the apparent inactivation kinetics of 3,4-methylenedioxyamphetamine, and found that the k_{inact} value ranged from 0.21 min^{-1} to 0.27 min^{-1} as the substrate concentration increased 10-fold from 2 to $20K_m$. The effect of substrate concentration on the K_I value was much more dramatic. It increased from 3.9 to 9.2 μM with increasing marker substrate concentration (122).

The spontaneous, time-dependent loss in enzyme activity in the absence of inhibitor must be taken into account when analyzing the data from mechanism-based inhibition studies (Fig. 14). For instance, the baseline loss of CYP3A4 activity in human liver microsomes incubated in the presence of NADPH and 400-μM alprazolam (which is a substrate for—but not a mechanism-based inhibitor of CYP3A4) has been reported to be from 0.0037 to 0.0039 min^{-1} (123,124). In other words, about 1% of CYP3A4 activity is lost every three minutes. Therefore, over the course of a 30-minute preincubation, one would expect to lose at least 10% of the initial CYP3A4 activity. To account for this loss, vehicle-control samples should be included and they should match all of the time points for the drug candidate. Even in the absence of solvent or substrate, there is a certain spontaneous loss of enzyme activity that appears to be different for each CYP. We have found that the loss of activity after a 30-minute preincubation with NADPH is approximately 30% for both CYP1A2 and CYP2D6, from 8% to 14% for CYP2A6, CYP2B6, and CYP3A4, and that there is little or no loss in activity or even an apparent increase in activity for CYP2C9, CYP2C19 or CYP2E1. Therefore, for certain CYP enzymes, significant error could be introduced if the data are not normalized for this spontaneous loss of activity during the preincubation stage. Normalization of the data can be accomplished by first calculating the decrease in activity over time for each concentration of drug candidate, including 0 μM (i.e., the solvent control):

$$\text{Activity over time} = \ln\left(\left(\frac{\text{Remaining activity}_t}{\text{Control activity}_t}\right) \times 100\right) \qquad (4)$$

where the remaining activity at time (t) is for a given concentration of drug candidate, and the control activity is the rate of reaction for the vehicle control at the corresponding preincubation time (rather than the zero-time control). By using the rate of reaction for the vehicle control at each time point (rather than the 0-minute vehicle control), the data are normalized for the spontaneous loss in CYP activity (an alternative approach is described below). Note that the natural log must be used (rather than the $\log_{10}$); otherwise the k_{inact} and K_I values will be off by a factor of 2.3.

Further data analysis initially consists of performing linear regression for each line defined by the natural log transformed data at each concentration of drug candidate. If a large dilution factor and saturating concentrations of marker

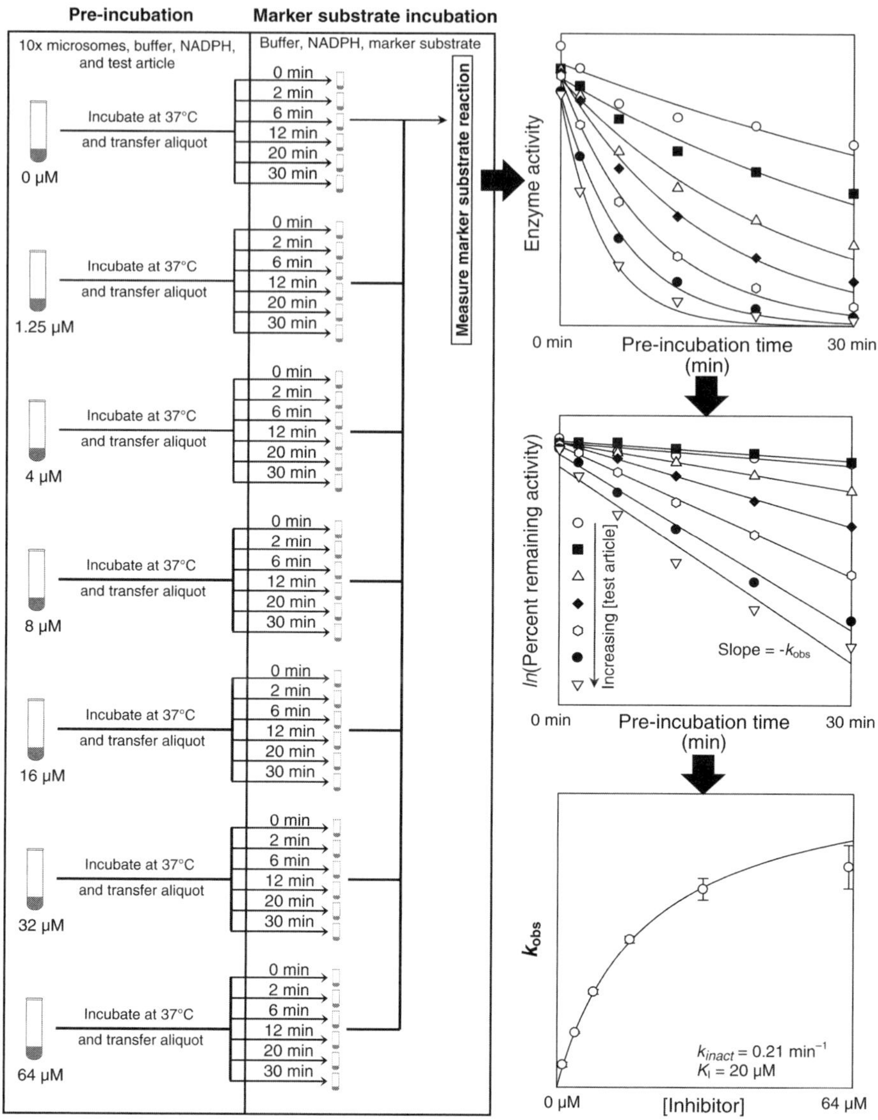

Figure 14 Design and graphical representation of irreversible or quasi-irreversible metabolism-dependent inhibition—determination of k_{inact} and K_i values. Actual data obtained for CYP2C8 inhibition by gemfibrozil glucuronide are shown. To determine the K_I and k_{inact} values for the inactivation of CYP2C8, various concentrations of gemfibrozil glucuronide (0.25 to 64 μM) were incubated for 2 to 30 minutes with pooled human liver microsomes (0.1 mg/mL) at 37°C. After the preincubation, an aliquot (40 μL) was transferred to another incubation tube (final volume 400 μL) containing paclitaxel (10 μM, due to limiting solubility) and an NADPH-generating system in order to measure residual

substrate have been used, these lines should intersect close to the "100% activity" value represented by the zero-minute vehicle control, as shown in Figure 14. The negative slope of the line is equal to k_{obs}, which represents the inactivation rate constant for that particular inhibitor concentration. These rates are then plotted against the inhibitor concentration and the data are fitted by nonlinear regression to the following equation:

$$k_{obs} = \frac{k_{inact} \cdot [I]}{K_I \cdot [I]} \tag{5}$$

Alternatively, if the individual data have not been normalized for the spontaneous loss of CYP activity, such a correction can be applied at this step. In this approach, the control activity in the above equation is always the zero-minute control for the solvent, rather than the solvent control at each time point. The apparent inactivation rate constant for the vehicle control, $k_{obs[I]=0}$ is then accounted for in the nonlinear regression according to the following equation (124):

$$k_{obs} + k_{obs[I]=0} + \frac{k_{inact} \cdot [I]}{K_I \cdot [I]} \tag{6}$$

These hyperbolic equations are analogous to the Michaelis-Menten equation. Nonlinear regression is preferable to the method proposed in the 1960s by Kitz and Wilson, which necessitates a double-reciprocal linear transformation of the data (analogous to a Lineweaver-Burk plot) that can bias the estimates of k_{inact} and K_I.

d. Evaluation of MI complex formation. The formation of MI complexes can be followed spectrophotometrically based on formation of an ~455-nm absorbing chromophore. In such an experiment, recombinant human CYP enzymes are usually the system of choice so that high concentrations of the CYP enzyme in question are achieved. Alternatively, if an individual human liver microsomal sample with very high levels of the CYP enzyme in question is available, this test system can also be used. The microsomes are diluted in 50 mM potassium phosphate (pH 7.4) or another appropriate buffer and divided

←

CYP2C8 activity. This procedure diluted the microsomes to 0.01 mg/mL and diluted gemfibrozil glucuronide to one-tenth its original concentration to minimize any direct inhibitory effects on CYP2C8. The incubations were then continued for two minutes to allow for conversion of paclitaxel to 6α-hydroxypaclitaxel. A short incubation with paclitaxel was used to minimize additional inactivation of CYP2C8 when measuring residual CYP2C8 activity. The data are plotted with incubation time on the *x* axis and enzymatic rates on the *y* axis. Subsequently, for each inhibitor concentration, the preincubation time (*x* axis) was plotted against the natural log of the percentage of remaining enzyme activity (*y* axis) (*middle graph*). The inhibitor concentration was then plotted against the *initial* rates of inactivation of the enzyme (negative slope of the lines in the middle graph). Nonlinear regression was then performed as described in the text (*bottom graph*).

equally between the sample and reference cuvettes to give approximately 0.2 to 1.0 nmol P450/mL. Baseline scans are recorded from 380 to 520 nm. The drug candidate is then added to the sample cuvette in 10 μL at a concentration known to produce maximal inhibition of the CYP enzyme. A corresponding volume of the solvent is added to the reference cuvette (another scan at this stage may reveal formation of a type I, type II, or reverse type I spectrum). The reaction is initiated with NADPH, which is added to both cuvettes to give a final concentration of 1 mM. Continuous scans (e.g., from 380 to 520 nm) are then recorded every minute or so for up to 30 minutes or more, depending on the spectral changes observed. The absorbance increase at approximately 455 nm, which is indicative of MI complex formation, can be used to estimate the amount of CYP complexed based on an extinction coefficient of 65 ($mM^{-1} \cdot cm^{-1}$). Positive controls should be used whenever possible. Figure 6 shows the MI complex spectra recorded when troleandomycin (50 μM) was incubated with NADPH-fortified human liver microsomes. Some MI complexes produce two absorbance maxima, one at approximately 455 nm [due to a complex with ferrous (Fe^{2+}) iron] and one at ~430 nm [due to a complex with ferric (Fe^{3+}) iron]. The ratio of these absorbance maxima is pH-dependent: alkaline conditions favor the 455-nm chromophore and acidic conditions favor the 430-nm chromophore.

e. Evaluation of covalent binding. Several methods have been developed to investigate the covalent binding of drug candidates to microsomal protein. The method used in our laboratory relies on the use of radiolabeled compound and is based on the method described by Munns et al. (125). To evaluate the ability of a drug candidate to bind covalently to protein, human liver microsomes (e.g., 2 mg protein/mL) are incubated at (37 ± 1)°C in 1-mL incubation mixtures containing potassium phosphate (50 mM, pH 7.4), $MgCl_2$ (3 mM), EDTA (1 mM), and a mixture of labeled and unlabeled candidate drug solution at varying concentrations in the presence and absence of an NADPH-generating system. Incubation times should be chosen to maximize binding (e.g., 30 to 120 minutes or more). Reactions are stopped at the designated time by the addition of 1 mL of 2% (w/v) sodium dodecyl sulfate (SDS) to solubilize the microsomal membrane (to reduce non-specific binding). Protein is then precipitated by the addition of 4-mL ice-cold acetone. Samples are kept on ice for 30 to 120 minutes, and then centrifuged at 920 × *g* for 10 minutes at 4°C to recover precipitated protein, and the amount of radioactivity in the supernatant fraction (1-mL aliquot) is determined by liquid scintillation counting. A 1-mL aliquot of supernatant fraction may be retained and stored at −80°C for potential future analysis. Precipitated protein is redissolved in 1 mL of 1% (w/v) SDS, followed by the addition of 4 mL of 1:2 (v/v) chloroform: methanol and a 30- to 60-minute incubation on ice to reprecipitate the protein. Precipitated protein is removed by centrifugation as above, after which the supernatant fraction is analyzed by liquid scintillation counting. The precipitated protein (the protein pellet) is then washed three times with neat methanol to remove traces of unbound drug candidate, with each wash step being followed by centrifugation at 920 × *g* for 10 minutes at 4°C [and by analysis of each supernatant (wash) fraction by liquid scintillation counting]. Following the methanol washes, additional extraction procedures with water or hexane may be performed to evaluate the ability of different solvents to remove unbound radioactivity from the precipitated protein. Following the final wash, the protein is redissolved in 2 mL of 4 M urea containing 1% (w/v) SDS and a 1-mL aliquot is then analyzed by liquid scintillation counting. A second 1-mL aliquot is used to determine the final protein concentration of each sample using a bicinchoninic acid protein assay kit. Values (bound radioactivity) from zero-time incubations are used to correct for nonspecific binding, and values from zero-cofactor incubations are used to correct for binding not mediated by CYP (or other NADPH-dependent mechanisms).

The general method described above can be combined with other techniques to further evaluate covalent binding of candidate drugs to CYP enzymes. For instance, the precipitated, washed protein can be heated for five minutes at 90°C and reconstituted with a loading buffer and subjected to SDS-PAGE (e.g., 10% polyacrylamide), as described for the binding of halothane to human liver microsomes (126). The gel can be stained with Coomassie Blue to locate proteins and then desiccated. The desiccated gel is then exposed to an appropriate film (e.g., Hyperfilm-βmax) and autoradiography is conducted for several days or even weeks to establish which protein or proteins are bound with radioactive material. In some cases it may also be possible to conduct in-gel proteolysis and LC/MS/MS to identify the specific protein to which the radioactive drug has become covalently attached.

D. Interpretation of CYP Inhibition Results

The primary goal of definitive in vitro CYP inhibition studies is to establish whether a drug can inhibit one or more major drug-metabolizing CYP enzymes in human liver, in order to assess the potential for clinically relevant drug-drug interactions. The importance of drug-drug interactions that contribute to adverse drug events (ADEs) cannot be understated given that Lazarou et al. estimated that, in 1994, over 2 million hospitalized patients had serious ADEs and 106,000 had fatal outcomes, which places ADEs (from all causes) only behind heart disease, cancer, stroke, and pulmonary disease as a leading cause of death (127). If there is little or no in vitro inhibition of CYP enzymes in a properly designed CYP inhibition study, at concentrations that are 10- to 100-fold higher than the known in vivo plasma concentrations, the interpretation is generally accepted to be straightforward (i.e., there will be no clinically significant CYP inhibition). However, as is discussed below, false negatives can occur. This is especially onerous considering that a negative result from an in vitro CYP inhibition study is often the basis for deciding not to conduct a clinical drug-drug interaction study. Furthermore, such negative results can appear in the labeling for the drug stipulating, a lack of inhibition of a given CYP enzyme. When substantial inhibition of one or more CYP enzymes is observed, this information must be put into context in order to make appropriate decisions about which clinical drug-drug interaction studies to conduct or even whether the drug should be developed further. Such decisions are based on in vitro to in vivo extrapolations. Great strides have been made in recent years in the *quantitative* prediction of metabolic drug interactions involving direct inhibition of CYP enzymes, which is the topic of chapter 12. Mechanism-based inhibition can complicate attempts to predict the clinical outcome, which is the topic of chapter 11. Although the FDA's recent draft guidance notes that "quantitative predictions of in vivo drug-drug interactions from in vitro studies are not possible," a rank-order approach can be used to prioritize clinical evaluations of drug-drug interactions.

1. FDA Perspective on Direct Inhibition

The FDA has adopted a conservative approach to interpreting in vitro CYP inhibition data given that there is often uncertainty regarding the concentration of drug at the enzyme's active site and the extent of first-pass drug metabolism (presystemic clearance). The FDA further acknowledges that an experimentally determined K_i value may vary with incubation conditions and individual drugs due to factors such as nonspecific binding of the drug to protein or other incubation components. Because of these uncertainties, the FDA recommends that $[I]/K_i$ values be calculated with [I] representing the mean C_{max} value at steady state for total drug (i.e., bound plus unbound) determined after the highest proposed clinical dose and that the results be interpreted as follows:

$[I]/K_i > 1.0$: Clinically significant inhibition is probable.

$[I]/K_i = 0.1$ to 1.0: Clinically significant inhibition is possible.

$[I]/K_i < 0.1$: Clinically significant inhibition is unlikely.

When the value of $[I]/K_i$ is >1.0, an in vivo drug-drug interaction study is recommended. If the value is <0.1, the likelihood of an interaction is "remote" and an in vivo drug-drug interaction study may not be necessary (1,2). Given the fact that a negative result in an in vitro CYP inhibition study can potentially allow a drug to be marketed without the conduct of in vivo CYP-based drug-drug interaction studies, great care should be taken to avoid false negatives. It is partially for this reason that some pharmaceutical companies routinely choose to have definitive GLP-compliant CYP inhibition studies performed for lead compounds even though the FDA does not require that in vitro studies be conducted in a strict GLP-compliant manner (20). The details on experimental design in section II.C illustrated how to avoid false negatives in CYP inhibition studies and how to obtain reliable measurements of a drug candidate's inhibitory potential.

When a drug's clearance is completely dependent on a single CYP enzyme (i.e., $f_{m(CYP)} = 1.0$), an $[I]/K_i$ ratio of 0.1 would theoretically result in only a 10% increase in AUC of that drug due to inhibition of the relevant CYP enzyme. Therefore, this criterion for possible exclusion from a clinical drug-drug interaction study is more conservative than the upper limit of the bioequivalence goalpost of 125% (see "Introduction"). In actual practice, it is uncommon to encounter a drug with an $f_{m(CYP)} = 1.0$. For instance, the CYP2C9 substrates tolbutamide and *S*-warfarin have $f_{m(CYP)}$ values of 0.78 and 0.91, respectively, whereas the $f_{m(CYP)}$ values for the CYP2C19 substrate omeprazole, the CYP2D6 substrate desipramine, and the CYP1A2 substrate theophylline are 0.87, 0.9, and 0.8, respectively (119). The CYP3A4 substrates buspirone, lovastatin, and simvastatin have reported $f_{m(CYP)}$ values of 0.99 (128), but only the CYP2C19 substrate, *S*-mephenytoin, is considered to have an $f_{m(CYP)}$ value of 1.0 (119) (although even *S*-mephenytoin is eventually eliminated in CYP2C19 PMs). The CYP2D6 substrate dextromethorphan is also estimated to have an $f_{m(CYP)}$ value of 1.0 based on the dextromethorphan/dextrorphan ratio, meaning that, for all practical purposes, CYP2D6 is the only enzyme that catalyzes the *O*-demethylation of dextromethorphan to dextrorphan (119). If a drug that was determined to be a CYP2C9 inhibitor with an $[I]/K_i$ value of 0.1 were coadministered with tolbutamide ($f_{m(CYP)} = 0.78$), the inhibitory drug would theoretically cause only a nominal increase in tolbutamide AUC (i.e., approximately 7.6%), assuming that no other pathways of elimination were concurrently affected (e.g., renal clearance). More than twice this modest increase in tolbutamide exposure is obtained when normal doses of the OTC H_2-receptor antagonist, cimetidine, are coadministered with tolbutamide, which is deemed acceptable even though this small interaction is associated with mild adverse effects in some patients (e.g., increased hunger and lethargy) (129). However, the FDA is rightly concerned about the possibility of in vitro false negatives from a perspective of public safety and has illustrated this possibility with a review of several NDAs, some of which showed that the actual magnitude of the in vivo interaction was grossly underpredicted by the in vitro evaluation. In some cases, such as fluvoxamine, the cause of the underestimation is not known (130). In other cases, in vitro results underestimate clinical outcomes because the clinical interaction involves transporter-mediated interactions (131) or, in the case of the interaction between gemfibrozil and cerivastatin, partly because CYP inhibition is caused by a metabolite formed by glucuronidation, not by CYP-dependent metabolism (11).

Drug candidates examined for their potential to inhibit CYP enzymes in vitro may inhibit multiple CYP enzymes with estimated $[I]/K_i$ values greater than the "cut-off value" of 0.1. In such situations, to minimize the conduct of many unnecessary clinical drug-drug interaction studies, the FDA notes that "although quantitative predictions of in vivo drug-drug interactions from in vitro studies are not possible, rank order across the different CYP enzymes for the same drug may help prioritize in vivo drug-drug interaction evaluations" (1,2). With this approach, an investigator would identify the enzyme that is most potently inhibited in vitro (with an $[I]/K_i$ value > 0.1) and conduct a clinical drug-drug-interaction study with a selective in vivo probe substrate for that enzyme. If there were a significant interaction (i.e., a significant increase in exposure to the probe substrate), then the next most potently inhibited enzyme would be examined in vivo, and so on. Criteria have not yet been developed to guide decision making as to when the next most potently inhibited enzyme does *not* need to be examined in vivo. For instance, if CYP2C9 and CYP2D6 were inhibited in vitro with $[I]/K_i$ values of 0.16 and 0.14, respectively, and if a subsequent in vivo study established there was no clinically significant inhibition of CYP2C9, the question may remain as to whether a clinical study for CYP2D6 interactions would be required. Industry perspectives on the rank-order approach have been published, which attempt to define criteria that ultimately prevent false negatives (119,120). These perspectives on the rank-order approach will be covered in detail in section II.D.3.

2. Interpretation of Metabolism-dependent Inhibition

A number of methods to predict in vivo inhibition from quasi-irreversible or irreversible inhibition have been reported, including the following equation which relates changes in substrate AUC (AUC_i-to-AUC_{ui} ratio) to the fraction of dose metabolized by the inhibited enzyme:

$$\frac{AUC_i}{AUC_{ui}} = \frac{1}{f_{m(CYP)}/\left(1 + k_{inact} \cdot [I]/K_I \cdot k_{deg}\right) + \left(1 - f_{m(CYP)}\right)} \tag{7}$$

where [I] is the in vivo inactivator concentration, k_{inact} and K_I are the kinetic parameters (as described earlier), $f_{m(CYP)}$ represents the fraction of total clearance of the victim drug to which the affected CYP enzyme contributes, and k_{deg} is the first-order rate constant of in vivo degradation of the affected enzyme (124). Note that various "surrogate" concentrations can be used for [I]. The estimated $C_{max,u,inlet}$ (estimated unbound steady-state C_{max} at the inlet to the liver) can be used in this equation in an attempt to approximate the actual unbound concentration in the liver, as described by Kanamitsu et al. (132). The estimated $C_{max,u,inlet}$ is higher than the unbound systemic concentration, but less than the total systemic concentration. This relationship relies on certain assumptions including (1) the conditions of the well-stirred pharmacokinetic model are met, (2) the substrate exhibits linear pharmacokinetics and is metabolized only in the liver, and (3) the complete absorption from the gastrointestinal tract occurs (123). The rate of enzyme degradation has a dramatic impact on the predictions made from this equation. These values are difficult to determine in humans. Approximate estimates of the rate of CYP3A4 degradation have ranged from 14 to 35 hours (derived from various rat or Caco-2 cells). The rate of human CYP3A4 degradation has also been estimated in vitro from liver slices, and in vivo based on the rate of return to basal activity following induction by either carbamazepine, ritonavir, or rifampin in clinical studies. The clinical studies reflected significant interindividual variability, with estimates of CYP3A4 half-life ranging from approximately one to six days (i.e., $k_{deg} = 0.00008$ to $0.0005\ min^{-1}$) and an average value of three days (i.e., $0.00016\ min^{-1}$), which is consistent with an estimate from liver slices of 79 hours (128). The k_{deg} for intestinal CYP3A4 has been estimated to be $0.000481\ min^{-1}$ following recovery after inactivation by grapefruit juice (124). CYP3A5 appears to have a shorter half-life (e.g., 35 hours for CYP3A5 in liver slices) (128). The k_{deg} for CYP2D6 has been estimated to be $0.000226\ min^{-1}$ based on the rate of return to baseline CYP2D6 activity after discontinuation of paroxetine (20 mg/day for 10 days) (124). The k_{deg} for CYP1A2 has been estimated to be $0.000296\ min^{-1}$ based on the time course of deinduction following smoking cessation (124).

As is the case for direct inhibition, the FDA takes a conservative stance on interpreting time-dependent inhibition. The FDA states that "any time-dependent and concentration-dependent loss of initial product formation rate indicates mechanism-based inhibition" and that this finding should be followed up with human in vivo studies (1,2).

3. The Rank-Order Approach

Although a proven method to consistently predict drug-drug interactions in a quantitative manner from in vitro studies has not been rigorously validated to date, a properly designed in vitro CYP inhibition study should at least establish which CYP enzymes are more affected than others. This assumption is made because any in vivo phenomena that can lead to discrepancies between in vitro and in vivo data (e.g., active hepatic uptake, free fraction, etc.) should be the same for a given drug regardless of the particular CYP enzyme that is affected (119,120). If these assumptions are true, then clinical drug-drug interaction studies need only be conducted for the enzymes that are most potently inhibited in vitro. According to the rank-order approach, if less than a twofold increase in AUC is observed in the first clinical drug-drug interaction study (which is conducted with an in vivo marker for the CYP that was most potently inhibited in vitro), then no further clinical drug-drug interaction studies need to be performed. On the other hand, if $>$ 2-fold interaction is observed, the next most potently inhibited CYP enzyme should be examined in vivo and so on until a clinical study establishes at what point in the rank order there ceases to be clinically significant CYP inhibition.

To assess whether or not the rank-order of inhibitory potency is the same both in vitro and in vivo, Obach et al. examined data for 21 drugs (119,120). Taking a very conservative approach, the authors found that the rank-order approach worked as expected in 18 of 21 cases. Thus, it would seem that the rank-order approach can be effectively applied, and false negatives avoided, in the vast majority of cases. However, some caution is warranted, and exceptions to the rule will be highlighted below. In the study by Obach et al., three interactions of $>$ 2-fold would have been missed had the rank-order approach been strictly followed, and these were with cimetidine, fluvoxamine, and troleandomycin. For cimetidine, CYP2D6 is the most potently inhibited enzyme in vitro, followed by CYP3A4, CYP2C9, and CYP1A2. A clinical drug-drug interaction study with desipramine as a clinical probe for CYP2D6 showed only a 56% increase in AUC by cimetidine, so the clinical studies would have stopped, thereby missing the ability of cimetidine to cause a reported 120% increase in AUC of the CYP3A4 substrate triazolam. For fluvoxamine, the in vitro inhibitory potency is CYP1A2 $>$ CYP2C19 $>$ CYP2D6 $\approx$ CYP2C9 $>$ CYP3A4, but the clinical study cascade would have stopped with CYP2D6 (14% increase in desipramine AUC), thereby missing the reported ability of fluvoxamine to cause a 140% increase in the AUC of the CYP3A4 substrate buspirone. In the case of troleandomycin, the in vitro inhibitory potency is CYP3A4 $>$ CYP2C19 $>$ CYP1A2, and the clinically relevant CYP3A4 interaction would be found following the rank-order approach. However, troleandomycin caused only a 26% increase in the AUC of the CYP2C19 substrate omeprazole, which would have resulted in missing the reported 100% increase in AUC of the CYP1A2 substrate, theophylline. In all three cases, however, other clinical drug-drug interaction studies have been performed that demonstrate $<$ 2-fold interactions with either the same or alternative in vivo probe substrates, so these exceptions do not seriously undermine the rank-order approach. It should also be mentioned that when the impact on intestinal CYP3A4 inhibition (discussed further below) by fluvoxamine is taken into account, the fluvoxamine-buspirone interaction can be quantitatively predicted, so the failure of the rank-order approach in this instance may be largely related to the inhibition of intestinal CYP enzymes. Some other reasons for the potential limitations of the rank-order approach, which relies on in vitro hepatic CYP inhibition data, include the following: (1) a metabolite that does not form in routine, in vitro CYP inhibition studies is ultimately responsible for a clinically relevant drug-drug interaction and (2) clinically relevant interactions are mediated by one or more transporters.

a. Exceptions to the rank-order approach: CYP2C8 inhibition by gemfibrozil glucuronide. The severe interaction between the antilipemic fibrate, gemfibrozil (perpetrator), and the cholesterol-lowering statin, cerivastatin (victim), which led to the withdrawal of cerivastatin from the market, illustrates the first scenario listed above in which the rank-order approach fails to predict the clinical outcome. On the basis of postmarketing adverse event reports, a labeling change highlighting this drug-drug interaction was made in January 1999, approximately 18 months after cerivastatin's approval in the United States (June 1997). The manufacturer voluntarily withdrew cerivastatin in August 2001, citing 31 deaths in the United States due to rhabdomyolysis (a side effect of statins). Of these deaths, 39% involved concomitant use of cerivastatin with gemfibrozil. Later investigation by the European Agency for the Evaluation of Medicinal Products (EMEA) found 546 worldwide reports of cerivastatin-induced rhabdomyolysis, 55% of which involved concomitant administration of gemfibrozil (8). The EMEA found a total of 99 fatal cases, 36.4% of which involved concomitant administration of cerivastatin with gemfibrozil. The EMEA concluded that this interaction "could not have been predicted based on what is currently known about the metabolism of these drugs" (8).

The metabolism of cerivastatin was relatively well characterized by 2002, as indicated by the final cerivastatin label issued in May 2001: "In vitro studies show that the hepatic CYP enzyme system catalyzes the cerivastatin biotransformation reactions. Specifically, two P450 enzyme sub-classes are involved. The first is CYP2C8, which leads predominately to the major active metabolite, M23, and to a lesser extent, the other active metabolite, M1. The second is CYP3A4, which primarily contributes to the formation of the less abundant metabolite, M1." On the other hand, the in vitro inhibitory potential of gemfibrozil was not well characterized in 2002, although it was reported by that time that gemfibrozil was a more potent inhibitor of CYP2C9 than CYP2C8 in vitro (133,134). In a subsequent in vitro study, we evaluated gemfibrozil as an inhibitor of several CYP enzymes under similar incubation conditions for all CYP enzymes, and showed that the in vitro rank-order of IC_{50} values for CYP inhibition by gemfibrozil was CYP2C9 (30 μM) < CYP1A2 (99 μM) ≈ CYP2C19 (100 μM) < CYP2C8 (120 μM) < CYP2B6, CYP2D6, and CYP3A4 (>300 μM) (11). Had the rank-order approach been taken based on these in vitro data, a clinical drug-drug interaction would have been performed first for CYP2C9, which would have tested negative. Coadministration of gemfibrozil with the CYP2C9 substrate warfarin does not increase the plasma concentrations of either *R*- or *S*-warfarin (in fact, it actually decreases them slightly) (135). No clinically relevant interactions between gemfibrozil and drugs that are primarily metabolized by CYP1A2 or CYP2C19, the next most potently inhibited CYP enzymes, have been reported. Therefore, based on the rank-order approach, gemfibrozil is predicted not to inhibit CYP2C8. However, contrary to prediction, there have been several reports of clinically significant interactions (i.e., 2- to 8-fold increases in AUC) between gemfibrozil and cerivastatin, repaglinide, rosiglitazone, and pioglitazone, which are predominantly metabolized by CYP2C8, the fourth most potently inhibited CYP enzyme in vitro (9,136–138).

An important clue to explaining why gemfibrozil is a more potent inhibitor of CYP2C9 than CYP2C8 in vitro but is a more potent inhibitor of CYP2C8 than CYP2C9 in vivo was provided by Shitara et al. (10), who demonstrated that gemfibrozil 1-*O*-β-glucuronide is a more potent inhibitor of CYP2C8 than is gemfibrozil. These same authors demonstrated that gemfibrozil glucuronide inhibits in vitro the CYP2C8-mediated metabolism of cerivastatin as well as the OATP1B1-mediated uptake of cerivastatin. We later showed that not only is gemfibrozil glucuronide a more potent inhibitor of CYP2C8 than the aglycone, but that it is also a potent irreversible mechanism-based inhibitor of CYP2C8 (11). In simulations of the in vivo effect of gemfibrozil glucuronide based on the k_{inact} and K_I values determined in vitro, we predicted an 8- to 19-fold increase in the AUC of a concomitantly administered drug whose clearance was 95% dependent on metabolism by CYP2C8, which varied depending on which reported in vivo concentration of gemfibrozil glucuronide was utilized. The AUC ratio for repaglinide, cerivastatin, pioglitazone, and rosiglitazone when coadministered with gemfibrozil, is $\sim$8.0 (range 5.5–15), 4.4 (range 1.1–8.0), 3.2 (range 2.3–6.5), and 2.3 (range 1.5–2.8), respectively (9,136–138). However, since the fm(CYP2C8) values for these drugs are $<$0.95, it is likely that mechanisms other than CYP2C8 inhibition are also involved in the interaction.

In the case of gemfibrozil, at least two metabolic events are necessary to observe the clinically significant mechanism-based inhibition of CYP2C8, and the first one does not involve CYP enzymes. First, gemfibrozil must be converted by UGT2B7 (and possibly other UGT enzymes) to the acyl glucuronide (139). This is a reaction that will not normally occur during in vitro CYP inhibition studies with human liver microsomes. However, pretreatment of microsomes with alamethicin (a UGT activator) and inclusion of UDP-glucuronic acid (UDPGA) (the cofactor for UGT enzymes) will allow glucuronidation to occur without altering CYP activity. We demonstrated that such a technique can be used to form the glucuronide of gemfibrozil in human liver microsomal incubations, thus increasing the in vitro inhibitory potency of gemfibrozil toward CYP2C8 (Fig. 15) (11). The next metabolic event is oxidation of gemfibrozil glucuronide by CYP2C8, which is the basis for the mechanism-based inhibition of CYP2C8. This example illustrates that xenobiotic oxidation can occur after conjugation. CYP2C8 has also been found to catalyze the oxidation of two other glucuronides, namely the acyl glucuronide of diclofenac and the 17β-glucuronide of 17β-estradiol (140,141). In both cases, hydroxylation occurs well away from the glucuronide moiety (4′-hydroxylation in the case of diclofenac glucuronide and 2-hydroxylation in the case of estradiol-17β-glucuroinide). These examples establish that conjugates can be substrates for further metabolism by CYP enzymes, and that this metabolism can be clinically relevant. In fact, a recent report suggests that direct glucuronidation with subsequent oxidation (and therefore a combination of UGT2B7 and CYP2C8 in humans) may be the major determinants of diclofenac clearance in humans (possibly as high as 75%) and monkeys ($>$90%), as opposed to earlier in vivo data suggesting that oxidative metabolism is the major determinant (142). Prueksaritanont et al. (142) further note that there are no clinical reports that implicate pharmacokinetic interactions between diclofenac and potent CYP2C9 inhibitors or inducers. Taken together, these observations suggest that the CYP-mediated oxidation of glucuronide metabolites has implications not only for the prediction of in

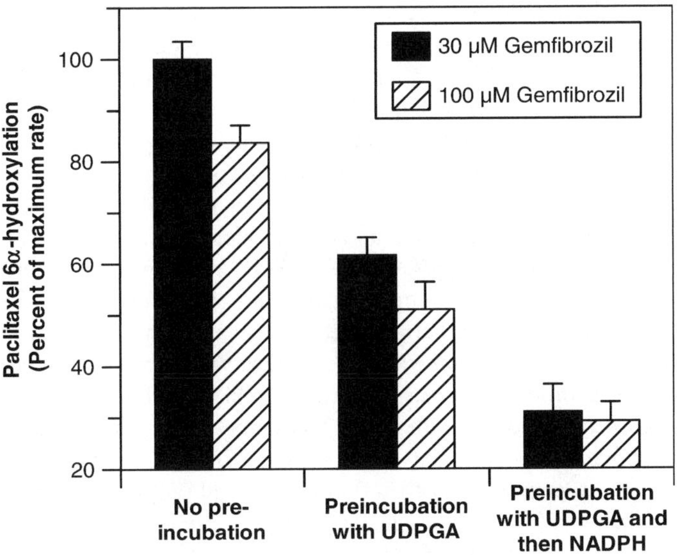

Figure 15 Two-stage activation of gemfibrozil by human liver microsomes in the presence of UDPGA and alamethicin (the glucuronidation step) and then NADPH (the CYP step). Gemfibrozil (30 or 100 μM) was preincubated for 0 or 60 minutes with pooled human liver microsomes (pretreated with alamethicin), UDPGA, and an additional 0 or 30 minutes with NADPH prior to measuring residual CYP2C8 activity. It should be noted that the preincubated samples were diluted 10-fold prior to measuring CYP2C8 activity; hence, the final concentration of gemfibrozil was 10 μM or less. Samples without preincubation served as controls. Values represent the average of triplicate determinations. *Abbreviations*: UDPGA, UDP-glucuronic acid; CYP, cytochrome P450.

vivo drug-drug interactions from in vitro data (i.e., gemfibrozil) but also for the prediction of in vivo clearance (i.e., diclofenac). These data further raise the concern that there may be a certain proportion of drugs that are rapidly and directly conjugated in vivo to such an extent that, if administered in high doses (i.e., gemfibrozil is given at 600 mg b.i.d.), inhibition of CYP enzymes by these conjugates may be clinically relevant, and therefore be exceptions to the standard rank-order approach.

b. Exceptions to the rank-order approach: Transporter inhibition by gemfibrozil and its glucuronide. The clinically significant interactions between gemfibrozil and the cholesterol-lowering statin drugs simvastatin and lovastatin illustrate the second scenario listed above in which the rank-order approach fails to predict clinical outcome (i.e., because clinically relevant interactions are mediated by one or more transporters). CYP3A4 is the major enzyme involved in the oxidative metabolism of simvastatin and lovastatin (which is not significantly inhibited by gemfibrozil or its glucuronide), and CYP2C8 does not contribute significantly to the overall metabolism of these drugs (although it does contribute to the hydroxylation of the acid forms of these drugs to some extent). Most statins have been shown to be substrates for OATP1B1 or one or more other transporters (e.g., OATP1B3, OATP2B1, OAT3, BCRP, MDR1, MRP2 etc.) (143). OATP1B1 in particular mediates the hepatic uptake of the acid form of statins. Gemfibrozil and its glucuronide are known to inhibit this transporter, and accordingly, gemfibrozil (600 mg b.i.d.) increases the AUC of the acid form of simvastatin and lovastatin by two- to threefold. Gemfibrozil also increases the AUC of atorvastatin, pravastatin, rosuvastatin, and pitavastatin, which suggests that OATP1B1 partly mediates these interactions because CYP2C8 plays only a minor role, if any, in the metabolism of these statins (143). Thus, the interactions between gemfibrozil and either simvastatin or lovastatin would not have been found through application of the rank-order approach because neither gemfibrozil nor its glucuronide inhibit CYP3A4 in vitro, so a clinical drug-drug interaction study with a CYP3A4 probe would not have been performed after a negative interaction study with warfarin. The important lesson from this example is that in vitro studies of CYP inhibition will only predict drug-drug interactions that involve CYP inhibition: they will not predict inhibition of other drug-metabolizing enzymes or transporters.

4. Route of Administration and Intestinal First-Pass Metabolism

The discussion above focuses on the rank-order approach for the prevention of false negatives rather than on a quantitative prediction of the in vivo magnitude of CYP inhibition from in vitro data, which will be discussed in greater detail in chapters 11 and 12. However, it is worth mentioning that there are some cases for which the rank-order of CYP inhibition in vitro and in vivo may be identical, but the magnitude of the actual interaction may differ drastically depending on the route of administration of the victim drug. CYP3A4 is the major CYP enzyme expressed in the mucosal enterocytes of the human small intestine, accounting for approximately 80% of the total immunoquantified CYP (144). Some drugs have low oral bioavailability because of extensive first-pass metabolism by CYP3A4 in the intestine (as well as the liver), including cyclosporine, midazolam, verapamil, nifedipine, and tirilazad. When intestinal first-pass metabolism is extensive, concomitantly administered CYP3A4 inhibitors cause a much greater increase in AUC when drugs are administered orally compared with intravenous administration. This process is illustrated by the interactions between either cyclosporine or tirilazad and the CYP3A4 inhibitor, ketoconazole, and between midazolam and either of the CYP3A4 inhibitors, itraconazole or fluconazole. The percent increase in the AUC of these victim drugs when administered orally ranged from 2.5- to 5-fold higher than the increases observed with intravenous dosing (145). Theoretically, the in vivo inhibition of intestinal CYP3A4 could be far greater than that predicted from in vitro K_i values and plasma inhibitor concentrations if the inhibitor and substrate are administered orally at the same time and if the concentration of the inhibitor in the gut lumen greatly exceeds those in plasma.

III. IDENTIFICATION OF CYP ENZYMES INVOLVED IN A GIVEN REACTION: REACTION PHENOTYPING

The in vitro technique of reaction phenotyping (also known as enzyme mapping) is the process of identifying which enzyme or enzymes are largely responsible for metabolizing a drug candidate (146). Although the experimental approaches described here focus largely on CYP enzymes, similar approaches can be used for other enzyme systems. A description of the experimental approaches to reaction phenotyping is preceded by an overview of the FDA's and PhRMA's perspective.

A. Guidelines for In Vitro Reaction Phenotyping Studies

The primary purpose of determining which enzymes are involved in the metabolism of a drug candidate in vitro is to determine its victim potential before advancing a candidate drug to a late stage of development. As noted in the Introduction, several examples of victim drugs have been withdrawn from the market because a single CYP enzyme largely mediates their clearance from the body so that inhibition of that enzyme by a perpetrator drug causes a large increase in the exposure to the victim drug. The development of victim drugs that are largely metabolized by a polymorphically expressed CYP enzyme is often halted during clinical development due to a high incidence of adverse events in PMs. In the United States, perhexilene and debrisoquine did not receive regulatory approval because they caused a high incidence of adverse effects in CYP2D6 PMs. On the other hand, there are very few cases of clinically significant drug-drug interactions related to drug-metabolizing enzymes other than CYP enzymes. Because of this trend, the importance of identifying the major CYP enzymes involved in the metabolism of a drug cannot be understated. Identification of the CYP enzymes involved in a drug's metabolism can also help predict the impact of CYP polymorphisms on the disposition of the drug. As detailed in the section II, the FDA issued a draft guidance entitled *Drug Interaction Studies—Study Design, Data Analysis, and Implications for Dosing and Labeling* (2). This guidance document along with the PhRMA perspective (5) provide the industry with a framework for the design of in vitro studies that can elucidate the enzymes involved in the metabolism of a drug or other xenobiotic. These guidelines will be briefly reviewed in the following sections.

1. Regulatory Perspective

The regulatory perspective will be covered in greater detail in chapter 16. This section will briefly highlight the latest recommendations regarding in vitro reaction phenotyping studies provided by the FDA (1–3). The FDA notes that one way to approach such studies is to first determine the metabolic profile of a drug and estimate the relative importance of CYP enzymes. It is recommended that preliminary experiments be conducted with human hepatocytes (or liver slices) followed by LC/MS/MS analysis to directly characterize the metabolites formed, and their relative importance. The relative importance of CYP enzymes to the formation of such metabolites can then be assessed in microsomes and/or hepatocytes in the presence and absence of NADPH (required for oxidation by CYP enzymes), 1-aminobenzotriazole (ABT) (a broad CYP enzyme inhibitor), and/or pretreatment of microsomes at 45°C to inactivate flavin monooxygenase (FMO). If the results of such preliminary experiments with human-derived systems (in vitro data) or clinical studies (in vivo data) indicate that CYP enzymes contribute significantly to a drug's clearance (i.e., >25%), then identification of the individual CYP enzymes involved in the metabolism of a drug is necessary, even if oxidative reactions are followed by conjugation. The guidance document also contains a table of preferred and acceptable chemical inhibitors to be used in reaction phenotyping experiments (sec. III.B). The remaining recommendations pertaining to reaction phenotyping studies in the guidance document can be summarized as follows:

1. Pooled human liver microsomes or individual human liver microsomal samples should be used for experiments designed to examine the effects of CYP-selective chemical inhibitors or selective inhibitory antibodies.
2. A bank of at least 10 individual human liver microsomal samples that have each been characterized for the drug-metabolizing CYP enzyme activities should be used for correlation analysis. These enzyme activities should be determined with appropriate marker substrates and experimental conditions. Furthermore, the variation in activity for each CYP among the individual samples should be sufficient to ensure adequate statistical power. Correlation analysis results should be considered as suspect if the regression line is unduly influenced by a single outlying data point or if the regression line does not pass near the origin.
3. Drug concentrations should be based on kinetic experiments whenever possible so that the concentration is $\leq K_m$ for a given reaction, and the incubations should be carried out under initial rate conditions.
4. Reliable analytical methods should be developed to quantify each metabolite that is produced by the individual CYP enzymes selected for identification.
5. Individual enantiomers of racemic drugs should be investigated separately.
6. Chemical inhibitors should be utilized at concentrations that maintain selectivity for a given CYP enzyme with adequate potency. A range of inhibitor concentrations can be used.
7. If a mechanism-based inhibitor is used, a 15- to 30-minute preincubation period with NADPH should be incorporated.
8. In experiments with recombinant human CYP enzymes, the rate of formation of a metabolite by multiple CYP enzymes does not provide adequate information on the relative importance of the individual pathways.
9. If CYP-selective inhibitory antibodies are used, multiple concentrations should be employed to establish a titration curve.

2. PhRMA Perspective

Major recommendations contained in the PhRMA paper (which add to the FDA documents) as they relate specifically to the design of in vitro reaction phenotyping studies can be summarized as follows:

1. If human metabolism studies with radiolabeled drug have not been performed prior to the conduct of reaction phenotyping studies, the initial experiments should use as "complete" an in vitro test system as possible, depending on the drug (e.g., tissue homogenates, liver slices, hepatocytes, etc.).
2. CYP enzyme reaction phenotyping can be performed with human liver microsomes.
3. Measurement of metabolite formation is preferred over substrate-depletion approaches, and linearity with incubation time and protein must be ensured.
4. If human in vivo concentrations of the test drug are not known, the substrate concentration should be $< K_m$.
5. The CYP enzymes of major or emerging importance include CYP1A2, CYP2B6, CYP2C8, CYP2C9, CYP2C19, CYP2D6, CYP3A4, and CYP3A5. Reaction phenotyping should be applied to these enzymes depending on the class of the drug, but should be applied to the major enzymes at a minimum (i.e., CYP1A2, CYP2C9, CYP2C19, CYP2D6, and CYP3A4).
6. Determining K_m values with recombinant human CYP enzymes can be used to differentiate high and low affinity enzymes involved in the metabolism of a drug candidate.
7. If one or more major circulating metabolites contribute significantly to the pharmacological action of a drug or if there are safety issues associated with such metabolites, reaction phenotyping for the individual metabolites should be considered.

B. Multiple Approaches to CYP Reaction Phenotyping

Four in vitro approaches have been developed for CYP reaction phenotyping. Each has its advantages and disadvantages, and each approach can provide incomplete or, on occasion, very misleading information. Therefore, a combination of approaches is highly recommended to identify which human CYP enzyme or enzymes are responsible for metabolizing a drug candidate. The four approaches to reaction phenotyping are:

1. Correlation analysis, which involves measuring the rate of xenobiotic metabolism by several samples of human liver microsomes and correlating reaction rates with the variation in the level or activity of the individual CYP enzymes in the same microsomal samples. This approach is successful because the levels of the CYP enzymes in human liver microsomes vary enormously from sample to sample (up to 100-fold), but

with judicious selection of individual samples, they can vary independently from each other.

2. Chemical inhibition, which involves an evaluation of the effects of known CYP enzyme inhibitors on the metabolism of a drug candidate by human liver microsomes. Chemical inhibitors of CYP must be used cautiously because most of them can inhibit more than one CYP enzyme and some chemicals can inhibit one enzyme but activate another. Some chemical inhibitors are mechanism-based inhibitors that require biotransformation to a metabolite that inhibits or inactivates CYP.
3. Antibody inhibition, which involves an evaluation of the effects of inhibitory antibodies against selected CYP enzymes on the metabolism of a drug candidate by human liver microsomes. Because of the ability of antibodies to inhibit selectively and noncompetitively, this method alone can potentially establish which human CYP enzyme is responsible for metabolizing a drug candidate. Unfortunately, the utility of this method is limited by the availability of specific inhibitory antibodies.
4. Metabolism by purified or recombinant (cDNA-expressed) human CYP enzymes, which can establish whether a particular CYP enzyme can or cannot metabolize a drug candidate, but it does not address whether that CYP enzyme contributes substantially to reactions catalyzed by human liver microsomes. The information obtained with purified or recombinant human CYP enzymes can be improved by taking into account large differences in the extent to which the individual CYP enzymes are expressed in human liver microsomes, which is summarized in Table 8, and by determining the

Table 8 Concentration of Individual P450 Enzymes in Human Liver and Intestinal Microsomes

	Liver				Intestine
	Specific content (pmol/mg protein)				
CYP enzyme	Source (1)	Source (2)	Source (3)	Source (4)	Source (5)
CYP1A2	52	45	42	15	
CYP2A6	36	68	42	12	
CYP2B6	11	39	1.0	3.0	
CYP2C8	24	64			
CYP2C9	73	96			11
CYP2C18	1	<2.5			
CYP2C19	14	19			2.1
CYP2D6	8	10	5.0	15	0.7
CYP2E1	61	49	22		
CYP3A4	155	108	98	40	58
CYP3A5	68	1.0			16
Total	503	534	344		

Source: (1) from Refs. 218, 219; (2–4) are from Ref. 193; (5) is from Ref. 144.

in vitro intrinsic clearance (V_{max}/K_m) of the drug candidate by each recombinant enzyme, which can be used to predict the contribution of each enzyme to metabolism of the drug candidate by human liver microsomes. The commercially available recombinant human CYP enzymes include some that are not expressed in human liver microsomes. For example, CYP1A1 and CYP1B1 are often included in a panel of recombinant CYP enzymes for reaction phenotyping even though these enzymes are expressed at such low levels in human liver microsomes that they do not contribute significantly to the hepatic metabolism of drugs. However, they may contribute to their metabolism in extrahepatic tissues. Other CYP enzymes are expressed in some but not all livers. For example, CYP3A5 is expressed in ~25% of human livers from Caucasian donors.

A combination of at least three approaches (correlation analysis, recombinant human CYP enzymes, and either chemical or antibody inhibition) is highly recommended to identify which human CYP enzyme or enzymes are responsible for metabolizing a drug candidate.

C. Stepwise Method for Reaction Phenotyping

1. Step 1

a. Selecting the appropriate in vitro test system. The two factors that frequently compromise a reaction phenotyping study are (1) the inappropriate selection of the in vitro test system and (2) the use of an inappropriate concentration of substrate (drug candidate), which is discussed in the next section.

Before conducting a CYP reaction phenotyping study, it is important to evaluate whether the disposition of the drug candidate is likely to be determined by its rate of oxidative metabolism by CYP enzymes. With the advent of combinatorial chemistry and high-throughput screening procedures, the selection of drug candidates is often biased against their interaction with CYP. Drug candidates are often selected for (1) good aqueous solubility and/or a high rate of dissolution in aqueous media, (2) their metabolic stability when incubated with NADPH-fortified human liver microsomes, and (3) their lack of inhibitory effect on CYP2D6 and CYP3A4 or possibly all the major drug-metabolizing CYP enzymes (CYP1A2, CYP2B6, CYP2C8, CYP2C9, CYP2C19, CYP2D6, and CYP3A4). Drug candidates that emerge from this selection process often contain exposed functional groups that can be metabolized by conjugating enzymes or other non-CYP enzymes. In such cases, a reaction phenotyping study designed to elucidate the role of CYP enzymes in the metabolism of the drug candidate may not produce clinically meaningful results.

The lipophilicity of the drug candidate is informative because drug candidates with a log $D_{7.4}$ greater than 1 are likely to require biotransformation to more polar metabolites to facilitate their elimination in urine or bile. The chemical structure of the drug candidate provides important information on the potential for enzymes other than CYP to be involved in its metabolism. Table 9 shows a variety

Table 9 Common Chemical Groups and Enzymes Possibly Involved in Their Metabolism

Chemical group	Enzyme(s)	Reaction(s)	Chemical group	Enzyme(s)	Reaction(s)
Alkane $R-CH_2-R$	CYP	Hydroxylation, dehydrogenation	Aldehyde 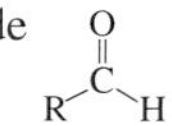	CYP, ALDH	Oxidative deformylation, oxidation to carboxylic acid
Alkene 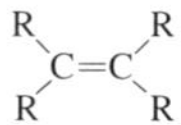	CYP, GST	Epoxidation, glutathione adduct formation	Amide $R-C(=O)-NH-R$	Amidase (esterase)	Hydrolysis
Alkyne $R-C\equiv C-R$	CYP	Oxidation to carboxylic acid	Aniline 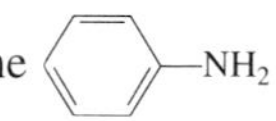	CYP, NAT, UGT, peroxidase, SULT	N-Hydroxylation, N-acetylation, N-glucuronidation, N-oxidation, N-sulfonation
Aliphatic alcohol $R-CH_2-OH$	CYP, ADH, catalase, UGT, SULT	Oxidation, glucuronidation and sulfonation	Aromatic azaheterocycles 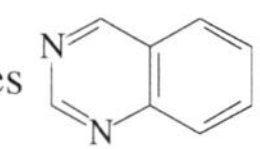	UGT, CYP, aldehyde oxidase	N-Glucuronidation, hydroxylation, N-oxidation, ring cleavage, oxidation
Aliphatic amine $R-NH_2$	CYP, FMO, MAO, UGT, SULT, MT, NAT, peroxidase	N-Dealkylation, N-oxidation, deamination, N-glucuronidation, N-carbamoyl glucuronidation, N-sulfonation, N-methylation, N-acetylation	Carbamate $R-NH-C(=O)-O-R$	CYP, esterase	Oxidative cleavage, hydrolysis

Amidine $HN{=}CR{-}NH_2$	CYP	N-Oxidation	Ester $R{-}CH_2{-}O{-}C({=}O){-}R$	CYP, esterase	Oxidative cleavage, hydrolysis
Arene 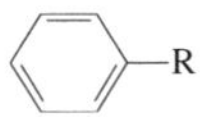	CYP	Hydroxylation and epoxidation	Ether $R{-}CH_2{-}O{-}CH_2{-}R$	CYP	O-Dealkylation
Carboxylic acid $R{-}COOH$	UGT, Amino acid transferases	Glucuronidation, amino acylation	Ketone $R{-}CH_2{-}C({=}O){-}R$	CYP, SDR, AKR	Baeyer-Villiger oxidation, reduction
Epoxide 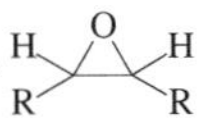	Epoxide hydrolase, GST	Hydrolysis, glutathione adduct formation	Phenol 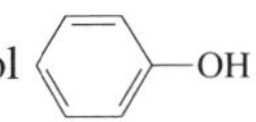	CYP, UGT, SULT, MT	Ipso-substitution, glucuronidation, sulfonation, methylation
Lactone 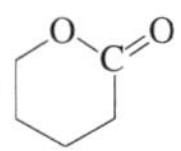	Lactonase (paraoxonase)	Hydrolysis (ring opening)	Thioether $R{-}CH_2{-}S{-}CH_2{-}R$	CYP, FMO	S-Dealkylation, S-oxidation

Abbreviations: ADH, alcohol dehydrogenase; ALDH, aldehyde dehydrogenase; AKR, aldo-keto reductases; GST, glutathione *S*-transferase; MT, methyltransferase; SDR, short-chain dehydrogenases/reductases; SULT, sulfotransferase; UGT, UDP-glucuronosyltransferase.

Source: Adapted from Ref. 146.

of chemical groups (but by no means all the chemical groups present in drug candidates) together with the enzymes that are typically involved in their metabolism. From Table 9 it is apparent that CYP enzymes can metabolize a very wide variety of functional groups. However, drug candidates with such functional groups as an ester, amide, or lactone may undergo hydrolysis whereas UGT, sulfotransferases (SULT), and other conjugating enzymes may directly conjugate those drug candidates containing such functional groups as a primary, secondary, or tertiary amine ($R\text{-}NH_2$, $R_1R_2\text{-}NH$, and $R_1R_2R_3\text{-}N$), an aliphatic hydroxyl group (an alcohol, R-OH), an aromatic hydroxyl group (phenol, R-OH), a free sulfhydryl group (R-SH), or a carboxylic acid (R-COOH).

Even when a drug candidate undergoes extensive metabolism by CYP, it is possible that the rate-limiting step in its systemic clearance is receptor-mediated hepatic uptake, not its rate of metabolism by CYP, as in the case of bosentan (Tracleer™) (147). Such drug candidates typically have a high liver-to-plasma ratio (which can be determined in laboratory animals) and their plasma clearance may be impaired by cyclosporine (a broad and potent inhibitor of several hepatic uptake transporters) rather than ABT (an inactivator of most drug-metabolizing CYP enzymes) (147,148). When the structure of a drug candidate leaves some doubt as to whether CYP alone will play the key role in its elimination, the selection of an appropriate in vitro test system can be guided by information on the routes of metabolism of the drug candidate in laboratory animals (and in humans, of course, although information from clinical studies is not always available when reaction phenotyping studies are conducted in the course of drug development). Rats treated with the general CYP inhibitor ABT can provide valuable information on the overall importance of CYP (both hepatic and intestinal) to the disposition of a drug candidate, as demonstrated recently by Strelevitz et al. for midazolam (148). CYP is unlikely to play a major role in the disposition of a drug candidate whose pharmacokinetic profile in rats is unaffected by treatment with ABT. A similar approach can be taken with human hepatocytes: CYP is unlikely to play a major role in the disposition of a drug candidate whose metabolic stability in suspended human hepatocytes is unaffected by a 30- or 60-minute preincubation with 100-μM ABT. The mechanism of CYP inactivation by ABT and experimental approaches to exploit the broad inactivating effect of ABT on most drug-metabolizing CYP enzymes are discussed below (in this section).

Studies with various subcellular fractions are useful to ascertain which enzyme systems are involved in the metabolism of a drug candidate. In the absence of added cofactors, oxidative reactions such as oxidative deamination that are supported by mitochondria or by liver microsomes contaminated with mitochondria membranes (as is the case with microsomes prepared from frozen liver samples) are likely catalyzed by monoamine oxidase (MAO), whereas oxidative reactions supported by cytosol are likely catalyzed by aldehyde oxidase and/or xanthine oxidase (a possible role for these enzymes in the metabolism of a drug candidate can be deduced in large part from the structure of the drug candidate).

Aldehyde oxidase is a molybdozyme present in the cytosolic fraction of liver and other tissues of several mammalian species (149). It catalyzes both oxidation reactions (e.g., aldehydes to carboxylic acids, azaheterocyclic aromatic compounds to lactams) and reduction reactions (e.g., nitrosoaromatics to hydroxylamines, isoxazoles to keto alcohols). In contrast to CYP, aldehyde oxidase does not require a pyridine nucleotide cofactor; hence, it catalyzes reactions without the need to add a cofactor. In general, xenobiotics that are good substrates for aldehyde oxidase are poor substrates for CYP and vice versa (150). Naphthalene (with no nitrogen atoms) is oxidized by CYP but not by aldehyde oxidase, whereas the opposite is true of pteridine (1,3,5,8-tetraazanaphthalene), which contains four nitrogen atoms. The intermediate structure, quinazolone (1,3-diazanaphthalene) is a substrate for both enzymes. This complementarity in substrate specificity reflects the opposing preference of the two enzymes for oxidizing carbon atoms; CYP prefers to oxidize carbon atoms with high electron density, whereas aldehyde oxidase (and xanthine oxidase) prefers to oxidize carbon atoms with low electron density. Aldehyde oxidase plays an important role in the metabolism of various drugs such as the hypnotic agent zaleplon (151–153), the antiepileptic agent zonisamide (154), and the antipsychotic agent ziprasidone (155). Thus, it is important to evaluate whether the metabolism of a drug is catalyzed only by CYP enzymes or if non-CYP enzymes such as aldehyde oxidase are also involved. If the latter is the case, it is useful to assess the relative contribution of CYP enzymes and non-CYP enzymes. One method to assess the involvement of CYP enzymes in the metabolism of a drug candidate is to incubate the drug candidate with test systems in the presence of ABT.

As a general inactivator of CYP, ABT provides useful information on the role of CYP in the in vitro disposition of a drug candidate. It can also be used in laboratory animals. Recently, ABT was used as an in vivo CYP inhibitor in rats to investigate the role of intestinal and hepatic metabolism in the disposition of midazolam. When rats were treated orally, ABT inhibited both intestinal and hepatic CYP enzymes involved in midazolam metabolism. However, only the hepatic CYP enzymes were inhibited when ABT was administered to rats intravenously (148). The mechanism of CYP inactivation by ABT involves the destruction of the heme moiety by benzyne, as illustrated in Figure 16. Although benzyne would be expected to bind to the heme moiety and inactivate all CYP enzymes, we have observed differential inactivation of CYP enzymes when NADPH-fortified human liver microsomes are incubated for 30 minutes with 2 mM ABT, as shown in Figure 17. Although ABT caused extensive inactivation (>80%) of CYP3A4, CYP1A2, CYP2A6, and CYP2E1, it only caused a 50–60% loss of CYP2C9 activity, and it caused only an ~25% decrease in CYP4A11 activity (unpublished results). Consequently, a reaction that is primarily catalyzed by CYP2C9 may be only partially inhibited by ABT.

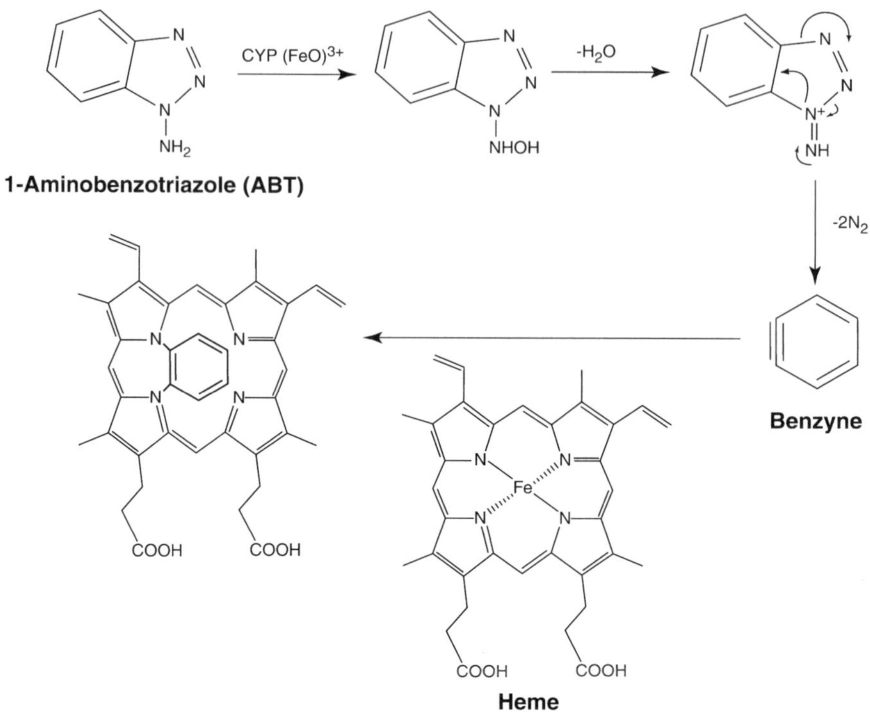

Figure 16 Mechanism of inhibition of the CYP enzyme inactivator ABT. *Abbreviations*: CYP, cytochrome P450; ABT, 1-aminobenzotriazole.

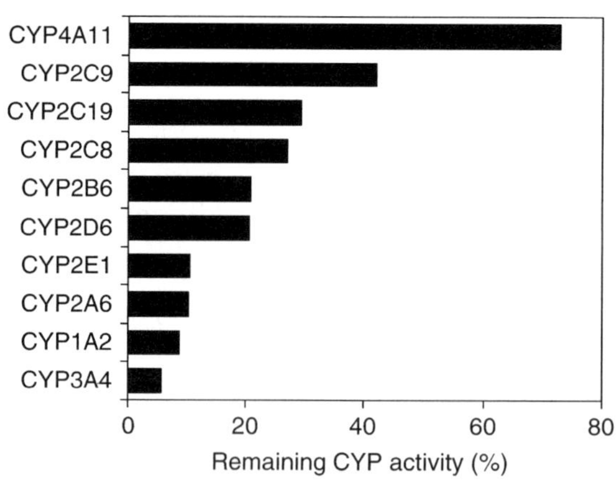

Figure 17 Effect of the CYP inactivator 1-aminobenzotriazle (1-ABT, 2 mM) on CYP activities in human liver microsomes (see text for details).

Typical experimental procedures are as follows: The test drug candidate is incubated with pooled human liver microsomes (e.g., 1 mg protein/mL) that were previously preincubated with ABT (1 or 2 mM) for 30 minutes at (37 ± 1)°C in the presence of an NADPH-generating system. Incubations of the drug candidate in the absence of ABT serve as controls. For hepatocytes, suspensions of freshly isolated or cryopreserved hepatocytes (1×10^6 cells/mL) are preincubated with 100-μM ABT for 30 minutes in 0.25 mL of Krebs-Henseleit buffer or Waymouth's medium (without phenol red) supplemented with FBS (4.5%), insulin (5.6 μg/mL), glutamine (3.6 mM), sodium pyruvate (4.5 mM), and dexamethasone (0.9 μM) at the final concentrations indicated. After the preincubation, the drug candidate is added to the incubation and the rate of metabolism of the drug candidate is compared in hepatocytes or microsomes with and without ABT treatment. A marked difference in metabolism caused by ABT is evidence that CYP plays a prominent role in the metabolism of the drug candidate.

Like CYP, FMO refers to a family of microsomal enzymes that require NADPH and oxygen (O_2) to catalyze the oxidative metabolism of drugs. Many of the reactions catalyzed by FMO can also be catalyzed by CYP enzymes (150,156). FMO enzymes catalyze N- and S-oxidation reactions but not C-oxidation reactions. Therefore, CYP but not FMO would be suspected of converting a drug candidate to a metabolite whose formation involves aliphatic or aromatic hydroxylation, epoxidation, dehydrogenation, or heteroatom dealkylation (N-, O- or S-dealkylation), whereas both enzymes would be suspected of forming metabolites by N- or S-oxygenation. Therefore, if NADPH-fortified human liver microsomes convert a nitrogen-containing drug candidate to an *N*-oxide or *N*-hydroxy metabolite, or if they convert a sulfur-containing drug candidate to a sulfoxide or sulfone, it is advisable to determine the relative contribution of FMO and CYP enzymes.

CYP and FMO enzymes can be distinguished by their differential sensitivity to inactivation with detergent or heat, as illustrated in Figures 18 and 19 for the metabolism of benzydamine, which undergoes N-oxygenation by FMO (to form an *N*-oxide) and N-demethylation by CYP (principally CYP3A4) to form *N*-desmethyl-benzydamine. As shown in Figure 18, treatment of human liver microsomes with the nonionic detergent Triton X-100 [final concentration 1% (v/v)] completely abolishes the CYP-dependent N-demethylation of benzydamine but only partially inhibits the FMO-dependent N-oxygenation. The converse is observed with heat inactivation, as shown in Figure 19. Incubating human liver microsomes at 50°C for one or two minutes causes extensive (>80%) loss of FMO activity with little loss (<20%) of CYP activity. Detergent and heat inactivation provide simple means to assess the relative contribution of CYP and FMO enzymes to oxidative reactions catalyzed by NADPH-fortified human liver microsomes. The general CYP inactivator ABT (see Fig. 17) can also be used to distinguish the role of CYP and FMO enzymes to reactions catalyzed by human liver microsomes or hepatocytes. Methimazole has been used to selectively inhibit FMO activity in liver

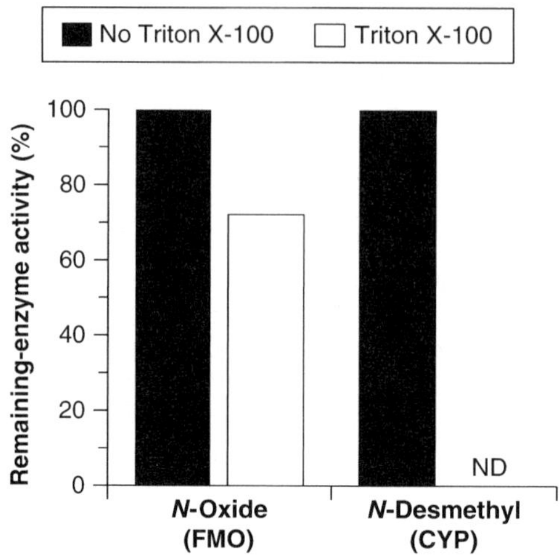

Figure 18 Effect of nonionic detergent (Triton X-100) on benzydamine N-oxygenation (FMO) and N-demethylation (CYP) by human liver microsomes. Benzydamine (500 μM) was incubated with pooled human liver microsomes (1.0 mg protein/mL) in tricine buffer (50 mM, pH 8.5 at 37°C) with or without Triton X-100 [1% (v/v)]. Reactions were initiated by the addition of an NADPH-generating system and stopped after 10 minute by the addition of an equal volume (500 μL) of methanol. Precipitated protein was removed by centrifugation, and an aliquot (25 μL) of the supernatant fraction was analyzed by HPLC with fluorescence detection. *Abbreviations*: FMO, flavin monooxygenase; CYP, cytochrome P450.

microsomes, but it can also inhibit CYP enzymes. Recombinant human FMO1, FMO3 [the most active form in human liver microsomes (157)], and FMO5 are commercially available and can be used to ascertain whether FMO enzymes are capable of catalyzing a given reaction.

Another important experimental consideration regarding FMO enzymes is that, in contrast to CYP enzymes, they cannot be inhibited by polyclonal antibodies, even when these antibodies recognize the FMO protein on a Western immunoblot.

b. Selecting the appropriate substrate concentration. CYP reaction phenotyping in vitro is not always carried out with pharmacologically relevant substrate concentrations. As a result, the CYP enzyme that appears to be largely responsible for metabolizing the drug candidate in vitro may not be the CYP enzyme responsible for its metabolism in vivo.

In general, CYP and other drug-metabolizing enzymes tend to catalyze reactions with relatively low affinity (with K_m values in the high micromolar to low millimolar range), as might be expected for enzyme systems with broad substrate specificities. Consequently, in most cases, the concentration of drugs in the plasma or liver is well below the K_m of the enzymes that metabolize them, although there

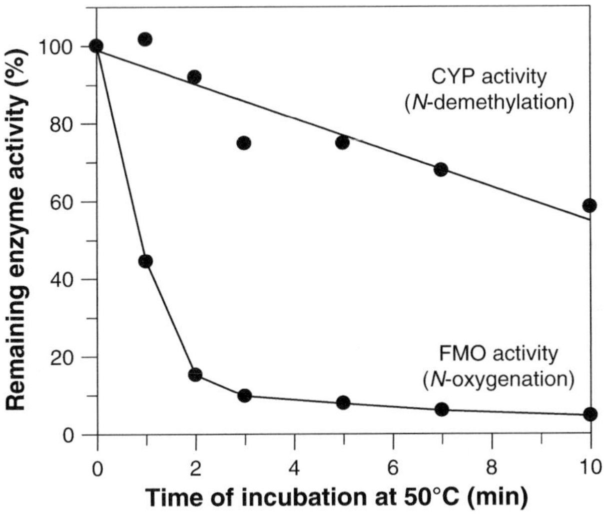

Figure 19 Time course of heat inactivation (50°C) of benzydamine N-oxygenation (FMO) and N-demethylation (CYP) by human liver microsomes. Human liver microsomes were preincubated at 50°C for 0, 1, 2, 3, 5, 7, and 10 minutes prior to measuring FMO and CYP activity toward benzydamine as outlined in Figure 18. *Abbreviations*: FMO, flavin monooxygenase; CYP, cytochrome P450.

are exceptions to this rule. Under these conditions (i.e., $[S] < K_m$), the rate of drug metabolism conforms to a first-order process so that a constant fraction of the remaining drug is cleared during each unit of time (and the unit of time required to clear 50% of the remaining drug is called half-life). When the substrate concentration exceeds K_m, as happens when someone drinks sufficient alcohol to saturate alcohol dehydrogenase, clearance becomes a zero-order process so that a constant amount of remaining drug, not a constant percentage of remaining drug, is cleared during each unit of time, until such time as the concentration of drug falls below K_m, at which point clearance conforms to a first-order process.

If two kinetically distinct enzymes are involved in a given reaction, there is a high probability that only the high affinity enzyme will contribute substantially to drug clearance (unless the drug is administered at sufficiently high doses to achieve hepatic drug levels that allow even the low affinity enzyme to contribute significantly to metabolite formation). If two kinetically distinct enzymes are involved in a given reaction, the sample-to-sample variation in metabolite formation by a panel of human liver microsomes can be determined at several drug concentrations to identify the enzyme that is more relevant at a given substrate concentrations. For example, the 5-hydroxylation of lansoprazole is mainly catalyzed by two CYP enzymes, CYP3A4 and CYP2C19 (158). At high substrate concentrations (~100 μM) the 5-hydroxylation of lansoprazole by human liver microsomes is dominated by CYP3A4, a low-affinity, high-capacity enzyme. However, at

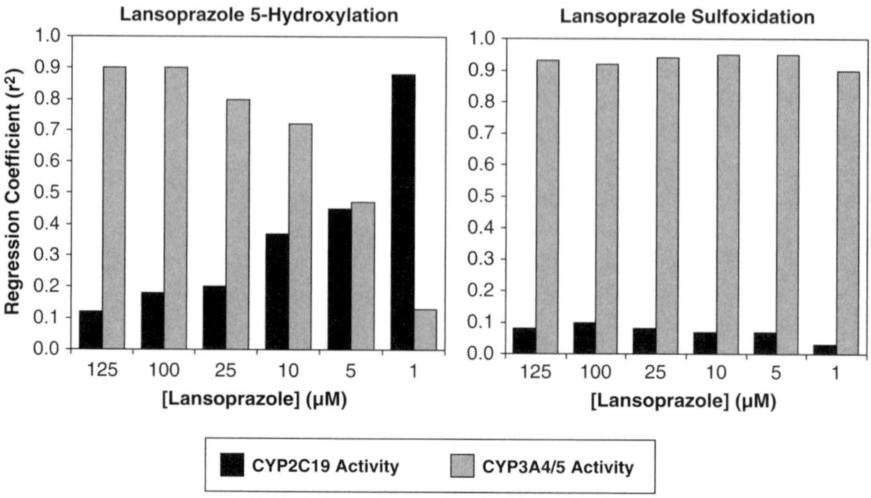

Figure 20 Relationship between the rates of *S*-mephenytoin 4′-hydroxylation (CYP2C19) or testosterone 6β-hydroxylation (CYP3A4/5) and the rates of lansoprazole 5-hydroxylation and lansoprazole sulfoxidation by human liver microsomes at substrate concentrations varying from 1 to 125 μM. Human liver microsomes were incubated with lansoprazole (1–125 μM) in the presence of an NADPH-generating system. The rates of lansoprazole 5-hydroxylation were compared with the rates of *S*-mephenytoin 4′-hydroxylation (a marker for CYP2C19 activity) or testosterone 6β-hydroxylation (a marker for CYP3A4/5 activity). Note that the correlation of CYP2C19 activity with lansoprazole 5-hydroxylation improved dramatically when the substrate concentration was lowered from 125 to 1 μM and, conversely, the correlation of the same activity with CYP3A4 progressively worsened. This is because, at low lansoprazole concentrations, lansoprazole 5-hydroxylation is catalyzed primarily by CYP2C19, but the contribution of CYP3A4/5 predominates at high concentrations of lansoprazole. In contrast, the correlation of lansoprazole sulfoxidation with CYP2C19 was poor and that with CYP3A4/5 was good, regardless of lansoprazole concentration. This is because CYP3A4/5 is the primary enzyme responsible for lansoprazole sulfoxidation in human liver microsomes. *Source*: Adapted from Ref. 158.

pharmacologically relevant concentrations (~1 μM), the 5-hydroxylation of lansoprazole is catalyzed primarily by CYP2C19, as it is in vivo. In the study by Pearce et al. (158), the correlation of CYP2C19 activity with lansoprazole 5-hydroxylation improved dramatically when the substrate concentration was lowered from 125 to 1 μM and, conversely, the correlation of the same activity with CYP3A4 progressively worsened (Fig. 20). The results in Figure 20 underscore an important principle: in vitro reaction phenotyping studies should be conducted with pharmacologically relevant drug concentrations. Unfortunately, in the case of drug candidates, clinical data may not be available; therefore, what constitutes a pharmacologically relevant concentration may not be known.

As a general rule of thumb, PhRMA recommended that for CYP reaction phenotyping the substrate concentration should be selected as follows: (1) if $K_m > C_{in\ vivo}$, then the substrate concentration for CYP reaction phenotyping should be $\leq K_m$; (2) if $K_m < C_{in\ vivo}$, then the substrate concentration selected should be approximately equal to $C_{in\ vivo}$; and (3) if $C_{in\ vivo}$ is unknown, then the substrate concentration selected should be $< K_m$. However, when $C_{in\ vivo}$ becomes available, it is fully justifiable to reassess the substrate concentration (and perhaps repeat the reaction phenotyping study) (5). It should be noted that these recommendations are based on a determination of K_m, and although correct in theory, the determination of K_m can be complicated when two enzymes participate in the formation of the metabolite of interest, one of which is high affinity and low capacity, the other of which is low affinity and high capacity [as in the case of lansoprazole 5-hydroxylation by CYP2C19 and CYP3A4 (Fig. 20)]. The in vitro results tend to be dominated by the latter; hence, the theoretical analysis could erroneously be based on a low-affinity enzyme rather than the high-affinity enzyme that is more likely to be important in vivo. For instance, the interaction between ethinyl estradiol–containing oral contraceptives and antibiotics, such as rifampin, is often attributed to the induction of CYP3A4, which is the major CYP involved in the oxidative metabolism of ethinyl estradiol (e.g., Ortho-Evra® prescribing information, 2005), but several lines of evidence suggest that induction of CYP3A4 is not the predominant mechanism by which rifampin increases the clearance of ethinyl estradiol. First, Li et al. reported that treatment of primary cultures of human hepatocytes with rifampin (33.3 μM) caused up to a 3.3-fold increase in ethinyl estradiol 3-*O*-sulfate formation (159). Second, CYP3A4 catalyzes the 2-hydroxylation of ethinyl estradiol with a K_m value of approximately 3.4 μM (160), whereas the average plasma concentrations are approximately 1/10,000th of this value. Third, SULTs 1A1, 1A2, 1A3, 1E1, and 2A1 catalyze the 3-O-sulfonation of ethinyl estradiol with K_m values ranging from 6.7 to 4500 nM, nearer the pharmacologically relevant concentrations (161). Finally, it is known that ethinyl estradiol is predominantly excreted in bile and urine as the 3-sulfate and, to a lesser extent, the 3-glucuronide (159), which suggests that 3-sulfonation is the major pathway of ethinyl estradiol metabolism. Taken together, these data suggest that induction of SULTs can be clinically relevant at least for low-dose drugs that can be sulfonated with high affinity.

With absent clinical information to guide the selection of a pharmacologically relevant substrate concentration, it is common practice to conduct reaction phenotyping experiments with 1-μM drug candidate. In most cases, this concentration is below K_m, which permits reaction phenotyping studies to be conducted under first order reaction kinetics and to identify, in most cases, the high affinity enzyme responsible for metabolizing the drug candidate. However, some drugs are metabolized by CYP with unusually high affinity. For example, the antimalarial drug candidate DB289 (2,5-bis(4-amidinophenyl)furan-bis-*O*-methylamidoxime) is metabolized by NADPH-fortified liver microsomes with a K_m of about 0.3 μM on the basis of measurements of substrate disappearance (a method that, if anything, leads to an overestimation of K_m) or formation of its O-demethylated metabolite M1 (162). Some drugs, such as many of the cholesterol-lowering statins (for which the liver is the therapeutic target) are actively transported into the liver. In some cases, the liver-to-plasma ratio is so high (an order of magnitude or more) that it is questionable whether the levels of drug in the systemic circulation provide a reliable estimate of the hepatic levels available to CYP and other drug-metabolizing enzymes. Finally, most acidic and sulfonamide-containing drugs bind extensively to plasma protein (in many cases their binding to plasma protein exceeds 99%), whereas such drugs do not bind extensively to microsomal protein (presumably because they are repelled by the negatively charged phosphate groups on the phospholipid membrane) (163). These few examples serve to illustrate the issues that can sometimes complicate the selection of a pharmacologically relevant concentration of drug candidate. It is for this reason that experiments with a range of substrate concentrations are conducted in order to determine the kinetic constants K_m and V_{max}, as outlined in section III.C.3. During the course of reaction phenotyping, there is one situation where the concentration of drug candidate is deliberately increased to a high level in order to support the formation of all possible metabolites. This is done to support the development of a suitable analytical method, which is the topic of the next section.

c. Development of the analytical procedure and its validation. A procedure must be developed to measure the rate of formation of metabolites of the drug candidate or possibly the disappearance of substrate (which is less sensitive and incapable of ascertaining whether different enzymes produce different metabolites). This typically involves incubating the appropriate test system with a range of substrate concentrations, some of which are not pharmacologically relevant but which support the formation of metabolites by both low- and high-affinity enzymes. The analysis of metabolites typically involves their chromatographic separation by HPLC with UV-VIS, fluorescent, radiometric, or mass spectrometric detection. Methods that have been developed for the analysis of the parent drug in formulations and blood (to support the analysis of clinical and toxicokinetic samples) are often unsuitable for reaction phenotyping because they are not designed to separate the parent drug from its metabolites, although they do provide a good starting point. The metabolites can be generated by incubating the parent drug with a pool of human liver microsomes in the presence of NADPH or an NADPH-generating system. These preliminary experiments are often conducted with a high concentration of microsomal protein (1–2 mg protein/mL) and drug candidate (1–100 μM) over extended incubation periods (up to 120 minutes) to facilitate the detection of all possible metabolites.

> Briefly, liver microsomes (e.g., 1 mg protein/mL) are incubated at (37 ± 1)°C in 0.25-mL incubation mixtures (final volume, target final pH 7.4) containing potassium phosphate buffer (50 mM), $MgCl_2$ (3 mM), EDTA

(1 mM), and the drug candidate (e.g., 1, 10, 100 μM) with and without an NADPH-generating system, at the final concentrations indicated. The NADPH-generating system consists of NADP (1 mM), glucose-6-phosphate (5 mM), and glucose-6-phosphate dehydrogenase (1 U/mL). If it is sufficiently water soluble, the drug is added to the incubation mixtures in water. Otherwise, the drug is added to each incubation in methanol [$\leq$1% (v/v)], DMSO [$\leq$0.1% (v/v)], or another suitable organic solvent (at the lowest concentration possible). Provided the drug candidate is not metabolized by non-CYP enzymes that do not require the addition of a cofactor (such as carboxylesterase or MAO), reactions are started by the addition of the NADPH-generating system. If metabolism by non-CYP enzymes occurs, reactions are started by the addition of the drug candidate. Reactions are stopped at designated times (typically up to 60 minutes in the preliminary experiment) by the addition of a stop reagent (e.g., organic solvent, acid, or base). Zero-time, zero-protein, zero-cofactor (no NADPH), and zero-substrate incubations serve as blanks. Precipitated protein is removed by centrifugation (typically 920 *g* for 10 minutes at 10°C), and an aliquot of the supernatant fraction is analyzed by HPLC or LC/MS/MS.

At this point, it is highly desirable to characterize by LC/MS/MS, the identity of the metabolites formed by NADPH-fortified human liver microsomes (or any other in vitro test system). Metabolite characterization is important to predict the enzyme system involved in a given reaction. For example, if a metabolite is formed by N-oxidation or S-oxygenation, FMO and CYP enzymes may both be involved. Alternatively, if a metabolite is a hydrolysis product of the parent drug, it points to involvement of carboxylesterases, especially if the reaction does not require NADPH. When liver microsomes are prepared from frozen liver tissue, microsomes are contaminated with the outer mitochondrial membrane where MAO activity is localized. MAO catalyzes oxidative deamination of primary, secondary, and tertiary amines. Oxidative deamination of a primary amine by MAO produces ammonia and an aldehyde ($R\text{-}CH_2\text{-}NH_2 + H_2O + O_2 \rightarrow R\text{-}CHO + NH_3 + H_2O_2$). Aldehydes are also formed during the N-, S-, and O-dealkylation of drugs by CYP (e.g., $R_1\text{-}CH_2\text{-}NH\text{-}R_2 + NADPH_2 + O_2 \rightarrow R_1\text{-}CHO + R_2\text{-}NH_2 + NADP + H_2O$), although in many cases the dealkylation reaction involves removal of a methyl or ethyl group, which is released as formaldehyde and acetaldehyde, respectively. In contrast to CYP, MAO does not require a pyridine nucleotide cofactor; therefore, if oxidation of an amine to an aldehyde is observed in microsomes, the reaction can likely be attributed to MAO if it proceeds in the absence of NADPH.

Aldehydes can also be formed by the sequential oxidation of a methyl group by CYP: $R\text{-}CH_3 \rightarrow R\text{-}CH_2OH \rightarrow R\text{-}CH(OH)_2 \rightarrow R\text{-}CHO$. Regardless of whether they are formed by MAO or CYP, aldehydes are usually further oxidized to the corresponding carboxylic acid ($R\text{-}CHO \rightarrow R\text{-}COOH$), although in some cases they are reduced to the corresponding alcohol ($R\text{-}CHO \rightarrow R\text{-}CH_2OH$). The conversion of aldehydes to carboxylic acids may be catalyzed by several enzymes including CYP, aldehyde dehydrogenase (present in cytosol and mitochondria), and aldehyde oxidase (present in cytosol). Therefore, formation of a carboxylic acid metabolite may start with CYP in the microsomal fraction (which generates the aldehyde) but the final step in its formation may require enzymes in other subcellular fractions. Benzylic methyl groups are often converted to carboxylic acids by CYP (as in the case of toluene being converted to benzoic acid). The changes in mass (atomic mass units, or amu) are as follows:

R-CH_3 → R-CH_2OH: +16 amu relative to the parent compound

R-CH_2OH → R-CHO: +14 amu relative to the parent compound (−2 amu relative to the precursor)

R-CHO → R-COOH: +30 amu relative to the parent compound (+16 amu relative to the precursor)

A metabolite with +16 amu is generally suspected of forming by hydroxylation (or by some other reaction involving the addition of oxygen). However, a metabolite with +14 amu is often suspected of forming by methylation (+CH_2), not by a combination of the addition of oxygen (+16) and dehydrogenation (−2). NADPH-fortified human liver microsomes cannot catalyze the methylation of drug candidates (such reactions are catalyzed by cytosolic enzymes in the presence of *S*-adenosylmethionine). However, methylation can sometimes occur as an artifact when mass spectrometry is conducted in the presence of methanol (164), and [M + 12] adducts can form from condensation reactions with formaldehyde, which is a microsomal metabolite of methanol (165). A metabolite with +30 amu is indicative of either formation of a carboxylic acid metabolite or a combination of hydroxylation (+16) and methylation (+14). Only the former can be catalyzed by NAPDH-fortified liver microsomes.

Mass spectrometry is widely used to characterize the structure of metabolites, and many instruments now come equipped with software to assist in this process, based on the fact that certain xenobiotic reactions are associated with discrete changes in mass. For example, the loss of 2 amu signifies dehydrogenation, whereas the loss of 14 amu usually signifies demethylation (-CH_2). Several reactions result in an increase in mass, including reduction (+2 amu = 2H), methylation (+14 amu = CH_2), oxidation (+16 amu = O), hydration (+18 amu = H_2O), acetylation (+42 amu = C_2H_2O), glucosylation (+162 amu = $C_6H_{10}O_5$), sulfonation (+80 amu = SO_3), glucuronidation (+176 amu = $C_6H_8O_6$), and conjugation with glutathione (+305 amu = $C_{10}H_{15}N_3O_6S$), glycine (+74 amu = $C_2H_4NO_2$), and taurine (+107 amu = $C_2H_6NO_3S$).

Occasionally, routine changes in mass can arise from unexpected reactions. For example, ziprasidone is converted to two metabolites, each of which involves an increase of 16 amu, which normally indicates addition of oxygen (e.g., hydroxylation, sulfoxidation, N-oxygenation). One of the metabolites is indeed formed by addition of oxygen to ziprasidone (sulfoxidation), as shown in Figure 21 (155). However, the other metabolite is formed by a combination of reduction (+2 amu) and methylation (+14 amu). The two pathways can be distinguished

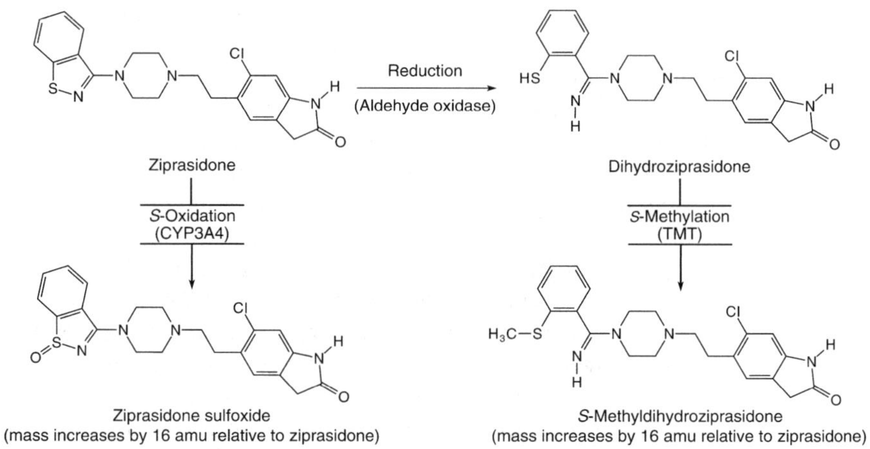

Figure 21 Conversion of ziprasidone to two different metabolites both involving a mass increase of 16 amu. *Abbreviation*: amu, atomic mass unit. *Source*: Adapted from Ref. 155.

by time-of-flight (TOF) LC/MS/MS analysis, which, in contrast to conventional LC/MS/MS, can distinguish the small difference in mass between the addition of CH_2 and the combination of the addition of oxygen and the loss of two hydrogen atoms. These few examples serve to underscore the point that care must be exercised in interpreting routes of metabolism based on changes in mass.

Mass spectrometry can typically provide information on which region of a molecule has undergone biotransformation, but it can seldom distinguish between several closely related possibilities. For example, based on mass spectrometry alone, it might be possible to ascertain that a certain phenyl group has been hydroxylated. However, analysis by nuclear magnetic resonance (NMR) is required to ascertain whether the hydroxylation occurred at the ortho, meta, or para position.

Once an analytical method (e.g., LC/MS/MS) is established, it is necessary to qualify or validate the procedure from a regulatory GLP perspective. The desired criteria for method validation/qualification include determining the lower and upper LOQ, inter- and intraday precision, specificity of the method, and linearity of the calibration curves (166). Validation/qualification must be performed in the presence of the representative biological matrix that will be used in reaction phenotyping. For CYP reaction phenotyping studies, the matrix of choice is a pool of human liver microsomes (166).

2. Step 2. Effect of Time and Protein

Step 2 involves an assessment of whether metabolite formation is proportional to incubation time and protein concentration under conditions that will be used for subsequent reaction phenotyping experiments. The goal is to establish in vitro conditions under which metabolites are formed under initial rate conditions. This procedure also helps establish whether the metabolites formed from the drug candidate are primary metabolites (no lag in formation) or secondary metabolites (lag in formation). For example, dextromethorphan is O-demethylated to dextrorphan by CYP2D6 and N-demethylated to 3-methoxymorphinan by CYP2B6 and CYP3A4 (167–169). Both dextrorphan and 3-methoxymorphinan are N-demethylated and O-demethylated, respectively, resulting in the formation of 3-hydroxymorphinan. In vitro formation of 3-hydroxymorphinan is always preceded by formation of dextrorphan or 3-methoxymorphinan and exhibits a time lag in its formation (170). On occasion, secondary metabolites are produced with no time delay, which may indicate that the primary metabolite is formed slowly by a relatively low-capacity and/or low-affinity enzyme, whereas the secondary metabolite is formed rapidly by a high-affinity and/or high-capacity enzyme. Alternatively, the lack of time delay in the formation of a secondary metabolite may indicate that the primary metabolite is not released from the enzyme active site but is converted immediately to the secondary metabolite. The antiangiogenic compound SU5416 provides an example of such concerted metabolism. A methyl group in SU5416 is hydroxylated by several human CYP enzymes ($R\text{-}CH_3 \rightarrow R\text{-}CH_2OH$) but only CYP1A2 further metabolizes the hydroxymethyl metabolite of SU5416 to the corresponding carboxylic acid ($R\text{-}CH_2OH \rightarrow R\text{-}COOH$). Recombinant CYP1A2 converts SU5416 to the acid metabolite but it does not release the hydroxymethyl metabolite, in contrast to several other CYP enzymes (which form the hydroxymethyl metabolite but not the carboxylic acid). Consequently, although formation of the hydroxymethyl metabolite by a bank of individual samples of human liver microsomes does not correlate well with CYP1A2, formation of the carboxylic metabolite correlates highly with CYP1A2 activity (171).

> The experimental design for evaluating the effects of incubation time and protein concentration on metabolite formation is often influenced by the results of the experiments to support the development of an analytical method (described in the preceding section), although the overall design often remains essentially the same. Unless there are reasons to do otherwise, a range of concentrations of the drug candidate (e.g., 1, 10, and 100 μM) is incubated with three concentrations of human liver microsomes (e.g., 0.125, 0.5, and 2.0 mg protein/mL) for a fixed time period (e.g., 0 and 15 minutes). Additionally, the drug candidate (e.g., 1, 10, and 100 μM) is incubated with a single concentration of human liver microsomes (e.g., 0.5 mg protein/mL) for multiple time periods (e.g., 0, 5, 10, 15, 20, 30, 45, 60 minutes). In addition to human liver microsomes and the drug candidate, the incubation mixture contains potassium phosphate (50 mM, pH 7.4), $MgCl_2$ (3 mM), EDTA (1 mM, pH 7.4), and an NADPH-generating system (1 mM NADP, 5 mM glucose-6-phosphate, and 1 U/mL glucose-6-phosphate dehydrogenase), at the final concentrations indicated. The remaining procedure is identical to that described previously.

If incubating as much as 100-μM drug candidate with liver microsomal protein for 120 minutes in the presence of NADPH results in no detectable metabolite formation, and if incubating as little as 1-μM drug candidate with liver microsomal protein for 120 minutes in the presence of NADPH results in no detectable loss of parent compound, it is reasonable to assume that the drug is minimally metabolized by CYP and FMO enzymes, unless other compelling data (such as in vivo pharmacokinetic data) are available that strongly suggest otherwise, in which case it would be prudent to examine microsomes from small intestine and other extrahepatic tissues as well as investigate the possibility that the drug candidate is metabolized by MAO, aldehyde oxidase, or another non-CYP enzyme.

3. Step 3. Determination of Kinetic Constants (K_m and V_{max})

If the goal of the in vitro study is to derive an estimate of in vitro intrinsic clearance (V_{max}/K_m) in order to predict in vivo clearance by a given enzymatic pathway, metabolite formation by the test system (e.g., human liver microsomes or hepatocytes) must be evaluated over a wide range of substrate concentrations. Such experiments must be designed carefully so that K_m and V_{max} are measured under appropriate kinetic conditions. It is important to verify that metabolite formation at all substrate concentrations (especially the lowest substrate concentration) is proportional to incubation time and protein concentration (i.e., that metabolite formation is measured under initial rate conditions). When kinetic parameters are determined with individual samples of human liver microsomes, V_{max} values generally vary enormously from one sample to the next, whereas K_m values remain relatively constant. The sample-to-sample variability in V_{max} values in a bank of human liver microsomes is related directly to the specific content of the given enzyme in the microsomal sample. However, the K_m value (the concentration of the substrate at which the reaction proceeds at one-half the maximum velocity) is independent of the specific content of the enzyme (although it may be seen to vary if those samples with a high V_{max} value result in over metabolism of the substrate so that initial rate conditions are not observed). For example, if the levels of a particular CYP enzyme vary 20-fold in a bank of human liver microsomes, then V_{max} values for a reaction catalyzed by that particular CYP enzyme would also be expected to vary 20-fold. However, K_m values would be expected to remain constant from one sample to the next because K_m is an intrinsic property of an enzyme and, as such, is not dependent on the amount of enzyme present. (A simple analogy will serve to underscore this point. Freezing point is an intrinsic property of liquids. Water, for example, freezes at 0°C, and it does so regardless of the amount of water being frozen, so ice cubes and icebergs freeze at the same temperature.)

Although K_m values would be expected to be constant, there are reports of K_m varying from one sample to the next. When K_m is found to increase with V_{max}, it is more than likely that the metabolism of the substrate was not determined under initial rate conditions. Therefore, sample-to-sample variation in K_m values, particularly when such variation coincides with the variation in V_{max} values, is usually an experimental artifact. For example, coumarin 7-hydroxylation is catalyzed by CYP2A6 in human liver microsomes and little sample-to-sample variability in the K_m for coumarin 7-hydroxylation was observed, which was approximately 0.5 μM regardless of whether the microsomal samples had high or low levels of CYP2A6 (112). However, it should be noted that great care was taken to measure initial rates of coumarin 7-hydroxylation. The percentage of substrate converted to 7-hydroxycoumarin ranged from less than 1% to about 15%. It was speculated that reports of higher K_m values for the 7-hydroxylation of coumarin by human liver microsomes, such as a K_m of 10 μM reported by Yamazaki et al. (172), stem from excessive metabolism of the substrate so that reaction rates did not reflect initial velocities.

The experiment designed to evaluate the effect of incubation time and protein concentration on the formation of metabolites (Step 2) provides the preliminary data necessary to select a range of substrate concentrations and experimental conditions to determine K_m and V_{max} for the metabolism of the drug candidate by human liver microsomes. A crude estimation of K_m can be obtained from the three substrate concentrations used in Step 2, provided rates of metabolite formation represents initial reaction velocities. K_m and V_{max} should be measured with a 100-fold range of substrate concentrations, one that ranges from one-tenth K_m to ten times K_m. However, this range of substrate concentrations may have to be expanded if metabolite formation is catalyzed by two kinetically distinct enzymes (one with low and one with high K_m). The kinetic constants (K_m and V_{max}) for a given reaction are usually determined with a pool of human liver microsomes as follows.

> Typically, the pool of human liver microsomes (single protein concentration) are incubated in triplicate for a specified time period with a drug candidate (e.g., $0.1K_m$, $0.2K_m$, $0.3K_m$, $0.4K_m$, $0.5K_m$, $0.6K_m$, $0.7K_m$, $0.8K_m$, $0.9K_m$, K_m, $1.25K_m$, $1.6K_m$, $2K_m$, $4K_m$, $7K_m$, and $10K_m$, where K_m is the crude estimate obtained from data generated in Step 2). For all substrate concentrations, the rate of reaction is measured under initial rate conditions; that is, the product formation is directly proportional to protein concentration and incubation time and the percentage metabolism of the substrate does not exceed 10%. However, with low K_m (i.e., high affinity) substrates, it may be difficult to limit the percentage metabolism of the substrate to $<10\%$ at low substrate concentrations. Initial rate conditions may be achieved by varying the incubation time or the protein concentration, if necessary.

Note that, in some cases, poor substrate solubility may prevent metabolite formation being measured at high substrate concentrations (especially at $10K_m$). Alternatively, low analytical sensitivity may impede the detection of metabolite formed at substrate concentrations well below K_m (especially one-tenth K_m).

Enzyme kinetic constants are calculated by nonlinear regression analysis with computer software, such as *GraFit* (Erithacus Software Limited,

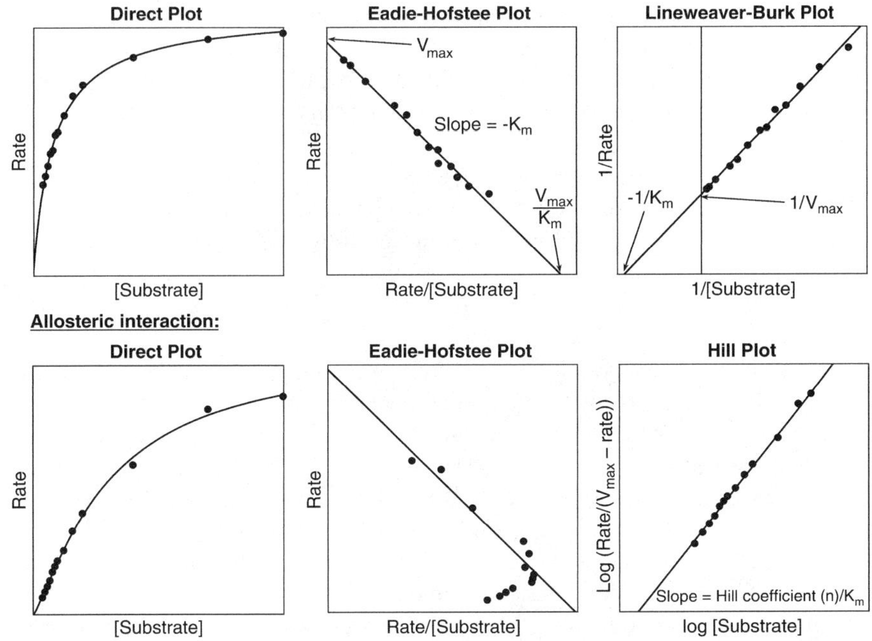

Figure 22 Examples of enzyme kinetic plots used for determination of K_m and V_{max} for a normal and an allosteric enzyme: Direct plot [(substrate) vs. initial rate of product formation] and various transformations of the direct plot (i.e., Eadie-Hofstee, Lineweaver-Burk, and/or Hill plots) are depicted for an enzyme exhibiting traditional Michaelis-Menten kinetics (coumarin 7-hydroxylation by CYP2A6) and one exhibiting allosteric substrate activation (testosterone 6β-hydroxylation by CYP3A4/5). The latter exhibits an S-shaped direct plot and a "hook"-shaped Eadie-Hofstee plot; such plots are frequently observed with CYP3A4 substrates. K_m and V_{max} are Michaelis-Menten kinetic constants for enzymes. K' is a constant that incorporates the interaction with the two (or more) binding sites but that is not equal to the substrate concentration that results in half-maximal velocity, and the symbol "n" (the Hill coefficient) theoretically refers to the number of binding sites. See the sec. III.C.3 for additional details.

London, U.K.), and the data are plotted on an Eadie-Hofstee plot (see Fig. 22). It should be noted that, at times, nonlinear regression lines represent the data points on an Eadie-Hofstee plots very poorly because the data reflect the contribution from two kinetically distinct enzymes whereas the computer software attempts to fit all data to an equation appropriate for a single enzyme. A relatively high standard error associated with the estimate of K_m suggests that the nonlinear regression did not fit the data very well, and it is possible that a two enzyme model or perhaps an atypical enzyme kinetics model needs to be selected. When K_m values are estimated by extrapolating data beyond the concentration range studied, the K_m values should be treated as estimates only. If the standard error associated with the K_m value is large (>25%) and/or if the K_m value falls outside the range of substrate concentrations studied, it is prudent to repeat this experiment with a new range of substrate concentrations that bracket the estimated K_m value.

When the Eadie-Hofstee plot suggests the involvement of two kinetically distinct enzymes in the formation of a particular metabolite, the data should be fitted to a dual-enzyme model according to the following equation:

$$v_{total} = v_1 + v_2 = \frac{V_{max_1} \cdot [S]}{K_{m1} + [S]} + \frac{V_{max_2} \cdot [S]}{K_{m2} + [S]} \qquad (8)$$

where v_{total} is the overall rate of metabolite formation at substrate [S], V_{max_1} and V_{max_2} are the maximal velocities of the reaction, and K_{m1} and K_{m2} are the Michaelis-Menten constants for enzyme 1 and enzyme 2, respectively. For simplicity, the following discussion assumes that enzyme 1 is the high-affinity (low-K_m) enzyme and that enzyme 2 is the low-affinity (high-K_m) enzyme. It further assumes that K_{m1} and K_{m2} differ by at least an order of magnitude and that the range of substrate concentrations extended well below K_{m1} and up to or above K_{m2}. Under such conditions, enzyme 2, the high-K_m enzyme, contributes negligibly to v_{total} at low substrate concentrations, and the range of substrate concentrations where this is largely true can be identified by visual inspection of the Eadie-Hofstee plot; (Fig. 23). Under these conditions, v_{total} is $\approx v_1$. These "enzyme 1" data are plotted on an Eadie-Hofstee plot to obtain K_{m1} and V_{max_1}. Subsequently, v_2 (which equals $v_{total} - v_1$) is calculated, and the data are plotted on an Eadie-Hofstee plot to obtain K_{m2} and V_{max_2}. As a rule of thumb, only data points for which v_2 is greater than $0.2v_{total}$ should be included in the latter determination because the experimental error associated with determination of v_{total} can give highly erroneous values for v_2. When K_{m1} and K_{m2} differ by less than an order of magnitude, or when the range of substrate concentrations does not bracket both K_{m1} and K_{m2}, it may not be possible to determine the kinetic constants of the individual enzymes.

Simply because a reaction fits the single-enzyme model well (i.e., the data conform to a straight line on an Eadie-Hofstee plot), it cannot be concluded that the reaction is catalyzed by only a single enzyme, although this is one possibility. Two enzymes with similar K_m values toward the same substrate have frequently been observed, and these will result in an Eadie-Hofstee plot consistent with single-enzyme kinetics. Applying the dual-enzyme model for such situations will not help; instead, reaction-phenotyping data must be used to tease out the role of the two enzymes. Some CYP enzymes (most notably CYP3A4) have been shown to exhibit kinetics consistent with allosteric interaction of the substrate with the enzyme, which is also known as *homotropic* or *substrate activation* (38,173). These result in an S-shaped curve on a (substrate) versus rate graph and a "hook"-shaped line graph on an Eadie-Hofstee plot. When allosteric interactions are

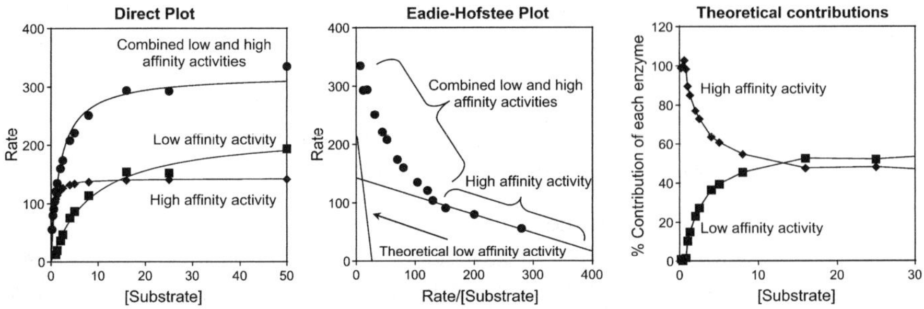

Figure 23 Depictions of a reaction catalyzed by two kinetically distinct enzymes. Theoretical data illustrate the method used to determine the kinetic constants when two enzymes are involved in the same reaction. Note that the direct plot (*left*) does not effectively indicate that two enzymes might be involved in a given reaction. However, this is readily achieved by a concave-appearing Eadie-Hofstee plot (*middle graph*). The kinetic constants (K_m and V_{max}) of the high-affinity (low-K_m) enzyme are determined using the initial rates observed at low substrate concentrations (*solid line* in the *middle graph*). Then, the contribution of the low-K_m enzyme is subtracted and the kinetic constants for the high-K_m enzyme are determined (*dotted line* in the *middle graph*). The theoretical contributions of the individual enzymes are shown (*right*). It is evident that the relative contribution of the high-K_m enzyme increases (and that of the low-K_m enzymes decreases) as the substrate concentration is increased.

observed, the Hill equation and a Hill plot can be used to calculate kinetic constants (109,174,175) (Fig. 22). The Hill equation is:

$$v = \frac{V_{max} \cdot [S]^n}{S_{50} + [S]^n} \tag{9}$$

where S_{50} is analogous to (but not identical to) K_m (i.e., it is the substrate concentration supporting half-maximal enzyme velocity) and incorporates the interaction of substrate with the two (or more) binding sites, and the symbol "n" (the Hill coefficient) theoretically refers to the number of binding sites. When n is greater than 1, it indicates positive cooperativity (substrate activation); when n is less than 1, it indicates negative cooperativity (substrate inhibition) (109). It should be noted that n need not be an integer. A Hill coefficient of 2 implies the presence of two discrete (nonoverlapping) substrate-binding sites on the enzyme, whereas a Hill coefficient of, say, 1.3 would indicate that there are two largely overlapping substrate-binding sites.

In addition to being prone to homotropic activation, CYP3A4 is also prone to heterotropic activation. The CYP1A2 inhibitor, α-naphthoflavone is an activator of certain CYP3A4-dependent reactions [a factor that complicates the use of this flavonoid in CYP inhibition studies (discussed later)]. CYP3A4-catalyzed reactions that exhibit homotropic activation (substrate activation) can sometimes exhibit typical Michaelis-Menten kinetics in the presence of α-naphthoflavone, as shown by Andersson et al. for the N-demethylation of diazepam (33).

4. Step 4. Correlation Analysis: Sample-to-Sample Variation in the Metabolism of the Drug Candidate

Correlation analysis is one of the four basic approaches to reaction phenotyping. It involves measuring the rate of drug metabolism by several samples of human liver microsomes (at least 10, according to the FDA) and correlating reaction rates with the variation in the level or activity of the individual CYP enzymes in the same bank of microsomal samples. This approach is successful because the levels of the CYP enzymes in human liver microsomes vary enormously from sample to sample (up to 100-fold), but with judicious selection of individual samples, they can vary independently from each other.

The experimental conditions for examining the in vitro metabolism of the drug candidate by a bank of human liver microsomes are based on the results of experiments described in Steps 2 and 3 (i.e., experiments designed to establish initial rate conditions and K_m and V_{max}). In order to obtain clinically relevant results, the metabolism of the drug candidate by human liver microsomes must be examined at pharmacologically relevant concentrations of the drug candidate, as illustrated for lansoprazole 5-hydroxylation in Figure 20.

In our laboratory, reaction phenotyping is carried out with a bank of human liver microsomal samples (e.g., $n = 16$) that has been analyzed to determine the sample-to-sample variation in the activity of the major drug-metabolizing CYP enzymes (namely, CYP1A2, CYP2A6, CYP2B6, CYP2C8, CYP2C9, CYPC19, CYP2D6, CYP2E1, CYP3A4/5, and CYP4A11) as well as FMO3. The marker substrates and reactions used to determine the sample-to-sample variation in CYP/FMO activity are shown in Figure 24, which also illustrates the extent of the variation in each enzyme activity. Banks of human liver microsomes intended for correlation analysis are commercially available as kits (e.g., Reaction Phenotyping Kit), and the manufacturers provide data on individual CYP enzyme and FMO3 activity (and perhaps data on the activity of UGT enzymes). For statistical purposes, it is important to select a bank of human liver microsomes (kit) in which the CYP enzyme activities do not correlate highly with each other. In other words, the independent variables (marker CYP enzyme activities supplied with the kits) must exhibit independent correlations. Differences in the rates of formation of the drug metabolites are compared with the sample-to-sample variation in CYP and FMO3 activity either by simple regression analysis (r^2 = coefficient of determination) or by Pearson's product moment correlation analysis (r = correlation coefficient), where the marker CYP/FMO enzyme activity is the independent variable and the rate of formation of drug metabolite is the dependent variable. The latter determination also provides a measure of the statistical significance of any correlations.

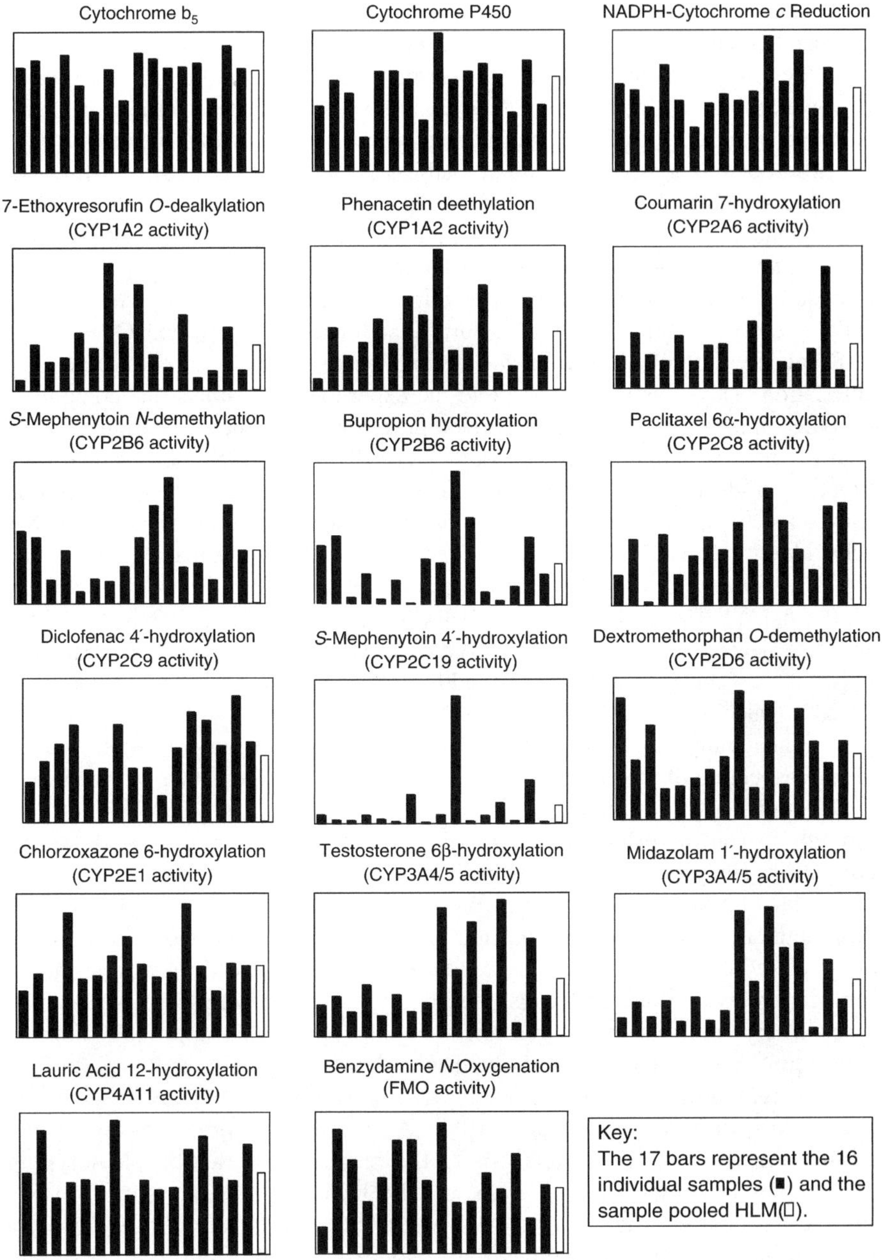

Figure 24 Sample-to-sample variation in CYP and FMO activities in a bank of 16 human liver microsomes. *Abbreviations*: CYP, cytochrome; FMO, flavin monooxygenase.

Correlation analysis provides valuable information on the extent to which the metabolism of a drug candidate will potentially vary from one subject to the next (i.e., it gives an estimate of pharmacokinetic variability in the clinic). However, when two or more enzymes contribute to metabolite formation, correlation analysis may lack the statistical power to establish the identity of each enzyme. For this reason, correlation analysis is often not conducted in favor of chemical or antibody inhibition studies and experiments with recombinant CYP incubations. However, of the four approaches to reaction phenotyping, correlation analysis generally provides the most reliable and clinically relevant information, provided the study is conducted under initial rate conditions with pharmacologically relevant concentrations of the drug candidate (and provided CYP and/or FMO plays a significant role in the disposition of the drug candidate) because correlation analysis is far less prone to the experimental artifacts that complicate all other approaches to reaction phenotyping. Furthermore, correlation analysis with a panel of human liver microsomes effectively assesses the potential contribution of all the CYP enzymes in human liver microsomes, whereas inhibition studies focus only those enzymes for which an inhibitory chemical or antibody has been identified or developed, and experiments with recombinant enzymes are typically conducted with the following panel of enzymes: CYP1A2, CYP2A6, CYP2B6, CYP2C8, CYP2C9, CYP2C19, CYP2D6, CYP2E1, CYP3A4, CYP3A5, and CYP4A11, and possibly CYP1A1, CYP1B1 and, rarely, CYP2C18. The importance of this principle is illustrated by results of studies on the metabolism of DB289, which are discussed later in section III.D.

Statistically significant correlations should always be confirmed with a visual inspection of the graph because there are two situations that can produce a misleadingly high correlation coefficient: (1) the regression line does not pass through or near the origin and (2) there is an outlying data point that skews the correlation analysis, as illustrated in Figure 25.

Correlation analysis works particularly well when a single enzyme dominates the formation of a particular metabolite. When two or more CYP enzymes contribute significantly to the metabolism of a drug at pharmacologically relevant concentrations, the identity of the enzymes involved can be assessed by multivariate regression analysis (176). This approach successfully identifies the enzymes involved when each enzyme contributes 25% or more to metabolite formation, but it will likely not identify an enzyme that contributes only approximately 10%. A graphical representation of the application of multivariate analysis to the results of a reaction phenotyping experiment is shown in Figure 26, on the basis of an examination of the sample-to-sample variation in the 1-hydroxylation of bufuralol (12 μM) by a panel of human liver microsomes. The 1-hydroxylation of bufuralol is widely used as a marker of CYP2D6 activity. However, experiments with recombinant or purified human CYP enzymes established that CYP1A2 and CYP2C19 can also catalyze this reaction (177,178). The sample-to-sample variation in bufuralol 1-hydroxylation correlates reasonably well with

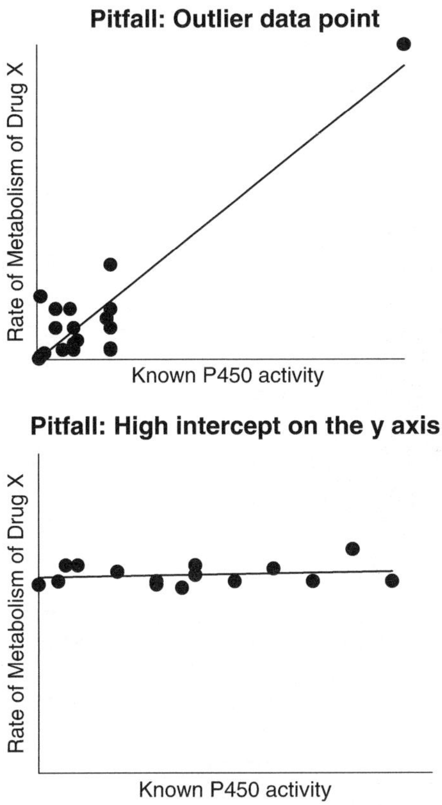

Figure 25 Common pitfalls in correlation analysis. Correlation analysis is suspected when the regression line is unduly affected by a single outlying data point, or when the regression line does not pass near the origin.

CYP2D6 activity measured by the O-demethylation of dextromethorphan ($r = 0.855$), but the correlation improves when the variation in CYP2C19 is taken into consideration ($r = 0.932$). The improvement in correlation coefficient only occurs when the third variable is CYP2C19 (measured by the 4′-hydroxylation of *S*-mephenytoin), which confirms the finding with recombinant enzymes that CYP2C19 is a potential contributor to the 1-hydroxylation of bufuralol by human liver microsomes. CYP1A2 appears to contribute negligibly to the 1-hydroxylation of bufuralol by human liver microsomes, but it does contribute significantly to the 4- and 6-hydroxylation of bufuralol (178). When two enzymes contribute significantly to metabolite formation, their identity and relative contribution can be established by performing correlation analysis in the presence and absence of an inhibitor of one of the participating enzymes (preferably the major contributor). This approach works even when one of the enzymes contributes substantially less than 25% to metabolite formation, as was demonstrated by

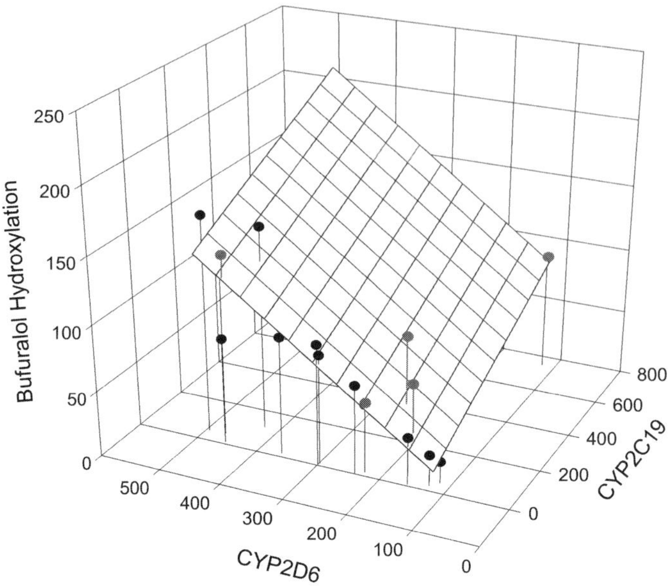

Figure 26 Multivariate correlation analysis of sample-to-sample variation in bufuralol 1′-hydroxylase activity in a bank of human liver microsomes with CYP2D6 and CYP2C19 activity. The sample-to-sample variation in the bufuralol 1′-hydroxylase activity in a bank of human liver microsomes was correlated with *S*-mephenytoin 4-hydroxylase (CYP2C19) and dextromethorphan *O*-demethylase (CYP2D6) activity by multivariate regression analysis. The regression coefficient of bufuralol 1′-hydroxylase activity improved from 0.855 (for CYP2D6 alone) to 0.932 when CYP2C19 activity was included in the analysis. Note that all points fall on the 3D plane best described by a combination of both CYP2C19 and CYP2D6 activity and that the bottom-right corner of the plane is very close to zero for bufuralol 1′-hydroxylase activity.

Yumibe et al. for the conversion of the antihistamine loratadine to desloratadine (179). The sample-to-sample variation in desloratadine formation by a panel of human liver microsomes was examined in the presence and absence of the CYP3A4 inhibitor ketoconazole (because studies with recombinant enzymes suggested that CYP3A4 was a major contributor to desloratadine formation). In the absence of ketoconazole, the sample-to-sample variation in desloratadine formation correlated highly with CYP3A4 (testosterone 6β-hydroxylase) activity ($r^2 = 0.96$), confirming a major role for CYP3A4 in loratadine metabolism. In the presence of ketoconazole, under conditions that caused extensive inhibition of CYP3A4, the residual activity (i.e., the uninhibited rate of desloratadine formation) correlated highly with CYP2D6 (dextromethorphan *O*-demethylase) activity ($r^2 = 0.81$), thereby establishing a minor role for CYP2D6 in loratadine metabolism. When a single enzyme dominates metabolite formation and correlation analysis is performed in the presence and absence of an inhibitor, the inhibited activity correlates well with both the residual activity and total activity, as illustrated in Figure 8 for chlorzoxazone 6-hydroxylase activity in the presence and absence of the CYP2E1 inhibitor DMSO.

5. Step 5. Chemical and Antibody Inhibition

Chemical and antibody inhibition represent the second and third approaches to reaction phenotyping. They typically involve an evaluation of the effects of known CYP enzyme inhibitors or inhibitory antibodies against selected CYP enzymes on the metabolism of a drug candidate by pooled human liver microsomes. As in the case of correlation analysis, chemical and antibody inhibition experiments must be conducted with pharmacologically relevant concentrations of the drug candidate in order to obtain clinically relevant results.

The FDA-approved and acceptable chemical inhibitors for reaction phenotyping are included in Table 2. Many of the inhibitors listed in Table 2 are metabolism-dependent inhibitors that, in order to inhibit CYP, require preincubation with NADPH-fortified human liver microsomes for 15 minutes or more. In the absence of the metabolism-dependent inhibitor, this preincubation of microsomes with NADPH can result in the partial, spontaneous loss of several CYP enzyme activities (see sec. II.C.7.c). Furthermore, the organic solvents commonly used to dissolve chemical inhibitors can themselves inhibit (or possibly activate) certain CYP enzymes, as discussed in section II.C.4. Therefore, appropriate solvent and preincubation controls should be included in all chemical inhibition experiments.

It is important to recognize that, in most cases, the specificity of chemical inhibitors is restricted to a particular concentration range, and that this concentration range can change with the concentration of microsomal protein (due to nonspecific binding), the concentration of substrate (which may compete with the inhibitor for binding to a particular CYP enzyme) and the type of substrate (because in some cases the degree of inhibition of a particular CYP enzyme is substrate dependent). For example, ketoconazole is a potent inhibitor of human CYP3A4 ($K_i < 20$ nM), but it is also capable of inhibiting several other CYP enzymes, including CYP1A1, CYP1B1, CYP2B6, CYP2C8, CYP2C9, CYP2C18, CYP2J2, CYP3A5, CYP4F2, and CYP4F12 (K_i in micromolar range) (109,180,181). Ticlopidine was initially proposed as a selective metabolism-dependent inhibitor of CYP2C19, but it was subsequently shown to be an even more effective metabolism-dependent inhibitor of CYP2B6 (182,183). The lack of specificity can complicate the interpretation of chemical inhibition experiments. It likely accounts for the majority of cases where the sum of the inhibitory effects of a panel of CYP inhibitors adds up to greater than 100%. The influence of protein concentration on chemical inhibition is particularly well illustrated in Figure 3, which shows that the inhibitory effect of montelukast on CYP2C8 activity declined almost 20-fold (based on IC_{50} values) when the concentration of microsomal protein was increased 20-fold (from 0.05 to 1 mg/mL), as previously reported by Walsky et al. (28).

If a drug candidate is metabolized by a high-affinity enzyme, the concentration of a competitive chemical inhibitor must be increased with increasing concentration of the drug candidate in order to achieve a high degree of inhibition. A good rule of thumb is to use multiples (generally up to 10-fold) of the lowest inhibitor concentration, which is calculated from the following equation:

$$\text{Lowest[Inhibitor]} = \frac{[\text{Drug}] \cdot K_{\text{i(inhibitor)}}}{K_{\text{m(Drug)}}} \tag{10}$$

where [Drug] is the intended final concentration of the drug candidate added to the microsomal incubation, K_i is the inhibition constant of the inhibitor for a given enzyme, and K_m is the Michaelis constant of the drug candidate (as determined in Step 3). This method of calculating of the lowest concentration of inhibitor is applicable to competitive inhibitors but not to noncompetitive or metabolism-dependent inhibitors. A range of inhibitor concentrations is recommended to demonstrate concentration dependence. For example, if the lowest concentration of the chemical inhibitor were calculated to be 1 μM (from the above equation), then the range of inhibitor concentration should span at least 10-fold (e.g., 1, 2, 5, and 10 μM).

One of the complicating factors with chemical inhibitors is that a chemical that inhibits one CYP enzyme may activate another enzyme. If both enzymes contribute to metabolite formation, the inhibitory effect of the chemical on one enzyme may be offset by its activating effect on the other enzyme. α-Naphthoflavone is an inhibitor of CYP1A2 but an activator of CYP3A4, whereas quinidine is an inhibitor of CYP2D6 but, in certain cases, an activator of CYP3A4 (184–186). α-Naphthoflavone and quinidine both appear on the list of FDA preferred and acceptable chemical inhibitors, so their ability to inhibit one enzyme but activate another are relevant to reaction phenotyping.

When chemical inhibition experiments are conducted with a relatively metabolically stable drug candidate (one that must be incubated with relatively high concentrations of human liver microsomes for a relatively long time in order to generate quantifiable levels of metabolite), it is important to take into account the metabolic stability of the inhibitors themselves. Lack of metabolic stability makes some compounds poor choices as chemical inhibitors despite their selectivity. For example, coumarin is a selective substrate of CYP2A6 (K_m ~0.25 to 0.5 μM) (111) and it would be a good selective competitive inhibitor of CYP2A6 if it were not metabolized so rapidly by human liver microsomes.

Finally, appropriate controls should be included in each chemical inhibition experiment to evaluate whether any of the chemical inhibitors interfere with the chromatographic analysis of the metabolites of interest and whether metabolite formation is inhibited by any of the organic solvents used to dissolve the chemical inhibitors.

Inasmuch as the selectivity of some chemical inhibitors is questionable or even variable depending on the incubation conditions, the use of selective inhibitory polyclonal, monoclonal, or antipeptide antibodies against individual CYP enzymes can be an alternative (or additional) approach to reaction phenotyping (187–189). CYP inhibition by antibodies is noncompetitive in nature and is therefore independent of the substrate concentration. Because of the ability of antibodies to inhibit selectively and noncompetitively, this method alone can potentially establish which human CYP enzyme is responsible for metabolizing a drug candidate. Unfortunately, the utility of this method is limited by the availability of specific inhibitory antibodies and by nonspecific effects associated with the addition of antiserum and ascites fluid to the microsomal incubation.

Although numerous antibodies against CYP enzymes are commercially available, there can be problems with cross-reactivity (lack of specificity), especially in the case of polyclonal or anti-peptide antibodies (the specificity of which can vary from lot to lot). The use of antiserum (for polyclonal antibodies) and ascites fluid (for monoclonal antibodies) rather than purified antibodies often necessitates adding a large amount of albumin and other proteins to the microsomal incubation. [The concentration of protein (mostly albumin) in serum and ascites fluid is ~70 and 30 mg/mL, respectively.] Adding albumin and other serum proteins to a microsomal incubation can exert several effects including (1) nonspecific binding of the drug candidate to serum proteins, which may artifactually decrease metabolite formation, (2) activation of CYP enzymes such as CYP2C9 by albumin (190,191), which may lead to increased metabolite formation or which mask the inhibitory effect of the antibody, and (3) metabolism of the drug candidate or its metabolites by serum enzymes, which may artifactually increase or decrease metabolite formation, respectively. For this reason, control (preimmune) serum and ascites fluid should be included as negative controls in antibody inhibition experiments. These issues are lessened when purified antibodies are used instead of antisera and ascites fluid. However, in the case of purified monoclonal antibodies, it may be necessary to include a large number of negative controls (perhaps one for each anti-CYP antibody) with different concentrations of "irrelevant" monoclonal antibodies (those prepared against enzymes other than CYP) and different antibody subtypes (IgM, IgG_1, IgG_2, etc.) to match each of the monoclonal antibodies used in the inhibition experiment.

As in the case of chemical inhibition, a lack of specificity can complicate the interpretation of antibody inhibition experiments. A lack of specificity and the nonspecific effects outlined above likely account for the majority of cases where the sum of the inhibitory effects of a panel of inhibitory antibodies adds up to greater than 100%. Another potential problem stems from the fact that many antibodies do not completely inhibit the activity of a microsomal CYP enzyme or the corresponding recombinant CYP enzyme. If an antibody inhibits the metabolism of a marker substrate by 80%, and if the same antibody inhibits the metabolism of drug candidate by 80%, there is uncertainty as to whether the inhibited enzyme contributes 80% or 100% to the metabolism of the drug candidate. This may seem like a trivial difference, but it has a large impact on the victim potential of the drug candidate. Genetic or drug-mediated loss of an enzyme that accounts for 80% of a drug's clearance will cause a fivefold increase in systemic exposure, whereas loss of an enzyme that accounts for 99% of a drug's clearance will cause a 100-fold increase in exposure.

6. Step 6. Recombinant Human CYP Enzymes

The fourth and final approach to reaction phenotyping involves the use of purified or recombinant (cDNA-expressed) human CYP enzymes, which can establish whether a particular CYP enzyme can or cannot metabolize a drug candidate, although it does not address whether that CYP enzyme contributes substantially to reactions catalyzed by human liver microsomes. Numerous human CYP enzymes have been cloned and expressed individually in various cell types. Microsomes from these cells, which contain a single human CYP enzyme with NADPH-CYP reductase with or without cytochrome b_5, are commercially available. The recombinant CYP enzymes differ in their catalytic competency, and they are not expressed in cells at concentrations that reflect their levels in human liver microsomes. Therefore, a simple evaluation of metabolism by a bank of recombinant human CYP enzymes does not establish the *extent* to which a CYP enzyme contributes to the metabolism of a particular drug candidate, only that a particular CYP enzyme *can* metabolize that drug candidate. Also, the recombinant CYP enzymes are usually expressed with much higher levels of NADPH-CYP reductase than those present in human liver microsomes. In many cases, the ratio of CYP to NADPH-CYP reductase in preparations of recombinant enzymes is more than an order of magnitude greater than that in human liver microsomes. This is a possible cause of artifacts. For example, the high levels of NADPH-CYP reductase that are used to reconstitute purified CYP enzymes, or expressed with recombinant CYP enzymes can potentially interfere with the metabolism of a drug candidate, as exemplified by the metabolism of 7-pentoxyresorufin by purified rat CYP2B1. When the molar ratio of NADPH-CYP reductase toCYP2B1 exceeds one-to-one, CYP2B1 loses its ability to catalyze the O-dealkylation of 7-pentoxyresorufin because the excess NADPH-CYP reductase reduces 7-pentoxyresorufin to a metabolite that is no longer O-dealkylated by CYP2B1. This does not occur in liver microsomes because the molar ratio of NADPH-CYP reductase to total CYP is considerably less than one-to-one (in microsomes there are 10–20 molecules of CYP for each molecule of NADPH-CYP reductase). The high levels of NADPH-CYP reductase expressed with recombinant CYP enzymes could also conceivably cause artificially high enzyme activity, as exemplified by studies with rat CYP2A2 (also known as P450m), which is expressed only in adult male rats. When purified and reconstituted with high levels of NADPH-CYP reductase, CYP2A2 catalyzes the 15α-hydroxylation of testosterone and does so at one-half the rate at which purified CYP2A1 (P450a) catalyzes the 7α-hydroxylation of testosterone. However, rat liver microsomes do not form these metabolites in a 2:1 ratio. In fact, rat liver microsomes produce very low levels of 15α-hydroxytestosterone (192), and the rate of testosterone 15α-hydroxylation by liver microsomes from adult male rats (which contain CYP2A2) is comparable to that catalyzed by liver microsomes from adult female rats and immature rats (which contain no detectable CYP2A2). Therefore, high levels of NADPH-CYP reductase can potentially lead to an overestimation of CYP activity (as in the case of CYP2A2) or an underestimation of its activity (as in the case of CYP2B1).

Cytochrome b_5 affects the kinetics of drug metabolism by certain CYP enzymes; hence, coexpression of this microsomal hemoprotein (together with NADPH-CYP reductase) can affect the catalytic efficiency of certain recombinant CYP enzymes (76,109). For example, the presence of cytochrome b_5 tends to increase V_{max} for reactions catalyzed by CYP3A4, whereas it tends to decrease K_m for reactions catalyzed by CYP2E1. In both cases, cytochrome b_5 increases V_{max}/K_m, which is a measure of in vitro intrinsic clearance. The fact that some commercially available recombinant CYP enzymes are expressed with cytochrome b_5 while others are not complicates the interpretation of results of studies performed with recombinant human CYP enzymes.

To facilitate a comparison of one recombinant CYP enzyme with another, Rodrigues (193) has proposed that recombinant human CYP enzymes should be used to measure in vitro intrinsic clearance based on a measurement of K_m and V_{max}. The kinetic constants are only determined for those enzymes that were shown in preliminary experiments to be capable of metabolizing the drug candidate. Care must be taken in the determination of kinetic constants, as described previously in Step 3, and the methodology used to determine K_m and V_{max} with recombinant CYP enzymes is very similar to that described previously. The V_{max} value obtained with each recombinant CYP enzyme (expressed as pmol product formed/min/pmol of P450) is multiplied by its average specific content in human liver microsomes (values for which are given in Table 8) (193), which provides an estimate of the V_{max} value in an average (or a pooled) sample of human liver microsomes. These estimates generally overestimate V_{max} values in microsomes as the catalytic activity of the recombinant CYP enzymes is artificially high because of the presence of artificially high levels of NADPH-CYP reductase. Nevertheless, this method forms the basis for evaluating the relative contribution of all the recombinant CYP enzymes that can metabolize the drug candidate. The assessment of relative contribution can be improved further by comparing in vitro intrinsic clearance (V_{max}/K_m) rather than V_{max} values, where the V_{max}/K_m values are again corrected for the specific content of each CYP enzyme in human liver microsomes. Unfortunately, this method is complicated by the empirical observation that K_m, in addition to V_{max}, can differ between a recombinant CYP enzyme and the same enzyme expressed in human liver microsomes. Some examples are given in Table 4. In the case of dextromethorphan O-demethylation by CYP2D6 and diclofenac 4′-hydroxylation by CYP2C9, the K_m for the recombinant enzyme is roughly one-fifth that of the microsomal enzyme. If the same difference in K_m were observed with a drug candidate, the estimate of in vitro intrinsic clearance by recombinant CYP would be at least five times greater than that in liver microsomes. Despite these difficulties, normalizing rates of drug metabolism by recombinant CYP enzymes by taking into account their specific content in human liver microsomes is one approach to assessing the relative contribution of CYP enzymes to the metabolism of a drug candidate.

An alternative approach to normalizing rates of drug metabolism by recombinant CYP enzymes is the application of a "relative activity factor (RAF)," in which the correction is not based on specific content but on specific activity, which requires a comparison of the rate of metabolism of a selective marker substrate by each recombinant CYP enzyme and human liver microsomes (75,194). The RAF is then multiplied by the observed rates of drug metabolism by each recombinant CYP enzyme before assessing the relative contribution of each enzyme to the metabolism of the drug. This approach has not been well validated. For example, it is not known whether the relative activity factor remains constant for several marker substrate reactions catalyzed by the same CYP enzyme. If the relative activity factor varies in a substrate-dependent manner, it would be difficult to know which RAF value to apply to the drug candidate under investigation. Another limitation of this approach is that the relative activity factor must be empirically determined for each lot of recombinant CYP enzyme (and preferably each batch of pooled human liver microsomes).

The FDA guidance document recognizes the difficulty of extrapolating the results obtained with recombinant enzymes to the situation in liver microsomes. Experiments with recombinant CYP enzymes provide valuable information on which CYP enzymes can and which ones cannot convert a drug candidate to a particular metabolite, and this information alone is particularly valuable in guiding the design or interpretation of correlation analysis, chemical inhibition, and antibody inhibition experiments.

D. The Relative Merits of the Four Approaches to Reaction Phenotyping

Many of the potential pitfalls and advantages or disadvantages of the four approaches to reaction phenotyping have been mentioned in the preceding sections, and they are summarized in Table 10. Additional potential pitfalls in reaction phenotyping do not apply simply to any one approach but apply to all of

Table 10 Advantages and Disadvantages of the Various Approaches to In Vitro CYP Reaction Phenotyping

Procedure	Attributes (advantages and disadvantages)
Correlation analysis with a bank of selected and well characterized human liver microsomes ($n = 10$ or more)	Contains all liver microsomal CYP enzymes with physiological levels of cytochrome b_5 and NADPH-CYP reductase. Establishes the degree of inter-individual variation in metabolic formation or substrate disappearance. A strong correlation clearly establishes the identity of the CYP enzyme responsible in metabolite formation. An outlying data point or a regression line that does not intersect near the origin can produce misleading results. Metabolite formation by high activity samples may violate initial rate conditions (<10% substrate loss). Even with multivariate analysis, correlation analysis may not positively identify multiple enzymes responsible for metabolite formation (unless correlation analysis is conducted in the presence and absence of a CYP-selective inhibitor).
Chemical inhibition with CYP-selective direct-acting and metabolism-dependent inhibitors and *antibody inhibition* with CYP-selective polyclonal or monoclonal antibodies (antiserum, ascites fluid or purified antibody)	Conducted with pooled human liver microsomes, which contains all the relevant CYP enzymes and which are generally used to establish initial rate conditions as well as K_m and V_{max}. Lack of specificity and inhibition values that total more than 100% complicate the interpretation of results. Lack of antibody specificity may reflect cross-reactivity or an artifact of adding albumin (present at high concentrations in serum and ascites fluid). Chemical inhibitors of one CYP enzyme can activate another CYP enzyme. This is also true of the albumin present in antiserum and ascites fluid. The inhibitory effect of some chemical inhibitors may be substrate-dependent. Solvent controls for all chemical inhibitors and preincubation controls for metabolism-dependent inhibitors must be included. Preimmune serum and ascites (or irrelevant antibodies) should be included as negative controls for antibody inhibition studies. The potency of chemical inhibitors often varies with the concentration of microsomal protein and substrate (drug candidate).

Continued

Table 10 *Continued*

Procedure	Attributes (advantages and disadvantages)
	Incomplete antibody inhibition complicates interpretation. If an antibody causes 80% inhibition both with a maker substrate and a drug candidate, does the targeted CYP enzyme contribute 80% or 100% to drug candidate metabolism?
Recombinant human enzymes (individual CYP and FMO enzymes expressed in mammalian cells, insect cells, bacteria or yeast)	Establishes whether a given CYP enzyme can or cannot metabolize the drug candidate or form a particular metabolite. Useful to study drug candidate metabolism by allelic variants of CYP enzymes. The panel of recombinant enzymes examined often fails to include all the CYP enzymes present in human liver microsomes (CYP2C18, 2J2, 3A7, 4F2, 4F3a, 4F3b are often omitted, but CYP1A1 and 1B1 are often included even though they are not expressed in human liver microsomes). The absence of cytochrome b_5 and/or artificially high levels of NADPH-CYP reductase can alter the kinetics of CYP enzymes and/or introduce artifacts. Compared with their microsomal counterparts, recombinant CYP enzymes may have a different K_m for marker substrates or a different K_i for chemical inhibitors. Rates of metabolism by recombinant CYP enzymes can be normalized based on specific content or relative activity factor, but neither method reliably predicts the relative contribution of CYP enzymes to reactions catalyzed by human liver microsomes.

Abbreviations: CYP, cytochrome P450; FMO, flavin monooxygenase.

the experimental approaches to identify which CYP enzyme is primarily responsible for metabolizing a drug candidate. The four most common errors are as follow:

1. An inappropriate test system is used to study the metabolism of the drug candidate. The preceding sections describe steps that can be taken to evaluate whether metabolism by CYP is likely to play an important role in the disposition of a drug candidate, and such information should be obtained prior to conducting a CYP reaction phenotyping study, especially if the drug candidate emerged from a screening program that favored the selection of drug candidates that do not interact with CYP (by selecting

those chemicals that cause little or no CYP inhibition, that are metabolically stable in NADPH-fortified human liver microsomes, and/or that have relatively high aqueous solubility or dissolution rates). In addition, although in vitro drug metabolism studies generally focus on hepatic metabolism, some drugs are extensively metabolized by enzymes in the intestine and other extrahepatic tissues (195–197).

2. The metabolism of the drug candidate is not measured under initial rate conditions. Prior to initiating reaction phenotyping studies, a pool of human liver microsomes should always be used to establish initial rate conditions (i.e., conditions under which metabolite formation is proportional to protein concentration and incubation time), and total amount of substrate consumed should be less than 10%.
3. The metabolism of the drug candidate is not measured at pharmacologically relevant concentrations. When reaction phenotyping studies are carried out with high, nonpharmacologically relevant substrate concentrations, metabolism of the drug candidate may be dominated by a low-affinity, high-capacity enzyme, but this enzyme may contribute negligibly to the metabolism of the drug candidate in the clinic (where drug concentrations tend to be low so that metabolism is dominated by high-affinity enzymes). This important principle is illustrated in Figure 20 for the 5-hydroxylation of lansoprazole (described in sec.III.C.1.b).
4. On a case-by-case basis, each of the four approaches to reaction phenotyping can give misleading results, which is why the FDA advocates a combination of approaches. There is a growing trend toward using two approaches to reaction phenotyping (chemical inhibition and recombinant CYP enzymes) rather than the three fundamentally different approaches (chemical inhibition, recombinant CYP enzymes, and correlation analysis). We have conducted a large number of reaction phenotyping studies, and we consider it unadvisable to exclude correlation analysis from a reaction phenotyping study. The following is but one example to support our conviction that correlation analysis should always be part of a reaction phenotyping study, along with chemical or antibody inhibition experiments and experiments with a large panel of recombinant human CYP enzymes.

The reliability of correlation analysis and the potential short-comings of other approaches to reaction phenotyping are illustrated by recent studies of DB289, a prodrug that is O-demethylated by CYP to M1 and subsequently converted by several steps to the active metabolite DB75, an aromatic dicationic antiparasitic agent that is effective against a broad range of pathogens, including African trypanosomiasis (African sleeping sickness) (162). Reaction phenotyping with the usual panel of recombinant human CYP enzymes (see above) identified members of the CYP1 family (namely, CYP1A1,

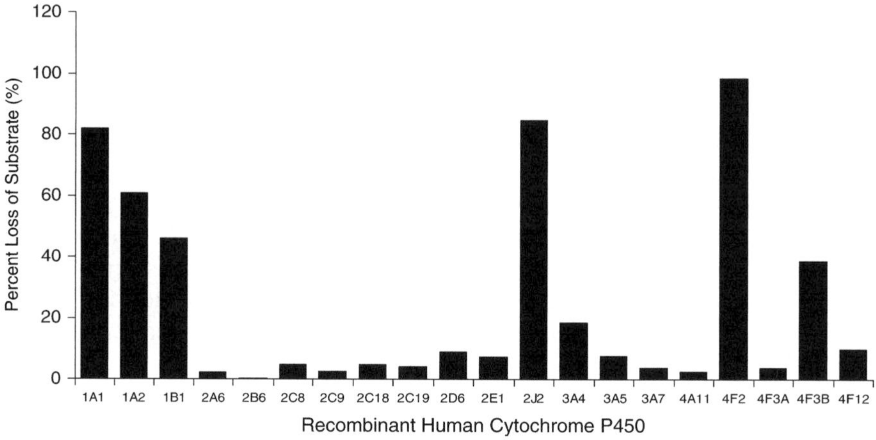

Figure 27 Metabolism of DB289 (antiparasitic prodrug) by a panel of recombinant human CYP enzymes. *Abbreviation*: CYP, cytochrome P450.

CYP1B1, and CYP1A2) as candidates for the microsomal enzyme that catalyzes the O-demethylation of DB289 to M1. However, because CYP1A1 and CYP1B1 are extrahepatic CYP enzymes that are not expressed in human liver microsomes (198,199), the study with recombinant CYP enzymes effectively identified only CYP1A2 as the enzyme responsible for the high rate of conversion of DB289 to the primary metabolite M1 by human liver microsomes. However, correlation analysis revealed no correlation between M1 formation and CYP1A2 activity (or any of the other major CYP enzyme activities). It was assumed (correctly) that the results of the experiments with recombinant CYP enzymes were incomplete (as opposed to assuming the results of the correlation analysis were incorrect). Inhibition studies with α-naphthoflavone (a direct-acting inhibitor of CYP1A2), furafylline (a metabolism-dependent inhibitor of CYP1A2), and an inhibitory antibody against CYP1A2, all confirmed that CYP1A2 was not the microsomal enzyme responsible for converting DB289 to M1. Although correlation analysis established that CYP1A2 was not the microsomal enzyme responsible for converting DB289 to M1, it did not positively identify which enzyme or enzymes were responsible, which illustrates both the strength and weakness of correlation analysis. The enzymes responsible for converting DB289 to M1 were identified when the panel of recombinant CYP enzymes was expanded to include CYP2C18, CYP2J2, CYP3A7, CYP4F2, CYP4FA, CYP4F3B, and CYP4F12. As illustrated in Figure 27, recombinant CYP2J2, CYP4F2, and CYP4F3B, all metabolized DB289, hence, these enzymes were investigated for their contribution to M1 formation in human liver microsomes. The results of experiments with additional chemical inhibitors and inhibitory antibodies established that CYP4F2 and CYP4F3b are the major enzymes in human liver microsomes responsible for converting DB289 to M1 (162). The studies with DB289 underscore the importance of using at least three approaches to reaction phenotyping, namely correlation analysis, studies with chemical or antibody inhibitors, and experiments with recombinant enzymes. They also underscore the value of evaluating a wide selection of recombinant human CYP enzymes, including those whose role in the metabolism of drugs has only recently come to light (200).

IV. IDENTIFICATION OF THE UGT ENZYMES INVOLVED IN CONJUGATION REACTIONS

The role of UGT enzymes in drug-drug interactions is discussed in more detail in chapter 4. This section will briefly highlight the application (and limitations) of reaction phenotyping methods to the identification of the UGT enzymes involved in the conjugation of a drug candidate. Like CYP-catalyzed oxidations, UGT-catalyzed reactions involving UDP-glucuronic acid (and occasionally UDP-glucose, UDP-xylose, and UDP-galactose) are a major pathway of drug metabolism in humans. The site of glucuronidation is generally an electron-rich nucleophilic O, N, or S heteroatom. Therefore, substrates for glucuronidation contain such functional groups as aliphatic alcohols and phenols (which form *O*-glucuronide ethers), carboxylic acids (which form *O*-glucuronide esters, also known as acyl glucuronides), primary and secondary aromatic and aliphatic amines (which form *N*-glucuronides), and free sulfhydryl groups (which form *S*-glucuronides). Certain xenobiotics, such as phenylbutazone, sulfinpyrazone, suxibuzone, ethchlorvynol, Δ^6-tetrahydrocannabinol, and feprazone contain carbon atoms that are sufficiently nucleophilic to form *C*-glucuronides. The C-glucuronidation of the enolic form of phenylbutazone is catalyzed specifically by UGT1A9 (201).

In contrast to the situation with CYP enzymes, there are fewer clinically relevant inhibitory drug-drug interactions caused by inhibition of UGT enzymes, and AUC increases are rarely greater than twofold (202), whereas dramatic AUC increases have been reported for CYP enzymes, such as the 190-fold increase in AUC reported for the CYP1A2 substrate ramelteon (Rozerem™) upon coadministration of fluvoxamine (Rozerem prescribing information, 2005). For instance, plasma levels of indomethacin are increased approximately twofold upon coadministration of diflunisal, and in vitro studies indicate that this interaction is due in part to inhibition of indomethacin glucuronidation in the intestine (203,204). Valproic acid coadministration increases the AUC of lorazepam and lamotrigine by 20% and 160%, respectively (202). Drug-drug interactions due to induction of UGT enzymes have also been observed. Rifampin coadministration increases mycophenolic acid clearance by 30% and increases the AUC of its acyl glucuronide (formed by UGT2B7) and its 7-*O*-glucuronide (formed by various UGT1 enzymes) by more than 100% and 20%, respectively (205). In spite of the limited number of reports of UGT-mediated drug-drug interactions, UGT enzymes are increasingly involved in the metabolism of new drug candidates. The apparent increased involvement of UGT enzymes is due in part to the selection process for new drug candidates, which is often biased against chemicals that interact with CYP enzymes. The selection of chemicals with little potential of inhibiting CYP enzymes, chemicals that are metabolically stable in NADPH-fortified human liver microsomes, and chemicals with high aqueous solubility are chemicals that are more likely to be metabolized by UGT than CYP. The FDA acknowledges that "there are few documented cases of clinically significant drug-drug interactions related to non-CYP enzymes...," but goes on to say that "the identification of drug metabolizing enzymes of this kind (i.e., glucuronosyltransferases, sulfotransferases, and *N*-acetyl transferases) is encouraged" (2).

Pooled and individual human liver microsomes with characterization data on the activity of the major UGT enzymes are commercially available. Figure 28 shows the variation in UGT activity toward five substrates in a bank of human liver microsomes. The activities were determined in native microsomes and in microsomes treated with the zwitterionic detergent, CHAPS, or the pore-forming peptide, alamethicin (treatments that stimulate UGT activity). In addition, individual recombinant UGT enzymes are commercially available. However, reaction phenotyping studies designed to identify the UGT enzymes involved in conjugating a new drug candidate must overcome certain challenges that are not evident (for the most part) in a CYP enzyme reaction phenotyping study. For example, selective UGT inhibitors have only been characterized for UGT1A4 (hecogenin) and UGT2B7 (fluconazole) (206), and there are relatively few specific inhibitory antibodies that are commercially available. Correlation analysis can be performed for UGT enzyme activities in individual human liver microsomes, but, as shown in Table 11, commercially available and specific probe drugs have been identified only for certain human UGT enzymes, including UGT1A1 (17β-estradiol, 3-glucuronidation, and bilirubin), UGT1A3 (hexafluoro-1α,25-trihydroxyvitamin D3), UGT1A4 (trifluoperazine), UGT1A6 (serotonin and 1-naphthol), UGT1A9 (propofol), UGT2B7 [morphine 6-glucuronidation and zidovudine (AZT)], and UGT2B15 (*S*-oxazepam) (206). Correlation analyses with most of these probe substrates are somewhat limited by the fact that there is often only a three- to fivefold difference from the minimum to the maximum rate of reaction among samples of human microsomes. Also, the relative abundance of UGT enzymes in human liver microsomes is not known (to date, at least 22 human UGT enzymes have been identified. The current UGT nomenclature may be found at: http://som.flinders.edu.au/FUSA/ClinPharm/UGT/index.html). Some UGT enzymes (e.g., UGT1A7, UGT1A8, UGT1A10, and UGT2A1) are not expressed in human liver (Table 11). Of the latter enzymes, UGT1A7, UGT1A8, and UGT1A10 are expressed in the gastrointestinal tract where they appear to be important for prehepatic elimination of various orally administered drugs. If screening assays with recombinant UGT enzymes indicate that one or more of these enzymes are predominantly involved in the conjugation of a drug candidate, pooled, or individual microsomal samples from extrahepatic tissue (e.g., intestine) should be evaluated for

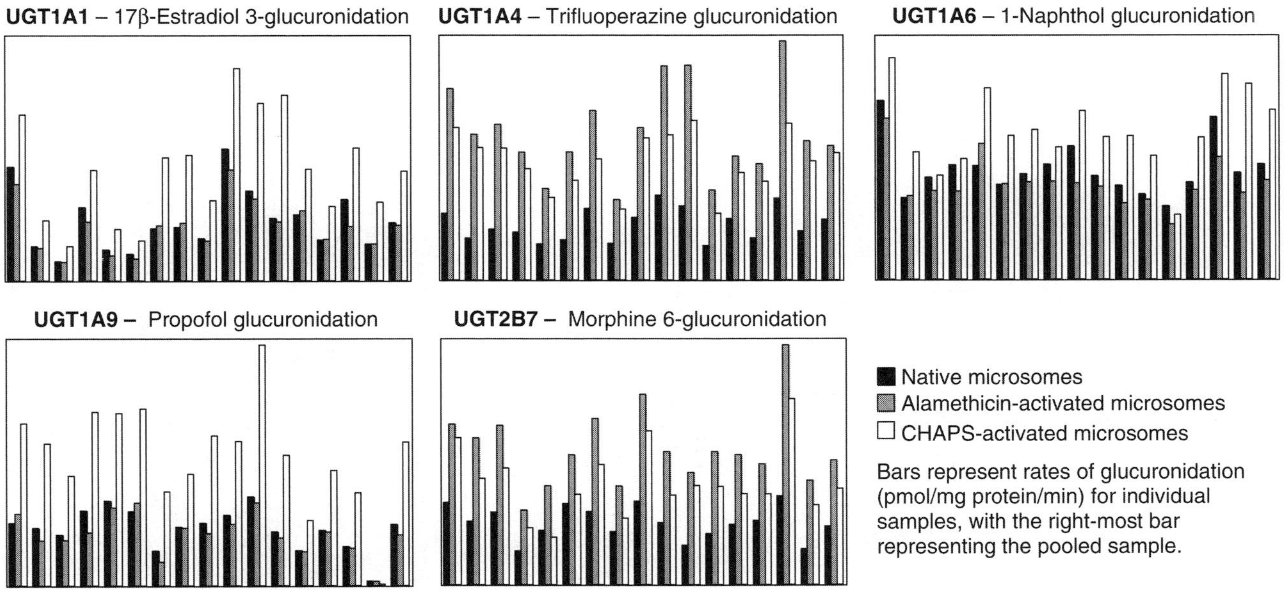

Figure 28 Probe UGT activities in a bank of human liver microsomes ($n = 16$). The following UGT activities were determined in the absence (native) or presence of the pore-forming peptide alamethicin or the zwitterionic detergent CHAPS: UGT1A1 (17β-estradiol 3-glucuronidation), UGT1A4 (trifluoperazine glucuronidation), UGT1A6 (1-naphthol glucuronidation), UGT1A9 (propofol glucuronidation), and UGT2B7 (morphine 6-glucuronidation). *Abbreviation*: UGT, UDP-glucuronosyltransferase.

Table 11 Tissue Distribution and Specific Substrates for Human UDP-Glucuronosyltransferase (UGT) Enzymes

UGT	Present in liver?	Tissue	Selective substrates
1A1	Yes	Liver, small intestine, colon	17β-estradiol (3-glucuronidation)
1A3	Yes	Liver, small intestine, colon	Hexafluoro-1α, 25-trihydroxyvitamin D3
1A4	Yes	Liver, small intestine, colon	Trifluoperazine, tertiary amines, antihistamines, lamotrigine, amitriptyline, cyclobenzaprine, olanzapine
1A5	Yes	Liver	
1A6	Yes	Liver, small intestine, colon, stomach	1-Naphthol
1A7	No	Esophagus, stomach, lung	
1A8	No	Colon, small intestine, kidney	
1A9	Yes	Liver, colon, kidney	Propofol
1A10	No	Stomach, small intestine, colon	
2A1	No	Olfactory	
2A2	Unknown	Unknown	
2A3	Unknown	Unknown	
2B4	Yes	Liver, small intestine	
2B7	Yes	Kidney, small intestine, colon	Morphine (6-glucuronidation[a])
2B10	Yes	Liver, ileum, prostate	
2B11	Yes	Mammary, prostate, others	
2B15	Yes	Liver, small intestine, prostate	*S*-Oxazepam
2B17	Yes	Liver, prostate	
2B28	Yes	Liver, mammary	

[a]UGT2B7 also catalyzes the 3-glucuronidation of morphine, but this reaction is also catalyzed by other UGTs.
Source: Adapted from Refs. 206, 220, 202, and 221.

their ability to conjugate the drug candidate. Of the hepatically expressed UGT enzymes, UGT1A1, UGT1A3, UGT1A4, UGT1A6, UGT1A9, UGT2B7, and UGT2B15 are considered to be the UGT enzymes most important for hepatic drug metabolism because UGT1A5, UGT2B4, UGT2B10, UGT2B11, UGT2B17, and UGT2B28 are reported to have low or negligible activity toward xenobiotics (206).

Typical reactions catalyzed by UGT enzymes require the cofactor uridine diphosphate-glucuronic acid, UDPGA. However, the C-terminus of all UGT enzymes contains a membrane-spanning domain that anchors the enzyme in the endoplasmic reticulum, and the enzyme faces the lumen of the endoplasmic reticulum, where it is ideally placed to conjugate lipophilic xenobiotics and their metabolites generated by oxidation, reduction, or hydrolysis. The lumenal orientation of UGT poses a problem because UDPGA is a water-soluble cofactor synthesized in the cytoplasm. In order to measure in vitro UGT activity in microsomes, this "latency" of UGT enzymes must be overcome. In vitro, the glucuronidation of xenobiotics by liver microsomes can be stimulated by detergents (e.g., CHAPS, Brij-58), which disrupt the lipid bilayer of the endoplasmic reticulum and allow UGT enzymes free access to UDPGA. High concentrations of detergent can inhibit UGT enzymes, presumably by disrupting their interaction with phospholipids, which are important for catalytic activity. Detergents can inhibit certain UGT enzymes (207), hence, they can affect UGT activity directly in addition to their indirect effects through interaction with phospholipids. Consequently, detergents, especially nonionic detergents like Brij-58, can alter the kinetics of UGT reactions (208). Treatment of microsomes with detergents also virtually eliminates CYP activity; hence, detergents cannot be used in studies designed to investigate the possible coupling of oxidation reactions catalyzed by CYP, with conjugation reactions catalyzed by UGT. This is not a limitation of alamethicin, a pore-forming peptide that activates UGT activity, like detergent, but which does not inhibit CYP (208). Furthermore, in contrast to certain detergents, alamethicin appears to increase V_{max} without affecting K_m. From a review of the literature, Miners et al. report that alamethicin and nonionic detergents, such as Brij-58, generally result in the highest UGT activity but that alamethicin is the preferred activator because the effects of detergents are not reproducible between substrates (206). However, in incubations designed strictly to measure UGT activity (and not both UGT and CYP activity), the zwitterionic detergent, CHAPS, can activate certain UGT activities to a comparable or even greater extent than alamethicin, as shown in Figure 28.

The in vitro activity of UGT enzymes is highly dependent not only on the substance used to activate the microsomal membranes but also on incubation conditions. The kinetic properties of UGT enzymes have been demonstrated to vary with the concentration of cofactor, membrane composition, type of buffer, ionic strength, and pH (206). In vitro intrinsic clearance (CL_{int}) values (measured as V_{max}/K_m) for AZT glucuronidation in human liver microsomes were shown to vary sixfold depending on incubation conditions, but even under conditions that produced the greatest CL_{int}, the in vivo clearance rate was underpredicted by three- to fourfold (206). This in vitro underprediction of the in vivo rate of clearance of drugs that are glucuronidated is typical when kinetic data based on experiments with human liver microsomal are used to assess CL_{int} and is likely due to a number of factors, including the presence or absence of albumin, nonspecific binding, atypical in vitro kinetics, active uptake into hepatocytes, and significant extrahepatic expression of various UGT enzymes (206). The prediction of the in vivo clearance of drugs that are glucuronidated by hepatocytes appears to be more accurate than for predictions made with microsomes, but underprediction is still the likely outcome. The case of AZT appears to be an exception to this rule, as in vivo clearance values were well-predicted with hepatocytes (206). The use of either microsomes or recombinant UGT2B7 also appears to underpredict the in vivo magnitude of the inhibitory interaction between fluconazole and AZT by 5- to 10-fold (206). However, when 2% bovine serum albumin (BSA) is added to either system, there is an 85% decrease in the K_i value, which results in a much improved prediction of the in vivo interaction (206). The effect of BSA is not due to nonspecific binding, but the exact mechanism remains to be elucidated.

The kinetic properties of UGT enzymes are possibly influenced by the formation of homo- and heterodimers among UGT enzymes and by the formation of heterodimers with other microsomal enzymes such as various CYP enzymes or epoxide hydrolase (206). For instance, it has been demonstrated that the ratio of morphine 3-glucuronide to morphine 6-glucuronide formed by UGT2B7 is altered by the presence of CYP3A4 (206). It is not known whether dimerization between UGT2B7 and CYP3A4, if it occurs, also alters the substrate specificity of other individual UGT enzymes, which would have implications for studies designed to determine the substrate specificity of recombinant enzymes, which are invariably expressed individually. Additionally, there is currently no universally accepted method to quantify UGT enzymes in a recombinant preparation, which precludes the accurate determination of relative activity factors, as for recombinant CYP enzymes. Finally, pos-translational modifications to UGT enzymes that occur in vivo in humans (e.g., phsophorylation and N-glycosylation) may not occur in the cell expression system chosen to produce the recombinant UGT enzymes (i.e., bacterial systems), which can impact activity in a substrate-dependent manner (206). All of these findings suggest that the use of recombinant human UGT enzymes may not provide accurate indications of the extent to which a given UGT can glucuronidate a given drug candidate.

An additional consideration that must be made for UGT reaction phenotyping studies that does not need to be made for CYP reaction phenotyping studies is the potential for a drug candidate to be converted to an *N*-carbamoyl glucuronide, which cannot be observed in vitro under typical incubation conditions. Some primary amines, or the demethylated metabolites of secondary and tertiary amines, such as carvedilol, sertraline, varenicline, mofegiline, garenoxacin, tocainide, and sibutramine, among others, have been reported to be converted to *N*-carbamoyl glucuronides (209–215). However, marked species difference have been found in the formation of *N*-carbamoyl glucuronides, and humans have only been found to produce these conjugates from even fewer drugs, including varenicline, sertraline, and mofegiline. To form this type of conjugate in vitro, the incubation must be performed under a CO_2 atmosphere, and include a carbonate buffer. Although not directly demonstrated, it has been hypothesized that a transient carbamic acid intermediate is formed by the interaction of the amine with the dissolved CO_2, followed by its glucuronidation (215). Since the intermediate is not stable, the hypothesis that UGT also catalyzes the formation of the carbamic acid cannot be disproved. However, in the case of varenicline and sertraline, it is predominantly UGT2B7 that forms the *N*-carbamoyl glucuronide, which also conjugates various carboxylic acids

(214,215). Given that the in vitro formation of *N*-carbamoyl glucuronides occurs only under special incubation conditions that are not typically employed, it is possible that many other primary and secondary amines or their oxidative metabolites can be converted to such conjugates but have not been detected because of the unusual incubation conditions required to support their formation. If one or more *N*-carbamoyl glucuronide conjugates of a new drug candidate are discovered in vivo, it may be worthwhile to utilize incubation conditions that will support the formation of such glucuronides in vitro.

REFERENCES

1. Preliminary Concept Paper: Drug Interaction Studies—Study Design, Data Analysis, and Implications for Dosing and Labeling. Preliminary Concept Paper: Office of Clinical Pharmacology and Biopharmaceutics, Center for Drug Evaluation and Research, United States Food and Drug Administration, 2004.
2. Guidance for Industry: Drug Interaction Studies—Study Design, Data Analysis, and Implications for Dosing and Labeling. Draft Guidance. Rockville, MD: Food and Drug Administration, 2006.
3. Food and Drug Administration. Available at: http://www.fda.gov/cder/drug/drugInteractions/default.htm. Accessed August 2006.
4. Tucker GT, Houston JB, Huang SM. Optimizing drug development: strategies to assess drug metabolism/transporter interaction potential–toward a consensus. Pharmacol Res 2001; 18:1071–1080.
5. Bjornsson TD, Callaghan JT, Einolf HJ, et al. The conduct of in vitro and in vivo drug-drug interaction studies: a Pharmaceutical Research and Manufacturers of America (PhRMA) perspective. Drug Metab Dispos 2003; 31:815–832.
6. Bjornsson TD, Callaghan JT, Einolf HJ, et al. The conduct of in vitro and in vivo drug-drug interaction studies: a PhRMA perspective. J Clin Pharmacol 2003; 43:443–469.
7. Ozdemir O, Boran M, Gokce V, et al. A case with severe rhabdomyolysis and renal failure associated with cerivastatin-gemfibrozil combination therapy–a case report. Angiology 2000; 51:695–697.
8. http://www.emea.eu.int/pdfs/human/referral/Cerivastatin/081102en.pdf. Accessed August 2006.
9. Backman J, Kyrklund C, Neuvonen M, et al. Gemfibrozil greatly increases plasma concentrations of cerivastatin. Clin Pharmacol Ther 2002; 72:685–691.
10. Shitara Y, Hirano M, Sato H, et al. Gemfibrozil and its glucuronide inhibit the organic anion transporting polypeptide 2 (OATP2/OATP1B1:SLC21A6)-mediated hepatic uptake and CYP2C8-mediated metabolism of cerivastatin: analysis of the mechanism of the clinically relevant drug-drug interaction between cerivastatin and gemfibrozil. J Pharmacol Exp Ther 2004; 311:228–236.
11. Ogilvie BW, Zhang D, Li W, et al. Glucuronidation converts gemfibrozil to a potent, metabolism-dependent inhibitor of CYP2C8: implications for drug-drug interactions. Drug Metab Dispos 2006; 34:191–197.
12. Tsunoda SM, Velez RL, von Moltke LL, et al. Differentiation of intestinal and hepatic cytochrome P450 3A activity with use of midazolam as an vivo probe: effect of ketoconazole. Clin Pharmacol Ther 1999; 66:461–471.

13. Gorski J, Vannaprasaht S, Hamman M, et al. The effect of age, sex, and rifampin administration on intestinal and hepatic cytochrome P450 3A activity. Clin Pharmacol Ther 2003; 74:275–287.
14. Holtbecker N, Fromm MF, Kroemer HK, et al. The nifedipine-rifampin interaction. Evidence for induction of gut wall metabolism. Drug Metab Dispos 1996; 24:1121–1123.
15. Fromm MF, Busse D, Kroemer HK, et al. Differential induction of prehepatic and hepatic metabolism of verapamil by rifampin. Hepatology 1996; 24:796–801.
16. Pope LE, Khalil MH, Berg JE, et al. Pharmacokinetics of dextromethorphan after single or multiple dosing in combination with quinidine in extensive and poor metabolizers. J Clin Pharmacol 2004; 44:1132–1142.
17. Gorski JC, Huang SM, Pinto A, et al. The effect of echinacea (Echinacea purpurea root) on cytochrome P450 activity in vivo. Clin Pharmacol Ther 2004; 75:89–100.
18. Zineh I, Beitelshees AL, Gaedigk A, et al. Pharmacokinetics and CYP2D6 genotypes do not predict metoprolol adverse events or efficacy in hypertension. Clin Pharmacol Ther 2004; 76:536–544.
19. Majerus PWT, Douglas M. Blood conjugation and anticoagulant, thrombolytic, and antiplatelet drugs. In: Goodman LS, Gilman A, Brunton LL, eds. Goodman & Gilman's The Pharmacological Basis of Therapeutics. 11th ed. New York: McGraw-Hill, 2006:1200–1256.
20. Bajpai M, Esmay JD. In vitro studies in drug discovery and development: an analysis of study objectives and application of good laboratory practices (GLP). Drug Metab Rev 2002; 34:679–689.
21. Yuan R, Madani S, Wei XX, et al. Evaluation of cytochrome P450 probe substrates commonly used by the pharmaceutical industry to study in vitro drug interactions. Drug Metab Dispos 2002; 30:1311–1319.
22. Walsky RL, Obach RS. Validated assays for human cytochrome P450 activities. Drug Metab Dispos 2004; 32:647–660.
23. Lim HK, Duczak N, Brougham L, et al. Automated screening with confirmation of mechanism-based inactivation of CYP3A4, CYP2C9, CYP2C19, CYP2D6, and CYP1A2 in pooled human liver microsomes. Drug Metab Dispos 2005; 33:1211–1219.
24. Bachmann KA. Inhibition constants, inhibitor concentrations and the prediction of inhibitory drug drug interactions: pitfalls, progress and promise. Curr Drug Metab 2006; 7:1–14.
25. McGinnity DF, Riley RJ. Predicting drug pharmacokinetics in humans from in vitro metabolism studies. Biochem Soc Trans 2001; 29:135–139.
26. Moody GC, Griffin SJ, Mather AN, et al. Fully automated analysis of activities catalysed by the major human liver cytochrome P450 (CYP) enzymes: assessment of human CYP inhibition potential. Xenobiotica 1999; 29:53–75.
27. Chiba M, Xu X, Nishime JA, et al. Hepatic microsomal metabolism of montelukast, a potent leukotriene D4 receptor antagonist, in humans. Drug Metab Dispos 1997; 25:1022–1031.
28. Walsky RL, Obach RS, Gaman EA, et al. Selective inhibition of human cytochrome P4502C8 by montelukast. Drug Metab Dispos 2005; 33:413–418.
29. Kajosaari LI, Niemi M, Backman JT, et al. Telithromycin, but not montelukast, increases the plasma concentrations and effects of the cytochrome P450 3A4 and 2C8 substrate repaglinide. Clin Pharmacol Ther 2006; 79:231–242.

30. Jaakkola T, Backman JT, Neuvonen M, et al. Montelukast and zafirlukast do not affect the pharmacokinetics of the CYP2C8 substrate pioglitazone. Eur J Clin Pharmacol 2006; 62:503–509.
31. Hallifax D, Houston JB. Binding of drugs to hepatic microsomes: comment and assessment of current prediction methodology with recommendation for improvement. Drug Metab Dispos 2006; 34:724–726.
32. Andersson T, Cederberg C, Edvardsson G, et al. Effect of omeprazole treatment on diazepam plasma levels in slow versus normal rapid metabolizers of omeprazole. Clin Pharmacol Ther 1990; 47:79–85.
33. Andersson T, Miners JO, Veronese ME, et al. Diazepam metabolism by human liver microsomes is mediated by both S-mephenytoin hydroxylase and CYP3A isoforms. Br J Clin Pharmacol 1994; 38:131–137.
34. Margolis JM, Obach RS. Impact of nonspecific binding to microsomes and phospholipid on the inhibition of cytochrome P4502D6: implications for relating in vitro inhibition data to in vivo drug interactions. Drug Metab Dispos 2003; 31:606–611.
35. Nielsen TL, Rasmussen BB, Flinois JP, et al. In vitro metabolism of quinidine: the (3*S*)-3-hydroxylation of quinidine is a specific marker reaction for cytochrome P-4503A4 activity in human liver microsomes. J Pharmacol Exp Ther 1999; 289:31–37.
36. Galetin A, Clarke SE, Houston JB. Quinidine and haloperidol as modifiers of CYP3A4 activity: multisite kinetic model approach. Drug Metab Dispos 2002; 30:1512–1522.
37. Galetin A, Clarke SE, Houston JB. Multisite kinetic analysis of interactions between prototypical CYP3A4 subgroup substrates: midazolam, testosterone and nifedipine. Drug Metab Dispos 2003; 31:1108–1116.
38. Galetin A, Ito K, Hallifax D, et al. CYP3A4 substrate selection and substitution in the prediction of potential drug-drug interactions. J Pharmacol Exp Ther 2005; 314:180–190.
39. Houston JB, Galetin A. Modelling atypical CYP3A4 kinetics: principles and pragmatism. Arch Biochem Biophys 2005; 433:351–360.
40. Gibbs M, Kunze K, Howald W, et al. Effect of inhibitor depletion on inhibitory potency: tight binding inhibition of CYP3A by clotrimazole. Drug Metab Dispos 1999; 27:596–599.
41. Li XQ, Andersson TB, Ahlstrom M, et al. Comparison of inhibitory effects of the proton pump-inhibiting drugs omeprazole, esomeprazole, lansoprazole, pantoprazole, and rabeprazole on human cytochrome P450 activities. Drug Metab Dispos 2004; 32:821–827.
42. Wright JN, Slatcher G, Akhtar M. 'Slow-binding' sixth-ligand inhibitors of cytochrome P-450 aromatase. Studies with 19-thiomethyl- and 19-azido-androstenedione. Biochem J 1991; 273(pt 3):533–539.
43. Li C, Benet L. Mechanistic role of acyl glucuronides. In: Kaplowitz N, DeLeve L, eds. Drug-Induced Liver Disease. New York: Marcel Dekker, Inc., 2003:151–181.
44. Silverman RB. Mechanism-based enzyme inactivators. Methods Enzymol 1995; 249:240–283.
45. Fontana E, Dansette PM, Poli SM. Cytochrome P450 enzymes mechanism based inhibitors: common sub-structures and reactivity. Curr Drug Metab 2005; 6:413–454.

46. Ortiz de Montellano PR. The 1994 Bernard B. Brodie award lecture: structure, mechanism, and inhibition of cytochrome P450. Drug Metab Dispos 1995; 23:1181–1187.
47. Melet A, Assrir N, Jean P, et al. Substrate selectivity of human cytochrome P450 2C9: importance of residues 476, 365, and 114 in recognition of diclofenac and sulfaphenazole and in mechanism-based inactivation by tienilic acid. Arch Biochem Biophys 2003; 409:80–91.
48. Lopez-Garcia MP, Dansette PM, Mansuy D. Thiophene derivatives as new mechanism-based inhibitors of cytochromes P-450: inactivation of yeast-expressed human liver cytochrome P-450 2C9 by tienilic acid. Biochemistry 1994; 33:166–175.
49. Robin MA, Maratrat M, Le Roy M, et al. Antigenic targets in tienilic acid hepatitis. Both cytochrome P450 2C11 and 2C11-tienilic acid adducts are transported to the plasma membrane of rat hepatocytes and recognized by human sera. J Clin Invest 1996; 98:1471–1480.
50. Walgren JL, Mitchell MD, Thompson DC. Role of metabolism in drug-induced idiosyncratic hepatotoxicity. Crit Rev Toxicol 2005; 35:325–361.
51. Uetrecht J. Prediction of a new drug's potential to cause idiosyncratic reactions. Curr Opin Drug Discov Devel 2001; 4:55–59.
52. Dexter A, Hager L. Transient heme N-alkylation of chloroperoxidase by terminal alkenes and alkynes. J Am Chem Soc 1995; 117:817–818.
53. Blobaum AL. Mechanism-based inactivation and reversibility: is there a new trend in the inactivation of cytochrome P450 enzymes? Drug Metab Dispos 2006; 34:1–7.
54. Sharma U, Roberts ES, Hollenberg PF. Formation of a metabolic intermediate complex of cytochrome P4502B1 by clorgyline. Drug Metab Dispos 1996; 24:1247–1253.
55. Ohashi K, Sudo T, Sakamoto K, et al. The influence of pretreatment periods with diltiazem on nifedipine kinetics. J Clin Pharmacol 1993; 33:222–225.
56. Bertelsen KM, Venkatakrishnan K, von Moltke LL, et al. Apparent mechanism-based inhibition of human CYP2D6 in vitro by paroxetine: comparison with fluoxetine and quinidine. Drug Metab Dispos 2003; 31:289–293.
57. Hemeryck A, Lefebvre RA, De Vriendt C, et al. Paroxetine affects metoprolol pharmacokinetics and pharmacodynamics in healthy volunteers. Clin Pharmacol Ther 2000; 67:283–291.
58. Belle DJ, Ernest CS, Sauer JM, et al. Effect of potent CYP2D6 inhibition by paroxetine on atomoxetine pharmacokinetics. J Clin Pharmacol 2002; 42:1219–27.
59. Liston HL, DeVane CL, Boulton DW, et al. Differential time course of cytochrome P450 2D6 enzyme inhibition by fluoxetine, sertraline, and paroxetine in healthy volunteers. J Clin Psychopharmacol 2002; 22:169–173.
60. Crespi CL, Stresser DM. Fluorometric screening for metabolism-based drug–drug interactions. J Pharmacol Toxicol Methods 2000; 44:325–331.
61. Delaporte E, Slaughter DE, Egan MA, et al. The potential for CYP2D6 inhibition screening using a novel scintillation proximity assay-based approach. J Biomol Screen 2001; 6:225–231.
62. Yamamoto T, Suzuki A, Kohno Y. Application of microtiter plate assay to evaluate inhibitory effects of various compounds on nine cytochrome P450 isoforms and to estimate their inhibition patterns. Drug Metab Pharmacokinet 2002; 17:437–448.

63. Ghosal A, Hapangama N, Yuan Y, et al. Rapid determination of enzyme activities of recombinant human cytochromes P450, human liver microsomes and hepatocytes. Biopharm Drug Dispos 2003; 24:375–384.
64. Jenkins KM, Angeles R, Quintos MT, et al. Automated high throughput ADME assays for metabolic stability and cytochrome P450 inhibition profiling of combinatorial libraries. J Pharm Biomed Anal 2004; 34:989–1004.
65. Ansede JH, Thakker DR. High-throughput screening for stability and inhibitory activity of compounds toward cytochrome P450-mediated metabolism. J Pharm Sci 2004; 93:239–255.
66. Cali JJ, Ma D, Sobol M, et al. Luminogenic cytochrome P450 assays. Expert Opin Drug Metab Toxicol 2006; 2:629–645.
67. Anzenbacherova E, Veinlichova A, Masek V, et al. Comparison of "High throughput" micromethods for determination of cytochrome P450 activities with classical methods using HPLC for product identification. Biomed Pap Med Fac Univ Palacky Olomouc Czech Repub 2005; 149:353–355.
68. Rodrigues AD, Surber BW, Yao Y, et al. [O-ethyl 14C]phenacetin O-deethylase activity in human liver microsomes. Drug Metab Dispos 1997; 25:1097–1100.
69. Di Marco A, Marcucci I, Verdirame M, et al. Development and validation of a high-throughput radiometric CYP3A4/5 inhibition assay using tritiated testosterone. Drug Metab Dispos 2005; 33:349–358.
70. Di Marco A, Marcucci I, Chaudhary A, et al. Development and validation of a high-throughput radiometric CYP2C9 inhibition assay using tritiated diclofenac. Drug Metab Dispos 2005; 33:359–364.
71. Katoh M, Nakajima M, Shimada N, et al. Inhibition of human cytochrome P450 enzymes by 1,4-dihydropyridine calcium antagonists: prediction of in vivo drug-drug interactions. Eur J Clin Pharmacol 2000; 55:843–852.
72. Tang W, Stearns RA, Wang RW, et al. Roles of human hepatic cytochrome P450s 2C9 and 3A4 in the metabolic activation of diclofenac. Chem Res Toxicol 1999; 12:192–199.
73. Ngui JS, Tang W, Stearns RA, et al. Cytochrome P450 3A4-mediated interaction of diclofenac and quinidine. Drug Metab Dispos 2000; 28:1043–1050.
74. Crespi C, Miller V, Penman B. Microtiter plate assays for inhibition of human, drug-metabolizing cytochromes P450. Anal Biochem 1997; 248:188–190.
75. Crespi CL, Miller VP. The use of heterologously expressed drug metabolizing enzymes state of the art and prospects for the future. Pharmacol Ther 1999; 84:121–131.
76. Yamazaki H, Nakajima M, Nakamura M, et al. Enhancement of cytochrome P-450 3A4 catalytic activities by cytochrome b(5) in bacterial membranes. Drug Metab Dispos 1999; 27:999–1004.
77. Yamazaki H, Gillam EM, Dong MS, et al. Reconstitution of recombinant cytochrome P450 2C10(2C9) and comparison with cytochrome P450 3A4 and other forms: effects of cytochrome P450-P450 and cytochrome P450-b5 interactions. Arch Biochem Biophys 1997; 342:329–337.
78. Kumar V, Rock DA, Warren CJ, et al. Enzyme source effects on CYP2C9 kinetics and inhibition. Drug Metab Dispos 2006; 34:1903–1908.
79. McGinnity DF, Griffin SJ, Moody GC, et al. Rapid characterization of the major drug-metabolizing human hepatic cytochrome P-450 enzymes expressed in Escherichia coli. Drug Metab Dispos 1999; 27:1017–1023.

80. Williams AJ, Ring BJ, Cantrell VE, et al. Comparative metabolic capabilities of CYP3A4, CYP3A5, and CYP3A7. Drug Metab Dispos 2002; 30:883–891.
81. Goodman LS, Gilman A, Brunton LL, et al. Goodman & Gilman's the pharmacological basis of therapeutics. 11th ed. New York: McGraw-Hill, 2006.
82. Kobayashi K, Yamamoto T, Chiba K, et al. The effects of selective serotonin reuptake inhibitors and their metabolites on S-mephenytoin 4′-hydroxylase activity in human liver microsomes. Br J Clin Pharmacol 1995; 40:481–485.
83. von Moltke LL, Greenblatt DJ, Schmider J, et al. Midazolam hydroxylation by human liver microsomes in vitro: inhibition by fluoxetine, norfluoxetine, and by azole antifungal agents. J Clin Pharmacol 1996; 36:783–791.
84. Lemberger L, Rowe H, Bosomworth JC, et al. The effect of fluoxetine on the pharmacokinetics and psychomotor responses of diazepam. Clin Pharmacol Ther 1988; 43:412–419.
85. Hall J, Naranjo CA, Sproule BA, et al. Pharmacokinetic and pharmacodynamic evaluation of the inhibition of alprazolam by citalopram and fluoxetine. J Clin Psychopharmacol 2003; 23:349–357.
86. http://www.wyeth.com/content/getfile.asp?id=93. Accessed September, 2006.
87. Ohyama K, Nakajima M, Suzuki M, et al. Inhibitory effects of amiodarone and its N-deethylated metabolite on human cytochrome P450 activities: prediction of in vivo drug interactions. Br J Clin Pharmacol 2000; 49:244–253.
88. Gascon MP, Dayer P. In vitro forecasting of drugs which may interfere with the biotransformation of midazolam. Eur J Clin Pharmacol 1991; 41:573–578.
89. Kobayashi K, Nakajima M, Chiba K, et al. Inhibitory effects of antiarrhythmic drugs on phenacetin O-deethylation catalysed by human CYP1A2. Br J Clin Pharmacol 1998; 45:361–368.
90. Walker DK, Alabaster CT, Congrave GS, et al. Significance of metabolism in the disposition and action of the antidysrhythmic drug, dofetilide. In vitro studies and correlation with in vivo data. Drug Metab Dispos 1996; 24:447–455.
91. Polasek TM, Elliot DJ, Lewis BC, et al. Mechanism-based inactivation of human cytochrome P4502C8 by drugs in vitro. J Pharmacol Exp Ther 2004; 311:996–1007.
92. Houston JB. Utility of in vitro drug metabolism data in predicting in vivo metabolic clearance. Biochem Pharmacol 1994; 47:1469–1479.
93. Neal JM, Kunze KL, Levy RH, et al. KIIV, an in vivo parameter for predicting the magnitude of a drug interaction arising from competitive enzyme inhibition. Drug Metab Dispos 2003; 31:1043–1048.
94. Kunze K, Trager W. Warfarin-fluconazole III: a rational approach to management of a metabolically based drug interaction. Drug Metab Dispos 1996; 24:429–435.
95. Zomorodi K, Houston JB. Effect of omeprazole on diazepam disposition in the rat: in vitro and in vivo studies. Pharm Res 1995; 12:1642–1646.
96. Jones HM, Hallifax D, Houston JB. Quantitative prediction of the in vivo inhibition of diazepam metabolism by omeprazole using rat liver microsomes and hepatocytes. Drug Metab Dispos 2004; 32:572–580.
97. McGinnity DF, Berry AJ, Kenny JR, et al. Evaluation of time-dependent cytochrome P450 inhibition using cultured human hepatocytes. Drug Metab Dispos 2006; 34:1291–1300.
98. Martin H, Sarsat JP, de Waziers I, et al. Induction of cytochrome P450 2B6 and 3A4 expression by phenobarbital and cyclophosphamide in cultured human liver slices. Pharm Res 2003; 20:557–568.

99. Renwick AB, Ball SE, Tredger JM, et al. Inhibition of zaleplon metabolism by cimetidine in the human liver: in vitro studies with subcellular fractions and precision-cut liver slices. Xenobiotica 2002; 32:849–862.
100. Lake BG, Ball SE, Kao J, et al. Metabolism of zaleplon by human liver: evidence for involvement of aldehyde oxidase. Xenobiotica 2002; 32:835–847.
101. Ludden LK, Ludden TM, Collins JM, et al. Effect of albumin on the estimation, in vitro, of phenytoin Vmax and Km values: implications for clinical correlation. J Pharmacol Exp Ther 1997; 282:391–396.
102. Salonen JS, Nyman L, Boobis AR, et al. Comparative studies on the cytochrome P450-associated metabolism and interaction potential of selegiline between human liver-derived in vitro systems. Drug Metab Dispos 2003; 31:1093–1102.
103. Persson K, Ekehed S, Otter C, et al. Evaluation of human liver slices and reporter gene assays as systems for predicting the cytochrome P450 induction potential of drugs in vivo in humans. Pharm Res 2006; 23:56–69.
104. Wang RW, Newton DJ, Liu N, et al. Human cytochrome P-450 3A4: in vitro drug-drug interaction patterns are substrate-dependent. Drug Metab Dispos 2000; 28:360–366.
105. Stresser DM, Blanchard AP, Turner SD, et al. Substrate-dependent modulation of CYP3A4 catalytic activity: analysis of 27 test compounds with four fluorometric substrates. Drug Metab Dispos 2000; 28:1440–1448.
106. Schrag ML, Wienkers LC. Triazolam substrate inhibition: evidence of competition for heme-bound reactive oxygen within the CYP3A4 active site. Drug Metab Dispos 2001; 29:70–75.
107. Wienkers LC. Problems associated with in vitro assessment of drug inhibition of CYP3A4 and other P-450 enzymes and its impact on drug discovery. J Pharmacol Toxicol Methods 2001; 45:79–84.
108. Schrag ML, Wienkers LC. Interactions with CYP3A4 are substrate-dependent. ISSX Proceedings 1999; 15:394.
109. Clarke SE. In vitro assessment of human cytochrome P450. Xenobiotica 1998; 28:1167–1202.
110. Ekroos M, Sjögren T. Structural basis for ligand promiscuity in cytochrome P450 3A4. Proc Natl Acad Sci U S A 2006; 103:13682–13687.
111. Draper AJ, Madan A, Parkinson A. Inhibition of coumarin 7-hydroxylase activity in human liver microsomes. Arch Biochem Biophys 1997; 341:47–61.
112. Hickman D, Wang JP, Wang Y, et al. Evaluation of the selectivity of in vitro probes and suitability of organic solvents for the measurement of human cytochrome P450 monooxygenase activities. Drug Metab Dispos 1998; 26:207–215.
113. Chauret N, Gauthier A, Nicoll-Griffith DA. Effect of common organic solvents on in vitro cytochrome P450-mediated metabolic activities in human liver microsomes. Drug Metab Dispos 1998; 26:1–4.
114. Busby W, Ackermann J, Crespi C. Effect of methanol, ethanol, dimethyl sulfoxide, and acetonitrile on in vitro activities of cdna-expressed human cytochromes P-450. Drug Metab Dispos 1999; 27:246–249.
115. Tang C, Shou M, Rodrigues D. Substrate-dependent effect of acetonitrile on human liver microsomal cytochrome P450 2C9 (CYP2C9) activity. Drug Metab Dispos 2000; 28:567–572.

116. Paris BL, Marcum AE, Clarin JR, et al. Effects of common organic solvents on CYP2E1 activity in human liver microsomes: effects of order of addition and preincubation with NADPH. Drug Metab Rev 2003; 35(suppl 2):180.
117. Wynalda MA, Wienkers LC. Assessment of potential interactions between dopamine receptor agonists and various human cytochrome P450 enzymes using a simple in vitro inhibition screen. Drug Metab Dispos 1997; 25:1211–1214.
118. Gorski JC, Jones DR, Wrighton SA, et al. Contribution of human CYP3A subfamily members to the 6-hydroxylation of chlorzoxazone. Xenobiotica 1997; 27:243–256.
119. Obach RS, Walsky RL, Venkatakrishnan K, et al. The utility of in vitro cytochrome P450 inhibition data in the prediction of drug-drug interactions. J Pharmacol Exp Ther 2006; 316:336–348.
120. Obach RS, Walsky RL, Venkatakrishnan K, et al. In vitro cytochrome P450 inhibition data and the prediction of drug-drug interactions: qualitative relationships, quantitative predictions, and the rank-order approach. Clin Pharmacol Ther 2005; 78:582–592.
121. Koenigs LL, Peter RM, Thompson SJ, et al. Mechanism-based inactivation of human liver cytochrome P450 2A6 by 8-methoxypsoralen. Drug Metab Dispos 1997; 25:1407–1415.
122. Van LM, Heydari A, Yang J, et al. The impact of experimental design on assessing mechanism-based inactivation of CYP2D6 by MDMA (Ecstasy). J Psychopharmacol 2006; 20:834–841.
123. Mayhew BS, Jones DR, Hall SD. An in vitro model for predicting in vivo inhibition of cytochrome P450 3A by metabolic intermediate complex formation. Drug Metab Dispos 2000; 28:1031–1037.
124. Obach RS, Walsky RL, Venkatakrishnan K. Mechanism-based inactivation of human cytochrome P450 enzymes and the prediction of drug-drug interactions. Drug Metab Dispos 2007; 35:246–255.
125. Munns AJ, De Voss JJ, Hooper WD, et al. Bioactivation of phenytoin by human cytochrome P450: characterization of the mechanism and targets of covalent adduct formation. Chem Res Toxicol 1997; 10:1049–1058.
126. Madan A, Parkinson A. Characterization of the NADPH-dependent covalent binding of [14-C]-halothane to human liver microsomes: a role for CYP2E1 at low substrate concentrations. Drug Metab Dispos 1996; 24:1307–1313.
127. Lazarou J, Pomeranz BH, Corey PN. Incidence of adverse drug reactions in hospitalized patients: a meta-analysis of prospective studies. JAMA 1998; 279:1200–1205.
128. Galetin A, Burt H, Gibbons L, et al. Prediction of time-dependent CYP3A4 drug-drug interactions: impact of enzyme degradation, parallel elimination pathways, and intestinal inhibition. Drug Metab Dispos 2006; 34:166–175.
129. Cate EW, Rogers JF, Powell JR. Inhibition of tolbutamide elimination by cimetidine but not ranitidine. J Clin Pharmacol 1986; 26:372–377.
130. Yao C, Kunze KL, Kharasch ED, et al. Fluvoxamine-theophylline interaction: gap between in vitro and in vivo inhibition constants toward cytochrome P4501A2. Clin Pharmacol Ther 2001; 70:415–424.
131. Davit B, Reynolds K, Yuan R, et al. FDA evaluations using in vitro metabolism to predict and interpret in vivo metabolic drug-drug interactions: impact on labeling. J Clin Pharmacol 1999; 39:899–910.

132. Kanamitsu SI, Ito K, Sugiyama Y. Quantitative prediction of in vivo drug-drug interactions from in vitro data based on physiological pharmacokinetics: use of maximum unbound concentration of inhibitor at the inlet to the liver. Pharm Res 2000; 17:336–343.
133. Wen X, Wang JS, Backman JT, et al. Gemfibrozil is a potent inhibitor of human cytochrome P450 2C9. Drug Metab Dispos 2001; 29:1359–1361.
134. Wang JS, Neuvonen M, Wen X, et al. Gemfibrozil inhibits CYP2C8-mediated cerivastatin metabolism in human liver microsomes. Drug Metab Dispos 2002; 30:1352–1356.
135. Lilja JJ, Backman JT, Neuvonen PJ. Effect of gemfibrozil on the pharmacokinetics and pharmacodynamics of racemic warfarin in healthy subjects. Br J Clin Pharmacol 2005; 59:433–439.
136. Niemi M, Backman JT, Neuvonen M, et al. Effects of gemfibrozil, itraconazole, and their combination on the pharmacokinetics and pharmacodynamics of repaglinide: potentially hazardous interaction between gemfibrozil and repaglinide. Diabetologia 2003; 46:347–351.
137. Niemi M, Backman JT, Granfors M, et al. Gemfibrozil considerably increases the plasma concentrations of rosiglitazone. Diabetologia 2003; 46:1319–1323.
138. Jaakkola T, Backman JT, Neuvonen M, et al. Effects of gemfibrozil, itraconazole, and their combination on the pharmacokinetics of pioglitazone. Clin Pharmacol Ther 2005; 77:404–414.
139. Soars MG, Riley RJ, Findlay KA, et al. Evidence for significant differences in microsomal drug glucuronidation by canine and human liver and kidney. Drug Metab Dispos 2001; 29:121–126.
140. Kumar S, Samuel K, Subramanian R, et al. Extrapolation of diclofenac clearance from in vitro microsomal metabolism data: role of acyl glucuronidation and sequential oxidative metabolism of the acyl glucuronide. J Pharmacol Exp Ther 2002; 303:969–978.
141. Delaforge M, Pruvost A, Perrin L, et al. Cytochrome P450-mediated oxidation of glucuronide derivatives: example of estradiol-17β-glucuronide oxidation to 2-hydroxy-estradiol-17β-glucuronide by CYP 2C8. Drug Metab Dispos 2005; 33:466–473.
142. Prueksaritanont T, Li C, Tang C, et al. Rifampin induces the in vitro oxidative metabolism, but not the in vivo clearance of diclofenac in rhesus monkeys. Drug Metab Dispos 2006; 34:1806–1810.
143. Neuvonen PJ, Niemi M, Backman JT. Drug interactions with lipid-lowering drugs: Mechanisms and clinical relevance. Clin Pharmacol Ther 2006; 80:565–581.
144. Paine MF, Hart HL, Ludington SS, et al. The human intestinal cytochrome P450 "pie". Drug Metab Dispos 2006; 34:880–886.
145. Thummel KEK, Kent L, Shen DD. Metabolically-based drug-drug interactions: principles and mechanisms. In: Levy RH, ed. Metabolic Drug Interactions. Philadelphia: Lippincott Williams & Wilkins, 2000:3–19.
146. Williams JA, Hurst SI, Bauman J, et al. Reaction phenotyping in drug discovery: moving forward with confidence? Curr Drug Metab 2003; 4:527–534.
147. Treiber A, Schneiter R, Delahaye S, et al. Inhibition of organic anion transporting polypeptide-mediated hepatic uptake is the major determinant in the pharmacokinetic interaction between bosentan and cyclosporin A in the rat. J Pharmacol Exp Ther 2004; 308:1121–1129.

148. Strelevitz TJ, Foti RS, Fisher MB. In vivo use of the P450 inactivator 1-aminobenzotriazole in the rat: varied dosing route to elucidate gut and liver contributions to first-pass and systemic clearance. J Pharm Sci 2006; 95:1334–1341.
149. Beedham C. Molybdenum hydroxylases. In: Ioannides C, ed. Enzyme Systems that Metabolise Drugs and Other Xenobiotics. New York: Wiley, 2002:147–187.
150. Rettie AE, Fisher MB. Transformation enzymes: oxidative; non-P450. In: Woolf T, ed. Handbook of Drug Metabolism. New York: Marcel Dekker, Inc., 1999:131–151.
151. Kawashima K, Hosoi K, Naruke T, et al. Aldehyde oxidase-dependent marked species difference in hepatic metabolism of the sedative-hypnotic, zaleplon, between monkeys and rats. Drug Metab Dispos 1999; 27:422–428.
152. Lake BG, Ball SE, Kao J, et al. Metabolism of zaleplon by human liver: evidence for involvement of aldehyde oxidase. Xenobiotica 2002; 32:835–847.
153. Renwick AB, Ball SE, Tredger JM, et al. Inhibition of zaleplon metabolism by cimetidine in the human liver: in vitro studies with subcellular fractions and precision-cut liver slices. Xenobiotica 2002; 32:849–862.
154. Sugihara K, Kitamura S, Tatsumi K. Involvement of mammalian liver cytosols and aldehyde oxidase in reductive metabolism of zonisamide. Drug Metab Dispos 1996; 24:199–202.
155. Beedham C, Miceli JJ, Obach RS. Ziprasidone metabolism, aldehyde oxidase, and clinical implications. J Clin Psychopharmacol 2003; 23:229–232.
156. Cashman J. In vitro metabolism: FMO and related oxygenations. In: Woolf EJ, ed. Handbook of Drug Metabolism. New York: Macel Dekker, Inc., 1999:477–505.
157. Cashman JR, Zhang J. Human flavin-containing monooxygenases. Annu Rev Pharmacol Toxicol 2006; 46:65–100.
158. Pearce RE, Rodrigues AD, Goldstein JA, et al. Identification of the human P450 enzymes involved in lasoprazole metabolism. J Pharmacol Exp Ther 1996; 277:805–816.
159. Li AP, Hartman NR, Lu C, et al. Effects of cytochrome P450 inducers on 17alpha-ethinyloestradiol (EE2) conjugation by primary human hepatocytes. Br J Clin Pharmacol 1999; 48:733–742.
160. Shiraga T, Niwa T, Ohno Y, et al. Interindividual variability in 2-hydroxylation, 3-sulfation, and 3-glucuronidation of ethynylestradiol in human liver. Biol Pharm Bull 2004; 27:1900–1906.
161. Schrag ML, Cui D, Rushmore TH, et al. Sulfotransferase 1E1 is a low km isoform mediating the 3-O-sulfation of ethinyl estradiol. Drug Metab Dispos 2004; 32:1299–1303.
162. Wang MZ, Saulter JY, Usuki E, et al. CYP4F enzymes are the major enzymes in human liver microsomes that catalyze the O-demethylation of the antiparasitic prodrug DB289 [2,5-Bis(4-amidinophenyl)furan-bis-O-methylamidoxime]. Drug Metab Dispos 2006; 34:1985–1994.
163. Obach RS. Prediction of human clearance of twenty-nine drugs from hepatic microsomal intrinsic clearance data: an examination of in vitro half-life approach and nonspecific binding to microsomes. Drug Metab Dispos 1999; 27:1350–1359.
164. Li TL, Giang YS, Hsu JF, et al. Artifacts in the GC-MS profiling of underivatized methamphetamine hydrochloride. Forensic Sci Int 2006; 162:113–120.
165. Yin H, Tran P, Greenberg GE, et al. Methanol solvent may cause increased apparent metabolic instability in in vitro assays. Drug Metab Dispos 2001; 29:185–193.

166. Guidance for Industry: Bioanalytical Method Validation. Guidance. Rockville, MD: Food and Drug Administration, 2001.
167. Kronbach T. Bufuralol, dextromethorphan, and debrisoquine as prototype substrates for human P450IID6. Methods Enzymol 1991; 206:509–517.
168. Gorski JC, Jones DR, Wrighton SA, et al. Characterization of dextromethorphan N-demethylation by human liver microsomes. Contribution of the cytochrome P450 3A (CYP3A) subfamily. Biochem Pharmacol 1994; 48:173–182.
169. McIntyre CJ, Madan A, Parkinson A. Investigation of the role of CYP2B6 in the N-demethylation of dextromethorphan by human liver microsomes. ISSX Proceedings 1996; 10:231.
170. Kerry NL, Somogyi AA, Mikus G, et al. Primary and secondary oxidative metabolism of dextromethorphan. In vitro studies with female Sprague-Dawley and Dark Agouti rat liver microsomes. Biochem Pharmacol 1993; 45:833–839.
171. Ye C, Sweeny D, Sukbuntherng J, et al. Distribution, metabolism, and excretion of the anti-angiogenic compound SU5416. Toxicol In Vitro 2006; 20:154–162.
172. Yamazaki H, Mimura M, Sugahara C, et al. Catalytic roles of rat and human cytochrome P450 2A enzymes in testosterone 7alpha- and coumarin 7-hydroxylations. Biochem Pharmacol 1994; 48:1524–1527.
173. Shou M, Mei Q, Ettore MW, et al. Sigmoidal kinetic model for two co-operative substrate-binding sites in a cytochrome P450 3A4 active site: an example of the metabolism of diazepam and its derivatives. Biochem J 1999; 340:845–853.
174. Andersson T, Miners JO, Veronese ME, et al. Identification of human liver cytochrome P450 isoforms mediating omeprazole metabolism. Br J Clin Pharmacol 1993; 36:521–530.
175. Andersson T, Miners JO, Veronese ME, et al. Identification of human liver cytochrome P450 isoforms mediating secondary omeprazole metabolism. Br J Clin Pharmacol 1994; 37:597–604.
176. Sharer J, Wrighton S. Identification of the human hepatic cytochromes P450 involved in the in vitro oxidation of antipyrine. Drug Metab Dispos 1996; 24:487–494.
177. Mankowski DC. The Role of CYP2C19 in the metabolism of (+/−) Bufuralol, the prototypic substrate of CYP2D6. Drug Metab Dispos 1999; 27:1024–1028.
178. Yamazaki H, Guo Z, Persmark M, et al. Burfuralol hydroxylation by cytochrome P450 2D6 and 1A2 enzymes in human liver microsomes. Mol Pharmacol 1994; 46:568–577.
179. Yumibe N, Huie K, Chen KJ, et al. Identification of human liver cytochrome P450 enzymes that metabolize the nonsedating antihistamine loratadine: formation of descarboethoxyloratadine by CYP3A4 and CYP2D6. Biochem Pharmacol 1996; 51:165–172.
180. Newton DJ, Wang RW, Lu AY. Cytochrome P450 inhibitors. Evaluation of specificities in the in vitrometabolism of therapeutic agents by human liver microsomes. Drug Metab Dispos 1995; 23:154–158.
181. Stresser DM, Broudy MI, Ho T, et al. Highly selective inhibition of human CYP3A in vitro by azamulin and evidence that inhibition is irreversible. Drug Metab Dispos 2004; 32:105–112.
182. Turpeinen M, Nieminen R, Juntunen T, et al. Selective inhibition of CYP2B6-catalyzed bupropion hydroxylation in human liver microsomes in vitro. Drug Metab Dispos 2004; 32:626–631.

183. Richter T, Murdter TE, Heinkele G, et al. Potent mechanism-based inhibition of human CYP2B6 by clopidogrel and ticlopidine. J Pharmacol Exp Ther 2004; 308:189–197.
184. Chen Q, Tan E, Strauss JR, et al. Effect of quinidine on the 10-hydroxylation of R-warfarin: species differences and clearance projection. J Pharmacol Exp Ther 2004; 311:307–314.
185. Ngui JS, Chen Q, Shou M, et al. In vitro stimulation of warfarin metabolism by quinidine: increases in the formation of 4′- and 10-hydroxywarfarin. Drug Metab Dispos 2001; 29:877–886.
186. Ludwig E, Schmid J, Beschke K, et al. Activation of human cytochrome P-450 3A4-catalyzed meloxicam 5′-methylhydroxylation by quinidine and hydroquinidine in vitro. J Pharmacol Exp Ther 1999; 290:1–8.
187. Gelboin HV, Krausz KW, Gonzalez FJ, et al. Inhibitory monoclonal antibodies to human cytochrome P450 enzymes: a new avenue for drug discovery. Trends Pharmacol Sci 1999; 20:432–438.
188. Stresser DM, Kupfer D. Monospecific antipeptide antibody to cytochrome P-450 2B6. Drug Metab Dispos 1999; 27:517–525.
189. Mei Q, Tang C, Assang C, et al. Role of a potent inhibitory monoclonal antibody to cytochrome P-450 3A4 in assessment of human drug metabolism. J Pharmacol Exp Ther 1999; 291:749–759.
190. Ludden LK, Ludden TM, Collins JM, et al. Effect of albumin on the estimation, in vitro, of phenytoin Vmax and Km values: implications for clinical correlation. J Pharmacol Exp Ther 1997; 282:391–396.
191. Tang C, Lin Y, Rodrigues AD, et al. Effect of albumin on phenytoin and tolbutamide metabolism in human liver microsomes: an impact more than protein binding. Drug Metab Dispos 2002; 30:648–654.
192. Arlotto MP, Greenway DJ, Parkinson A. Purification of two isozymes of rat liver microsomal cytochrome P450 with testosterone 7 alpha-hydroxylase activity. Arch Biochem Biophys 1989; 270:441–457.
193. Rodrigues A. Integrated cytochrome P450 reaction phenotyping. Biochem Pharmacol 1999; 57:465–480.
194. Crespi CL, Code EL, Penman BW, et al. An activity based method for integrating metabolism data from cDNA expressed cytochrome P450 enzymes to the balance of enzymes in human liver microsomes. ISSX Proceedings 1995; 8:40.
195. Stevens JC, Melton RJ, Zaya MJ, et al. Expression and characterization of functional dog flavin-containing monooxygenase 1. Mol Pharmacol 2003; 63:271–275.
196. Furnes B, Schlenk D. Extrahepatic metabolism of carbamate and organophosphate thioether compounds by the flavin-containing monooxygenase and cytochrome P450 systems. Drug Metab Dispos 2005; 33:214–218.
197. Clement B, Mau S, Deters S, et al. Hepatic, extrahepatic, microsomal, and mitochondrial activation of the n-hydroxylated prodrugs benzamidoxime, guanoxabenz, and RO 48-3656 ([[1-[(2S)-2-[[4-[(hydroxyamino)iminomethyl]benzoyl] amino]-1-oxopropyl]-4-piperidinyl]oxy]-acetic acid). Drug Metab Dispos 2005; 33:1740–1747.
198. Guengerich FP. Human cytochrome P450 enzymes. In: Ortiz de Montellano P, ed. Cytochrome P450: Structure, Mechanism, and Biochemistry. New York: Plenum Press, 1995:473–535.

199. Wrighton SA, Stevens JC. The human hepatic cytochromes P450 involved in drug metabolism. Crit Rev Toxicol 1992; 22:1–21.
200. Parkinson A, Ogilvie BW. Biotransformation of xenobiotics. In: Klaassen CD, ed. Casarett and Doull's Toxicology: The Basic Science of Poisons. New York: McGraw-Hill Medical Pub. Division, 2007 (in press).
201. Nishiyama T, Kobori T, Arai K, et al. Identification of human UDP-glucuronosyltransferase isoform(s) responsible for the C-glucuronidation of phenylbutazone. Arch Biochem Biophys 2006; 454:72–79.
202. Williams JA, Hyland R, Jones BC, et al. Drug-drug interactions for UDP-glucuronosyltransferase substrates: a pharmacokinetic explanation for typically observed low exposure (AUCI/AUC) ratios. Drug Metab Dispos 2004; 32:1201–1208.
203. Gidal BE, Sheth R, Parnell J, et al. Evaluation of VPA dose and concentration effects on lamotrigine pharmacokinetics: implications for conversion to lamotrigine monotherapy. Epilepsy Res 2003; 57:85–93.
204. Mano Y, Usui T, Kamimura H. In vitro drug interaction between diflunisal and indomethacin via glucuronidation in humans. Biopharm Drug Dispos 2006; 27:267–273.
205. Naesens M, Kuypers DR, Streit F, et al. Rifampin induces alterations in mycophenolic acid glucuronidation and elimination: implications for drug exposure in renal allograft recipients. Clin Pharmacol Ther 2006; 80:509–521.
206. Miners JO, Knights KM, Houston JB, et al. In vitro-in vivo correlation for drugs and other compounds eliminated by glucuronidation in humans: pitfalls and promises. Biochem Pharmacol 2006; 71:1531–1539.
207. Visser TJ, Kaptein E, van Toor H, et al. Glucuronidation of thyroid hormone in rat liver: effects of in vivo treatment with microsomal enzyme inducers and in vitro assay conditions. Endocrinology 1993; 133:2177–2186.
208. Fisher M, Campanale K, Ackermann B, et al. In vitro glucuronidation using human liver microsomes and the pore-forming peptide alamethicin. Drug Metab Dispos 2000; 28:560–566.
209. Gipple KJ, Chan KT, Elvin AT, et al. Species differences in the urinary excretion of the novel primary amine conjugate: tocainide carbamoyl O-beta-D-glucuronide. J Pharm Sci 1982; 71:1011–1014.
210. Tremaine LM, Stroh JG, Ronfeld RA. Characterization of a carbamic acid ester glucuronide of the secondary amine sertraline. Drug Metab Dispos 1989; 17:58–63.
211. Beconi MG, Mao A, Liu DQ, et al. Metabolism and pharmacokinetics of a dipeptidyl peptidase IV inhibitor in rats, dogs, and monkeys with selective carbamoyl glucuronidation of the primary amine in dogs. Drug Metab Dispos 2003; 31:1269–1277.
212. Hayakawa H, Fukushima Y, Kato H, et al. Metabolism and disposition of novel desfluoro quinolone garenoxacin in experimental animals and an interspecies scaling of pharmacokinetic parameters. Drug Metab Dispos 2003; 31:1409–1418.
213. Link M, Hakala KS, Wsol V, et al. Metabolite profile of sibutramine in human urine: a liquid chromatography-electrospray ionization mass spectrometric study. J Mass Spectrom 2006; 41:1171–1178.
214. Obach RS, Reed-Hagen AE, Krueger SS, et al. Metabolism and disposition of varenicline, a selective alpha4beta2 acetylcholine receptor partial agonist, in vivo and in vitro. Drug Metab Dispos 2006; 34:121–130.

215. Obach RS, Cox LM, Tremaine LM. Sertraline is metabolized by multiple cytochrome P450 enzymes, monoamine oxidases, and glucuronyl transferases in human: an in vitro study. Drug Metab Dispos 2005; 33:262–270.
216. Ogilvie BW, Carrott PW, Yang C, et al. A convenient non-HPLC marker assay for measuring CYP2C8 activity in human liver microsomes: 5- and 6-chloromethylfluorescein diethyl ether (CMFDEE) O dealkylation. Drug Metab Rev 2001; 33 (suppl 1):230.
217. Ogilvie BW, Otradovec SM, Paris BP, et al. 10-(Imidazolyl)-decanoic Acid (10-IDA) is a selective and potent inhibitor of CYP4A11. Drug Metab Rev 2000; 32 (suppl 2):233.
218. Howgate EM, Rowland YK, Proctor NJ, et al. Prediction of in vivo drug clearance from in vitro data. I: impact of inter-individual variability. Xenobiotica 2006; 36:473–497.
219. Rowland-Yeo K, Rostami-Hodjean A, Tucker GT. Abundance of cytochromes P450 in human liver: a meta-analysis. Br J Clin Pharmacol 2004; 57:687.
220. Kiang TKL, Ensom MHH, Chang TKH. UDP-glucuronosyltransferases and clinical drug-drug interactions. Pharmacol Ther 2005; 106:97–132.
221. Fisher MB, VandenBranden M, Findlay K, et al. Tissue distribution and interindividual variation in human UDP-glucuronosyltransferase activity: relationship between UGT1A1 promoter genotype and variability in a liver bank. Pharmacogenetics 2000; 10:727–739.

8

The Role of P-Glycoprotein in Drug Disposition: Significance to Drug Development

Matthew D. Troutman
Pfizer Inc., Groton, Connecticut, U.S.A.

Gang Luo
Covance Laboratories Inc., Madison, Wisconsin, U.S.A.

Beverly M. Knight and Dhiren R. Thakker
The University of North Carolina at Chapel Hill, Chapel Hill, North Carolina, U.S.A.

Liang-Shang Gan
Biogen Idec, Inc., Cambridge, Massachusetts, U.S.A.

I. INTRODUCTION

The activity of transporters is now widely accepted as an important determinant of drug disposition. Certainly one major reason that transporters have become a key area of research that continues to grow involves the efflux pump, P-glycoprotein (P-gp). The transporter was initially discovered by Juliano and Ling as a transmembrane (TM) protein that was overexpressed in Chinese hamster ovary (CHO) cells treated with various chemotherapeutic agents that had become resistant to these cytotoxic drugs (1,2). In several cancerous tissues, overexpression of this protein is often associated with conferring the multidrug resistance (MDR) phenotype that involves insensitivity to a variety of structurally unrelated cytotoxic compounds, some of which the MDR cell had not been previously exposed to. The "P" in P-gp stands for permeability, as this efflux transporter was found to reduce the permeability of a wide variety of chemically unrelated cell permeable substrates. Subsequent to the recognition of P-gp's role in cancer, P-gp was found to be expressed in many normal tissues, namely, epithelial and endothelial barrier tissues (3,4). In this capacity, P-gp provides a biochemical mechanism to modulate the trafficking of endogenous compounds and drugs across these barriers, and this activity has been shown to influence the disposition of these compounds. Since the recognition of its role in limiting the oral absorption of certain drugs (5–10), P-gp has emerged as an important determinant of the oral bioavailability of drug molecules. Additionally, P-gp is known to be a critical determinant of the distribution of its substrates, particularly to organs protected by blood-tissue barriers such as the central nervous system (CNS) (11,12). For certain substrates, P-gp has been shown to be a determinant of elimination, playing a role in renal and biliary excretion (13,14). Recently, it has been shown that P-gp efflux activity can have a profound influence on the extent of metabolism (15–18). In addition to influencing a substrate's disposition, it has recently been demonstrated that changes to P-gp efflux can cause clinically significant drug-drug interactions (DDIs) (19,20). These findings and many others have clearly demonstrated the importance of P-gp in disposition. For these reasons, the elucidation of P-gp's role in disposition continues to be a key scientific goal in drug discovery and development and in the further understanding of clinically used therapies that are substrates for this important efflux transporter.

Extensive multidisciplinary studies have been conducted in an attempt to understand P-gp, and significant progress has been made. Furthermore, the knowledge gained from the study of P-gp has been invaluable in aiding the understanding of how other recently discovered transporters affect the disposition of their substrates. Salient knowledge key to understanding P-gp has been gained from the molecular level to the clinic. The scope of this chapter is to provide context and information about P-gp along this continuum. The purpose of this work is to facilitate the understanding of P-gp as it relates to drug disposition, i.e., absorption, distribution, metabolism, and excretion (ADME). Additionally, a review of the methodologies used to understand P-gp efflux and its role in disposition including DDI is provided.

II. P-gp AND OTHER EFFLUX TRANSPORTERS

A. P-Glycoprotein

P-gp was initially discovered by Juliano and Ling as a TM protein, which was overexpressed in CHO cells that had acquired resistance to the effects of cytotoxic drugs (1,2). In several cancerous tissues, overexpression of P-gp often is associated with conferring the MDR phenotype that involves cellular insensitivity to a variety of structurally unrelated cytotoxic compounds; many of which the cancerous cell has never been exposed to. Indeed, the substrate specificity of P-gp is quite broad and encompasses compounds from various chemical classes. More recently, it has been recognized that P-gp is constitutively expressed in many normal tissues, namely epithelial and endothelial barrier tissues, where it provides a biochemical mechanism to modulate the trafficking of endogenous compounds and xenobiotics across these barriers. The broad substrate specificity of P-gp, its location in endothelial and epithelial tissues within many organs, and its transport activity can potentially make P-gp an important determinant of the ADME of its substrates. The following sections provide information about P-gp (and other efflux transporters), which may aid in further understanding of how this important transporter can ultimately affect drug disposition in patients.

1. Nomenclature

Genetic analysis has revealed the existence of multiple MDR mammalian genes (21). Members of the MDR gene family can be divided into three classes (Table 1) on the basis of sequence homology (22). Two nomenclatures exist to describe these genes; the legacy system (MDR) and a more recent system implemented for ATP binding cassette (ABC) transporters (see Table 2). In humans, the genes are denoted as MDR1 (class I) (23,24) and MDR3 (class III), or ABCB1 and ABCB4, respectively (25,26). In mice, the genes are denoted as mdr3 (mdr1a, class I), mdr1

Table 1 Nomenclature and Function of Mammalian P-gp Gene Family

Species	Member	Function	References
Human	MDR1 (ABCB1)[a]	MDR	23,24
	MDR2/3 (ABCB4)[a]	Phosphatidylcholine translocation	25,26
Mouse	mdr3 (mdr1a)	MDR	27–29,35
	mdr1 (mdr1b)	MDR	
	mdr2	Phosphatidylcholine translocation	
Rat	mdr3	MDR	455
	mdr1	MDR	30
	mdr2	Phosphatidylcholine translocation	
Hamster	Pgp1	MDR	
	Pgp2	MDR	21,22,31,446
	Pgp3	Phosphatidylcholine translocation	

[a]In the new nomenclature system, human MDR1 and MDR2/3 are named ABCB1 and ABCB4, respectively.

Abbreviation: MDR, multidrug resistance.

Table 2 Comparison Between Human MDR1 and MRP1

	MDR	MRP1
Family member	MDR1 and MDR3	MRP1, MRP2, MRP3, MRP4, MRP5, MRP6
Protein chemistry		
Amino acid residues	1280	1531
Molecular weight (kDa)	~170	~190
Glycosylation sites	1	2
TM domains	12	17
Extracellular N-terminus	No	Yes
Molecular biology		
Locus on chromosome	7q21.1	16p13.12-13
Gene expression		
In normal tissues	Adrenal cortex, liver, kidney, intestine, brain, testes, placenta, lymphocytes	High in lung, bladder, spleen, thyroid, testes, adrenal gland, low in kidney, stomach, liver, colon
In tumor tissues	High in colon, renal and adrenal carcinomas, rarely in lung and gastric carcinomas	
Substrates and Inhibitors		
Calcium channel blockers (verapamil)	Yes	Yes
Immunosuppressants (cyclosporin A)	Yes	Yes
Anthracycline (doxorubincin)	Yes	Yes
Vinca alkaloids (vincristine)	Yes	Yes
Calmodulin antagonists (trifluoperazine)	Yes	Yes
Toxic peptides (valinomycin)	Yes	Yes
Steroids (tamoxifen)	Yes	Yes
Glucuronide conjugates	No	Yes
Glutathione conjugates	No	Yes
Sulfate conjugates	No	Yes
Others		
Colchine	Yes	No
Taxol	Yes	No
Digoxin	Yes	No
Heavy-metal oxyanions	No	Yes
Inhibitors	PSC833, GF120918	Genestein

Abbreviations: MDR, multidrug resistance; MRP, MDR-associated protein; TM, transmembrane.

(mdr1b, class II), and mdr2 (class III) (27–29). In rats, the genes are denoted as mdr3 (class I), mdr1 (class II), and mdr2 (class III) (30). In hamster, these genes are named pgp1 (class I), pgp2 (class II), and pgp3 (class III) (23,31). In pigs, five members of the P-gp superfamily have been identified—four class I genes and one class III gene (32). In addition, P-gp was also cloned from monkeys (Rhesus, Cynomolgus, and African green) and dogs, with amino acid residue identity as high as 96% and 90%, respectively. Their DNA and protein sequences are available from National Center for Biotechnology Information.

It has been shown experimentally that only the class I and class II (human MDR1, rodent mdr3 and mdr1) confer the MDR phenotype (24,27,28,33,34). The human MDR3 and rodent mdr2 genes encode a protein expressed in the bile canalicular membrane that translocates phosphatidylcholine from the inner to outer leaflet of this membrane (35,36). In this chapter, only the gene products conferring the MDR phenotype will be discussed as the activities of these proteins can influence drug disposition.

2. Endogenous Expression

P-gp is constitutively expressed in nearly all barrier tissues. Techniques involving Northern blots (37) or Western blots with monoclonal antibodies such as C219 (38) and MRK 16 (39) have been used extensively to determine the tissue distribution of P-gp. It is expressed in adrenal cortex, kidney, liver, intestine, and pancreas; endothelial cells at blood-tissue barriers, namely, the CNS, the testis, and in the papillary dermis (3,4,38,40,41). P-gp displays specific subcellular localization in cells with a polarized excretion or absorption function. More specifically, P-gp is found at the apical (AP) canalicular surface of hepatocytes, in the AP membrane of the columnar epithelial cells of colon and jejunum, and the AP brush border of the renal proximal tubule epithelium (3,4,40–42). In endothelial cells, P-gp is located in the luminal membrane (4,43).

3. Physiological Functions

The tissue-specific expression and cellular localization of P-gp has provided some insight regarding its physiological function and roles in pharmacology. Borst and Schinkel (44), Borst et al. (45), and Lum and Gosland (46) have postulated several likely physiological functions of P-gp. P-gp protects against the entry of exogenous toxins ingested with food, evidenced by expression in small intestine, colon, and blood-tissue barrier sites. It extrudes or precludes entry of toxic compounds from the CNS and testis (4,43). Indeed, literature is replete with examples of how P-gp-mediated efflux activity from barrier-forming cells affects the disposition of its substrates. P-gp excretes toxins or metabolites, as evidenced by its expression in liver canalicular membrane and kidney (47). Recently, evidence has been reported to show how P-gp-mediated efflux can make intestinal secretion a potential mechanism for drug elimination (5,9,10,48). P-gp is expressed in adrenal gland and it was demonstrated that it transports steroid hormones such as cortisol, corticosterone, and aldosterone (49). P-gp extrudes polypeptides and large macrocyclic molecules, as seen by the ability of P-gp to efflux cyclosporin A and tacrolimus (50,51). P-gp activates endogenous chloride channel activity. P-gp itself is not a volume-sensitive chloride channel; however, P-gp has been shown to play an indirect role in chloride channel activation. P-gp also enhances the ability of cells to downregulate their volume through modulation of volume sensitivity of the chloride channel in a manner independent of its ATPase activity (52,53).

B. Other Efflux Transporters Known to be Important to Drug Disposition

1. MDR-Associated Protein

In addition to P-gp, MDR-associated protein (MRP) also plays an important role in MDR of cancer therapy and in affecting the behavior of other drug substrates. The MRP proteins are a relatively large protein family consisting of at least nine members, MRP1, MRP2, MRP3, MRP4, MRP5, MRP6, MRP7, MRP8, and MRP9, each with diverse specificities, structure, and function (54–60). MRP1, encoded by ABCC1 gene, consists of 1531 amino acid residues, with a molecular weight of 190 kDa. Like P-gp, MRP1 is glycosylated posttranslationally (at two sites vs. one site seen for P-gp) and thus the actual molecular weights are greater than those predicted from the primary sequences of amino acid residues. The amino acid sequences for P-gp and MRP2 show only 15% similarity (61). Other differences in their protein structure include different numbers of TM domains (12 for MDR1 and 17 for MRP) and different orientation of their N-termini.

The differential expression of MDR1 and MRP in various tissues suggests that they may have different physiological functions and that they may play different roles in the pharmacology and toxicology of their substrates. P-gp is expressed in the AP membranes of certain normal human tissue cells and in tumor cells as described above. Pharmacologically, P-gp plays a role in preventing intestinal drug absorption, brain entry, and in eliminating drugs by excretion into bile and urine. MRP1 is extensively expressed in lung (bronchial epithelia), bladder, spleen, and testes (haploid spermatid), but to a lesser extent in kidney, stomach, liver, and colon (62–64). Unlike P-gp, MRP1 is found in the basolateral (BL) membranes of cells (65).

Both P-gp and MRP1 exhibit broad but different spectrums of substrate specificity. Generally speaking, P-gp transports hydrophobic compounds and MRP1 effluxes hydrophilic chemicals. For example, P-gp transports hydrophobic molecules that often possess a positive charge, a nitrogen group, and an aromatic group, whereas MRP1 has been shown to transport heavy metal oxyanions, glutathione conjugates, glucuronide conjugates, and sulfate conjugates (the reader is cautioned that these are very general characteristics regarding P-gp and MRP1 substrates and numerous deviations exist in each case). Despite their very different substrate selectivity, they do exhibit some overlap in their activity toward some substrates. As listed in Table 2, verapamil, cyclosporin A, doxorubicin, vincristine, and tamoxifen are examples of substrates for both P-gp and MRP1 protein.

MRP2, previously referred to as canalicular multispecific organic anion transporter (cMOAT), is an important ABC transporter and consists of 1541 amino acid residues, encoded by ABCC2 gene (66–68). This AP membrane bound transport protein is highly expressed in the canalicular membrane and is also present in intestine, kidney, and the blood-brain barrier (BBB) (66–70). MRP2 plays a critical role in the biliary excretion of certain endogenous anionic compounds and many xenobiotics. Bilirubin glucuronide, for example, is excreted into bile via MRP2. Many drugs and their glucuronide, glutathione, and sulfate conjugates are substrates of MRP2 (66–70). MRP2 deficiency causes hyperbilirubinemia in humans (Dubin-Johnson syndrome) as well as in TR^- rats and Eisai hyperbilirubinemic rats (EHBR) (67,70–72). Biliary excretion of methotrexate is markedly inhibited by MRP2 inhibitors such as probenecid (73) and becomes insignificant in EHBR (69). MK-571 is an inhibitor of MRP2, with IC_{50} in a range of 4 to 16 μM (74–76), which is frequently used as a probe for this transporter. Readers are cautioned that the inhibitory activity of MK-571 toward all MRP isoforms has not been elucidated, but there is no reason to believe that it is selective for MRP2 to the exclusion of other MRPs.

2. Breast Cancer Resistance Protein

Breast cancer resistance protein (BCRP), also known as placenta-specific ABC transporter (ABCP) and mitoxantrone resistance protein (MXR), is the product of the ABC half-transporter gene ABCG2. Initially, BCRP was isolated from a multidrug resistant human breast cancer cell line (MCF-7/Adr Vp) (77–81). The human ABCG2 gene is located on chromosome 4q22, contains 16 exons and 15 introns, and encodes 655 amino acid residues with molecular weight of 72 kDa (77,79,81,82). BCRP has been proposed to have six putative TM domains, four potential glycosylation sites, and one nucleotide-binding domain (NBD) (79,83,84). Furthermore, BCRP is only about half in size compared with other ABC transporters, and it is believed to function as a dimer bridged by disulfide bonds (85,86), which is different from other ABC transporters such as P-gp and MRP2.

BCRP expression is high in placenta and hematopoietic stem cells, while low in small intestine, colon, hepatic canalicular membrane, breast, venous, and capillary endothelium (82,84,87). Since BCRP is highly expressed in placental syncytiotrophoblast, it has been speculated that this transporter may play a significant role in protection of fetus from toxic xenobiotics (88). Notably, BCRP is expressed in lactating, but not virgin and nonlactating, mammary gland epithelia of mice, cows, and humans, suggesting that BCRP may play a critical role in secreting certain drugs into the human milk (89). For example, nitrofurantoin (an antibiotic commonly prescribed to lactating women suffering from urinary tract infection) is actively secreted into human milk, likely via BCRP-mediated transport (90).

The substrate specificity of BCRP is relatively broad, including some endogenous chemicals such as estrone sulfate, antitumor drugs such as mitoxantrone, topotecan, prazosin, and polyglutamated methotrexate, nucleoside reverse transcriptase inhibitors such as zidovudine and lamivudine, and others such as statins (84). Many chemicals are now reported to inhibit BCRP, including fumitremorgin C, Ko143, GF120918, cyclosporin A (84). Interestingly, BCRP shares some substrates as well as inhibitors with P-gp and MRPs. BCRP has been reported to affect oral absorption, distribution across BBB and blood-placenta barrier, and elimination via hepatobiliary as well as milk secretion [summary by Xia et al. (84)]. Furthermore, the polymorphisms of BCRP could significantly contribute to interindividual differences in response to some drugs in the clinic.

3. Bile Salt Export Pump

Bile salt export pump (BSEP) is a 160-kDa ABC transport protein encoded by ABCB11. Since it is closely related to P-gp, it is also known as sister of P-gp. Recent results have suggested that BSEP is the major canalicular BSEP expressed in mammalian liver, and therefore, the major determinant of bile salt dependent bile flow (91,92). The expression of BSEP [determined by real-time reverse transcriptase-polymerase chain reaction (RT-PCR)] is high in the liver, and significant in the brain gray cortex and large gut mucosa, but was not detected in the kidney or BBB (93). The subcellular distribution of BSEP in the liver (determined by immunofluorescence and immunogold labeling experiments) appears to be localized to the canalicular microvilli and to subcanalicular vesicles (94). BSEP appears to be important in the biliary secretion of taurocholate, taurochenodeoxycholate, tauroursodeoxycholate, glycholate, and cholate (94).

Although BSEP is related to P-gp, its substrate specificity is different. The actions of BSEP on several known P-gp substrates were examined by expressing BSEP cDNA in LLC-PK1 and MDCKII cells. Cells expressing BSEP displayed decreased uptake of taurocholate and vinblastine compared with control cells, and the accelerated efflux of vinblastine was observed in the cells (95). BSEP has no effect on the uptake of the P-gp substrates vincristine, daunomycin, paclitaxel, digoxin, and rhodamine 123, but did cause an efflux of calcein acetoxymethyl ester (calcein-AM). The transport of calcein-AM via BSEP-mediated efflux was not inhibited by the P-gp inhibitors cyclosporin A or reserpine, but was inhibited by ditekiren (a linear hexapeptide). The involvement of this protein in drug transport has only recently been explored and its role in drug elimination will become clearer as more studies are performed to address the significance of BSEP-mediated efflux of drugs.

Table 3 Examples of Overlapping Substrate and Inhibitor Specificity Among ABC Transporters

	P-gp	MRPs	BCRP	Reference
Substrate				
Doxorubicin	X	X	X	456
Pravastatin	X	X	X	102
Saquinavir	X	X		457
Vinblastine	X	X		456
Inhibitor				
Cyclosporin A	X	X	X	458
GF120918	X		X	84
MK-571	X	X		76

Abbreviations: ABC, ATP binding cassette; P-gp, P-glycoprotein; MRPs, MDR-associated proteins; BCRP, breast cancer resistance protein.

C. P-gp Substrates and Inhibitors

It is well established that the substrate specificity of P-gp is quite broad with respect to both chemical structure and size (Table 2). The structural diversity of P-gp substrates (and inhibitors) is so broad that it is difficult to define specific structural features that are required for the substrates/inhibitors of P-gp. However, some of the properties that are shared by many P-gp substrates include the presence of a nitrogen group, aromatic moieties, planar domains, molecular size $\geq$300 Da, presence of a positive charge at physiological pH, amphipathicity, and lipophilicity (96–100). Substrates and inhibitors of P-gp can be found within the same chemical classes (101). All P-gp substrates and modulators show moderate to high lipophilicity (have high membrane partition coefficients). However, significant differences have been shown for the membrane diffusion properties of substrates versus inhibitors (101). Although several structure-activity relationships have been constructed in attempts to elucidate structural requirements for P-gp substrates, the ability to predict P-gp substrate specificity or inhibitory potency a priori, remains limited. Notably, P-gp has been found to share a number of substrates and inhibitors with other ABC transporters (Table 3) as well as solute carriers such as organic anion-transporting polypeptides (OATPs) (102). Therefore, caution should be taken in experimental designs and data interpretation. The reader is directed to the following excellent references for further discussion of substrates and inhibitors (96–100,103).

D. P-gp Biochemistry

An exhaustive and thorough review of P-gp biochemistry is beyond the scope of this chapter and the reader is directed to the following excellent reviews on this subject for further information (104–110). It is however important to understand key aspects of P-gp and its function on the molecular level to better understand how P-gp may influence disposition. A discussion of these is presented in the following sections.

1. Protein Structure and Transport Mechanisms

P-gp is as a member of the superfamily of transporters known as ABC transporters (107). To date, more than 200 membrane transporters have been identified as members of the ABC transporter family, and this family includes the sodium potassium ATPase, the calcium ATPase, and the outwardly rectifying chloride ion channel, CFTR (107). The general architecture of ABC transporters includes four major domains, two membrane bound domains, each with six TM segments, and two cytosolic ATP-binding motifs commonly known as the Walker A and B domains, which bind and hydrolyze ATP (also known as the NBDs (Fig. 1) (104,106). There is a high degree of homology among many of the ABC transporters with certain structural features conserved between highly related members of the family, most notably the NBDs. Attempts to extrapolate the information known to other ABC transporters has not been successful and is likely due to the differences in substrate-binding domains.

All four domains of mammalian P-gp are encoded by one gene (106), and the human MDR1 gene product consists of 1280 amino acid residues with a large degree of homology between a carboxy half and an amino half (23,111). Both terminal halves must be present and act in a cooperative manner for maximal activity (112,113). Each homologous half contains a hydrophobic membrane-associated domain, consisting of approximately 300 amino acids, and a hydrophilic NBD also consisting of approximately 300 amino acids (104,111,114,115).

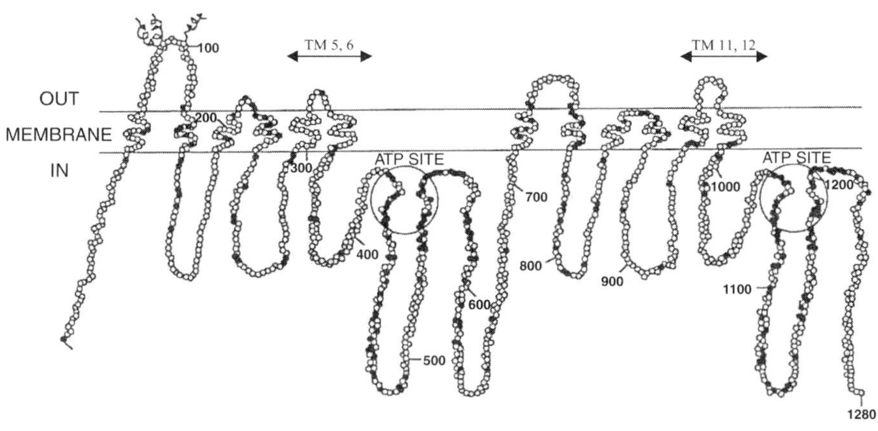

Figure 1 Model of human P-gp derived from sequence analysis (23). TM domains 5, 6, 11, and 12 are thought to compose the binding site(s) (104). *Abbreviations*: P-gp, P-glycoprotein; TM, transmembrane.

Furthermore, many chemotherapeutic drugs or chemosensitizers do not inhibit P-gp via prevention of ATP binding, which suggests that separate ATP- and drug-binding sites exist (Fig. 1) (111). The molecular weight of P-gp is 120 to 140 kDa preglycosylated and 160 to 180 kDa in glycosylated form; size depends on cell type and species (116,117). The first extracellular loop of P-gp contains N-linked carbohydrate moieties that are postulated to contribute to proper folding and routing and stability of P-gp within the plasma membrane rather than influencing function or ATPase activity (116–124). Mature P-gp is phosphorylated by multiple kinases, including protein kinases A and C and perhaps by serine threonine kinase at a cluster of maximally four serine residues located in a central cytosolic linker region connecting the homologous halves (125,126). It is currently poorly understood how the degree of P-gp phosphorylation may affect efflux activity (125–127). P-gp can exist as a monomer, dimer, or higher order oligomer (128).

Multiple studies have provided significant evidence that P-gp directly effluxes its substrates in the manner of a primary ATPase (107). The most convincing evidence comes from the studies with acetoxymethyl esters of fluorescent dyes, fluorescently labeled daunorubicin, and the measurement of structural changes associated with substrate efflux (111,129). Early hypotheses suggested that the broad substrate specificity of P-gp was the result of P-gp's ability to change physical parameters of the surrounding medium such as modifying pH or by influencing the osmotic gradient similar to a chloride ion channel or an ATP channel (130,131). This latter hypothesis has been refuted with clear experimental results (132).

Substrates bind to P-gp while they are associated with the plasma membrane; this process is possibly the most important aspect of P-gp-mediated efflux activity to appreciate. By using fluorescent dye esters, it was shown that P-gp interacts with its substrates within the plasma membrane. As these dye esters cross the membranes, esterases quickly hydrolyze the esters to their free acid form in the cytoplasm. Cells expressing P-gp showed no accumulation of the free acid dye in the cytoplasm clearly illustrating that P-gp can efflux substrates directly from the membrane (129). Additionally, P-gp can bind to substrates at the inner leaflet—cytosolic interface as demonstrated in studies with the P-gp substrate rhodamine 123 (133). It was shown that P-gp does not influence drug concentration in the exofacial leaflet (134), thus implying that P-gp only binds compounds from either within the inner leaflet or at the inner leaflet—cytosolic interface. These findings clearly show that the behavior of the substrate/inhibitor within the lipid barrier is likely to be a primary determinant of P-gp-mediated efflux activity. This separates P-gp from traditional transporters in which binding of the substrate to the active site in an enzyme-like fashion is the primary determinant of transport activity.

Two models have been developed to describe P-gp's unique transport activity. Higgins and Gottesman have postulated that P-gp acts as a hydrophobic vacuum cleaner, clearing the plasma membrane of substrates before they enter the cytoplasm (104,135). This hypothesis serves to explain two deviations from the classical transporter model. First, by acting to remove substrates directly from the membrane, the primary determinant of substrate specificity is the ability of the drug to interact with the plasma membrane, and the secondary determinant would be the ability of the drug to interact with the protein itself. This serves to explain the broad substrate specificity of P-gp and why nearly all P-gp substrates are lipophilic. The second deviation is often associated with the kinetics of transport. It is usually quite difficult to correlate the initial rate of drug efflux with drug concentration in MDR cells. The vacuum cleaner model hypothesizes that the actual concentration seen by the transporter would not correspond to the concentration of drug used in the experiment, but actually would depend on the ability of the drug to partition into the lipid bilayer as well as the lipid composition of the membrane (104,135). The second widely accepted model builds on the vacuum cleaner model to explain how P-gp actually translocates substrates. It has been proposed that P-gp acts like a flippase to ''flip'' substrates from the inner leaflet to the outer leaflet or aqueous space (135). According to this model, the concentration of the substrate in the outer leaflet and the extracellular space is in equilibrium. There also exists an equilibrium between the inner leaflet and the cytoplasm, and finally an equilibrium exists between the leaflets of the plasma membrane. The pump would create a gradient by flipping the substrate from the inner to outer leaflet, and thus force the substrate to partition from the outer leaflet into the extracellular space. Further rationale for this model is given by the large degree of homology (75%) seen between the MDR1 and three gene products, and postulation that these proteins may act similarly. MDR3 activity involves translocation of phospholipids between the membrane leaflets (25,136,137).

2. Binding Sites

Several studies have been performed to identify the specific regions of P-gp involved in drug transport. Photoaffinity labeling with azidopine or azidoprazosin and numerous mutational analyses have shown that P-gp's binding site is located in two regions, TM 5 and 6, and TM 11 and 12 (104,110,138–140). The amino acid sequence of TM segments 5, 6, 11, and 12, which compose the binding pocket, contain several aromatic side chains shown to be important in binding and transport of substrates (138). P-gp contains a high amount of aromatic amino acids compared with other ABC transporters, and these residues are highly conserved across species (105). The aromatic and hydrophobic amino acid residues in the binding region of P-gp are thought to comprise a hydrophobic channel that provides binding sites for substrates with P-gp (108,115). This channel reduces the interactions of the substrates of P-gp with the lipid bilayer, thus making substrate transport across the membrane more energetically favorable (141). The proposed involvement of several aromatic rich segments (located in the binding regions, TM 5, 6, 11, and 12) in drug binding and transport are thought to give P-gp the conformational flexibility needed to interact with several chemically unrelated substrates of various sizes and shapes (108). These results have led to the most widely accepted current hypothesis, which states that amino acid residues of both N- and C-terminal halves of P-gp interact and cooperate to form one major drug interaction pore capable of accommodating two small compounds to one large compound (110,115,138). This model allows multiple sites for drug recognition and rationalizes the findings that show different classes of drugs bind to different, possibly allosterically coupled, regions within P-gp (142–144). Indeed several pharmacological studies have shown differential drug interactions with P-gp with respect to binding affinity and capacity and ATP activity (145–148). Evidence has shown that P-gp transport activity toward certain compounds can be increased in the presence of other P-gp substrates, perhaps by some unknown allosteric mechanism (149). Using equilibrium and kinetic radioligand binding techniques, it has been shown that a minimum of four distinct drug-binding regions exist for P-gp—three sites were identified as transport sites and one was identified as a modulatory site (150,151). Findings around multiple binding sites along with the nature of the P-gp drug-binding TM domains have led Loo and Clarke to propose a ''substrate-induced fit'' mechanism that accounts for the complexity and near compound-dependent nature of P-gp-binding sites and resultant effects on transport activity (152). Rather than a static drug interaction pore, Loo and Clarke hypothesize that the substrate could create its own binding pocket via interaction with key residues in TM domains 5, 6, 11, and 12. In this model, a compound's affinity for P-gp would depend on the number and types of residues involved in its binding and requires that the TM domains be quite mobile to accommodate a wide range of chemotypes (110). These findings on the nature of P-gp compound binding and resulting activities highlight key points to consider with regards to P-gp's role in drug disposition and, in particular, potential DDI related to modulation of P-gp activity. The nature of an interaction between two P-gp substrates or a substrate and inhibitor may be unique. Therefore, caution must be exercised when trying to extrapolate how the substrate/inhibitor may interact to an untested substrate/inhibitor.

3. Mutations and Impact on (In Vitro) Function

A systemic screening for functional polymorphisms of the human Pgp was first carried out by Hoffmeyer et al. (153). So far, a total of 28 single nucleotide polymorphisms (SNPs) have been identified at 27 positions on the MDR1 (ABCB1) gene. Among them, polymorphism in wobble position of exon 26 (C3435T, a silent point mutation) and in exon 21 (G2677T/A, Ala893Ser/Thr) of MDR1 are believed to link to duodenal expression levels and function of P-gp. In homozygous (3435TT) individuals, a significantly lower duodenal P-gp expression has been linked to the higher plasma levels of digoxin (153,154) and tacrolimus (155,156), clopidogrel absorption, and thereby active metabolite formation (157), and nortriptyline-induced postural hypotension (158). In addition, C3435T is also reported to be a risk factor for certain diseases, including renal neoplasms (159), Parkinson disease (160–164), inflammatory bowel disease (165,166), refractory seizures (167), and HIV infection (168). The rate of Caucasian individuals who are homozygous carriers of 3435TT with functionally restrained P-gp is about 24–29% (153,169). Notably, this mutation is markedly influenced by ethnicity (170,171). However, the reports are not always consistent regarding the effect of C3435T mutation on subsequent expression level and the exposures of P-gp substrates. For example, some reports indicated that C3435T mutation did not affect P-gp expression level in the intestine, nor on the disposition of talinolol, loperamide, and fexofenadine in humans (172–175). Therefore, the haplotype analysis of the gene should be included, and the clinical trials must be designed properly to avoid misinterpretation (163). In addition to C3435T mutation, G2667T/A mutation may influence the risk of development of lung cancer (176).

4. The Role of the Plasma Membrane in P-gp-Mediated Efflux Activity

Unlike most transporters, the composition and physical state of the plasma membrane and the interaction of the substrate with the plasma membrane are important determinants of P-gp-mediated efflux activity. A discussion of these phenomena is helpful to aid in further understanding of the nature of P-gp efflux activity.

a. Relationship of P-gp-mediated efflux activity to substrate membrane permeability. The permeability of a substrate across a lipid bilayer occurs in three steps, all of which are determined by the structure of the plasma membrane bilayer and the structure of the substrate (101). The first step of permeability involves adsorption (partitioning) of the substrate within the interfacial region of the bilayer. Nearly all P-gp substrates and inhibitors have moderate to high lipophilicity/membrane partitioning coefficients (177,178). Although the complex processes underlying partitioning are not fully understood, several parameters that affect partitioning have been identified. These include the nature of the lipids (where composition of the headgroup and fatty acid structure are important), the physical state of the bilayer, and the composition of the aqueous buffer. The nature of the substrate, with regards to lipophilicity and charge, dictates where in the bilayer the substrate partitions (within the headgroup region or in the fatty acid region) (101). The site of substrate partitioning in the membrane may affect the access of specific binding sites on P-gp to the substrate (101). Several studies have shown that closely related steroids and 1,4-dihydropyridines noncompetitively interact with P-gp, clearly showing these compounds interact with different binding sites/regions of P-gp (148,179,180). The process of partitioning is further complicated in the case of charged and lipophilic substrates. For basic compounds, the protonated form of these compounds has particularly high partition coefficients because of the electrostatic interactions with zwitterionic or anionic lipids (181). Furthermore, two forms exist (protonated and unprotonated) for basic drugs, and each is likely to possess a unique partitioning ability into the membrane (182). The proportion of these forms at the membrane depends on the (microenvironment) pH and ionic composition of the aqueous phase, and also on the properties of the membrane, including the dielectric constant and surface potential (183). For compounds with a permanent positive charge, the electrostatic properties of the membrane bilayer suggest that the energetically favorable site of partitioning is at the interface (184,185).

The second step in permeability involves the energetically unfavorable process of TM movement of the substrate from one leaflet of the bilayer to the other through the hydrophobic core of the membrane. This step is rate limiting in permeability and has been shown to be markedly different for P-gp substrates versus inhibitors (177,186–188). Multilamellar vesicles and large unilamellar vesicles have been used to measure the transbilayer movement of both P-gp substrates (doxorubicin, rhodamine 123, vinblastine, taxol, and mitoxantrone) and inhibitors (verapamil, quinidine, quinine, trifuoroperazine, and progesterone) (177,188). Substrates were shown to diffuse across these membranes at much lower rates than the inhibitors. It was hypothesized that inhibitors act in a competitive manner to occupy P-gp by crossing the membrane as fast as or faster than efflux can occur. Further evidence for this hypothesis has been presented by the inverse correlation of the rates of diffusion of a series of rhodamine 123 derivatives through model membranes, with the accumulation of these compounds into cells expressing P-gp (186). These studies have provided some insight into how substrate membrane diffusion determines P-gp-mediated efflux activity.

Finally, the third step of substrate permeability across the plasma membrane involves partitioning of the substrate from the opposite interface (desorption). This process involves membrane partitioning and the same factors that determine adsorption also determine desorption; but for desorption versus adsorption, the relationships are reversed (101). It is important to note that because of membrane asymmetry (between inner and outer leaflets) present in all cells, the processes of adsorption and desorption may be vastly different depending on the direction of substrate transport (from external milieu to cytosol or vice versa). Consequently, differences in adsorption and desorption can lead to differences in substrate permeability across inner and outer leaflets, as shown for doxorubicin (189). Indeed, it has been hypothesized that direction of substrate transport may affect how P-gp effluxes its substrates (190).

b. Relationship of substrate binding to P-gp and ATPase activity to the composition and physical state of the plasma membrane. Although for most experimental systems in which P-gp is studied, the state of the plasma membrane remains constant, it is important to understand when differences in the composition and physical state of the plasma membrane can affect P-gp-mediated efflux activity. Differences in the lipid composition of plasma membranes have been shown to affect the binding characteristics of substrates to P-gp. The importance of the membrane environment on substrate specificity has been illustrated by transfection of P-gp into cells with dissimilar lipid composition (106). The relative ability of P-gp to efflux vinblastine and daunorubicin is reversed when the efflux pump is transfected in insect cells that have different membrane compositions than mammalian cells. The lipid environment directly affects the basal catalytic (ATPase) activity of P-gp and the degree of substrate stimulation (191–195). Contrary to other membrane transporters, P-gp binds ATP and its substrates with higher affinity when the membrane is in the gel state (108); most detergents (at nonsolubilizing concentrations) completely abolish both P-gp ATPase activity and drug binding to P-gp (196,197). The dependence of P-gp ATPase activity and substrate binding on the composition and physical state of the membrane has been clearly shown with studies in which P-gp was reconstituted into proteoliposomes of various lipid content (198). It was observed that the binding affinities of vinblastine, daunorubicin, and verapamil to P-gp were directly correlated to the substrate-lipid partition coefficients determined for each lipid system and that these compounds bound to P-gp with much greater affinity when each lipid membrane was in the gel phase versus the liquid crystalline phase. Stimulation or inhibition of P-gp ATPase activity observed for each lipid composition, in gel or liquid crystalline state, was highly correlated to the binding affinities. From these results, it was concluded that the effective concentration of the substrate in the membrane, determined by substrate-lipid partition coefficient and physical state of the membrane, are important for both the interaction of the substrate with P-gp and the ATPase activity of P-gp. When one system with constant plasma membrane composition is used, it is important to understand that agents that affect the physical state of the plasma membrane (i.e., detergents and membrane fluidizers such as benzyl alcohol, methanol, and ethanol) may significantly alter P-gp-mediated efflux activity.

5. Kinetics and Mechanisms of P-gp

Several reports have shown that the kinetics of P-gp transport activity can be sufficiently described by one-site Michaelis-Menten saturable kinetics (199–206). Where $J_{\text{P-gp}}$ is the flux mediated by P-gp transport activity, $J_{\max}$ is the maximal flux mediated by P-gp transport activity, K_m is the Michaelis-Menten constant, and C_t is the concentration of substrate present at the target (binding) site of P-gp. When donor concentration is used in place of C_t, apparent K_m and $J_{\max}$ values are obtained. Binding affinity of the substrate to P-gp and the catalytic (ATPase) activity of P-gp combine to determine K_m, and $J_{\max}$ is determined by the catalytic (ATPase) activity of P-gp and the expression of P-gp in the system (concentration of P-gp protein). It has recently been noted that since substrates must first partition or cross the membrane to access the binding site, accurate assessing of P-gp kinetics can be difficult (207). Furthermore, the requirement of first partitioning into the membrane has been shown to produce asymmetric apparent kinetics in polarized cells where AP and BL membrane compositions may be sufficiently different (206).

III. P-gp's ROLE IN DETERMINING DRUG DISPOSITION OF ITS SUBSTRATES

A. P-gp-Mediated Efflux Activity on the Cellular Level

Within the cell, P-gp can be expressed in several organelles and as such can influence the cellular distribution of its substrates. Studies with tumor cells have shown P-gp expression on the cell surface, in cytoplasmic vesicles, in Golgi apparatus, and in the nuclear envelope (208,209). Within vesicles and in Golgi apparatus, P-gp acts to sequester compounds as the transport is directed within the vesicle. At the nuclear membrane, P-gp acts to restrict access of substrates to the nucleus by directing transport in a cytoplasmic direction. This subcellular localization of P-gp can be an important consideration for P-gp substrates with intracellular targets (208).

The actions of P-gp located on the cell surface that act to restrict substrate access and to enhance elimination via efflux directed from cytoplasm to extracellular milieu are the most widely studied and understood. The remainder of this section will focus on the ramifications of the P-gp activity at the cell membrane level to disposition, which has been shown to be particularly relevant in barrier tissues such as the intestine, BBB, and blood-testes barrier, and in cells of eliminating organs such as hepatocytes and renal tubule cells.

B. Influence of the P-gp-Mediated Efflux Activity on ADME

Much of the information known about the role of P-gp in determining the pharmacokinetic profile of drugs has come from in vivo experimentation. These experiments can roughly be classified into two categories—studies performed in the P-gp-deficient mouse model as pioneered by Schinkel et al. and further findings contributed by other researchers using this elegant model (11,12,14,124,210–220), and those performed to determine the pharmacokinetic parameters of P-gp substrates in normal mice and man (19,48,221–231). These studies have helped to elucidate the overall importance of P-gp in affecting the ADME of its substrates. The following is a brief review of some of the important findings of these studies that have clearly shown how P-gp efflux activity can influence each aspect of drug disposition.

1. Absorption

All orally administered drugs must pass through the gastrointestinal tract to reach the blood and thus pass the barrier formed by the enterocytes in the intestine. For years, low first-pass bioavailability of a drug was attributed mainly to clearance via hepatic metabolism and biliary clearance or incomplete absorption in the intestine due to poor solubility or intrinsic permeability properties. Although these are important factors in determining the overall oral bioavailability of certain drugs, recent studies have shown that P-gp-mediated efflux may also play a role in attenuating oral absorption of many drug molecules (9,16,217,224,226,232–234). During absorption, P-gp-mediated efflux activity can potentially attenuate the overall bioavailability of its substrates by multiple mechanisms. It can attenuate the rate at which its substrates permeate from gut across intestinal enterocytes (where P-gp is located on AP membrane) into blood thus potentially delaying absorption time, reducing C_{max}, and possibly reducing total exposure [area under the curve (AUC)]. Additionally, P-gp efflux during intestinal absorption may enhance intestinal metabolic elimination, thus indirectly reducing the amount of compound able to reach the bloodstream (details of how P-gp can affect intestinal metabolism are discussed below) (16). In the liver, P-gp is expressed in the canalicular membrane of the hepatocyte and can potentially enhance first-pass biliary excretion of the compound.

It has been shown through studies with mdr1a (−/−) mice and in clinical trials that the mean absorption time, AUC, and maximal plasma concentration (C_{max}) following oral administration of P-gp substrates are affected by apically directed efflux activity of P-gp in the intestine (12,212,216,218,226). For example, the plasma AUC values determined for a 10-mg/kg dose of paclitaxel observed in mdr1a(−/−) mice were indeed several times higher following oral administrations compared with values obtained in wild-type mice (217). The oral bioavailability of paclitaxel was 35% for the mdr1a(−/−) mice versus 11% for the wild-type mice. Similar studies have been performed with other P-gp substrates such as cyclosporin A and fexofenadine; an increased oral absorption of all these substrates was observed in the P-gp-deficient mice (212,235). The effects of P-gp-mediated efflux activity have been shown in studies aimed at elucidating the oral disposition of β-adrenoceptor antagonists. The dose-normalized AUC was found to increase with dose, but the oral clearance was found to decrease with increasing dose (226). These findings were not compatible with the saturable first-pass effect attributable to metabolism. The t_{max} and mean absorption times of the orally administered P-gp substrate, talinolol, were significantly reduced with coadministration of verapamil. By using verapamil to alter the pharmacokinetic properties (specifically the intestinal absorption) of the β-adrenoceptor antagonist talinolol, it has been clearly shown in an intact model that the absorption of this drug is significantly affected by P-gp present in the intestine (226). Similarly, the nonlinear and limited bioavailability of celiprolol and pafenolol have been shown to be due to the actions of P-gp-mediated efflux activity in the intestine (5,236–238).

P-gp-mediated efflux activity is an important determinant of digoxin oral absorption, and this has been observed from DDIs between digoxin and quinidine, and digoxin and rifampicin. Digoxin absorption is not influenced by first-pass metabolism and any changes to digoxin absorption are thought to be due to changes in the actions of P-gp. The interaction between orally administered quinidine and digoxin results in a dramatic enhancement in digoxin C_{max} and AUC (239–242). Conversely, treatment with the P-gp inducer, rifampicin, has been shown to decrease digoxin C_{max} and AUC in humans (243). In fact, it was shown that intestinally expressed P-gp (induced by treatment with rifampicin) closely correlated with digoxin AUC in a negative fashion.

The examples above have clearly shown how P-gp efflux can influence absorption. However, it is important to note that P-gp efflux activity does not always dictate these outcomes to a compound's absorption profile. Absorption is a highly complex multifactorial process in which P-gp can play a part. The magnitude of the effect of P-gp efflux activity on a compound's absorption profile ultimately depends on the P-gp activity in combination with other critical factors such as solubility, permeability, and metabolism.

2. Distribution

In some instances, P-gp can significantly affect the profile of drug distribution from systemic circulation into organs and tissues, most notably those that possess a specialized blood-tissue barrier such as the brain. Experiments with mdr1a(–/–) mice have shown how P-gp affects the distribution of its substrates into certain tissues (11,12,124,212–219). A few examples are shown below to demonstrate the role played by P-gp in the tissue distribution of drugs.

Some of the most informative results came from a study involving altered distribution of P-gp substrates in mdr1a(–/–) mice. At moderate doses (1 mg/kg) of vinblastine, the concentrations of the parent drug in the heart, muscle, brain, and plasma were 3-, 7-, 20-, and 2-fold higher, respectively, in the mdr1a(–/–) mice compared with the normal mice (results summarized in Table 4) (12). The levels in the other tissues expressing the mdr1a P-gp were two- to threefold higher in mdr1a (–/–) mice (12). At a dose of 6 mg/kg, the differences in tissue distribution were still significant, but reduced, most likely due to saturation of P-gp (12). A 12-fold increase in brain concentration was seen at this dose; plasma and tissue differences of approximately 2-fold were seen (12). The concentration of ivermectin was found to be 87-fold higher in the brain of mdr1a(–/–) mice than that of the wild-type mice. Not surprisingly, compared with the wild-type mice, the mdr1a (–/–) mice displayed an increased sensitivity to ivermectin (100-fold) (12). The effect of P-gp efflux on opioid peptide pharmacodynamics was studied using mdr1a(–/–) mice. The brain tissue concentration of 2,5-D-penicillamine enkephalin (DPDPE) was found to be two-to fourfold higher in the mdr1a(–/–) mice and the dose required to elicit a comparable antinociception was nearly 30-fold lower in the mdr1a(–/–) mice (210). Similar studies have been performed with the P-gp substrates dexamethasone, digoxin, loperamide, and cyclosporin A (212,216,244). The differences seen in plasma and tissue concentrations between the mdr1a-deficient mice and the normal mice differ from drug to drug, but a common theme observed in the mdr1a-deficient mice was the increased tissue accumulation of these substrates (216).

Studies have been performed with normal mice to demonstrate that P-gp efflux activity that limits extravascular exposure is inhibitable. The treatment of mice with the P-gp inhibitor GF120918 resulted in a 13-fold and 3.3-fold increase in brain and cerebrospinal fluid concentration of amprenavir, respectively, over that in the vehicle-treated mice (245). Also, for loperamide, a 10-fold higher brain uptake was observed in mdr1a(−/−) compared with mdr1a(+/+) mice. Treatment with quinidine fully abolished this difference (246,247).

These findings of how P-gp limits the disposition of its substrates are critical to elucidate, particularly regarding potential toxicity and CNS exposure. With regards to the former, a change in P-gp efflux activity, either via saturation by substrate, disease state, or inhibition by another compound, can lead to a significant increase in drug concentration in organs normally protected by P-gp. In certain cases, as seen with ivermectin and vinblastine, this increased exposure can significantly increase toxicity that is related to exposure in these organs. With regards to CNS exposure, P-gp efflux liability can be a significant hurdle to the successful development of CNS active agents (11,248).

3. Metabolism

Recent evidence has shown that P-gp can play a role in determining the oxidative metabolism of its substrates that are also substrates of CYP3A. Several factors have led to the observation that P-gp and CYP3A4 may act in concert to determine the oral absorption and bioavailability of drugs. These barrier-forming proteins are colocalized to the AP region of the enterocytes that form the epithelial lining of the small intestine and also in the hepatocytes (249–251). P-gp and CYP3A4 can be induced by many of the same compounds, although it has recently been shown that these proteins are not coregulated (252). It is well known that there exists a large degree of overlap between the broad substrate specificities of P-gp and CYP3A4 (253). Given this fact, it seems reasonable that the combined actions of P-gp and CYP3A4 could account, in some part, for the low oral bioavailability determined for many of these dual substrates.

Until recently, intestinal metabolism via CYP3A4-mediated metabolic pathways was thought to be insignificant because of the lower levels of expression compared with that seen in the liver and slower metabolic rates measured for intestinal microsomes (224). However, similar K_m values have been reported for midazolam 1′-hydroxylation by microsomes obtained in the upper intestine and the liver (254,255). This correlation indicates that the upper intestine and hepatic CYP3A4 are functionally equivalent. Such findings further establish the importance of the intestine in the elimination of orally administered substrates for CYP3A4-mediated metabolic pathways. Additionally, coadministration of substrates/inhibitors that may alter the function of these proteins (induction, inhibition) could further be responsible for the variability in intestinal absorption (drug interactions) seen for some drugs.

The interplay between P-gp and CYP3A4 has been studied extensively, and the results from certain in vitro experiments have suggested the interesting possibility that these two proteins may act in concert (Fig. 2). Studies involving cyclosporin A transport across Caco-2 cell monolayers have shown how the

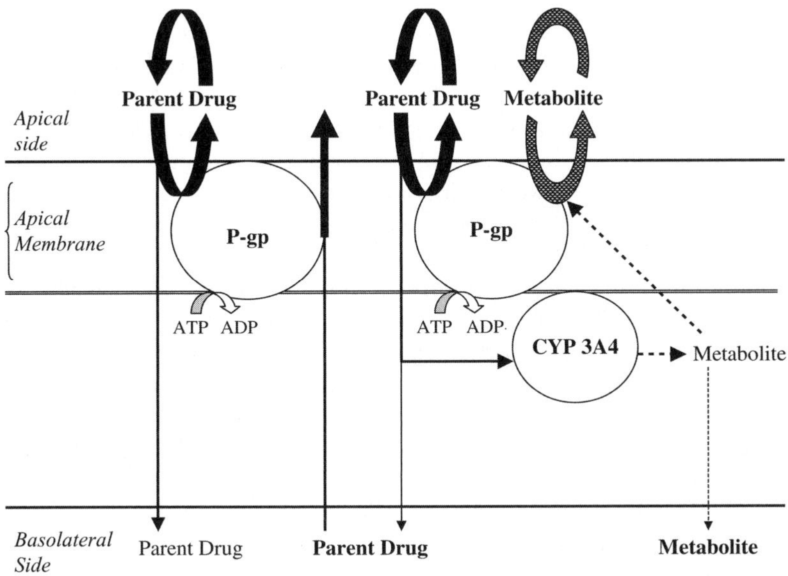

Figure 2 Apically directed P-gp-mediated efflux of drugs across intestinal epithelium and synergistic interactions of P-gp with CYP3A4 in attenuating the absorptive transport. Heavy arrows versus light arrows indicate relative magnitudes of the flux. This exemplifies an elimination mechanism that a dual substrate of P-gp and CYP3A4 may encounter in the enterocyte. Conceivably, the metabolite may or may not be a substrate for P-gp (as drawn, it is a substrate). *Abbreviation*: P-gp, P-glycoprotein.

actions of P-gp and CYP3A4 may act coordinately to enhance the attenuation of AP to BL transport of this drug. It was observed that cyclosporin A metabolism was much greater when the compound was transported in the AP to BL (absorptive) direction than in the BL to AP (secretory) direction (16). It was postulated that the reduction in the AP to BL flux of cyclosporin A caused by apically directed P-gp efflux enhanced the exposure of the compound to CYP3A4, and thus a greater amount of metabolism was achieved (16). It appears that P-gp may be able to increase the susceptibility of some compounds to CYP3A4-mediated metabolic pathways in the intestine (15,16,256–259).

Quantification of the primary metabolites formed in these experiments has also provided some interesting observations. The metabolites of cyclosporin A generated by a CYP3A4-like enzyme were preferentially transported to the AP side, indicating that these metabolites were also P-gp substrates (16). Similar results were obtained in experiments following the metabolism of midazolam as it diffused across Caco-2 cells induced to express CYP3A4 (257). Fisher et al. found that the distribution of 1′-hydroxymidazolam and 4-hydroxymidazolam, two primary metabolites of midazolam generated by the CYP3A4-mediated metabolic pathway, were preferentially transported to the AP compartment regardless of the transport direction of midazolam (257). This is interesting because midazolam is not subject to polarized efflux, but does interact with P-gp in an inhibitory fashion (260). These results suggest that CYP3A4-mediated metabolism of midazolam makes the P-gp-mediated efflux of this compound (via its metabolites) more efficient. The oxidative metabolism, mediated by CYP3A4, may possibly reduce the passive membrane permeability of the metabolites (e.g., 1′-hydroxymidazolam), thus allowing P-gp to more effectively establish a concentration gradient. Furthermore, the addition of oxygen may act to increase the affinity of P-gp for these metabolites with similar structures. An example of this is the drug-metabolite pair terfenadine and fexofenadine—fexofenadine is known to be a much better substrate for P-gp than terfenadine (261). Making the metabolite a better substrate of P-gp also has consequences for the catalytic activity of CYP3A4. If the efflux of primary metabolites is more efficient than that of the parent, the amount of competing secondary oxidative metabolism will be reduced, and thus the metabolism of the parent will be more complete (262).

In contrast to these findings, several pharmacokinetic models have been published, which fail to predict an increase in CYP3A metabolism due to P-gp efflux (263–266). In fact, some have even predicted that P-gp efflux could decrease the rate of metabolism by effectively decreasing the intracellular concentration of parent drug. This conflict has been experimental, and theoretical results are yet to be resolved; further experimentation is needed to characterize the nature of this interaction (267).

Converse to what has been observed for P-gp and CYP3A activity in the intestine, reports have shown that P-gp efflux activity can act to reduce CYP3A4 metabolism that is primarily hepatically mediated (268). This finding highlights an important caveat for dual substrates—how P-gp and CYP3A4 interact (in concert or opposed) may be tissue or organ specific. Indeed, the role P-gp plays in the intestine to reduce absorption rate is different from that in the liver where it acts as a competing clearance mechanism with CYP3A4, potentially reducing total CYP3A4-mediated metabolism (267).

These findings have raised several interesting questions regarding how these proteins may act in concert to maximize their protective activities. It is not well understood how key parameters (e.g., substrate affinity, protein expression, substrate permeability properties) combine to ultimately determine disposition and, furthermore, how the activity of each affects the other in intestine versus liver. These issues remain as key challenges for both furthering understanding of the determinants of bioavailability and prediction and understanding of potential DDIs.

4. Elimination

In addition to affecting absorption, distribution, and possibly the metabolism of drugs, P-gp can also play a role in hepatic, renal, and intestinal elimination of its substrates (13). The mechanisms of how P-gp acts to make the intestine an important route of elimination are only now being appreciated (221). Certain drugs administered by the intravenous route are indeed eliminated to a high degree in the intestine via a process other than biliary excretion (12,48,212,216–218,221,223). The enormous surface area of the intestine ($\sim$200 m^2 in adult man) allows the organ to act as a giant dialysis membrane for drugs as the concentrations in the plasma exceed those in the intestinal lumen, and passive diffusion across the mucosa into the gut lumen can occur (221). Some of the same driving forces that affect the intestinal absorption of drugs also exist for exsorption. Other biochemical and physiological factors that are likely to affect this process include protein binding, blood flow to gut, and specificity for intestinal P-gp-mediated efflux activity. P-gp can affect the rate at which drugs are eliminated from tissues and from the plasma via elimination through the liver, intestine, and/or kidney. The oral, systemic, and tissue clearances (rate of elimination) are affected by P-gp efflux, and thus the terminal half-lives of P-gp substrates may be related to the efflux activity seen in the organism.

The effect of P-gp-mediated efflux activity on excretion has been clearly shown through experiments with vinblastine and paclitaxel in mdr1a(−/−) mice. The results of these experiments have shown how P-gp-mediated efflux activity accelerates tissue clearances and also systemic clearances of its substrates. Additionally, these studies have highlighted the role of the intestine in elimination. While the role of intestinally expressed P-gp in limiting absorption is recognized, these experiments have helped elucidate its role in making the intestine a significant route of elimination.

In normal mice, the elimination of vinblastine in the feces within 24 hours of administration was determined to be approximately 25% of the dose as unchanged drug at two doses (1 and 6 mg/kg) (218). In the mdr1a-deficient mice, the amount of unchanged drug recovered in the feces was reduced to 9.4% for the 1-mg/kg dose and 3.4% for the 6-mg/kg dose (218). The amount of vinblastine remaining in the brain tissue of the P-gp-deficient mice was approximately 1000-ng/g tissue at the 6-mg/kg dose 24 hours after administration, whereas the amount of vinblastine remaining in the brain tissue of normal mice at the same dose was only 22 ng/g tissue (218). The normal mice showed much more rapid elimination of vinblastine from both the plasma and tissue than the mdr1a-deficient mice, and a significant reduction in terminal elimination half-life and reduced clearances of vinblastine were observed for each of these dose levels for the P-gp-deficient mice (12,218).

As seen with vinblastine, clearances of paclitaxel were reduced and elimination of half-life increased in the mdr1a(−/−) mice (217). Nearly 90% of the radioactivity following an IV dose of paclitaxel was recovered in the feces of the wild-type mice, mainly as unchanged drug or hydroxylated metabolites (217). For the mdr1a-deficient mice, a mere 1.5% of the dose was recovered in the feces and approximately 45% of the dose was recovered in the urine as unknown metabolites (217). Following an oral dose (10 mg/kg), 90% of the dose was recovered in feces of the wild-type mice compared with only 2% seen in the mdr1a-deficient mice (217). The levels of the hydroxylated metabolites excreted by the mdr1a-deficient mice were not dependent on the route of elimination, whereas in wild-type mice three times as much hydroxylated paclitaxel was collected following an IV dose (217).

The contributions of the mdr1a P-gp to the hepatic and intestinal clearances of paclitaxel, digoxin, vinblastine, and doxorubicin have been determined by comparing the amounts of biliary and intestinal secretion of each drug in wild-type and mdr1a(−/−) mice (Table 4). The amounts of biliary excretion of paclitaxel and the hydroxylated metabolites were not significantly different between the wild-type mice and the mdr1a-deficient mice (217). Further, when the biliary excretion into the intestinal lumen was blocked, nearly three times the amount of a 10 mg/kg IV dose was recovered in the lumen of the wild-type mice versus the mdr1a(−/−) mice within 90 minutes of administration (217). Like paclitaxel, absence of the mdr1a P-gp seems to have a minimal effect on the biliary secretion of digoxin and vinblastine, whereas the intestinal secretion of these compounds is significantly affected (219,269). An opposite situation exists for the intestinal and biliary secretion of doxorubicin. Nearly five times the amount of unchanged doxorubicin was secreted into the bile of the wild-type mice versus the mdr1a(−/−) mice, whereas the intestinal secretion of doxorubicin was approximately equal (~ 10% of the dose) in both sets of mice (219). These results illustrate that in mice, mdr1a P-gp is active in the intestinal excretion of paclitaxel, digoxin, and vinblastine, and that the mouse liver has the ability to utilize alternate pathways of elimination for these compounds. Conversely, the biliary excretion of doxorubicin in mice appears to be highly dependent on mdr1a P-gp-mediated efflux activity, whereas intestinal mdr1a P-gp plays less of a role in the intestinal excretion of doxorubicin (Table 4) (270).

IV. RAMIFICATIONS OF ALTERING P-gp ACTIVITY: EFFECTS ON DRUG DISPOSITION AND DDIs

As highlighted in section III, P-gp-mediated efflux activity can play a significant role in its substrates' disposition. Alteration of P-gp-mediated efflux activity can have consequences to these substrates' disposition. Furthermore, inter- and intrapatient variability in P-gp expression and function can be a potential source of pharmacokinetic variability. The following sections summarize findings following changes to P-gp-mediated efflux activity resulting from coadministration of P-gp substrates and inhibitors and those related to genetic variability in P-gp.

A. Outcomes from Alteration of P-gp Efflux Activity Mediated by Coadministration of Substrates and Inhibitors

The efflux activity mediated by P-gp is a saturable process and as such is subject to interactions between substrates and inhibitors. These interactions, however, can be quite complex. Lin notes that interactions between P-gp

Table 4 Relative Influence of P-gp on the Biliary and Intestinal Excretion of P-gp Substrates[a]

	Plasma level[b] (ng/mL)		Biliary excretion[c] (% of administered dose)		Intestinal excretion[c] (% of administered dose)		
Drug (dose) (mg/kg)	Wild type	mdr1a(−/−)	Wild type	mdr1a(−/−)	Wild type	mdr1a(−/−)	References
Paclitaxel (5)	289 ± 38	327 ± 44	25.7 ± 4.5	26.6 ± 2.90	4.63 ± .49	1.52 ± 0.05	217
Digoxin (0.2)	125 ± 10	216[d] ± 14	24.0 ± 4.8	15.8 ± 2.9	16.4 ± 2.6	2.2[d] ± 0.4	269
Vinblastine (1)	NA	NA	26.7 ± 1.3	28.9 ± 2.0	10.4 ± 0.4	6.8[d] ± 0.5	218
Doxorubicin (5)	150 ± 22	166 ± 23	13.3 ± 1.7	2.4[d] ± 0.3	10.5 ± 0.5	10.0 ± 0.4	219

[a]Biliary and intestinal excretion of total [^{3}H] label of the drugs (parent and metabolites) in the first 90 minutes after IV bolus administration was determined in the wild type and mdr1a(−/−) mice with a cannulated gallbladder,

[b]Plasma concentration at $t = 90$ minutes,

[c]Data (means ± S.E.) are represented as a percentage of the administered dose,

[d]$p < 0.05$ versus wild-type mice.

Abbreviations: P-gp, P-glycoprotein; NA, not available.

substrates can result in competitive inhibition, noncompetitive inhibition, and cooperative stimulation (271). Furthermore, these interactions can occur either at transport-binding sites or at ATP-binding domains, as seen with vanadate and cyclosporin A (interacts with transport and binding domains) (144,271–273). This multiplicity of mechanisms makes assessment of interaction difficult and further nearly impossible to predict a priori (271). While it is certain that coadministration of P-gp substrates results in changes to disposition, it is often unclear to what degree these changes may occur as a result of alteration of efflux. Indeed, the overlapping substrate specificity of P-gp and CYPs, namely CYP3A4, has confounded attempts to elucidate the ramifications of P-gp-related interactions (274). A method to deconvolute this relative contribution of interaction of P-gp from CYP3A involves comparing inhibitory potencies toward each protein and relating this to relevant clinical concentrations (274). Finally, P-gp expression can be induced, and this phenomenon has the potential to lead to interactions.

Although a precise understanding of the implications of P-gp interactions is yet to be realized, some insight into how coadministration of substrates and inhibitors interact can be gained from clinical findings. The first area involves attempts to reverse the MDR phenotype via P-gp inhibition. The second area involves reports of DDIs potentially mediated by P-gp. A summary of the findings in each area and conclusions around P-gp interactions that they provide is given below.

1. Clinical Trials with P-gp Modulators

A large area of research has involved determining the possible use of P-gp modulators to reverse the MDR phenotype associated with P-gp-mediated efflux in an attempt to improve the efficacy of chemotherapeutic agents and regimens. Clinical trials have been performed to assess the use of P-gp modulators (i.e., verapamil, cyclosporin A, etc.) to improve the intracellular delivery/efficacy of the chemotherapeutic agents (i.e., doxorubicin, vinblastine, and etoposide). However, the interpretation of the results of these clinical trials involving the use of P-gp inhibitors in an attempt to reverse the MDR phenotype has been complicated by unknown pharmacokinetic interactions between the target cytotoxic drug and the modulator (275). Results obtained from trials with first generation inhibitors have been somewhat disappointing; however, some promising results were obtained in hematolymphoid malignancies (275,276). There are several possible reasons why this line of therapy has not been successful. Difficulties in detecting the MDR1 phenotype in clinical practice, the inability to achieve target concentrations of the modulator, methodology of the trial, and the multifacial array of chemoresistance mechanisms could all act to confound the results of these trials (276). Importantly, a key reason why first- and second-generation inhibitors have not been successful is the change in systemic pharmacokinetics, which some have imparted to the chemotherapeutic agent; this change is often increased absorption or decreased elimination, leading to greater systemic concentrations and subsequent toxicity (277,278). Furthermore, many of the first- and second-generation inhibitors known to produce these changes in systemic pharmacokinetics have inhibitory potency against both P-gp and CYP3A4 (278). For this reason, third-generation inhibitors have been developed that more specifically interact with P-gp, and consequently these selective inhibitors have shown less propensity to alter the systemic pharmacokinetics of the chemotherapeutic agents (278).

a. First generation. These agents represent drugs in clinical use for other indications that had been shown to inhibit P-gp efflux through in vitro experiments. Because of the relatively low binding affinity of these compounds for P-gp and the need to increase the doses of these modulators to toxic levels, few of these agents have been further studied for use in clinical modulation of P-gp. However, early trials with these drugs have provided invaluable information regarding the consequences of inhibiting P-gp. These first-generation inhibitors include verapamil, cyclosporin A, tamoxifen, quinidine, and quinine.

i. Verapamil Many of the early trials aimed at reversing the MDR phenotype associated with overexpression of P-gp involved coadministration of the phenylalkamine voltage-dependent L-type calcium channel blocker verapamil. Racemic verapamil was shown to reverse P-gp-mediated resistance to vincristine and vinblastine in vitro and in vivo in P388 leukemia (279). These early findings and the fact that verapamil was a clinically used drug with an established record of safety provided the rationale for its use clinically as a P-gp modulator. The maximum tolerated dose of verapamil has been reported to be 480 mg/day orally (leading to blood levels of 1 mM) with the dose-limiting toxicity being hypotension (280). Dose escalation studies with intravenously administered verapamil found that for the dose range of 0.15–0.6 mg/kg/hr, cardiovascular toxicities may be seen along with edema and weight gain (281).

Oral verapamil has been shown to increase peak plasma levels, prolong the terminal half-life, and increase the volume of distribution at steady state of doxorubicin (282). Gigante et al. (283) performed similar studies in which the pharmacokinetics of doxorubicin in combination with verapamil given at high doses intravenously were followed for 17 patients with advanced neoplasms. The steady-state concentration and systemic and renal clearances were found to be statistically similar for various doses of verapamil and doxorubicin, and for doxorubicin administered alone.

Additional trials were designed to assess the usefulness of verapamil in improving the efficacy of chemotherapeutic regimens for the treatment of small cell lung cancer (284,285), refractory multiple myeloma (286), and breast cancer (287). The results of these trials showed that verapamil had only a modest positive effect on the overall effectiveness of the regimen.

ii. Cyclosporin A The immunosuppressive cyclic undecapeptide cyclosporin A has been used in several clinical trials as a modulator of P-gp.

Cyclosporin A readily inhibits CYP3A metabolism and may lead to significant pharmacokinetic interactions (288). Several studies have been performed using cyclosporin A as a P-gp modulator in combination with etoposide, doxorubicin, and paclitaxel as described below.

Background work with mice was performed to assess the feasibility of using cyclosporin A as a modulator of P-gp-mediated drug resistance. The AUC of doxorubicin in the liver, kidney, and adrenals increased nearly two- to threefold with respect to the levels measured in control animals 30 minutes after a single intraperitoneal injection of cyclosporin A (289). The serum levels of doxorubicin following cyclosporin A treatment were unchanged, indicating that cyclosporin A was altering the drug concentrations in the tumor without affecting its plasma concentration.

The effects of cyclosporin A on the pharmacokinetics of etoposide have been determined and were shown to be dose dependent. A variable range of cyclosporin A concentrations was obtained (297–5073 ng/mL), and it was observed that patients with higher cyclosporin A concentrations also had larger increases in etoposide AUC (290). Results from studies using clinically relevant plasma concentrations of cyclosporin A (1000–5000 ng/mL) as a P-gp inhibitor resulted in mean 48%, 52%, and 52% decreases in the systemic, renal, and nonrenal clearances of intravenously administered etoposide (232,290). Similar decreases in the systemic, renal, and nonrenal clearances of doxorubicin were observed with administration of cyclosporin A (232,291).

iii. Tamoxifen, Quinine, and Quinidine Quinine and quinidine are both alkaloid drugs (quinine is the S-diastereoisomer of quinidine) used as antiarrhythmic drugs. Both have been shown to modulate P-gp-mediated efflux in vitro with quinidine being the stronger inhibitor of the two (143,292). Few positive results have been seen with the use of these agents in reversing MDR in clinical trials (293–296). The relatively low affinity of each of these compounds has limited their use clinically to reverse MDR.

Tamoxifen is an estrogen receptor antagonist that weakly binds to P-gp and exerts inhibitory effects in vitro at concentrations above 1 μM (297). Tamoxifen is used clinically for the treatment of breast cancer and initial trials with this P-gp inhibitor have focused on using this drug to not only treat breast cancer but also to reverse P-gp-mediated MDR. In a dose escalation study, a vinblastine and tamoxifen combination proved to be neurotoxic (298). Neurotoxicity also occurred in a trial with high-dose tamoxifen and etoposide, and at this dose, the plasma concentration of tamoxifen was below the concentration reported to reverse etoposide resistance in P-gp-expressing cell lines (297,298). Tamoxifen has very complex pharmacokinetics, which are not fully understood presently. The drug exhibits high plasma protein binding (98%), enterohepatic recirculation, distribution into fatty tissue, and a long terminal half-life (299). Other trials with tamoxifen have been performed, all of which have reported adverse toxic effects without much success at reversing MDR (300,301). Because of these severe toxic effects of tamoxifen, such as dizziness, tremor, unsteady gait, grand mal seizure, and myelosuppression, no further trials have been conducted with this drug.

b. Second generation. These compounds represent a more focused attempt to develop potent P-gp modulators that would be much less toxic than first-generation inhibitors, so that adequate P-gp inhibitory concentrations can be achieved clinically without the risk of toxic effects. The second generation modulators include dexniguldipine (B8509-035), dexverapamil (*R*-verapamil), and PSC833 (valspodar).

The (−) isomer of the L-type calcium channel blocker (+)-niguldipine is dexniguldipine. This agent binds to an intracellular domain of P-gp with a K_i of 10 nm. In addition, this compound can block RNA synthesis at 5 μM and possesses some anticancer activity (302). Currently, only a few studies have been conducted to evaluate the use of this compound as a P-gp modulator. Definitive results are yet to be reported.

Dexverapamil is just as effective at blocking P-gp-mediated efflux as its enantiomer verapamil, but this compound is seven times less potent at inhibiting the contractile force of isolated human heart muscle tissue (303). This reduction in the dose-limiting factor of verapamil has led to clinical trials with dexverapamil as a possible P-gp-reversing agent. A clinical trial to evaluate the effect of dexverapamil in Hodgkin's and 154 non-Hodgkin's lymphoma refractory to EPOCH (etoposide, vincristine, doxorubicin, cyclophosphamide, and prednisone) chemotherapy was conducted. The combination therapy was well tolerated, but the results showed that the effect of dexverapamil in improving the EPOCH chemotherapeutic regimen was at best minimal (304,305). A trial involving combination therapy of dexverapamil and paclitaxel in heavily pretreated patients with metastatic breast cancer showed that the combination resulted in hematological toxicity that was greater than paclitaxel alone along with increased mean peak paclitaxel concentrations and delayed mean paclitaxel clearance (306).

Valspodar (PSC833) is an analog of cyclosporin D but has no immunosuppressive activity. Results from in vitro assays have shown that PSC833 may be a 20-fold more potent inhibitor of P-gp than cyclosporin A (307–309). Several clinical trials have been performed with PSC833 with some promising results. Patients with relapsed acute myelogenous leukemia (AML) were administered PSC833 with mitoxantrone and etoposide, and it was concluded that this regimen was tolerable and had antileukemic activity (310,311). Plasma concentrations of PSC833 shown to reverse P-gp-mediated efflux in vitro were achievable in patients treated with other P-gp substrates with no associated toxicity assigned to PSC833. However, the toxicity of the chemotherapeutic agents tends to be somewhat pronounced when coadministered with PSC833 (312,313). The effectiveness of PSC833 in increasing the efficacy of chemotherapeutic agents/regimens appears promising, but more trials must be performed to confirm these initial results.

c. Third generation. Like the second-generation modulators, these compounds represent further attempts to produce agents whose primary activity involves the inhibition of P-gp-mediated efflux with reduced toxic effects. Many of these compounds have been shown to possess low nanomolar potency as P-gp inhibitors in vitro. Additionally, many of these agents do not inhibit CYP3A4 and consequently have produced fewer of the adverse interactions mediated by gross changes to substrate pharmacokinetics. These compounds include GF120918 (Elacridar), CGP41251, S9788, and LY335979 (Zosuquidar).

GF120918 (GW918) is an acridonecarboxamide derivative that has been shown to inhibit P-gp with an EC_{50} of 20 nM, making it one of the most potent P-gp modulators reported (314). Initial trials were performed to assess the alteration in the pharmacokinetic profile of doxorubicin that may occur with coadministration of GF120918. The results indicate that plasma concentrations of GF120918 that modulate P-gp in vitro were obtainable in vivo and that at these concentrations the pharmacokinetics and pharmacodynamic toxicity (involving myelotoxicity) of doxorubicin appear to be unaltered by GF120918 (14). It has recently been shown that GF120918 coadministered with doxorubicin has minimal effects on doxorubicin pharmacokinetics and was found to be safe to use at doses producing P-gp efflux inhibition (315).

CGP41251 is the *N*-benzyl derivative of staurosporine and appears to have some affinity for protein kinase C along with an ability to inhibit P-gp-mediated efflux (299). There have been few clinical studies performed with this agent to date.

S9788 has been shown to be five times more potent than verapamil in inhibiting P-gp in vitro (316). The triazinodiaminopiperidine derivative S9788 represents one of the first attempts in the development of a high-affinity agent used specifically to reverse P-gp-mediated resistance. It is possible to achieve nontoxic plasma concentrations of S9788 that are known to reverse P-gp-mediated efflux in vitro (317). The adverse effects of this compound seem to involve cardiotoxic events, including AV blocks and QT prolongation, leading to ventricular arrhythmia and torsade de pointes, which occur at the maximum tolerated dose (96 mg/m^2) (317,318). In a preliminary study, coadministration of S9788 did not enhance the toxicity of doxorubicin, and the pharmacokinetic profile of doxorubicin was not altered by S9788 (318). The potency and safety of this compound has led to the initiation of further clinical trials with this compound as a P-gp modulator.

LY335979 was developed as a highly potent and specific inhibitor of MDR1, but does not inhibit MRP2, BCRP, or CYP enzymes (319). Interestingly, although LY335979 serum concentrations would predict complete inhibition of P-gp efflux, the pharmacokinetics of coadministered substrates doxorubicin, daunorubicin, and paclitaxel appeared minimally affected; slight changes noted were increases in plasma concentrations and decreases in clearance (320).

2. DDIs Related to P-gp Alteration

Although DDIs are typically associated with a change in a compound's metabolic profile, it has recently become apparent that interactions between P-gp substrates can also lead to significant alterations in the pharmacokinetic profiles of these drugs. The actions of transporters in the elimination of their substrates in the liver, kidney, and intestine (exsorption) have recently been elucidated. It is now known that primary active transport mechanisms contribute greatly to the biliary excretion of various cytotoxic agents, organic cations and anions, and compounds that have been conjugated via phase II metabolism (321). The elimination of organic cations by the kidney is highly dependent on active transport (322). It is known that intestinally expressed P-gp can act to limit the absorption of its substrates and, like in the liver and kidney, the presence of P-gp in the intestine can make it an efficient organ of elimination. Because of the extensive distribution and physiologically protective nature of P-gp, it is inevitable that DDIs between substrates of this pump will be seen, given the importance of P-gp in determining the absorption, distribution, and elimination of its substrates (323). Knowledge regarding the importance of these interactions is presently limited. Some examples of drug interactions caused by coadministration of compounds that affect P-gp-mediated efflux are given below.

a. Digoxin. The cardiac glycoside digoxin, widely used for the treatment of congestive heart failure, has a very narrow therapeutic window and any interactions that alter the blood concentration of this agent are potentially dangerous (324,325). Digoxin has been shown to be a substrate of P-gp both in vitro (240) and in vivo (326). Because of the strict monitoring of digoxin pharmacokinetics, valuable information regarding the interaction between this agent and other P-gp substrates has been elucidated. Digoxin absorption is known to be affected by P-gp efflux and additionally, in humans, digoxin is primarily eliminated unchanged renally with minimal metabolism; thus changes to digoxin disposition can to some degree be attributed to changes in P-gp-mediated efflux activity (327,328).

The ratio of renal clearance of digoxin to creatinine clearance decreased with the coadministration of clarithromycin (0.64 and 0.73), and was restored (1.30) after administration of clarithromycin had stopped (326). The role of P-gp efflux in this interaction was confirmed using an in vitro kidney epithelial cell line (326). The administration of itraconazole, a P-gp inhibitor, with digoxin resulted in an increased trough concentration and a decrease in the amount of renal clearance, possibly by an inhibition of the renal tubular secretion of digoxin via P-gp (329). The P-gp modulator verapamil has also been shown to decrease the renal clearance of digoxin (330).

It is well known that a DDI occurs between digoxin and quinidine. It has been shown that quinidine can alter the secretion of digoxin in the kidney and also in the intestine (234). The plasma concentrations of digoxin following intravenous injection increased twofold when quinidine (1 mg/hr) was coadministered. The total clearance decreased from 318 ± 19.3 to 167 ± 11.0 mL/hr. The coadministration of quinidine decreased the amount of digoxin appearing in the intestine by approximately 40%. The intestinal clearance also decreased from 28.8 ± 1.7 to 11.1 ± 1.6 mL/hr following quinidine coadministration. These studies demonstrate how quinidine can affect the absorption and secretion of digoxin.

In some cases of atrial fibrillation, both digoxin and verapamil are used (324,325). Observations from this coadministration have shown how P-gp modulation by verapamil altered the distribution and elimination of digoxin (214,331–335).

Dietary factors and herbal agents can also lead to drug interactions. The effects of St. John's wort (*Hypericum perforatum*), a widely used herbal antidepressant, on digoxin were examined in a single-blind placebo-controlled clinical trial, designed to study the changes in the pharmacokinetics of digoxin used in combination with this supplement (20). This herbal extract was shown to have significant effects on the pharmacokinetic profile of digoxin. The results of this study indicate that St. John's wort extract appears to increase the elimination of digoxin.

Another interaction that has been reported to affect digoxin pharmacokinetics involves the induction of P-gp (243). It has been shown clinically that the blood concentration of digoxin decreases significantly for patients receiving rifampin. A clinical trial was designed to confirm that this decrease was indeed due to induction of P-gp; the single-dose pharmacokinetics of digoxin (oral and IV) were determined before and after administration of rifampin. The rifampin treatment increased the level of P-gp in the intestine by 3.5-fold. The AUC of orally administered digoxin was significantly lower after administration of rifampin, whereas the decrease in intravenously administered digoxin was affected to a lesser degree. Additionally, the renal clearance and the half-life of digoxin were found to be unaltered by rifampin. These findings led the authors to postulate that the digoxin rifampin interaction occurs largely at the level of the intestine and seems to have a large effect on the absorption of digoxin. The ability of orally administered rifampin to induce intestinally expressed P-gp may have further consequences for the intestinal absorption of other P-gp substrates/inhibitors (336).

b. Cyclosporin A. Like digoxin, the plasma concentrations of cyclosporin A are strictly monitored. The determination of the effects of other agents on the pharmacokinetic profile of cyclosporin A has provided valuable information regarding possible DDIs involving P-gp-mediated efflux.

A toxic interaction between escalating doses of intravenously administered cyclosporin A (6–27 mg/kg/day, median: 19.5 mg/kg/day) and a standard chemotherapeutic regimen was observed in patients diagnosed with soft tissue sarcoma (323,337). The regimen consisted of courses of etoposide and ifosfamide (days 1 and 2) (VP16/Ifos cycles), alternating with courses of vincristine, dactinomycin, and cyclophosphamide (days 1 and 5) (VAC cycles) (337). The administration of cyclosporin A dramatically increased the systemic toxicity of the VAC cycle, but only mildly increased the systemic toxicity of the VP16/Ifos cycle (337). A possible mechanism for this increased toxicity was proposed to involve increases in serum concentrations (due to decreased elimination) of etoposide, vincristine, and dactinomycin, all of which are P-gp substrates, following the inhibition of P-gp by cyclosporin A (337).

An enhancement in the absorption of orally administered cyclosporin A (10 mg) was observed, as evidenced by an increase in the AUC, when a solution of vitamin E was given concomitantly with one of the cyclosporin A doses in a randomized trial (338). The levels of metabolites of cyclosporin were unchanged by oral administration of the vitamin E solution. This observation led the researchers to conclude that the vitamin E solution acted to either enhance the absorptive transport or decrease the counter transport of cyclosporin A in the intestine by inhibition of P-gp.

It has recently been proposed that the Biopharmaceutical Classification System (BCS) can be used to predict intestinal drug disposition with regards to efflux transport and metabolism (339). Furthermore, on the basis of the key substrate BCS-related properties, permeability, and solubility, the system may be used to predict potential interactions mediated through changes in efflux and/or metabolism at the level of the intestine.

B. Genetic-Related Differences in P-gp Function and Outcomes: P-gp Polymorphisms and Relation to Pharmacokinetics and Pharmacodynamics of P-gp Substrates

Since the first systemic screening for functional polymorphisms of the human P-gp, the impact of polymorphisms on pharmacokinetics and pharmacodynamics of P-gp substrates has attracted much attention. The readers are referred to recent review articles (340–342). In the evaluation of the impact, we should have some special considerations. First, P-gp polymorphisms only bring about twofold change in expression level but not the entire loss of function. This means that the impact on the pharmacokinetics and pharmacodynamics of P-gp substrates could be only moderate or even weak. The scenario is very different from that observed with the polymorphisms of CYP2D6. Second, the impact is substrate dependent, i.e., dependent on the role that P-gp plays in the in vivo disposition of the test drugs. Therefore, the polymorphisms may have quite different effects on different drugs. Third, the human P-gp polymorphisms may significantly affect absorption and distribution, but probably not clearance, of the test drugs. In addition, the design of clinical studies and the drugs that the subjects were concomitantly taking may greatly affect the results of the studies.

Digoxin, a classical P-gp substrate, is so far the best example that the human P-gp polymorphisms affect the pharmacokinetics profile of the test drugs. The C_{max} and AUC of digoxin after oral administration were significantly higher in the subjects with genotypes of 3435TT and 2677TT/3435TT than in the subjects with other genotypes (153,154,343,344). In the case of digoxin, two factors should be noted. First, P-gp is a major determinant in the intestinal absorption of digoxin. Second, the therapeutic dose of digoxin is low, and therefore the P-gp is very unlikely saturated in the studies. The human P-gp polymorphisms also impact the dose requirement and plasma concentrations of cyclosporin A and tacrolimus in kidney transplant patients. But the impact is relatively slight compared with that on digoxin because P-gp only plays a minor role in the disposition of these drugs (341). Another significant instance is nortriptyline. Nortriptyline, a drug used for the treatment of depression, may cause postural hypotension (a side effect) in some patients. Roberts et al. found that the occurrence rate of nortriptyline-induced postural hypotension was MDR1 genotype dependent—0% in patients with CC, 9% with CT, and 25% with TT, while there were no differences in the dose administration, efficacy (improvement of depression assessed by MADRS), and nortriptyline blood concentrations among the three genotypes (158). The authors believed that nortriptyline and its metabolites, as substrates of P-gp, accumulated in the brain more in the patients with TT genotype than those with CC or CT genotype (158). In addition, P-gp polymorphisms may affect the distribution of HIV protease inhibitors into the brain, and therefore enhance therapeutics and cause CNS toxicity because all the currently marketed HIV protease inhibitors are reported to be P-gp substrates (341). However, there have been no reports that demonstrate the significant impact of polymorphisms on pharmacokinetics, pharmacodynamics, and toxicity of protease inhibitors.

V. EXPERIMENTAL MODELS USED TO UNDERSTAND P-gp EFFLUX LIABILITY AND DDI POTENTIAL

The involvement of P-gp in the absorption and consequent distribution and elimination of orally administered drugs has been extensively studied in in vitro, in situ, and in vivo models. Some routinely used systems include cultured cell lines, isolated intestinal segments, everted sacs, and brush border membranes. Organ (brain, liver, and kidney) perfusion and gene knockout mice have also been used. Each of these models has certain advantages and disadvantages. A brief description of the models that are being used to evaluate the role of P-gp in the disposition of drug molecules is given below.

A. In Vitro Models Commonly Employed to Study P-gp

1. Tissue Culture Transport Models

a. Caco-2. Of the many cell types utilized to model drug behavior in the human intestine, the immortalized human colorectal carcinoma-derived cell line, Caco-2, is the most widely accepted in vitro model to date. This cell line has several advantages over others that make it the cell line of choice in both academia and in the pharmaceutical industry (345–351). Perhaps the most attractive feature of the Caco-2 cell line is the spontaneous differentiation into mature enterocytes that occurs after plating the cells on porous membranes and the ability to maintain the cells under normal culturing conditions. Accompanying this differentiation is the expression of several biochemical and anatomical features common to normal enterocytes. Caco-2 monolayers become polarized and display a well-defined brush border membrane located in the AP domain. Because of the various enzyme and transport activities associated with the brush border membrane, the expression of a fully defined brush border membrane in cell lines used to model enterocytes is critical. The brush border membrane contains several transporters, metabolic enzymes, and efflux pumps, such as P-gp, whose expression is both stable and functional (7,50,352,353). The expression of P-gp has been demonstrated by Western blot analysis and by polarized transport of P-gp substrates, such as cyclosporin A, which is reversed (i.e., polarity is abolished) by P-gp inhibitors such as verapamil (7,50,348,352,354,355). Recently it has been shown that P-gp expression in Caco-2 cells is nearly identical to P-gp expression in normal adult intestinal tissue (356). The kinetics of P-gp-mediated efflux activity in Caco-2 cells are equivalent to those observed in rat intestine (based on P-gp-mediated efflux of digoxin) (357).

The functional activity of P-gp in Caco-2 cells has been extensively evaluated with respect to various methodological factors such as culture time and passage number (355). Western blot analysis demonstrated that P-gp was expressed as early as day 7 of culturing. The absorptive transport of cyclosporin A was relatively constant from day 5 of culturing (treatment with the P-gp inhibitor verapamil significantly increased absorptive permeability, consistent with inhibition of polarized efflux mechanism). The secretory transport of cyclosporin A increased until day 17, at which time this permeability became constant. The reduced barrier function observed before day 17 is most likely due to incomplete monolayer differentiation or incomplete P-gp expression versus that observed at day 17. Passage number has also been considered as a variable that may affect the amount of P-gp present in the AP brush border membrane (354). Caco-2 cells of lower passage numbers (~22) have been shown to have a shorter doubling time than those of higher passage number (~72), resulting in an increased number of cells per monolayer and thus an increased amount of membrane protein. However, several reports have stated that Caco-2 cells at higher passage numbers (>90) contain significantly more P-gp than those at lower passage numbers. P-gp expression in the Caco-2 cells has been shown to be stable, and this allows relatively accurate comparison of data from various monolayers as long as they represent a relatively narrow range of passage numbers.

Expression of specific proteins can be induced in Caco-2 cells using simple culturing techniques. For example, the induction and overexpression of cytochrome P450 3A4 (CYP3A4) was achieved by culturing the cells with 1α,25-dihydroxyvitamin D3 beginning at confluence, and this overexpression was shown to be dose and duration dependent (18). Overexpression of P-gp can also be achieved in the Caco-2 cell line by culturing with vinblastine, verapamil, and celiprolol (358,359). No morphological differences were noticed for vinblastine cultured cells with respect to appearance, formation of tight monolayers, and the corresponding transepithelial resistance (359).

b. Madine-Darby canine kidney. Examples of studies involving P-gp-mediated efflux of therapeutic compounds in immortalized Madine-Darby canine kidney (MDCK) cells are less numerous than those utilizing the Caco-2 cell line;

however, this model and the transfected MDR1-MDCK variant are important tools for the study of P-gp efflux activity. Both have been used to follow the passive diffusion of compounds across monolayers. The most significant advantage the MDCK cell line has over the Caco-2 cell line is the much shorter culture time because of the enhanced growth rate of MDCK cells (360,361). Studies by Simons et al. have shown that these cells are polarized and contain a well-defined AP brush border membrane with a membrane composition similar to that of the intestine (362,363). The spontaneous differentiation of MDCK into polarized cell monolayers with defined AP and BL domains make study of the actions of transporters expressed in a polarized fashion facile. In addition, this cell line can be readily transfected and other drug effluxing transporters (expressed in either AP or BL domain) that have been incorporated into these cells to study their effects on altering the flux of a compound as it crosses a polarized monolayer (364).

Although there is a widespread perception that wild-type MDCK cells contain insignificant levels of P-gp to affect substrate transport, it has been demonstrated that this is not the case. It was shown that the transport of vinblastine sulfate across MDCK monolayers was indeed apically polarized (203). These results were duplicated by Hirst et al. using the same test compound, verapamil, in two different strains of MDCK cells. The transport profiles of verapamil showed polarity in both a high-resistance strain [TEER $\sim$2000 $\Omega\cdot\text{cm}^2$ and a low-resistance strain (TEER $<$ 200 $\Omega\cdot\text{cm}^2$] (365). Recently, parallel studies were performed measuring the transport of a novel peptide, KO2, across both MDCK and Caco-2 cells (364). The results showed nearly identical profiles for the AP to BL and BL to AP transport of this agent in both cell types. Although it is unlikely that all P-gp substrates will behave identically in both cell lines, these studies indicate that there is sufficient P-gp expression in MDCK cells to affect transport studies. Thus, MDCK cells can be used to evaluate the transport of compounds that are suspected to be substrates of P-gp.

The human MDR1 gene has been successfully transfected into MDCK cells (366). The expression of MDR1 gene product in these MDCK cells was shown to be nearly 10-fold higher than that seen in Caco-2 cells (as determined by Western blot analysis) (364). The high expression of P-gp and the short culturing time make these transfected MDCK cells an attractive model to study how P-gp-mediated efflux activity alters substrate transport across polarized epithelium. The model's considerable advantages have led to it being increasingly used as the model of choice to screen for P-gp efflux liability.

c. Brain microvessel endothelial cells. The delivery of therapeutic agents into the CNS poses a particularly difficult problem because transport of compounds across the very formidable barrier formed by the specialized endothelial cells lining the capillaries that perfuse the brain, the BBB is not facile (367,368). The BBB is a blood-tissue barrier within the CNS that regulates the transport of nutrients into the brain and limits exposure of the brain to toxic compounds via mechanisms such as P-gp. As is the case with the intestinal epithelium, P-gp plays an important role in limiting the transport of drugs across the BBB (213,369). Because the primary pharmacological targets of many drugs are receptors within the CNS and many of these drugs have been shown to be substrates for P-gp in other organs and in various in vitro systems, investigation of the processes surrounding the transport of compounds across the BBB, and specifically, susceptibility of compounds to P-gp-mediated efflux in the BBB remains an important area of research.

A frequently used in vitro model to study drug behavior at the BBB is cultured brain microvessel endothelial cells (BMECs), a primary culture that forms confluent monolayers 9 to 12 days after initial seeding (370). These cultured cells have been shown to retain many morphological and biochemical properties of their in vivo counterparts, including distinguishable luminal and abluminal membrane domains that are functionally and biochemically distinct (371–381). One of the major advantages of BMECs is that these cells can be grown on collagen-coated or fibronectin-treated polycarbonate membranes, and thus this system can be used to study transport across the monolayer by various mechanisms (i.e., passive diffusion, transcytosis, endocytosis, inwardly directed carrier proteins, polarized efflux, and uptake in both luminal and abluminal directions) (370). One limitation of the system is that the tight junctional complexes of BMECs are not as developed as those seen in vivo, and thus the contribution of paracellular permeability to the overall permeability of a compound is much greater in this in vitro system than what would be seen for a compound crossing the BBB in vivo (382). The comparable leakiness of the system can also make it difficult to quantify differences in transport that may be mediated by transporter activity.

Both functional assays [vincristine transport (381) and rhodamine 123 transport (383)] and biochemical assays involving immunohistochemical analysis (381,384) have confirmed the expression of P-gp in the luminal membrane of BMECs cultured on polycarbonate membranes. Additionally, immunohistochemical methods showed the expression of P-gp in BMEC to be constant and at a high level in five- to seven-day-old old primary cultures (384). Like many other barrier-forming cells, BMECs appear to express other efflux proteins, for example, RT-PCR and immunoblot analysis have shown the presence of MRP1 in rat BMECs (385,386). Functional evidence has also been presented to confirm the expression of MRP1 in BMECs (387).

BMECs have been used to study various aspects of the P-gp-mediated efflux of compounds from the endothelial cells that comprise the BBB. Several examples have demonstrated the usefulness of this system to study polarized efflux via P-gp. For example, the influence of P-gp expressed in brain capillary endothelial cells on the transport of cyclosporin A (388,389), vincristine (381), protease inhibitors (amprenavir, saquinavir, and indinavir) (245,390), rhodamine 123 (211,383), opioid peptides (211,391,392), and the β-blocking agent bunitrolol (393) have all been determined using this system.

2. Experimental Methods Used with Tissue Culture Transport Models to Study P-gp Efflux

The use of appropriate experimental design can provide definitive evidence that P-gp-mediated efflux is altering the transport of a compound and can provide further mechanistic information regarding the transport of a compound. Recently it has been appreciated that P-gp efflux can be a potential source for drug interactions and in vitro experimentation can be very helpful to understand potential liability. The techniques described in this section can be used with any tissue culture transport model.

a. Transport studies used to understand P-gp efflux. Transport across cell monolayers can be easily determined using a bicameral system, such as the Transwell® system, in which the compartments are separated by the polarized cell monolayer (attached to a porous filter support). The AP domain interacts with one compartment and the BL domain interacts with the other. The flux of the test compound can be measured in absorptive (AP to BL) or secretory (BL to AP) direction, and from the flux the permeability can be determined. One of the most significant advantages of this experimental system is that the appearance rather than the disappearance of the compound can be easily quantified to yield a permeability value. The most direct way to positively identify substrates for P-gp-mediated efflux activity in polarized epithelium is to measure permeability (in either transport direction) in the presence of a specific P-gp inhibitor such as GW918 (314) or antibodies such as MRK16 (48,365). Comparison of the permeability values provides a true measure of how P-gp affects the transport of the substrate across polarized epithelium and correctly identifies if the transport is subject to P-gp-mediated efflux activity (vs. some efflux mediated by another transporter). This experimental format allows an assessment of how significantly P-gp efflux attenuates or enhances absorptive versus secretory transport, respectively (394). The absorptive quotient (AQ) and secretory quotient (SQ) are metrics that have been created to quantify P-gp's effects on absorptive and secretory transport, respectively (394).

Another well-established metric used to identify P-gp substrates is the efflux ratio, in which secretory permeability is compared with absorptive permeability. An efflux ratio greater than one can imply apically directed transport polarity, suggesting that the compound is a substrate for efflux transport (395). It is important to note that apically directed transport as determined by efflux ratio does not provide unambiguous evidence that P-gp is responsible for the efflux of the compound (transporters other than or in combination with P-gp may be responsible for transport polarity). For these reasons, the wild-type cells that show low to insignificant transporter activity, corresponding to transfected cell systems such as MDR1-MDCK and MDR1-LLC-PK1, are often used to generate these efflux ratios with higher confidence of correctly assessing P-gp efflux liability. Although the efflux ratio can, under the proper experimental construct, be useful to identify

P-gp substrates, it does not quantify the functional activity of P-gp and furthermore cannot be used to understand how P-gp affects absorptive or secretory transport (206,394,396). For example, although digoxin and rhodamine 123 have similar efflux ratios, P-gp affects these compounds in much different ways; P-gp efflux affects digoxin absorption, but not rhodamine 123 absorption, and affects rhodamine 123 secretion greatly but digoxin secretion modestly (396). Using the unidirectional approach of studying transport under normal conditions and in the presence of a P-gp inhibitor and subsequently quantifying the effects of P-gp efflux via AQ and SQ clearly elucidated this difference in digoxin and rhodamine 123 absorptive versus secretory transport (394,396).

There is one major caveat of using the tissue culture transport experiment to study P-gp efflux that cannot be overlooked—P-gp efflux is not directly determined in this experiment. Rather, the effects of P-gp-mediated efflux activity and changes to this activity are inferred from the resulting overall transport data. Particularly with regards to substrate identification, there is the potential for false negatives. For a compound to be affected by P-gp-mediated efflux, it must reach P-gp's binding site that is within the cell. Compounds with poor membrane (transcellular) permeability are not likely to be identified as substrates (395,397). Conversely, compounds with very high passive membrane permeability can saturate P-gp efflux at low micromolar concentrations and are often not identified as substrates (206,395,397). The tissue culture transport study is a powerful tool, but the reasons listed above make it an absolute necessity to incorporate proper controls while performing and making conclusions from these studies.

b. Methods used to understand DDI potential. Increasingly, efforts are being made to quantify the inhibitory potency of new molecular entities against P-gp-mediated efflux using interaction studies performed in vitro. In particular, several efforts have specifically focused on determination of inhibitory potency against P-gp efflux of digoxin, a substrate with a narrow therapeutic window with kinetics known to be determined in part by P-gp (398,399). The transport study using a probe substrate such as digoxin, verapamil, or taxol can be conducted in the presence of a test compound over a series of concentrations to determine the inhibitory potency of the test compound (199,201,359,398,399). A comparison of this inhibitory potency to expected systemic concentrations can provide some insight into potential interactions that may be seen following coadministration of the compounds of interest.

Fluorescent dyes such as calcein-AM and rhodamine derivatives have been demonstrated to be P-gp substrates (400–407). These compounds can be used in any competition assay in which the test compound is added with these dyes. Any reduction in the dye efflux would be indicative of the inhibitory properties of the test compounds toward P-gp. Both rhodamine 123 and calcein-AM have been used in high-throughput assays, including the NCI assay, to screen large numbers of compounds as inhibitors of P-gp in several cell types. Calcein-AM itself is a weakly fluorescent molecule. When the acetoxymethyl ester group is cleaved by intracellular esterases, the fluorescent intensity of the metabolite calcein increases significantly (401,404,406). The amount of P-gp inhibition can be directly correlated with the amount of intracellular fluorescence. This is because calcein-AM is transported via P-gp, and thus the efflux pump attenuates its intracellular accumulation, unless it is inhibited by another P-gp inhibitor. However, calcein-AM is not significantly transported by P-gp because of the negative charge and subsequent lack of binding to membranes; thus it accumulates in the cytoplasm when formed by hydrolysis of intracellular calcein-AM (401,404,406). These probe substrates are also applicable to tissue culture transport studies. Rhodamine 123 has been used in conjunction with cell monolayers grown on polycarbonate membranes to detect the presence of P-gp in the AP cell membrane and to assess its inhibition by a variety of compounds in a competition style assay (405,407).

It is important to note that P-gp inhibition by a compound for the efflux of any of these ligands does not directly correlate with the ability of P-gp to efflux the compound of interest (177). Such is the case with paclitaxel, which is considered to be an excellent P-gp substrate but a poor inhibitor as determined by the dye-efflux method. The converse is seen with progesterone, which is a good inhibitor of P-gp-mediated efflux and yet is a poor substrate. It is important to note that P-gp inhibition can occur in several ways—competitively, noncompetitively, and via inhibition of ATP hydrolysis at the Walker A and B motifs (271). Furthermore, the false negatives due to poor permeability noted for transport assays can also produce false negatives in these interaction assays.

3. Other In Vitro Models

a. Membrane vesicles. Membrane vesicles are typically formed from intact cells and require some skill for their preparation. Given this relative limitation, the use of membrane vesicles as a rapid screen for P-gp efflux activity has not been extensive and has proven a better tool for studying the microscopic aspects of P-gp-mediated efflux.

Rat liver canalicular membrane vesicles (CMV) have been used to examine the mechanisms of uptake of P-gp substrates such as daunomycin, daunorubicin, and vinblastine, whose biliary excretion is extensive (47,137, 408,409). Early work with plasma membrane vesicles, partially purified from MDR human KB carcinoma cells that accumulated [^{3}H]vinblastine in an ATP-dependent manner, definitively showed how P-gp can act to efflux substrates from cancer cells (410). Additionally, these vesicles have been used to study microscopic aspects of P-gp-mediated efflux, such as the relationship of P-gp function to the membrane fluidity (137).

Brush border membrane vesicles (BBMV), prepared from rat intestine, were used to elucidate the function of P-gp in this organ and to show that the subcellular distribution of P-gp is localized to the AP membrane (411). The differences in P-gp-mediated efflux seen in the ileum, jejunum, and duodenum of rat intestine were studied by preparing BBMV from each of these distinct regions and then determining the Michaelis-Menten parameters, K_m and V_{max}, associated with the P-gp-mediated efflux of several substrates and inhibitors and the corresponding ATPase activity associated with efflux (412). Renal BBMV have been used to show P-gp's actions on its substrates in the kidney (413).

Membrane vesicles, prepared from CHO cells, have been used to determine the kinetic parameters associated with P-gp efflux (97,98). Factors such as the ATP hydrolysis rate associated with transport of various substrates have been studied along with the Michaelis-Menten parameters of efflux for various substrates.

b. Isolated intestinal segments and everted gut sacs. In the intestinal segment study, the intestine is removed and either mounted in a diffusion apparatus (Ussing chamber) or everted to make an everted sac (234,414–416). Factors affecting the transport of drugs (i.e., metabolism and efflux) can be studied by determining the fate of the test compound as it crosses the intestinal epithelium.

The transport characteristics of verapamil were determined for each region of the intestine as well as the colon with this model system. The duodenum and jejunum showed the most P-gp activity followed by lower activity in the colon and, surprisingly, none in the ileum (416). Polarized transport of quinidine due to P-gp efflux was demonstrated by using intestinal segments mounted in Ussing chambers (414). Further studies using everted sacs showed that P-gp inhibition by quinidine caused an altered drug absorption of digoxin and explained the interaction seen with coadministration of these agents (234). Metabolism and P-gp-mediated efflux of the macrolide antibiotic tacrolimus were studied in perfusion studies and in everted sacs (415). It was shown that inhibiting P-gp with miconazole (a P-gp inhibitor) greatly increased the amount of tacrolimus in the tissue (415). The results of these experiments provided evidence that P-gp is active in limiting tissue exposure to drugs and also that the intestinal metabolism of certain compounds can be significant.

c. Expression systems. The availability of full-length cDNA for functional mammalian MDR genes has made it possible to evaluate protein structure and structure-activity relationship and determine substrate-binding affinity through the in vitro P-gp expression system. Presently, MDR1 gene has been successfully expressed in *Escherichia coli* (417,418), in Sf9 cells using a recombinant baculovirus (120,123), in *Xenopus* oocytes (419), and in yeast (121,420,421). P-gp, expressed in these in vitro systems, is thought to function normally (analogous to function seen in in vivo systems) even though the former lacks glycosylation at N-terminal. Despite the normal functional activity of P-gp, researchers found it difficult to use P-gp expressed in *E. coli* for functional assay because many drugs cannot penetrate the cell walls. To solve this problem, Beja and Bibi developed a method to express P-gp in ''leaky'' *E. coli* cells (417). The results of these assays may be significantly different than those obtained in studies performed with mammalian cells due to differences that exist between bacteria, the insect cells, and mammalian cells.

P-gp associated ATPase is vanadate sensitive. A membrane product prepared from baculovirus infected insect cells containing this activity is now commercially available from Gentest Corp. (Woburn, Massachusetts, U.S.). Substrates of P-gp, such as verapamil, have been demonstrated to stimulate this vanadate-sensitive membrane ATPase (123). By determination of inorganic phosphate liberated in the reaction containing a P-gp preparation and a test compound, in the presence and absence of vanadate, one can determine if the test compound is a substrate/inhibitor of P-gp (123,422). Any compound that binds to P-gp would stimulate the magnesium-dependent ATPase, and thus, this method cannot distinguish between a substrate and inhibitor of P-gp.

B. In Situ and In Vivo Models

Whereas in vitro models are the tool of choice to identify P-gp substrates and to specifically study molecular aspects of P-gp-mediated efflux activity, extrapolation of these data to predict relevance in vivo can sometimes be difficult. Indeed, P-gp-mediated efflux activity is often one of a multitude of parameters that ultimately combine to confer substrate disposition; these exact relationships between key parameters are complex and remain to be resolved. For these reasons, models with greater complexity, more specifically those in which more key factors are present such as in situ and in vivo models, are essential to gain insight into the overall relevance of P-gp efflux for substrate disposition. Furthermore, the complexity underlying drug interactions involving P-gp and CYP3A4 are beginning to be appreciated, and in vivo models are being increasingly employed to study these interactions in combination with in vitro–derived inhibitory potency data (423). The following section summarizes the respective strengths and weaknesses of in situ and in vivo models currently used to study P-gp efflux.

1. In Situ Studies and Models

Some efforts have been made to determine the effect P-gp has on its substrates by use of in situ perfusion methods, including intestinal perfusion, liver perfusion, kidney perfusion, and brain perfusion. These experiments allow the researcher to study the transport of compounds in a physiologically relevant environment in which the integrity of the organ is preserved with regards to cell polarity and representation of all cell types seen in the organ. Furthermore, the reduction in complexity of in situ models versus in vivo studies facilitates the conduct of complex studies and allows more definitive conclusions to be made regarding the role P-gp may play in disposition.

In situ intestinal perfusion studies are typically done with live animals in which a perfusion loop has been inserted into the intestine (233,424). Depending on the experimental protocol, the system can offer a relatively unbiased view of intestinal transport with respect to normal expression of transporters in healthy animals. One limitation of this protocol is that the disappearance rather than the appearance of a compound is often determined (appearance can be determined by collection of blood in the vessels perfusing the section of intestine studied, a process requiring significant surgical skill). Estimates of the polarity of transport imparted by P-gp are difficult to assess and typically can only be determined by using an inhibitor or antibody to P-gp. Often the animal is anesthetized, and the anesthetizing agent can further affect the results (altered membrane fluidity, possible inhibitory effects on P-gp-mediated efflux activity) (187). There are some other obvious limitations. Using the intact intestine adds more levels of complexity that can further confound studies meant to elucidate the role of transporters, which act on the cellular level. However, this complexity can be a strength to the role P-gp plays in concert with other key factors that influence absorption and can be studied in parallel. It is possible that results will differ for intestinal region and also due to the presence of Peyer's patches that have different physiological roles from enterocytes (414,416). Furthermore, these studies suffer from an interspecies variability (rats are typically the test subjects). Despite certain disadvantages, if these studies are conducted with appropriate controls involving known P-gp substrates, it can provide valuable insights on how to correlate the effect of P-gp observed in cellular transport studies to that expressed in the absorption of drugs in vivo.

By measuring the intestinal absorption from small intestine of rat in situ, Saitoh et al. studied the differences between the oral bioavailabilities of methylprednisolone, prednisolone, and hydrocortisone, three structurally related glucocorticoids (233). Compared with prednisolone and hydrocortisone, methylprednisolone absorption was significantly retarded in jejunum and ileum by an intestinal efflux system. In the presence of verapamil and quinidine, the attenuation in the absorption of methylprednisolone was reversed, suggesting that P-gp is responsible for the unique features of methylprednisolone absorption. This study provides a good example of the usefulness of an intestinal perfusion experiment in further determining the regional differences in intestinal drug absorption modulated by P-gp that would otherwise be difficult to deduce in experiments performed with cell culture models or performed with whole animal systems.

The isolated perfused rat liver has been extensively used because of the minimal surgical manipulation needed due to its size and because the organ is less than 25 g, the perfusate used can be hemoglobin-free while ensuring adequate oxygen delivery at the flow rates used in these experiments (425). The isolated perfused liver system provides an excellent model for studying the hepatobiliary disposition of compounds without confounding influences that may be seen in vivo, such as influences on hepatic metabolism and additional metabolism or excretion by other organs of clearance (270,425). The isolated perfused rat liver can be used to study biochemical regulation of hepatic metabolism, synthetic function of liver, and mechanism of bile formation and secretion (270).

This model has provided important results regarding the influence of MDR modulators on hepatobiliary disposition of chemotherapeutic agents (426,427).

The effects of the P-gp inhibitor, GF120918, on the hepatobiliary disposition (biliary excretion) of doxorubicin were determined using a perfused rat liver system (270). Biliary excretion is the rate-limiting process for doxorubicin elimination. In the presence of GF120918, the biliary excretion of doxorubicin and its major metabolite, doxorubicinol, was decreased significantly without alterations in doxorubicin perfusate concentrations or doxorubicin and doxorubicinol liver concentrations. In a similar study on the hepatic elimination of other P-gp substrates, including vincristine and daunorubicin, it was reported that canalicular P-gp plays a significant role in the biliary secretion of these compounds (428,429).

Because of the kidney's involvement in the excretion of hydrophilic compounds and because most of the substrates of P-gp are hydrophobic compounds that are likely to be cleared mainly by biliary excretion or intestinal secretion, comparably fewer studies have been performed with the isolated perfused kidney. The isolated perfused rat kidney model was used to demonstrate that digoxin is actively secreted by P-gp located on the luminal membrane of renal tubular epithelial cells and that clinically important interactions with quinidine and verapamil are caused by the inhibition of P-gp activity in the kidney (332). These results provide an excellent example of how the isolated perfused kidney model can be used to definitively conclude that P-gp-mediated efflux is involved in the renal excretion of a compound and also to elucidate possible DDIs that might arise in the kidney following coadministration of P-gp substrates/inhibitors.

The brain perfusion system has been used to study the disposition of several compounds across a functionally intact BBB, which has been shown to possess nearly identical structural and functional features as those seen in the BBB in vivo, including the presence of multiple tight junctional complexes between cells and P-gp (43,430–432). This in situ technique involves stopping the heart and perfusing the brain via the carotid artery at a flow rate that does not alter the integrity of the BBB (432,433). The brain capillary endothelium, the choroid plexus epithelium, and the arachnoid membrane, which comprise the functional BBB in vivo, are all present in this technique and this provides a major advantage over in vitro models used to study the BBB (e.g., BMEC). One major advantage this technique has over an in vivo experiment involves the perfusion fluid used in the experiment. The composition of the solution can be controlled with respect to test compounds, plasma proteins, nutrients, and metabolic cofactors (432). However, the use of a perfusate solution can also be a disadvantage as it may not be possible to provide all the necessary nutrients or metabolic cofactors that would be present in vivo and, thus, may lead to incorrect conclusions (430). The major disadvantages of the model with respect to in vitro models include the lack of control of the extracellular fluid concentration for studies of drug efflux from the brain and a greater complexity that the brain matrix provides. As with other perfusion systems, this technique requires anesthesia and thereby may act to confound results.

Some of the more notable applications of this in situ model system in the study of CNS drug disposition have involved the determination of drug permeability across the BBB, drug uptake kinetics, transport mechanisms (uptake and efflux), elucidation of the CNS metabolic pathways (the drug has no access to peripheral metabolism), and the effects of plasma protein binding (430,434, 435). This model has been used to study the effects of P-gp-mediated efflux in the BBB on antibacterial agents (436), colchicines (437,438), and vinblastine (438), to evaluate a prodrug strategy for increasing doxorubicin uptake into the brain (439) and to determine the brain uptake parameter, log PS (permeability surface area product), a value that more accurately determines rate of brain uptake (435). The system has also been used to determine the effects of P-gp modulators such as verapamil (440) and PSC833 (441) on the BBB transport of P-gp substrates. Recently, the system has been adopted and validated for use in the gene knockout mdr1a(−/−) mice, and results obtained from this model compared with those from experiments performed in wild-type mice can be used to gauge the overall effect of P-gp-mediated efflux on the transport of P-gp substrates across the BBB (244).

These in situ techniques can be powerful tools to gauge the actual extent of P-gp efflux that can be expected in vivo. There are confounding factors that must be addressed when interpreting data obtained from these studies, and as with all biological models, the appropriate controls must be used to ensure that the observed effect appears to be due to P-gp-mediated efflux activity.

2. In Vivo Models

The major advantages to in vivo models are that they provide a method to understand relevance on an organism level and that these models have been used successfully to predict outcomes in humans. The obvious disadvantages of these models are their limitations with regards to study designs and sampling, reduced ability to deconvolute complex processes, and the need for animal experimentation. For that reason, the in vivo model is a tool more suitable for aiding the understanding of the ramifications of P-gp efflux liability for gross disposition processes. A great deal of understanding around how P-gp affects disposition has come from in vivo models. Below is a brief summary of key models and findings.

Schinkel et al. have generated mice with disruption of individual mdr1a, mdr1b, or mdr2 genes and furthermore, they have generated a double knockout in which both mdr1a and mdr1b are disrupted (12,36,44,212–216,442,443). In mice, mdr1a and mdr1b genes encode two separate P-gp proteins that are analogous to MDR1 gene product expressed in humans (12). The mdr1a RNA is found abundantly in the brain, intestine, liver, and testis (444), while mdr1b RNA is usually associated with the adrenal cortex, placenta, ovaries, and uterus (445). Both gene products are expressed in the kidney, heart, lung, thymus, and spleen (12,444). The relative sequence identity of the human P-gp with the mouse mdr1a P-gp is 82% (227,446,447). The greatest homology of the two proteins is seen in ATP-binding regions, the second, fourth, and eleventh TM domains, and the first and second intracytoplasmic loops in each half of the molecule (31,227,448). The proteins show the least homology in the first extracellular loop, the connecting region between the homologous halves, and at both terminal ends (31,227,448). It was concluded that mdr1 P-gp has no essential physiological function, since no gross disturbance in corticosteroid metabolism during pregnancy and in bile formation was observed in mdr1a (−/−) mice. However, lack of mdr1 P-gp significantly altered the disposition profile of P-gp substrates. In P-gp gene knockout mice, the absorption was increased, the elimination was decreased, and the concentration of certain substrates in key organs, such as the brain, testes, and heart, was increased dramatically (12). Although the mouse mdr1a P-gp is not totally homologous to the human P-gp, mice that are dominant negative for the mdr1a gene continue to provide an excellent in vivo tool to probe the effects of P-gp on the ADME of drugs that is extensively used in industry and academia (11,210,248,449).

A transgenic mouse model involving MDR1 has been used to study the function of P-gp. A transgenic system was developed to express human MDR1 gene in the marrow of mice leading to bone marrow that is resistant to the cytotoxic effect of anticancer drugs, which are substrates of P-gp (450–453). When exposed to anticancer agents, the transgenic mice showed normal peripheral white blood cell counts implying that the MDR1 P-gp protects the marrow (451). When the efflux activity of the MDR1 P-gp expressed in these mice was inhibited with other P-gp substrates or MRK16, an antibody to an external epitope of P-gp, the mice became sensitized to cytotoxic drug therapy that manifested in a drop in the white blood counts. This model has seen widespread use to evaluate safety of chemotherapeutic agents. However, this and other transgenic models have not been widely employed in the evaluation of the effects of P-gp on drug pharmacokinetics.

C. In Vivo/In Vitro Correlations

In vitro models have provided invaluable information about properties of compounds that affect their in vivo transport and absorption. Regardless of how closely in vitro systems model in vivo conditions, they cannot completely represent what may be seen in vivo by virtue of their reduced nature. For that reason, it is important to consider that a focused endpoint generated using an in vitro model will only correlate to a much more complex parameter like absorption when that endpoint is a major determinant of the complex parameter. The lack of in vitro/in vivo correlation does not necessarily implicate a failure of the model, but rather that the endpoint may not be sufficient to describe the in vivo process. Furthermore, the in vivo data used for these correlations are rarely precise or granular enough to gauge differences that may be related to P-gp efflux. Pharmacokinetic parameters commonly used for in vivo correlations like $C_{\max}$, AUC, and systemic and oral clearance are gross parameters that are determined by a multitude of factors and typically only describe the central compartment. For any number of reasons above, attempts to elucidate a quantitative in vivo/in vitro correlation for P-gp efflux have been difficult and have had limited success. However, recent efforts to generate qualitative understandings have shown some utility.

Despite our inability to predict quantitatively the influence P-gp may have on the in vivo transport of substrates in normal tissues with respect to other processes, in vitro experiments remain the best means of demonstrating that a compound is a substrate for polarized efflux. Nearly all experiments designed to study the extent of P-gp efflux of test compounds in vivo require adequate in vitro data to support the hypothesis (48,217,226,454). In vitro studies on P-gp substrates such as vinblastine, paclitaxel, cyclosporin A, talinolol, acebutolol, and digoxin have provided a good indication of the effect of P-gp on the in vivo pharmacokinetic behavior of these compounds. These studies show that results from the in vitro studies provide a qualitative estimate of the influence of P-gp on its in vivo pharmacokinetic behavior. Findings such as these give confidence that results from in vitro experiments can be extrapolated to explain modulation of drug disposition by P-gp efflux.

Recently, classification systems have been proposed that give further refinement to the understanding of the potential role of P-gp efflux in vivo. Substrate transport across polarized epithelium can utilize various routes, and P-gp efflux does not affect each in the same manner. A system has been proposed that uses a metric created to quantify the functional activity of P-gp (absorptive and secretory quotients) coupled with substrate transport pathway across the cell in order to give further clarity regarding the mechanism of P-gp efflux that may be seen during various disposition processes (394). A system has been proposed that utilizes the BCS to predict how and when transporter activity may be important for disposition (339). In brief, it has been proposed that the importance of efflux and influx transport can be correlated with the BCS class to predict the extent to which the transport activity will affect disposition. Particularly for class II compounds, where permeability is high but solubility is low, efflux is predicted to play a role in disposition. Additionally, this system has been used to predict potential DDI resulting from transporter and/or metabolism inhibition. As for disposition, class II BCS compounds that are dual efflux and metabolism substrates are predicted to have the greatest potential for significant DDI. These qualitative relationships highlight the advances that have been made in understanding efflux and its effects on disposition and, furthermore, show how knowledge of disposition and mechanisms can be used to gain ability to predict possible outcomes in vivo.

VI. COMMENTARY ON FEDERAL DRUG ADMINISTRATION GUIDANCE ON THE USE OF IN VITRO MODELS TO DETERMINE P-gp-RELATED IN VIVO DDI POTENTIAL OF DISCOVERY/DEVELOPMENT CANDIDATES

Having developed guidelines for using in vitro metabolism studies to assess drug metabolism (specifically cytochrome P450)-mediated potential DDIs, the Food and Drug Administration (FDA) has initiated an effort to develop similar guidelines for transporter-mediated DDIs (http://www.fda.gov/cder/guidance/index.htm). In what appears to be an initial attempt at developing a more comprehensive guideline, the FDA has chosen to focus on P-gp-mediated DDIs. One must assume that this reflects the availability of extensive literature and industry data on the role of P-gp in such interactions rather than greater importance of this transporter over others in causing DDIs.

For details of the guideline and decision trees developed by the FDA, the reader is referred to the Web site http://www.fda.gov/cder/guidance/index.htm. The salient features of the guideline to identify P-gp substrates for in vivo drug interactions studies include (1) examining bidirectional transport of the test compounds in Caco-2 or MDR-MDCK cell monolayers, (2) selecting likely substrates based on an efflux ratio of >2, (3) confirming P-gp substrate activity by showing that specific inhibitors of P-gp decrease the efflux ratio, and (4) selecting compounds for in vivo P-gp-related interactions studies if transport of a test compound with an efflux ratio of >2 in these test systems is inhibited by P-gp inhibitors. The salient features of the guideline to identify P-gp inhibitors for in vivo drug interactions include (1) examining bidirectional transport of P-gp probe substrates across Caco-2 or MDR-MDCK cell monolayers, (2) determining the ability of the test compound to reduce the magnitude of the efflux ratio of P-gp probe substrates, (3) determining K_i or $[I]/IC_{50}$ of the test compounds, and (4) selecting compounds with K_i or $IC_{50}/[I] < 0.1$ for in vivo drug interaction studies.

While it is prudent to start with relatively simple and limited guidelines, such simplicity also pose significant risk by oversimplifying the real behavior of test compounds and arriving at misleading or false conclusions about the potential of compounds to cause in vivo drug interactions. First and foremost, the guideline implies that P-gp is much more important than other transporters in causing drug interactions, clearly this has not been established by definitive studies. Specifically, the proposed experimental scheme and decision trees will likely lead to too many unnecessary clinical drug interaction studies for the simple reason that disposition of compounds with efflux ratio of ~ 2 is not going to be significantly affected across a P-gp competent epithelial or endothelial tissue; this is an unrealistically wide, ''catch-all'' guidance rather than selecting potent P-gp substrates likely to have serious drug interactions. The guidance for identifying P-gp inhibitors is equally unrealistic and also quite ambiguous. For example, it is extremely difficult, if not impossible, to achieve $IC_{50}/[I]$ of <0.1 in the cellular test systems proposed. Further, it is not clear what the [I] represents—e.g., in vivo plasma concentrations, intraintestinal concentrations! Also, it is not clear whether these test systems are designed to assess in vivo drug interactions at the absorption site or at the bile canalicular site or at the BBB site. The in vitro models, experimental design, and screening parameters are expected to be significantly different to predict P-gp-related drug interactions at each of these sites. Finally, the guidance for identifying P-gp inducers is even more premature than the guidance for identifying P-gp substrates and inhibitors that might cause drug interactions. Clearly, much work needs to be done before a comprehensive and meaningful guidance can be developed for conducting in vitro studies to identify test compounds that should be tested for transporter-related in vivo drug interactions.

VII. CONCLUSIONS

Originally discovered as an adaptive response of cancer cells that are exposed to high concentrations of toxic drugs, P-gp is now recognized as a widely distributed constitutive protein that plays a pivotal role in the systemic disposition of a wide variety of hormones, drugs, and other xenobiotics. Furthermore, recent investigations have uncovered a large family of efflux proteins, with diverse and overlapping substrate specificities, which play critical roles in the disposition of therapeutic agents. The scope of the biochemical, cellular, physiological, and clinical implications of these proteins is just beginning to be recognized. An exhaustive review of this vast and complex area of emerging research is beyond the scope of this chapter. Furthermore, an exhaustive review of the research specifically focusing on P-gp would be prohibitive. Instead, we have focused on P-gp efflux with a bias toward its role in drug disposition. The studies presented here have demonstrated the dual role played by P-gp in minimizing the systemic and tissue/organ exposure to foreign agents—it acts as a biochemical barrier in preventing the entry (absorption) of drugs across epithelial or endothelial tissues, and it provides a driving force for excretion of drugs and metabolites by mediating their active secretion into the excretory organs. By virtue of its presence in epithelial and endothelial cells, P-gp can also play a decisive role in the tissue and organ distribution of a drug. The most notable example of this is the role played by P-gp (as a component of the BBB) in attenuating the access of drugs to brain tissues. P-gp, when colocalized with metabolic enzymes in certain tissues (e.g., CYP3A in intestinal epithelium), can modulate the metabolic transformation of some drugs markedly, both at the cellular and tissue/organ level. The importance of P-gp efflux as a potential mechanism underlying DDIs has been recently appreciated. Hence, in designing drugs with optimal pharmacokinetic profiles, it is imperative that the role of P-gp (and other efflux proteins) in the ADME of the drug candidates is elucidated. It is equally important to recognize that other factors—e.g., coadministered drug(s), diet, disease—can significantly affect the disposition of a given therapeutic agent by modulating the activity of P-gp (and other efflux proteins), resulting in serious incidents of therapeutic failure or unexpected toxicity. Although much progress has been made, it is still difficult to predict how and when P-gp efflux may affect disposition and how this may change in the context of the factors listed above. Elucidation of these relationships is a critical goal certain to advance our knowledge and predictive ability. However, the complexity underlying these relationships is likely to require technological advancements and a multidisciplinary approach to solve. Investigation of P-gp and other efflux proteins promises to be a very fertile area of research in the years to come across a wide array of scientific disciplines.

REFERENCES

1. Juliano RL, Ling V. A surface glycoprotein modulating drug permeability in Chinese hamster ovary cell mutants. Biochim Biophys Acta 1976; 455(1):152–162.
2. Kartner N, Riordan JR, Ling V. Cell surface P-glycoprotein associated with multidrug resistance in mammalian cell lines. Science 1983; 221(4617):1285–1288.
3. Thiebaut F, Tsuruo T, Hamada H, et al. Cellular localization of the multidrug-resistance gene product P-glycoprotein in normal human tissues. Proc Natl Acad Sci U S A 1987; 84(21):7735–7738.
4. Thiebaut F, Tsuruo T, Hamada H, et al. Immunohistochemical localization in normal tissues of different epitopes in the multidrug transport protein P170: evidence for localization in brain capillaries and crossreactivity of one antibody with a muscle protein. J Histochem Cytochem 1989; 37(2):159–164.
5. Gramatte T, Oertel R, Terhaag B, et al. Direct demonstration of small intestinal secretion and site-dependent absorption of the beta-blocker talinolol in humans. Clin Pharmacol Ther 1996; 59(5):541–549.
6. Hunter J, Hirst BH, Simmons NL. Epithelial secretion of vinblastine by human intestinal adenocarcinoma cell (HCT-8 and T84) layers expressing P-glycoprotein. Br J Cancer 1991; 64(3):437–444.
7. Hunter J, Hirst BH, Simmons NL. Drug absorption limited by P-glycoprotein-mediated secretory drug transport in human intestinal epithelial Caco-2 cell layers. Pharm Res 1993; 10(5):743–749.
8. Ince P, Elliott K, Appleton DR, et al. Modulation by verapamil of vincristine pharmacokinetics and sensitivity to metaphase arrest of the normal rat colon in organ culture. Biochem Pharmacol 1991; 41(8):1217–1225.
9. Meyers MB, Scotto KW, Sirotnak FM. P-glycoprotein content and mediation of vincristine efflux: correlation with the level of differentiation in luminal epithelium of mouse small intestine. Cancer Commun 1991; 3(5):159–165.
10. Wetterich U, Spahn-Langguth H, Mutschler E, et al. Evidence for intestinal secretion as an additional clearance pathway of talinolol enantiomers: concentration- and dose-dependent absorption in vitro and in vivo. Pharm Res 1996; 13(4):514–522.
11. Chen C, Liu X, Smith BJ. Utility of Mdr1-gene deficient mice in assessing the impact of P-glycoprotein on pharmacokinetics and pharmacodynamics in drug discovery and development. Curr Drug Metab 2003; 4(4):272–291.
12. Schinkel AH, Smit JJ, van Tellingen O, et al. Disruption of the mouse mdr1a P-glycoprotein gene leads to a deficiency in the blood-brain barrier and to increased sensitivity to drugs. Cell 1994; 77(4):491–502.

13. Chandra P, Brouwer KL. The complexities of hepatic drug transport: current knowledge and emerging concepts. Pharm Res 2004; 21(5):719–735.
14. Sparreboom A, Planting AS, Jewell RC, et al. Clinical pharmacokinetics of doxorubicin in combination with GF120918, a potent inhibitor of MDR1 P-glycoprotein. Anticancer Drugs 1999; 10(8):719–728.
15. Cummins CL, Jacobsen W, Benet LZ. Unmasking the dynamic interplay between intestinal P-glycoprotein and CYP3A4. J Pharmacol Exp Ther 2002; 300(3): 1036–1045.
16. Gan LS, Moseley MA, Khosla B, et al. CYP3A-like cytochrome P450-mediated metabolism and polarized efflux of cyclosporin A in Caco-2 cells. Drug Metab Dispos 1996; 24(3):344–349.
17. Pan L, Ho Q, Tsutsui K, et al. Comparison of chromatographic and spectroscopic methods used to rank compounds for aqueous solubility. J Pharm Sci 2001; 90(4):521–529.
18. Schmiedlin-Ren P, Thummel KE, Fisher JM, et al. Expression of enzymatically active CYP3A4 by Caco-2 cells grown on extracellular matrix-coated permeable supports in the presence of 1alpha,25-dihydroxyvitamin D3. Mol Pharmacol 1997; 51(5):741–754.
19. Fromm MF, Busse D, Kroemer HK, et al. Differential induction of prehepatic and hepatic metabolism of verapamil by rifampin. Hepatology 1996; 24(4):796–801.
20. Johne A, Brockmoller J, Bauer S, et al. Pharmacokinetic interaction of digoxin with an herbal extract from St John's wort (Hypericum perforatum). Clin Pharmacol Ther 1999; 66(4):338–345.
21. Juranka PF, Zastawny RL, Ling V. P-glycoprotein: multidrug-resistance and a superfamily of membrane-associated transport proteins. FASEB J 1989; 3(14):2583–2592.
22. Ng WF, Sarangi F, Zastawny RL, et al. Identification of members of the P-glycoprotein multigene family. Mol Cell Biol 1989; 9(3):1224–1232.
23. Chen CJ, Chin JE, Ueda K, et al. Internal duplication and homology with bacterial transport proteins in the mdr1 (P-glycoprotein) gene from multidrug-resistant human cells. Cell 1986; 47(3):381–389.
24. Fardel O, Lecureur V, Guillouzo A. The P-glycoprotein multidrug transporter. Gen Pharmacol 1996; 27(8):1283–1291.
25. Smith AJ, Timmermans-Hereijgers JL, Roelofsen B, et al. The human MDR3 P-glycoprotein promotes translocation of phosphatidylcholine through the plasma membrane of fibroblasts from transgenic mice. FEBS Lett 1994; 354 (3):263–266.
26. van der Bliek AM, Kooiman PM, Schneider C, et al. Sequence of mdr3 cDNA encoding a human P-glycoprotein. Gene 1988; 71(2):401–411.
27. Devault A, Gros P. Two members of the mouse mdr gene family confer multidrug resistance with overlapping but distinct drug specificities. Mol Cell Biol 1990; 10 (4):1652–1663.
28. Gros P, Croop J, Housman D. Mammalian multidrug resistance gene: complete cDNA sequence indicates strong homology to bacterial transport proteins. Cell 1986; 47(3):371–380.
29. Gros P, Raymond M, Bell J, et al. Cloning and characterization of a second member of the mouse mdr gene family. Mol Cell Biol 1988; 8(7):2770–2778.
30. Deuchars KL, Duthie M, Ling V. Identification of distinct P-glycoprotein gene sequences in rat. Biochim Biophys Acta 1992; 1130(2):157–165.

31. Devine SE, Hussain A, Davide JP, et al. Full length and alternatively spliced pgp1 transcripts in multidrug-resistant Chinese hamster lung cells. J Biol Chem 1991; 266 (7):4545–4555.
32. Childs S, Ling V. Duplication and evolution of the P-glycoprotein genes in pig. Biochim Biophys Acta 1996; 1307(2):205–212.
33. Schurr E, Raymond M, Bell JC, et al. Characterization of the multidrug resistance protein expressed in cell clones stably transfected with the mouse mdr1 cDNA. Cancer Res 1989; 49(10):2729–2733.
34. Ueda K, Cardarelli C, Gottesman MM, et al. Expression of a full-length cDNA for the human ''MDR1'' gene confers resistance to colchicine, doxorubicin, and vinblastine. Proc Natl Acad Sci U S A 1987; 84(9):3004–3008.
35. Buschman E, Gros P. The inability of the mouse mdr2 gene to confer multidrug resistance is linked to reduced drug binding to the protein. Cancer Res 1994; 54 (18):4892–4898.
36. Oude Elferink RP, Ottenhoff R, van Wijland M, et al. Regulation of biliary lipid secretion by mdr2 P-glycoprotein in the mouse. J Clin Invest 1995; 95(1):31–38.
37. Fojo AT, Ueda K, Slamon DJ, et al. Expression of a multidrug-resistance gene in human tumors and tissues. Proc Natl Acad Sci U S A 1987; 84(1):265–269.
38. Cordon-Cardo C, O'Brien JP, Boccia J, et al. Expression of the multidrug resistance gene product (P-glycoprotein) in human normal and tumor tissues. J Histochem Cytochem 1990; 38(9):1277–1287.
39. Hamada H, Tsuruo T. Functional role for the 170- to 180-kDa glycoprotein specific to drug-resistant tumor cells as revealed by monoclonal antibodies. Proc Natl Acad Sci U S A 1986; 83(20):7785–7789.
40. Sugawara I, Kataoka I, Morishita Y, et al. Tissue distribution of P-glycoprotein encoded by a multidrug-resistant gene as revealed by a monoclonal antibody, MRK 16. Cancer Res 1988; 48(7):1926–1929.
41. Sugawara I, Nakahama M, Hamada H, et al. Apparent stronger expression in the human adrenal cortex than in the human adrenal medulla of Mr 170,000–180,000 P-glycoprotein. Cancer Res 1988; 48(16):4611–4614.
42. Lieberman DM, Reithmeier RA, Ling V, et al. Identification of P-glycoprotein in renal brush border membranes. Biochem Biophys Res Commun 1989; 162(1): 244–252.
43. Cordon-Cardo C, O'Brien JP, Casals D, et al. Multidrug-resistance gene (P-glycoprotein) is expressed by endothelial cells at blood-brain barrier sites. Proc Natl Acad Sci U S A 1989; 86(2):695–698.
44. Borst P, Schinkel AH. What have we learnt thus far from mice with disrupted P-glycoprotein genes? Eur J Cancer 1996; 32A(6):985–990.
45. Borst P, Schinkel AH, Smit JJ, et al. Classical and novel forms of multidrug resistance and the physiological functions of P-glycoproteins in mammals. Pharmacol Ther 1993; 60(2):289–299.
46. Lum BL, Gosland MP. MDR expression in normal tissues. Pharmacologic implications for the clinical use of P-glycoprotein inhibitors. Hematol Oncol Clin North Am 1995; 9(2):319–336.
47. Kamimoto Y, Gatmaitan Z, Hsu J, et al. The function of Gp170, the multidrug resistance gene product, in rat liver canalicular membrane vesicles. J Biol Chem 1989; 264(20):11693–11698.

48. Terao T, Hisanaga E, Sai Y, et al. Active secretion of drugs from the small intestinal epithelium in rats by P-glycoprotein functioning as an absorption barrier. J Pharm Pharmacol 1996; 48(10):1083–1089.
49. Ueda K, Okamura N, Hirai M, et al. Human P-glycoprotein transports cortisol, aldosterone, and dexamethasone, but not progesterone. J Biol Chem 1992; 267 (34):24248–24252.
50. Augustijns PF, Bradshaw TP, Gan LS, et al. Evidence for a polarized efflux system in CACO-2 cells capable of modulating cyclosporin A transport. Biochem Biophys Res Commun 1993; 197(2):360–365.
51. Pourtier-Manzanedo A, Boesch D, Loor F. FK-506 (fujimycin) reverses the multidrug resistance of tumor cells in vitro. Anticancer Drugs 1991; 2(3):279–283.
52. Miwa A, Ueda K, Okada Y. Protein kinase C-independent correlation between P-glycoprotein expression and volume sensitivity of Cl-channel. J Membr Biol 1997; 157(1):63–69.
53. Valverde MA, Bond TD, Hardy SP, et al. The multidrug resistance P-glycoprotein modulates cell regulatory volume decrease. EMBO J 1996; 15(17):4460–4468.
54. Borst P, Evers R, Kool M, et al. The multidrug resistance protein family. Biochim Biophys Acta 1999; 1461(2):347–357.
55. Borst P, Evers R, Kool M, et al. A family of drug transporters: the multidrug resistance-associated proteins. J Natl Cancer Inst 2000; 92(16):1295–1302.
56. Borst P, Kool M, Evers R. Do cMOAT (MRP2), other MRP homologues, and LRP play a role in MDR? Semin Cancer Biol 1997; 8(3):205–213.
57. Borst P, Zelcer N, van de Wetering K, et al. On the putative co-transport of drugs by multidrug resistance proteins. FEBS Lett 2006; 580(4):1085–1093.
58. Kool M, de Haas M, Scheffer GL, et al. Analysis of expression of cMOAT (MRP2), MRP3, MRP4, and MRP5, homologues of the multidrug resistance-associated protein gene (MRP1), in human cancer cell lines. Cancer Res 1997; 57(16):3537–3547.
59. Kool M, van der Linden M, de Haas M, et al. Expression of human MRP6, a homologue of the multidrug resistance protein gene MRP1, in tissues and cancer cells. Cancer Res 1999; 59(1):175–182.
60. Kool M, van der Linden M, de Haas M, et al. MRP3, an organic anion transporter able to transport anti-cancer drugs. Proc Natl Acad Sci U S A 1999; 96(12): 6914–6919.
61. Kusuhara H, Suzuki H, Sugiyama Y. The role of P-glycoprotein and canalicular multispecific organic anion transporter in the hepatobiliary excretion of drugs. J Pharm Sci 1998; 87(9):1025–1040.
62. Flens MJ, Zaman GJ, van der Valk P, et al. Tissue distribution of the multidrug resistance protein. Am J Pathol 1996; 148(4):1237–1247.
63. Stride BD, Valdimarsson G, Gerlach JH, et al. Structure and expression of the messenger RNA encoding the murine multidrug resistance protein, an ATP-binding cassette transporter. Mol Pharmacol 1996; 49(6):962–971.
64. Zaman GJ, Versantvoort CH, Smit JJ, et al. Analysis of the expression of MRP, the gene for a new putative transmembrane drug transporter, in human multidrug resistant lung cancer cell lines. Cancer Res 1993; 53(8):1747–1750.
65. Evers R, Zaman GJ, van Deemter L, et al. Basolateral localization and export activity of the human multidrug resistance-associated protein in polarized pig kidney cells. J Clin Invest 1996; 97(5):1211–1218.

66. Gerk PM, Vore M. Regulation of expression of the multidrug resistance-associated protein 2 (MRP2) and its role in drug disposition. J Pharmacol Exp Ther 2002; 302 (2):407–415.
67. Ito K, Suzuki H, Hirohashi T, et al. Molecular cloning of canalicular multispecific organic anion transporter defective in EHBR. Am J Physiol 1997; 272(1 pt 1): G16–G22.
68. Xiong H, Turner KC, Ward ES, et al. Altered hepatobiliary disposition of acetaminophen glucuronide in isolated perfused livers from multidrug resistance-associated protein 2-deficient TR(-) rats. J Pharmacol Exp Ther 2000; 295(2):512–518.
69. Masuda M, I'Izuka Y, Yamazaki M, et al. Methotrexate is excreted into the bile by canalicular multispecific organic anion transporter in rats. Cancer Res 1997; 57 (16):3506–3510.
70. Paulusma CC, Bosma PJ, Zaman GJ, et al. Congenital jaundice in rats with a mutation in a multidrug resistance-associated protein gene. Science 1996; 271 (5252):1126–1128.
71. Iyanagi T, Emi Y, Ikushiro S. Biochemical and molecular aspects of genetic disorders of bilirubin metabolism. Biochim Biophys Acta 1998; 1407(3):173–184.
72. Jansen PL, Peters WH, Lamers WH. Hereditary chronic conjugated hyperbilirubinemia in mutant rats caused by defective hepatic anion transport. Hepatology 1985; 5(4):573–579.
73. Furst DE. Practical clinical pharmacology and drug interactions of low-dose methotrexate therapy in rheumatoid arthritis. Br J Rheumatol 1995; 34(suppl 2):20–25.
74. Hagmann W, Schubert J, Konig J, et al. Reconstitution of transport-active multidrug resistance protein 2 (MRP2; ABCC2) in proteoliposomes. Biol Chem 2002; 383 (6):1001–1009.
75. Leveque D, Wisniewski S, Renault C, et al. The effect of rifampin on the pharmacokinetics of vinorelbine in the micropig. Anticancer Res 2003; 23(3B): 2741–2744.
76. Yamazaki M, Li B, Louie SW, et al. Effects of fibrates on human organic anion-transporting polypeptide 1B1-, multidrug resistance protein 2- and P-glycoprotein-mediated transport. Xenobiotica 2005; 35(7):737–753.
77. Allikmets R, Schriml LM, Hutchinson A, et al. A human placenta-specific ATP-binding cassette gene (ABCP) on chromosome 4q22 that is involved in multidrug resistance. Cancer Res 1998; 58(23):5337–5339.
78. Bates SE, Robey R, Miyake K, et al. The role of half-transporters in multidrug resistance. J Bioenerg Biomembr 2001; 33(6):503–511.
79. Doyle LA, Yang W, Abruzzo LV, et al. A multidrug resistance transporter from human MCF-7 breast cancer cells. Proc Natl Acad Sci U S A 1998; 95(26): 15665–15670.
80. Litman T, Brangi M, Hudson E, et al. The multidrug-resistant phenotype associated with overexpression of the new ABC half-transporter, MXR (ABCG2). J Cell Sci 2000; 113(pt 11):2011–2021.
81. Miyake K, Mickley L, Litman T, et al. Molecular cloning of cDNAs which are highly overexpressed in mitoxantrone-resistant cells: demonstration of homology to ABC transport genes. Cancer Res 1999; 59(1):8–13.
82. Maliepaard M, Scheffer GL, Faneyte IF, et al. Subcellular localization and distribution of the breast cancer resistance protein transporter in normal human tissues. Cancer Res 2001; 61(8):3458–3464.

83. Gottesman MM, Fojo T, Bates SE. Multidrug resistance in cancer: role of ATP-dependent transporters. Nat Rev Cancer 2002; 2(1):48–58.
84. Xia CQ, Yang JJ, Gan LS. Breast cancer resistance protein in pharmacokinetics and drug-drug interactions. Expert Opin Drug Metab Toxicol 2005; 1(4):595–611.
85. Kage K, Tsukahara S, Sugiyama T, et al. Dominant-negative inhibition of breast cancer resistance protein as drug efflux pump through the inhibition of S-S dependent homodimerization. Int J Cancer 2002; 97(5):626–630.
86. Xu J, Liu Y, Yang Y, et al. Characterization of oligomeric human half-ABC transporter ATP-binding cassette G2. J Biol Chem 2004; 279(19):19781–19789.
87. Zhou S, Schuetz JD, Bunting KD, et al. The ABC transporter Bcrpl/ABCG2 is expressed in a wide variety of stem cells and is a molecular determinant of the side-population phenotype. Nat Med 2001; 7(9):1028–1034.
88. Young AM, Allen CE, Audus KL. Efflux transporters of the human placenta. Adv Drug Deliv Rev 2003; 55(1):125–132.
89. Jonker JW, Merino G, Musters S, et al. The breast cancer resistance protein BCRP (ABCG2) concentrates drugs and carcinogenic xenotoxins into milk. Nat Med 2005; 11(2):127–129.
90. Merino G, Jonker JW, Wagenaar E, et al. The breast cancer resistance protein (BCRP/ABCG2) affects pharmacokinetics, hepatobiliary excretion, and milk secretion of the antibiotic nitrofurantoin. Mol Pharmacol 2005; 67(5):1758–1764.
91. Geier A, Wagner M, Dietrich CG, et al. Principles of hepatic organic anion transporter regulation during cholestasis, inflammation and liver regeneration. Biochim Biophys Acta 2007; 1773(3):283–308.
92. Muller M, Jansen PL. Molecular aspects of hepatobiliary transport. Am J Physiol 1997; 272(6 pt 1):G1285–1303.
93. Torok M, Gutmann H, Fricker G, et al. Sister of P-glycoprotein expression in different tissues. Biochem Pharmacol 1999; 57(7):833–835.
94. Gerloff T, Stieger B, Hagenbuch B, et al. The sister of P-glycoprotein represents the canalicular bile salt export pump of mammalian liver. J Biol Chem 1998; 273 (16):10046–10050.
95. Lecureur V, Sun D, Hargrove P, et al. Cloning and expression of murine sister of P-glycoprotein reveals a more discriminating transporter than MDR1/P-glycoprotein. Mol Pharmacol 2000; 57(1):24–35.
96. Etievant C, Schambel P, Guminski Y, et al. Requirements for P-glycoprotein recognition based on structure-activity relationships in the podophyllotoxin series. Anticancer Drug Des 1998; 13(4):317–336.
97. Litman T, Zeuthen T, Skovsgaard T, et al. Structure-activity relationships of P-glycoprotein interacting drugs: kinetic characterization of their effects on ATPase activity. Biochim Biophys Acta 1997; 1361(2):159–168.
98. Litman T, Zeuthen T, Skovsgaard T, et al. Competitive, non-competitive and cooperative interactions between substrates of P-glycoprotein as measured by its ATPase activity. Biochim Biophys Acta 1997; 1361(2):169–176.
99. Scala S, Akhmed N, Rao US, et al. P-glycoprotein substrates and antagonists cluster into two distinct groups. Mol Pharmacol 1997; 51(6):1024–1033.
100. Seelig A. A general pattern for substrate recognition by P-glycoprotein. Eur J Biochem 1998; 251(1-2):252–261.

101. Ferte J. Analysis of the tangled relationships between P-glycoprotein-mediated multidrug resistance and the lipid phase of the cell membrane. Eur J Biochem 2000; 267(2):277–294.
102. Shitara Y, Horie T, Sugiyama Y. Transporters as a determinant of drug clearance and tissue distribution. Eur J Pharm Sci 2006; 27(5):425–446.
103. McDevitt CA, Callaghan R. How can we best use structural information on P-glycoprotein to design inhibitors? Pharmacol Ther 2007; 113(2):429–441.
104. Gottesman MM, Pastan I. Biochemistry of multidrug resistance mediated by the multidrug transporter. Annu Rev Biochem 1993; 62:385–427.
105. Gottesman MM, Pastan I. Modulation of the Multidrug Resistance Phenotype. Cellular Pharmacology 1993; 1 (suppl 1):S111–S112.
106. Higgins CF. ABC transporters: from microorganisms to man. Annu Rev Cell Biol 1992; 8:67–113.
107. Sharom FJ. The P-glycoprotein efflux pump: how does it transport drugs? J Membr Biol 1997; 160(3):161–175.
108. Sharom FJ. The P-glycoprotein multidrug transporter: interactions with membrane lipids, and their modulation of activity. Biochem Soc Trans 1997; 25(3):1088–1096.
109. Stein WD. Kinetics of the P-glycoprotein, the multidrug transporter. Exp Physiol 1998; 83(2):221–232.
110. Loo TW, Clarke DM. Recent progress in understanding the mechanism of P-glycoprotein-mediated drug efflux. J Membr Biol 2005; 206(3):173–185.
111. Leveille-Webster CR, Arias IM. The biology of the P-glycoproteins. J Membr Biol 1995; 143(2):89–102.
112. Currier SJ, Ueda K, Willingham MC, et al. Deletion and insertion mutants of the multidrug transporter. J Biol Chem 1989; 264(24):14376–14381.
113. Loo TW, Clarke DM. Reconstitution of drug-stimulated ATPase activity following coexpression of each half of human P-glycoprotein as separate polypeptides. J Biol Chem 1994; 269(10):7750–7755.
114. Endicott JA, Ling V. The biochemistry of P-glycoprotein-mediated multidrug resistance. Annu Rev Biochem 1989; 58:137–171.
115. Germann UA. P-glycoprotein–a mediator of multidrug resistance in tumour cells. Eur J Cancer 1996; 32A(6):927–944.
116. Greenberger LM, Lothstein L, Williams SS, et al. Distinct P-glycoprotein precursors are overproduced in independently isolated drug-resistant cell lines. Proc Natl Acad Sci U S A 1988; 85(11):3762–3766.
117. Richert ND, Aldwin L, Nitecki D, et al. Stability and covalent modification of P-glycoprotein in multidrug-resistant KB cells. Biochemistry 1988; 27(20):7607–7613.
118. Beck WT, Cirtain MC. Continued expression of vinca alkaloid resistance by CCRF-CEM cells after treatment with tunicamycin or pronase. Cancer Res 1982; 42 (1):184–189.
119. Buschman E, Arceci RJ, Croop JM, et al. mdr2 encodes P-glycoprotein expressed in the bile canalicular membrane as determined by isoform-specific antibodies. J Biol Chem 1992; 267(25):18093–18099.
120. Germann UA, Willingham MC, Pastan I, et al. Expression of the human multidrug transporter in insect cells by a recombinant baculovirus. Biochemistry 1990; 29 (9):2295–2303.
121. Kuchler K, Thorner J. Functional expression of human mdr1 in the yeast Saccharomyces cerevisiae. Proc Natl Acad Sci U S A 1992; 89(6):2302–2306.

122. Ling V, Kartner N, Sudo T, et al. Multidrug-resistance phenotype in Chinese hamster ovary cells. Cancer Treat Rep 1983; 67(10):869–874.
123. Sarkadi B, Price EM, Boucher RC, et al. Expression of the human multidrug resistance cDNA in insect cells generates a high activity drug-stimulated membrane ATPase. J Biol Chem 1992; 267(7):4854–4858.
124. Schinkel AH, Kemp S, Dolle M, et al. N-glycosylation and deletion mutants of the human MDR1 P-glycoprotein. J Biol Chem 1993; 268(10):7474–7481.
125. Chambers TC, Germann UA, Gottesman MM, et al. Bacterial expression of the linker region of human MDR1 P-glycoprotein and mutational analysis of phosphorylation sites. Biochemistry 1995; 34(43):14156–14162.
126. Orr GA, Han EK, Browne PC, et al. Identification of the major phosphorylation domain of murine mdr1b P-glycoprotein. Analysis of the protein kinase A and protein kinase C phosphorylation sites. J Biol Chem 1993; 268(33):25054–25062.
127. Scala S, Dickstein B, Regis J, et al. Bryostatin 1 affects P-glycoprotein phosphorylation but not function in multidrug-resistant human breast cancer cells. Clin Cancer Res 1995; 1(12):1581–1587.
128. Poruchynsky MS, Ling V. Detection of oligomeric and monomeric forms of P-glycoprotein in multidrug resistant cells. Biochemistry 1994; 33(14):4163–4174.
129. Homolya L, Hollo Z, Germann UA, et al. Fluorescent cellular indicators are extruded by the multidrug resistance protein. J Biol Chem 1993; 268(29):21493–21496.
130. Boyum R, Guidotti G. Effect of ATP binding cassette/multidrug resistance proteins on ATP efflux of Saccharomyces cerevisiae. Biochem Biophys Res Commun 1997; 230(1):22–26.
131. Higgins CF. P-glycoprotein and cell volume-activated chloride channels. J Bioenerg Biomembr 1995; 27(1):63–70.
132. Tominaga M, Tominaga T, Miwa A, et al. Volume-sensitive chloride channel activity does not depend on endogenous P-glycoprotein. J Biol Chem 1995; 270 (46):27887–27893.
133. Altenberg GA, Vanoye CG, Horton JK, et al. Unidirectional fluxes of rhodamine 123 in multidrug-resistant cells: evidence against direct drug extrusion from the plasma membrane. Proc Natl Acad Sci U S A 1994; 91(11):4654–4657.
134. Chen Y, Pant AC, Simon SM. P-glycoprotein does not reduce substrate concentration from the extracellular leaflet of the plasma membrane in living cells. Cancer Res 2001; 61(21):7763–7769.
135. Higgins CF, Gottesman MM. Is the multidrug transporter a flippase? Trends Biochem Sci 1992; 17(1):18–21.
136. van Helvoort A, Smith AJ, Sprong H, et al. MDR1 P-glycoprotein is a lipid translocase of broad specificity, while MDR3 P-glycoprotein specifically translocates phosphatidylcholine. Cell 1996; 87(3):507–517.
137. Sinicrope FA, Dudeja PK, Bissonnette BM, et al. Modulation of P-glycoprotein-mediated drug transport by alterations in lipid fluidity of rat liver canalicular membrane vesicles. J Biol Chem 1992; 267(35):24995–25002.
138. Bruggemann EP, Currier SJ, Gottesman MM, et al. Characterization of the azidopine and vinblastine binding site of P-glycoprotein. J Biol Chem 1992; 267 (29):21020–21026.

139. Loo TW, Clarke DM. Functional consequences of phenylalanine mutations in the predicted transmembrane domain of P-glycoprotein. J Biol Chem 1993; 268 (27):19965–19972.
140. Loo TW, Clarke DM. Mutations to amino acids located in predicted transmembrane segment 6 (TM6) modulate the activity and substrate specificity of human P-glycoprotein. Biochemistry 1994; 33(47):14049–14057.
141. Pawagi AB, Wang J, Silverman M, et al. Transmembrane aromatic amino acid distribution in P-glycoprotein. A functional role in broad substrate specificity. J Mol Biol 1994; 235(2):554–564.
142. Ferry DR, Malkhandi PJ, Russell MA, et al. Allosteric regulation of [^{3}H]vinblastine binding to P-glycoprotein of MCF-7 ADR cells by dexniguldipine. Biochem Pharmacol 1995; 49(12):1851–1861.
143. Ferry DR, Russell MA, Cullen MH. P-glycoprotein possesses a 1,4-dihydropyridine-selective drug acceptor site which is alloserically coupled to a vinca-alkaloid-selective binding site. Biochem Biophys Res Commun 1992; 188(1):440–445.
144. Tamai I, Safa AR. Azidopine noncompetitively interacts with vinblastine and cyclosporin A binding to P-glycoprotein in multidrug resistant cells. J Biol Chem 1991; 266(25):16796–16800.
145. Ayesh S, Shao YM, Stein WD. Co-operative, competitive and non-competitive interactions between modulators of P-glycoprotein. Biochim Biophys Acta 1996; 1316 (1):8–18.
146. Malkhandi J, Ferry DR, Boer R, et al. Dexniguldipine-HCl is a potent allosteric inhibitor of [3H]vinblastine binding to P-glycoprotein of CCRF ADR 5000 cells. Eur J Pharmacol 1994; 288(1):105–114.
147. Martin C, Berridge G, Higgins CF, et al. The multi-drug resistance reversal agent SR33557 and modulation of vinca alkaloid binding to P-glycoprotein by an allosteric interaction. Br J Pharmacol 1997; 122(4):765–771.
148. Pascaud C, Garrigos M, Orlowski S. Multidrug resistance transporter P-glycoprotein has distinct but interacting binding sites for cytotoxic drugs and reversing agents. Biochem J 1998; 333(pt 2):351–358.
149. Shapiro AB, Ling V. Positively cooperative sites for drug transport by P-glycoprotein with distinct drug specificities. Eur J Biochem 1997; 250(1):130–137.
150. Martin C, Berridge G, Higgins CF, et al. Communication between multiple drug binding sites on P-glycoprotein. Mol Pharmacol 2000; 58(3):624–632.
151. Martin C, Berridge G, Mistry P, et al. Drug binding sites on P-glycoprotein are altered by ATP binding prior to nucleotide hydrolysis. Biochemistry 2000; 39 (39):11901–11906.
152. Loo TW, Clarke DM. Identification of residues within the drug-binding domain of the human multidrug resistance P-glycoprotein by cysteine-scanning mutagenesis and reaction with dibromobimane. J Biol Chem 2000; 275(50):39272–39278.
153. Hoffmeyer S, Burk O, von Richter O, et al. Functional polymorphisms of the human multidrug-resistance gene: multiple sequence variations and correlation of one allele with P-glycoprotein expression and activity in vivo. Proc Natl Acad Sci U S A 2000; 97(7):3473–3478.
154. Johne A, Kopke K, Gerloff T, et al. Modulation of steady-state kinetics of digoxin by haplotypes of the P-glycoprotein MDR1 gene. Clin Pharmacol Ther 2002; 72(5): 584–594.

155. Fredericks S, Moreton M, Reboux S, et al. Multidrug resistance gene-1 (MDR-1) haplotypes have a minor influence on tacrolimus dose requirements. Transplantation 2006; 82(5):705–708.
156. Macphee IA, Fredericks S, Tai T, et al. Tacrolimus pharmacogenetics: polymorphisms associated with expression of cytochrome p4503A5 and P-glycoprotein correlate with dose requirement. Transplantation 2002; 74(11):1486–1489.
157. Taubert D, von Beckerath N, Grimberg G, et al. Impact of P-glycoprotein on clopidogrel absorption. Clin Pharmacol Ther 2006; 80(5):486–501.
158. Roberts RL, Joyce PR, Mulder RT, et al. A common P-glycoprotein polymorphism is associated with nortriptyline-induced postural hypotension in patients treated for major depression. Pharmacogenomics J 2002; 2(3):191–196.
159. Siegsmund M, Brinkmann U, Schaffeler E, et al. Association of the P-glycoprotein transporter MDR1(C3435T) polymorphism with the susceptibility to renal epithelial tumors. J Am Soc Nephrol 2002; 13(7):1847–1854.
160. Drozdzik M, Bialecka M, Mysliwiec K, et al. Polymorphism in the P-glycoprotein drug transporter MDR1 gene: a possible link between environmental and genetic factors in Parkinson's disease. Pharmacogenetics 2003; 13(5):259–263.
161. Furuno T, Landi MT, Ceroni M, et al. Expression polymorphism of the blood-brain barrier component P-glycoprotein (MDR1) in relation to Parkinson's disease. Pharmacogenetics 2002; 12(7):529–534.
162. Lee CG, Tang K, Cheung YB, et al. MDR1, the blood-brain barrier transporter, is associated with Parkinson's disease in ethnic Chinese. J Med Genet 2004; 41(5):e60.
163. Sakaeda T, Nakamura T, Okumura K. Pharmacogenetics of MDR1 and its impact on the pharmacokinetics and pharmacodynamics of drugs. Pharmacogenomics 2003; 4(4):397–410.
164. Tan EK, Chan DK, Ng PW, et al. Effect of MDR1 haplotype on risk of Parkinson disease. Arch Neurol 2005; 62(3):460–464.
165. Onnie CM, Fisher SA, Pattni R, et al. Associations of allelic variants of the multidrug resistance gene (ABCB1 or MDR1) and inflammatory bowel disease and their effects on disease behavior: a case-control and meta-analysis study. Inflamm Bowel Dis 2006; 12(4):263–271.
166. Schwab M, Schaeffeler E, Marx C, et al. Association between the C3435T MDR1 gene polymorphism and susceptibility for ulcerative colitis. Gastroenterology 2003; 124(1):26–33.
167. Hung CC, Tai JJ, Lin CJ, et al. Complex haplotypic effects of the ABCB1 gene on epilepsy treatment response. Pharmacogenomics 2005; 6(4):411–417.
168. Fellay J, Marzolini C, Meaden ER, et al. Response to antiretroviral treatment in HIV-1-infected individuals with allelic variants of the multidrug resistance transporter 1: a pharmacogenetics study. Lancet 2002; 359(9300):30–36.
169. Cascorbi I, Gerloff T, Johne A, et al. Frequency of single nucleotide polymorphisms in the P-glycoprotein drug transporter MDR1 gene in white subjects. Clin Pharmacol Ther 2001; 69(3):169–174.
170. Ameyaw MM, Regateiro F, Li T, et al. MDR1 pharmacogenetics: frequency of the C3435T mutation in exon 26 is significantly influenced by ethnicity. Pharmacogenetics 2001; 11(3):217–221.
171. Schaeffeler E, Eichelbaum M, Brinkmann U, et al. Frequency of C3435T polymorphism of MDR1 gene in African people. Lancet 2001; 358(9279):383–384.

172. Drescher S, Schaeffeler E, Hitzl M, et al. MDR1 gene polymorphisms and disposition of the P-glycoprotein substrate fexofenadine. Br J Clin Pharmacol 2002; 53 (5):526–534.
173. Goto M, Masuda S, Saito H, et al. C3435T polymorphism in the MDR1 gene affects the enterocyte expression level of CYP3A4 rather than Pgp in recipients of living-donor liver transplantation. Pharmacogenetics 2002; 12(6):451–457.
174. Pauli-Magnus C, Feiner J, Brett C, et al. No effect of MDR1 C3435T variant on loperamide disposition and central nervous system effects. Clin Pharmacol Ther 2003; 74(5):487–498.
175. Siegmund W, Ludwig K, Giessmann T, et al. The effects of the human MDR1 genotype on the expression of duodenal P-glycoprotein and disposition of the probe drug talinolol. Clin Pharmacol Ther 2002; 72(5):572–583.
176. Gervasini G, Carrillo JA, Garcia M, et al. Adenosine triphosphate-binding cassette B1 (ABCB1) (multidrug resistance 1) G2677T/A gene polymorphism is associated with high risk of lung cancer. Cancer 2006; 107(12):2850–2857.
177. Eytan GD, Regev R, Oren G, et al. The role of passive transbilayer drug movement in multidrug resistance and its modulation. J Biol Chem 1996; 271(22):12897–12902.
178. Seydel JK, Coats EA, Cordes HP, et al. Drug membrane interaction and the importance for drug transport, distribution, accumulation, efficacy and resistance. Arch Pharm (Weinheim) 1994; 327(10):601–610.
179. Boer R, Ulrich WR, Haas S, et al. Interaction of cytostatics and chemosensitizers with the dexniguldipine binding site on P-glycoprotein. Eur J Pharmacol 1996; 295 (2-3):253–260.
180. Orlowski S, Mir LM, Belehradek J Jr., et al. Effects of steroids and verapamil on P-glycoprotein ATPase activity: progesterone, desoxycorticosterone, corticosterone and verapamil are mutually non-exclusive modulators. Biochem J 1996; 317: pt 2515–522.
181. Limbacher HP Jr., Blickenstaff GD, Bowen JH, et al. Multiequilibrium binding of a spin-labeled local anesthetic in phosphatidylcholine bilayers. Biochim Biophys Acta 1985; 812(1):268–276.
182. Avdeef A, Box KJ, Comer JE, et al. pH-metric logP 10. Determination of liposomal membrane-water partition coefficients of ionizable drugs. Pharm Res 1998; 15 (2):209–215.
183. Fernandez MS, Fromherz P. Lipoid pH indicators as probes of electrical potential and polarity in micelles. J Phys Chem 1977; 81:1755–1761.
184. Flewelling RF, Hubbell WL. The membrane dipole potential in a total membrane potential model. Applications to hydrophobic ion interactions with membranes. Biophys J 1986; 49(2):541–552.
185. Franklin JC, Cafiso DS, Flewelling RF, et al. Probes of membrane electrostatics: synthesis and voltage-dependent partitioning of negative hydrophobic ion spin labels in lipid vesicles. Biophys J 1993; 64(3):642–653.
186. Eytan GD, Regev R, Oren G, et al. Efficiency of P-glycoprotein-mediated exclusion of rhodamine dyes from multidrug-resistant cells is determined by their passive transmembrane movement rate. Eur J Biochem 1997; 248(1):104–112.
187. Regev R, Assaraf YG, Eytan GD. Membrane fluidization by ether, other anesthetics, and certain agents abolishes P-glycoprotein ATPase activity and modulates efflux from multidrug-resistant cells. Eur J Biochem 1999; 259(1-2):18–24.

188. Regev R, Eytan GD. Flip-flop of doxorubicin across erythrocyte and lipid membranes. Biochem Pharmacol 1997; 54(10):1151–1158.
189. Speelmans G, Staffhorst RW, de Kruijff B, et al. Transport studies of doxorubicin in model membranes indicate a difference in passive diffusion across and binding at the outer and inner leaflets of the plasma membrane. Biochemistry 1994; 33 (46):13761–13768.
190. Stein WD, Cardarelli C, Pastan I, et al. Kinetic evidence suggesting that the multidrug transporter differentially handles influx and efflux of its substrates. Mol Pharmacol 1994; 45(4):763–772.
191. Doige CA, Yu X, Sharom FJ. The effects of lipids and detergents on ATPase-active P-glycoprotein. Biochim Biophys Acta 1993; 1146(1):65–72.
192. Sharom FJ. Characterization and functional reconstitution of the multidrug transporter. J Bioenerg Biomembr 1995; 27(1):15–22.
193. Urbatsch IL, Sankaran B, Bhagat S, et al. Both P-glycoprotein nucleotide-binding sites are catalytically active. J Biol Chem 1995; 270(45):26956–26961.
194. Urbatsch IL, Sankaran B, Weber J, et al. P-glycoprotein is stably inhibited by vanadate-induced trapping of nucleotide at a single catalytic site. J Biol Chem 1995; 270(33):19383–19390.
195. Urbatsch IL, Senior AE. Effects of lipids on ATPase activity of purified Chinese hamster P-glycoprotein. Arch Biochem Biophys 1995; 316(1):135–140.
196. Callaghan R, Berridge G, Ferry DR, et al. The functional purification of P-glycoprotein is dependent on maintenance of a lipid-protein interface. Biochim Biophys Acta 1997; 1328(2):109–124.
197. Orlowski S, Selosse MA, Boudon C, et al. Effects of detergents on P-glycoprotein ATPase activity: differences in perturbations of basal and verapamil-dependent activities. Cancer Biochem Biophys 1998; 16(1-2):85–110.
198. Romsicki Y, Sharom FJ. The membrane lipid environment modulates drug interactions with the P-glycoprotein multidrug transporter. Biochemistry 1999; 38 (21):6887–6896.
199. Doppenschmitt S, Langguth P, Regardh CG, et al. Characterization of binding properties to human P-glycoprotein: development of a [3H]verapamil radioligand-binding assay. J Pharmacol Exp Ther 1999; 288(1):348–357.
200. Doppenschmitt S, Spahn-Langguth H, Regardh CG, et al. Role of P-glycoprotein-mediated secretion in absorptive drug permeability: an approach using passive membrane permeability and affinity to P-glycoprotein. J Pharm Sci 1999; 88 (10):1067–1072.
201. Gao J, Murase O, Schowen RL, et al. A functional assay for quantitation of the apparent affinities of ligands of P-glycoprotein in Caco-2 cells. Pharm Res 2001; 18 (2):171–176.
202. Ho NF, Burton PS, Conradi RA, et al. A biophysical model of passive and polarized active transport processes in Caco-2 cells: approaches to uncoupling apical and basolateral membrane events in the intact cell. J Pharm Sci 1995; 84(1):21–27.
203. Horio M, Chin KV, Currier SJ, et al. Transepithelial transport of drugs by the multidrug transporter in cultured Madin-Darby canine kidney cell epithelia. J Biol Chem 1989; 264(25):14880–14884.
204. Jang SH, Wientjes MG, Au JL. Kinetics of P-glycoprotein-mediated efflux of paclitaxel. J Pharmacol Exp Ther 2001; 298(3):1236–1242.

205. Michelson S, Slate D. A mathematical model of the P-glycoprotein pump as a mediator of multidrug resistance. Bull Math Biol 1992; 54(6):1023–1038.
206. Troutman MD, Thakker DR. Efflux ratio cannot assess P-glycoprotein-mediated attenuation of absorptive transport: asymmetric effect of P-glycoprotein on absorptive and secretory transport across Caco-2 cell monolayers. Pharm Res 2003; 20(8):1200–1209.
207. Bentz J, Tran TT, Polli JW, et al. The steady-state Michaelis-Menten analysis of P-glycoprotein mediated transport through a confluent cell monolayer cannot predict the correct Michaelis constant Km. Pharm Res 2005; 22(10):1667–1677.
208. Molinari A, Calcabrini A, Meschini S, et al. Subcellular detection and localization of the drug transporter P-glycoprotein in cultured tumor cells. Curr Protein Pept Sci 2002; 3(6):653–670.
209. Molinari A, Cianfriglia M, Meschini S, et al. P-glycoprotein expression in the Golgi apparatus of multidrug-resistant cells. Int J Cancer 1994; 59(6):789–795.
210. Chen C, Pollack GM. Altered disposition and antinociception of [D-penicillamine (2,5)] enkephalin in mdr1a-gene-deficient mice. J Pharmacol Exp Ther 1998; 287 (2):545–552.
211. Letrent SP, Polli JW, Humphreys JE, et al. P-glycoprotein-mediated transport of morphine in brain capillary endothelial cells. Biochem Pharmacol 1999; 58(6):951–957.
212. Schinkel AH. Pharmacological insights from P-glycoprotein knockout mice. Int J Clin Pharmacol Ther 1998; 36(1):9–13.
213. Schinkel AH. P-Glycoprotein, a gatekeeper in the blood-brain barrier. Adv Drug Deliv Rev 1999; 36(2-3):179–194.
214. Schinkel AH, Mol CA, Wagenaar E, et al. Multidrug resistance and the role of P-glycoprotein knockout mice. Eur J Cancer 1995; 31A(7-8):1295–1298.
215. Schinkel AH, Roelofs EM, Borst P. Characterization of the human MDR3 P-glycoprotein and its recognition by P-glycoprotein-specific monoclonal antibodies. Cancer Res 1991; 51(10):2628–2635.
216. Schinkel AH, Wagenaar E, van Deemter L, et al. Absence of the mdr1a P-Glycoprotein in mice affects tissue distribution and pharmacokinetics of dexamethasone, digoxin, and cyclosporin A. J Clin Invest 1995; 96(4):1698–1705.
217. Sparreboom A, van Asperen J, Mayer U, et al. Limited oral bioavailability and active epithelial excretion of paclitaxel (Taxol) caused by P-glycoprotein in the intestine. Proc Natl Acad Sci U S A 1997; 94(5):2031–2035.
218. van Asperen J, Schinkel AH, Beijnen JH, et al. Altered pharmacokinetics of vinblastine in Mdr1a P-glycoprotein-deficient mice. J Natl Cancer Inst 1996; 88 (14):994–999.
219. van Asperen J, van Tellingen O, Beijnen JH. The role of mdr1a P-glycoprotein in the biliary and intestinal secretion of doxorubicin and vinblastine in mice. Drug Metab Dispos 2000; 28(3):264–267.
220. Washington CB, Wiltshire HR, Man M, et al. The disposition of saquinavir in normal and P-glycoprotein deficient mice, rats, and in cultured cells. Drug Metab Dispos 2000; 28(9):1058–1062.
221. Arimori K, Nakano M. Drug exsorption from blood into the gastrointestinal tract. Pharm Res 1998; 15(3):371–376.
222. Aszalos A, Ross DD. Biochemical and clinical aspects of efflux pump related resistance to anti-cancer drugs. Anticancer Res 1998; 18(4C):2937–2944.

223. Ferry DR. Testing the role of P-glycoprotein expression in clinical trials: applying pharmacological principles and best methods for detection together with good clinical trials methodology. Int J Clin Pharmacol Ther 1998; 36(1):29–40.
224. Hebert MF. Contributions of hepatic and intestinal metabolism and P-glycoprotein to cyclosporine and tacrolimus oral drug delivery. Adv Drug Deliv Rev 1997; 27(2-3):201–214.
225. Ling V. Multidrug resistance: molecular mechanisms and clinical relevance. Cancer Chemother Pharmacol 1997; 40(suppl):S3–S8.
226. Spahn-Langguth H, Baktir G, Radschuweit A, et al. P-glycoprotein transporters and the gastrointestinal tract: evaluation of the potential in vivo relevance of in vitro data employing talinolol as model compound. Int J Clin Pharmacol Ther 1998; 36 (1):16–24.
227. van de Vrie W, Marquet RL, Stoter G, et al. In vivo model systems in P-glycoprotein-mediated multidrug resistance. Crit Rev Clin Lab Sci 1998; 35(1):1–57.
228. Wang Q, Yang H, Miller DW, et al. Effect of the p-glycoprotein inhibitor, cyclosporin A, on the distribution of rhodamine-123 to the brain: an in vivo microdialysis study in freely moving rats. Biochem Biophys Res Commun 1995; 211(3):719–726.
229. Beaumont K, Harper A, Smith DA, et al. The role of P-glycoprotein in determining the oral absorption and clearance of the NK2 antagonist, UK-224,671. Eur J Pharm Sci 2000; 12(1):41–50.
230. Chiou WL, Chung SM, Wu TC, et al. A comprehensive account on the role of efflux transporters in the gastrointestinal absorption of 13 commonly used substrate drugs in humans. Int J Clin Pharmacol Ther 2001; 39(3):93–101.
231. Fromm MF. P-glycoprotein: a defense mechanism limiting oral bioavailability and CNS accumulation of drugs. Int J Clin Pharmacol Ther 2000; 38(2):69–74.
232. Relling MV. Are the major effects of P-glycoprotein modulators due to altered pharmacokinetics of anticancer drugs? Ther Drug Monit 1996; 18(4):350–356.
233. Saitoh H, Hatakeyama M, Eguchi O, et al. Involvement of intestinal P-glycoprotein in the restricted absorption of methylprednisolone from rat small intestine. J Pharm Sci 1998; 87(1):73–75.
234. Su SF, Huang JD. Inhibition of the intestinal digoxin absorption and exsorption by quinidine. Drug Metab Dispos 1996; 24(2):142–147.
235. Cvetkovic M, Leake B, Fromm MF, et al. OATP and P-glycoprotein transporters mediate the cellular uptake and excretion of fexofenadine. Drug Metab Dispos 1999; 27(8):866–871.
236. Gramatte T, Oertel R. Intestinal secretion of intravenous talinolol is inhibited by luminal R-verapamil. Clin Pharmacol Ther 1999; 66(3):239–245.
237. Lennernas H, Regardh CG. Dose-dependent intestinal absorption and significant intestinal excretion (exsorption) of the beta-blocker pafenolol in the rat. Pharm Res 1993; 10(5):727–731.
238. Riddell JG, Shanks RG, Brogden RN. Celiprolol. A preliminary review of its pharmacodynamic and pharmacokinetic properties and its therapeutic use in hypertension and angina pectoris. Drugs 1987; 34(4):438–458.
239. Angelin B, Arvidsson A, Dahlqvist R, et al. Quinidine reduces biliary clearance of digoxin in man. Eur J Clin Invest 1987; 17(3):262–265.
240. de Lannoy IA, Silverman M. The MDR1 gene product, P-glycoprotein, mediates the transport of the cardiac glycoside, digoxin. Biochem Biophys Res Commun 1992; 189(1):551–557.

241. Doering W, König E. [The influence of quinidine on serum digoxin concentrations (author's transl)]. Med Klin 1978; 73(30):1085–1088.
242. Leahey EB Jr., Reiffel JA, Drusin RE, et al. Interaction between quinidine and digoxin. JAMA 1978; 240(6):533–534.
243. Greiner B, Eichelbaum M, Fritz P, et al. The role of intestinal P-glycoprotein in the interaction of digoxin and rifampin. J Clin Invest 1999; 104(2):147–153.
244. Dagenais C, Rousselle C, Pollack GM, et al. Development of an in situ mouse brain perfusion model and its application to mdr1a P-glycoprotein-deficient mice. J Cereb Blood Flow Metab 2000; 20(2):381–386.
245. Polli JW, Jarrett JL, Studenberg SD, et al. Role of P-glycoprotein on the CNS disposition of amprenavir (141W94), an HIV protease inhibitor. Pharm Res 1999; 16(8):1206–1212.
246. Dagenais C, Graff CL, Pollack GM. Variable modulation of opioid brain uptake by P-glycoprotein in mice. Biochem Pharmacol 2004; 67(2):269–276.
247. Kalvass JC, Graff CL, Pollack GM. Use of loperamide as a phenotypic probe of mdr1a status in CF-1 mice. Pharm Res 2004; 21(10):1867–1870.
248. Doran A, Obach RS, Smith BJ, et al. The impact of P-glycoprotein on the disposition of drugs targeted for indications of the central nervous system: evaluation using the MDR1A/1B knockout mouse model. Drug Metab Dispos 2005; 33(1):165–174.
249. Cummins CL, Mangravite LM, Benet LZ. Characterizing the expression of CYP3A4 and efflux transporters (P-gp, MRP1, and MRP2) in CYP3A4-transfected Caco-2 cells after induction with sodium butyrate and the phorbol ester 12-O-tetradecanoylphorbol-13-acetate. Pharm Res 2001; 18(8):1102–1109.
250. Hall SD, Thummel KE, Watkins PB, et al. Molecular and physical mechanisms of first-pass extraction. Drug Metab Dispos 1999; 27(2):161–166.
251. Kolars JC, Lown KS, Schmiedlin-Ren P, et al. CYP3A gene expression in human gut epithelium. Pharmacogenetics 1994; 4(5):247–259.
252. Seree E, Villard PH, Hever A, et al. Modulation of MDR1 and CYP3A expression by dexamethasone: evidence for an inverse regulation in adrenals. Biochem Biophys Res Commun 1998; 252(2):392–395.
253. Wacher VJ, Wu CY, Benet LZ. Overlapping substrate specificities and tissue distribution of cytochrome P450 3A and P-glycoprotein: implications for drug delivery and activity in cancer chemotherapy. Mol Carcinog 1995; 13(3):129–134.
254. Paine MF, Khalighi M, Fisher JM, et al. Characterization of interintestinal and intraintestinal variations in human CYP3A-dependent metabolism. J Pharmacol Exp Ther 1997; 283(3):1552–1562.
255. Paine MF, Hart HL, Ludington SS, et al. The human intestinal cytochrome P450 "pie". Drug Metab Dispos 2006; 34(5):880–886.
256. Benet LZ, Izumi T, Zhang Y, et al. Intestinal MDR transport proteins and P-450 enzymes as barriers to oral drug delivery. J Control Release 1999; 62(1–2):25–31.
257. Fisher JM, Wrighton SA, Watkins PB, et al. First-pass midazolam metabolism catalyzed by 1alpha,25-dihydroxy vitamin D3-modified Caco-2 cell monolayers. J Pharmacol Exp Ther 1999; 289(2):1134–1142.
258. Wacher VJ, Silverman JA, Zhang Y, et al. Role of P-glycoprotein and cytochrome P450 3A in limiting oral absorption of peptides and peptidomimetics. J Pharm Sci 1998; 87(11):1322–1330.

259. Cummins CL, Salphati L, Reid MJ, et al. In vivo modulation of intestinal CYP3A metabolism by P-glycoprotein: studies using the rat single-pass intestinal perfusion model. J Pharmacol Exp Ther 2003; 305(1):306–314.
260. Kim RB, Wandel C, Leake B, et al. Interrelationship between substrates and inhibitors of human CYP3A and P-glycoprotein. Pharm Res 1999; 16(3):408–414.
261. Tahara H, Kusuhara H, Fuse E, et al. P-glycoprotein plays a major role in the efflux of fexofenadine in the small intestine and blood-brain barrier, but only a limited role in its biliary excretion. Drug Metab Dispos 2005; 33(7):963–968.
262. Watkins PB. The barrier function of CYP3A4 and P-glycoprotein in the small bowel. Adv Drug Deliv Rev 1997; 27(2-3):161–170.
263. Johnson BM, Charman WN, Porter CJ. Application of compartmental modeling to an examination of in vitro intestinal permeability data: assessing the impact of tissue uptake, P-glycoprotein, and CYP3A. Drug Metab Dispos 2003; 31(9):1151–1160.
264. Johnson BM, Chen W, Borchardt RT, et al. A kinetic evaluation of the absorption, efflux, and metabolism of verapamil in the autoperfused rat jejunum. J Pharmacol Exp Ther 2003; 305(1):151–158.
265. Tam D, Sun H, Pang KS. Influence of P-glycoprotein, transfer clearances, and drug binding on intestinal metabolism in Caco-2 cell monolayers or membrane preparations: a theoretical analysis. Drug Metab Dispos 2003; 31(10):1214–1226.
266. Tam D, Tirona RG, Pang KS. Segmental intestinal transporters and metabolic enzymes on intestinal drug absorption. Drug Metab Dispos 2003; 31(4):373–383.
267. Knight B, Troutman M, Thakker DR. Deconvoluting the effects of P-glycoprotein on intestinal CYP3A: a major challenge. Curr Opin Pharmacol 2006; 6(5):528–532.
268. Lan LB, Dalton JT, Schuetz EG. Mdr1 limits CYP3A metabolism in vivo. Mol Pharmacol 2000; 58(4):863–869.
269. Mayer U, Wagenaar E, Beijnen JH, et al. Substantial excretion of digoxin via the intestinal mucosa and prevention of long-term digoxin accumulation in the brain by the mdr 1a P-glycoprotein. Br J Pharmacol 1996; 119(5):1038–1044.
270. Booth CL, Brouwer KR, Brouwer KL. Effect of multidrug resistance modulators on the hepatobiliary disposition of doxorubicin in the isolated perfused rat liver. Cancer Res 1998; 58(16):3641–3648.
271. Lin JH. Drug-drug interaction mediated by inhibition and induction of P-glycoprotein. Adv Drug Deliv Rev 2003; 55(1):53–81.
272. Ford JM. Experimental reversal of P-glycoprotein-mediated multidrug resistance by pharmacological chemosensitisers. Eur J Cancer 1996; 32A(6):991–1001.
273. Ramachandra M, Ambudkar SV, Chen D, et al. Human P-glycoprotein exhibits reduced affinity for substrates during a catalytic transition state. Biochemistry 1998; 37(14):5010–5019.
274. Wandel C, Kim RB, Kajiji S, et al. P-glycoprotein and cytochrome P-450 3A inhibition: dissociation of inhibitory potencies. Cancer Res 1999; 59(16):3944–3948.
275. Kaye SB. Multidrug resistance: clinical relevance in solid tumours and strategies for circumvention. Curr Opin Oncol 1998; 10(suppl 1):S15–S19.
276. Leveque D, Jehl F. P-glycoprotein and pharmacokinetics. Anticancer Res 1995; 15 (2):331–336.
277. Modok S, Mellor HR, Callaghan R. Modulation of multidrug resistance efflux pump activity to overcome chemoresistance in cancer. Curr Opin Pharmacol 2006; 6(4):350–354.

278. Sparreboom A, Nooter K. Does P-glycoprotein play a role in anticancer drug pharmacokinetics? Drug Resist Updat 2000; 3(6):357–363.
279. Tsuruo T, Iida H, Tsukagoshi S, et al. Overcoming of vincristine resistance in P388 leukemia in vivo and in vitro through enhanced cytotoxicity of vincristine and vinblastine by verapamil. Cancer Res 1981; 41(5):1967–1972.
280. Cantwell B, Buamah P, Harris AL. Phase I and II study of oral verapamil and intravenous vindesine. Proceedings of the American Society of Clinical Oncology 1985; 42:161.
281. Pennock GD, Dalton WS, Roeske WR, et al. Systemic toxic effects associated with high-dose verapamil infusion and chemotherapy administration. J Natl Cancer Inst 1991; 83(2):105–110.
282. Kerr DJ, Graham J, Cummings J, et al. The effect of verapamil on the pharmacokinetics of adriamycin. Cancer Chemother Pharmacol 1986; 18(3):239–242.
283. Gigante M, Toffoli G, Boiocchi M. Pharmacokinetics of doxorubicin co-administered with high-dose verapamil. Br J Cancer 1995; 71(1):134–136.
284. Millward MJ, Cantwell BM, Munro NC, et al. Oral verapamil with chemotherapy for advanced non-small cell lung cancer: a randomised study. Br J Cancer 1993; 67 (5):1031–1035.
285. Milroy R. A randomised clinical study of verapamil in addition to combination chemotherapy in small cell lung cancer. West of Scotland Lung Cancer Research Group, and the Aberdeen Oncology Group. Br J Cancer 1993; 68(4):813–818.
286. Dalton WS, Crowley JJ, Salmon SS, et al. A phase III randomized study of oral verapamil as a chemosensitizer to reverse drug resistance in patients with refractory myeloma. A Southwest Oncology Group study. Cancer 1995; 75(3):815–820.
287. Taylor CW, Dalton WS, Mosley K, et al. Combination chemotherapy with cyclophosphamide, vincristine, adriamycin, and dexamethasone (CVAD) plus oral quinine and verapamil in patients with advanced breast cancer. Breast Cancer Res Treat 1997; 42(1):7–14.
288. Kronbach T, Fischer V, Meyer UA. Cyclosporine metabolism in human liver: identification of a cytochrome P-450III gene family as the major cyclosporine-metabolizing enzyme explains interactions of cyclosporine with other drugs. Clin Pharmacol Ther 1988; 43(6):630–635.
289. Colombo T, Zucchetti M, D'Incalci M. Cyclosporin A markedly changes the distribution of doxorubicin in mice and rats. J Pharmacol Exp Ther 1994; 269(1):22–27.
290. Lum BL, Kaubisch S, Yahanda AM, et al. Alteration of etoposide pharmacokinetics and pharmacodynamics by cyclosporine in a phase I trial to modulate multidrug resistance. J Clin Oncol 1992; 10(10):1635–1642.
291. Bartlett NL, Lum BL, Fisher GA, et al. Phase I trial of doxorubicin with cyclosporine as a modulator of multidrug resistance. J Clin Oncol 1994; 12(4):835–842.
292. Tsuruo T, Iida H, Kitatani Y, et al. Effects of quinidine and related compounds on cytotoxicity and cellular accumulation of vincristine and adriamycin in drug-resistant tumor cells. Cancer Res 1984; 44(10):4303–4307.
293. Fardel O, Lecureur V, Guillouzo A. Regulation by dexamethasone of P-glycoprotein expression in cultured rat hepatocytes. FEBS Lett 1993; 327(2):189–193.
294. Solary E, Caillot D, Chauffert B, et al. Feasibility of using quinine, a potential multidrug resistance-reversing agent, in combination with mitoxantrone and cytarabine for the treatment of acute leukemia. J Clin Oncol 1992; 10(11):1730–1736.

295. Verrelle P, Meissonnier F, Fonck Y, et al. Clinical relevance of immunohistochemical detection of multidrug resistance P-glycoprotein in breast carcinoma. J Natl Cancer Inst 1991; 83(2):111–116.
296. Wishart GC, Plumb JA, Morrison JG, et al. Adequate tumour quinidine levels for multidrug resistance modulation can be achieved in vivo. Eur J Cancer 1992; 28 (1):28–31.
297. Stuart NS, Philip P, Harris AL, et al. High-dose tamoxifen as an enhancer of etoposide cytotoxicity. Clinical effects and in vitro assessment in p-glycoprotein expressing cell lines. Br J Cancer 1992; 66(5):833–839.
298. Trump DL, Smith DC, Ellis PG, et al. High-dose oral tamoxifen, a potential multidrug-resistance-reversal agent: phase I trial in combination with vinblastine. J Natl Cancer Inst 1992; 84(23):1811–1816.
299. Ferry DR, Traunecker H, Kerr DJ. Clinical trials of P-glycoprotein reversal in solid tumours. Eur J Cancer 1996; 32A(6):1070–1081.
300. Samuels BL, Hollis DR, Rosner GL, et al. Modulation of vinblastine resistance in metastatic renal cell carcinoma with cyclosporine A or tamoxifen: a cancer and leukemia group B study. Clin Cancer Res 1997; 3(11):1977–1984.
301. Smith DC, Trump DL. A phase I trial of high-dose oral tamoxifen and CHOPE. Cancer Chemother Pharmacol 1995; 36(1):65–68.
302. Wilisch A, Haussermann K, Noller A, et al. MDR modulating and antineoplastic effects of B859-035, the (−)isomer of niguldipine. J Cancer Res Clin Oncol 1991; 117:S110.
303. Ferry DR, Glossmann H, Kaumann AJ. Relationship between the stereoselective negative inotropic effects of verapamil enantiomers and their binding to putative calcium channels in human heart. Br J Pharmacol 1985; 84(4):811–824.
304. Wilson WH, Bates SE, Fojo A, et al. Modulation of multidrug resistance by dexverapamil in EPOCH-refractory lymphomas. J Cancer Res Clin Oncol 1995; 121 (suppl 3):R25–R29.
305. Wilson WH, Jamis-Dow C, Bryant G, et al. Phase I and pharmacokinetic study of the multidrug resistance modulator dexverapamil with EPOCH chemotherapy. J Clin Oncol 1995; 13(8):1985–1994.
306. Tolcher AW, Cowan KH, Solomon D, et al. Phase I crossover study of paclitaxel with r-verapamil in patients with metastatic breast cancer. J Clin Oncol 1996; 14 (4):1173–1184.
307. Glisson B, Gupta R, Hodges P, et al. Cross-resistance to intercalating agents in an epipodophyllotoxin-resistant Chinese hamster ovary cell line: evidence for a common intracellular target. Cancer Res 1986; 46(4 pt 2):1939–1942.
308. te Boekhorst PA, van Kapel J, Schoester M, et al. Reversal of typical multidrug resistance by cyclosporin and its non-immunosuppressive analogue SDZ PSC 833 in Chinese hamster ovary cells expressing the mdr1 phenotype. Cancer Chemother Pharmacol 1992; 30(3):238–242.
309. Twentyman PR, Bleehen NM. Resistance modification by PSC-833, a novel non-immunosuppressive cyclosporin [corrected]. Eur J Cancer 1991; 27(12):1639–1642.
310. Advani R, Visani G, Milligan D, et al. Treatment of poor prognosis AML patients using PSC833 (valspodar) plus mitoxantrone, etoposide, and cytarabine (PSC-MEC). Adv Exp Med Biol 1999; 457:47–56.

311. Kornblau SM, Estey E, Madden T, et al. Phase I study of mitoxantrone plus etoposide with multidrug blockade by SDZ PSC-833 in relapsed or refractory acute myelogenous leukemia. J Clin Oncol 1997; 15(5):1796–1802.
312. Giaccone G, Linn SC, Welink J, et al. A dose-finding and pharmacokinetic study of reversal of multidrug resistance with SDZ PSC 833 in combination with doxorubicin in patients with solid tumors. Clin Cancer Res 1997; 3(11):2005–2015.
313. Lee EJ, George SL, Caligiuri M, et al. Parallel phase I studies of daunorubicin given with cytarabine and etoposide with or without the multidrug resistance modulator PSC-833 in previously untreated patients 60 years of age or older with acute myeloid leukemia: results of cancer and leukemia group B study 9420. J Clin Oncol 1999; 17(9):2831–2839.
314. Hyafil F, Vergely C, Du Vignaud P, et al. In vitro and in vivo reversal of multidrug resistance by GF120918, an acridonecarboxamide derivative. Cancer Res 1993; 53 (19):4595–4602.
315. Planting AS, Sonneveld P, van der Gaast A, et al. A phase I and pharmacologic study of the MDR converter GF120918 in combination with doxorubicin in patients with advanced solid tumors. Cancer Chemother Pharmacol 2005; 55(1):91–99.
316. Dhainaut A, Regnier G, Atassi G, et al. New triazine derivatives as potent modulators of multidrug resistance. J Med Chem 1992; 35(13):2481–2496.
317. Punt CJ, Voest EE, Tueni E, et al. Phase IB study of doxorubicin in combination with the multidrug resistance reversing agent S9788 in advanced colorectal and renal cell cancer. Br J Cancer 1997; 76(10):1376–1381.
318. Tranchand B, Catimel G, Lucas C, et al. Phase I clinical and pharmacokinetic study of S9788, a new multidrug-resistance reversal agent given alone and in combination with doxorubicin to patients with advanced solid tumors. Cancer Chemother Pharmacol 1998; 41(4):281–291.
319. Dantzig AH, Law KL, Cao J, et al. Reversal of multidrug resistance by the P-glycoprotein modulator, LY335979, from the bench to the clinic. Curr Med Chem 2001; 8(1):39–50.
320. Takara K, Sakaeda T, Okumura K. An update on overcoming MDR1-mediated multidrug resistance in cancer chemotherapy. Curr Pharm Des 2006; 12(3):273–286.
321. Yamazaki M, Suzuki H, Sugiyama Y. Recent advances in carrier-mediated hepatic uptake and biliary excretion of xenobiotics. Pharm Res 1996; 13(4):497–513.
322. Adams DJ, Knick VC. P-glycoprotein mediated resistance to 5′-nor-anhydro-vinblastine (Navelbine). Invest New Drugs 1995; 13(1):13–21.
323. Yu DK. The contribution of P-glycoprotein to pharmacokinetic drug-drug interactions. J Clin Pharmacol 1999; 39(12):1203–1211.
324. Cobbe SM. Using the right drug. A treatment algorithm for atrial fibrillation. Eur Heart J 1997; 18(suppl C):C33–C39.
325. Mackstaller LL, Alpert JS. Atrial fibrillation: a review of mechanism, etiology, and therapy. Clin Cardiol 1997; 20(7):640–650.
326. Wakasugi H, Yano I, Ito T, et al. Effect of clarithromycin on renal excretion of digoxin: interaction with P-glycoprotein. Clin Pharmacol Ther 1998; 64(1):123–128.
327. Kovarik JM, Rigaudy L, Guerret M, et al. Longitudinal assessment of a P-glycoprotein-mediated drug interaction of valspodar on digoxin. Clin Pharmacol Ther 1999; 66(4):391–400.
328. Lin JH. Transporter-mediated drug interactions: clinical implications and in vitro assessment. Expert Opin Drug Metab Toxicol 2007; 3(1):81–92.

329. Alderman CP, Allcroft PD. Digoxin-itraconazole interaction: possible mechanisms. Ann Pharmacother 1997; 31(4):438–440.
330. Ito S, Woodland C, Harper PA, et al. P-glycoprotein-mediated renal tubular secretion of digoxin: the toxicological significance of the urine-blood barrier model. Life Sci 1993; 53(2):L25–L31.
331. Brody TM, Larner J, Minneman RG. Human Pharmacology: Molecular to Clinical. Mosby-Year Book, Inc.St. Louis1998.
332. Hori R, Okamura N, Aiba T, et al. Role of P-glycoprotein in renal tubular secretion of digoxin in the isolated perfused rat kidney. J Pharmacol Exp Ther 1993; 266 (3):1620–1625.
333. Kelly RA, Smith TW. Pharmacological treatment of heart failure. In: Gilman AG, Limbird LE, Molinoff PB, et al., eds. Goodman & Gilman's The Pharmacological Basis of Therapeutics. McGraw-Hill Company1996.
334. Koren G, Woodland C, Ito S. Toxic digoxin-drug interactions: the major role of renal P-glycoprotein. Vet Hum Toxicol 1998; 40(1):45–46.
335. Pedersen KE. Digoxin interactions. The influence of quinidine and verapamil on the pharmacokinetics and receptor binding of digitalis glycosides. Acta Med Scand Suppl 1985; 697:1–40.
336. Matheny CJ, Ali RY, Yang X, et al. Effect of prototypical inducing agents on P-glycoprotein and CYP3A expression in mouse tissues. Drug Metab Dispos 2004; 32(9):1008–1014.
337. Theis JG, Chan HS, Greenberg ML, et al. Increased systemic toxicity of sarcoma chemotherapy due to combination with the P-glycoprotein inhibitor cyclosporin. Int J Clin Pharmacol Ther 1998; 36(2):61–64.
338. Chang T, Benet LZ, Hebert MF. The effect of water-soluble vitamin E on cyclosporine pharmacokinetics in healthy volunteers. Clin Pharmacol Ther 1996; 59 (3):297–303.
339. Wu CY, Benet LZ. Predicting drug disposition via application of BCS: transport/absorption/elimination interplay and development of a biopharmaceutics drug disposition classification system. Pharm Res 2005; 22(1):11–23.
340. Dey S. Single nucleotide polymorphisms in human P-glycoprotein: its impact on drug delivery and disposition. Expert Opin Drug Deliv 2006; 3(1):23–35.
341. Eichelbaum M, Fromm MF, Schwab M. Clinical aspects of the MDR1 (ABCB1) gene polymorphism. Ther Drug Monit 2004; 26(2):180–185.
342. Sakurai A, Tamura A, Onishi Y, et al. Genetic polymorphisms of ATP-binding cassette transporters ABCB1 and ABCG2: therapeutic implications. Expert Opin Pharmacother 2005; 6(14):2455–2473.
343. Kurata Y, Ieiri I, Kimura M, et al. Role of human MDR1 gene polymorphism in bioavailability and interaction of digoxin, a substrate of P-glycoprotein. Clin Pharmacol Ther 2002; 72(2):209–219.
344. Verstuyft C, Schwab M, Schaeffeler E, et al. Digoxin pharmacokinetics and MDR1 genetic polymorphisms. Eur J Clin Pharmacol 2003; 58(12):809–812.
345. Artursson P. Epithelial transport of drugs in cell culture. I: A model for studying the passive diffusion of drugs over intestinal absorptive (Caco-2) cells. J Pharm Sci 1990; 79(6):476–482.
346. Artursson P. Cell cultures as models for drug absorption across the intestinal mucosa. Crit Rev Ther Drug Carrier Syst 1991; 8(4):305–330.

347. Cogburn JN, Donovan MG, Schasteen CS. A model of human small intestinal absorptive cells. 1. Transport barrier. Pharm Res 1991; 8(2):210–216.
348. Gan LS, Thakker DR. Applications of the Caco-2 model in the design and development of orally active drugs: elucidation of biochemical and physical barriers posed by the intestinal epithelium. Adv Drug Deliv Rev 1997; 23(1-3):77–98.
349. Hidalgo IJ, Raub TJ, Borchardt RT. Characterization of the human colon carcinoma cell line (Caco-2) as a model system for intestinal epithelial permeability. Gastroenterology 1989; 96(3):736–749.
350. Hilgers AR, Conradi RA, Burton PS. Caco-2 cell monolayers as a model for drug transport across the intestinal mucosa. Pharm Res 1990; 7(9):902–910.
351. Wilson G, Hassan I, Dix C, et al. Transport and permeability properties of human Caco-2 cells: an in vitro model of the intestinal epithelial cell barrier. J Control Release 1990; 11:25–40.
352. Burton PS, Conradi RA, Hilgers AR, et al. Evidence for a polarized efflux system for peptides in the apical membrane of Caco-2 cells. Biochem Biophys Res Commun 1993; 190(3):760–766.
353. Chantret I, Barbat A, Dussaulx E, et al. Epithelial polarity, villin expression, and enterocytic differentiation of cultured human colon carcinoma cells: a survey of twenty cell lines. Cancer Res 1988; 48(7):1936–1942.
354. Briske-Anderson MJ, Finley JW, Newman SM. The influence of culture time and passage number on the morphological and physiological development of Caco-2 cells. Proc Soc Exp Biol Med 1997; 214(3):248–257.
355. Hosoya KI, Kim KJ, Lee VH. Age-dependent expression of P-glycoprotein gp170 in Caco-2 cell monolayers. Pharm Res 1996; 13(6):885–890.
356. Paine MF, Leung LY, Lim HK, et al. Identification of a novel route of extraction of sirolimus in human small intestine: roles of metabolism and secretion. J Pharmacol Exp Ther 2002; 301(1):174–186.
357. Stephens RH, O'Neill CA, Warhurst A, et al. Kinetic profiling of P-glycoprotein-mediated drug efflux in rat and human intestinal epithelia. J Pharmacol Exp Ther 2001; 296(2):584–591.
358. Anderle P, Niederer E, Rubas W, et al. P-Glycoprotein (P-gp) mediated efflux in Caco-2 cell monolayers: the influence of culturing conditions and drug exposure on P-gp expression levels. J Pharm Sci 1998; 87(6):757–762.
359. Doppenschmitt S, Spahn-Langguth H, Regardh CG, et al. Radioligand-binding assay employing P-glycoprotein-overexpressing cells: testing drug affinities to the secretory intestinal multidrug transporter. Pharm Res 1998; 15(7):1001–1006.
360. Cho MJ, Adson A, Kezdy FJ. Transepithelial transport of aliphatic carboxylic acids studied in Madin Darby canine kidney (MDCK) cell monolayers. Pharm Res 1990; 7(4):325–331.
361. Cho MJ, Thompson DP, Cramer CT, et al. The Madin Darby canine kidney (MDCK) epithelial cell monolayer as a model cellular transport barrier. Pharm Res 1989; 6(1):71–77.
362. Simons K, Fuller SD. Cell surface polarity in epithelia. Annu Rev Cell Biol 1985; 1:243–288.
363. Simons K, van Meer G. Lipid sorting in epithelial cells. Biochemistry 1988; 27 (17):6197–6202.

364. Zhang Y, Benet LZ. Characterization of P-glycoprotein mediated transport of K02, a novel vinylsulfone peptidomimetic cysteine protease inhibitor, across MDR1-MDCK and Caco-2 cell monolayers. Pharm Res 1998; 15(10):1520–1524.
365. Hunter J, Hirst BH, Simmons NL. Transepithelial secretion, cellular accumulation and cytotoxicity of vinblastine in defined MDCK cell strains. Biochim Biophys Acta 1993; 1179(1):1–10.
366. Pastan I, Gottesman MM, Ueda K, et al. A retrovirus carrying an MDR1 cDNA confers multidrug resistance and polarized expression of P-glycoprotein in MDCK cells. Proc Natl Acad Sci U S A 1988; 85(12):4486–4490.
367. Bradbury MW. The blood-brain barrier. Exp Physiol 1993; 78(4):453–472.
368. Pardridge WM. CNS drug design based on principles of blood-brain barrier transport. J Neurochem 1998; 70(5):1781–1792.
369. Tsuji A. Blood-brain barrier function of P-glycoprotein. Adv Drug Deliv Rev 1997; 25(2,3):287–298.
370. Audus KL, Ng L, Wang W, et al. Brain microvessel endothelial cell culture systems. Pharm Biotechnol 1996; 8:239–258.
371. Audus KL, Borchardt RT. Characteristics of the large neutral amino acid transport system of bovine brain microvessel endothelial cell monolayers. J Neurochem 1986; 47(2):484–488.
372. Audus KL, Borchardt RT. Bovine brain microvessel endothelial cell monolayers as a model system for the blood-brain barrier. Ann N Y Acad Sci 1987; 507:9–18.
373. Betz AL, Firth JA, Goldstein GW. Polarity of the blood-brain barrier: distribution of enzymes between the luminal and antiluminal membranes of brain capillary endothelial cells. Brain Res 1980; 192(1):17–28.
374. Borges N, Shi F, Azevedo I, et al. Changes in brain microvessel endothelial cell monolayer permeability induced by adrenergic drugs. Eur J Pharmacol 1994; 269 (2):243–248.
375. Drewes LD, Lidinsky WA. Studies of cerebral capillary endothelial membrane. Adv Exp Med Biol 1980; 131:17–27.
376. Guillot FL, Audus KL. Angiotensin peptide regulation of fluid-phase endocytosis in brain microvessel endothelial cell monolayers. J Cereb Blood Flow Metab 1990; 10 (6):827–834.
377. Joo F. The cerebral microvessels in culture, an update. J Neurochem 1992; 58(1): 1–17.
378. Joo F. The blood-brain barrier in vitro: the second decade. Neurochem Int 1993; 23 (6):499–521.
379. Miller DW, Audus KL, Borchardt RT. Application of cultured endothelial cells of the brain microvasculature in the study of the blood-brain barrier. J Tiss Cult Meth 1992; 14:217–224.
380. Raub TJ, Audus KL. Adsorptive endocytosis and membrane recycling by cultured primary bovine brain microvessel endothelial cell monolayers. J Cell Sci 1990; 97 (pt 1):127–138.
381. Tsuji A, Terasaki T, Takabatake Y, et al. P-glycoprotein as the drug efflux pump in primary cultured bovine brain capillary endothelial cells. Life Sci 1992; 51 (18):1427–1437.
382. Crone C. The blood-brain barrier as a tight epithelium: where is information lacking? Ann N Y Acad Sci 1986; 481:174–185.

383. Fontaine M, Elmquist WF, Miller DW. Use of rhodamine 123 to examine the functional activity of P-glycoprotein in primary cultured brain microvessel endothelial cell monolayers. Life Sci 1996; 59(18):1521–1531.
384. Lechardeur D, Scherman D. Functional expression of the P-glycoprotein mdr in primary cultures of bovine cerebral capillary endothelial cells. Cell Biol Toxicol 1995; 11(5):283–293.
385. Regina A, Koman A, Piciotti M, et al. Mrp1 multidrug resistance-associated protein and P-glycoprotein expression in rat brain microvessel endothelial cells. J Neurochem 1998; 71(2):705–715.
386. Seetharaman S, Barrand MA, Maskell L, et al. Multidrug resistance-related transport proteins in isolated human brain microvessels and in cells cultured from these isolates. J Neurochem 1998; 70(3):1151–1159.
387. Huai-Yun H, Secrest DT, Mark KS, et al. Expression of multidrug resistance-associated protein (MRP) in brain microvessel endothelial cells. Biochem Biophys Res Commun 1998; 243(3):816–820.
388. Shirai A, Naito M, Tatsuta T, et al. Transport of cyclosporin A across the brain capillary endothelial cell monolayer by P-glycoprotein. Biochim Biophys Acta 1994; 1222(3):400–404.
389. Tsuji A, Tamai I, Sakata A, et al. Restricted transport of cyclosporin A across the blood-brain barrier by a multidrug transporter, P-glycoprotein. Biochem Pharmacol 1993; 46(6):1096–1099.
390. Glynn SL, Yazdanian M. In vitro blood-brain barrier permeability of nevirapine compared to other HIV antiretroviral agents. J Pharm Sci 1998; 87(3):306–310.
391. Egleton RD, Abbruscato TJ, Thomas SA, et al. Transport of opioid peptides into the central nervous system. J Pharm Sci 1998; 87(11):1433–1439.
392. Henthorn TK, Liu Y, Mahapatro M, et al. Active transport of fentanyl by the blood-brain barrier. J Pharmacol Exp Ther 1999; 289(2):1084–1089.
393. Matsuzaki J, Yamamoto C, Miyama T, et al. Contribution of P-glycoprotein to bunitrolol efflux across blood-brain barrier. Biopharm Drug Dispos 1999; 20(2): 85–90.
394. Troutman MD, Thakker DR. Novel experimental parameters to quantify the modulation of absorptive and secretory transport of compounds by P-glycoprotein in cell culture models of intestinal epithelium. Pharm Res 2003; 20(8):1210–1224.
395. Polli JW, Wring SA, Humphreys JE, et al. Rational use of in vitro P-glycoprotein assays in drug discovery. J Pharmacol Exp Ther 2001; 299(2):620–628.
396. Troutman MD, Thakker DR. Rhodamine 123 requires carrier-mediated influx for its activity as a P-glycoprotein substrate in Caco-2 cells. Pharm Res 2003; 20(8): 1192–1199.
397. Lentz KA, Polli JW, Wring SA, et al. Influence of passive permeability on apparent P-glycoprotein kinetics. Pharm Res 2000; 17(12):1456–1460.
398. Keogh JP, Kunta JR. Development, validation and utility of an in vitro technique for assessment of potential clinical drug-drug interactions involving P-glycoprotein. Eur J Pharm Sci 2006; 27(5):543–554.
399. Rautio J, Humphreys JE, Webster LO, et al. In vitro p-glycoprotein inhibition assays for assessment of clinical drug interaction potential of new drug candidates: a recommendation for probe substrates. Drug Metab Dispos 2006; 34(5):786–792.

400. Bucana CD, Giavazzi R, Nayar R, et al. Retention of vital dyes correlates inversely with the multidrug-resistant phenotype of adriamycin-selected murine fibrosarcoma variants. Exp Cell Res 1990; 190(1):69–75.
401. Dhar S, Nygren P, Liminga G, et al. Relationship between cytotoxic drug response patterns and activity of drug efflux transporters mediating multidrug resistance. Eur J Pharmacol 1998; 346(2-3):315–322.
402. Efferth T, Lohrke H, Volm M. Reciprocal correlation between expression of P-glycoprotein and accumulation of rhodamine 123 in human tumors. Anticancer Res 1989; 9(6):1633–1637.
403. Lee JS, Paull K, Alvarez M, et al. Rhodamine efflux patterns predict P-glycoprotein substrates in the National Cancer Institute drug screen. Mol Pharmacol 1994; 46 (4):627–638.
404. Liminga G, Nygren P, Larsson R. Microfluorometric evaluation of calcein acetoxymethyl ester as a probe for P-glycoprotein-mediated resistance: effects of cyclosporin A and its nonimmunosuppressive analogue SDZ PSC 833. Exp Cell Res 1994; 212(2):291–296.
405. Nagai J, Takano M, Hirozane K, et al. Specificity of p-aminohippurate transport system in the OK kidney epithelial cell line. J Pharmacol Exp Ther 1995; 274 (3):1161–1166.
406. Tiberghien F, Loor F. Ranking of P-glycoprotein substrates and inhibitors by a calcein-AM fluorometry screening assay. Anticancer Drugs 1996; 7(5):568–578.
407. Yumoto R, Murakami T, Nakamoto Y, et al. Transport of rhodamine 123, a P-glycoprotein substrate, across rat intestine and Caco-2 cell monolayers in the presence of cytochrome P-450 3A-related compounds. J Pharmacol Exp Ther 1999; 289(1):149–155.
408. Bohme M, Jedlitschky G, Leier I, et al. ATP-dependent export pumps and their inhibition by cyclosporins. Adv Enzyme Regul 1994; 34:371–380.
409. Kwon Y, Kamath AV, Morris ME. Inhibitors of P-glycoprotein-mediated daunomycin transport in rat liver canalicular membrane vesicles. J Pharm Sci 1996; 85 (9):935–939.
410. Horio M, Gottesman MM, Pastan I. ATP-dependent transport of vinblastine in vesicles from human multidrug-resistant cells. Proc Natl Acad Sci U S A 1988; 85 (10):3580–3584.
411. Hsing S, Gatmaitan Z, Arias IM. The function of Gp170, the multidrug-resistance gene product, in the brush border of rat intestinal mucosa. Gastroenterology 1992; 102(3):879–885.
412. Makhey VD, Guo A, Norris DA, et al. Characterization of the regional intestinal kinetics of drug efflux in rat and human intestine and in Caco-2 cells. Pharm Res 1998; 15(8):1160–1167.
413. Dutt A, Heath LA, Nelson JA. P-glycoprotein and organic cation secretion by the mammalian kidney. J Pharmacol Exp Ther 1994; 269(3):1254–1260.
414. Emi Y, Tsunashima D, Ogawara K, et al. Role of P-glycoprotein as a secretory mechanism in quinidine absorption from rat small intestine. J Pharm Sci 1998; 87 (3):295–299.
415. Hashimoto Y, Sasa H, Shimomura M, et al. Effects of intestinal and hepatic metabolism on the bioavailability of tacrolimus in rats. Pharm Res 1998; 15 (10):1609–1613.

416. Saitoh H, Aungst BJ. Possible involvement of multiple P-glycoprotein-mediated efflux systems in the transport of verapamil and other organic cations across rat intestine. Pharm Res 1995; 12(9):1304–1310.
417. Beja O, Bibi E. Functional expression of mouse Mdr1 in an outer membrane permeability mutant of Escherichia coli. Proc Natl Acad Sci U S A 1996; 93 (12):5969–5974.
418. Evans GL, Ni B, Hrycyna CA, et al. Heterologous expression systems for P-glycoprotein: E. coli, yeast, and baculovirus. J Bioenerg Biomembr 1995; 27(1): 43–52.
419. Castillo G, Vera JC, Yang CP, et al. Functional expression of murine multidrug resistance in Xenopus laevis oocytes. Proc Natl Acad Sci U S A 1990; 87(12):4737–4741.
420. Raymond M, Gros P, Whiteway M, et al. Functional complementation of yeast ste6 by a mammalian multidrug resistance mdr gene. Science 1992; 256(5054):232–234.
421. Saeki T, Shimabuku AM, Azuma Y, et al. Expression of human P-glycoprotein in yeast cells–effects of membrane component sterols on the activity of P-glycoprotein. Agric Biol Chem 1991; 55(7):1859–1865.
422. Drueckes P, Schinzel R, Palm D. Photometric microtiter assay of inorganic phosphate in the presence of acid-labile organic phosphates. Anal Biochem 1995; 230 (1):173–177.
423. Marathe PH, Rodrigues AD. In vivo animal models for investigating potential CYP3A- and P-gp-mediated drug-drug interactions. Curr Drug Metab 2006; 7(7): 687–704.
424. Lennernas H. Human jejunal effective permeability and its correlation with preclinical drug absorption models. J Pharm Pharmacol 1997; 49(7):627–638.
425. Brouwer KL, Thurman RG. Isolated perfused liver. Pharm Biotechnol 1996; 8: 161–192.
426. Booth CL, Pollack GM, Brouwer KL. Hepatobiliary disposition of valproic acid and valproate glucuronide: use of a pharmacokinetic model to examine the rate-limiting steps and potential sites of drug interactions. Hepatology 1996; 23(4):771–780.
427. Smit JW, Duin E, Steen H, et al. Interactions between P-glycoprotein substrates and other cationic drugs at the hepatic excretory level. Br J Pharmacol 1998; 123(3): 361–370.
428. Hayes JH, Soroka CJ, Rios-Velez L, et al. Hepatic sequestration and modulation of the canalicular transport of the organic cation, daunorubicin, in the Rat. Hepatology 1999; 29(2):483–493.
429. Watanabe T, Miyauchi S, Sawada Y, et al. Kinetic analysis of hepatobiliary transport of vincristine in perfused rat liver. Possible roles of P-glycoprotein in biliary excretion of vincristine. J Hepatol 1992; 16(1-2):77–88.
430. Smith QR. Brain perfusion systems for studies of drug uptake and metabolism in the central nervous system. In: Borchardt RT, ed. Models for Assessing Drug Absorption and Metabolism. New York: Plenum Press, 1996.
431. Smith QR, Takasato Y. Kinetics of amino acid transport at the blood-brain barrier studied using an in situ brain perfusion technique. Ann N Y Acad Sci 1986; 481:186–201.
432. Takasato Y, Rapoport SI, Smith QR. An in situ brain perfusion technique to study cerebrovascular transport in the rat. Am J Physiol 1984; 247(3 pt 2):H484–H493.

433. Smith QR, Nagura H, Takada Y, et al. Facilitated transport of the neurotoxin, beta-N-methylamino-L-alanine, across the blood-brain barrier. J Neurochem 1992; 58 (4):1330–1337.
434. Liu X, Smith BJ, Chen C, et al. Use of a physiologically based pharmacokinetic model to study the time to reach brain equilibrium: an experimental analysis of the role of blood-brain barrier permeability, plasma protein binding, and brain tissue binding. J Pharmacol Exp Ther 2005; 313(3):1254–1262.
435. Liu X, Tu M, Kelly RS, et al. Development of a computational approach to predict blood-brain barrier permeability. Drug Metab Dispos 2004; 32(1):132–139.
436. Murata M, Tamai I, Kato H, et al. Efflux transport of a new quinolone antibacterial agent, HSR-903, across the blood-brain barrier. J Pharmacol Exp Ther 1999; 290 (1):51–57.
437. Drion N, Lemaire M, Lefauconnier JM, et al. Role of P-glycoprotein in the blood-brain transport of colchicine and vinblastine. J Neurochem 1996; 67(4):1688–1693.
438. Drion N, Risede P, Cholet N, et al. Role of P-170 glycoprotein in colchicine brain uptake. J Neurosci Res 1997; 49(1):80–88.
439. Rousselle C, Clair P, Lefauconnier JM, et al. New advances in the transport of doxorubicin through the blood-brain barrier by a peptide vector-mediated strategy. Mol Pharmacol 2000; 57(4):679–686.
440. Chikhale EG, Burton PS, Borchardt RT. The effect of verapamil on the transport of peptides across the blood-brain barrier in rats: kinetic evidence for an apically polarized efflux mechanism. J Pharmacol Exp Ther 1995; 273(1):298–303.
441. Lemaire M, Bruelisauer A, Guntz P, et al. Dose-dependent brain penetration of SDZ PSC 833, a novel multidrug resistance-reversing cyclosporin, in rats. Cancer Chemother Pharmacol 1996; 38(5):481–486.
442. Groen AK, Van Wijland MJ, Frederiks WM, et al. Regulation of protein secretion into bile: studies in mice with a disrupted mdr2 p-glycoprotein gene. Gastroenterology 1995; 109(6):1997–2006.
443. Smit JJ, Schinkel AH, Oude Elferink RP, et al. Homozygous disruption of the murine mdr2 P-glycoprotein gene leads to a complete absence of phospholipid from bile and to liver disease. Cell 1993; 75(3):451–462.
444. Croop JM, Raymond M, Haber D, et al. The three mouse multidrug resistance (mdr) genes are expressed in a tissue-specific manner in normal mouse tissues. Mol Cell Biol 1989; 9(3):1346–1350.
445. Arceci RJ, Croop JM, Horwitz SB, et al. The gene encoding multidrug resistance is induced and expressed at high levels during pregnancy in the secretory epithelium of the uterus. Proc Natl Acad Sci U S A 1988; 85(12):4350–4354.
446. Endicott JA, Sarangi F, Ling V. Complete cDNA sequences encoding the Chinese hamster P-glycoprotein gene family. DNA Seq 1991; 2(2):89–101.
447. Silverman JA, Raunio H, Gant TW, et al. Cloning and characterization of a member of the rat multidrug resistance (mdr) gene family. Gene 1991; 106(2):229–236.
448. Hsu SI, Lothstein L, Horwitz SB. Differential overexpression of three mdr gene family members in multidrug-resistant J774.2 mouse cells. Evidence that distinct P-glycoprotein precursors are encoded by unique mdr genes. J Biol Chem 1989; 264 (20):12053–12062.
449. Lin JH, Chiba M, Balani SK, et al. Species differences in the pharmacokinetics and metabolism of indinavir, a potent human immunodeficiency virus protease inhibitor. Drug Metab Dispos 1996; 24(10):1111–1120.

450. Galski H, Sullivan M, Willingham MC, et al. Expression of a human multidrug resistance cDNA (MDR1) in the bone marrow of transgenic mice: resistance to daunomycin-induced leukopenia. Mol Cell Biol 1989; 9(10):4357–4363.
451. Mickisch GH, Merlino GT, Galski H, et al. Transgenic mice that express the human multidrug-resistance gene in bone marrow enable a rapid identification of agents that reverse drug resistance. Proc Natl Acad Sci U S A 1991; 88(2):547–551.
452. Mickisch GH, Pastan I, Gottesman MM. Multidrug resistant transgenic mice as a novel pharmacologic tool. Bioessays 1991; 13(8):381–387.
453. Mickisch GH, Rahman A, Pastan I, et al. Increased effectiveness of liposome-encapsulated doxorubicin in multidrug-resistant-transgenic mice compared with free doxorubicin. J Natl Cancer Inst 1992; 84(10):804–805.
454. Walle UK, Walle T. Taxol transport by human intestinal epithelial Caco-2 cells. Drug Metab Dispos 1998; 26(4):343–346.
455. Elferink RP, Tytgat GN, Groen AK. Hepatic canalicular membrane 1: the role of mdr2 P-glycoprotein in hepatobiliary lipid transport. FASEB J 1997; 11(1):19–28.
456. Cascorbi I. Role of pharmacogenetics of ATP-binding cassette transporters in the pharmacokinetics of drugs. Pharmacol Ther 2006; 112(2):457–473.
457. Deeley RG, Cole SP. Substrate recognition and transport by multidrug resistance protein 1 (ABCC1). FEBS Lett 2006; 580(4):1103–1111.
458. Xia CQ, Liu N, Miwa GT, et al. Interactions of cyclosporin a with breast cancer resistance protein. Drug Metab Dispos 2007; 35(4):576–582.

9

Cytochrome P450 Protein Modeling and Ligand Docking

K. Anton Feenstra, Chris de Graaf, and Nico P. E. Vermeulen

Department of Pharmacochemistry,
Vrije Universiteit, Amsterdam, The Netherlands

I. INTRODUCTION

Cytochromes P450 (CYPs) constitute the most important family of biotransformation enzymes and play an important role in the disposition of drugs and their pharmacological and toxicological effects. Early consideration of ADME-properties (absorption disposition, metabolism, and excretion) is increasingly seen as essential for efficient discovery and development of new drugs and drug candidates (1,2). Apart from the application of in vitro tools, this necessitates the application of novel in silico tools that can accurately predict ADME-properties of drug candidates already in early stages of the lead finding and optimization process (3–5).

CYPs generally detoxify potentially harmful xenobiotic compounds; however, in a number of cases, nontoxic compounds are also bioactivated to toxic reactive intermediates and procarcinogens into ultimate carcinogens (6). Furthermore, CYPs catalyze key reactions in the formation of endogenous compounds such as hormones and steroids. The catalytic activities of CYPs can be divided into (1) monooxygenase activity, usually resulting in the incorporation of an oxygen atom into the substrate, (2) oxidase activity, resulting in the formation of superoxide anion radicals or hydrogen peroxide (i.e., uncoupling of the catalytic cycle), and (3) substrate reductase activity, usually producing free radical intermediates under anaerobic conditions. For more details we refer to the review by Guengerich (7). Apart from the incorporation of oxygen into substrates, other reactions can be performed by CYPs, e.g., desaturation, dehydrogenation, ring formation, and dehalogenation. CYPs can be classified according to the electron transfer chain that delivers the electrons for the one-electron reduction reactions from NAD(P)H: class I CYPs are found in bacteria and in eukaryotic mitochondrial membranes and require a flavin adenine dinucleotide (FAD) containing reductase and an iron-sulfur protein (putidaredoxin); class II CYPs are bound to the endoplasmic reticulum and interact directly with a CYP reductase containing FAD and flavin mononucleotide (FMN) (8).

In all known CYP crystal structures available today, the same general three-dimensional (3D) fold is present with a conserved core region containing the C-terminal half of helix I, helices L, E, K, and K′, and the heme coordination region (9). A model CYP structure is given in Figure 1 that shows a recent homology model of CYP2D6. The 3D structure of these regions is well conserved amongst the CYP2 family despite a low sequence homology. Other regions (e.g., the active site region containing the B′ helix, the loops between helices C and D, the region spanning helices F and G, and most of the β-strands) are more variable. The largest structural variations are found in the B′ helix and helices F and G and their connecting loop, which makes sense, since these regions are known to be involved in substrate access and recognition (8). Six substrate recognition sites (SRSs) have been assigned on the basis of mutagenesis and sequence alignment studies with representative members of the CYP2 family and CYP101 (*cam*) (10). Figure 1 illustrates the organization of SRSs and secondary structure elements in CYP2D6, which shares all major features of CYPs.

The state-of-the-art computational methods used in CYP modeling vary considerably, as does the reliability of the results obtained. Protein homology modeling in combination with automated docking and molecular dynamics (MD) simulations (11,12) have been used successfully for the rationalization and prediction of metabolite formation by several CYP isoenzymes. In recent years, the lack of detailed structural knowledge from crystallography has largely been overcome by the prediction of CYP structures using computer-aided homology modeling techniques in combination with pharmacophore or small molecule models and mechanism-based ab initio and semiempirical calculations (13). It has been concluded that not one computational approach is capable of rationalizing and reliably predicting metabolite formation by CYPs, but that it is rather the combination of various complimentary computational approaches (14).

The primary aim of this chapter is to describe recent advances in CYP homology modeling methods and methods for modeling of CYP protein-ligand interactions. Section II deals with crystal structures, homology models, and various ways in which this structural information is used. Section III covers issues as binding orientation, ligand dynamics, and binding affinity. Finally, a general summary and conclusions section is presented, and an outlook of expected developments in the field. Two comprehensive tables are supplied that list the most important homology models (Table 1) and the most commonly used software for CYP molecular modeling (Table 2).

(*text continues on page 442*)

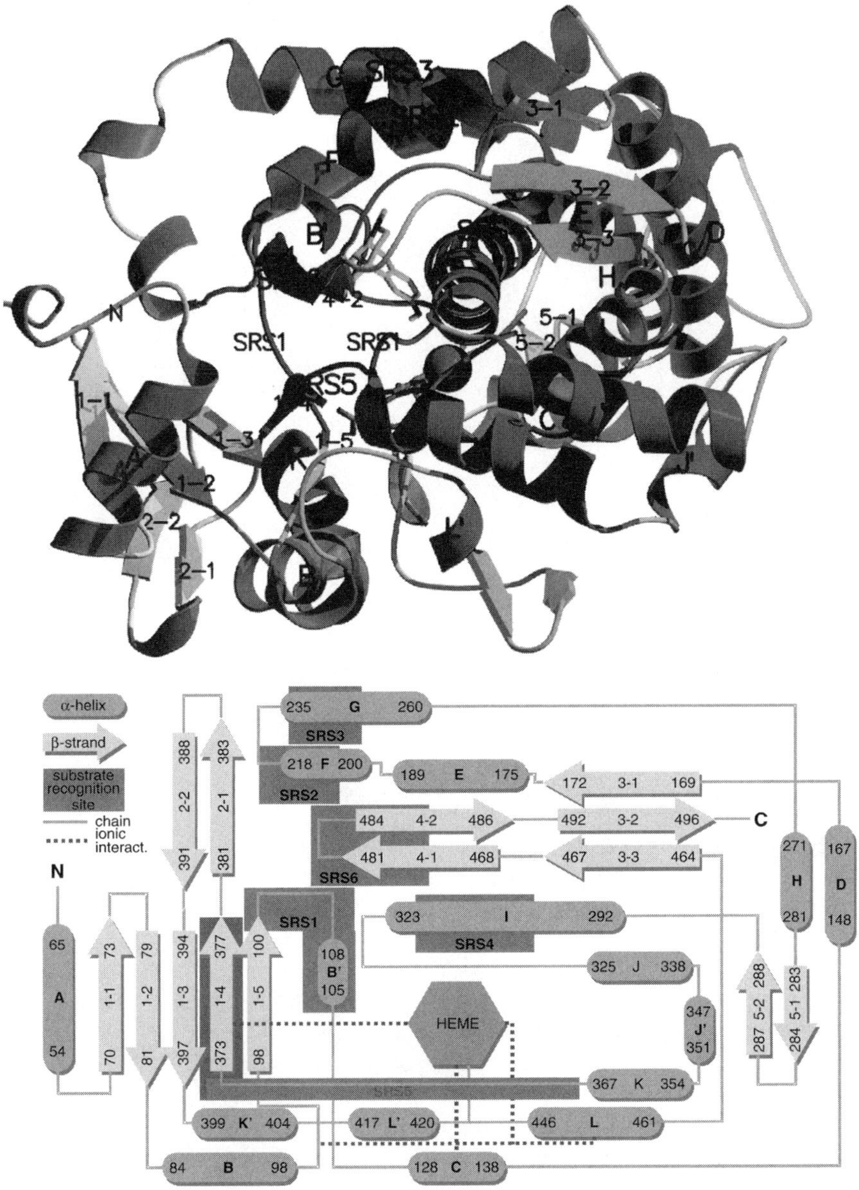

Figure 1 Three-dimensional structure (*top*) and topology and tertiary organization (*bottom*) of a homology model of CYP2D6 (de Graaf et al., Current Drug Metabolism, 2007; 8:59–77). The SRSs (after 10) are indicated in dark grey shading. Central in the structure is the heme group with the iron as a space-filling sphere. The ligand dextromethorphan is shown in the binding pocket. The figure was generated using Molscript (119) and Raster3D (120). *Abbreviations*: CYP, cytochrome P450; SRSs, substrate recognition sites.

Table 1 Selection of 52 of the 120 CYP Homology Models Built So Far Based on Available CYP Crystal Structures

CYP isoform	CYP template(s)[a]	Predicted property[b]	Method of refinement and validation[c]	Remarks	Reference[d]
1A1	101				122
	2C5	CSP[e], K_m	MD, SM		113
1A2	102	CSP[e]	SM		123
	102	CSP, mutagenicity	AD, P (Lozano, 2000), SM	Evaluation of alignment strategies	124
	101/102/107A/108	CSP, IS	AD	Comparison with 2D6 and 3A4. Suggestion for membrane attachment	42
	2C5	CSP[e]		Rationalization of inhibition mechanisms	125
1B1	2C5	CSP[e], K_m[f]	SM		126
2A1/2A4/2A5	102	CSP[e], K_m[f], IS	SM		127
2A6	102	CSP[e], K_m[f]	SM		128
	2C5	CSP[e], K_m[f], IS	SM		129
2B1	101	CSP[e]	SM		130
	102	CSP[e]		Suggestion for membrane attachment	131
2B1/2B2	101/102/108	CSP[e]	SM [132]	2B2 model built from 2B1 template	133,134
2B1/2B4/2B5	2C5	IS	SM		54
2B4	101/102/107A1/108		SM, MD	Suggestion for binding regions with redox partners	135
2B4	2C5/2C9	K_s	SM, MD, QM	Comparison of models derived from 101, 2C5 or 2C9	49

2B6	101/102/107A1/108	CSP, IS	SM		136
	102/2C5	CSP	AD, MD	Bound and unbound forms of 102 as additional templates for the F-G-loop	56
	2C5	CSP, K_m	AD, P		81
2C8/2C9/2C18/2C19	2C5	CSP, IS	AD, P	Comparison of 2C active sites	55
2C9	101		SM		137
	102/2C5	CSP[e]	P	B′ region from 102	13
2C18/2C19	102	CSP[e], IS	MD		138
2D6	101	CSP[e]		Preliminary model of active site regions (11 segments)	35
	101/102/108	CSP[e]	P		36,40,41
	101/102/108	CSP		Refinement and validation using NMR-derived ligand-heme distance restraints	28,29
	101/102/107A1/108	CSP, IS	AD	Comparison with 1A2 and 3A4. Includes suggestion for membrane attachment	42
	101/102/2C5	CSP, K_m	AD, P		139
	101/102/107A1/108/2C5	CSP	AD, SM	Evaluation of alignment strategies	44,46
	2C5	CSP, IS	AD, MD	Comparison with rat 2D1/2/3/4 models built on 2D6	11

Continued

Table 1 *Continued*

CYP isoform	CYP template(s)[a]	Predicted property[b]	Method of refinement and validation[c]	Remarks	Reference[d]
2E1	102	CSP[e], IS	SM		140
	2C5/2C9	CSP	AD, MD, QM		82
3A4	101	CSP[e]			141
	101/102/107A1/108	CSP	AD, MD, QM	Active site chemical probe studies	83,32, 142–144
	101/102/107A1/108	CSP, IS	AD	Comparison with 1A2 and 2D6. Suggestion for membrane attachment	42
4A1/4A4/4A11	102	CSP[e]	SM		145
4A11	102	CSP[e]	MD		146
6B1	102	CSP	AD, SM		84,147
11A (SCC)	101			Suggestion for membrane attachment	148
	102		AD, MD, SM	Protein-protein docking with adrenodoxin redox-partner	149
17 (17α)	101	CSP[e]	MD, SM		150
	108	CSP[e]	MD		151
19 (arom)	101	CSP[e]	SM		152
	102	CSP[e]	SM		153
	2C5	CSP[e]	SM		154
51 (14α)	101		SM		155
	101/102/107A1	CSP[e]	MD		156
27A1	102		SM		157

				Suggestion for membrane attachment. F-G-loop from Arc Repressor	
73A1	101/102/107A1/108	CSP[e]	SM	Refinement and validation using NMR-derived ligand-heme distance restraints	30,158
105C1A1/B1	101	CSP[e]		Models with different alignments.	159
105C1	101/102/107A1/108		MD		160
119	101/102/107A1/108/55A1		MD		161

For each CYP isoform, in addition to the earliest published model, only the most recent and most properly refined and validated homology models are included.
[a]Key references for substrate-free crystal structures for CYP isoforms 101 (162), 102 (163), 107A (59), 108 (164), 55A1 (165), 2C5 (50), and 2C9 (15).
[b]Predicted property: ''Affinity'' (Michaelis-Menten constant K_m or spectral dissociation constant K_s); CSP, catalytic site prediction; IS, isoenzyme specificity.
[c]Method of prediction and validation: Automated docking (AD) of substrates in homology model to predict site of catalysis; molecular dynamics (MD) simulations to predict site of catalysis or affinity (LIE method); combined with pharmacophore model (P); site-directed mutagenesis (SM) studies used to refinement and validation of sequence alignment and/or homology model; quantum mechanics (QM) calculations to predict site of catalysis. References are given for applications of methods in separate studies.
[d]First published homology models for each isoform are underlined.
[e]Catalytic site prediction using manual docking in combination with energy minimizations and/or MD simulations.
[f]Affinity prediction using manual docking in combination with energy minimizations and 1D-QSARs.

II. HOMOLOGY MODEL BUILDING

A. Introduction

A wealth of detailed information on CYP enzyme action and substrate or inhibitor interactions with the binding cavity becomes accessible when a structure of the CYP protein is available. The main source for these structures is X-ray crystallography, but an even larger number of CYP structures have been built on the basis of these crystal structures using homology modeling techniques, which are summarized in Table 1. The progression in fields of X-ray crystallography and homology model building, as well as their mutual complementarity, will be illustrated and discussed in the following sections. The commonly used software for CYP homology model building and CYP protein structure analysis is listed in Table 2.

B. Crystal Structures

Crystal structures have been resolved for several cytosolic bacterial ($n = 75$), cytosolic eukaryotic ($n = 23$), and, recently, also of membrane-bound mammalian ($n = 6$) CYP oxygenases. In addition to these 104 CYP oxygenase structures, including 3 CYP102 (*BM3*) oxygenase-reductase complexes, 6 individual CYP reductase structures have been resolved. In addition, several new structures have been announced on the Protein Data Bank (PDB), such as human CYP3A4 and inhibitor-bound CYP2B4. The progression in CYP crystallography has recently culminated in the first structure of a human CYP2C9 (15). However, the distance of the bound *S*-warfarin substrate to the heme iron is large (~10 Å), which makes it unlikely that this crystal structure corresponds to a catalytically active state. In line with this suggestion, Arg105 and Arg108, implicated in the formation of putative anionic-binding sites by mutagenesis, pharmacophore and homology modeling studies both point away from the active site. Previously, in the substrate-bound CYP102 (*BM3*) structures, unreasonably long substrate-heme distances were found (16,17). It has been suggested that protein conformational changes driven by electron transfer trigger the movement of these substrates into effective positions for hydroxylation (18). In crystal structures of CYP101 (*cam*) (19–22) and CYP2C5 (23), multiple ligand-binding modes were observed. In CYP107A (*EryF*) (24), homotropic cooperativity of multiple ligands was observed. These effects probably are a result of ligand-induced conformational changes of the active site and the relatively large volume of the active site as compared with the volume occupied by the ligand.

Although crystallographical protein structures are firmly based on experimental data, it must be borne in mind that for resolutions of around 2 Å and worse, the electron density maps are not sufficiently detailed to resolve individual atoms. In order to circumvent this problem, molecular modeling techniques are often applied and usually yield reliable structural models that best represent the measured diffraction patterns (25). However, the likelihood of errors in the structure like misthreading or misplacement of secondary structure elements increases rapidly with diminishing resolution. In some cases, it is more appropriate to characterize these structures as crystallographic protein *models* to emphasize the distinction with atomic-resolution crystallographic *structures*.

C. Homology-based Protein Models

1. Methods Background

Like X-ray crystal structures, CYP protein homology model structures can be a valuable source of detailed information on binding cavity characteristics and other details of the ligand-protein interaction. In addition, a variety of biochemical, spectroscopical, and mutational data may be incorporated during the model building process, providing a powerful way to consolidate and rationalize the different experimental insights into the structure and function of the CYP enzyme. The commonly used software for CYP homology model building is listed in Table 2.

Building a homology model of a protein is a highly iterative process involving (often many) cycles of sequence or structural alignments and model building, analysis, and validation, as is depicted in the flowchart in Figure 2. Starting with the sequence of the target CYP protein, homologous sequences are found by scanning against sequences of proteins from the PDB using one of the many search algorithms available, such as PSI-BLAST (26); more are listed in Table 2. This is step 1 in the top cycle in Figure 2. Often, first a multiple sequence alignment is performed against homologous sequences of the CYP subfamily, as available from sequence databases, which is step 2 in this cycle. The sequence alignment of templates must be validated, e.g., using mutational studies and isoenzyme specificity to verify that the location of essential amino acids is correct, which is step 3 in this cycle, and if necessary it can be refined by structural alignment, step 4 in the top cycle in Figure 2. Although all CYPs share the same fold, variations in length and position of secondary structure elements between isoforms make structural alignments less than straightforward. Alignments can be carried out in two different ways (8). The first method is to align the sequence according to a superposition of available protein (crystal) structures. Using this approach, Hasemann et al. made the first CYP structural sequence alignment using crystal structures of CYP101 (*cam*), 102 (*BM3*), and 108 (*ter*) (9). The second method focuses on local structural similarities between equivalent structure elements. Jean et al. developed an algorithm to define the so-called ''common structure building blocks'' (CSBs) by comparing torsion angles of protein backbones (27). From the same three CYP crystal structures, 15 CSBs were derived, consisting of most CYP secondary structure elements except helices A and B′ (see also Fig. 1).

Once the templates have been aligned, step 4 in the top and step 1 in the central cycle in Figure 2, the amino acid sequence of the target protein is aligned with them, step 2 in the central cycle, and this alignment must be validated using

Table 2 Commercially Available Computational Molecular Modeling Software Used in the Studies Described in This Chapter

Program	Description	Reference	Web site
Homology modeling			
ExPASy	Proteomics web-server for the analysis of protein sequences and structures	166	www.expasy.org
(PSI-)BLAST	Algorithm for searching of nucleotide and protein databases	26	www.ncbi.nlm.nih.gov/BLAST/
ClustalW	Sequence alignment program/web server	167	www.ebi.ac.uk/clustalw/
Dali	Automatic structure comparison program for structural sequence alignments	168	www.ebi.ac.uk/dali/
GCG	Sequence alignment program	169	
JPRED	Secondary structure prediction and sequence alignment web server	170	www.compbio.dundee.ac.uk/~www-jpred/
T-COFFEE	Sequence alignment program	171	www.ch.embnet.org/software/TCoffee.html
NNPREDICT	Secondary structure prediction web server	172	www.cmpharm.ucsf.edu/~nomi/nnpredict.html
Prosa	Algorithm for searching of nucleotide and protein databases		
PSIPRED	Automatic sequence alignment, protein fold recognition, and secondary structure prediction web server	173	bioinf.cs.ucl.ac.uk/psipred/
Consensus	Comparative homology modeling program	174	www.accelrys.com/insight/consensus.html
Modeller	Comparative homology modeling program	175	salilab.org/modeller/modeller.html
SWISS-MODEL	Automated comparative protein-modeling server	176	expasy.hcuge.ch/swissmod/SWISS-MODEL.html
PredictProtein	Automatic sequence alignment, protein fold recognition, and secondary structure prediction web server	177	www.embl-heidelberg.de/predictprotein/
Protein structure analysis			
ERRAT	Program to evaluate the amino acid side-chain environment (packing quality) of protein structures	178	www.doe-mbi.ucla.edu/Services/ERRAT/
PROCHECK	Program to evaluate the stereochemical quality of protein structures	179	www.biochem.ucl.ac.uk/~roman/procheck/procheck.html

PROSA II	Program to evaluate the fold of protein structures	180	www.came.sbg.ac.at/Services/prosa.html
PROVE	Program for protein volume evaluation	181	www.ucmb.ulb.ac.be/SCMBB/PROVE/
Verify3D	Program to evaluate the amino acid side-chain environment (solvent accessibility) of protein structures	182	www.doe-mbi.ucla.edu/Services/Verify_3D/
Molecular dynamics simulations			
AMBER	MD simulation engine and force field	93	amber.ch.ic.ac.uk
CHARMM	MD simulation engine and force field	94	www.charmm.org
GROMACS	MD simulation engine and force field	96	www.gromacs.org
GROMOS	MD simulation engine and force field	183	www.igc.ethz.ch/gromos/
Automated docking			
AutoDock	Automated ligand-protein docking program with genetic algorithm and regression-based scoring function	67	www.scripps.edu/pub/olson-web/doc/autodock/
Chemscore	Regression-based scoring function	184	
Cscore	Suite of scoring functions (ChemScore, DOCK, FlexX, GOLD, PMF) and consensus scoring tool	185	www.tripos.com/sciTech/inSilicoDisc/virtualScreening/cscore.html
DOCK	Automated docking program with incremental construction and random search algorithm and force field–based scoring function	66	dock.compbio.ucsf.edu
FlexX	Automated docking program with incremental construction algorithm and regression-based scoring function	65	www.biosolveit.de/FlexX/
GLIDE	Automated docking program with exhaustive search algorithm and regression-based scoring function	186	www.schrodinger.com/Products/glide.html
GOLD	Automated ligand-protein docking program with genetic algorithm and force field–based scoring function	64	www.ccdc.cam.ac.uk/products/life_sciences/gold/

Continued

Table 2 *Continued*

Program	Description	Reference	Web site
PMF	Knowledge-based scoring function	187	
SCORE	Regression-based scoring function	188	
SLIDE	Automated docking program with multilevel and mean field algorithm and regression-based scoring function	189	www.bch.msu.edu/labs/kuhn/web/projects/slide/home.html
X-SCORE	Suite of three regression-based scoring functions and consensus scoring tool	190	sw16.im.med.umich.edu/software/xtool/
General modeling programs			
Insight II	Program with modules for homology modeling, protein structure analysis, pharmacophore/3D-QSAR modeling, MD		www.accelrys.com/insight/Insight2.html
QUANTA	Program with modules for homology modeling, protein structure analysis, MD		www.accelrys.com/quanta/
Sybyl	Program with modules for homology modeling, protein structure analysis, pharmacophore/3D-QSAR modeling, MD, automated docking		www.tripos.com/sciTech/inSilicoDisc/moleculeModeling/sybyl.html
WHAT IF	Program with modules for homology modeling, protein structure analysis, MD		www.cmbi.kun.nl/whatif/

Abbreviation: MD, molecular dynamics.

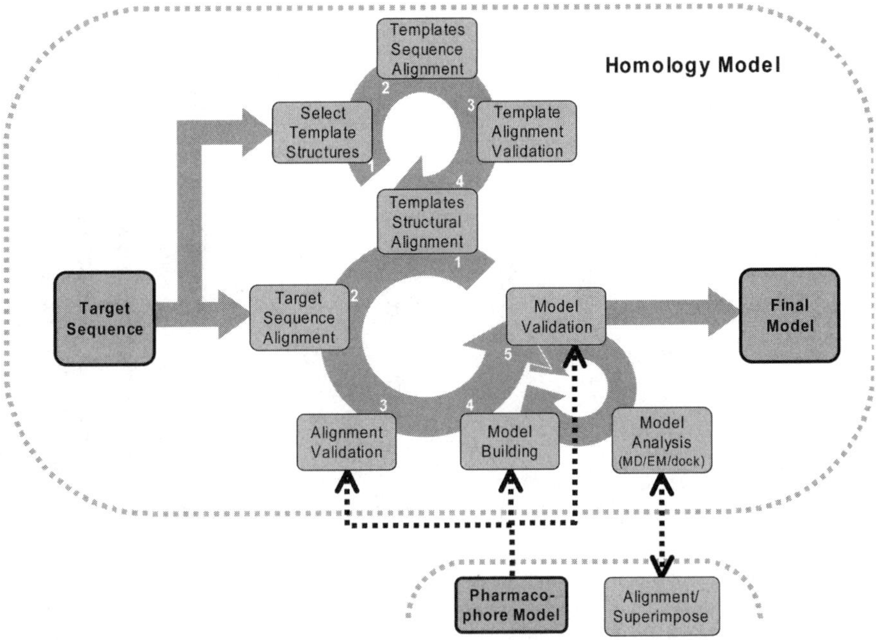

Figure 2 Flowchart for homology modeling. The highly iterative processes in alignment, model building, and analysis and validation involving three connected cycles is clearly visible. Details from a pharmacophore model can be used to optimize sequence alignment, to choose arrangements of certain important groups during model building, and for final validation. *Source*: Modified from Ref. 121.

mutational studies to verify that the location of essential amino acids is correct, step 3 in this cycle. For the parts of the target protein that have the best alignment with the template structure, or one of the template structures, the coordinates of the backbone of the target are taken from the homologous parts of the backbone of the template. Side-chain coordinates are transferred as well using the ''maximum overlap'' principle to keep the coordinates of all atoms from the template residue side chain that have topologically corresponding atoms in the target. Missing side-chain atoms are added and conformations are generated for inserted residues and loops and, finally, the structure can be optimized using, e.g., energy minimization, simulated annealing, or ''room temperature'' MD simulations. On the basis of basic protein structure quality parameters (e.g., Ramachandran distributions, packing quality), one of several generated model structures is selected. Together this constitutes the ''model building'' step 4 in the central cycle in Figure 2.

The 3D homology model of the protein structure is validated as much as possible with experimental data, such as mutagenesis data, NMR spin-relaxation measurements (28–30), active site chemical probe studies (31,32), and predictions of CYP substrate- and/or inhibitor-binding and regioselective metabolism and

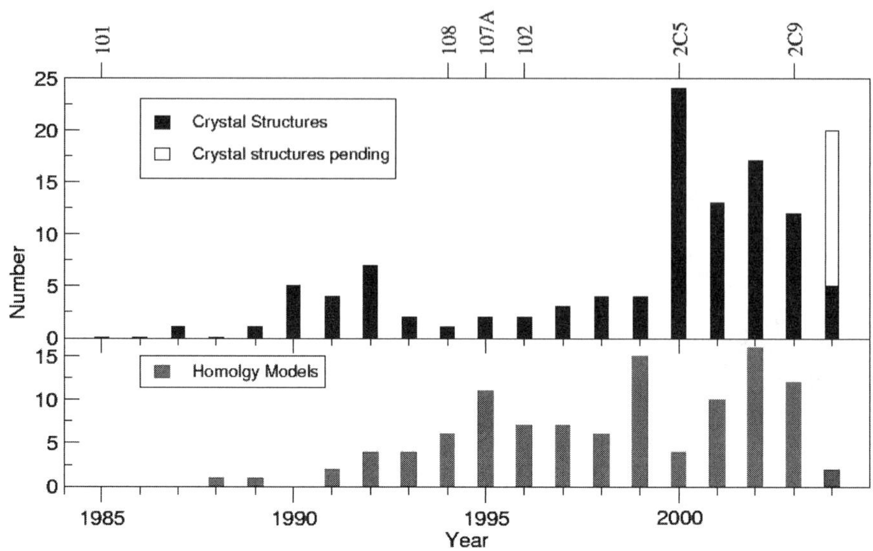

Figure 3 Development of the number of new crystal structures and homology models published over the years. The overall steady increase is clearly visible, as is the increase in homology and pharmacophore models during (temporary) declines in number of crystal structures, illustrating the complementary nature of experimental structure determination and model building. Publication of structures for isoforms that have been extensively used for homology model building are indicated at the top.

isoenzyme specificity, using, e.g., automated docking (sec. III.B) and MD simulations (sec. III.C), which is step 5 in the central cycle in Figure 2. Depending on the quality of the model as indicated by the validations used, it will be refined by additional cycles of structure optimization and repeated model building, which is the bottom-right cycle in Figure 2 and also involves steps 3 and 4 of the central cycle, as well as by improving the alignment with the template sequence(s) and/or structure(s) and repeating all steps in the central cycle.

Table 1 shows a selection of the many homology models built so far based on available CYP crystal structures (middle, 2004). Especially in the last few years, the use of homology modeling has increased considerably (Fig. 3). Therefore, for each CYP isoform, in addition to the earliest published model, only the most recent and most properly refined and validated homology models are included in this review. A more extensive overview of the earlier CYP homology models until 2000 is available in an earlier review by Ter Laak et al. (33) and a more recent review by us (34). The dependency of homology model building on the availability of (quality) crystal structures is obvious, the correlated growth in published crystal structures and homology models is clearly visible in Figure 3. In total, we counted 120 homology models published, of which 52 are included in Table 1. The progress in homology modeling is demonstrated below by the case of CYP2D6.

2. Applications

The first preliminary homology model for the human CYP2D6, containing only active site regions of the protein (11 segments) indicated Asp301 as an important amino acid residue for catalytic activity (35). For one of the first complete CYP2D6 homology models, a structural sequence alignment of the crystal structures of bacterial CYP101, 102, and 108 isoenzymes (9) was aligned with a multiple sequence alignment for members of the CYP2 family using site-directed mutagenesis data (36). Three known substrates (debrisoquine, dextromethorphan, and GBR 12909) and one inhibitor (ajmalicine) were docked into the active site of the CYP2D6 model (36) indicating that the protein model was able to accommodate large substrates and inhibitors. The docked conformations indicated a crucial H-bond with the amino acid Asp301. Two amino acids for which site-directed mutagenesis data were available, namely, Asp301 (37) and Val374 (38,39), were part of the active sites of the protein model (36). Later, a cluster of 51 manually superimposed and energy-minimized substrates were docked and evaluated in a refined model explaining 72 metabolic pathways catalyzed by CYP2D6 (40,41). It appeared that this model could predict correctly six out of eight metabolites observed in a test set of compounds. A CYP2D6 model was published together with models of CYP1A2 and CYP3A4 based on four bacterial crystal structures (42). In total, 14 CYP2D6 substrates and four nonspecific substrates known to be metabolized by 2D6 were successfully automatically docked into the active site. Almost all substrates had important Van der Waals interactions with Val307, Phe483, and Leu484, whereas Asp301 was always involved in charge-reinforced H-bonds with the protonated nitrogen atom of the substrates. Modi et al. introduced an original approach for the construction of CYP2D6 protein models (28,29). They used NMR spin-relaxation rate-derived distance restraints for codeine and MPTP to guide the homology modeling process.

Recently, four different sets of comparative models of CYP2D6 [including the first CYP2D6 homology models based on rabbit CYP2C5, the first mammalian CYP crystal structure (43)] were constructed using different structural alignments of combinations of four bacterial and/or CYP2C5 crystal structure templates (44). Selection of the best model was based on stereochemical quality and side-chain environment. The resulting model was further refined and validated by automated docking in combination with NMR-derived distance restraints mentioned above (28,29) and GRID/GOLPE analysis (45) of the active site. This latter analysis suggested that both Asp301 and Glu216 are required for metabolism of basic substrates, as Glu216 was identified as a key determinant in the binding of the basic moiety of substrates, while Asp301 might exert indirect effects on protein-ligand binding by movement of the B′-C loop and creating a net negative charge together with Glu216. This hypothesis was later confirmed by site-direct mutagenesis studies, which showed that neutralizing both Glu216 and Asp301 alters CYP2D6 substrate recognition (46).

More recently, a new homology model of CYP2D6 based on rabbit CYP2C5 was reported (11). The model was validated on its ability to (1) accommodate codeine in the binding orientation determined by Modi et al. (28), (2) to reproduce the binding orientations corresponding to metabolic routes of 10 substrates with rigid docking studies, (3) to remain stable during unrestrained MD at 300 K, and (4) inhibition characteristics of 11 CYP2D6 model ligands. Analogous homology models were generated for the rat CYP2D isoforms, 2D1/2/3/4 (11). Electrostatic potential calculations on the binding pocket residues showed large differences in the negative charge of active sites of the different isoenzymes, as is shown in Figure 4. These differences correlated well with observed IC_{50}s (11). This observation provided novel insights into differences in active site topology of human and rat 2D isoforms and on the validity of the rat as a model for human CYP2D6 activity. Interestingly, in addition to Asp301 and Glu216, this study also indicated Phe120 as a key interaction residue. Importance of Phe120 in CYP2D6 metabolism was later confirmed by site-directed mutagenesis studies (47,48).

In summary, protein homology models of CYP2D6 have provided a detailed insight in the overall protein structure as well as in the active site topology. The availability of the first crystal structure of a membrane bound CYP, i.e., the rabbit CYP2C5, as well as more advanced computational modeling techniques, e.g., MD simulations, have significantly improved the predictive quality of the resulting protein models. The most recent human CYP2C9 crystal structure has so far only been used in the construction of a new CYP2B4 model (49). Homology models proved to be very suitable to predict specificity and sites of metabolism for CYP substrates, but the quantitative prediction of (binding) affinities and turnover rates remains a challenge for the near future.

D. Conclusions: Homology Model Building

The first human CYP crystal structure [i.e., CYP2C9 (15)] together with some other crystal structures of mammalian CYPs [rabbit CYP2C5 (23,50,51) and human CYP2B4 (52) and CYP2C8 (53)] provide a still growing basis for modeling work. When using crystal structures for CYP-modeling studies, special attention should be given to the state in which the CYP was crystallized, as this may not correspond to a catalytically active state. It may be necessary to combine such a structure with a relevant active site loop structure from a related CYP crystal structure in which the loop is known to be in the active state. Although the resulting model is sometimes classified as "only" a homology model, the model can be much more relevant for rationalizing and predicting enzyme activity than the respective crystal structure.

As shown in Table 1, the trend in CYP homology model building is clearly toward a multi-integrated approach using multiple sequence alignments (multiple), structural alignments in combination with mutational, spectroscopic and enzyme kinetic experimental data and new or available pharmacophore models,

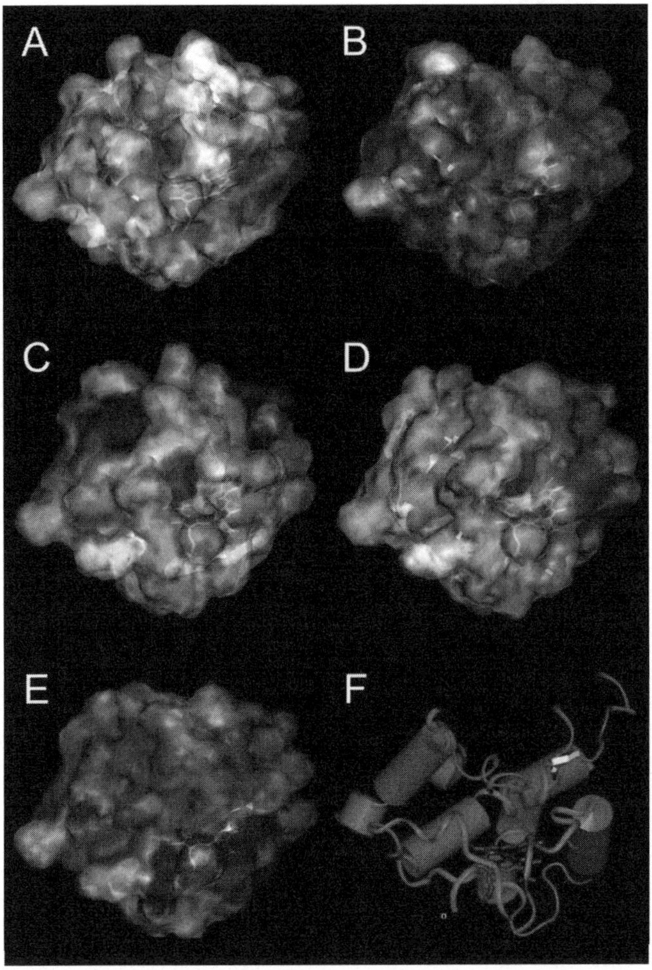

Figure 4 Electronic potential of the active site of (**A**) CYP2D1, (**B**) CYP2D2, (**C**) CYP2D3, (**D**) CYP2D4, and (**E**) CYP2D6. Grey, black, and white indicate negative, positive, and neutral electrostatic potentials, respectively. The electrostatic potential surfaces are shown in the same orientation as the enzyme shown in (**F**), with the heme group at the bottom of the active site. *Source*: Modified from Ref. 11.

automated docking, and MD simulations. Mutational data and experimental knowledge of SRSs are very frequently used to optimize sequence and structural alignments and validate final homology models. Recent homology models show good agreement with site-directed mutagenesis studies, especially in the case of CYP2D6, where Phe120, Glu216, and Asp301 were indicated as key determinants of selectivity and regiospecificity of substrate binding and catalysis by CYP2D6. Several cases are known in which CYP homology models are combined with pharmacophore models (i.e., CYP1A2, 2B6, 2C8/9/18/19, and 2D6) or QM calculations (i.e., CYP2E1 and 3A4). In recent studies, for nearly all isoforms, predictions of site(s) of catalysis in substrates, substrate selectivity, isoenzyme specificity, e.g., CYP1A2, 2D6, and 3A4 (42), 2B1/4/5 (54), and 2C8/9/18/19 (55), and ligand (binding) affinity by automated docking and MD simulations have been used during the model building process to evaluate intermediate stages, as well as for a first validation of the final model. An upcoming and potentially powerful approach is the use of MD simulation techniques for optimization of the final model, e.g., for CYP2B6 (56) and other oxygenase enzymes (12). Although CYP homology models are rarely used for quantitative binding affinity predictions (e.g., CYP1A1, as described in sec. III.C.2), CYP homology models for many isoforms have been used successfully for the prediction of substrate selectivity and site of metabolism by automated docking and MD (see sec. III.B and III.C, respectively).

III. LIGAND-PROTEIN BINDING

A. Introduction

One of the critical issues in predicting substrate metabolism is the combination of statistic, orientational, and energetic factors that each contributes to various degrees in different CYP enzymes. A first step is made by automated docking programs that aim to predict energetically favorable binding conformations of ligands in the active site cavity. Further exploration of binding conformations and relative probabilities can be done using MD simulations, which can also be used for predictions of binding affinities. The commonly used software for CYP ligand-binding predictions is listed in Table 2.

Comparison between different ligand-bound crystal structures of CYP101 reveals small perturbations at the binding site and for some ligands result in important alterations in the recognition pattern for binding (57). Comparison between ligand-bound and ligand-free crystal structures of other CYP isoenzymes, i.e., 102 (*BM3*) (58), 107A (*EryF*) (59), 119 (60), 51 (*14α*) (61), and 2C5 (23,51), showed ligand-induced changes in size, shape, and hydration of the active site. This conformational flexibility of the active site is likely to underlie the capacity of many CYP isoenzymes to metabolize structurally diverse ligands of different sizes and to bind ligands in multiple binding modes (generating different metabolic products).

B. Automated Ligand-Protein Docking

1. Methods Background

Automated ligand-protein docking methods predict energetically favorable conformations and orientations of ligands (or substrates) in the binding pocket of a protein. Algorithms to generate different poses (docking) are combined with scoring functions that consider the tightness of the protein-ligand interactions (62,63). When a structural model of a CYP isoenzyme is available, from X-ray crystallography or homology model building, docking methods can be used for predictions of the catalytic site and binding affinity for CYP substrates and to identify potential CYP substrates in a chemical database (protein-based virtual screening). In addition, automated docking can be used to generate input structures for pharmacophore modeling and MD simulations (see sec. III.C''). The heme prosthetic group, the presence of water in the active site, the relatively large size and large conformational flexibility of the active site, the broad range of substrates and many catalytic sites in most substrates, as described in section II.B, make CYPs rather difficult docking targets.

Several docking algorithms and scoring functions have been described in the past few years, the most commonly used for CYPs are GOLD (64), FlexX (65), DOCK (66), and AutoDock (67), and the scoring functions from C-Score (TRIPOS Inc., St. Louis, Missouri, U.S.); more are listed in Table 2. Docking accuracy, i.e., structure prediction, and scoring accuracy, i.e., prediction of binding free energy, of docking-scoring combinations vary considerably with the selected target protein and physicochemical details of target-ligand interactions (68,69) and even depend on fine details of the CYP protein structure, as shown for CYP101 (*cam*) (70). Therefore, a docking-scoring strategy must be tailored to the system of interest on the basis of a test set of ligand-bound protein crystal structures. For CYP crystal structures such a methodological evaluation was performed recently (71) and so far has not been explored for CYP homology models. Water molecules can play an essential role in ligand-protein binding (72–74), and water-mediated ligand-protein interactions have been observed in crystal structures of most CYP isoforms, e.g., rabbit CYP2C5 (23,51) and human CYP2C9 (15). Most docking algorithms, however, ignore the possibility of water-mediated interactions between protein and ligand (75,76). Ligand-induced changes in the active sites of CYPs can lead to important alterations in ligand recognition patterns and are accounted for in some new methods (75,77,78), but the inclusion of full protein flexibility in automated ligand-protein docking gives an enormous increase in complexity (62,75), and until now no such method has been validated extensively (76,79).

2. Applications

Automated docking has primarily been applied to refine and validate CYP pharmacophore models (55,80) and protein homology models (see Table 1), and manual docking has been extensively used as well in the past. Validation is achieved by comparing predicted binding modes of substrates with reported metabolic product(s) and was used during the homology modeling process of CYP1A2 (42), 2B6 (56,81), the 2C subfamily (55), 2D6 (11,42,44), 2E1 (82), 3A4 (42,83), and 6B1 (84). In Figure 5, dextromethorphan is shown in its docked binding orientation in the binding cavity of the CYP2D6 homology model.

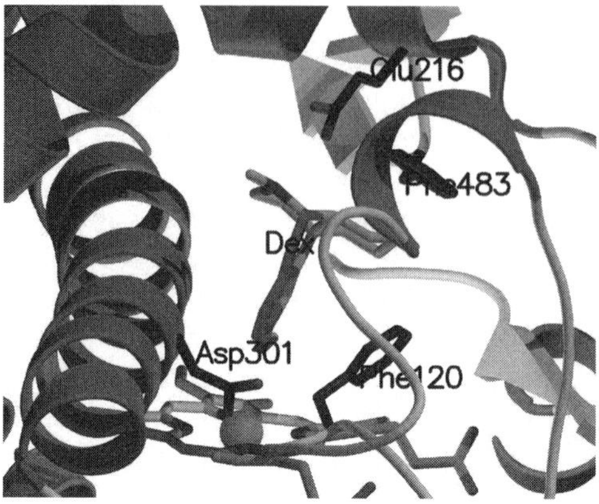

Figure 5 Example of substrate docking. Dextromethorphan docked in the active site of the homology model of CYP2D6. Key interacting side chains (Phe120, Glu216, Asp301, and Phe483) in the CYP2D6 active site binding cavity are shown as stick models, the rest of the active site is shown in cartoons. The figure was generated using Molscript and Raster3D. *Source*: From Ref. 119 and 120.

Experimentally determined binding affinities of 11 different CYP101 (*cam*)-ligand complexes showed no clear correlation with values from different C-Score scoring functions on CYP101(*cam*)-ligand crystal structures or on complexes produced by the automated docking program FlexX (85). Nevertheless, the docking and scoring could still be used to prioritize in silico screening hits of a chemical database. For wild-type CYP101 (*cam*) and the L244A mutant, DOCK was used for in silico screening to identify potential substrates (86,87) and imidazole inhibitors (88).

The automated docking program GOLD was recently used to predict binding of octane, octanoic, and lauric acid to CYP102 (*BM3*) (89). Predicted binding modes were further analyzed with MD simulations and QM calculations on substrate reactivity, as described in more detail in section III.C. Marked differences were seen in the probabilities of different binding modes in wild-type CYP102 and three mutants, and these trends were comparable to experimental product formation.

Recently, three popular automated docking programs AutoDock, FlexX, and GOLD have been evaluated on their ability to predict the binding mode of ligands from 19 crystallized complexes of CYPs and ligands, either without water, with crystal waters, or with water in positions based on predictions by GRID (71). Although the individual programs only showed mediocre performance, pooling and rescoring of all solutions improved the prediction accuracy significantly. Binding mode predictions were strongly improved by including either crystal or predicted water, and even with higher deviation from the reference crystal structure, correct prediction of the site of catalysis was still possible.

C. Molecular Dynamics

1. Methods Background

The role of CYPs as biotransformation enzymes and the broad substrate selectivity of many CYPs are reflected by the size of the active site, which is generally large compared with the size of the substrates. Furthermore, many substrates and inhibitors exhibit high degrees of internal flexibility, and large conformational changes of the CYP protein structure on ligand binding are known to play an important role in the CYP reaction cycle. By using MD simulations, these processes associated with CYP-ligand interaction can be described and understood in atomic detail (89–91).

MD simulations provide a detailed insight in the behavior of molecular systems in both space and time, with ranges of up to nanometers and nanoseconds attainable for a system of the size of a CYP enzyme in solution. However, MD simulations are based on empirical molecular mechanics (MM) force field descriptions of interactions in the system, and therefore depend directly on the quality of the force field parameters (92). Commonly used MD programs for CYPs are AMBER (93), CHARMM (94), GROMOS (95), and GROMACS (96), and results seem to be comparable between methods (also listed in Table 2). For validation, direct comparisons between measured parameters and parameters calculated from MD simulations are possible, e.g., for fluorescence (97) and NMR (cross-relaxation) (98,99). In many applications where previously only energy minimization would be applied, it is now common to perform one or several MD simulations, as Ludemann et al. and Winn et al. (100–102) performed in studies of substrate entrance and product exit.

Starting conformations of MD simulations are generally obtained from automated docking methods (see also the previous sec. III.B). Conversely, structural snapshots from MD runs can be used, with or without energy minimization, to include protein flexibility into docking simulations. Because of limited timescales of simulations, results can remain dependent on the starting conformations. The state of the art is to solvate the simulated system in explicit water in a periodic box, which although computationally more expensive than continuum solvent or limited solvation schemes, avoids many of the possible pitfalls associated with them. Furthermore, several methodological developments to increase the efficiency of simulations of proteins (and ligand) in water are coming available (103,104). Various analysis methods are available to estimate statistics of ligand binding and orientation and energies of ligand-protein interactions or free energies of binding (105).

2. Applications

Many applications of MD can be found in refinement and optimization of homology models, as highlighted in section II.C and Table 1. This type of use of MD simulations is rapidly expanding. In some cases, the dynamics of the ligand is included and distance of oxidation site in the substrate to the heme iron during the simulation is taken as a quality measure for the homology model being constructed, e.g., CYP2B6 (56) and CYP2D6 (11).

Extensive MD simulations were used to identify key conformational changes upon ligand binding for CYP102 (*BM3*) (91,106). MD simulations were also used to study substrate access and product exit, as described in section III. C.2, by random expulsion simulations and related methods for CYP101 (*cam*), 102 (*BM3*), 107A (*EryF*), (100–102,107). Free energy of hydration of the active site of CYP101 has been studied using thermodynamic integration (TI) (108,109). Putative protein domain boundaries of CYP102 (*BM3*) were identified using dynamical cross-correlation (DCMM) analysis (110). On the basis of MD simulations of substrates in the bound state in the enzyme and free state in solution, substrate binding affinities have been predicted with the LIE method for CYP101 (111,112) and CYP1A1 (113), FEP for CYP101 (114) and normal modes, and Poisson-Boltzmann for CYP2B4 (49).

For the prediction of regiospecific substrate oxidation, QM calculations and MD simulations have been combined to account for differences in reactivity in the substrate as well as distributions of multiple binding conformations, as applied for CYPs 101 (*cam*) (111,115,116), 107A (*EryF*) (117), and 2E1 (82). For CYP102 (*BM3*), probabilities for substrate orientation based on statistics of substrate binding from MD simulations were combined with probabilities for substrate reaction based on QM calculations of H-atom abstraction by a hydroxyl radical (89). The experimentally known preference for subterminal, distal hydroxylation of octane, octanoic acid, and lauric acid are thus reproduced.

D. Conclusions: Ligand-Protein Binding

Manual docking has been extensively used in the past, and automated docking approaches are nowadays successfully applied for the prediction of the site of catalysis in substrates, the refinement and validation of CYP homology models (i.e., CYP1A2, 2B6, the 2C subfamily, 2D6, 2E1, 3A4, and 6B1, see Table 1 and Figure 2), and the construction of pharmacophore models (i.e., CYPs of the 2C subfamily). Docking algorithms are successful in determining binding conformations and orientations of CYP-ligand complexes, but docking scoring functions are still not suitable for accurate prediction of ligand-binding affinity. Despite the fact that database docking methods have been successfully applied to crystal structures (69) as well as protein homology models (118), structure-based in silico screening studies on CYPs have so far only been reported for CYP101 (*cam*).

New automated docking strategies, considering explicit water molecules, partial protein flexibility, and (consensus) rescoring of docking poses have already been found to improve binding mode prediction of CYP ligands. It is expected that these improved docking strategies will also be used more and more for CYP in silico screening studies in the future. The combination of docking with MD simulations to improve docking predictions and explore conformational flexibility of substrates and CYP enzymes, as well as the explicit inclusion of water, is seen in an increasing number of studies and shows much promise.

MD simulations are used in various ways to study CYP-ligand interactions. As shown in Table 1, applications for homology model optimization and validation of model stability and the prediction of sites of catalysis in substrates are becoming common practice. Prediction of substrate and inhibitor binding affinity and orientation have been reliable in the cases of CYP101 (*cam*), 2B4, and 1A1, and combined with QM calculations on the substrate for predictions of product formation for CYP101 (*cam*), 102 (*BM3*), 107A (*EryF*), and 2E1.

Currently, most MD studies have been limited to short (~100 picoseconds, ps) or even very short (~10 ps) simulation times and only a single simulation. A notable exception is the study by Winn et al. (102), where 17 simulations of several hundreds of picoseconds to a few nanoseconds are used to describe substrate entrance to and product exit from CYP101 (*cam*), 102 (*BM3*), and 107A (*EryF*). For a proper description of ligand dynamics and conformational changes in the enzyme, simulation times exceeding hundreds of nanoseconds may well be needed. Applications of MD simulations for homology model optimization and validation are becoming common practice. Prediction of substrate and inhibitor binding affinity and orientation have been reliable in the cases of CYP101 (*cam*) and CYP1A1, and combined with QM calculations on the substrate for predictions of product formation for CYP101 (*cam*), 102 (*BM3*), 107A (*EryF*), 2E1, and 2B4.

A remaining hurdle in the accurate calculation of CYP ligand binding affinity is the prediction of multiple binding modes of substrates, reflected in a flexible binding cavity that is large in relation to the size of the substrates, and, subsequently, for many substrates the binding mode is not as strictly defined as is the case for many other more substrate-specific enzymes. This restricts the application of MD-based methods that can provide the most accurate free-energy calculations.

IV. CONCLUSIONS AND PERSPECTIVES

Early consideration of properties of the CYP family of biotransformation enzymes involved in drug metabolism is increasingly seen as essential for efficient discovery and development of new drugs and drug candidates. Apart from the application of in vitro tools, this necessitates the application of in silico tools that can accurately rationalize and predict metabolic properties and products of drug candidates already in early stages of the lead finding and optimization process. The primary aim of this chapter was to describe recent advances in CYP homology model-building methods and methods for modeling of CYP-ligand interactions.

The growth in available CYP protein structures from X-ray crystallography, presented in Figure 3, has inspired several investigators to detailed analysis of CYP structure and of CYP-ligand interaction, and it has been the basis for many homology modeling work, see Table 1 for homology structures of CYPs and Table 2 for most commonly used software for CYP molecular modeling. Figure 1 shows the 3D structure of a CYP2D6, which shares all major features of CYPs, together with the organization of SRSs and secondary structure elements. In this respect, homology structures have been complementary to crystal structures, especially for the mammalian CYPs now that mammalian crystal structures have recently become available. Several CYP isoforms have only been crystallized in a nonactive ligand-bound state, and for some it is not known whether the crystal structure corresponds to an active state or not. It is important to realize that this may make a crystallized structure of a certain CYP isoenzyme not directly applicable to, e.g., explaining observed substrate selectivity and product formation of the enzyme. For several CYPs, ligand-induced changes in the size, shape, and hydration of the active sites as well as multiple binding modes for some ligands can be observed. These effects are likely to underlie the capacity of many CYPs to metabolize structurally diverse substrates of different sizes and to generate different metabolic products from these substrates.

Automated docking approaches are successfully applied for prediction of substrate orientations and binding and of the site of catalysis in substrates for many CYP isoforms, for the refinement and optimization of homology models, and for the construction of pharmacophore models. Docking is still not suitable for quantitative prediction of ligand binding affinity, which may be attributed at least in part to the lack of proper consideration of active site water and protein flexibility, and structure-based in silico screening studies on CYPs have so far been reported only for CYP101 (*cam*). MD-based methods for prediction of binding affinities, for finding a more comprehensive description of multiple sites of catalysis in substrates, for exploring ligand binding, and for exploring protein flexibility of homology models in more detail have shown to be more reliable.

Generally speaking, it is clear that computational approaches in parallel with high(er) throughput experimental technologies are among the newer and fastest developing approaches in drug discovery, drug metabolism, and toxicology. New links with other current developments, such as in neural network computing, genomics, proteomics, and bioinformatics, are within reach, and successful exploitation of these links will allow significant progress toward in silico prediction of drug activity, drug metabolism, and toxicity. Future developments toward a fully in silico prediction of ADME(T) will need more accurate models for the prediction of drug absorption (A), disposition (D), and elimination (E), in addition to the models for prediction of metabolism (M), and toxicology (T). Furthermore, in general, a better understanding of the key issues in drug activity should be developed. This type of preexperimental prediction of nonpharmacological drug properties with in silico ADME(T) screening is likely to be one of the challenging developments with a great scientific and practical impact.

It has been shown for CYP2D6 and CYP2C9 that good protein homology models can be built for CYPs and can be used to obtain useful information concerning amino acids important for substrate and/or inhibitor binding. Because of relatively low sequence and structure homology in the ligand-binding site region of some CYPs, these predictions, however, have to be experimentally verified and critically considered. For elucidation of the role of specific amino acids in the catalytic activity of CYPs, homology models can be used as well to rationalize observed differences; however, the turnover activities cannot yet be predicted accurately. Indications of substrate selectivity can be obtained qualitatively on the basis of homology models, although these predictions should always be experimentally verified and critically considered as well.

Evaluation of different scoring functions, docking algorithms and pooling of docking results from different docking algorithms, has recently been shown to be critical in obtaining good results for ligand docking in CYP crystal structures and also to make docking into homology model structures feasible. Much additional gain is expected from these methodological developments. Moreover, addition of water bound at specific (experimentally derived or predicted) locations in the active sites is the topic of current investigations (71,99) and is expected to significantly improve the quality of docking results. Another important factor is the existence of substrate-bound crystal structures where the substrate apparently is not in a catalytically active position with respect to the heme iron. This indicates the importance of the kinetics and dynamics of substrate binding and oxygen binding and reduction.

CYPs generally have broad substrate selectivities and often substrates are found in multiple binding modes, instead of in shape-complementary fits that are common for other enzymes with high specificities (and some CYPs as well). This makes statistical considerations like relative weights of different binding modes crucial and in addition increases the flexibility of ligands within each binding mode. Energetic consideration of detailed ligand-protein interactions gives insight in the (relative) importance of different sites of interaction between the ligand and the protein. Finally, the inclusion of explicit water molecules into the active site allows the elucidation of the (mediating) role of water in ligand binding. A remaining hurdle in the accurate calculation of CYP ligand binding affinity is the prediction of multiple binding modes for substrates, reflected in a flexible binding cavity that is large compared to the size of the substrates and, subsequently, for many substrates the binding mode is not as strictly defined as is the case for many other more substrate selective enzymes. This restricts the application of MD-based methods that can provide the most accurate free-energy calculations.

A trend in CYP homology model building is clearly toward an integrated approach of multiple sequence and structural alignments, combination with new or available pharmacophore models, automated docking, MD simulations and validation with mutational, spectroscopic, structural, and enzyme kinetic data. Predictions of substrate selectivities and sites of metabolism have been successful for homology models of several CYP isoforms, and, for example, recent models of CYP2D6 and 2B and 2C enzymes show very good agreement with site-directed mutation studies. Furthermore, the use of homology modeling methods to enrich X-ray crystallographical structures with other experimental data, e.g., biochemical or spectroscopical, on activities and binding orientations of substrates, has been successful in some applications and is expected to become important in the near future.

It has been shown that the structure-based modeling of CYP enzymes has successfully added to our understanding of CYP structure and function in a way that is complementary to experimental studies. Another clear trend in many recent computational studies is to combine several modeling techniques to arrive at meaningful rationalization of experimental data, and interpretations of CYP function. Indeed, much of the recent progress in this area stems from combining computational methods with solid experimental data, a development that will prove to be critical for the success of CYP research as a whole.

REFERENCES

1. Clark DE, Grootenhuis PDJ. Progress in computational methods for the prediction of ADMET properties. Curr Opin Drug Discov Dev 2002; 5:382–390.
2. Hou TJ, Xu XJ. Recent development and application of virtual screening in drug discovery: an overview. Curr Pharm Des 2004; 10:1011–1033.
3. Li AP. Screening for human ADME/Tox drug properties in drug discovery. Drug Discov Today 2001; 6:357–366.
4. van de Waterbeemd H, Gifford E. ADMET in silico modelling: towards prediction paradise? Nat Rev Drug Discov 2003; 2:192–204.
5. Yu H, Adedoyin A. ADME-Tox in drug discovery: integration of experimental and computational technologies. Drug Discov Today 2003; 8:852–861.
6. Vermeulen NPE. Role of metabolism in chemical toxicity. In: Ioannides C, ed. Cytcochromes P450: Metabolic and Toxicological Aspects. Boca Raton, FL: CRC Press, 1996:29–53.
7. Guengerich FP. Common and uncommon cytochrome P450 reactions related to metabolism and chemical toxicity. Chem Res Toxicol 2001; 14:611–650.
8. Li H. In: Messerschmidt A. ed. Handbook of Metalloproteins. Chichester, UK: Wiley, 2001:267–282.
9. Hasemann CA, Kurumbail RG, Boddupalli SS, et al. Structure and function of cytochromes P450: a comparative analysis of three crystal structures. Structure 1995; 3:41–62.
10. Gotoh O. Substrate recognition sites in cytochrome P450 family 2 (CYP2) proteins inferred from comparative analyses of amino acid and coding nucleotide sequences. J Biol Chem 1992; 267:83–90.
11. Venhorst J, ter Laak AM, Commandeur JN, et al. Homology modeling of rat and human cytochrome P450 2D (CYP2D) isoforms and computational rationalization of experimental ligand-binding specificities. J Med Chem 2003; 46:74–86.
12. Feenstra KA, Hofstetter K, Bosch R, et al. Enantioselective substrate binding in a monooxygenase protein model by molecular dynamics and docking. *Biophys J* 2006; 91:3206–3216.

13. de Groot MJ, Alex AA, Jones BC. Development of a combined protein and pharmacophore model for cytochrome P4502C9. J Med Chem 2002; 45: 1983–1993.
14. Vermeulen NPE. Prediction of drug metabolism: the case of cytochrome P450 2D6. Curr Top Med Chem 2003; 3:1227–1239.
15. Williams PA, Cosme J, Ward A, et al. Crystal structure of human cytochrome P450 2C9 with bound warfarin. Nature 2003; 424:464–468.
16. Li HY, Poulos TL. The structure of the cytochrome p450BM-3 haem domain complexed with the fatty acid substrate, palmitoleic acid. Nat Struct Biol 1997; 4:140–146.
17. Haines DC, Tomchick DR, Machius M, et al. Pivotal role of water in the mechanism of P450BM-3. Biochemistry 2001; 40:13456–13465.
18. Modi S, Sutcliffe MJ, Primrose WU, et al. The catalytic mechanism of cytochrome P450 BM3 involves a 6 A movement of the bound substrate on reduction. Nat Struct Biol 1996; 3:414–417.
19. Raag R, Poulos TL. Crystal structures of cytochrome P-450CAM complexed with camphane, thiocamphor, and adamantane: factors controlling P-450 substrate hydroxylation. Biochemistry 1991; 30:2674–2684.
20. Schlichting I, Jung C, Schulze H. Crystal structure of cytochrome P-450cam complexed with the (1S)-camphor enantiomer. FEBS Lett 1997; 415:253–257.
21. Bell SG, Chen X, Sowden RJ, et al. Molecular recognition in (+)-alpha-pinene oxidation by cytochrome P450cam. J Am Chem Soc 2003; 125:705–714.
22. Strickler M, Goldstein BM, Maxfield K, et al. Crystallographic studies on the complex behavior of nicotine binding to P450cam (CYP101). Biochemistry 2003; 42:11943–11950.
23. Wester MR, Johnson EF, Marques-Soares C, et al. Structure of a substrate complex of mammalian cytochrome P450 2C5 at 2.3 A resolution: evidence for multiple substrate binding modes. Biochemistry 2003; 42:6370–6379.
24. Cupp-Vickery J, Anderson R, Hatziris Z. Crystal structures of ligand complexes of P450eryF exhibiting homotropic cooperativity. Proc Natl Acad Sci U S A 2000; 97:3050–3055.
25. Brunger AT, Adams PD. Molecular dynamics applied to X-ray structure refinement. Acc Chem Res 2002; 35:404–412.
26. Altschul SF, Madden TL, Schaffer AA, et al. Gapped BLAST and PSI-BLAST: a new generation of protein database search programs. Nucleic Acids Res 1997; 25:3389–3402.
27. Jean P, Pothier J, Dansette PM, et al. Automated multiple analysis of protein structures: application to homology modeling of cytochromes P450. Proteins 1997; 28:388–404.
28. Modi S, Paine MJ, Sutcliffe MJ, et al. A model for human cytochrome P450 2D6 based on homology modeling and NMR studies of substrate binding. Biochemistry 1996; 35:4540–4550.
29. Modi S, Gilham DE, Sutcliffe MJ, et al. 1-Methyl-4-phenyl-1,2,3,6-tetrahydropyridine as a substrate of cytochrome P450 2D6: allosteric effects of NADPH-cytochrome P450 reductase. Biochemistry 1997; 36:4461–4470.
30. Schoch GA, Attias R, Le Ret M, et al. Key substrate recognition residues in the active site of a plant cytochrome P450, CYP73A1. Homology guided site-directed mutagenesis. Eur J Biochem 2003; 270:3684–3695.

31. Demontellano PRO. Arylhydrazines as probes of hemoprotein structure and function. Biochimie 1995; 77:581–593.
32. Yamaguchi Y, Khan KK, He YA, et al. Topological changes in the CYP3A4 active site probed with phenyldiazene: effect of interaction with NADPH-cytochrome P450 reductase and cytochrome B5 and of site-directed mutagenesis. Drug Metab Dispos 2004; 32:155–161.
33. Ter Laak AM, Vermeulen NP, de Groot MJ. Molecular modeling approaches to predicting drug metabolism and toxicity. In: Rodrigues AD, ed. Drug-Drug Interactions. New York: Marcel Dekker, Inc., 2002:505–548.
34. de Graaf C, Vermeulen NP, Feenstra KA. Cytochrome p450 in silico: an integrative modeling approach. J Med Chem 2005; 48:2725–2755.
35. Koymans LM, Vermeulen NP, Baarslag A, et al. A preliminary 3D model for cytochrome P450 2D6 constructed by homology model building. J Comput Aided Mol Des 1993; 7:281–289.
36. de Groot MJ, Vermeulen NP, Kramer JD, et al. A three-dimensional protein model for human cytochrome P450 2D6 based on the crystal structures of P450 101, P450 102, and P450 108. Chem Res Toxicol 1996; 9:1079–1091.
37. Ellis SW, Hayhurst GP, Smith G, et al. Evidence that aspartic acid 301 is a critical substrate-contact residue in the active site of cytochrome P450 2D6. J Biol Chem 1995; 270:29055–29058.
38. Ellis SW, Rowland K, Ackland MJ, et al. Influence of amino acid residue 374 of cytochrome P-450 2D6 (CYP2D6) on the regio- and enantio-selective metabolism of metoprolol. Biochem J 1996; 316(pt 2):647–654.
39. Ellis SW, Rowland K, Harlow JR, et al. Regioselective and enantioselective metabolism of metoprolol by 2 human cDNA-derived Cyp2d6 proteins. Br J Pharmacol 1994; 112:U124–U124.
40. de Groot MJ, Ackland MJ, Horne VA, et al. A novel approach to predicting P450 mediated drug metabolism. CYP2D6 catalyzed N-dealkylation reactions and qualitative metabolite predictions using a combined protein and pharmacophore model for CYP2D6. J Med Chem 1999; 42:4062–4070.
41. de Groot MJ, Ackland MJ, Horne VA, et al. Novel approach to predicting P450-mediated drug metabolism: development of a combined protein and pharmacophore model for CYP2D6. J Med Chem 1999; 42:1515–1524.
42. De Rienzo F, Fanelli F, Menziani MC, et al. Theoretical investigation of substrate specificity for cytochromes P450 IA2, P450 IID6 and P450 IIIA4. J Comput Aided Mol Des 2000; 14:93–116.
43. Cosme J, Johnson EF. Engineering microsomal cytochrome P450 2C5 to be a soluble, monomeric enzyme. Mutations that alter aggregation, phospholipid dependence of catalysis, and membrane binding. J Biol Chem 2000; 275:2545–2553.
44. Kirton SB, Kemp CA, Tomkinson NP, et al. Impact of incorporating the 2C5 crystal structure into comparative models of cytochrome P450 2D6. Proteins 2002; 49: 216–231.
45. Cruciani G, Watson KA. Comparative molecular field analysis using GRID force-field and GOLPE variable selection methods in a study of inhibitors of glycogen phosphorylase b. J Med Chem 1994; 37:2589–2601.
46. Paine MJ, McLaughlin LA, Flanagan JU, et al. Residues glutamate 216 and aspartate 301 are key determinants of substrate specificity and product regioselectivity in cytochrome P450 2D6. J Biol Chem 2003; 278:4021–4027.

47. Flanagan JU, Marechal JD, Ward R, et al. Phe120 contributes to the regiospecificity of cytochrome P450 2D6: mutation leads to the formation of a novel dextromethorphan metabolite. Biochem J 2004; 380:353–360.
48. Keizers PH, Lussenburg BM, de Graaf C, et al. Influence of phenylalanine 120 on cytochrome P450 2D6 catalytic selectivity and regiospecificity: crucial role in 7-methoxy-4-(aminomethyl)-coumarin metabolism. Biochem Pharmacol 2004; 68:2263–2271.
49. Harris DL, Park JY, Gruenke L, et al. Theoretical study of the ligand-CYP2B4 complexes: effect of structure on binding free energies and heme spin state. Proteins 2004; 55:895–914.
50. Williams PA, Cosme J, Sridhar V, et al. Mammalian microsomal cytochrome P450 monooxygenase: structural adaptations for membrane binding and functional diversity. Mol Cell 2000; 5:121–131.
51. Wester MR, Johnson EF, Marques-Soares C, et al. Structure of mammalian cytochrome P450 2C5 complexed with diclofenac at 2.1 A resolution: evidence for an induced fit model of substrate binding. Biochemistry 2003; 42:9335–9345.
52. Scott EE, He YA, Wester MR, et al. An open conformation of mammalian cytochrome P450 2B4 at 1.6-A resolution. Proc Natl Acad Sci U S A 2003; 100: 13196–13201.
53. Schoch GA, Yano JK, Wester MR, et al. Structure of human microsomal cytochrome P450 2C8. Evidence for a peripheral fatty acid binding site. J Biol Chem 2004; 279:9497–9503.
54. Spatzenegger M, Wang QM, He YQ, et al. Amino acid residues critical for differential inhibition of CYP2B4, CYP2B5, and CYP2B1 by phenylimidazoles. Mol Pharmacol 2001; 59:475–484.
55. Ridderstrom M, Zamora I, Fjellstrom O, et al. Analysis of selective regions in the active sites of human cytochromes P450, 2C8, 2C9, 2C18, and 2C19 homology models using GRID/CPCA. J Med Chem 2001; 44:4072–4081.
56. Bathelt C, Schmid RD, Pleiss J. Regioselectivity of CYP2B6: homology modeling, molecular dynamics simulation, docking. J Mol Model 2002; 8:327–335.
57. Fradera X, De La Cruz X, Silva CH, et al. Ligand-induced changes in the binding sites of proteins. Bioinformatics 2002; 18:939–948.
58. Li H, Poulos TL. Conformational dynamics in cytochrome P450-substrate interactions. Biochimie 1996; 78:695–699.
59. Cupp-Vickery JR, Garcia C, Hofacre A, et al. Ketoconazole-induced conformational changes in the active site of cytochrome P450eryF. J Mol Biol 2001; 311: 101–110.
60. Park SY, Yamane K, Adachi S, et al. Thermophilic cytochrome P450 (CYP119) from Sulfolobus solfataricus: high resolution structure and functional properties. J Inorg Biochem 2002; 91:491–501.
61. Podust LM, Poulos TL, Waterman MR. Crystal structure of cytochrome P450 14alpha -sterol demethylase (CYP51) from Mycobacterium tuberculosis in complex with azole inhibitors. Proc Natl Acad Sci U S A 2001; 98:3068–3073.
62. Taylor RD, Jewsbury PJ, Essex JW. A review of protein-small molecule docking methods. J Comput Aided Mol Des 2002; 16:151–166.
63. Bohm HJ, Stahl M. The use of scoring function in drug discovery applications. In: Lipkowitz KB, Boyd DB, eds. Reviews in Computational Chemistry, vol 18. Weinheim, Germany: Wiley VCH, 2002:41–87.

64. Jones G, Willett P, Glen RC, et al. Development and validation of a genetic algorithm for flexible docking. J Mol Biol 1997; 267:727–748.
65. Rarey M, Kramer B, Lengauer T, et al. A fast flexible docking method using an incremental construction algorithm. J Mol Biol 1996; 261:470–489.
66. Ewing TJ, Makino S, Skillman AG, et al. DOCK 4.0: search strategies for automated molecular docking of flexible molecule databases. J Comput Aided Mol Des 2001; 15:411–428.
67. Morris GM, Goodsell DS, Halliday RS, et al. Automated docking using a Lamarckian genetic algorithm and an empirical binding free energy function. J Comput Chem 1998; 19:1639–1662.
68. Paul N, Rognan D. ConsDock: a new program for the consensus analysis of protein-ligand interactions. Proteins 2002; 47:521–533.
69. Bissantz C, Folkers G, Rognan D. Protein-based virtual screening of chemical databases. 1. Evaluation of different docking/scoring combinations. J Med Chem 2000; 43:4759–4767.
70. Kramer B, Rarey M, Lengauer T. Evaluation of the FLEXX incremental construction algorithm for protein-ligand docking. Proteins 1999; 37:228–241.
71. de Graaf C, Pospisil P, Pos W, et al. Binding mode prediction of cytochrome p450 and thymidine kinase protein-ligand complexes by consideration of water and rescoring in automated docking. J Med Chem 2005; 48:2308–2318.
72. Poornima CS, Dean PM. Hydration in drug design. 1. Multiple hydrogen-bonding features of water molecules in mediating protein-ligand interactions. J Comput Aided Mol Des 1995; 9:500–512.
73. Poornima CS, Dean PM. Hydration in drug design. 2. Influence of local site surface shape on water binding. J Comput Aided Mol Des 1995; 9:513–520.
74. Poornima CS, Dean PM. Hydration in drug design. 3. Conserved water molecules at the ligand-binding sites of homologous proteins. J Comput Aided Mol Des 1995; 9:521–531.
75. Carlson HA, McCammon JA. Accommodating protein flexibility in computational drug design. Mol Pharmacol 2000; 57:213–218.
76. McConkey BJ, Sobolev V, Edelman M. The performance of current methods in ligand-protein docking. Curr Sci 2002; 83:845–856.
77. Halperin I, Ma B, Wolfson H, et al. Principles of docking: an overview of search algorithms and a guide to scoring functions. Proteins 2002; 47:409–443.
78. Kairys V, Gilson MK. Enhanced docking with the mining minima optimizer: acceleration and side-chain flexibility. J Comput Chem 2002; 23:1656–1670.
79. Birch L, Murray CW, Hartshorn MJ, et al. Sensitivity of molecular docking to induced fit effects in influenza virus neuraminidase. J Comput Aided Mol Des 2002; 16:855–869.
80. Afzelius L, Zamora I, Ridderstrom M, et al. Competitive CYP2C9 inhibitors: enzyme inhibition studies, protein homology modeling, and three-dimensional quantitative structure-activity relationship analysis. Mol Pharmacol 2001; 59:909–919.
81. Wang Q, Halpert JR. Combined three-dimensional quantitative structure-activity relationship analysis of cytochrome P450 2B6 substrates and protein homology modeling. Drug Metab Dispos 2002; 30:86–95.
82. Park JY, Harris D. Construction and assessment of models of CYP2E1: predictions of metabolism from docking, molecular dynamics, and density functional theoretical calculations. J Med Chem 2003; 46:1645–1660.

83. Szklarz GD, Halpert JR. Molecular modeling of cytochrome P450 3A4. J Comput Aided Mol Des 1997; 11:265–272.
84. Baudry J, Li W, Pan L, et al. Molecular docking of substrates and inhibitors in the catalytic site of CYP6B1, an insect cytochrome p450 monooxygenase. Protein Eng 2003; 16:577–587.
85. Keseru GM. A virtual high throughput screen for high affinity cytochrome P450cam substrates. Implications for in silico prediction of drug metabolism. J Comput Aided Mol Des 2001; 15:649–657.
86. Devoss JJ, Demontellano PRO. Computer-assisted, structure-based prediction of substrates for cytochrome P450 (Cam). J Am Chem Soc 1995; 117:4185–4186.
87. DeVoss JJ, Sibbesen O, Zhang ZP, et al. Substrate docking algorithms and prediction of the substrate specificity of cytochrome P450(cam) and its L244A mutant. J Am Chem Soc 1997; 119:5489–5498.
88. Verras A, Kuntz ID, de Montellano PRO. Computer-assisted design of selective imidazole inhibitors for cytochrome p450 enzymes. J Med Chem 2004; 47: 3572–3579.
89. Feenstra KA, Starikov EB, Urlacher VB, et al. Combining substrate dynamics, binding statistics, and energy barriers to rationalize regioselective hydroxylation of octane and lauric acid by CYP102A1 and mutants. Protein Sci 2007; 16:420–431.
90. Wade RC, Gabdoulline RR, Ludemann SK, et al. Electrostatic steering and ionic tethering in enzyme-ligand binding: insights from simulations. Proc Natl Acad Sci U S A 1998; 95:5942–5949.
91. Chang YT, Loew GH. Molecular dynamics simulations of P450 BM3–examination of substrate-induced conformational change. J Biomol Struct Dyn 1999; 16:1189–1203.
92. van Gunsteren WF, Berendsen HJC. Computer simulation of molecular dynamics: methodology, applications and perspectives in chemistry. Angew Chem Int Ed English 1990; 29:992–1023.
93. Pearlman DA, Case DA, Caldwell JW, et al. Amber, a package of computer-programs for applying molecular mechanics, normal-mode analysis, molecular-dynamics and free-energy calculations to simulate the structural and energetic properties of molecules. Comput Phys Commun 1995; 91:1–41.
94. MacKerell AD Jr., Brooks B, Brooks CL III, et al., eds. CHARMM: The Energy Function and Its Parameterization with an Overview of the Program. Chichester, U.K.: John Wiley & Sons 1998.
95. Oostenbrink C, Villa A, Mark AE, et al. A biomolecular force field based on the free enthalpy of hydration and solvation: the GROMOS force-field parameter sets 53A5 and 53A6. J Comput Chem 2004; 25:1656–1676.
96. Lindahl E, Hess B, van der Spoel D. GROMACS 3.0: a package for molecular simulation and trajectory analysis. J Mol Model 2001; 7:306–317.
97. van den Berg PAW, Feenstra KA, Mark AE, et al. Dynamic conformations of flavin adenine dinucleotide: simulated molecular dynamics of the flavin cofactor related to the time- resolved fluorescence characteristics. J Phys Chem B 2002; 106:8858–8869.
98. Feenstra KA, Peter C, Scheek RM, et al. A comparison of methods for calculating NMR cross-relaxation rates (NOESY and ROESY intensities) in small peptides. J Biomol NMR 2002; 23:181–194.

99. Keizers PH, de Graaf C, de Kanter FJ, et al. Metabolic regio- and stereoselectivity of cytochrome P450 2D6 towards 3,4-methylenedioxy-N-alkylamphetamines: in silico predictions and experimental validation. J Med Chem 2005; 48:6117–6127.
100. Ludemann SK, Lounnas V, Wade RC. How do substrates enter and products exit the buried active site of cytochrome P450cam? 1. Random expulsion molecular dynamics investigation of ligand access channels and mechanisms. J Mol Biol 2000; 303:797–811.
101. Ludemann SK, Lounnas V, Wade RC. How do substrates enter and products exit the buried active site of cytochrome P450cam? 2. Steered molecular dynamics and adiabetic mapping of substrate pathways. J Mol Biol 2000; 303:813–830.
102. Winn PJ, Ludemann SK, Gauges R, et al. Comparison of the dynamics of substrate access channels in three cytochrome P450s reveals different opening mechanisms and a novel functional role for a buried arginine. Proc Natl Acad Sci U S A 2002; 99:5361–5366.
103. Feenstra KA, Hess B, Berendsen HJC. Improving efficiency of large ime-scale molecular dynamics simulations of hydrogen-rich systems. J Comput Chem 1999; 20:786–798.
104. Oostenbrink C, van Gunsteren WF. Single-step perturbations to calculate free energy differences from unphysical reference states: limits on size, flexibility, and character. J Comput Chem 2003; 24:1730–1739.
105. Hansson T, Oostenbrink C, van Gunsteren W. Molecular dynamics simulations. Curr Opin Struct Biol 2002; 12:190–196.
106. Paulsen MD, Ornstein RL. Dramatic differences in the motions of the mouth of open and closed cytochrome P450bm-3 by molecular-dynamics simulations. Proteins, 1995; 21:237–243.
107. Ludemann SK, Carugo O, Wade RC. Substrate access to cytochrome P450cam: a comparison of a thermal motion pathway analysis with molecular dynamics simulation data. J Mol Model 1997; 3:369–374.
108. Helms V, Wade RC. Thermodynamics of water mediating protein-ligand interactions in cytochrome P450cam: a molecular dynamics study. Biophys J 1995; 69:810–824.
109. Helms V, Wade RC. Hydration energy landscape of the active site cavity in cytochrome P450cam. Proteins 1998; 32:381–396.
110. Arnold GE, Ornstein RL. Molecular dynamics study of time-correlated protein domain motions and molecular flexibility: cytochrome P450BM-3. Biophys J 1997; 73:1147–1159.
111. Jones JP, Trager WF, Carlson TJ. The binding and regioselectivity of reaction of (R)-nicotine and (S)-nicotine with cytochrome-P-450cam - parallel experimental and theoretical-studies. J Am Chem Soc 1993; 115:381–387.
112. Paulsen MD, Ornstein RL. Binding free energy calculations for P450cam-substrate complexes. Protein Eng 1996; 9:567–571.
113. Szklarz GD, Paulsen MD. Molecular modeling of cytochrome P450 1A1: enzyme-substrate interactions and substrate binding affinities. J Biomol Struct Dyn 2002; 20:155–162.
114. Helms V, Wade RC. Computational alchemy to calculate absolute protein-ligand binding free energy. J Am Chem Soc 1998; 120:2710–2713.
115. Harris D, Loew G. Prediction of regiospecific hydroxylation of camphor analogs by cytochrome-P450(Cam). J Am Chem Soc 1995; 117:2738–2746.

116. Keseru GM, Kolossvary I, Bertok B. Cytochrome P-450 catalyzed insecticide metabolism. Prediction of regio- and stereoselectivity in the primer metabolism of carbofuran: a theoretical study. J Am Chem Soc 1997; 119:5126–5131.
117. Harris DL, Loew GH. Investigation of the proton-assisted pathway to formation of the catalytically active, ferryl species of P450s by molecular dynamics studies of P450eryF. J Am Chem Soc 1996; 118:6377–6387.
118. Bissantz C, Bernard P, Hibert M, et al. Protein-based virtual screening of chemical databases. II. Are homology models of G-protein coupled receptors suitable targets? Proteins 2003; 50:5–25.
119. Kraulis PJ. MOLSCRIPT a program to produce both detailed and schematic plots of protein structures. J Appl Cryst 1991; 24:946–950.
120. Merritt EA, Murphy MEP. Raster3D version 2.0: a program for photorealistic molecular graphics. BTacd 1994; 50:869–873.
121. Kirton SB, Baxter CA, Sutcliffe MJ. Comparative modelling of cytochromes P450. Adv Drug Deliv Rev 2002; 54:385–406.
122. Zvelebil MJ, Wolf CR, Sternberg MJ. A predicted three-dimensional structure of human cytochrome P450: implications for substrate specificity. Protein Eng 1991; 4:271–282.
123. Lewis DF, Lake BG. Molecular modelling of CYP1A subfamily members based on an alignment with CYP102: rationalization of CYP1A substrate specificity in terms of active site amino acid residues. Xenobiotica 1996; 26:723–753.
124. Lozano JJ, Lopez-de-Brinas E, Centeno NB, et al. Three-dimensional modelling of human cytochrome P450 1A2 and its interaction with caffeine and MeIQ. J Comput Aided Mol Des 1997; 11:395–408.
125. Cho US, Park EY, Dong MS, et al. Tight-binding inhibition by alpha-naphthoflavone of human cytochrome P450 1A2. Biochim Biophys Acta 2003; 1648:195–202.
126. Lewis DF, Gillam EMJ, Everett SA, et al. Molecular modelling of human CYP1B1 substrate interactions and investigation of allelic variant effects on metabolism. Chem Biol Interact 2003; 145:281–295.
127. Lewis DF, Lake BG. Molecular modelling of members of the P4502A subfamily: application to studies of enzyme specificity. Xenobiotica 1995; 25:585–598.
128. Lewis DF. 3-Dimensional models of human and other mammalian microsomal P450s constructed from an alignment with P450102 (P450(Bm3)). Xenobiotica 1995; 25:333–366.
129. Lewis DF, Lake BG, Dickins M, et al. Homology modelling of CYP2A6 based on the CYP2C5 crystallographic template: enzyme-substrate interactions and QSARs for binding affinity and inhibition. Toxicol In Vitro 2003; 17:179–190.
130. Szklarz GD, Ornstein RL, Halpert JR. Application of 3-dimensional homology modeling of cytochrome P450 2B1 for interpretation of site-directed mutagenesis results. J Biomol Struct Dyn 1994; 12:61–78.
131. Dai R, Pincus MR, Friedman FK. Molecular modeling of cytochrome P450 2B1: mode of membrane insertion and substrate specificity. J Protein Chem 1998; 17:121–129.
132. Kobayashi Y, Fang XJ, Szklarz GD, et al. Probing the active site of cytochrome P450 2B1: metabolism of 7-alkoxycoumarins by the wild type and five site-directed mutants. Biochemistry 1998; 37:6679–6688.
133. Strobel SM, Halpert JR. Reassessment of cytochrome P450 2B2: catalytic specificity and identification of four active site residues. Biochemistry 1997; 36:11697–11706.

134. Szklarz GD, He YA, Halpert JR. Site-directed mutagenesis as a tool for molecular modeling of cytochrome P450 2B1. Biochemistry 1995; 34:14312–14322.
135. Chang YT, Stiffelman OB, Vakser IA, et al. Construction of a 3D model of cytochrome P450 2B4. Protein Eng 1997; 10:119–129.
136. Domanski TL, Schultz KM, Roussel F, et al. Structure-function analysis of human cytochrome P-4502B6 using a novel substrate, site-directed mutagenesis, and molecular modeling. J Pharmacol Exp Ther 1999; 290:1141–1147.
137. Korzekwa KR, Jones JP. Predicting the cytochrome P450 mediated metabolism of xenobiotics. Pharmacogenetics 1993; 3:1–18.
138. Payne VA, Chang YT, Loew GH. Homology modeling and substrate binding study of human CYP2C18 and CYP2C19 enzymes. Proteins 1999; 37:204–217.
139. Snyder R, Sangar R, Wang JB, et al. Three-dimensional quantitative structure activity relationship for CYP2D6 substrates. Quant Struct-Act Relat 2002; 21: 357–368.
140. Lewis DF, Bird MG, Parke DV. Molecular modelling of CYP2E1 enzymes from rat, mouse and man: an explanation for species differences in butadiene metabolism and potential carcinogenicity, and rationalization of CYP2E substrate specificity. Toxicology 1997; 118:93–113.
141. Ferenczy GG, Morris GM. The active site of cytochrome P-450 nifedipine oxidase: a model-building study. J Mol Graph 1989; 7:206–211.
142. Xue L, Wang HF, Wang Q, et al. Influence of P450 3A4 SRS-2 residues on cooperativity and/or regioselectivity of aflatoxin B(1) oxidation. Chem Res Toxicol 2001; 14:483–491.
143. Kuhn B, Jacobsen W, Christians U, et al. Metabolism of sirolimus and its derivative everolimus by cytochrome P450 3A4: insights from docking, molecular dynamics, and quantum chemical calculations. J Med Chem 2001; 44:2027–2034.
144. Torimoto N, Ishii I, Hata M, et al. Direct interaction between substrates and endogenous steroids in the active site may change the activity of cytochrome P450 3A4. Biochemistry 2003; 42:15068–15077.
145. Lewis DF, Lake BG. Molecular modelling of CYP4A subfamily members based on sequence homology with CYP102. Xenobiotica 1999; 29:763–781.
146. Chang YT, Loew GH. Homology modeling and substrate binding study of human CYP4A11 enzyme. Proteins 1999; 34:403–415.
147. Pan LP, Wen ZM, Baudry J, et al. Identification of variable amino acids in the SRS1 region of CYP6B1 modulating furanocoumarin metabolism. Arch Biochem Biophys 2004; 422:31–41.
148. Vijayakumar S, Salerno JC. Molecular modeling of the 3-D structure of cytochrome P-450scc. Biochim Biophys Acta 1992; 1160:281–286.
149. Usanov SA, Graham SE, Lepesheva GI, et al. Probing the interaction of bovine cytochrome P450scc (CYP11A1) with adrenodoxin: evaluating site-directed mutations by molecular modeling. Biochemistry 2002; 41:8310–8320.
150. Laughton CA, Zvelebil MJ, Neidle S. A detailed molecular model for human aromatase. J Steroid Biochem Mol Biol 1993; 44:399–407.
151. Schappach A, Holtje HD. Molecular modelling of 17 alpha-hydroxylase-17,20-lyase. Pharmazie 2001; 56:435–442.
152. Graham-Lorence S, Khalil MW, Lorence MC, et al. Structure-function relationships of human aromatase cytochrome P-450 using molecular modeling and site-directed mutagenesis. J Biol Chem 1991; 266:11939–11946.

153. Auvray P, Nativelle C, Bureau R, et al. Study of substrate specificity of human aromatase by site directed mutagenesis. Eur J Biochem 2002; 269:1393–1405.
154. Chen S, Zhang F, Sherman MA, et al. Structure-function studies of aromatase and its inhibitors: a progress report. J Steroid Biochem Mol Biol 2003; 86:231–237.
155. Ishida N, Aoyama Y, Hatanaka R, et al. A single amino acid substitution converts cytochrome P450(14DM) to an inactive form, cytochrome P450SG1: complete primary structures deduced from cloned DNAs. Biochem Biophys Res Commun 1988; 155:317–323.
156. Holtje HD, Fattorusso C. Construction of a model of the Candida albicans lanosterol 14-alpha-demethylase active site using the homology modelling technique. Pharm Acta Helv 1998; 72:271–277.
157. Murtazina D, Puchkaev AV, Schein CH, et al. Membrane-protein interactions contribute to efficient 27-hydroxylation of cholesterol by mitochondrial cytochrome P450 27A1. J Biol Chem 2002; 277:37582–37589.
158. Schoch GA, Attias R, Belghazi M, et al. Engineering of a water-soluble plant cytochrome P450, CYP73A1, and NMR-based orientation of natural and alternate substrates in the active site. Plant Physiol 2003; 133:1198–1208.
159. Braatz JA, Bass MB, Ornstein RL. An evaluation of molecular models of the cytochrome P450 Streptomyces griseolus enzymes P450SU1 and P450SU2. J Comput Aided Mol Des 1994; 8:607–622.
160. Chang YT, Loew GH. Construction and evaluation of a three-dimensional structure of cytochrome P450choP enzyme (CYP105C1). Protein Eng 1996; 9:755–766.
161. Chang YT, Loew G. Homology modeling, molecular dynamics simulations, and analysis of CYP119, a P450 enzyme from extreme acidothermophilic archaeon Sulfolobus solfataricus. Biochemistry 2000; 39:2484–2498.
162. Poulos TL, Finzel BC, Howard AJ. Crystal structure of substrate-free Pseudomonas putida cytochrome P-450. Biochemistry 1986; 25:5314–5322.
163. Sevrioukova IF, Li H, Zhang H, et al. Structure of a cytochrome P450-redox partner electron-transfer complex. Proc Natl Acad Sci U S A 1999; 96:1863–1868.
164. Hasemann CA, Ravichandran KG, Peterson JA, et al. Crystal structure and refinement of cytochrome P450terp at 2.3 A resolution. J Mol Biol 1994; 236: 1169–1185.
165. Shimizu H, Park SY, Shiro Y, et al. X-ray structure of nitric oxide reductase (cytochrome P450nor) at atomic resolution. Acta Crystallogr D 2002; 58:81–89.
166. Gasteiger E, Gattiker A, Hoogland C, et al. ExPASy: The proteomics server for in-depth protein knowledge and analysis. Nucleic Acids Res 2003; 31:3784–3788.
167. Thompson JD, Higgins DG, Gibson TJ. CLUSTAL W: improving the sensitivity of progressive multiple sequence alignment through sequence weighting, position-specific gap penalties and weight matrix choice. Nucleic Acids Res 1994; 22:4673–4680.
168. Holm L, Sander C. Dali/FSSP classification of three-dimensional protein folds. Nucleic Acids Res 1997; 25:231–234.
169. Butler BA. Sequence analysis using GCG. Methods Biochem Anal 1998; 39:74–97.
170. Cuff JA, Clamp ME, Siddiqui AS, et al. JPred: a consensus secondary structure prediction server. Bioinformatics 1998; 14:892–893.
171. Notredame C, Higgins DG, Heringa J. T-Coffee: a novel method for fast and accurate multiple sequence alignment. J Mol Biol 2000; 302:205–217.
172. Kneller DG, Cohen FE, Langridge R. Improvements in protein secondary structure prediction by an enhanced neural network. J Mol Biol 1990; 214:171–182.

173. McGuffin LJ, Bryson K, Jones DT. The PSIPRED protein structure prediction server. Bioinformatics 2000; 16:404–405.
174. Havel TF, Snow ME. A new method for building protein conformations from sequence alignments with homologues of known structure. J Mol Biol 1991; 217:1–7.
175. Sali A, Blundell TL. Comparative protein modelling by satisfaction of spatial restraints. J Mol Biol 1993; 234:779–815.
176. Schwede T, Kopp J, Guex N et al, SWISS-MODEL: an automated protein homology-modeling server. Nucleic Acids Res 2003; 31:3381–3385.
177. Rost B, Liu J. The PredictProtein server. Nucleic Acids Res 2003; 31:3300–3304.
178. Colovos C, Yeates TO. Verification of protein structures: patterns of nonbonded atomic interactions. Protein Sci 1993; 2:1511–1519.
179. Laskowski RA, Macarthur MW, Moss DS, et al. Procheck - a program to check the stereochemical quality of protein structures. J Appl Crystallogr 1993; 26:283–291.
180. Sippl MJ. Recognition of errors in three-dimensional structures of proteins. Proteins 1993; 17:355–362.
181. Pontius J, Richelle J, Wodak SJ. Deviations from standard atomic volumes as a quality measure for protein crystal structures. J Mol Biol 1996; 264:121–136.
182. Luthy R, Bowie JU, Eisenberg D. Assessment of protein models with three-dimensional profiles. Nature 1992; 356:83–85.
183. Scott WRP, Hunenberger PH, Tironi IG, et al. The GROMOS biomolecular simulation program package. J Phys Chem A 1999; 103:3596–3607.
184. Eldridge MD, Murray CW, Auton TR, et al. Empirical scoring functions: I. The development of a fast empirical scoring function to estimate the binding affinity of ligands in receptor complexes. J Comput Aided Mol Des 1997; 11:425–445.
185. Clark RD, Strizhev A, Leonard JM, et al. Consensus scoring for ligand/protein interactions. J Mol Graph Model 2002; 20:281–295.
186. Friesner RA, Banks JL, Murphy RB, et al. Glide: a new approach for rapid, accurate docking and scoring. 1. Method and assessment of docking accuracy. J Med Chem 2004; 47:1739–1749.
187. Muegge I, Martin YC. A general and fast scoring function for protein-ligand interactions: a simplified potential approach. J Med Chem 1999; 42:791–804.
188. Wang RX, Liu L, Lai LH, et al. SCORE: A new empirical method for estimating the binding affinity of a protein-ligand complex. J Mol Model 1998; 4:379–394.
189. Schnecke V, Kuhn LA. Virtual screening with solvation and ligand-induced complementarity. Perspect Drug Disc Des 2000; 20:171–190.
190. Wang RX, Lai LH, Wang SM. Further development and validation of empirical scoring functions for structure-based binding affinity prediction. J Comput Aided Mol Des 2002; 16:11–26.

10

Role of the Gut Mucosa in Metabolically Based Drug-Drug Interactions

Kenneth E. Thummel, Danny D. Shen, and Nina Isoherranen
*University of Washington,
Seattle, Washington, U.S.A.*

I. INTRODUCTION

A. Background

The gastrointestinal mucosa represents a major physical and metabolic barrier to the systemic availability of orally ingested drug molecules. A critical component of that barrier is a collection of drug-metabolizing enzymes localized primarily at the apical aspect of the intestinal epithelium. The liver is generally considered to be the dominant site of drug metabolism, but it is now clear that for a number of drug molecules (e.g., midazolam, nifedipine, and verapamil), biotransformation at the intestinal mucosa contributes significantly to their first-pass removal. Indeed, some prodrugs have been developed that take advantage of the enzymatic activity of the intestinal mucosa (e.g., carboxyesterases and cytochromes P450) to promote the absorption and subsequent release of pharmacologically active drug into the hepatic portal circulation.

In addition to drug metabolism, it is recognized that efflux drug transporters, such as P-glycoprotein (P-gp, gene product of *ABCB1* or *MDR1*) and breast cancer resistance protein (BCRP, gene product of *ABCG2*), present at the apical membranes of enterocytes, promote the efflux of drugs from intracellular sites into the gut lumen. Thus, for some orally administered drugs (e.g., cyclosporine and digoxin), extensive apically directed drug efflux effectively reduces their intestinal permeability and bioavailability. It has also been suggested that a functional interaction occurs between P-gp and metabolic enzymes at the apical interface of the mucosa that increases the residence time of a drug molecule and results in enhanced first-pass metabolism at the intestine.

It stands to reason that modulation of intestinal drug-metabolizing enzyme and efflux transporter function could constitute a mechanism of drug-drug interaction. In this chapter, we review the expression and localization of intestinal enzymes and transporters that have been implicated in metabolically based drug-drug interactions and the pharmacokinetic characteristics of those interaction events.

B. Pharmacokinetic Principles

If a metabolically based drug-drug interaction is to have clinical significance, the affected process of drug metabolism must represent an appreciable part of the overall drug elimination scheme. In the case of intestinal metabolism, it is the fraction of a dose metabolized by the gut mucosa on first pass (E_{gm}) that is most relevant. In general, intestinal mucosal enzymes that contribute significantly to the first-pass metabolism of a drug have a much lower contribution to the systemic clearance of the same molecule because of the relatively low blood flow to enterocytes that express drug-metabolizing enzymes (1,2). Thus, important drug interactions involving gut metabolism will generally be associated with drugs that have an appreciable first-pass intestinal extraction.

The involvement of gut wall metabolism in a drug-drug interaction is usually inferred from the difference in the magnitude of AUC change between oral and intravenous dosing of the affected drug, except in the case of a very high extraction drug with blood flow rate–limited systemic clearance. Because most intestinal enzymes are also found in the liver, an overall change in AUC may reflect an interaction involving both hepatic and intestinal first-pass metabolism. The oral AUC is a function of the systemic clearance (CL) of the drug, the product of the fractions available following passage through the gut (F_{gm}) and the liver (F_h), and the fraction released from the dosage form that is absorbed (F_a).

$$\mathrm{AUC}_{\mathrm{po}} = \frac{\left(F_a \cdot F_{gm} \cdot F_h\right) \cdot \mathrm{Dose}_{\mathrm{po}}}{\mathrm{CL}} \tag{1}$$

Each term with a metabolic component (F_{gm}, F_h, CL) is a function of unbound fraction in blood (f_u) and intrinsic clearance (i.e., V_{max}/K_m in biochemical terms) and can be modified by an enzyme/transporter inducer or inhibitor. If drug is completely absorbed and elimination occurs exclusively in the liver, the systemic AUC observed in the presence of a metabolic modulator would directly reflect modification of intrinsic hepatic clearance in the absence of a change in f_u (3,4), as indicated in Eq. (2) by an asterisk (*):

$$\frac{\mathrm{AUC}^{*}{}_{\mathrm{po}}}{\mathrm{AUC}_{\mathrm{po}}} = \frac{f_u \cdot \mathrm{CL}_h^{\mathrm{int}}}{f_u \cdot \mathrm{CL}_h^{\mathrm{int}*}} \tag{2}$$

When both intestine and liver contribute significantly to the metabolic elimination of an orally administered drug, the resulting mathematical relationship between the oral AUC for the affected drug and organ intrinsic clearance will be more complex. If we consider the pharmacokinetic model depicted in Figure 1 for a drug that is metabolized in the liver and intestine but is not subject to intestinal or hepatic efflux transport, a series of equations can be derived with the following simplifying assumptions:

- Complete absorption of the oral dose
- Sequential gut mucosal and liver first-pass metabolic extraction
- Liver is the exclusive site of systemic clearance (CL); i.e., no significant renal or intestinal contributions to systemic clearance
- Flow-limited, well-stirred model for organ extraction
- First-order liver and intestinal metabolism

$$\frac{\mathrm{Dose}_{\mathrm{po}}}{\mathrm{AUC}_{\mathrm{po}}} = \frac{\mathrm{CL}_h}{F_{\mathrm{gm}} \cdot F_h} \tag{3}$$

Recognizing that,

$$\mathrm{CL}_h = \frac{Q_h \cdot f_u \cdot \mathrm{CL}_h^{\mathrm{int}}}{f_u \cdot \mathrm{CL}_h^{\mathrm{int}} + Q_h} \tag{4}$$

$$F_{\mathrm{gm}} = \frac{Q_{\mathrm{gm}}}{f_u \cdot \mathrm{CL}_{\mathrm{gm}}^{\mathrm{int}} + Q_{\mathrm{gm}}} \tag{5}$$

$$F_h = \frac{Q_h}{f_u \cdot \mathrm{CL}_h^{\mathrm{int}} + Q_h} \tag{6}$$

Substituting Eqs. (4–6) into (3), we obtain

$$\frac{\mathrm{Dose}_{\mathrm{po}}}{\mathrm{AUC}_{\mathrm{po}}} = \frac{f_u \cdot \mathrm{CL}_h^{\mathrm{int}}}{Q_{\mathrm{gm}}/\left(f_u \cdot \mathrm{CL}_{\mathrm{gm}}^{\mathrm{int}} + Q_{\mathrm{gm}}\right)} \tag{7}$$

where $\mathrm{CL}_{\mathrm{gm}}^{\mathrm{int}}$ and $\mathrm{CL}_h^{\mathrm{int}}$ are the unbound intrinsic clearances of the gastrointestinal mucosa and liver, respectively, and Q_{gm} and Q_h are blood flows to the gastrointestinal mucosa and liver, respectively. In the context of first-pass metabolism after oral administration, it is important to define in what region of the gastrointestinal tract the majority of the drug dose will be absorbed when assigning

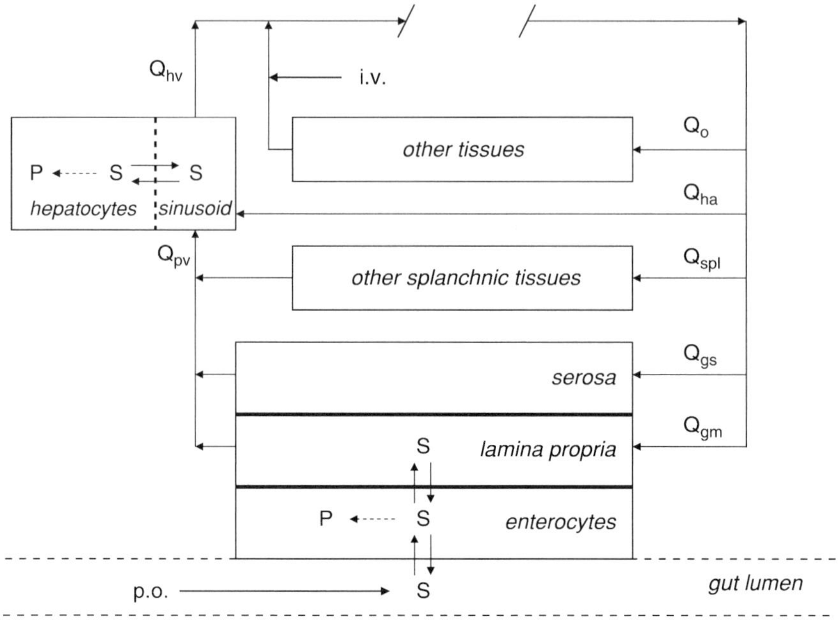

Figure 1 Physiological model for sequential intestinal and hepatic first-pass metabolism. Blood flow to the small intestine is functionally divided into mucosal (Q_{gm}) and serosal (Q_{gs}) blood flow. Mucosal blood flow in the lamina propria perfuses the enterocyte epithelium. Portal blood flow (Q_{pv}), which perfuses the liver is comprised of blood leaving the small intestine and other splanchnic organs such as the stomach and spleen. Blood flow leaving the liver (Q_{hv}) represents the sum of hepatic arterial flow (Q_{ha}) and Q_{pv}. First-pass metabolism of an orally administered substrate (S) to product (P) may occur in the enterocyte or hepatocyte.

values to the mucosal blood flow and unbound intrinsic clearance. For example, many CYP3A substrates are absorbed predominantly in the small intestine and, thus, one can define the total unbound intrinsic clearance for mucosal tissue of the small intestine and the blood flow that perfuses that tissue.

The assumption that intestinal metabolism does not contribute significantly to systemic clearance is based on studies of acetaminophen, enalapril, morphine, and (−)-aminocarbovir disposition in perfused, rat small intestine [see recent review by Pang (2)], and one human study on intestinal metabolism of midazolam during the anhepatic phase of liver transplantation (5). It should also be noted that Eq. (5) for F_{gm} assumes that permeability at either the apical or basolateral membrane of the epithelium (or both) is not rate limiting (i.e., permeability being much higher than the intrinsic clearance and mucosal flow). Mizuma et al. have proposed a more complex model that features the joint actions of permeability and metabolism (6,7). The theoretical considerations presented in this section hold true qualitatively even if drug translocation across the mucosal

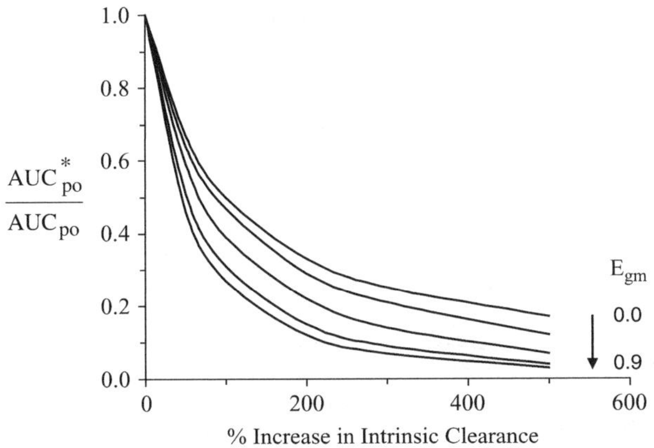

Figure 2 Simulation of the effect of enzyme induction on oral midazolam AUC. Hepatic extraction in the absence of an inducer was set at 0.44. The initial intestinal extraction was varied from 0.0 to 0.9. The inducer was assumed to cause an equivalent change in hepatic and intestinal intrinsic clearance. Mucosal and hepatic plasma flows were assumed to be 240 and 780 mL/min. Simulations were obtained from Eq. (6), assuming an initial intestinal extraction ratio of 0.00, 0.07, 0.27, 0.43, 0.60, and 0.88.

barriers was to be rate limiting; if anything, the role of intestinal metabolism during first-pass intestinal extraction would be enhanced (5).

When the above equations are applied to the context of drug-drug interaction, the AUC ratio in the absence and presence of a modulating drug can be expressed as the following:

$$\frac{\mathrm{AUC}^*{}_{\mathrm{po}}}{\mathrm{AUC}_{\mathrm{po}}} = \left(\frac{Q_{\mathrm{gm}} + f_{\mathrm{u}} \cdot \mathrm{CL}_{\mathrm{gm}}^{\mathrm{int}}}{Q_{\mathrm{gm}} + f_{\mathrm{u}} \cdot \mathrm{CL}_{\mathrm{gm}}^{\mathrm{int}^*}}\right) \cdot \left(\frac{f_{\mathrm{u}} \cdot \mathrm{CL}_{\mathrm{h}}^{\mathrm{int}}}{f_{\mathrm{u}} \cdot \mathrm{CL}_{\mathrm{h}}^{\mathrm{int}^*}}\right) \tag{8}$$

The * denotes inhibited conditions.

The above equation features the multiplicative effect that simultaneous changes in hepatic and intestinal intrinsic clearance can have on the systemic AUC, as illustrated by the simulations for midazolam presented in Figures 2 and 3. For example, a fourfold increase in hepatic and mucosal intrinsic clearances, as a consequence of enzyme induction, can lead to as much as a 94% or 16-fold reduction in systemic AUC, depending on the initial mucosal extraction ratio (Fig. 2). Without mucosal first-pass extraction, the reduction in systemic AUC proportionately reflects the increase in hepatic intrinsic clearance (i.e., a fourfold change). Viewed in another way, the lowering in systemic AUC is four times greater for a drug that undergoes extensive sequential intestinal and liver first-pass extraction compared with the one that only undergoes liver first-pass extraction; i.e., comparing the scenarios represented by Eqs. (2) and (8).

If we now turn to inhibitory interactions, fourfold reductions in hepatic and intestinal intrinsic clearance may cause up to a 16-fold increase in systemic AUC

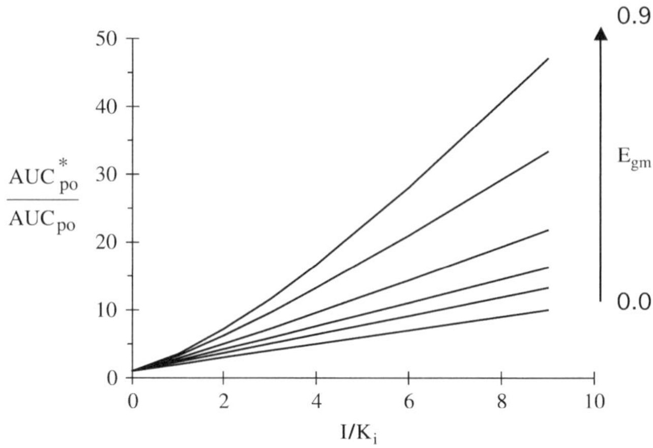

Figure 3 Simulation of the effect of enzyme inhibition on oral midazolam AUC. Hepatic extraction in the absence of an inhibitor was set at 0.44. The initial intestinal extraction was varied from 0.0 to 0.9. The inhibitor/K_i ratio was assumed to be equivalent for inhibition of hepatic and intestinal metabolism (i.e., $I_u = I_{gm}$). Mucosal and hepatic plasma flows were assumed to be 240 and 780 mL/min. Simulations were obtained from Eq. (8), assuming an initial intestinal extraction ratio of 0.00, 0.27, 0.43, 0.60, 0.78, and 0.88.

compared with control, which is illustrated in Figure 3 for midazolam. In the case of rapidly reversible enzyme inhibitors, the change in intrinsic clearance in the liver and intestine can be expressed as a function of the in vivo K_i and the respective unbound inhibitor concentration in the liver and intestine (I_u and I_{gm}):

$$\frac{f_u \cdot CL_h^{int}}{f_u \cdot CL_h^{int*}} = 1 + \frac{I_u}{K_i} \tag{9}$$

$$\frac{f_u \cdot CL_{gm}^{int}}{f_u \cdot CL_{gm}^{int*}} = 1 + \frac{I_{gm}}{K_i} \tag{10}$$

Although we have assumed that there is no mucosal diffusional barrier for the CYP3A substrate, the unbound inhibitor concentration in the intestinal epithelia (I_{gm}) may or may not be equivalent to that in plasma and the liver (I_u). I_{gm} may exceed the unbound portal plasma concentration during the inhibitor absorption phase or be less than the unbound portal concentration postabsorption if there is not rapid equilibrium between the intracellular and portal plasma compartments (i.e., a basolateral membrane diffusional barrier exists). This obviously makes it challenging to anticipate the quantitative effect of an inhibitor on intestinal first-pass metabolism.

For competitive inhibitors, Eqs. (9) and (10) are quantitatively accurate only when the substrate concentration is below its respective K_m. Saturation of metabolic enzymes by the substrate is generally not an issue for hepatic elimination of most drugs, but may influence the extent of inhibition for intestinal metabolism as the enterocytes may face concentrations of drugs that greatly exceed those observed in plasma. Competition for the metabolic enzyme between the substrate and inhibitor will decrease the overall inhibition observed. Substitution of Eqs. (9) and (10) into (8) yields an expression (Eq. 11) that illustrates the multiplicative effect that an enzyme/transporter inhibitor can have on the systemic exposure to an orally administered drug:

$$\frac{\mathrm{AUC}^*_{\mathrm{po}}}{\mathrm{AUC}_{\mathrm{po}}} = \left(\frac{Q_{\mathrm{gm}} + f_{\mathrm{u}} \cdot \mathrm{CL}_{\mathrm{gm}}^{\mathrm{int}}}{Q_{\mathrm{gm}} + \left\{f_{\mathrm{u}} \cdot \mathrm{CL}_{\mathrm{gm}}^{\mathrm{int}} / \left[1 + \left(I_{\mathrm{gm}}/K_{\mathrm{i}}\right)\right]\right\}}\right) \cdot \left(1 + \frac{I_{\mathrm{u}}}{K_{\mathrm{i}}}\right) \tag{11}$$

Note that the effect of the inhibitor in the liver is independent of blood flow. This is not the case for the intestine, where the relative magnitude of mucosal blood flow compared with the baseline mucosal intrinsic clearance and the apparent intrinsic clearance in the presence of inhibitor must be considered. When the baseline mucosal intestinal intrinsic clearance is negligible compared to mucosal blood flow (i.e., negligible mucosal extraction), Eq. (11) will collapse into a much simpler and better recognized equation for a hepatic inhibitory interaction (3).

$$\frac{\mathrm{AUC}^*_{\mathrm{po}}}{\mathrm{AUC}_{\mathrm{po}}} = 1 + \frac{I_{\mathrm{u}}}{K_{\mathrm{i}}} \tag{12}$$

For an orally administered drug that is completely absorbed from the gastrointestinal tract and subject to first-pass intestinal extraction, one can assign a lower limit to the impact of change in mucosal extraction ratio on systemic AUC. For example, complete inhibition of an intestinal extraction at 25% in the control state would result in a 1.33-fold increase in AUC, independent of any changes in hepatic metabolism. The lower the mucosal extraction ratio, the more the interaction will be defined by the liver. For interactions involving induction of intestinal processes, a lower limit of significance for the initial mucosal extraction is more difficult to deduce. However, given the magnitude of change in enzyme/transporter expression commonly observed in a clinical setting (four- to sixfold), induction of intestinal first-pass metabolism that has an extraction of $<10\%$ of the oral dose in the uninduced state is unlikely to have an appreciable effect on the AUC of the affected drug. Again, when the initial mucosal extraction ratio is moderate to high, both induction and inhibition of mucosal intrinsic clearance will have a pronounced effect on the oral AUC (Eqs. 8 and 11).

It is possible to have a drug-drug interaction that is confined to the intestine during first-pass extraction. Regular strength grapefruit juice appears to selectively inhibit CYP3A in the intestinal mucosa, but not in the liver. As discussed below, the magnitude of AUC changes observed in vivo are dependent on the initial intestinal extraction ratio for any given individual. The change in oral AUC can be quite remarkable when the intestinal first-pass extraction is extensive. Conversely, if E_{gm} is already low, grapefruit juice will cause little change to the intestinal availability and to the oral AUC.

Another important consideration for understanding metabolically based drug-drug interactions is that the level of exposure of the liver and intestinal mucosa to an inhibitor or inducer need not be identical (as discussed above), particularly during the periabsorptive phase, when modulator concentration at the intestinal mucosa may be much greater than that in the portal blood. It is also important to recognize that the intracellular mechanism underlying an interaction (e.g., transcriptional activation for induction) may not simultaneously take effect in the intestine and the liver, or if it were to occur in both organs, they do not follow the same temporal course or to the same degree. Consequently, the extent of induction or inhibition at each site of metabolism/transport following acute or chronic administration of an interacting drug could be quite different (e.g., for induction, differential transcriptional activation because of a different spectrum of expressed coactivators and corepressors).

In the remainder of this chapter, we review the characteristic features of drug interactions that involve modulation of the first-pass intestinal metabolism of orally administered drugs. A full complement of drug-metabolizing enzymes is expressed in the human intestinal epithelium. The most important phase 1 enzymes in the context of drug-drug interactions are the cytochrome P450 enzymes. The specific P450 content of microsomes isolated from mucosal epithelium of the human proximal small intestine is roughly one-sixth to one-eighth of that found in liver microsomes (8,9). However, some P450 isozymes, such as CYP3A4, CYP3A5, and CYP2C, are expressed more prominently than others (10). A number of phase 2 enzymes are also present in human intestinal mucosa, of which sulfotransferases (SULTs) and UDP-glucuronosyltransferases (UGTs) are the two families of conjugating enzymes most involved in clinically significant drug-drug interactions (11–13). Notable aspects of drug interactions involving each of the aforementioned enzyme classes will be presented in the following sections.

II. CYP3A

A. Localization and Function

CYP3A4 is expressed at high levels within columnar epithelial cells lining the gastrointestinal tract and hepatic parenchymal cells (14). In 1987, Watkins et al. first reported the identification of functionally active CYP3A4 enzyme in the human gastrointestinal tract (15), and it is the dominant cytochrome P450 enzyme present in the small intestine (16,17). Subsequent studies revealed the expression of another member of the CYP3A family—CYP3A5 in the mucosa of the small intestine. Expression of CYP3A5 protein is variable because of genetic polymorphisms. Individuals with variant CYP3A5 alleles (viz. *CYP3A5*3*, *CYP3A5*6*, and *CYP3A5*7*) have very low to nondetectable CYP3A5 protein in the small intestine, whereas CYP3A5 levels comparable to CYP3A4 are observed in individuals carrying the *CYP3A5*1* wild-type allele (8,18,19). There is considerable overlap in the substrate selectivity of CYP3A4 and CYP3A5; they do differ in product regioselectivity and turnover kinetics and, more interestingly, in the potency of competitive (K_i) and mechanism-based inhibitors (K_I and k_{inact}) (20–22).

The expression of CYP3A along the gastrointestinal tract is not uniform. Mucosal enzyme concentration is greatest within the duodenal and jejunal sections of the small intestine and declines distally and proximally. In an early study by DeWaziers et al. (10), mean microsomal CYP3A content reported as percent of mean hepatic microsomal CYP3A content was 2.5% in esophagus, 4.3% in stomach, 48% in duodenum, 25% in jejunum, 15% in ileum, and 1.5% in colon. In a later study of 20 full-length intestines and livers from organ donors, Paine et al. (8) reported a median value of 70 (4–262), 31 (<2–91), 23 (<2–98), and 17 (<2–60) pmol/mg protein in microsomes isolated from the liver, duodenum, jejunum and ileum, respectively. A similar heterogeneous pattern was described for CYP3A-catalyzed midazolam 1′-hydroxylation activity.

von Richter et al. (23) recently reported CYP3A4 protein expression in paired biopsy specimens of duodenum or proximal jejunum and liver from 15 patients undergoing gastrointestinal surgery. A villous fraction consisting of mainly mature enterocytes was isolated from the intestinal samples. CYP3A4 protein content was found to be about threefold higher in the enterocyte homogenate than in the liver homogenate, i.e., 76 versus 23.6 pmol/mg protein, respectively. The enterocyte CYP3A4 content when expressed per milligram of microsomal protein was much higher than corresponding values for mucosal scrapings from the duodenum reported by Paine et al. (210 pmol/mg vs. 31 pmol/mg) (8), although later calculations by Yang et al. (24) indicated that the total content of CYP3A4 in the small intestine for the two studies are comparable. The data from von Richter et al. suggest that the abundance of CYP3A4 in the apical aspect of enterocytes may be higher than that in hepatocytes, which further argues for the important role of intestinal CYP3A4 in first-pass metabolism.

The relative expression of CYP3A4 and CYP3A5 also varies along the length of the gastrointestinal tract. CYP3A4 expression predominates in the small intestine. In the colon and stomach, mucosal expression of CYP3A5 mRNA and protein is more prominent than corresponding CYP3A4 measures (16,17). For example, Gervot et al. (25) detected CYP3A5 protein, but not CYP3A4 protein, in colonic mucosa from 40 different uninduced tissue donors. CYP3A5 also appears to be the dominant CYP3A isozyme that is expressed at relatively low levels in various epithelial cell lines derived from the colon (25,26). In this respect, these cell lines may represent an excellent model for xenobiotic metabolism in the human colon and its role in chemical-induced mutagenesis or cytotoxicity. However, enhancement of CYP3A4 expression, such as treatment of the Caco-2 cell line by vitamin D_3 (26) or stable cDNA transfection (27), would be needed if the colonic cell line is to be used as a model for first-pass metabolism in the small intestine.

Total CYP3A content within a defined region of small bowel varies considerably between individuals. Lown et al. (18) found an 8-fold difference in CYP3A4 mRNA and 11-fold difference in immunoreactive CYP3A protein in an analysis of duodenal pinch biopsies obtained from 20 "normal" volunteers. Variability in CYP3A4 and CYP3A5 mRNA expression was confirmed in two recent studies with human duodenal biopsies (28,29). Paine et al. (8) described even greater variability (>50-fold) in an analysis of CYP3A protein content in duodenal and jejunal mucosal scrapings from 20 organ donors. Although some of this extreme variability in the latter study could be the result of events preceding the procurement of tissue (i.e., reduced nutritional intake, antibiotic administration, treatment with known CYP3A inducers, and brain death), it does suggest a remarkably dynamic system and that enzyme expression that may respond to a variety of dietary, therapeutic, and pathophysiological conditions. For example, exposure of volunteers to the known hepatic inducer rifampin causes an increase in duodenal CYP3A4 content (30). In contrast, acute or chronic ingestion of grapefruit juice reduces duodenal content of CYP3A4 (31).

Despite several years of effort, there is still only limited knowledge about the regulation of intestinal CYP3A4 by constitutive factors. It has been suggested that endogenous ligands of human pregnane-X-receptor (hPXR) may regulate both basal and inducible expression of hepatic CYP3A4 (32). Accordingly, the high level of hPXR expression in the intestinal mucosa suggests that basal intestinal CYP3A4 expression could be mediated by activation of hPXR. However, in addition to hPXR, studies with Caco-2 and LS-180 cells suggest a possible role for the vitamin D receptor (VDR) in the regulation of intestinal CYP3A. Treatment of these cells with 1α,25-dihydroxy vitamin D_3 stimulates CYP3A4 expression and its associated midazolam 1′-hydroxylation activity (26). The effect on CYP3A4 expression is ligand- and VDR-dependent and involves binding of the activated VDR/RXR (RXR, retinoid X receptor) heterodimer to the same CYP3A4 response elements that mediate enzyme induction by hPXR ligands (e.g., rifampin) (33,34). Two different studies have shown that administration of 1α,25-dihydroxy vitamin D_3 (or a precursor) to rodents activates CYP3A genes that contain the appropriate ER6 or DR3 5′-flanking response elements (35,36). It has also been shown that several endogenous bile acids (lithocholic acid and its keto derivatives) also bind to VDR and activate CYP3A4 transcription and, thus, may also play a role in regulating intestinal CYP3A4 expression (35,37).

1α,25-Dihydroxy vitamin D_3 exerts an important physiological role in calcium homeostasis by a process that includes VDR-mediated transcriptional activation of genes coding for the intestinal luminal calcium transporter, TRPV6 (38), and the intracellular calcium binding protein, calbindin D9k (39). In addition, it has recently been shown that CYP3A4 metabolizes 1α,25-dihydroxy vitamin D_3 to less active products (36). Thus, induction of intestinal CYP3A4 by 1α,25-dihydroxy vitamin D_3 may provide self-limiting control of the transcriptional effects of the hormone. It also provides a plausible mechanism for the osteomalacia associated with chronic administration of potent PXR-mediated CYP3A4 inducers, such as phenytoin, carbamazepine, and barbiturates to adults (40).

Mucosal homogenates and microsomes from human intestine have been shown to catalyze the metabolism of a number of CYP3A substrates, including the oxidation of cyclosporine (15,41), erythromycin (15), ethinyl estradiol (42), flurazepam (43), indinavir (44,45), irinotecan (46), L-α-acetylmethadol (47), lovastatin (48), methadone (47), midazolam (49), quazepam (50), rifabutin (51), ritonavir (45), saquinavir (52,53), sirolimus (54), tacrolimus (55), terfenadine (52), and testosterone (56). However, corresponding evidence for CYP3A-mediated intestinal first-pass metabolism in vivo is only available for a small subset of these drug substrates: buspirone (57), cyclosporine (58), felodipine (59), midazolam (5), racemic verapamil (60,61), nifedipine (62), tirilazad (63), and triazolam (64). Given the inaccessibility of the intestinal site of metabolism, it is understandable that most of the in vivo evidence is indirect in nature. The study by Paine et al. (5) was a rare exception, which reported direct measurement of intestinal extraction after both intraduodenal and intravenous administration of midazolam through arterial and hepatoportal venous blood sampling during the anhepatic phase of liver transplantation surgery. All the other studies cited above relied on pharmacokinetic estimation of in vivo intestinal extraction based on oral and intravenous AUC data. The usual approach is to assume that systemic clearance represents hepatic clearance for drugs that are extensively metabolized, and that intestinal metabolism does not contribute to systemic clearance. Hence, hepatic extraction can be estimated according to the well-stirred clearance model as follows:

$$F_h = 1 - \frac{CL_{IV}}{Q_h} = 1 - \frac{Dose_{IV}/AUC_{IV}}{Q_h} \tag{13}$$

Recognizing that,

$$F = F_a \cdot F_{gm} \cdot F_h \tag{14}$$

and substituting Eq. (13) into (14) yields,

$$F_a \cdot F_{gm} = \frac{F}{F_h} = \frac{F}{1 + CL_{IV}/Q_h} = \frac{(AUC_{po}/Dose_{po})/(AUC_{IV}/Dose_{IV})}{1 + (Dose_{IV}/AUC_{IV})/Q_h} \tag{15}$$

If F_a is known, F_{gm} can be estimated from AUC_{po} and AUC_{IV} on the basis of Eq. (15). For most drugs that are readily released from the oral formulation and have high intestinal permeability, F_a is often assumed to be unity. The assumption that intestinal mucosa does not contribute to systemic clearance (i.e., negligible access of drug in systemic circulation to the intestinal mucosa) may be true under basal conditions, but might be inappropriate after treatment with the inducer, particularly if the inducer exerts a more profound effect on the gut wall enzyme compared with hepatic enzyme. Ignoring a possible contribution of the intestinal mucosal metabolism to systemic clearance would lead to an underestimation of the true intestinal first-pass extraction.

In the case of terfenadine (65) and saquinavir (66), the remarkably low oral bioavailability of these two drugs (<4%) has been attributed to extensive sequential first-pass metabolism by intestinal and hepatic CYP3A, although hard evidence is lacking. In fact, a recent study by Mouly et al. (67) questioned the quantitative importance of intestinal CYP3A-mediated metabolism of saquinavir during its first-pass extraction. It is possible that conventional oral doses of saquinavir saturates (i.e. auto inhibits) intestinal CYP3A enzymes and most of the dose escapes the gut unchanged.

Significant first-pass intestinal extraction occurs despite the fact that total CYP3A content in the human small intestine is much less than total hepatic CYP3A: 70 nmol versus 5490 nmol (8). However, the entire dose released from the dosage form is exposed to the mucosal enzymes during its obligatory passage through the intestinal epithelium and, thus, the more relevant comparison is the intracellular enzyme concentration and intrinsic clearance (i.e., V_{max}/K_m) of the enterocytes versus hepatocytes. In studies that offer comparative metabolic kinetics between hepatic and intestinal (duodenal or jejunal) microsomes, mean intrinsic clearance for intestinal microsomes varied from 20% to as high as 200% of that for liver microsomes; the list of drugs include cyclosporine (41), erythromycin (15), indinavir (44,45), irinotecan (46), lovastatin (48), midazolam (8), quazepam (50), rifabutin (51), saquinavir (53), and tacrolimus (55). For many of these drug substrates, mucosal intrinsic clearances are comparable to the corresponding mean hepatic intrinsic clearance.

Whether there will be a similar intestinal and hepatic first-pass extraction for the aforementioned drugs is more difficult to predict since intestinal first-pass extraction is dependent on a number of other factors, including total oral dose, rate of drug absorption, enzyme saturability (K_m), the absorptive region, mucosal barrier permeability, and binding to blood components. For example, should the dose be high enough to cause enzyme saturation, it is possible that a drug with a high hepatic and intestinal intrinsic clearance could largely escape intestinal first-pass extraction but not hepatic extraction. Also, if the basolateral membrane represents the rate-limiting barrier to the passage of drug across the intestinal epithelium, the residence time of the drug substrate within the enterocytes would increase and result in a greater metabolic first-pass loss than a comparable substrate with better permeability (6,68).

B. Induction of Intestinal CYP3A4

Induction of intestinal CYP3A4 by drugs has been demonstrated at both the biochemical and functional level. In their initial characterization of human intestinal CYP3A4, Kolars et al. (30) noted its inducibility in healthy volunteers following seven days of oral rifampin administration at 600 mg/day. CYP3A4 mRNA contents in biopsies of the duodenal mucosa were elevated five- to eightfold compared with untreated control biopsies. A later study conducted by Greiner et al. (69) showed a 4.4-fold increase of immunoreactive CYP3A protein in duodenal biopsy taken after 10 days of rifampin pretreatment. Yet another study evaluating shed human enterocytes sampled through a multilumen perfusion catheter placed in the small intestine indicated an approximate threefold increase in mucosal CYP3A4 mRNA and protein (70). Finally, in a recent study by Zhou et al. (71), the time course of duodenal CYP3A4 induction following rifampin treatment (150 mg, q6h) showed a 6-fold increase in CYP3A4 mRNA 2 days after the initiation of treatment and that it actually declined slightly to 3.5-fold above control after 14 days of drug administration. One possible explanation for this finding is that a concomitant increase in 1α,25-dihydroxy vitamin D_3 metabolism resulted in negative feedback suppression of duodenal CYP3A4 expression, or that there was some other compensatory response, so that the steady-state effect was lower than the initial peak response.

The transcriptional effect of intestinal CYP3A4 inducers is likely to be mediated by hPXR. This nuclear receptor is expressed prominently in the human duodenal and jejunal mucosa (71), and studies with LS-180 and Caco-2 cells indicate that its presence is essential for activation of CYP3A4 transcription by classic hPXR agonists (34,71). However, there may be ligand- and gene-specific differences in the magnitude of CYP3A4 induction in liver and small intestine. For example, some inducers such as the tocotrienols cause preferential induction of CYP3A4 in hepatocytes, compared with intestinal LS-180 cells, whereas the same compounds preferentially induce ABCB1 (MDR1/P-gp) in intestinal cells compared with hepatocytes. In contrast, rifampin induces CYP3A4 equally well in both cell types, but preferentially induces ABCB1 transcription in LS-180 cells. These differences have been attributed in part to a greater level of expression of a corepressor protein, NCoR, in intestinal cells and its gene-specific effects on hPXR-mediated transcriptional activation (72). This finding has some in vivo parallels. For example, efavirenz appears to be a much more effective inducer of CYP3A4 in the liver than in the small intestine (73).

Conflicting data have been reported on the inducibility of CYP3A5. Burk et al. (74) demonstrated induction of CYP3A5 mRNA in human hepatocytes exposed to 10 μM rifampin, although to a much more modest maximum degree than the highest degree of CYP3A4 induction (8-fold vs. 700-fold). More interestingly, these investigators also showed selective increase in the level of intestinal CYP3A5 mRNA following rifampin treatment in three of eight subjects who were all carriers of the CYP3A5*1 wide-type allele. The other five subjects were homozygotes of the variant CYP3A5*3 allele, who had very low basal expression of CYP3A5 transcript and did not respond to rifampin treatment. The investigators also showed that the effect was most likely mediated through binding of ligand bound hPXR/RXR to a proximal ER6 motif similar to that found in the CYP3A4 gene. These data indicate that individuals who are carriers of CYP3A5*1 allele may show a greater response to intestinal microsomal enzyme inducers because of simultaneous induction of CYP3A4 and CYP3A5.

In vivo, rifampin profoundly reduces the AUC of orally administered CYP3A substrates that exhibit a modest or low systemic availability because of intestinal and hepatic first-pass metabolism. Insights into the pharmacokinetics of rifampin induction have mostly been gained through studies with the prototype CYP3A substrate—midazolam. A number of studies have shown that induction of midazolam elimination is highly route dependent (75–77). For example, Gorski et al. (77) observed a 90% reduction in oral midazolam AUC when healthy subjects were pretreated with 600 mg/day of rifampin for seven days. This corresponds to a 22-fold increase in the apparent oral clearance (Cl/*F*). Although systemic clearance (Cl) of midazolam was also induced by rifampin, the effect was modest by comparison; i.e., at a 2.2-fold increase. This remarkable route dependency in the induction of midazolam clearance can be explained by induction of sequential first pass at the intestinal mucosa and the liver after oral administration, whereas only hepatic extraction is operative and inducible after intravenous administration. It was estimated that treatment with rifampin decreased F_h from 0.71 to 0.38 and F_{gm} from 0.47 to 0.15, which together resulted in a reduction in overall systemic availability (*F*) from 0.32 to 0.038. It should also be appreciated that the increase in systemic or hepatic clearance of midazolam following rifampin is limited to some extent by the ceiling of liver blood flow, as hepatic extraction of midazolam is increased from around 0.3 to as high as 0.6. Gorski et al. (77) also noted a significant inverse relationship between the fold increase in hepatic intrinsic clearance and the fold increase in intestinal intrinsic clearance. The inductive effects at the two sites appeared not to be concordant; in fact, the extent of induction was high at either the hepatic or intestinal site, but not both. Other orally administered CYP3A4 substrates whose AUCs are affected to an equally remarkable extent by rifampin and, therefore, suggestive of induction of intestinal CYP3A can be found in Table 1. Of note, all the drugs listed for which unambiguous data are available exhibit incomplete oral bioavailability. In addition, where both intravenous and oral administration have been studied, rifampin appears to increase the extent of intestinal first-pass metabolism and decrease intestinal bioavailability substantially; for example, alfentanil, $F_{gm} = 0.52$ and 0.16 under control and rifampin treatment; for nifedipine, $F_{gm} = 0.78$ and 0.24, respectively; and for *S*-verapamil, $F_{gm} = 0.52$ and 0.08, respectively.

Other CYP3A4 inducers have also been reported to exert a pronounced effect on oral midazolam AUC (78), presumably through induction of intestinal and hepatic CYP3A4. For example, administration of phenytoin and carbamazepine led to a 94% reduction in midazolam AUC compared with an untreated control population. Induction of CYP3A by St. John's wort, a widely used herbal supplement for the treatment of mild to moderate depression, has also attracted considerable interest. The major bioactive ingredient of St. John's wort, hyperforin, is a very potent in vitro activator of hPXR (79,80). Several groups have investigated induction of midazolam clearance by St. John's wort (81–83). As in the case of rifampin, clearance of oral midazolam

Table 1 Effect of Rifampin on the Oral AUC of CYP3A Substrates

CYP3A substrate	Baseline oral bioavailability[a] (%)	Rifampin dosage[b]	Decrease in AUC (%)[b]
Alfentanil	43	600 mg, q.d. (bedtime), × 5 days	96
Buspirone	4	600 mg, q.d. (8 pm), × 5 days	90
Gefetinib	60	600 mg, q.d., × 10 days	83
LAAM	48	600 mg, q.d. (evening), × 5 days	94
Midazolam	30	600 mg, q.d. (evening), × 7 days	90
Nifedipine	50	600 mg, q.d., × 7 days	92
Simvastatin	≤5	600 mg, q.d., × 9 days	91
Tamoxifen	–	600 mg, q.d. (8 pm), × 5 days	86
Toremifene	–	600 mg, q.d. (8 pm), × 5 days	87
Triazolam	44	600 mg, q.d. (8 pm), × 5 days	95
Verapamil	22	600 mg, q.d., × 12 days	97

Source: [a]Oral bioavailability data abstracted from Ref. 86, except for, alfentanil (205), midazolam (49), LAAM (206), triazolam (207).

[b]Oral rifampin dose and AUC changes abstracted from: alfentanil (205), buspirone (208), gefitinib (209), L-α-acetylmethadol (210), midazolam (77); nifedipine (62), simvastatin (211), tamoxifen (212), toremifene (212), triazolam (213), and verapamil (60).

was increased to a larger extent than that of intravenous midazolam; except, the induction of oral midazolam clearance was more modest than that with rifampin, about 1.5- to 3-fold. Estimation of F_{gm} and F_h revealed that the effect of St. John's wort on intestinal extraction was slightly greater than that on hepatic extraction.

It is important to point out that the effect of an hPXR agonist on in vivo CYP3A4 activity is highly variable and does not always correspond to the induction potency of the compounds in vitro. This can be explained in part by differences in systemic exposure to the hPXR agonist. For example, although hyperforin is a potent CYP3A4 inducer in cultured human hepatocytes (EC_{50} = 0.5 μM) (84), its effects on CYP3A4 activity in vivo are much more modest than that of rifampin, which is an equally potent inducer of CYP3A4 in vitro (EC_{50} = 0.5–1 μM) (85). The difference can be attributed to the much higher peak circulating rifampin concentrations (8 μM) (86), compared with that of hyperforin (0.28 μM) (87), following standard treatment doses. Indeed, the relatively poor hPXR agonist phenobarbital (hepatocytes EC_{50} = 100–200 μM) (85) is an effective CYP3A4 inducer in vivo simply because its therapeutic use results in very high systemic blood concentrations (50–250 μM) (86). Conversely, compounds such as lovastatin that are relatively effective inducers of CYP3A4 in vitro (EC_{50} = 1 μM) (85) are not associated with induction of CYP3A4 activity in vivo most likely because of the very low circulating blood concentrations (0.15 μM), following standard drug therapy. Admittedly, during the absorption period, intestinal concentrations of a CYP3A4 inducer could be higher than those seen systemically and result in greater induction of intestinal enzyme than what might be expected based on circulating concentration. However, the reasonableness of the in vitro–in vivo predictions suggests that intraenterocyte concentrations are not much greater than portal or systemic concentrations, which is what one might expect for a highly permeable and rapidly absorbed CYP3A4 inducer.

Many CYP3A substrates are also subject to efflux transport by P-gp at the intestinal mucosa (88–90). As discussed above, P-gp is inducible via activation of nuclear orphan receptors by the same inducers of CYP enzymes (91). Accordingly, for those drugs which are substrates of both CYP and P-gp, reduction in intestinal availability following treatment with known microsomal enzyme inducers could reflect the joint effects of increased mucosal metabolism and apical efflux.

There is still one other complication that should be recognized with induction of intestinal CYP enzymes. There is evidence indicating that microsomal enzyme inducers can simultaneously act as inhibitors. For example, Bidstrup et al. (92) recently suggested that rifampin acted as a dual inducer and inhibitor of repaglinide metabolism, most likely mediated by CYP2C8 and CYP3A4. The AUC of a single oral dose of repaglinide was measured before and after daily rifampin treatment for seven days, either concurrent with the last dose of rifampin or at 24 hours afterward. When repaglinide was given with the last dose of rifampin, a 50% reduction in the median AUC compared to the baseline was observed. In comparison, the reduction in repaglinide AUC was 80% when repaglinide was given 24 hours after the last dose. These investigators proposed that the inductive effect of rifampin was masked by its simultaneous inhibition of repaglinide metabolism when repaglinide clearance was assessed immediately after concurrent rifampin administration when the circulating concentration of rifampin was high. The inductive effect of rifampin was fully revealed after the washout of rifampin by 24 hours. Similarly, Xie and Kim (93) have proposed that St. John's wort can acutely inhibit and chronically induce CYP enzymes and P-gp. Ritonavir is also a potent inducer of CYP3A4 (94), as well as other CYP enzymes (95). However, it is also a very potent mechanism-based inhibitor of CYP3A4, so that its net effect on the intestinal and hepatic clearance of CYP3A substrates appears to be inhibitory (96). It stands to reason that the masking of an inductive effect by simultaneous inhibition is more likely to occur with intestinal first-pass metabolism; however, supportive evidence is lacking. The interplay between induction and inhibition also means that the outcomes of interaction studies with enzyme inducers may depend on study design; that is the relative timing of the inducer and substrate administrations.

C. Inhibition of Intestinal CYP3A

Considerable research has been directed toward understanding the effect of potent CYP3A inhibitors on the first-pass extraction of drugs at both the intestinal mucosa and the liver. The impetus for this research began in the early 1990s when it was discovered that ketoconazole profoundly elevated the AUC of orally administered terfenadine, resulting in a prolonged QT interval that could lead to a life-threatening cardiac arrhythmia. It was estimated that oral terfenadine AUC increased by 16- to 73-fold following chronic ketoconazole administration (97). Although some of the pharmacokinetic changes observed were surely the result of an interaction in the liver, it is likely that the enzyme/transporter barrier at the intestinal mucosa was also affected by ketoconazole. We now recognize that ketoconazole is capable of inhibiting CYP3A-dependent first-pass metabolism and P-gp-mediated active efflux processes at the intestinal mucosa. Drug-drug interactions involving inhibition of intestinal first-pass metabolism have since been identified for a number of other CYP3A substrates. These interactions can conveniently be grouped according to the mechanism of inhibition, namely those involving reversible (i.e., competitive or non-competitive) inhibition and those involving irreversible (i.e., mechanism-based) inhibition.

The commonly used azole antifungals—ketoconazole, fluconazole, voriconazole, and itraconazole—are potent, reversible CYP3A inhibitors and have been shown to inhibit the intestinal first-pass metabolism of several CYP3A drug substrates with very low oral bioavailability. For example, in an earlier study of the interaction between ketoconazole and tirilazad, Fleishaker et al. (63) reported a 1.7- and 3.1-fold increase in AUC after a single intravenous and oral dose of tirilazad, respectively. Pharmacokinetic analysis using Eqs. (13–15) presented in the preceding section suggested that tirilazad underwent sequential intestinal and hepatic first-pass metabolism. Moreover, ketoconazole inhibited intestinal and hepatic extraction to nearly the same extent; the hepatic availability (F_h) increased from 0.44 to 0.66 and the intestinal availability (F_{gm}) increased from 0.20 to 0.32. Olkkola et al. (98) found that itraconazole has a profound inhibitory effect on intestinal CYP3A, as evident by an increase in oral midazolam bioavailability from 39% to 96% following six days of oral itraconazole (200 mg/day) administration. Estimated F_h increased from 0.59 to 0.87. More remarkable is the increase in F_{gm} from 0.66 to near 1.0, suggesting that intestinal extraction was completely abolished by itraconazole treatment. Ketoconazole and itraconazole are potent reversible inhibitors of CYP3A4 in vitro, with respective unbound microsomal K_i values of 20 nM (20) and 6 nM (99). Thus, their ability to inhibit intestinal CYP3A-mediated drug metabolism is relatively predictable.

Studies in recent years have revealed a number of remarkable drug interactions with irreversible or mechanism-based inhibitors of CYP3A, many of which can be attributed to inhibition of sequential intestinal and hepatic first-pass metabolism. Mechanism-based inhibition involves the metabolism of an inhibitor to a reactive metabolite, which either forms a slowly reversible metabolic-intermediate (MI) complex with the heme moiety or inactivates the enzyme irreversibly via covalent binding to the enzyme catalyzing the last step in the bioactivation sequence. As a result, mechanism-based inhibition is both concentration and time dependent in manner. It should be noted that mechanism-based inhibitors also act as competitive inhibitors. The best-known mechanism-based inhibitors of CYP3A are some of the macrolide antibiotics (erythromycin, clarithromycin, troleandomycin, roxithromycin), HIV-protease inhibitors (ritonavir, saquinavir, and delavirdine), and the calcium channel antagonists—diltiazem and mibefradil (100).

The effects of the aforementioned mechanism-based inhibitors on intestinal first-pass metabolism have mostly been investigated using midazolam as the model CYP3A substrate that undergoes sequential intestinal and hepatic first-pass elimination. For example, clarithromycin is a modest reversible inhibitor of CYP3A4, with a K_i (10–28 μM) that is much higher than the peak plasma concentrations achieved after standard doses of the antibiotic (<5 μM) (51,101). Mayhew et al. (102) observed spectral formation of an MI complex when clarithromycin was incubated with human liver microsomes in the presence of NADPH. Moreover, they and Tinel et al. (103) showed time-dependent inactivation of microsomal CYP3A by clarithromycin. Accordingly, clarithromycin has been shown to be an effective inhibitor of both hepatic and intestinal midazolam metabolism in vivo. In a study with healthy volunteers, Gorski et al. (104) found that administration of clarithromycin (500 mg, b.i.d.) for seven days increased the hepatic midazolam availability (F_h) from 0.74 to 0.90 and the intestinal availability (F_{gm}) from 0.42 to 0.83. Overall, the inhibition of intestinal metabolism by clarithromycin had a much greater impact on the systemic availability of midazolam than it did on the inhibition of hepatic midazolam metabolism. A recent follow-up study by the same investigative group (105) showed that homogenates of duodenal biopsies taken from healthy subjects before and after clarithromycin treatment showed a 74% reduction in the rate of CYP3A-mediated 1′-hydroxylation of midazolam, which is consistent with the doubling of intestinal midazolam availability observed earlier. There was no change in the duodenal expression of CYP3A4 and CYP3A5 mRNA and protein. They observed a strong positive correlation between the decrease in duodenal CYP3A activity and baseline catalytic activity. In addition, CYP3A5 expressors (i.e., carriers of *CYP3A5*1* allele), which exhibited greater baseline intestinal CYP3A activity, showed greater inhibition of CYP3A activity after clarithromycin treatment compared with CYP3A5 nonexpressors.

Similarly, five days of saquinavir administration (1200 mg, t.i.d.) inhibited the metabolic clearance of intravenous and oral midazolam (106). The oral bioavailability of midazolam increased from 41% to 90%. The hepatic availability fraction was estimated to have increased from 0.64 to 0.84 and the intestinal availability fraction increased from 0.64 to about 1.0. Thus, saquinavir treatment resulted in a near complete inhibition of first-pass intestinal extraction and a lesser inhibition of hepatic extraction of midazolam.

Wang et al. (107) characterized the mechanism-based inhibition of testosterone 6β-hydroxylation (another CYP3A phenotypic marker) by verapamil enantiomers and each of their CYP3A metabolites—norverapamil and

N-desalkyl-verapamil in pooled human liver microsomes and recombinant CYP3A4 with coexpressed cytochrome b_5. The in vitro CYP3A inactivation experiments yielded two salient kinetic parameters, namely k_{inact}, the maximum first-order rate constant for inactivation, and $K_{I,u}$, the unbound concentration of the inhibitor that results in a first-order inactivation constant that is 50% of maximum. Estimates of the inactivation parameters, the average steady-state plasma unbound concentration of the verapamil enantiomers and metabolites, and literature values of the turnover half-life of CYP3A were incorporated into a sequential first-pass intestinal and hepatic extraction model to predict the increase in the bioavailability of orally administered midazolam. Backman et al. (108) had previously reported a threefold increase in the AUC of an oral dose of midazolam (15 mg) following three times daily treatment of oral racemic verapamil (80 mg) for two days, prior to full attainment of steady-state plasma verapamil concentration. On the basis of the model developed by Wang et al. (107), a 3.2- to 4.4-fold increase in oral midazolam AUC is expected during verapamil treatment at steady state if F_{gm} is increased from a baseline of around 0.5 to a maximum of 1.0. Again, this example illustrates the ability of a mechanism-based inhibitor to achieve near complete inactivation of intestinal CYP3A. Recently, Galetin et al. (109) extended the prediction of the Wang model to 37 in vivo cases of mechanism-based inhibition by erythromycin, clarithromycin, azithromycin, and diltiazem. The model predictions were within twofold of the in vivo observations for 26 cases involving the macrolide antibiotics. There were 11 overpredictions, curiously all belonging to interactions involving diltiazem and especially notable for the substrates cyclosporine and buspirone. Apparently, inactivation of intestinal CYP3A by diltiazem may not be as extensive or complete as in the case of the macrolide antibiotics. It is important to keep in mind that the Wang model does not take into account the possible confounding effects of CYP3A induction and modulation of intestinal P-gp that have been observed with some macrolide antibiotics. In addition, like all of the models for a mechanism-based interaction, it requires knowledge of the endogenous in vivo enzyme half-life (i.e., k_{deg}). There have been no direct measurements of this parameter in humans and, thus, one must use either values obtained from studies of pulse-labeled rodent CYP3A turnover, CYP3A turnover in human cell (Caco-2 or LS-180) culture experiments, or characterization of the time course of change in the oral clearance of a CYP3A substrate in vivo following treatment with a probe inducer or inhibitor. All of these alternative methods are subject to error, yield only approximations of the true enzyme half-life in the human enterocyte (or hepatocyte) in vivo and, thus, must be used with that limitation in mind.

One final consideration of drug interactions involving intestinal CYP3A is the fact that differential modulation of intestinal and hepatic CYP3A is possible. Simultaneous inhibition of intestinal first-pass metabolism and stimulation or induction of hepatic first-pass metabolism has been reported for the interaction between the herbal supplement echinacea (*Echinacea purpurea* root) and midazolam. Gorski et al. (110) reported that after a short course of echinacea (400 mg q.i.d. for 8 days), clearance of intravenous midazolam increased by 34%, whereas no significant alteration in oral midazolam AUC was observed, which means that the oral bioavailability must have increased by the same extent. Estimation of intestinal and hepatic availability from the IV and oral midazolam data indicated that echinacea treatment nearly doubled F_{gm} from 0.33 to 0.61, while F_h decreased from 0.72 to 0.61. These investigators suggested that echinacea extract when given orally is capable of inhibiting intestinal CYP3A and inducing hepatic CYP3A. The masking of a strong inhibitory effect on intestinal extraction by an opposing effect on hepatic extraction or systemic clearance certainly complicates the interpretation of interaction data involving sequential first-pass metabolism at the two sites.

D. Interactions with Grapefruit Juice

The ability of grapefruit juice to potently inhibit intestinal CYP3A-dependent first-pass metabolism has been recognized since the discovery in 1991 that grapefruit juice consumption led to a remarkable increase in C_{max} and AUC of the calcium channel antagonists—felodipine and nifedipine (111). Since then, there has been a flood of studies documenting the effect of grapefruit juice on the bioavailability of CYP3A drug substrates that exhibit significant first-pass metabolic extraction (Table 2); these include calcium channel antagonists, benzodiazepines, HMG-CoA reductase inhibitors, antimalarials, and other miscellaneous drugs [see reviews by Bailey et al. (112), Fuhr (113), and Saito et al. (114)]

An interesting aspect of the grapefruit juice effect is that it appears to be highly variable in a given population. Some subjects/patients will experience significant changes, whereas the change for others is minor or nonexistent (115–117). It has been suggested that the magnitude of the grapefruit juice interaction depends on the basal level of intestinal CYP3A expression (31). Higher levels of intestinal CYP3A are associated with a greater magnitude of interaction. In addition, the magnitude of the grapefruit juice interaction clearly depends on the strength of the juice and the frequency of administration. Repeated ingestion of 200 mL of double-strength grapefruit juice, three times a day, for two days, caused a ninefold increase in mean oral buspirone AUC (118), and a 15-fold and 16-fold increase in mean oral lovastatin (119) and simvastatin (120) AUC, respectively. In comparison, a subsequent study on the consumption of 250 mL of regular-strength juice, once every morning, for three days, had a much more modest effect on lovastatin AUC (30% increase) when lovastatin was dosed on the evening of the last day of grapefruit juice consumption, mimicking, the authors state, a more "typical" pattern of juice consumption and statin administration (121). Likewise, the effect of 200 mL of normal strength juice taken once a day in the morning for two days caused only a 3.3- and 3.6-fold increase in simvastatin and simvastatin acid AUC, respectively, when the drug was taken simultaneously with the last morning dose of juice (122). Indeed, where comparative data are available (e.g., atorvastatin, lovastatin,

Table 2 Effect of Grapefruit Juice on the Oral AUC of CYP3A Substrates

CYP3A substrate	Baseline oral bioavailability[a] (%)	GFJ dosage[b,c]	Mean increase in AUC (%)[c]
Alfentanil	43	240 mL NS at bedtime, 90 mL DS on study day (−2 hr)	61
Amiodarone	46	300 mL NS, 2 doses	50
Artemether	–	350 mL DS, SD	144
Artemether	–	350 mL DS, q.d., × 5 days	251
Atorvastatin	12	250 mL NS, t.i.d., × 3 days	33 (acid), 63 (lactone)
Atorvastatin	12	200 mL DS, t.i.d., × 3 days	150 (acid), 230 (lactone)
Buspirone	4	200 mL DS, t.i.d., × 3 days	823
Carbamazepine	–	300 mL NS, q.d., × 2 days	41
Cisapride	40–50	250 mL NS, SD	39
Cisapride	40–50	200 mL DS, t.i.d., × 2 days	144
Cyclosporine	28	250 mL NS, 2 doses	60
Diltiazem	38	250 mL NS, SD	20
Erythromycin	35	300 mL NS, SD	49
Ethinyl estradiol	42	100–200 mL, NS, ≤ 5 doses	28
Felodipine	15	240 mL NS, SD	104
Felodipine	15	240 mL NS, t.i.d., × 6 days	211
Lovastatin	≤5	250 mL NS, q.d., × 2+ days	57 (acid), 94 (lactone)
Lovastatin	≤5	200 mL DS, t.i.d., × 2+ days	400 (acid), 1430 (lactone)
Midazolam	30	240 mL NS, q.d., × 3 days	141
Midazolam	30	240 mL DS, t.i.d., × 3 days	495
Nifedpine	50	200 mL DS, 2 doses	58
Saquinavir	4	200 mL NS, 2 doses	50
Simvastatin	≤5	200 mL NS, q.d., × 2+ days	234 (acid), 255 (lactone)
Simvastatin	≤5	200 mL DS, t.i.d., × 2+ days	580 (acid), 1510 (lactone)
Triazolam	44	250 mL NS, SD	48
Triazolam	44	200 mL DS, t.i.d., × 3 days	143
Verapamil	22	250 mL, q.i.d. × 3 days	43

Source: [a]Bioavailability data abstracted from Ref. 86, except alfentanil (205), cisapride (214), ethinyl estradiol (215), midazolam (49), triazolam (207), saquinavir (216).

[b]NS and DS, normal-strength and double-strength grapefruit juice, respectively. A listing of two or more doses indicates that multiple doses of juice were given on the same pharmacokinetic study day.

[c]Oral grapefruit juice dose information and AUC changes abstracted from the following: alfentanil (205), amiodarone (217), artemether (218), atorvastatin (219,220), buspirone (120), carbamazepine (221), cisapride (221,222), cyclosporine (116), diltiazem (223), erythromycin (224), ethinyl estradiol (225), felodipine (31), lovastatin (121), midazolam (125), nifedipine (124), triazolam (226,227), saquinavir (228), simvastatin (120,122), verapamil (229).

midazolam, simvastatin), consumption of double-strength grapefruit juice, multiple times a day always results in a more profound change in the AUC of the CYP3A substrate than once a day consumption of normal strength juice (Table 2).

The ability of grapefruit juice to selectively elevate the bioavailability of CYP3A drug substrates that are subject to extensive first-pass metabolism indicates an inhibition of intestinal first-pass metabolism. Moreover, there is good evidence that normal consumption (one glass of regular strength juice a day) appears to alter only the function of intestinal CYP3A and not hepatic CYP3A. For example, the AUC of midazolam is increased only after its oral administration but not after intravenous administration (123). A similar observation was made with cyclosporine (116) and nifedipine (124). A recent study by Veronese et al. (125) showed that consumption of one glass of either regular- or double-strength grapefruit juice increased oral midazolam C_{max} and AUC (twofold), with little effect on midazolam elimination half-life and erythromycin breath test administered concurrently. The erythromycin breath test is a measure of hepatic CYP3A activity. In contrast, consumption of double-strength grapefruit juice thrice daily for three days significantly increased oral midazolam C_{max} (2.5-fold), AUC (6-fold), and elimination half-life (2-fold), and caused a significant reduction in the recovery of exhaled radiolabeled CO_2 in the erythromycin breath test. The prolongation in midazolam elimination half-life and change in erythromycin breath test indicates additional inhibition of hepatic CYP3A by absorbable grapefruit juice ingredients. These findings suggest that regular-strength grapefruit juice can be used to selectively probe for the presence of CYP3A-dependent intestinal metabolism of a drug molecule.

Lown et al. (31) were the first investigators to show that grapefruit juice consumption leads to a loss of intestinal CYP3A protein without a corresponding decrease in mRNA. There is strong evidence that grapefruit juice ingredients act as mechanism-based inhibitors of intestinal CYP3A (126), which in turn should accelerate mucosal turnover of CYP3A protein. The identity of the inhibitory components of grapefruit juice has been clarified in recent years [see recent review by Saito et al. (114)]. Early studies suggested that the inhibitory components of grapefruit juice might be flavonoids, specifically naringin or its aglycone naringenin, and quercetin (112,113). However, feeding of commercially available pure naringin showed little effect on nisoldipine or nifedipine pharmacokinetics (127,128). More recent studies in vitro and in vivo implicated furanocoumarins, namely 6′,7′-dihydroxybergamottin (129–131), bergamottin (132–135), and dimers of furanocoumarins (133,136). These grapefruit juice ingredients have been shown to inhibit intestinal CYP3A in vitro by both reversible and suicide inactivation mechanisms. Moreover, Paine et al (126) recently showed that removal of furanocoumarins from grapefruit juice abolished its inhibitory effect on oral felodipine clearance, establishing their importance in the interaction in vivo.

The maximum inhibitory effect appears to require consumption of more than one glass of normal- or double-strength juice. Moreover, inactivation of intestinal CYP3A by grapefruit juice means that recovery of intestinal CYP3A function depends on the de novo synthesis of CYP3A in the intestinal mucosa. Several studies have investigated the time it takes for the inhibitory effect of grapefruit juice to dissipate (137–140). The most recent study by Greenblatt et al. (139) on the time course of recovery of CYP3A function (oral midazolam clearance) after a single 300 mL of regular-strength grapefruit juice suggested a recovery half-life of 23 hours. Full recovery of intestinal CYP3A function was attained within three days.

E. Coupling of P-gp and CYP3A

Many of the substrates for CYP3A are subjected to P-gp-mediated cell efflux (141,142). P-gp is expressed prominently on the apical (luminal) membrane of intestinal epithelia. As discussed in Chapters 8 and 12, it can function to delay or limit the oral absorption of its substrates, depending on the relative magnitude of the secretory and diffusional clearances. In addition, Benet and collaborators have suggested that P-gp can work synergistically with CYP3A to reduce the intestinal availability of a common substrate drug beyond what would be expected if each operated independently (142–145,147). Specifically, they suggest that P-gp-mediated cell efflux increases the probability of a drug being exposed to the biotransformation enzyme through successive cycles of absorption and efflux, which increases intracellular residence time and enhances the probability that it will undergo first-pass metabolism. P-gp activity may also reduce the likelihood of product inhibition of CYP3A, if the metabolites are substrates for the transporter. In addition, P-gp efflux activity may ensure that intracellular concentrations of the P-gp/CYP3A substrate remain below the CYP3A K_m and that first-order kinetics are operative during the drug absorption period (146). Accordingly, inhibitors or inducers of intestinal P-gp may affect the extent of CYP3A-dependent first-pass metabolism. For example, Cummins et al. (147) reported that treatment of CYP3A4-induced Caco-2 cells with the P-gp inhibitor GG918 increased the intracellular concentration of a dual CYP3A and P-gp substrate, K77, and decreased the extent of apical to basolateral first-pass metabolism in the model system. Similarly, Hochman et al. (146) showed that inhibition of P-gp with cyclosporine also led to a decrease in the extent of CYP3A-dependent first-pass indinavir metabolism in the same Caco-2 cell model. Although cyclosporine is also a substrate/inhibitor of CYP3A4, the authors reported that its effect on CYP3A4 function was minimal at the concentration employed in the cell culture study.

The extent to which P-gp activity influences intestinal CYP3A-mediated first-pass extraction in vivo is still unclear. However, Johnson et al. (148) did show that selective inhibition of P-gp mediated verapamil efflux with PSC833 reduced the extent of intestinal first-pass metabolism in an autoperfused rat jejunum model. Interestingly, their kinetic analysis of the data indicated that the effect was mediated primarily by a disproportionate increase in the rate constant for verapamil transport from tissue into mesenteric blood, with no change in the apparent rate constant for verapamil metabolism. The authors proposed that saturation of intracellular verapamil binding as a consequence of P-gp inhibition and buildup in enterocyte concentration led to the change in verapamil transport. Why the apparent rate constant for intracellular metabolism would not also be affected is unclear if the two processes draw from the same pool of unbound substrate. Indeed, Tam et al. (149) have suggested through model simulations that changes in the apical secretion of a dual P-gp/CYP3A substrate would not affect the extent of metabolism at infinite time in the Caco-2 cell monolayer system, so long as first-order transport and metabolic conditions exist. However, the model employed by these authors represents a closed system and 100% metabolism will eventually occur. In vivo, drug that is absorbed into the basolateral (i.e., vascular) compartment will be removed by villous blood drainage. In other words, there is a competition between apical cycling via P-gp and absorptive loss in vivo. The autoperfused jejunum model of Johnson et al. maintains mesenteric perfusion, together with sampling of mesenteric blood draining the jejunal mucosal tissue. In addition, the cultured cell monolayer studies of Hochman and Cummins involved evaluation of the extraction process over a fixed (180 minutes) period of time. Although the latter experiments are not an ideal representation of the actual in vivo absorption and first-pass metabolism process, the combined data suggests some type of functional interrelationship between P-gp and CYP3A that is susceptible to modulation by P-gp inhibition.

III. OTHER INTESTINAL CYP ENZYMES (CYP2C9, CYP2C19, CYP2D6, AND CYP1A1)

Four other intestinal P450 isozymes that merit consideration from the perspective of oral drug bioavailability are CYP2C9, CYP2C19, CYP2D6, and CYP1A1. As pointed out by Paine et al. (150), these and CYP2J2 are the next most abundant P450 isozymes in the small intestine after CYP3A. Although DeWaziers et al. (10) reported the detection of what was described as CYP2C8-10 in mucosal microsomes, it was found only in the small intestine and preferentially in the proximal region. Subsequent studies indicate that CYP2C9 and CYP2C19 are the major CYP2C isoforms expressed in the human small intestine (70,151–153). The recent study of Läpple et al. examined paired intestinal and liver biopsies from a panel of 15 patients. CYP2C9 protein was measured in 12 of 15 intestinal biopsies, ranging from 0.5 to 3.4 pmol/mg protein. Mean CYP2C9 protein content in the intestinal biopsies was only 10% of that in liver biopsies (2.1 pmol/mg vs. 21 pmol/mg); likewise, mean clearance for diclofenac 4′-hydroxylation (marker of CYP2C9 activity) in intestinal microsomes was 16% of that in liver microsomes. CYP2C19 protein content was measured in 10 of 15 intestinal specimens, ranging from 0.3 to 3.2 pmol/mg protein. Mean CYP2C19 protein in the intestine was 69% of that in the liver (1.5 pmol/protein vs. 2.2 pmol/mg). Mean CYP2C19 activity as measured by *S*-mephenytoin 4-hydroxylation clearance in intestinal microsomes was 46% that in the corresponding liver microsomes. There was no significant correlation in the expression or activity of either CYP2C isoforms between the paired biopsy specimens, which suggests that these two CYP2C isoforms are independently regulated in the human intestine and liver. In their study, Lapple et al. also probed for CYP2C8 protein and activity. They were not able to measure CYP2C8 protein because of its low levels. However, using N-dealkylation of verapamil (i.e., formation of D-703), they found that the CYP2C8 activity in the intestinal microsomes was about 13% that in the paired liver microsomes. In contrast, Glaeser et al. (154) were able to quantitate CYP2C8 in duodenal biopsy tissue and found basal levels similar to that of CYP2C9. Paine and Thummel (153) reported similar results on the level of expression of CYP2C9 and CYP2C19 and probe substrate activity in duodenal mucosa. In addition, they showed that the pattern of CYP2C9/19 expression along the length of the small intestine was remarkably similar to that of CYP3A4, with much higher levels of both enzymes in the proximal small intestine compared with the ileum, confirming the earlier observation of DeWaziers et al. (10).

In view of their incomplete oral bioavailability, several CYP2C substrates may undergo significant first-pass intestinal metabolism, including the CYP2C9 substrates verapamil (155), losartan (156), fluvastatin (157), and diclofenac (158), and the CYP2C19 substrates (*S*)-mephenytoin (159) and omeprazole (160). However, there is no firm evidence to date to indicate that these intestinal enzymes are involved in drug-drug interactions. Interestingly, Glaeser et al. (154) recently reported that rifampin treatment induced intestinal CYP2C9 expression and ex vivo CYP2C9 activity toward verapamil. Schuetz and colleagues have also found that duodenal CYP2C9 mRNA is significantly increased by daily rifampin treatment (Erin Schuetz, St. Jude Children's Research Center, personal communication). This increase raises the possibility that some of the effects of rifampin on the disposition of CYP2C9 substrates, such as fluvastatin (161) and celecoxib (162), may involve an induction of intestinal first-pass metabolism.

DeWaziers et al. first described the identification of CYP2D6 in human intestinal microsomes (10). Like CYP3A4, it is localized within mucosal enterocytes and most concentrated within the proximal small intestine. The mean specific enzyme content of duodenal and jejunal microsomes was reported to be approximately 20% of hepatic CYP2D6 microsomal content. However, it was undetected in ileum and colon. Other investigators have confirmed the expression of CYP2D6 in the human gastrointestinal tract. In separate studies, Prueksaritanont et al. detected CYP2D6 in mucosal microsomes from several human donors (163,164). In addition, they reported observing microsomal 1′-hydroxylation activity toward the CYP2D6 substrate

(+)-bufuralol in all preparations. Further, the activity was largely inhibited by the known CYP2D6 inhibitors, quinidine and ajmaline, as well as anti-CYP3A1 IgG (163).

In a more comprehensive study, Madani et al. (165) quantified CYP2D6 protein in 20 human jejunum and 31 human livers. They found that the median microsomal-specific CYP2D6 content was less than 8% of the hepatic microsomal content (0.85 vs. 12.8 pmol/mg) and that there was extensive interindividual variability in protein content for both tissues. These investigators also characterized the catalytic activity of the same jejunal microsomes toward the recognized CYP2D6 substrate metoprolol and found that α-hydroxylation reaction rate was significantly correlated with CYP2D6 protein content ($r = 0.75$).

Although there are many CYP2D6 substrates that undergo extensive first-pass metabolism and are involved in many clinically significant inhibitory drug interactions, it is not likely that the gut wall contributes significantly to the elimination process. For example, duodenal and jejunal microsomal intrinsic clearances for metoprolol oxidation reactions were found to be only a fraction of the hepatic intrinsic clearance (165). On the basis of the well-stirred model and assuming villous blood-flow limited absorption, the first-pass intestinal and hepatic extraction ratios for metoprolol were predicted to be 2% and 61%, respectively. Thus, any potential inhibition of intestinal CYP2D6 activity should have little impact on the systemic bioavailability of the drug. However, recently, Mizuma et al. (7) argued that permeability limitation at either the apical or basolateral membrane (or both) could enhance the intestinal extraction of metoprolol and, hence, increase the possibility of drug-drug interactions at the intestinal site. Definitive evidence for interactions involving CYP2D6 is still lacking.

CYP1A1 is an extrahepatic isoform expressed in human intestine as well as the lung, skin, larynx, and placenta but not in the human liver, whereas CYP1A2 is expressed in the liver and in the duodenum. It is unclear to what extent CYP1A enzymes are constitutively expressed in the intestine. In a recent study by Paine et al. (150), CYP1A1 protein was detected in mucosal scrapings of 3 of 31 organ donors. In contrast, in a study by McDonnell et al. (166) in which endoscopic biopsy specimens were obtained from various tissues including duodenum and colon of six healthy volunteers before and one week after taking 20 mg of omeprazole daily, CYP1A1 mRNA and ethoxyresorufin activity were present constitutively in the duodenum of each volunteer. The omeprazole treatment induced CYP1A1 mRNA and enzymatic activity in five of the six volunteers. Consumption of chargrilled meat diet resulted in unequivocal induction of CYP1A enzymes in the liver and small intestine of healthy adults fed a diet enriched with chargrilled meat for seven days (167). At present, the in vivo significance of intestinal CYP1A1 with respect to first-pass drug metabolism remains unknown.

IV. INTESTINAL SULFOTRANSFERASES

Cytosolic sulfotransferases (ST) catalyze the transfer of sulfate, donated by 3′-phosphoadenosine-5′-phosphosulfate (PAPS), to a phenolic (OH) group on an acceptor substrate. Aside from the liver, SULTs are expressed in the intestinal mucosa and other extrahepatic tissues. Drugs that are subject to sulfonation in the human small intestine include isoproterenol, salicylamide, acetaminophen, ethinyl estradiol, terbutaline, salbutamol, minoxidil, apomorphine, and budesonide (168,169). At least four isoforms of sulfotransferase have been identified in the human intestinal mucosa (13,170): simple phenolic ST (SULT1A1), monoamine phenolic ST (SULT1A3), estrogen phenolic ST (SULT1E1), and hydroxysteroid ST (SULT2A1). Chen et al. (170) conducted a thorough study of the distribution of STs along the human intestine using four ST isoform-selective substrates: 2-naphthol (SULT1A1), dopamine (SULT1A3), estradiol (SULT1E1), and dehydroepiandrosterone (SULT2A1). Small intestine has the highest activity for all four substrates compared with the stomach and colon. Dopamine sulfonation was three times as high as that of human liver. Estrogen sulfonation was about comparable between small intestine and liver. In contrast, intestinal sulfonation rate of 2-naphthol and dihydroepiandrosterone were one-half and one-fifth that of human liver, respectively. Western blot results generally agree with the catalytic activity measurements. High variability in the activity of all four ST isoforms was noted.

Although data is limited, it has been suggested that gut wall SULTs contribute to the first-pass metabolism of the β_2-agonists, isoproteranol, terbutaline (171,172), and ethinyl estradiol (42). Likewise, evidence for the involvement of human intestinal SULTs in drug-drug interactions is limited and, in some cases, circumstantial. For example, first-pass sulfonation of isoproterenol in the dog can be reduced by coadministration of competitive substrates, salicylamide (173) and ascorbic acid (174). Also, both oral acetaminophen (175) and ascorbate (176) administration increase the bioavailability of ethinyl estradiol through an inhibition of sulfotransferase activity. Inhibitors of SULTs may exert their effect by competition for the enzyme or the limited pool of PAPS. The effects of acetaminophen and ascorbate, both given in gram doses, are attributed to a reduction in first-pass intestinal ethinyl estradiol sulfonation, via depletion of the mucosal sulfate pool. Recently, mefenamic acid was identified as a potent, selective competitive inhibitor of human liver phenolic sulfotransferases (SULT1A1), with an IC_{50} of 20 nM (177,178). It was further shown that the IC_{50} values of mefenamic acid for the sulfonation of salbutamol and apomorphine in cytosolic fraction prepared from human duodenum were four orders of magnitude higher (i.e., $IC_{50} > 160\ \mu M$) than that from human liver. A similar differential inhibition of SULT1A1 activity between duodenal and liver cytosol fraction was also observed with salicylic acid. The in vivo significance of these in vitro results on the inhibition of SULT1A1 awaits further investigation.

V. INTESTINAL GLUCURONOSYLTRANSFERASES

The human UGTs fall into three subfamilies, UGT1A, UGT2A, and UGT2B. Like the CYP enzymes, several UGT isoforms are expressed within the human intestinal mucosa (179–182) and could contribute toward first-pass drug metabolism. Messenger RNA expression of UGT1A1, UGT1A3, UGT1A4, UGT1A6, UGT2B4, UGT2B7, UGT2B10 (ileum only), UGT2B15, and UGT2B28 in the small human intestine has been reported. Most notably, UGT1A8 and UGT1A10 are expressed exclusively in the gastrointestinal tract, including the intestine (180). UGT1A7 is expressed in the upper gastrointestinal tract, but not in the intestine. However, evidence for expression of UGT proteins and associated catalytic activities in the human intestine is incomplete or conflicting. Radominska-Pandya et al. (11) failed to detect UGT2B4 in the human intestine, but a UGT enzyme putatively identified as UGT2B7 was detected. UGT2B7 has been proposed to be responsible for the glucuronidation of ezetimibe in intestinal microsomes (183). However, the rate of 3-glucuronidation of morphine, a UGT2B7-selective reaction, was very low in intestinal microsomes in comparison to liver, suggesting minimal expression of UGT2B7 protein in the intestine (184). Similarly, acetaminophen-*O*-glucuronide formation rates, reflecting UGT1A6 activity, were low (184). In a study by Paine and Fisher (182), significant UGT1A1 protein expression was confirmed in the small intestine, whereas UGT1A6 expression was undetectable and UGT2B7 immunoreactivity was faint to detectable. As such, one may conclude that UGT1A1, UGT1A8, and UGT1A10 proteins are present in the intestine at appreciable levels that may influence oral drug bioavailability, whereas other UGT isoforms, such as UGT1A6, UGT2B4, and UGT2B7 do not play a role in first pass glucuronidation reactions. One reason for the discrepancies between studies may be related to the polymorphic expression of most UGT isoforms in the human intestine (185,186). The mechanisms of polymorphic expression are unclear, except for UGT1A1. The UGT1A1*28 allelic variant contains an additional TA repeat in the TATA-box promoter that has been linked to decreased expression of the enzyme (187,188).

Microsomes isolated from human intestine display appreciable glucuronidation activity toward several drugs, including estradiol and 17β-estradiol, ethinyl estradiol (42,189), acetaminophen, propofol (190), amitriptyline, desipramine, imipramine, ibuprofen (12,191), raloxifene (192), resveratrol (193), ezetimibe (183), and troglitazone (194). Using recombinant enzymes, it has been shown that the two intestine-specific UGT isoforms, UGT1A8 and UGT1A10, glucuronidate numerous compounds, including raloxifene troglitazone, resveratrol, opioids, valproic acid, clofibrate, furosemide, acetaminophen, propranolol, and probenecid (192–195), which raises the possibility that intestinal metabolism may limit the oral bioavailability of these drugs. We have also observed that the 3- and 17-glucuronidation of estradiol by UGT1A8 and UGT1A10 is as efficient as that observed by UGT1A1 (Isoherranen and Thummel, unpublished observations), suggesting that the intestine may be a more significant site of estradiol glucuronidation than the liver. It is likely that many of the substrates for the intestinal UGTs will undergo at least some first-pass metabolism at the intestinal mucosa. However, the quantitative importance of this process compared with hepatic extraction remains to be elucidated.

Despite the fact that UGTs contribute significantly to the elimination of numerous drugs, clinically significant inhibitory drug-drug interactions involving hepatic or intestinal UGTs are relatively rare. There may be two reasons for this. First, UGT enzymes are typically low-affinity enzymes, and K_m and K_i values of ligands are in the high-micromolar range and above the circulating inhibitor concentrations. Second, considerable substrate overlap between UGT enzymes exists, which attenuates the impact of isoform-specific interactions (196). These characteristics have also made the UGTs difficult to study as very few isoform-specific probes or inhibitors have been identified. Nevertheless, intestinal UGTs appear to be involved in drug interactions. The best example is perhaps the clinically observed interaction between mycophenolate mofetil and tacrolimus. UGT1A8 and UGT1A10 along with UGT1A9 (major) and UGT1A1 (minor) catalyze the 7-O-glucuronidation of mycophenolic acid (195,197,198), the active metabolite produced from ester hydrolysis of mycophenolate mofetil. In intestinal microsomes, the UGT1A9 probe propofol decreased mycophenolic acid glucuronidation by 47%. Tacrolimus is reportedly a good inhibitor of mycophenolic acid conjugation, both in vitro (199) and in vivo (200). Thus, it is possible that the drug-drug interaction that occurs in patients is, in part, a consequence of the inhibition of first-pass intestinal UGT1A8/9/10 activity.

An interesting aspect of drug-drug interactions involving UGTs is the impairment of enterohepatic circulation (EHC). During EHC, a drug is generally glucuronidated in the liver, its conjugate metabolite is excreted to the bile, cleaved in the gut lumen by the gut microflora β-glucuronidases, and the liberated aglycone reabsorbed back into the portal circulation (201). Theoretically, inhibition of the β-glucuronidases in gut microflora will decrease the exposure to drugs that are subject to significant EHC. An example of such interaction would be the effect of antibiotics on oral contraceptive (OC) pharmacokinetics. Epidemiological evidence suggests that oral contraceptive failures are associated with the use of oral antibiotics. However, prospective pharmacokinetics studies to date have failed to demonstrate a clear acceleration of OC estrogen or progestin clearance after oral antibiotic administration (203). Case reports have suggested that some women have significantly reduced concentrations of ethinyl estradiol when taken in combination with tetracyclines and penicillin derivatives (202). It is likely that the interaction, if it exists, occurs only in selected individuals who are poor metabolizers for the nonconjugative pathways (e.g., CYP3A) and have a significant pool of conjugate metabolites undergoing EHC, among other factors (204).

Inductive interactions can be of a concern with UGTs. Similar to some of the P450 enzymes, the nuclear hormone receptors CAR and PXR regulate induction of UGT1A1, one of the intestinal UGT isoforms and may contribute to the regulation of others as well. Numerous studies have documented induction of glucuronidation activity in vivo by rifampin using various drug substrates, but questions regarding which isoform(s) of UGT are induced and the site of induction have not been addressed.

In conclusion, the involvement of intestinal CYP3A in metabolically based drug interactions is reasonably well characterized. In contrast, the role of other intestinal drug-metabolizing enzymes in drug interactions remains speculative or controversial. The major obstacle appears to be our poor understanding of the functional importance of other CYP and conjugating enzymes in intestinal first-pass drug metabolism. Future progress in this area will require a concerted effort in developing appropriate in vitro cellular systems and conducting rigorous human studies to elucidate in vivo function and regulation of intestinal drug-metabolizing enzymes.

REFERENCES

1. Thummel KE, Kunze KL, Shen DD. Enzyme-catalyzed processes of first-pass hepatic and intestinal drug extraction. Adv Drug Deliv Rev 1997; 27:99–127.
2. Pang KS. Modeling of intestinal drug absorption: roles of transporters and metabolic enzymes (for the Gillette Review Series). Drug Metab Dispos 2003; 31(12): 1507–1519.
3. Rowland M, Matin SB. Kinetics of drug-drug interactions. J Pharmacokinet Biopharm 1973; 1:553–567.
4. Shaw PN, Houston JB. Kinetics of drug metabolism inhibition: use of metabolite concentration-time profiles. J Pharmacokinet Biopharm 1987; 15:497–510.
5. Paine MF, Shen DD, Kunze KL, et al. First-pass metabolism of midazolam by the human intestine. Clin Pharmacol Ther 1996; 60:14–24.
6. Mizuma T. Kinetic impact of presystemic intestinal metabolism on drug absorption: experiment and data analysis for the prediction of in vivo absorption from in vitro data. Drug Metab Pharmacokinet 2002; 17(6):496–506.
7. Mizuma T, Tsuji A, Hayashi M. Does the well-stirred model assess the intestinal first-pass effect well? J Pharm Pharmacol 2004; 56(12):1597–1599.
8. Paine MF, Khalighi M, Fisher JM, et al. Characterization of inter- and intra-intestinal differences in human CYP3A-dependent metabolism. J Pharmacol Exp Ther 1997; 283:1552–1562.
9. Zhang QY, Dunbar D, Ostrowska A, et al. Characterization of human small intestinal cytcochromes P-450. Drug Metab Dispos 1999; 27:804–809.
10. DeWaziers I, Cugnenc PH, Yang CS, et al. Cytochrome P450 isoenzymes, epoxide hydrolase and glutathione transferases in rat and human hepatic and extrahepatic tissues. J Pharmacol Exp Ther 1990; 253:387–394.
11. Radominska-Pandya A, Little J, Pandya J, et al. UDP-glucuronosyltransferases in human intestinal mucosa. Biochimica et Biophysica Acta 1998; 1394:199–208.
12. Strassburg C, Nguyen N, Manns M, et al. UDP-glucuronosyltransferase activity in human liver and colon. Gastroenterology 1999; 116:149–160.

13. Her C, Szumlanski C, Aksoy I, et al. Human jejunal estrogen sulfotransferase and dehydroepiandrosterone sulfotransferase. Immunochemical characterization of individual variation. Drug Metab Dispos 1996; 24:1328–1335.
14. Murray GI, Barnes TS, Sewell HF, et al. The immunocytochemical localisation and distribution of cytochrome P-450 in normal hepatic and extrahepatic tissues with a monoclonal antibody to human cytcohrome P-450. Br J Clin Pharmacol 1988; 25: 465–475.
15. Watkins PB, Wrighton SA, Schuetz EG, et al. Identification of glucocorticoid-inducible cytochromes P-450 in the intestinal mucosa of rats and man. J Clin Invest 1987; 80:1029–1036.
16. Kolars JC, Lown KS, Schmiedlin-Ren P, et al. CYP3A gene expression in human gut epithelium. Pharmacogenetics 1994; 4:247–259.
17. McKinnon R, Burgess W, Hall PDLM, et al. Characterization of *CYP3A* gene subfamily expression in human gastrointestinal tissues. Gut 1995; 36:259–267.
18. Lown KS, Kolars JC, Thummel KE, et al. Interpatient heterogeneity in expression of CYP3A4 and CYP3A5 in small bowel: lack of prediction by the erythromycin breath test. Drug Metab Dispos 1994; 22:947–955.
19. Lin YS, Dowling AL, Quigley SD, et al. Co-regulation of CYP3A4 and CYP3A5 and contribution to hepatic and intestinal midazolam metabolism. Mol Pharmacol 2002; 62(1):162–172.
20. Gibbs MA, Thummel KE, Shen DD, et al. Inhibition of cytochrome P-450 3A (CYP3A) in human intestinal and liver microsomes: comparison of Ki values and impact of CYP3A5 expression. Drug Metab Dispos 1999; 27(2):180–187.
21. Huang W, Lin YS, McConn DJ II, et al. Evidence of significant contribution from CYP3A5 to hepatic drug metabolism. Drug Metab Dispos 2004; 32(12): 1434–1445.
22. McConn DJ, Lin YS, Allen KE, et al. Differences in the inhibition of cytochromes P450 3A4 and 3A5 by metabolite-inhibitor complex-forming drugs. Drug Metab Dispos 2004; 32(10):1083–1091.
23. von Richter O, Burk O, Fromm MF, et al. Cytochrome P450 3A4 and P-glycoprotein expression in human small intestinal enterocytes and hepatocytes: a comparative analysis in paired tissue specimens. Clin Pharmacol Ther 2004; 75(3): 172–183.
24. Yang J, Tucker GT, Rostami-Hodjegan A. Cytochrome P450 3A expression and activity in the human small intestine. Clin Pharmacol Ther 2004; 76(4):391.
25. Gervot L, Carriére V, Costet P, et al. CYP3A5 is the major cytochrome P450 3A expressed in human colon and colonic cell lines. Environ Toxicol Pharmacol 1996; 2:381–388.
26. Schmiedlin-Ren P, Thummel KE, Fisher JM, et al. Expression of enzymatically active CYP3A4 by Caco-2 cells grown on extracellular matrix-coated permeable supports in the presence of 1a,25-dihydroxyvitamin D_3. Mol Pharmacol 1997; 51: 741–754.
27. Hu M, Li Y, Davitt CM, et al. Transport and metabolic characterization of Caco-2 cells expressing CYP3A4 and CYP3A4 plus oxidoreductase [in process citation]. Pharm Res 1999; 16(9):1352–1359.
28. Lindell M, Karlsson MO, Lennernas H, et al. Variable expression of CYP and Pgp genes in the human small intestine. Eur J Clin Invest 2003; 33(6):493–499.

29. Thorn M, Finnstrom N, Lundgren S, et al. Cytochromes P450 and MDR1 mRNA expression along the human gastrointestinal tract. Br J Clin Pharmacol 2005; 60(1): 54–60.
30. Kolars JC, Schmiedlin-Ren P, Schuetz JD, et al. Identification of rifampin-inducible P450IIIA4 (CYP3A4) in human small bowel enterocytes. J Clin Invest 1992; 90:1871–1878.
31. Lown KS, Bailey DG, Fontana RJ, et al. Grapefruit juice increases felodipine oral availability in humans by decreasing intestinal CYP3A protein expression. J Clin Invest 1997; 99:2545–2553.
32. Blumberg B, Kang H, Bolado J Jr., et al. BXR, an embryonic orphan nuclear receptor activated by a novel class of endogenous benzoate metabolites. Genes Dev 1998; 12(9):1269–1277.
33. Thompson PD, Jurutka PW, Whitfield GK, et al. Liganded VDR induces CYP3A4 in small intestinal and colon cancer cells via DR3 and ER6 vitamin D responsive elements. Biochem Biophys Res Commun 2002; 299(5):730–738.
34. Thummel KE, Brimer C, Yasuda K, et al. Transcriptional control of intestinal cytochrome P-4503A by 1alpha,25-dihydroxy vitamin D(3). Mol Pharmacol 2001; 60(6):1399–1406.
35. Makishima M, Lu TT, Xie W, et al. Vitamin D receptor as an intestinal bile acid sensor. Science 2002; 296(5571):1313–1316.
36. Xu Y, Iwanaga K, Zhou C, et al. Selective induction of intestinal CYP3A23 by 1alpha,25-dihydroxyvitamin D3 in rats. Biochem Pharmacol 2006; 72(3):385–392.
37. Jurutka PW, Thompson PD, Whitfield GK, et al. Molecular and functional comparison of 1,25-dihydroxyvitamin D(3) and the novel vitamin D receptor ligand, lithocholic acid, in activating transcription of cytochrome P450 3A4. J Cell Biochem 2005; 94(5):917–943.
38. Meyer MB, Watanuki M, Kim S, et al. The human transient receptor potential vanilloid type 6 distal promoter contains multiple vitamin D receptor binding sites that mediate activation by 1,25-dihydroxyvitamin D3 in intestinal cells. Mol Endocrinol 2006; 20(6):1447–1461.
39. Fleet JC, Eksir F, Hance KW, et al. Vitamin D-inducible calcium transport and gene expression in three Caco-2 cell lines. Am J Physiol Gastrointest Liver Physiol 2002; 283(3):G618–G625.
40. Valsamis HA, Arora SK, Labban B, et al. Antiepileptic drugs and bone metabolism. Nutr Metab (Lond) 2006; 3:36.
41. Lampen A, Christians U, Bader A, et al. Drug interactions and interindividual variability of ciclosporin metabolism in the small intestine. Pharmacology 1996; 52 (3):159–168.
42. Rogers SM, Back DJ, Orme ML. Intestinal metabolism of ethinyloestradiol and paracetamol in vitro: studies using Ussing chambers. Br J Clin Pharmacol 1987; 23 (6):727–734.
43. Mahon WA, Inaba T, Stone RM. Metabolism of flurazepam by the small intestine. Clin Pharmacol Ther 1977; 22(2):228–233.
44. Chiba M, Hensleigh M, Lin JH. Hepatic and intestinal metabolism of indinavir, an HIV protease inhibitor, in rat and human microsomes. Major role of CYP3A. Biochem Pharmacol 1997; 53(8):1187–1195.
45. Koudriakova T, Iatsimirskaia E, Utkin I, et al. Metabolism of the human immunodeficiency virus protease inhibitors indinavir and ritonavir by human intestinal

microsomes and expressed cytochrome P4503A4/3A5: mechanism-based inactivation of cytochrome P4503A by ritonavir. Drug Metab Dispos 1998; 26(6):552–561.
46. Fujita K, Ando Y, Narabayashi M, et al. Gefitinib (Iressa) inhibits the CYP3A4-mediated formation of 7-ethyl-10-(4-amino-1-piperidino)carbonyloxycamptothecin but activates that of 7-ethyl-10-[4-N-(5-aminopentanoic acid)-1-piperidino] carbonyloxycamptothecin from irinotecan. Drug Metab Dispos 2005; 33(12): 1785–1790.
47. Oda Y, Kharasch ED. Metabolism of methadone and levo-alpha-acetylmethadol (LAAM) by human intestinal cytochrome P450 3A4 (CYP3A4): potential contribution of intestinal metabolism to presystemic clearance and bioactivation. J Pharmacol Exp Ther 2001; 298(3):1021–1032.
48. Jacobsen W, Kirchner G, Hallensleben K, et al. Small intestinal metabolism of the 3-hydroxy-3-methylglutaryl-coenzyme A reductase inhibitor lovastatin and comparison with pravastatin. J Pharmacol Exp Ther 1999; 291(1):131–139.
49. Thummel KE, O'Shea D, Paine MF, et al. Oral first-pass elimination of midazolam involves both gastrointestinal and hepatic CYP3A-mediated metabolism. Clin Pharmacol Ther 1996; 59:491–502.
50. Miura M, Ohkubo T. In vitro metabolism of quazepam in human liver and intestine and assessment of drug interactions. Xenobiotica 2004; 34(11–12):1001–1011.
51. Iatsimirskaia E, Tulebaev S, Storozhuk E, et al. Metabolism of rifabutin in human enterocyte and liver microsomes: kinetic paramters, identification of enzyme systems, and drug interactions with macrolides and antifungal agents. Clin Pharmacol Ther 1997; 61:554–562.
52. Fitzsimmons ME, Collins JM. Selective biotransformation of the human immunodeficiency virus protease inhibitor saquinavir by human small-intestinal cytochrome P4503A4. Drug Metab Dispos 1997; 25:256–266.
53. Eagling VA, Wiltshire H, Whitcombe IW, et al. CYP3A4-mediated hepatic metabolism of the HIV-1 protease inhibitor saquinavir in vitro. Xenobiotica 2002; 32(1):1–17.
54. Lampen A, Zhang Y, Hackbarth I, et al. Metabolism and transport of the macrolide immunosuppressant sirolimus in the small intestine. J Pharmacol Exp Ther 1998; 285(3):1104–1112.
55. Lampen A, Christians U, Guengerich FP, et al. Metabolism of the immunosuppressant tacrolimus in the small intestine: cytochrome P450, drug interactions, and interindividual variability. Drug Metab Dispos 1995; 23:1315–1324.
56. Obach RS, Zhang QY, Dunbar D, et al. Metabolic characterization of the major human small intestinal cytochrome p450s. Drug Metab Dispos 2001; 29(3): 347–352.
57. Mahmood I, Sahajwalla C. Clinical pharmacokinetics and pharmacodynamics of buspirone, an anxiolytic drug. Clin Pharmacokinet 1999; 36(4):277–287.
58. Hebert MF, Roberts JP, Prueksaritanont T, et al. Bioavailability of cyclosporine with concomitant rifampin administration is markedly less than predicted by hepatic enzyme induction. Clin Pharmacol Ther 1992; 52:453–457.
59. Lundahl J, Regardh CG, Edgar B, et al. Effects of grapefruit juice ingestion–pharmacokinetics and haemodynamics of intravenously and orally administered felodipine in healthy men. Eur J Clin Pharmacol 1997; 52(2):139–145.
60. Fromm MF, Busse D, Kroemer HK, et al. Differential induction of prehepatic and hepatic metabolism of verapamil by rifampin. Hepatology 1996; 24:796–801.

61. von Richter O, Greiner B, Fromm MF, et al. Determination of in vivo absorption, metabolism, and transport of drugs by the human intestinal wall and liver with a novel perfusion technique. Clin Pharmacol Ther 2001; 70(3):217–227.
62. Holtbecker N, Fromm M, Kroemer HK, et al. The nifedipine-rifampin interaction. Evidence for induction of gut wall metabolism. Drug Metab Dispos 1996; 24: 1121–1123.
63. Fleishaker JC, Pearson PG, Wienkers LC, et al. Biotransformation of tirilazad in humans: 2. Effect of ketoconazole on tirilazad clearance and oral bioavailability. J Pharmacol Exp Ther 1996; 277:991–998.
64. Masica AL, Mayo G, Wilkinson GR. In vivo comparisons of constitutive cytochrome P450 3A activity assessed by alprazolam, triazolam, and midazolam. Clin Pharmacol Ther 2004; 76(4):341–349.
65. Garteiz DA, Hook RH, Walker BJ, et al. Pharmacokinetics and biotransformation studies of terfenadine in man. Arzneimittelforschung 1982; 32(9a):1185–1190.
66. Williams PEO, Muirhead GJ, Madigan MJ, et al. Disposition and bioavailability of the HIV-proteinase inhibitor, Ro 31-8959, after single doses in healthy volunteers. Br J Clin Pharmacol 1992; 34:155P–156P.
67. Mouly SJ, Matheny C, Paine MF, et al. Variation in oral clearance of saquinavir is predicted by CYP3A5*1 genotype but not by enterocyte content of cytochrome P450 3A5. Clin Pharmacol Ther 2005; 78(6):605–618.
68. Gwilt P, Comer S, Chaturvedi P, et al. The influence of diffusional barriers on presystemic gut elimination. Drug Metab Dispos 1988; 16:521–526.
69. Greiner B, Eichelbaum M, Fritz P, et al. The role of intestinal P-glycoprotein in the interaction of digoxin and rifampin. J Clin Invest 1999; 104(2):147–153.
70. Glaeser H, Drescher S, Eichelbaum M, et al. Influence of rifampicin on the expression and function of human intestinal cytochrome P450 enzymes. Br J Clin Pharmacol 2005; 59(2):199–206.
71. Zhou C, Assem M, Tay JC, et al. Steroid and xenobiotic receptor and vitamin D receptor crosstalk mediates CYP24 expression and drug-induced osteomalacia. J Clin Invest 2006; 116(6):1703–1712.
72. Zhou C, Tabb MM, Sadatrafiei A, et al. Tocotrienols activate the steroid and xenobiotic receptor, SXR, and selectively regulate expression of its target genes. Drug Metab Dispos 2004; 32(10):1075–1082.
73. Mouly S, Lown KS, Kornhauser D, et al. Hepatic but not intestinal CYP3A4 displays dose-dependent induction by efavirenz in humans. Clin Pharmacol Ther 2002; 72(1):1–9.
74. Burk O, Koch I, Raucy J, et al. The induction of cytochrome P450 3A5 (CYP3A5) in the human liver and intestine is mediated by the xenobiotic sensors pregnane X receptor (PXR) and constitutively activated receptor (CAR). J Biol Chem 2004; 279(37):38379–38385.
75. Backman JT, Olkkola KT, Neuovnen PJ. Rifampin drastically reduces plasma concentrations and effects of oral midazolam. Clin Pharmacol Ther 1996; 59:7–13.
76. Kharasch ED, Russell M, Mautz D, et al. The role of cytochrome P450 3A4 in alfentanil clearance. Implications for interindividual variability in disposition and perioperative drug interactions. Anesthesiology 1997; 87:36–50.
77. Gorski JC, Vannaprasaht S, Hamman MA, et al. The effect of age, sex, and rifampin administration on intestinal and hepatic cytochrome P450 3A activity. Clin Pharmacol Ther 2003; 74(3):275–287.

78. Backman JT, Olkkola KT, Ojala M, et al. Concentrations and effects of oral midazolam are greatly reduced in patients treated with carbamazepine or phenytoin. Epilepsia 1996; 37:253–257.
79. Moore LB, Goodwin B, Jones SA, et al. St. John's wort induces hepatic drug metabolism through activation of the pregnane X receptor. Proc Natl Acad Sci U S A 2000; 97(13):7500–7502.
80. Wentworth JM, Agostini M, Love J, et al. St John's wort, a herbal antidepressant, activates the steroid X receptor. J Endocrinol 2000; 166(3):R11–R16.
81. Dresser GK, Schwarz UI, Wilkinson GR, et al. Coordinate induction of both cytochrome P4503A and MDR1 by St John's wort in healthy subjects. Clin Pharmacol Ther 2003; 73(1):41–50.
82. Wang Z, Gorski JC, Hamman MA, et al. The effects of St John's wort (Hypericum perforatum) on human cytochrome P450 activity. Clin Pharmacol Ther 2001; 70 (4):317–326.
83. Xie R, Tan LH, Polasek EC, et al. CYP3A and P-glycoprotein activity induction with St. John's Wort in healthy volunteers from 6 ethnic populations. J Clin Pharmacol 2005; 45(3):352–356.
84. Komoroski BJ, Zhang S, Cai H, et al. Induction and inhibition of cytochromes P450 by the St. John's wort constituent hyperforin in human hepatocyte cultures. Drug Metab Dispos 2004; 32(5):512–518.
85. LeCluyse EL. Human hepatocyte culture systems for the in vitro evaluation of cytochrome P450 expression and regulation. Eur J Pharm Sci 2001; 13(4):343–368.
86. Brunton LL, Lazo JS, Parker KL. Goodman & Gilman's The Pharmacological Basis of Therapeutics. 11th ed. New York: McGraw-Hill; 2006.
87. Biber A, Fischer H, Romer A, et al. Oral bioavailability of hyperforin from hypericum extracts in rats and human volunteers. Pharmacopsychiatry 1998; 31 (suppl 1):36–43.
88. Chan LM, Lowes S, Hirst BH. The ABCs of drug transport in intestine and liver: efflux proteins limiting drug absorption and bioavailability. Eur J Pharm Sci 2004; 21(1):25–51.
89. Fricker G, Miller DS. Relevance of multidrug resistance proteins for intestinal drug absorption in vitro and in vivo. Pharmacol Toxicol 2002; 90(1):5–13.
90. Van Asperen J, Van Tellingen O, Beijnen JH. The pharmacological role of P-glycoprotein in the intestinal epithelium. Pharmacol Res 1998; 37(6):429–435.
91. Xu C, Li CY, Kong AN. Induction of phase I, II and III drug metabolism/transport by xenobiotics. Arch Pharm Res 2005; 28(3):249–268.
92. Bidstrup TB, Stilling N, Damkier P, et al. Rifampicin seems to act as both an inducer and an inhibitor of the metabolism of repaglinide. Eur J Clin Pharmacol 2004; 60(2):109–114.
93. Xie HG, Kim RB. St John's wort-associated drug interactions: short-term inhibition and long-term induction? Clin Pharmacol Ther 2005; 78(1):19–24.
94. Luo G, Cunningham M, Kim S, et al. CYP3A4 induction by drugs: correlation between a pregnane X receptor reporter gene assay and CYP3A4 expression in human hepatocytes. Drug Metab Dispos 2002; 30(7):795–804.
95. Faucette SR, Wang H, Hamilton GA, et al. Regulation of CYP2B6 in primary human hepatocytes by prototypical inducers. Drug Metab Dispos 2004; 32(3): 348–358.

96. Yeh RF, Gaver VE, Patterson KB, et al. Lopinavir/ritonavir induces the hepatic activity of cytochrome P450 enzymes CYP2C9, CYP2C19, and CYP1A2 but inhibits the hepatic and intestinal activity of CYP3A as measured by a phenotyping drug cocktail in healthy volunteers. J Acquir Immune Defic Syndr 2006; 42(1): 52–60.
97. Honig PK, Wortham DC, Zamani K, et al. The terfenadine-ketoconazole interaction: pharmacokinetic and electrocardiographic consequences. JAMA 1993; 269: 1513–1518.
98. Olkkola KT, Ahonen J, Neuvonen PJ. The effect of systemic antimycotics, itraconazole and fluconazole, on the pharmacokinetics and pharmacodynamics of intravenous and oral midazolam. Anesth Analg 1996; 82:511–516.
99. Isoherranen N, Kunze KL, Allen KE, et al. Role of itraconazole metabolites in CYP3A4 inhibition. Drug Metab Dispos 2004; 32(10):1121–1131.
100. Zhou S, Yung Chan S, Cher Goh B, et al. Mechanism-based inhibition of cytochrome P450 3A4 by therapeutic drugs. Clin Pharmacokinet 2005; 44(3):279–304.
101. Jurima-Romet M, Crawford K, Cyr T, et al. Terfenadine metabolism in human liver. In vitro inhibition by macrolide antibiotics and azole antifungals. Drug Metab Dispos 1994; 22:849–857.
102. Mayhew BS, Jones DR, Hall SD. An in vitro model for predicting in vivo inhibition of cytochrome P450 3A4 by metabolic intermediate complex formation. Drug Metab Dispos 2000; 28(9):1031–1037.
103. Tinel M, Descatoire V, Larrey D, et al. Effects of clarithromycin on cytochrome P-450. Comparison with other macrolides. J Pharmacol Exp Ther 1989; 250:746–751.
104. Gorski JC, Jones DR, Haehner-Daniels BD, et al. The contribution of intestinal and hepatic CYP3A to the interaction between midazolam and clarithromycin. Clin Pharmacol Ther 1998; 64:133–143.
105. Pinto AG, Wang YH, Chalasani N, et al. Inhibition of human intestinal wall metabolism by macrolide antibiotics: effect of clarithromycin on cytochrome P450 3A4/5 activity and expression. Clin Pharmacol Ther 2005; 77(3):178–188.
106. Palkama VJ, Ahonen J, Neuvonen PJ, et al. Effect of saquinavir on the pharmacokinetics and pharmacodynamics of oral and intravenous midazolam. Clin Pharmacol Ther 1999; 66(1):33–39.
107. Wang YH, Jones DR, Hall SD. Prediction of cytochrome P450 3A inhibition by verapamil enantiomers and their metabolites. Drug Metab Dispos 2004; 32(2): 259–266.
108. Backman JT, Olkkola KT, Aranko K, et al. Dose of midazolam should be reduced during diltiazem and verapamil treatments. Br J Clin Pharmacol 1994; 37: 221–225.
109. Galetin A, Burt H, Gibbons L, et al. Prediction of time-dependent CYP3A4 drug-drug interactions: impact of enzyme degradation, parallel elimination pathways, and intestinal inhibition. Drug Metab Dispos 2006; 34(1):166–175.
110. Gorski JC, Huang SM, Pinto A, et al. The effect of echinacea (Echinacea purpurea root) on cytochrome P450 activity in vivo. Clin Pharmacol Ther 2004; 75(1): 89–100.
111. Bailey DG, Spence JD, Munoz C, et al. Interaction of citrus juices with felodipine and nifedipine. Lancet 1991; 377:268–269.
112. Bailey DG, Malcolm J, Arnold O, et al. Grapefruit juice-drug interactions. Br J Clin Pharmacol 1998; 46(2):101–110.

113. Fuhr U. Drug interactions with grapefruit juice. Extent, probable mechanism and clinical relevance. Drug Saf 1998; 18(4):251–272.
114. Saito M, Hirata-Koizumi M, Matsumoto M, et al. Undesirable effects of citrus juice on the pharmacokinetics of drugs: focus on recent studies. Drug Saf 2005; 28 (8):677–694.
115. Gross AS, Goh YD, Addison RS, et al. Influence of grapefruit juice on cisapride pharmacokinetics. Clin Pharmacol Ther 1999; 65:395–401.
116. Ducharme MP, Warbasse LH, Edwards DJ. Disposition of intravenous and oral cyclosporine after administration with grapefruit juice. Clin Pharmacol Ther 1995; 57:485–491.
117. Bailey DG, Arnold JMO, Spence JD. Grapefruit juice and drugs. How significant is the interaction? Clin Pharmacokinet 1994; 26:91–98.
118. Lilja JJ, Kivisto KT, Backman JT, et al. Grapefruit juice substantially increases plasma concentrations of buspirone. Clin Pharmacol Ther 1998; 64(6):655–660.
119. Kantola T, Kivisto KT, Neuvonen PJ. Grapefruit juice greatly increases serum concentrations of lovastatin and lovastatin acid. Clin Pharmacol Ther 1998; 63 (4):397–402.
120. Lilja JJ, Kivisto KT, Neuvonen PJ. Grapefruit juice-simvastatin interaction: effect on serum concentrations of simvastatin, simvastatin acid, and HMG-CoA reductase inhibitors. Clin Pharmacol Ther 1998; 64(5):477–483.
121. Rogers JD, Zhao J, Liu L, et al. Grapefruit juice has minimal effects on plasma concentrations of lovastatin-derived 3-hydroxy-3-methylglutaryl coenzyme A reductase inhibitors [in process citation]. Clin Pharmacol Ther 1999; 66(4): 358–366.
122. Lilja JJ, Neuvonen M, Neuvonen PJ. Effects of regular consumption of grapefruit juice on the pharmacokinetics of simvastatin. Br J Clin Pharmacol 2004; 58(1): 56–60.
123. Kupferschmidt HHT, Ha HR, Ziegler WH, et al. Interaction between grapefruit juice and midazolam in humans. Clin Pharmacol Ther 1995; 58:20–28.
124. Rashid TJ, Martin U, Clarke H, et al. Factors affecting the absolute bioavailability of nifedipine. Br J Clin Pharmacol 1995; 40(1):51–58.
125. Veronese ML, Gillen LP, Burke JP, et al. Exposure-dependent inhibition of intestinal and hepatic CYP3A4 in vivo by grapefruit juice. J Clin Pharmacol 2003; 43 (8):831–839.
126. Paine MF, Widmer WW, Hart HL, et al. A furanocoumarin-free grapefruit juice establishes furanocoumarins as the mediators of the grapefruit juice-felodipine interaction. Am J Clin Nutr 2006; 83(5):1097–1105.
127. Bailey DG, Arnold JM, Munoz C, et al. Grapefruit juice–felodipine interaction: mechanism, predictability, and effect of naringin. Clin Pharmacol Ther 1993; 53 (6):637–642.
128. Bailey DG, Arnold JM, Strong HA, et al. Effect of grapefruit juice and naringin on nisoldipine pharmacokinetics. Clin Pharmacol Ther 1993; 54(6):589–594.
129. Edwards DJ, Bellevue FH III, Woster PM. Identification of 6′,7′-dihydroxybergamottin, a cytochrome P450 inhibitor, in grapefruit juice. Drug Metab Dispos 1996; 24:1287–1290.
130. Schmiedlin-Ren P, Edwards DJ, Fitzsimmons ME, et al. Mechanisms of enhanced oral availability of CYP3A4 substrates by grapefruit constituents: decreased

enterocyte CYP3A4 concentration and mechanism-based inactivation by furanocoumarins. Drug Metab Dispos 1997; 25:1228–1233.

131. Edwards DJ, Fitzsimmons ME, Schuetz EG, et al. 6′,7′-Dihydroxybergamottin in grapefruit juice and Seville orange juice: effects on cyclosporine disposition, enterocyte CYP3A4, and P- glycoprotein. Clin Pharmacol Ther 1999; 65(3):237–244.
132. Bailey DG, Kreeft JH, Munoz C, et al. Grapefruit juice-felodipine interaction: effect of naringin and 6′,7′-dihydroxybergamottin in humans. Clin Pharmacol Ther 1998; 64(3):248–256.
133. Guo LQ, Taniguchi M, Xiao YQ, et al. Inhibitory effect of natural furanocoumarins on human microsomal cytochrome P450 3A activity. Jpn J Pharmacol 2000; 82(2): 122–129.
134. Bailey DG, Dresser GK, Bend JR. Bergamottin, lime juice, and red wine as inhibitors of cytochrome P450 3A4 activity: comparison with grapefruit juice. Clin Pharmacol Ther 2003; 73(6):529–537.
135. Kakar SM, Paine MF, Stewart PW, et al. 6′7′-Dihydroxybergamottin contributes to the grapefruit juice effect. Clin Pharmacol Ther 2004; 75(6):569–579.
136. Fukuda K, Ohta T, Oshima Y, et al. Specific CYP3A4 inhibitors in grapefruit juice: furocoumarin dimers as components of drug interaction. Pharmacogenetics 1997; 7: 391–396.
137. Lundahl J, Regardh CG, Edgar B, et al. Relationship between time of intake of grapefruit juice and its effect on pharmacokinetics and pharmacodynamics of felodipine in healthy subjects. Eur J Clin Pharmacol 1995; 49(1–2):61–67.
138. Takanaga H, Ohnishi A, Murakami H, et al. Relationship between time after intake of grapefruit juice and the effect on pharmacokinetics and pharmacodynamics of nisoldipine in healthy subjects. Clin Pharmacol Ther 2000; 67(3):201–214.
139. Greenblatt DJ, von Moltke LL, Harmatz JS, et al. Time course of recovery of cytochrome p450 3A function after single doses of grapefruit juice. Clin Pharmacol Ther 2003; 74(2):121–129.
140. Lilja JJ, Kivisto KT, Neuvonen PJ. Duration of effect of grapefruit juice on the pharmacokinetics of the CYP3A4 substrate simvastatin. Clin Pharmacol Ther 2000; 68(4):384–390.
141. Kim RB, Wandel C, Leake B, et al. Interrelationship between substrates and inhibitors of human CYP3A and P-glycoprotein. Pharm Res 1999; 16(3):408–414.
142. Wacher VJ, Wu CY, Benet LZ. Overlapping substrate specificities and tissue distribution of cytochrome P450 3A and P-glycoprotein: implications for drug delivery and activity in cancer chemotherapy. Mol Carcinog 1995; 13(3):129–134.
143. Lown KS, Mayo RR, Leichtman AB, et al. Role of intestinal P-glycoprotein (mdr1) in interpatient variation in the oral bioavailability of cyclosporine. Clin Pharmacol Ther 1997; 62:248–260.
144. Wacher VJ, Silverman JA, Zhang Y, et al. Role of P-glycoprotein and cytochrome P450 3A in limiting oral absorption of peptides and peptidomimetics. J Pharm Sci 1998; 87(11):1322–1330.
145. Watkins PB. The barrier function of CYP3A4 and P-glycoprotein in the small bowel. Adv Drug Deliv Rev 1997; 27(2–3):161–170.
146. Hochman JH, Chiba M, Nishime J, et al. Influence of P-glycoprotein on the transport and metabolism of indinavir in Caco-2 cells expressing cytochrome P-450 3A4. J Pharmacol Exp Ther 2000; 292(1):310–318.

147. Cummins CL, Jacobsen W, Benet LZ. Unmasking the dynamic interplay between intestinal P-glycoprotein and CYP3A4. J Pharmacol Exp Ther 2002; 300(3): 1036–1045.
148. Johnson BM, Chen W, Borchardt RT, et al. A kinetic evaluation of the absorption, efflux, and metabolism of verapamil in the autoperfused rat jejunum. J Pharmacol Exp Ther 2003; 305(1):151–158.
149. Tam D, Sun H, Pang KS. Influence of P-glycoprotein, transfer clearances, and drug binding on intestinal metabolism in Caco-2 cell monolayers or membrane preparations: a theoretical analysis. Drug Metab Dispos 2003; 31(10):1214–1226.
150. Paine MF, Hart HL, Ludington SS, et al. The human intestinal cytochrome P450 ''pie''. Drug Metab Dispos 2006; 34(5):880–886.
151. Klose TS, Blaisdell JA, Goldstein JA. Gene structure of CYP2C8 and extrahepatic distribution of the human CYP2Cs. J Biochem Mol Toxicol 1999; 13(6):289–295.
152. Lapple F, von Richter O, Fromm MF, et al. Differential expression and function of CYP2C isoforms in human intestine and liver. Pharmacogenetics 2003; 13(9): 565–575.
153. Paine M, Thummel K. Role of intestinal cytochromes P450 in drug disposition. Lee J, Obach SR, Fisher M, eds. Drug Metabolizing Enzymes. Lausanne, Switzerland: Fontis Media, 2003:421–451.
154. Glaeser H, Drescher S, Hofmann U, et al. Impact of concentration and rate of intraluminal drug delivery on absorption and gut wall metabolism of verapamil in humans. Clin Pharmacol Ther 2004; 76(3):230–238.
155. Busse D, Cosme J, Beaune P, et al. Cytochromes of the P450 2C subfamily are the major enzymes involved in the O-demethylation of verapamil in humans. Naunyn Schmiedebergs Arch Pharmacol 1995; 353(1):116–121.
156. Stearns RA, Chakravarty PK, Chen R, et al. Biotransformation of losartan to its active carboxylic acid metabolite in human liver microsomes. Role of cytochrome P4502C and 3A subfamily members. Drug Metab Dispos 1995; 23(2):207–215.
157. Kirchheiner J, Kudlicz D, Meisel C, et al. Influence of CYP2C9 polymorphisms on the pharmacokinetics and cholesterol-lowering activity of (−)-3S,5R-fluvastatin and (+)-3R,5S-fluvastatin in healthy volunteers. Clin Pharmacol Ther 2003; 74(2): 186–194.
158. Leemann T, Kondo M, Zhao J, et al. [The biotransformation of NSAIDs: a common elimination site and drug interactions]. Schweiz Med Wochenschr 1992; 122 (49):1897–1899.
159. Goldstein JA, Faletto MB, Romkes-Sparks M, et al. Evidence that CYP2C19 is the major (S)-mephenytoin 4′-hydroxylase in humans. Biochemistry 1994; 33(7): 1743–1752.
160. Lasker JM, Wester MR, Aramsombatdee E, et al. Characterization of CYP2C19 and CYP2C9 from human liver: respective roles in microsomal tolbutamide, S-mephenytoin, and omeprazole hydroxylations. Arch Biochem Biophys 1998; 353(1): 16–28.
161. Scripture CD, Pieper JA. Clinical pharmacokinetics of fluvastatin. Clin Pharmacokinet 2001; 40(4):263–281.
162. Jayasagar G, Krishna Kumar M, Chandrasekhar K, et al. Influence of rifampicin pretreatment on the pharmacokinetics of celecoxib in healthy male volunteers. Drug Metabol Drug Interact 2003; 19(4):287–295.

163. Prueksaritanont T, Dwyer LM, Cribb AE. (+)-Bufuralol 1′-hydroxylation activity in human and rhesus monkey intestine and liver. Biochem Pharmacol 1995; 50(9):1521–1525.
164. Prueksaritanont T, Gorham LM, Hochman JH, et al. Comparative studies of drug-metabolizing enzymes in dog, monkey, and human small intestines, and in Caco-2 cells. Drug Metab Dispos 1996; 24(6):634–642.
165. Madani S, Paine MF, Lewis L, et al. Comparison of CYP2D6 content and metoprolol oxidation between microsomes isolated from human livers and small intestines [in process citation]. Pharm Res 1999; 16(8):1199–1205.
166. McDonnell WM, Scheiman JM, Traber PG. Induction of cytochrome P450IA genes (CYP1A) by omeprazole in the human alimentary tract. Gastroenterology 1992; 103 (5):1509–1516.
167. Fontana RJ, Lown KS, Paine MF, et al. Effects of a chargrilled meat diet on expression of CYP3A, CYP1A, and P-glycoprotein levels in healthy volunteers. Gastroenterology 1999; 117(1):89–98.
168. Mulder GJ, Jakoby WB. Sulfation. In: Mulder GJ, ed. Conjugation Reactions in Drug Metabolism. New York: Taylor and Francis, 1990:107–161.
169. Kauffman FC. Sulfonation in pharmacology and toxicology. Drug Metab Rev 2004; 36(3–4):823–843.
170. Chen G, Zhang D, Jing N, et al. Human gastrointestinal sulfotransferases: identification and distribution. Toxicol Appl Pharmacol 2003; 187(3):186–197.
171. Conolly ME, Davies DS, Dollery CT, et al. Metabolism of isoprenaline in dog and man. Br J Pharmacol 1972; 46(3):458–472.
172. Pacifici GM, Eligi M, Giuliani L. (+) and (−) terbutaline are sulphated at a higher rate in human intestine than in liver. Eur J Clin Pharmacol 1993; 45(5):483–487.
173. Bennett PN, Blackwell E, Davies DS. Competition for sulphate during detoxification in the gut wall. Nature 1975; 258(5532):247–248.
174. Houston JB, Wilkens HJ, Levy G. Potentiation of isoproterenol effect by ascorbic acid. Res Commun Chem Pathol Pharmacol 1976; 14(4):643–650.
175. Rogers SM, Back DJ, Stevenson PJ, et al. Paracetamol interaction with oral contraceptive steroids: increased plasma concentrations of ethinyloestradiol. Br J Clin Pharmacol 1987; 23(6):721–725.
176. Back DJ, Breckenridge AM, MacIver M, et al. Interaction of ethinyloestradiol with ascorbic acid in man. Br Med J (Clin Res Ed) 1981; 282(6275):1516.
177. Vietri M, Pietrabissa A, Spisni R, et al. Differential inhibition of hepatic and duodenal sulfation of (−)-salbutamol and minoxidil by mefenamic acid. Eur J Clin Pharmacol 2000; 56(6–7):477–479.
178. Vietri M, Vaglini F, Pietrabissa A, et al. Sulfation of R(−)-apomorphine in the human liver and duodenum, and its inhibition by mefenamic acid, salicylic acid and quercetin. Xenobiotica 2002; 32(7):587–594.
179. de Wildt SN, Kearns GL, Leeder JS, et al. Cytochrome P450 3A: ontogeny and drug disposition. Clin Pharmacokinet 1999; 37(6):485–505.
180. Gregory PA, Lewinsky RH, Gardner-Stephen DA, et al. Regulation of UDP glucuronosyltransferases in the gastrointestinal tract. Toxicol Appl Pharmacol 2004; 199(3):354–363.
181. Kiang TK, Ensom MH, Chang TK. UDP-glucuronosyltransferases and clinical drug-drug interactions. Pharmacol Ther 2005; 106(1):97–132.

182. Paine MF, Fisher MB. Immunochemical identification of UGT isoforms in human small bowel and in caco-2 cell monolayers. Biochem Biophys Res Commun 2000; 273(3):1053–1057.
183. Ghosal A, Hapangama N, Yuan Y, et al. Identification of human UDP-glucuronosyltransferase enzyme(s) responsible for the glucuronidation of ezetimibe (Zetia). Drug Metab Dispos 2004; 32(3):314–320.
184. Fisher MB, Vandenbranden M, Findlay K, et al. Tissue distribution and interindividual variation in human UDP-glucuronosyltransferase activity: relationship between UGT1A1 promoter genotype and variability in a liver bank. Pharmacogenetics 2000; 10(8):727–739.
185. Strassburg CP, Kneip S, Topp J, et al. Polymorphic gene regulation and interindividual variation of UDP-glucuronosyltransferase activity in human small intestine. J Biol Chem 2000; 275(46):36164–36171.
186. Strassburg CP, Nguyen N, Manns MP, et al. Polymorphic expression of the UDP-glucuronosyltransferase UGT1A gene locus in human gastric epithelium. Mol Pharmacol 1998; 54(4):647–654.
187. Monaghan G, Ryan M, Seddon R, et al. Genetic variation in bilirubin UPD-glucuronosyltransferase gene promoter and Gilbert's syndrome. Lancet 1996; 347 (9001):578–581.
188. Raijmakers MT, Jansen PL, Steegers EA, et al. Association of human liver bilirubin UDP-glucuronyltransferase activity with a polymorphism in the promoter region of the UGT1A1 gene. J Hepatol 2000; 33(3):348–351.
189. Back DJ, Bates M, Breckenridge AM, et al. The in vitro metabolism of ethinyloestradiol, mestranol and levonorgestrel by human jejunal mucosa. Br J Clin Pharmacol 1981; 11(3):275–278.
190. Raoof AA, van Obbergh LJ, de Ville de Goyet J, et al. Extrahepatic glucuronidation of propofol in man: possible contribution of gut wall and kidney. Eur J Clin Pharmacol 1996; 50(1–2):91–96.
191. Cappiello M, Giuliani L, Pacifici GM. Distribution of UDP-glucuronosyltransferase and its endogenous substrate uridine 5′-diphosphoglucuronic acid in human tissues. Eur J Clin Pharmacol 1991; 41(4):345–350.
192. Jeong EJ, Liu Y, Lin H, et al. Species- and disposition model-dependent metabolism of raloxifene in gut and liver: role of UGT1A10. Drug Metab Dispos 2005; 33 (6):785–794.
193. Watanabe Y, Nakajima M, Yokoi T. Troglitazone glucuronidation in human liver and intestine microsomes: high catalytic activity of UGT1A8 and UGT1A10. Drug Metab Dispos 2002; 30(12):1462–1469.
194. Sabolovic N, Humbert AC, Radominska-Pandya A, et al. Resveratrol is efficiently glucuronidated by UDP-glucuronosyltransferases in the human gastrointestinal tract and in Caco-2 cells. Biopharm Drug Dispos 2006; 27(4):181–189.
195. Cheng Z, Radominska-Pandya A, Tephly TR. Studies on the substrate specificity of human intestinal UDP-glucuronosyltransferases 1A8 and 1A10 [in process citation]. Drug Metab Dispos 1999; 27(10):1165–1170.
196. Williams JA, Hyland R, Jones BC, et al. Drug-drug interactions for UDP-glucuronosyltransferase substrates: a pharmacokinetic explanation for typically observed low exposure (AUCi/AUC) ratios. Drug Metab Dispos 2004; 32(11): 1201–1208.

197. Mojarrabi B, Mackenzie PI. The human UDP glucuronosyltransferase, UGT1A10, glucuronidates mycophenolic acid. Biochem Biophys Res Commun 1997; 238(3): 775–778.
198. Picard N, Ratanasavanh D, Premaud A, et al. Identification of the UDP-glucuronosyltransferase isoforms involved in mycophenolic acid phase II metabolism. Drug Metab Dispos 2005; 33(1):139–146.
199. Zucker K, Tsaroucha A, Olson L, et al. Evidence that tacrolimus augments the bioavailability of mycophenolate mofetil through the inhibition of mycophenolic acid glucuronidation. Ther Drug Monit 1999; 21(1):35–43.
200. Zucker K, Rosen A, Tsaroucha A, et al. Augmentation of mycophenolate mofetil pharmacokinetics in renal transplant patients receiving Prograf and CellCept in combination therapy. Transplant Proc 1997; 29(1–2):334–336.
201. Roberts MS, Magnusson BM, Burczynski FJ, et al. Enterohepatic circulation: physiological, pharmacokinetic and clinical implications. Clin Pharmacokinet 2002; 41(10):751–790.
202. Dickinson BD, Altman RD, Nielsen NH, et al. Drug interactions between oral contraceptives and antibiotics. Obstet Gynecol 2001; 98(5 pt 1):853–860.
203. Bauer KL, Wolf D, Patel M, et al. Clinical inquiries. Do antibiotics interfere with the efficacy of oral contraceptives? J Fam Pract 2005; 54(12):1079–1080.
204. Back DJ, Orme ML. Pharmacokinetic drug interactions with oral contraceptives. Clin Pharmacokinet 1990; 18(6):472–484.
205. Kharasch ED, Walker A, Hoffer C, et al. Intravenous and oral alfentanil as in vivo probes for hepatic and first-pass cytochrome P450 3A activity: noninvasive assessment by use of pupillary miosis. Clin Pharmacol Ther 2004; 76(5):452–466.
206. Walsh SL, Johnson RE, Cone EJ, et al. Intravenous and oral l-alpha-acetylmethadol: pharmacodynamics and pharmacokinetics in humans. J Pharmacol Exp Ther 1998; 285(1):71–82.
207. Garzone PD, Kroboth PD. Pharmacokinetics of the newer benzodiazepines. Clin Pharmacokinet 1989; 16(6):337–364.
208. Lamberg TS, Kivisto KT, Neuvonen PJ. Concentrations and effects of buspirone are considerably reduced by rifampicin. Br J Clin Pharmacol 1998; 45(4):381–385.
209. Swaisland HC, Ranson M, Smith RP, et al. Pharmacokinetic drug interactions of gefitinib with rifampicin, itraconazole and metoprolol. Clin Pharmacokinet 2005; 44(10):1067–1081.
210. Kharasch ED, Whittington D, Hoffer C, et al. Paradoxical role of cytochrome P450 3A in the bioactivation and clinical effects of levo-alpha-acetylmethadol: importance of clinical investigations to validate in vitro drug metabolism studies. Clin Pharmacokinet 2005; 44(7):731–751.
211. Chung E, Nafziger AN, Kazierad DJ, et al. Comparison of midazolam and simvastatin as cytochrome P450 3A probes. Clin Pharmacol Ther 2006; 79(4):350–361.
212. Kivistö KT, Villikka K, Nyman L, et al. Tamoxifen and toremifene concentrations in plasma are greatly decreased by rifampin. Clin Pharmacol Ther 1998; 64(6): 648–654.
213. Villikka K, Kivisto KT, Backman JT, et al. Triazolam is ineffective in patients taking rifampin. Clin Pharmacol Ther 1997; 61(1):8–14.
214. McCallum RW, Prakash C, Campoli-Richards DM, et al. Cisapride. A preliminary review of its pharmacodynamic and pharmacokinetic properties, and therapeutic use

as a prokinetic agent in gastrointestinal motility disorders. Drugs 1988; 36(6): 652–681.

215. Back DJ, Breckenridge AM, Crawford FE, et al. An investigation of the pharmacokinetics of ethynylestradiol in women using radioimmunoassay. Contraception 1979; 20(3):263–273.
216. Perry CM, Noble S. Saquinavir soft-gel capsule formulation. A review of its use in patients with HIV infection. Drugs 1998; 55(3):461–486.
217. Libersa CC, Brique SA, Motte KB, et al. Dramatic inhibition of amiodarone metabolism induced by grapefruit juice. Br J Clin Pharmacol 2000; 49(4):373–378.
218. van Agtmael MA, Gupta V, van der Wosten TH, et al. Grapefruit juice increases the bioavailability of artemether. Eur J Clin Pharmacol 1999; 55(5):405–410.
219. Fukazawa I, Uchida N, Uchida E, et al. Effects of grapefruit juice on pharmacokinetics of atorvastatin and pravastatin in Japanese. Br J Clin Pharmacol 2004; 57(4):448–455.
220. Lilja JJ, Kivisto KT, Neuvonen PJ. Grapefruit juice increases serum concentrations of atorvastatin and has no effect on pravastatin. Clin Pharmacol Ther 1999; 66(2): 118–127.
221. Garg SK, Kumar N, Bhargava VK, et al. Effect of grapefruit juice on carbamazepine bioavailability in patients with epilepsy. Clin Pharmacol Ther 1998; 64(3): 286–288.
222. Kivisto KT, Lilja JJ, Backman JT, et al. Repeated consumption of grapefruit juice considerably increases plasma concentrations of cisapride. Clin Pharmacol Ther 1999; 66(5):448–453.
223. Christensen H, Asberg A, Holmboe AB, et al. Coadministration of grapefruit juice increases systemic exposure of diltiazem in healthy volunteers. Eur J Clin Pharmacol 2002; 58(8):515–520.
224. Kanazawa S, Ohkubo T, Sugawara K. The effects of grapefruit juice on the pharmacokinetics of erythromycin. Eur J Clin Pharmacol 2001; 56(11):799–803.
225. Weber A, Jager R, Borner A, et al. Can grapefruit juice influence ethinylestradiol bioavailability? Contraception 1996; 53(1):41–47.
226. Hukkinen SK, Varhe A, Olkkola KT, et al. Plasma concentrations of triazolam are increased by concomitant ingestion of grapefruit juice. Clin Pharmacol Ther 1995; 58(2):127–131.
227. Lilja JJ, Kivisto KT, Backman JT, et al. Effect of grapefruit juice dose on grapefruit juice-triazolam interaction: repeated consumption prolongs triazolam half-life. Eur J Clin Pharmacol 2000; 56(5):411–415.
228. Kupferschmidt HH, Fattinger KE, Ha HR, et al. Grapefruit juice enhances the bioavailability of the HIV protease inhibitor saquinavir in man. Br J Clin Pharmacol 1998; 45(4):355–359.
229. Fuhr U, Muller-Peltzer H, Kern R, et al. Effects of grapefruit juice and smoking on verapamil concentrations in steady state. Eur J Clin Pharmacol 2002; 58(1):45–53.

11

Mechanism-Based Inhibition of Human Cytochromes P450: In Vitro Kinetics and In Vitro–In Vivo Correlations

Xin Zhang, David R. Jones, and Stephen D. Hall
Indiana University School of Medicine,
Indianapolis, Indiana, U.S.A.

I. INTRODUCTION

Over the past decade there has been a substantial improvement in the ability to predict metabolism-based in vivo drug interactions from kinetic data obtained in vitro. This advance has been most evident for interactions that occur at the level of cytochrome P450 (CYP)-catalyzed oxidation and reflects the availability of human tissue samples, cDNA-expressed CYPs, and well-defined substrates and inhibitors of individual enzymes. The most common paradigm in the prediction of in vivo drug interactions has been first to determine the enzyme selectivity of a suspected inhibitor and subsequently to estimate the constant that quantifies the potency of reversible inhibition in vitro. This approach has been successful in identifying clinically important potent competitive inhibitors, such as quinidine, fluoxetine, and itraconazole. However, there is a continuing concern that a number of well-established and clinically important CYP-mediated drug interactions are not predictable from the classical approach that assumes reversible mechanisms of inhibition are ubiquitous.

Irreversible inhibition is an additional mechanism that may reduce the catalytic activity of an enzyme in vitro and in vivo. This mechanism has been extensively characterized in vitro and is particularly common for CYP-mediated biotransformations, in part because of the high-energy intermediates that are characteristic of these reactions. A seminal illustration of the importance of an irreversible mechanism of inhibition is provided by erythromycin, the widely used macrolide antibiotic. Steady-state plasma concentrations of erythromycin are below the in vitro estimated constant for competitive inhibition of CYP3A4 (1,2), and consequently no in vivo drug interactions are expected with CYP3A4 substrates. However, in clinical practice, erythromycin is a well-established inhibitor of CYP3A-mediated biotransformation (1). This is not surprising in view of the ample evidence demonstrating that both human and animal CYP3A enzymes convert erythromycin to a metabolite that complexes with heme to cause inactivation (3). Thus, the goal of this text is to describe the scope of irreversible inhibition of drug metabolizing enzymes and to indicate how the prediction of in vivo drug interactions can be incorporated into this phenomenon.

II. CHARACTERISTICS OF IRREVERSIBLE INHIBITORS

In general, three types of CYP inhibition have been described (4). The most common type of inhibition is displayed by agents that reversibly bind to CYP and is displayed by all substrates of an enzyme at sufficiently high concentrations. A second type of inhibition occurs when substrates or their metabolites form quasi-irreversible complexes with the prosthetic heme; this is typified by the inhibition of CYP3A enzymes by macrolide antibiotics. The third type of inhibition occurs when a substance binds irreversibly to structural motifs of the CYP apoprotein or to the prosthetic heme group, or accelerates the degradation of the prosthetic heme group. The latter two modes of inhibition are most commonly displayed by inhibitors that are dependent on the enzyme itself to reveal their inhibition, and they are therefore commonly referred to as mechanism-based inhibitors (5). A mechanism-based inhibitor must first bind and then become catalytically activated by the enzyme. The activated species irreversibly alters the enzyme and removes it permanently from the pool of active enzyme. For a substance to be classified as a direct mechanism-based inhibitor, it should meet the following rigorous criteria proposed by Silverman (5):

1. Under conditions that support catalysis, a time-dependent loss of enzyme activity is observed.
2. The rate of enzyme inactivation is proportional to low inactivator concentration but is independent at high inactivator concentration [Eq. (1)].
3. The rate of inactivation is slower in the presence of a competing substrate than in its absence.
4. Enzyme activity does not return upon physical removal of inactivator, e.g., by dialysis, filtration, or centrifugation.

5. A catalytic step for the conversion of inactivator to a reactive intermediate can be proposed.
6. There is no lag time for inactivation; the presence of exogenous nucleophiles has no effect on the inactivation rate; following inactivation, a second, equal addition of enzyme results in the same rate of inactivation as the first addition in the absence of inactivator and cofactor depletion.

III. TYPES OF IRREVERSIBLE INHIBITORS

A. Compounds That Covalently Bind to the Protein

Examples of xenobiotics that bind to proteins and fall into this class of mechanism-based inhibitor include tienilic acid, cannabidiol, chloramphenicol, secobarbital, some psoralens, spironolactone, mifepristone, and grapefruit juice.

Tienilic acid is oxidized by CYP2C9 to form metabolites that appear to covalently bind to the protein at the active site, thus rendering the enzyme inactive (6,7). Evidence suggests that an electrophilic sulfoxide metabolite of tienilic acid is the reactive species. When tienilic acid was incubated with CYP2C9 and NADPH, three protein species were detected: native CYP2C9, a monoadduct of CYP2C9 and tienilic acid, and a diadduct that incorporated two molecules of tienilic acid in CYP2C9. Further evidence suggested that each tienilic acid that was covalently adducted to CYP2C9 contained a hydroxyl group, which is consistent with initial ring oxidation and/or with initial sulfoxide formation, provided the attached sulfoxide does not dehydrate (8).

Preincubation of human liver microsomes (HLMs) with cannabidiol decreased the formation of all detectable metabolites of cyclosporine, a substrate of CYP3A (9). Cannabidiol is metabolized by CYP3A to form a cannabidiol hydroxyquinone. This metabolite binds to the apoprotein of CYP3A and renders it inactive (10).

Chloramphenicol and secobarbital exhibit properties similar to those of tienilic acid, but they have not been studied in humans (11). Oxidative dechlorination of chloramphenicol with formation of reactive acyl chlorides appears to be an important metabolic pathway for irreversible inhibition of CYP. Chloramphenicol binds to CYP, and subsequent substrate hydroxylation and product release are not impaired. The inhibition of CYP oxidation and the inhibition of endogenous NADPH oxidase activity suggest that some modification of the CYP has taken place, which inhibits its ability to accept electrons from the CYP reductase (11). Secobarbital completely inactivates rat CYP2B1 functionally, with partial loss of the heme chromophore. Isolation of the N-alkylated secobarbital heme adduct and the modified CYP2B1 protein revealed that the metabolite partitioned between heme N-alkylation, CYP2B1 protein modification, and epoxidation. A small fraction of the prosthetic heme modifies the protein and contributes to the CYP2B1 inactivation (12).

Psoralens, e.g., 8-methoxypsoralen, are a family of furanocoumarin derivatives that have been used in part to treat diseases such as psoriasis and cutaneous

T-cell lymphoma. Additionally, 8-methoxypsoralen has been shown to inhibit CYP2A6 (13). The mechanism of inhibition by this compound appears to be an initial oxidation to generate an epoxide that reacts with a nucleophilic amino acid at the active site (14).

CYP inactivation by spironolactone is due to a reactive species that binds covalently to the protein and/or modifies the heme group (15). However, this has not been investigated in human tissue. More recently, mifepristone (RU486) was characterized as a mechanism-based inhibitor of CYP3A4 (16). Mifepristone irreversibly modified the CYP3A4 apoprotein at the active site. The proposed mechanism of inactivation involved addition of reactive oxygen to the carbon-carbon triple bond of mifepristone to yield a highly reactive ketene intermediate that reacts with a nucleophilic residue at the enzyme active site (16). Components of grapefruit juice have also been shown to inactivate CYP3A4, and the mechanism is partly through modifying the apoprotein (see sec. VII).

B. Compounds That Quasi-Irreversibly Coordinate to the Prosthetic Heme

These compounds are catalytically oxidized to intermediates or products that coordinate tightly to the prosthetic heme of the CYP. This coordination can only be displaced under nonphysiological experimental conditions (e.g., potassium ferricyanide). Many nitrogen-containing compounds, usually amines, are found in this group. Primary amines are required for the metabolic intermediate complex (MIC) formation, although secondary and tertiary amines are appropriate precursors. The primary amines are hydroxylated and then further oxidized to a nitroso group that appears to chelate to the heme, which results in a more stable (ferrous) state of iron. This ferrous state exhibits a spectrum with an absorbance maximum of 445–455 nm (17). Nonnitrogenous compounds, composed primarily of methylenedioxybenzene derivatives, also form MIC and exhibit absorbance peaks at ~430 nm and 455 nm when the iron is in the ferrous state. However, the MIC formed by these compounds can also be observed at ~437 nm when the iron is in the ferric state. The MIC formed by the nonnitrogenous compounds may be a result of metabolism at the methylene carbon (17).

C. Compounds That Covalently Bind to the Prosthetic Heme

This class of compounds irreversibly inactivates CYP by the covalent attachment of the inhibitor, or a derivative of the inhibitor, to the prosthetic heme group. Compounds that fall into this class are terminal acetylenes, e.g., gestodene (18) and ethynyl estradiol (19), which selectively inactivate CYP3A; furafylline, which selectively inactivates CYP1A1/2 (20); hydrazines, e.g., phenelzine (21), and other xenobiotics, e.g., griseofulvin (22); and phencyclidine, which has been shown to be a substrate and inhibitor of CYP3A (23). Phenelzine and griseofulvin have exhibited mechanism-based inhibition in mouse or rat liver microsomes but have not been investigated with human tissue.

D. Compounds That Degrade the Prosthetic Heme Group

Certain CYPs undergo mechanism-based inactivation as a result of conversion of their prosthetic heme groups to products that irreversibly bind to the protein. Hydrogen peroxide and cumene hydroperoxide partially degrade the prosthetic heme to monopyrrole and dipyrrole fragments that bind to the protein (24). Presently, no drugs have been shown to fall into this class.

IV. KINETICS OF MECHANISM-BASED INHIBITION

Scheme 1 is the simplest one that is consistent with the inactivation of an enzyme while a drug is metabolized (25). As with conventional enzyme kinetics, there is an initial, reversible step that combines the inhibitor and free enzyme to form an enzyme-inhibitor complex.

In the absence of catalysis, the inhibitor concentration and the ratio of k_1 to k_{-1}, the equilibrium association constant, will define the fraction of the enzyme bound with inhibitor at a given enzyme concentration. The enzyme-inhibitor complex proceeds to transform the inhibitor to an intermediate that may decompose to form a metabolite or react with the enzyme to form an inactive complex. First-order rate constants k_2, k_3, and k_4 determine the rates of these reactions and the concentration of intermediate at a given concentration of inhibitor and enzyme.

Most commonly, the rate of formation of the inactivated enzyme, under steady-state conditions, can be described by the rectangular hyperbolic function often associated with the traditional Henri-Michaelis-Menten function (25,26):

$$\text{Rate of inactive enzyme formation} = I_{\text{max}} \cdot \frac{I}{K_{\text{I}} + I} = k_{\text{inact}} \cdot E \cdot \frac{I}{K_{\text{I}} + I} \qquad (1)$$

where I is the concentration of inhibitor or inactivator, K_{I} is the inhibitor concentration that supports half the maximal rate of inactivation, and I_{max} is the maximal rate of inactivation (when $I \gg K_{\text{I}}$). The symbol K_{I} is employed in the context of inactivation kinetics to distinguish it from the equilibrium inhibition constant K_{i} (see Chap. 2) that is commonly used in the description of reversible enzyme inhibition (5). The maximal rate of inactivation, I_{max}, will occur when inhibitor binds to all of the available enzymes:

$$\text{Maximal rate of formation of inactive enzyme} = E \cdot k_{\text{inact}} \qquad (2)$$

Thus, k_{inact} is the first-order rate constant that relates the maximal rate of formation of inactive enzyme to the active enzyme concentration. Tatsunami et al. demonstrated that under steady-state conditions the following relationships exist for the reaction displayed in Scheme 1 (27):

$$K_{\text{I}} = \frac{k_{-1} + k_2}{k_1} \cdot \frac{k_3 + k_4}{k_2 + k_3 + k_4} \qquad (3)$$

$$k_{\text{inact}} = \frac{k_2 \cdot k_4}{k_2 + k_3 + k_4} \qquad (4)$$

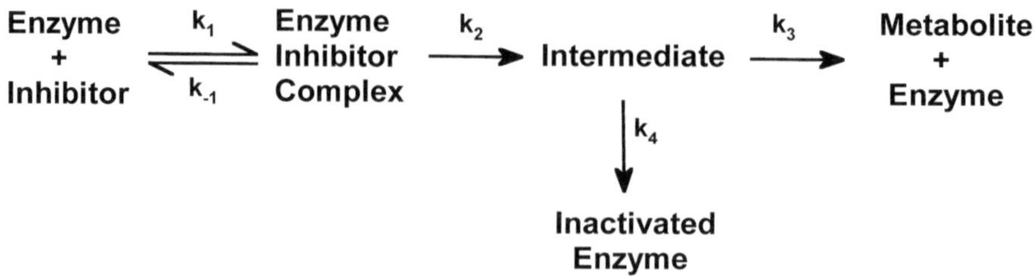

Scheme 1

It is clear from Eqs. (3) and (4) that K_I and k_{inact} are complex functions of several microrate constants. It is important to note that only under restrictive conditions can k_{inact} be equated with k_2, e.g., when k_4 is much greater than k_2 plus k_3. Similarly, K_I cannot simply be equated with the inverse of the equilibrium association constant for inhibitor and free enzyme.

Analogous relationships exist for the rate of metabolite formation in this enzymatic scheme:

$$\text{Rate of metabolite formation} = V_{max} \cdot \frac{I}{K_I + I} = k_{cat} \cdot E \cdot \frac{I}{K_I + I} \tag{5}$$

where V_{max} is the maximal rate of metabolite formation (when $I \gg K_I$) and K_I is the inhibitor concentration that supports half the maximal rate of metabolite formation and is exactly the same as the constant defined in Eqs. (1) and (3). k_{cat} is the first-order rate constant that relates maximal rate of metabolite formation to E and that is analogous to k_{inact} and can be defined as a function of the microrate constants:

$$k_{cat} = \frac{k_2 \cdot k_3}{k_2 + k_3 + k_4} \tag{6}$$

In the context of mechanism-based inhibition, k_{cat} does not have the same definition as that commonly used in metabolite formation kinetics; k_{cat} is not equivalent to k_2 unless k_3 greatly exceeds k_2 and k_4, which may occur when the inactivation pathway is minor in comparison to the formation of metabolite.

A useful index of the propensity for an enzyme to undergo inactivation, as opposed to metabolite formation, is the partition ratio, r (28), defined as the ratio of the rate of metabolite formation to the rate of inactive enzyme formation. Thus, by combining Eqs. (1) and (5):

$$r = \frac{k_{cat}}{k_{inact}} \tag{7}$$

Furthermore from the relationships in Eqs. (4) and (6) that include the microrate constants of Scheme 1:

$$r = \frac{k_3}{k_4} \tag{8}$$

From Eqs. (7) and (8) it is clear that, in the context of the current model, r is independent of inhibitor concentration. The value of r varies from infinity, when the inactivation reaction is a rare event, to a value of zero, where inactivation of enzyme occurs during every catalytic cycle.

It should be noted that the mechanism depicted in Scheme 1 is the simplest that is consistent with mechanism-based inhibition. The mechanism for a given inhibitor and enzyme may be considerably more complex due to (a) multiple intermediates [e.g., MIC formation often involves four or more intermediates (29)], (b) detectable metabolite that may be produced from more than one intermediate, and (c) the fact that enzyme-inhibitor complex may produce a metabolite that is mechanistically unrelated to the inactivation pathway. Events such as these will necessitate alternate definitions for k_{inact}, K_{I}, and r in terms of the microrate constants of the appropriate model. The hyperbolic relationship between rate of inactivation and inhibitor concentration will, however, remain, unless nonhyperbolic kinetics characterize this interaction. Silverman discussed this possibility from the perspective of an allosteric interaction between inhibitor and enzyme (5). Nonhyperbolic kinetics has been observed for the interaction of several drugs with members of the CYPs (30).

V. DETERMINATION OF ENZYME CONSTANTS IN VITRO

Characterization of a mechanism-based inhibitor may involve the estimation of the constants described in section IV, namely, k_{inact}, K_{I}, k_{cat}, and r. The most common approach has been to incubate inhibitor, enzyme, and cofactors together and to determine the decline in enzyme activity with time (26). In practice, this approach often employs the measurement of residual enzyme activity in a subsequent incubation with a specific substrate under conditions that limit further inactivation and competitive inhibition by the inactivator, usually by an appropriate dilution (10-fold or greater) of the original incubate (5).

On the basis of foregoing discussion, the rate of change of enzyme activity in the presence of an inhibitor concentration I is given by:

$$\frac{\mathrm{d}E_{(t)}}{\mathrm{d}t} = -k_{\text{inact}} \cdot I \cdot \frac{E_{(t)}}{K_{\text{I}} + I} \tag{9}$$

where $E_{(t)}$ is the enzyme concentration at some time t. This expression can be integrated to provide a relationship that has been widely used to estimate the desired parameters:

$$E_{(t)} = E_{(0)} \cdot \mathrm{e}^{-[k_{\text{inact}} \cdot I/(K_{\text{I}}+I)] \cdot t} \tag{10}$$

where $E_{(0)}$ is the initial enzyme concentration or activity. Thus, a plot of $E_{(t)}$ against time can be generated for a range of inhibitor concentrations that encompass K_I to estimate k_{inact} and K_I. Alternatively, by taking the natural log of Eq. (10),

$$\ln\left(\frac{E_{(t)}}{E_{(0)}}\right) = -\left(k_{inact} \cdot \frac{I}{K_I + I}\right) \cdot t \tag{11}$$

Equation (11) indicates that a plot of log fractional enzyme activity against time will be linear, and the negative of the slope will be equivalent to $k_{inact} \cdot I/(K_I + I)$ (26). The family of curves obtained by varying inhibitor concentration should share the same value of $\ln(E_{(t)}/E_{(0)}) = 1$ at $t = 0$, unless the experiment is confounded by the occurrence of significant competitive inhibition. The relationship between the slope of these plots and inhibitor concentration can be analyzed by nonlinear regression [see Eq. (1)] or double reciprocal plots to estimate k_{inact} and K_I [Figs. 1 (inset) and 2]. Many mechanism-based inhibitors of CYP3A4 (Table 1) and other drug metabolizing CYPs (Table 2) have been characterized in this manner and the corresponding k_{inact} and K_I values estimated. These approaches assume a constant inhibitor concentration equal to the starting concentration and that loss of enzyme activity is due only to the specific effect of the inhibitor. Preliminary experiments are indicated to verify these assumptions.

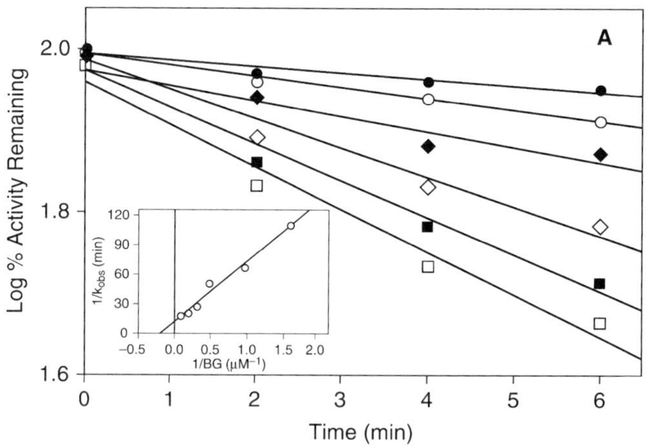

Figure 1 Time- and concentration-dependent inactivation of the catalytic activity of P450 2B6 by bergamottin. Inactivation of the EFC O-deethylation activity of P450 2B6 in the reconstituted system incubated with 0.6 (●), 1 (○), 2 (◆), 3 (◇), 5 (■), and 10 (□) μM bergamottin. Aliquots were removed at the indicated time and assayed for residual activity. The insets show the double reciprocal plots of the initial rates of inactivation as a function of the bergamottin concentrations. The kinetic constants K_I, k_{inact}, and $t_{1/2}$ were determined from this plot. The data shown represent the average of three experiments that did not differ by more than 10%. *Source*: From Ref. 72.

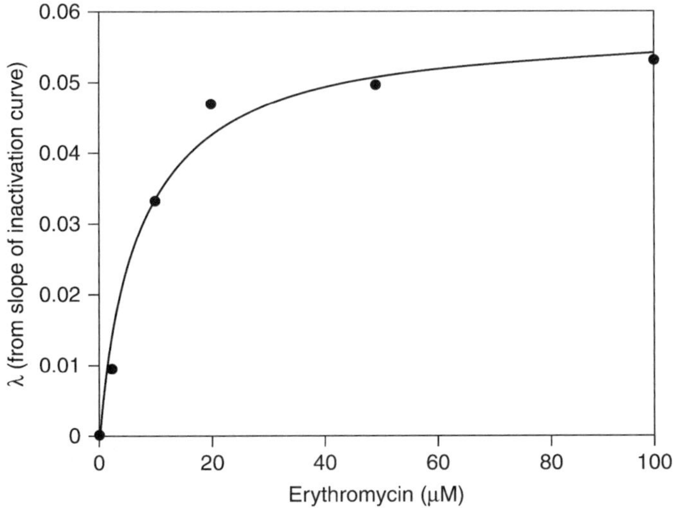

Figure 2 Determination of irreversible inhibition constants for erythromycin and CYP3A4 microsomes. Microsomes (1 mg/mL microsomal protein) were incubated with erythromycin (0–100 μM) in the presence of NADPH for differing incubation times. The pseudo-first-order rate constant for enzyme inactivation was plotted versus erythromycin concentration to estimate K_I and k_{inact} (14.4 μM and 0.045 min^{-1}, respectively). The curve represents the line of best fit. *Source*: From Ref. 32.

In some cases the rate of enzyme inactivation can be quantified without an assay for enzyme activity. For example, inactivation of CYPs due to MIC formation can be directly quantified spectrophotometrically, which avoids the potential artifacts introduced by the measurement of catalytic activity. Microsomes, or purified enzymes, are incubated with a substrate and NADPH and monitored for MIC formation over time in a spectrophotometer. An example of MIC formation by diltiazem in human liver microsomes (HLMs) is shown in Figure 3 (36). The MIC exhibits an absorbance maximum between 448 nm and 456 nm when the heme iron is in the reduced state (17). Extinction coefficients of MIC are approximately 64mM/cm (79). Thus, MIC formation by diltiazem in the example is 59% of the total CYP, which would be consistent with inactivation of most of the CYP3A in the microsomes.

The value of k_{cat} and K_I can also be estimated by quantifying the rate of metabolite formation from the inhibitor either simultaneously with the decline in enzyme activity or under the same incubation conditions. The rate of change of metabolite, $dM_{(t)}/dt$, is given by

$$\frac{dM(t)}{dt} = \frac{k_{cat} \cdot I}{K_I + I} \cdot E_{(0)} \cdot e^{-[k_{inact} \cdot I/(K_I + I)] \cdot t} \tag{12}$$

Table 1 Mechanism-Based Inhibitors of CYP3A4 and Corresponding Estimates of In Vitro Constants

Drug	k_{inact} (min^{-1})	K_I (μM)	MIC[a] formation	References
Antibiotics				
Clarithromycin	0.072	5.49	Y	31
Erythromycin	0.046	10.9	Y	32
Troleandomycin	0.15	0.18	Y	33
Dirithromycin	ND	ND	Y	34
Isoniazid	0.042	48.6		35
Antihypertensive agents				
Diltiazem	0.11	2	Y	36
Verapamil	0.09	1.7	Y	37
Nicardipine	0.06	1.29	Y	32
Amlodipine	0.35[b]	2.6[b]		Unpublished data
Dihydralazine	0.05	35		38
Mibefradil	0.4	2.3		39
Anti-HIV agents				
Ritonavir	0.4[b]	0.17[b]	Y	40
Indinavir	ND	ND	Y	41
Lopinavir	0.11	1	Y	40
Nelfinavir	0.22	1	Y	40
Amprenavir	0.59[b]	0.37[b]	Y	40
Saquinavir	0.31[b]	0.2[b]		40
Delavirdine	0.44	9.5		41
Anticancer agents				
Tamoxifen	0.04[b]	0.2[b]		42
Irinotecean	0.06[b]	24[b]		43
SN-38	0.1[c]	26[c]		43
Antidepressants				
Fluoxetine	0.017	5.26		31
Nortriptyline	0.04[b]	2.1[b]		Unpublished data
Fluvoxamine	0.05[b]	3.7[b]		Unpublished data
Nefazodone	0.27[b]	12.5[b]		Unpublished data
Dietary Supplements				
Bergamottin (grapefruit juice)	0.3	4.2		44
6′,7′-dihydroxybergamottin (grapefruit juice)	0.16	59		45
Glabridin (licorice extract)	0.14	7		46
Oleuropein (olive oil)	0.09	22.2		47
Resveratrol (red wine)	0.2	20		48
Limonin (grapefruit oil)	0.266	23.2		49
Rutaecarpine (evodia fruit)	0.387	107.7		49
Kaempferol glycosides (Indonesian medicinal plant *Zingiber aromaticum*)	0.23–0.65	2.21–27.01		50

Continued

Table 1 *Continued*

Drug	k_{inact} (min^{-1})	K_I (μM)	MIC[a] formation	References
(−)-Hydrastine (goldenseal extract)	0.23	110		57
Steroids				
17α-Ethynylestradiol	0.04	18		51
Mifepristone	0.4	2.3		16
Gestodene	0.4	46		52
Raloxifene	0.16	9.9		53
Antidiabetics				
Troglitazone	0.0335	5		54
Rosiglitazone	0.0195	11.9		54
Pioglitazone	0.0112	10.4		54
Others				
Midazolam	0.15	5.8		55
Diclofenac	0.25	1640		56
4-Ipomeanol	0.15	20		58
Tabimorelin	0.08	4.7		59
K11002	0.026	0.5		60
K11777	0.054	0.06		60

[a]Compounds for which MIC formation has been observed (Y).
[b]Parameters estimated using recombinant enzyme. Otherwise, parameters were estimated using HLMs.
Abbreviations: ND, not determined; MIC, metabolic intermediate complex.

If the rate of metabolite formation can be determined over a time period that is sufficiently short that significant enzyme inactivation does not occur ($k_{cat} > k_{inact}$), then the exponential term in Eq. (12) approaches unity and may be ignored. Equation (12) illustrates that the apparent V_{max} for formation of a metabolite will decline as the incubation time increases when simultaneous enzyme inactivation occurs (Fig. 4).

The partition ratio can be obtained from estimates of k_{inact} and k_{cat} or can be determined directly. This is achieved by simultaneously quantifying the moles of enzyme inactivated and the moles of metabolite formed for given incubation conditions. Clearly, if any two parameters from k_{inact}, k_{cat}, and r are known, then the third can be calculated.

Under conditions where it is not possible to approximate the steady state, i.e., constant inactivator concentration, it is possible to estimate k_{inact} and K_I, if the inactivator concentration and residual enzyme activity are quantified simultaneously. If a fixed quantity of enzyme and inactivator are combined under

Table 2 Mechanism-Based Inhibitors of CYPs (other than CYP3A4) and Corresponding Estimates of In Vitro Constants

Drug	k_{inact} (min^{-1})	K_I (μM)	References
CYP 1A1			
Rapontigenin	0.06	0.09	61
Bergamottin	ND	ND	62
CYP 1A2			
Zileuton	0.035	117	63
Rofecoxib	0.07	4.6	64
trans-Resveratrol	0.28	8.5	65
Oltipraz	0.19	9	66
Clorgyline	0.15	6.8	35
CYP 2A6			
Valproic acid	0.048	9150	67
Nicotine	ND	ND	68
β-Nicotyrine	ND	ND	68
CYP 2B6			
TA[a]	0.09	5.1	69
TioTEPA	0.16	3.8	70
Clopidogrel	0.35	0.5	71
Ticlopidine	0.5	0.2	71
Bergamottin	0.09	5	72
CYP 2C8			
Nortriptyline	0.036	49.9	73
Verapamil	0.065	17.5	73
Fluoxetine	0.083	294	73
Amiodarone	0.079	1.5	73
Isoniazid	0.042	374	73
Phenelzine	0.243	1.2	73
Gemfibrizol	0.21	20–52	74
CYP 2E1			
Phenethyl isothiocyanate	0.339	9.98	75
CYP 2D6			
Paroxetine	0.17	4.85	76
MDMA (Ecstacy)	0.2–0.58	2.22–4.86	77
Serpentine	0.09	0.148	78

[a]*N*-(3,5-dichloro-4-pyridyl)-4-methoxy-3-(prop-2-ynyloxy)benzamide.
Abbreviation: ND, not determined.

non-steady-state conditions that support catalysis, then the rate of formation of inactive enzyme at some time t, $dE_{(t)}/dt$, is given by

$$\frac{dE_{(t)}}{dt} = k_{\text{inact}} \cdot E_{(t)} \cdot \frac{I_{(t)}}{K_I + I} \tag{13}$$

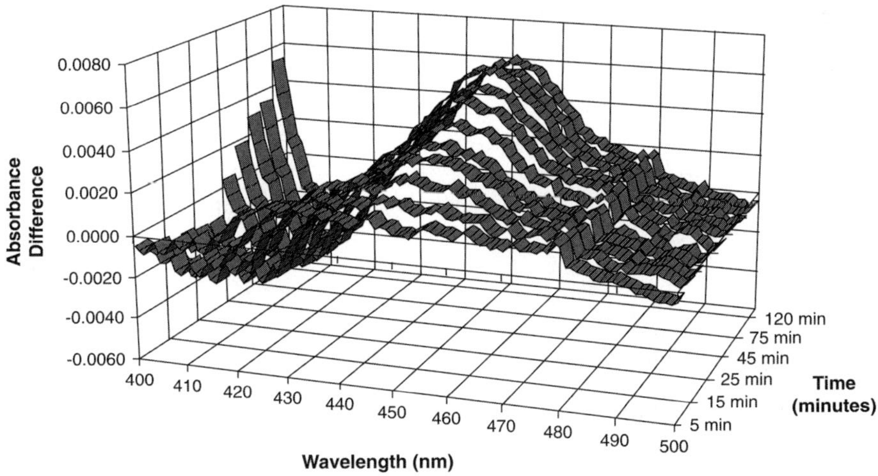

Figure 3 MIC formation by diltiazem (5 μM) in HLMs. The sample cuvette contained HLMs, diltiazem, and NADPH, whereas the reference cuvette contained HLMs, buffer, and NADPH. The ribbons represent the change in absorbance difference for scans from 5 to 120 minutes. *Abbreviations*: MIC, metabolic intermediate complex; HLMs, human liver microsomes. *Source*: From Ref. 37.

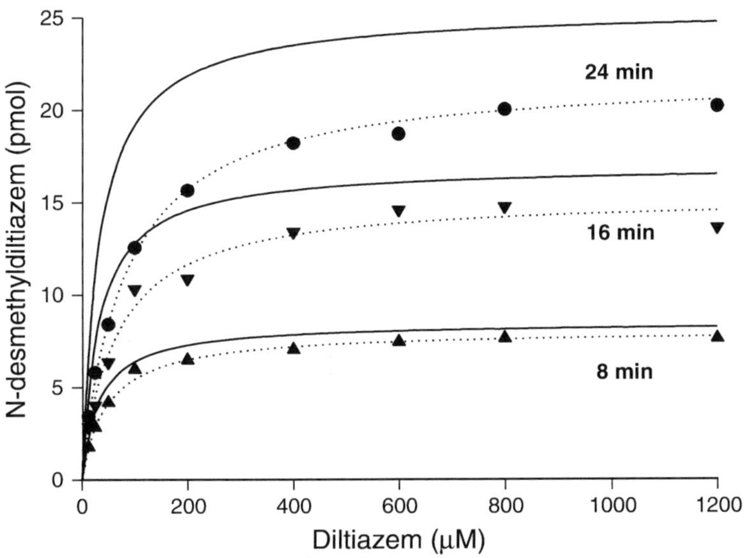

Figure 4 Effect of incubation time on the formation of *N*-desmethyldiltiazem (MA) in HLMs. HLMs (50 μg) were incubated with diltiazem (12.5–1200 μM) and NADPH (1 mM) at 37° for 8 (▲), 16 (▼), and 24 (●) minutes. The dashed line is the line of best fit of the data with the Michaelis-Menten equation. The solid line is the line that represents the predicted MA formation at the corresponding time using instantaneous formation rates. *Abbreviations*: HLMs, human liver microsomes; MA, *N*-desmethyldiltiazem. *Source*: From unpublished data.

The corresponding decline in inhibitor concentration at some time t, $-\mathrm{d}I_{(t)}/\mathrm{dt}$, will be given by the rate of metabolite formation plus the rate of inactive enzyme formation:

$$-\frac{\mathrm{d}I_{(t)}}{\mathrm{dt}} = (k_{\mathrm{inact}} + k_{\mathrm{cat}}) \cdot E_{(t)} \cdot \frac{I_{(t)}}{K_{\mathrm{I}} + I} \tag{14}$$

Equation (14) can also be written to include the partition ratio:

$$-\frac{\mathrm{d}I_{(t)}}{\mathrm{dt}} = (1 + r) \cdot k_{\mathrm{inact}} \cdot E_{(t)} \cdot \frac{I_{(t)}}{K_{\mathrm{I}} + I} \tag{15}$$

VI. PREDICTION OF DRUG INTERACTIONS IN VIVO

A. Extent of Interaction

When one drug has the capability to inactivate an enzyme, the elimination of a second drug that relies on that enzyme may be impaired. The net effect of exposure to an enzyme inactivator is to enhance the rate of degradation of active enzyme from the endogenous pool. Under baseline conditions the rate of change of active enzyme concentration, $\mathrm{d}E_{(t)}/\mathrm{dt}$, is determined by the balance between the rate of de novo synthesis and the rate of degradation. Enzyme synthesis rate is generally assumed to be a zero-order process, whereas the rate of degradation is a first-order process (80):

$$\frac{\mathrm{d}E_{(t)}}{\mathrm{dt}} = R_0 - k_{\mathrm{E}} \cdot E_{(t)} \tag{16}$$

where R_0 is the rate of enzyme synthesis and k_{E} is the endogenous degradation rate constant. Therefore, at steady state ($\mathrm{d}E_{(t)}/\mathrm{dt} = 0$), the enzyme concentration, E_{ss}, is given by

$$E_{\mathrm{ss}} = \frac{R_0}{k_{\mathrm{E}}} \tag{17}$$

In turn the steady-state enzyme concentration in the liver determines the baseline hepatic intrinsic clearance, $\mathrm{CL}_{\mathrm{int}}$, for the metabolism of a drug substrate by the enzyme. When substrate concentration, S, is low relative to the Michaelis constant, K_{m}, for a particular biotransformation,

$$\mathrm{CL}_{\mathrm{int}} = \frac{V_{\max}}{K_{\mathrm{m}}} = E_{\mathrm{ss}} \cdot \frac{k_{\mathrm{cat}}}{K_{\mathrm{m}}} \tag{18}$$

where $V_{\max}$ is the maximal rate of substrate metabolism and k_{cat} is the first-order rate constant that relates $V_{\max}$ to E_{ss} In the presence of an inactivator of the enzyme, the rate of change of active enzyme, $\mathrm{d}E'_{(t)}/\mathrm{dt}$, is given by:

$$\frac{\mathrm{d}E'_{(t)}}{\mathrm{dt}} = R_0 - k_{\mathrm{E}} \cdot E_{(t)} - k_{\mathrm{I}} \cdot E_{(t)} \tag{19}$$

where k_I is the rate constant for inactivation of enzyme. Consequently, the steady-state enzyme concentration in the presence of inactivator, E'_{ss}, is reduced:

$$E'_{ss} = \frac{R_0}{k_E + k_I} \tag{20}$$

The inactivator will therefore produce a corresponding reduction in intrinsic clearance to CL'_{int}.

$$CL'_{int} = E'_{ss} \cdot \frac{k_{cat}}{K_m} \tag{21}$$

The ratio of the intrinsic clearances in the absence and presence of an inactivator is given by

$$\frac{CL_{int}}{CL'_{int}} = \frac{E_{ss}}{E'_{ss}} = \frac{k_E + k_I}{k_E} = 1 + \frac{k_I}{k_E} \tag{22}$$

For a drug that is eliminated exclusively by the liver and that is completely absorbed following oral administration, the intrinsic clearance can be related to the area under the plasma concentration–time curve (AUC_{po}) if the well-stirred model of hepatic elimination is assumed (81,82):

$$CL_{int} \cdot f_u = \frac{Dose_{po}}{AUC_{po}} \tag{23}$$

where AUC_{po} is obtained from time zero to infinity following a single oral dose or over a dosing interval when the drug is administered orally in the steady state; f_u is the fraction of drug unbound in plasma. Thus, for a drug that is eliminated from the body by a single hepatic pathway that is the target of an inactivator, the following relationship describes the predicted increase AUC_{po} from the baseline state to the inactivated state AUC'_{po}:

$$\frac{AUC'_{po}}{AUC_{po}} = \frac{CL_{int}}{CL'_{int}} = 1 + \frac{k_I}{k_E} \tag{24}$$

If the inactivated pathway is only one of multiple elimination pathways in the liver, then the predictive model of Eq. (24) becomes

$$\frac{AUC'_{po}}{AUC_{po}} = \frac{1}{[f_m/(1 + k_I/k_E)] + 1 - f_m} \tag{25}$$

where f_m represents the fraction of the total hepatic elimination at baseline that is due to the pathway that is susceptible to inactivation.

From Eqs. (24) and (25) it is clear that in order to predict the effect of an inactivator on the AUC_{po} of a coadministered drug, the determinants of k_I must be understood. From our earlier discussion, the rate of formation of inactive enzyme is given by

$$\text{Rate inactive enzyme formation} = k_{\text{inact}} \cdot E \cdot \frac{I}{K_{\text{I}} + I} \tag{26}$$

From Eq. (19) the rate of inactivation of enzyme is also given by $k_{\text{t}} \cdot E$; therefore, when $K_{\text{I}} \gg I$,

$$k_{\text{I}} = k_{\text{inact}} \cdot \frac{I}{K_{\text{I}}} \tag{27}$$

Consequently, the predicted effect of inactivator on AUC_{po} is given by

$$\frac{\text{AUC}'_{\text{po}}}{\text{AUC}_{\text{po}}} = 1 + \left(k_{\text{inact}} \cdot \frac{I}{K_{\text{I}} \cdot k_{\text{E}}} \right) \tag{28}$$

An analogous expression can readily be derived from Eq. (22). Thus estimates of k_{inact} and K_{I} determined in vitro can be combined with estimates of baseline enzyme turnover ($1/k_{\text{E}}$) and in vivo concentration of inhibitor to predict the extent (fold increase in AUC) of an interaction. This expression is reminiscent of the model used to predict interactions involving reversible, competitive inhibition (see chap. 1), with the substitution of $k_{\text{inact}}/(K_{\text{I}} \cdot k_{\text{E}})$ for $1/K_{\text{i}}$ (83). The ratio of k_{inact} to K_{I} is a useful parameter that can be considered the intrinsic efficiency of inactivation independent of inhibitor concentration. The concentration of inhibitor that should be used in this predictive model is the concentration at the enzyme, but in practice plasma concentrations are often used as a surrogate. However, if the assumption that $K_{\text{I}} \gg$ I is not appropriate, the function that is nonlinear with respect to inhibitor concentration [Eq. (26)] must be employed and a time-average concentration would not be appropriate. In this nonlinear system, the effect of an inactivator on steady-state enzyme concentrations can be predicted by iteratively solving the differential equations that describe the rate of change of enzyme and inactivator concentration.

B. Time Course of Inactivation

An important characteristic of inhibition of drug metabolism by an inactivator is the time dependence of both the onset and offset of the effect. The time course of the change in enzyme concentration from the baseline, E_{ss}, to that in the presence of inactivator, E'_{ss}, is given by

$$E_{(t)} = E_{\text{ss}} \cdot \text{e}^{-kt} + E'_{\text{ss}} \tag{29}$$

where $E_{(t)}$ is the enzyme concentration at some time t. This relationship indicates that the half-life of the decline in enzyme concentration ($0.693/k_{\text{I}}$) is dependent on k, which in turn is a function of k_{inact}, K_{I}, and I [Eq. (27)]. Therefore, the greater the potency of the inactivator, the faster the onset of the interaction. In contrast, the rate of offset of the interaction is given by

$$E_{(t)} = E'_{\mathrm{ss}} + R_0 \cdot \left(1 - \mathrm{e}^{-k_E \cdot t}\right) \tag{30}$$

In this case the half-life for the return to baseline enzyme concentration ($0.693/k_{\mathrm{E}}$) is controlled by the turnover of the enzyme. Thus, the offset of the interaction is independent of the properties of a given inactivator.

C. Intestinal Wall Metabolism

For many drugs, particularly those that are substrates for CYP3A4, it is inappropriate to assume that all metabolism occurs in the liver, because intestinal enterocytes also contribute to first-pass elimination of these substrates (see Chap. 10). When intestinal wall metabolism contributes to the "first-pass" effect, the prediction of the effect of inactivation on $\mathrm{AUC_{po}}$ must be modified as follows (84):

$$\frac{\mathrm{AUC}'_{\mathrm{po}}}{\mathrm{AUC}_{\mathrm{po}}} = \left(\frac{F'_{\mathrm{G}}}{F_{\mathrm{G}}}\right) \cdot \left[1 + \left(k_{\mathrm{inact}} \cdot \frac{I}{K_{\mathrm{I}} \cdot k_{\mathrm{E}}}\right)\right] \tag{31}$$

where F_{G} and F'_{G} are the intestinal wall availabilities in the absence and presence of inhibitor, respectively. In turn F_{G} is a function of the intrinsic clearance at the intestinal wall, $\mathrm{CL_{int,G}}$ (85):

$$F_{\mathrm{G}} = \frac{A}{A + \mathrm{CL_{int,G}}} \tag{32}$$

where A is the absorption constant which may be a function of epithelial permeability or intestinal blood flow. The ratio $F'_{\mathrm{G}}/F_{\mathrm{G}}$ can therefore be estimated from the relative change in $\mathrm{CL_{int,G}}$ caused by the inactivator at the concentration in contact with the enterocytes. This assumes that A is unaffected by the presence of inactivator.

VII. DRUG-DRUG INTERACTIONS DUE TO IRREVERSIBLE INHIBITION

A. Evidence for In Vivo Drug-Drug Interactions Caused by Mechanism-Based Inhibition

Direct evidence that irreversible inhibition is the principle mechanism underlying in vivo drug-drug interactions (DDIs) is often lacking because of the requirement for either direct tissue sampling to reveal inactivated enzyme or in vivo inhibition of activity after drug is essentially eliminated from the body. Nevertheless the steady-state plasma concentrations of several clinically important CYP inhibitors are well below the in vitro estimated competitive inhibition constant, K_{i}. This suggests that competitive inhibition is unlikely to occur in vivo, yet these compounds inhibit CYP activity in a time and concentration-dependant manner when cDNA-expressed CYPs or HLMs are used as an enzyme source. Therefore, the mechanism underlying the observed DDIs in vivo has been attributed to mechanism-based inhibition. Although the conclusion that irreversible inhibition is determinant in these examples is reasonable, caution must be exercised because it is often not possible to rule out a role for reversible inhibition arising from high tissue drug concentrations and/or the presence of inhibitory metabolites.

Erythromycin has a K_i of ~148 μM with HLMs and an average serum concentration of 5.4 μM at steady state after 500 mg was administered three times a day. Therefore, erythromycin would not be expected to cause significant reversible inhibition of CYP3A activity in vivo (2). However, clinically important drug interactions between erythromycin and midazolam (1), dextromethorphan (86), cyclosporin A (87), alfentanil (88), triazolam (89), alprazolam (90), and carbamazepine (91) have been observed clinically. Erythromycin and troleandomycin (TAO) are well-established mechanism-based inhibitors of CYP3A4 via the formation of MIC in vitro (34). In a key clinical study, liver specimens were obtained by surgical biopsy of six patients receiving erythromycin propionate (2 g daily for 7 days) (3). MIC formation in vivo was detected by ultraviolet absorbance difference spectra, and an increased CYP concentration was observed in microsomes prepared from these liver specimens (3). Similarly, liver microsomes were prepared from liver biopsy samples in six patients given TAO (2 g daily for 7 days) and six untreated patients (92); compared with the control group, the total CYP concentration was increased (0.39 ± 0.11 to 0.69 ± 0.15 nmol/mg protein) but 33% of the total P450 was complexed by a TAO metabolite, leading to a 45% decrease in antipyrine clearance in the treatment group (92). These seminal studies established the potential for MIC formation to represent a functionally irreversible mode of CYP inhibition in vivo. The increased concentration of CYP that accompanied this inhibition may represent stabilization of complexed enzyme and/or transcriptional upregulation of CYP synthesis.

Endoscopic biopsy samples of small intestinal enterocytes have also provided direct evidence of irreversible inhibition of CYP3A4 in vivo. Biopsies were obtained before and after the consumption of 8 oz of grapefruit juice three times a day for six days in 10 healthy men (93). The concentration of CYP3A4 in enterocytes, determined by immunoblotting, was decreased by 62% with no change in CYP3A4 mRNA levels. This is consistent with the in vitro finding that grapefruit juice inactivates CYP3A4 by modifying the apoprotein (44) and initiating rapid proteasomal degradation following ubiquitination (94). Furanocoumarins characteristic of grapefruit juice, including bergamottin and 6′,7′-dihydroxybergamottin, are mechanism-based inhibitors of CYP3A4 in vitro and when used alone reproduce the effect exhibited by grapefruit juice (95). The reduction in intestinal wall CYP3A4 resulted in an increase in maximal plasma concentration of felodipine taken with grapefruit juice (93). Subsequently, the oral bioavailability of more than 20 CYP3A4 substrates has been shown to be increased by grapefruit juice consumption (96).

As expected for irreversible inhibition, there is a delay in the return of CYP3A4 activity to baseline upon discontinuation of inhibitor administration. When an oral dose of midazolam was given before and 2, 26, 50, and 74 hours after a single administration of 300-mL regular-strength grapefruit juice, the oral midazolam AUC was increased by 1.65-fold at 2 hours after grapefruit juice treatment, and 1.29-, 1.29-, and 1.06-fold at 26, 50, and 74 hours, respectively (97). The time course of recovery indicated an average recovery half-life of 23 hours and is consistent with enzyme regeneration after mechanism-based inhibition. Similarly, there was a delayed recovery of CYP2D6 activity in vivo (dextromethorphan metabolic ratio) after discontinuation of paroxetine [20 mg/day for 10 days (98)]. Paroxetine, a selective serotonin reuptake inhibitor, is a mechanism-based inhibitor of CYP2D6 in vitro (76) with a relatively short serum half-life (16–24 hours) and no inhibitory metabolites (99). Mechanism-based inhibition probably contributes to the persistent inhibition of in vivo CYP2D6 activity. With a deconvolution approach, where CYP2D6 half-life and paroxetine steady-state half-life could be distinguished, Venkatakrishnan and Obach estimated a k_E of 0.000226 min^{-1}, translating into a degradation half-life of 51 hours for CYP2D6 (100).

Clarithromycin is characterized by a K_i value in vitro of 10 μM and average serum concentrations of 0.9 μM (101), but precipitates clinically significant interactions with CYP3A substrates such as midazolam (102), carbamazepine (103), ritonavir (104), and cyclosporine (105). As with grapefruit juice, intestinal biopsy studies have been employed to obtain direct evidence of mechanism-based inhibition in vivo for clarithromycin. After clarithromycin 500 mg was given twice daily for seven days, a 74% decrease in intestinal biopsy homogenate 1′-hydroxylation of midazolam was observed without a significant change in CYP3A mRNA or protein expression (106). This significant reduction in CYP3A activity cannot be explained by competitive inhibition because the measured free inhibitor concentration in intestinal homogenate was 1.2 ± 0.7 μM, which is approximately 10,000-fold lower than the reported K_i (10 μM) for competitive inhibition of clarithromycin in vitro (101). In common with erythromycin and TAO, clarithromycin forms an MIC with CYP3A4 and HLMs in vitro, and this is presumed to be the mechanism of inhibition in vivo (106). In contrast to grapefruit juice, clarithromycin inhibits the intestinal and hepatic activities of CYP3A4 as reflected in the significant increases in systemic and oral clearances of midazolam (106,107). CYP3A activity recovered over a seven-day period following discontinuation of clarithromycin, whereas clarithromycin in serum was undetectable within two days (107).

A similar phenomenon has been observed with the well-established mechanism-based inhibitor, diltiazem (108). Diltiazem (Cardizem SR®, 120 mg b.i.d. for 7 days) caused a decrease in small bowel CYP3A activity of 62% with no corresponding change in intestinal CYP3A mRNA or protein expression. Many clinical studies have shown that diltiazem, a calcium channel blocker, inhibits the metabolism of CYP3A substrates, such as triazolam (109), midazolam (110), and lovastatin (84), but the in vitro estimated K_i is ~60 μM and the steady-state serum concentration is ~0.3 μM for diltiazem after regular doses (111,112).

The demonstration of mechanism-based inhibition in vitro should not be assumed to result in significant inhibition in vitro. This is illustrated by two mechanism-based inhibitors of CYP3A, gestodene and 17α-ethynyl estradiol (17-EE), that are commonly administered in oral contraceptive pills. Gestodene exhibits a low partition ratio of ~9 in vitro using HLMs, but the normal dosage is 75 μg/day (113). Guengerich calculated that gestodene would inactivate about 3% of total CYP3A4 per day if the in vitro partition ratio was the same in vivo (18,19); this appears consistent with the lack of interaction with CYP3A substrates in vivo. However, the partition ratio approach to in vivo prediction is not recommended because an in vivo partition ratio that accounts for competing routes of elimination is generally not available and the contribution of new enzyme synthesis is ignored. 17-EE is also a mechanism-based inhibitor of CYP3A4 ($k_{inact} = 0.04\ min^{-1}$, $K_I = 18$ μM) and the mechanism of inhibition involves heme destruction and the irreversible modification of the apoprotein at the active site (51). However, daily dosing of Ovral® (50-μg 17-EE/500-μg norgestrel) for 10 days resulted in no change in the oral or intravenous clearance of midazolam (114). This lack of effect reflects the fact that peak plasma concentration of 17-EE are 0.6 to 0.7 nM after a 50 μg dose and consequently the rate of inactivation is insignificant relative to the endogenous rate of CYP3A degradation. Thus, no change in CYP3A activity is predicted (Eq. 8) (115,116).

B. In Vitro–In Vivo Prediction of Mechanism-Based Inhibition

Quantitative predictions of in vivo DDIs involving mechanism-based inhibition have been successful for several inhibitors. Approaches to predicting in vivo DDIs involving mechanism-based inhibition have ranged from a relatively simplistic method using a single inhibitor concentration to a more complicated, physiologically based pharmacokinetic (PBPK) model, considering the change of inhibitor and substrate concentrations (unbound or total in blood or tissue) with time. However, the core interaction model used in these approaches was described earlier (Eqs. 16–22) for modeling the changes in the amount of active enzyme in the presence of a mechanism-based inhibitor, which in turn, determines the nonlinear elimination of inhibitors and the corresponding DDIs. The inactivation parameters k_{inact} and K_I are estimated using either recombinant enzyme or HLMs in vitro. The degradation rate constant, k_E, is characterized by considerable uncertainty due to the difficulties estimating the value in vivo; estimates of k_E using a variety of approaches are presented in Table 3.

The in vivo interactions between midazolam and several macrolides were predicted using PBPK modeling of drug concentration and the time course of active enzyme concentration in the liver (123). Using a value for k_E of $0.0005\ min^{-1}$, the AUC of midazolam after oral administration was predicted to increase 2.9- or 3.0-fold following pretreatment with erythromycin (500 mg t.i.d.

Table 3 Estimated Values of Endogenous CYP Degradation Rate Constant (k_E) and Half-Life ($0.693/k_E$)

CYPs	k_{deg} (min^{-1})	$t_{1/2}$ (hr)	Methods	References
1A2	0.000296	39	Time course of de-induction in vivo	117
2D6	0.000226	51	Time course of recovery following inactivation in vivo	100
3A4	0.000825	14	Turnover in CYP3A4-expressing Caco-2 cells	94
3A4	0.000263	44	Turnover in primary human hepatocytes	80
3A4	0.00015	79	Turnover in liver slices	118
3A4	0.000481	24	Time course of recovery following inactivation by grapefruit juice (GFJ) in vivo	97
3A4	0.000321	36	Time course of de-induction (rifampin on verapamil) in vivo	119
3A4	0.00014	85	Autoinduction following ritonavir in vivo	120
3A4	0.00008–0.0005	20–146	Time course of de-induction following carbamazepine in vivo	121
Rat CYPs	0.0012–0.00058	10–20	Turnover in rats	122

Abbreviation: CYP, cytochrome P450.

for 5 or 6 days, respectively) and 2.1- or 2.5-fold by clarithromycin (250 mg b.i.d. for 5 days or 500 mg b.i.d. for 7 days, respectively), whereas azithromycin (500 mg once a day for 3 days) was predicted to have little effect on midazolam AUC. These results agreed well with the reported in vivo observations.

A similar model was employed to predict the DDIs between 5-fluorouracil and sorivudine (124). Sorivudine is converted by gut flora to (E)-5-(2-bromovinyl) uracil, which inactivates dihydropyrimidine dehydrogenase and impairs the metabolism of 5-fluorouracil by this enzyme. This interaction led to 15 deaths in Japan from 5-fluorouracil toxicity due to elevated exposure to the drug. Using a k_E of 0.00018 min^{-1}, a fivefold increase in the AUC of 5-fluorouracil was predicted after administration of sorivudine (150 mg/day for 5 days), which was close to the observed data in patients.

Mayhew et al. predicted that therapeutic unbound plasma concentrations (0.1 mM) of fluoxetine, clarithromycin, and the primary metabolite of diltiazem, *N*-desmethyl diltiazem, would reduce the steady-state concentration of hepatic CYP3A4 to approximately 72%, 39%, or 21% of untreated levels, respectively

(31). These reductions correspond to 1.4-, 2.6-, or 4.7-fold increases, respectively, in the AUC of a coadministered drug that is eliminated exclusively by hepatic CYP3A4 metabolism. The value of k_E used in this study was 0.000825 min^{-1} ($t_{1/2}$ = 14 hours). This basic approach was also used to predict the inhibitory effect of verapamil enantiomers and their major metabolites, norverapamil and *N*-desalkyl verapamil (D617), on midazolam clearance. The enantiomers of verapamil and metabolites inhibited CYP3A in a time- and concentration-dependent manner by using the cDNA-expressed CYP3A4 and pooled HLMs (37). Combining the values of k_{inact} and K_I estimated with the cDNA-expressed CYP3A4, the unbound inhibitor concentrations, k_E of 0.000769 min^{-1} to 0.00128 min^{-1}, and a fraction metabolized by CYP3A (f_m) of 0.9, the model predicted oral midazolam AUC was increased by 3.2- to 4.5-fold at steady state after chronic verapamil dosing. The predicted results agreed well with the in vivo drug interaction data.

The extent of CYP2D6 inactivation by paroxetine was predicted from in vitro inactivation kinetics (k_{inact} of 0.17 min^{-1} and unbound K_I of 0.315 μM), in vivo inhibitor concentrations, and an estimated CYP2D6 degradation half-life of 51 hours, using the mathematical model of mechanism-based inhibition (100). The model predicted the nonlinear disposition of paroxetine with an accumulation ratio that is fivefold higher than the expected accumulation if linear kinetics were assumed; this was in excellent agreement with the observed five- to sixfold greater accumulation. Extent of interactions between paroxetine (20–30 mg/day) with the CYP2D6 substrates desipramine, risperidone, perphenazine, atomoxetine, (*S*)-metoprolol, and (*R*)-metoprolol were estimated and the predicted fold-increases in victim drug AUC were five-, six-, five-, six-, four-, and sixfold, respectively. These predictions are in reasonable agreement with observed data but the incorporation of in vitro microsomal binding was essential for good predictive accuracy.

Quantitative prediction of irreversible inhibition at the level of the gut wall remains challenging because of the added uncertainty in the effective inhibitor concentration at this site. For strong inhibitors, it is reasonable to assume complete inhibition of gut wall CYP3A, and this was useful in predicting the interactions between verapamil and midazolam (37). However, a model that successfully predicted the grapefruit juice–felodipine interaction at the gut wall has been described and used to recommend consumption strategies for minimizing the severity of the interaction, although prospective evaluations of the predictions were not described (125).

An attractive strategy for predicting the clinical significance of irreversible inhibition is to use human hepatocytes wherein the "natural" turnover of enzymes might be preserved and in vivo cellular concentrations of inhibitors and metabolites would be achieved. Zhao et al. demonstrated time-dependent inactivation of CYP3A in cryopreserved hepatocytes for amprenavir, diltiazem, erythromycin, raloxifene, and TAO (126). Except for TAO, significant differences in inactivation efficiency potency between hepatocytes and HLMs were noted; hepatocytes experienced greater inactivation for raloxifene, amprenavir, and erythromycin (observed IC_{50} values were 6.2-, 55-, and 7.8-fold higher, respectively, than HLMs), whereas diltiazem caused fourfold less inactivation in hepatocytes compared with HLMs. However, using cultured primary human hepatocytes, McGinity et al. found that for tienilic acid (CYP2C9), erythromycin, TAO, and fluoxetine (CYP3A4), there was a good agreement between the k_{inact} and K_I values derived using hepatocytes, cDNA-expressed CYPs, and HLMs (127).

VIII. CONCLUSION

Many drugs are mechanism-based inhibitors of CYP. This property could affect a drug's own metabolism or the metabolism of coadministered drugs, which could lead to serious drug interactions. Even though in vitro K_i values have been determined for a number of drugs and have been used to predict an in vivo interaction, the effect of mechanism-based inhibitors can be observed at in vivo concentrations below these K_i values. This effect can be predicted if in vitro estimates of kinetic constants (e.g., K_I and k_{inact}) for mechanism-based inhibitors are known. A theoretical basis and application have been presented that applies in vitro estimates of mechanism-based inhibitors to accurately predict in vivo drug interactions.

REFERENCES

1. Olkkola K, Aranko K, Luurila H, et al. A potentially hazardous interaction between erythromycin and midazolam. Clin Pharmacol Ther 1993; 53:298–305.
2. Gascon MP, Dayer P. In vitro forcasting of drugs which may interfere with the biotransformation of midazolam. Eur J Clin Pharmacol 1991; 41:573–578.
3. Larrey D, Funck-Brentano C, Breil P, et al. Effects of erythromycin on hepatic drug-metabolizing enzymes in humans. Biochem Pharmacol 1983; 32(6):1063–1068.
4. Ortiz de Montellano P, Correia M. Inhibition of cytochrome P450 Enzymes. New York: Plenum Press, 1995.
5. Silverman R. Mechanism-Based Enzyme Inactivation: Chemistry and Enzymology. Boca Raton, Florida: CRC Press, 1988.
6. Lopez-Garcia M, Dansette P, Mansuy D. Thiophene derivatives as new mechanism-based inhibitors of cytochromes P-450: inactivation of yeast-expressed human liver cytochrome450 2C9 by tienilic acid. Biochem Pharmacol 1994; 33:166–175.
7. Jean P, Lopez-Garcia P, Dansette P, et al. Oxidation of tienilic acid by human yeast-expressed cytochromes P-450 2C8, 2C9, 2C18 and 2C19: evidence that this drug is a mechanism-based inhibitor specific for cytochrome P-450 2C9. Eur J Biochem 1996; 241:797–804.
8. Koenigs L, Peter R, Hunter A, et al. Electrospray ionization mass spectrometric analysis of intact cytochrome P450: identification of tienilic acid adducts to P450 2C9. Biochem Pharmacol 1999; 38:2312–2319.
9. Jaeger W, Benet L, Bornheim L. Inhibition of cyclosporine and tetrahydrocan nabinol metabolism by cannabidiol in mouse and human microsomes. Xenobiotica 1996; 26:275–284.

10. Bornheim L, Grille M. Characterization of cytochrome P450 3A inactivation by cannabidiol: possible involvement of cannabidiol-hydroxyquinone as a P450 inactivator. Chem Res Toxicol 1990; 11:1209–1216.
11. Halpert J, Naslund B, Betner I. Suicide inactivation of rat liver cytochrome P-450 by chloramphenicol in vivo and in vitro. Mol Pharmacol 1983; 23:445–452.
12. He K, Falick A, Chen B, et al. Identification of the heme adduct and an active site peptide modified during mechanism-based inactivation of rat liver cytochrome P450 2B1 by secobarbital. Chem Res Toxicol 1996; 9:614–622.
13. Koenigs L, Peter R, Thompson S, et al. Mechanism-based inactivation of human liver cytochrome P450 2A6 by 8-methoxypsoralen. Drug Metab Dispos 1997; 25: 1407–1415.
14. Koenigs L, Trager W. Mechanism-based inactivation of P450 2A6 by furanocoumarins. Biochem Pharmacol 1998; 37:10047–10061.
15. Decker C, Rashed M, Baillie T, et al. Oxidative metabolism of spironolactone: evidence for the involvement of electrophilic thiosteroid species in drug-mediated destruction of rat hepatic cytochrome P450. Biochem Pharmacol 1989; 28:5128–5136.
16. He K, Woolf TF, Hollenberg PF. Mechanism-based inactivation of cytochrome P-450-3A4 by mifepristone (RU486). J Pharmacol Exp Ther 1999; 288(2):791–797.
17. Franklin R. Inhibition of mixed-function oxidations by substrates forming reduced cytochrome P-450 metabolic-intermediate complexes. Pharmacol Ther 1977; 2: 227–245.
18. Guengerich F. Mechanism-based inactivation of human liver microsomal cytochrome P-450 IIIA4 by gestodene. Chem Res Toxicol 1990; 3:363–371.
19. Guengerich F. Oxidation of 17 α-ethynylestradiol by human liver cytochrome P-450. Mol Pharmacol 1988; 33:500–508.
20. Tassaneeyakul W, Birkett D, Veronese M, et al. Direct characterization of the selectivity of furafylline as an inhibitor of human cytochromes P450 1A1 and 1A2. Pharmacogenetics 1994; 4:281–284.
21. Muakkassah S, Yang W. Mechanism of the inhibitory action of phenelzine on microsomal drug metabolism. J Pharmacol Exp Ther 1981; 219:147–155.
22. De Mateis F, Gibbs A. Drug-induced conversion of liver haem into modified porphyrins. Evidence for two classes of products. Biochem J 1980; 187:285–288.
23. Laurenzana E, Owens S. Metabolism of phencyclidine by human liver microsomes. Drug Metab Dispos 1997; 25:557–563.
24. Schaefer W, Harris T, Guengerich F. Characterization of the enzymatic and non-enzymatic peroxidative degradation of iron porphyrins and cytochrome P-450 heme. Biochem J 1985; 24:3254–3263.
25. Waley S. Kinetics of suicide substrates. Biochem J 1980; 185:771–773.
26. Waley S. Kinetics of suicide substrates. Practical procedures for determining parameters. Biochem J 1985; 227:843–849.
27. Tatsunami S, Yago N, Hosoe M. Kinetics of suicide substrates. Steady-state treatments and computer-aided exact solutions. Biochim Biophys Acta 1981; 662: 226–235.
28. Walsh C, Cromartie T, Marcotte P, et al. Suicide substrates for flavoprotein enzymes. Meth Enzymol 1978; 53:437–448.
29. Bensoussan C, Delaforge M, Mansuy D. Particular ability of cytochromes P450 3A to form inhibitory P450-iron-metabolite complexes upon metabolic oxidation of aminodrugs. Biochem Pharmacol 1995; 49:591–602.

30. Korzekwa K, Krishnamachary N, Shou M, et al. Evaluation of atypical cytochrome P450 kinetics with two-substrate models: evidence that multiple substrates can simultaneously bind to cytochrome P450 active sites. Biochem J 1998; 37:4137–4147.
31. Mayhew BS, Jones DR, Hall SD. An in vitro model for predicting in vivo inhibition of cytochrome P450 3A4 by metabolic intermediate complex formation. Drug Metab Dispos 2000; 28(9):1031–1037.
32. McConn DJ II, Lin YS, Allen K, et al. Differences in the inhibition of cytochromes P450 3A4 and 3A5 by metabolite-inhibitor complex-forming drugs. Drug Metab Dispos 2004; 32(10):1083–1091.
33. Tinel M, Descatoire V, Larrey D, et al. Effects of clarithromycin on cytochrome P-450. Comparison with other macrolides. J Pharmacol Exp Ther 1989; 250(2): 746–751.
34. Lindstrom T, Hanssen B, Wrighton S. Cytochrome P-450 complex formation by dirithromycin and other macrolides in rat and human livers. Antimicrob Agents Chemother 1993; 37:265–269.
35. Polasek TM, Elliot DJ, Somogyi AA, et al. An evaluation of potential mechanism-based inactivation of human drug metabolizing cytochromes P450 by monoamine oxidase inhibitors, including isoniazid. Br J Clin Pharmacol 2006; 61(5):570–584.
36. Jones DR, Gorski JC, Hamman MA, et al. Diltiazem inhibition of cytochrome P-450 3A activity is due to metabolite intermediate complex formation. J Pharmacol Exp Ther 1999; 290(3):1116–1125.
37. Wang YH, Jones DR, Hall SD. Prediction of cytochrome P450 3A inhibition by verapamil enantiomers and their metabolites. Drug Metab Dispos 2004; 32(2): 259–266.
38. Masubuchi Y, Horie T. Mechanism-based inactivation of cytochrome P450s 1A2 and 3A4 by dihydralazine in human liver microsomes. Chem Res Toxicol 1999; 12 (10):1028–1032.
39. Prueksaritanont T, Ma B, Tang C, et al. Metabolic interactions between mibefradil and HMG-CoA reductase inhibitors: an in vitro investigation with human liver preparations. Br J Clin Pharmacol 1999; 47(3):291–298.
40. Ernest CS II, Hall SD, Jones DR. Mechanism-based inactivation of CYP3A by HIV protease inhibitors. J Pharmacol Exp Ther 2005; 312(2):583–591.
41. Voorman RL, Maio SM, Hauer MJ, et al. Metabolism of delavirdine, a human immunodeficiency virus type-1 reverse transcriptase inhibitor, by microsomal cytochrome P450 in humans, rats, and other species: probable involvement of CYP2D6 and CYP3A. Drug Metab Dispos 1998; 26(7):631–639.
42. Zhao XJ, Jones DR, Wang YH, et al. Reversible and irreversible inhibition of CYP3A enzymes by tamoxifen and metabolites. Xenobiotica 2002; 32(10):863–878.
43. Hanioka N, Ozawa S, Jinno H, et al. Interaction of irinotecan (CPT-11) and its active metabolite 7-ethyl-10-hydroxycamptothecin (SN-38) with human cytochrome P450 enzymes. Drug Metab Dispos 2002; 30(4):391–396.
44. He K, Iyer KR, Hayes RN, et al. Inactivation of cytochrome P450 3A4 by bergamottin, a component of grapefruit juice. Chem Res Toxicol 1998; 11(4):252–259.
45. Schmiedlin-Ren P, Edwards DJ, Fitzsimmons ME, et al. Mechanisms of enhanced oral availability of CYP3A4 substrates by grapefruit constituents. Decreased enterocyte CYP3A4 concentration and mechanism-based inactivation by furanocoumarins. Drug Metab Dispos 1997; 25(11):1228–1233.

46. Kent UM, Aviram M, Rosenblat M, et al. The licorice root derived isoflavan glabridin inhibits the activities of human cytochrome P450S 3A4, 2B6, and 2C9. Drug Metab Dispos 2002; 30(6):709–715.
47. Stupans I, Murray M, Kirlich A, et al. Inactivation of cytochrome P450 by the food-derived complex phenol oleuropein. Food Chem Toxicol 2001; 39(11):1119–1124.
48. Piver B, Berthou F, Dreano Y, et al. Inhibition of CYP3A, CYP1A and CYP2E1 activities by resveratrol and other non volatile red wine components. Toxicol Lett 2001; 125(1–3):83–91.
49. Iwata H, Tezuka Y, Kadota S, et al. Mechanism-based inactivation of human liver microsomal CYP3A4 by rutaecarpine and limonin from Evodia fruit extract. Drug Metab Pharmacokinet 2005; 20(1):34–45.
50. Usia T, Watabe T, Kadota S, et al. Mechanism-based inhibition of CYP3A4 by constituents of Zingiber aromaticum. Biol Pharm Bull 2005; 28(3):495–459.
51. Lin HL, Kent UM, Hollenberg PF. Mechanism-based inactivation of cytochrome P450 3A4 by 17 alpha-ethynylestradiol: evidence for heme destruction and covalent binding to protein. J Pharmacol Exp Ther 2002; 301(1):160–167.
52. Guengerich F. Inhibition of oral contraceptive steroid-metabolizing enzymes by steroids and drugs. Am J Obstet Gynecol 1990; 163(6 pt 2):2159–2163.
53. Chen Q, Ngui JS, Doss GA, et al. Cytochrome P450 3A4-mediated bioactivation of raloxifene: irreversible enzyme inhibition and thiol adduct formation. Chem Res Toxicol 2002; 15(7):907–914.
54. Lim H-K, Nicholas Duczak J, Brougham L, et al. Automated screening with confirmation of mechanism-based inactivation of CYP3A4, CYP2C9, CYP2C19, CYP2D6, and CYP1A2 in pooled human liver microsomes. Drug Metab Dispos 2005; 33(8):1211–1219.
55. Khan KK, He YQ, Domanski TL, et al. Midazolam oxidation by cytochrome P450 3A4 and active-site mutants: an evaluation of multiple binding sites and of the metabolic pathway that leads to enzyme inactivation. Mol Pharmacol 2002; 61(3): 495–506.
56. Masubuchi Y, Ose A, Horie T. Diclofenac-induced inactivation of CYP3A4 and its stimulation by quinidine. Drug Metab Dispos 2002; 30(10):1143–1148.
57. Chatterjee P, Franklin MR. Human cytochrome p450 inhibition and metabolic-intermediate complex formation by goldenseal extract and its methylenedioxyphenyl components. Drug Metab Dispos 2003; 31(11):1391–1397.
58. Alvarez-Diez TM, Zheng J. Mechanism-based inactivation of cytochrome P450 3A4 by 4-ipomeanol. Chem Res Toxicol 2004; 17(2):150–157.
59. Zdravkovic M, Olsen AK, Christiansen T, et al. A clinical study investigating the pharmacokinetic interaction between NN703 (tabimorelin), a potential inhibitor of CYP3A4 activity, and midazolam, a CYP3A4 substrate. Eur J Clin Pharmacol 2003; 58(10):683–688.
60. Jacobsen W, Christians U, Benet LZ. In vitro evaluation of the disposition of A novel cysteine protease inhibitor. Drug Metab Dispos 2000; 28(11):1343–1351.
61. Chun YJ, Ryu SY, Jeong TC, et al. Mechanism-based inhibition of human cytochrome P450 1A1 by rhapontigenin. Drug Metab Dispos 2001; 29(4 pt 1):389–393.
62. Cai Y, Baer-Dubowska W, Ashwood-Smith MJ, et al. Mechanism-based inactivation of hepatic ethoxyresorufin O-dealkylation activity by naturally occurring coumarins. Chem Res Toxicol 1996; 9(4):729–736.

63. Lu P, Schrag ML, Slaughter DE, et al. Mechanism-based inhibition of human liver microsomal cytochrome P450 1A2 by zileuton, a 5-lipoxygenase inhibitor. Drug Metab Dispos 2003; 31(11):1352–1360.
64. Karjalainen MJ, Neuvonen PJ, Backman JT. Rofecoxib is a potent, metabolism-dependent inhibitor of CYP1A2: implications for in vitro prediction of drug interactions. Drug Metab Dispos 2006; 34(12):2091–2096.
65. Chang TK, Chen J, Lee WB. Differential inhibition and inactivation of human CYP1 enzymes by trans-resveratrol: evidence for mechanism-based inactivation of CYP1A2. J Pharmacol Exp Ther 2001; 299(3):874–882.
66. Langouet S, Furge LL, Kerriguy N, et al. Inhibition of human cytochrome P450 enzymes by 1,2-dithiole-3-thione, oltipraz and its derivatives, and sulforaphane. Chem Res Toxicol 2000; 13(4):245–252.
67. Wen X, Wang JS, Kivisto KT, et al. In vitro evaluation of valproic acid as an inhibitor of human cytochrome P450 isoforms: preferential inhibition of cytochrome P450 2C9 (CYP2C9). Br J Clin Pharmacol 2001; 52(5):547–553.
68. Denton TT, Zhang X, Cashman JR. Nicotine-related alkaloids and metabolites as inhibitors of human cytochrome P-450 2A6. Biochem Pharmacol 2004; 67(4): 751–756.
69. Fan PW, Gu C, Marsh SA, et al. Mechanism-based inactivation of cytochrome P450 2B6 by a novel terminal acetylene inhibitor. Drug Metab Dispos 2003; 31(1):28–36.
70. Richter T, Schwab M, Eichelbaum M, et al. Inhibition of human CYP2B6 by N,N′, N″-triethylenethiophosphoramide is irreversible and mechanism-based. Biochem Pharmacol 2005; 69(3):517–524.
71. Richter T, Murdter TE, Heinkele G, et al. Potent mechanism-based inhibition of human CYP2B6 by clopidogrel and ticlopidine. J Pharmacol Exp Ther 2004; 308(1):189–197.
72. Lin HL, Kent UM, Hollenberg PF. The grapefruit juice effect is not limited to cytochrome P450 (P450) 3A4: evidence for bergamottin-dependent inactivation, heme destruction, and covalent binding to protein in P450s 2B6 and 3A5. J Pharmacol Exp Ther 2005; 313(1):154–164.
73. Polasek TM, Elliot DJ, Lewis BC, et al. Mechanism-based inactivation of human cytochrome P4502C8 by drugs in vitro. J Pharmacol Exp Ther 2004; 311(3):996–1007.
74. Ogilvie BW, Zhang D, Li W, et al. Glucuronidation converts gemfibrozil to a potent, metabolism-dependent inhibitor of CYP2C8: implications for drug-drug interactions. Drug Metab Dispos 2006; 34(1):191–197.
75. Nakajima M, Yoshida R, Shimada N, et al. Inhibition and inactivation of human cytochrome P450 isoforms by phenethyl isothiocyanate. Drug Metab Dispos 2001; 29(8):1110–1113.
76. Bertelsen KM, Venkatakrishnan K, Von Moltke LL, et al. Apparent mechanism-based inhibition of human CYP2D6 in vitro by paroxetine: comparison with fluoxetine and quinidine. Drug Metab Dispos 2003; 31(3):289–293.
77. Van LM, Heydari A, Yang J, et al. The impact of experimental design on assessing mechanism-based activation of CYP2D6 by MDMA (Ecstasy). J Psychopharmacol 2006; 20(6):834–841.
78. Usia T, Watabe T, Kadota S, et al. Cytochrome P450 2D6 (CYP2D6) inhibitory constituents of Catharanthus roseus. Biol Pharm Bull 2005; 28(6):1021–1024.
79. Pershing L, Franklin MR. Cytochrome P-450 metabolic-intermediate complex formation and induction by macrolide antibiotics: a new class of agents. Xenobiotica 1982; 12:687–699.

80. Pichard L, Fabre I, Daujat M, et al. Effect of corticosteroids on the expression of cytochromes P450 and on cyclosporin A oxidase activity in primary cultures of human hepatocytes. Mol Pharmacol 1992; 41(6):1047–1055.
81. Wilkinson G, Shand D. A physiological approach to hepatic drag clearance. Clin Pharmacol Ther 1975; 18:377–390.
82. Houston J. Drug metabolite kinetics. Pharmacol Ther 1982; 15:521–552.
83. Jacqz E, Hall S, Branch R, et al. Polymorphic metabolism of mepheny toin in man: pharmacokinetic interaction with a co-regulated substrate, mephobarbital. Clin Pharmacol Ther 1986; 6:646–653.
84. Azie N, Brater D, Becker P, et al. The interaction of diltiazem with lovastatin and pravastatin. Clin Pharmacol Ther 1998; 64:369–377.
85. Tozer T. Concepts basic to pharmacokinetics. Pharmacol Ther 1981; 12:109–131.
86. Jones D, Gorski J, Haehner B, et al. Determination of cytochrome P450 3A4/5 activity in vivo with dextromethorphan iV-demethylation. Clin Pharmacol Ther 1996; 60:374–384.
87. Freeman J, Martell R, Carrufhers S, et al. Cyclosporin-erythromycin interaction in normal subjects. Br J Clin Pharmacol 1987; 23:776–778.
88. Bartkowski P, Goldberg M, Larijani G, et al. Inhibition of alfentanil metabolism by erythromycin. Clin Pharmacol Ther 1989; 46:99–102.
89. Phillips J, Antal E, Smith R. A pharmacokinetic drug interaction between erythromycin and triazolam. J Clin Psychopharmacol 1986; 6:297–299.
90. Gorski J, Jones D, Hamman M, et al. Biotransformation of alprazolam by members of the human cytochrome P450 3A subfamily. Xenobiotica 1999; 29:931–944.
91. Wong YY, Ludden TM, Bell RD. Effect of erythromycin on carbamazepine kinetics. Clin Pharmacol Ther 1983; 33:460–464.
92. Pessayre D, Larrey D, Vitaux J, et al. Formation of an inactive cytochrome P-450 Fe(II)-metabolite complex after administration of troleandomycin in humans. Biochem Pharmacol 1982; 31(9):1699–704.
93. Lown KS, Bailey DG, Fontana RJ, et al. Grapefruit juice increases felodipine oral availability in humans by decreasing intestinal CYP3A protein expression. J Clin Invest 1997; 99(10):2545–2553.
94. Malhotra S, Schimiedlin-Ren P, Paine MF, et al. The furocoumarin 6′, 7′-dihydroxybergamottin (DHB) accelarates CYP3A4 degradation via the ubiquitin-proteasomal pathway. Drug Metab Rev 2001; 33:97.
95. Guo LQ, Fukuda K, Ohta T, et al. Role of furanocoumarin derivatives on grapefruit juice-mediated inhibition of human CYP3A activity. Drug Metab Dispos 2000; 28 (7):766–771.
96. Arayne MS, Sultana N, Bibi Z. Grape fruit juice-drug interactions. Pak J Pharm Sci 2005; 18(4):45–57.
97. Greenblatt DJ, von Moltke LL, Harmatz JS, et al. Time course of recovery of cytochrome p450 3A function after single doses of grapefruit juice. Clin Pharmacol Ther 2003; 74(2):121–129.
98. Liston HL, DeVane CL, Boulton DW, et al. Differential time course of cytochrome P450 2D6 enzyme inhibition by fluoxetine, sertraline, and paroxetine in healthy volunteers. J Clin Psychopharmacol 2002; 22(2):169–173.
99. Sindrup SH, Brosen K, Gram LF, et al. The relationship between paroxetine and the sparteine oxidation polymorphism. Clin Pharmacol Ther 1992; 51(3):278–287.

100. Venkatakrishnan K, Obach RS. In vitro-in vivo extrapolation of CYP2D6 inactivation by paroxetine: prediction of nonstationary pharmacokinetics and drug interaction magnitude. Drug Metab Dispos 2005; 33(6):845–852.
101. Jurima-Romet M, Crawford K, Cyr T, et al. Terfenadine metabolism in human liver. In vitro inhibition by macrolide antibiotics and azole antifungals. Drug Metab Dispos 1994; 22(6):849–857.
102. Gorski JC, Jones DR, Haehner-Daniels BD, et al. The contribution of intestinal and hepatic CYP3A to the interaction between midazolam and clarithromycin. Clin Pharmacol Ther 1998; 64(2):133–143.
103. Albani F, Riva R, Baruzzi A. Clarithromycin-carbamazepine interaction: a case report. Epilepsia 1993; 34:161–162.
104. Ouellet D, Hsu A, Granneman G, et al. Pharmacokinetic interaction between ritonavir and clarithromycin. Clin Pharmacol Ther 1998; 64:355–362.
105. Sketris I, Wright M, ML W. Possible role of the intestinal P-450 enzyme system in a cyclosporine-clarithromycin interaction. Pharmacotherapy 1996; 16:301–305.
106. Pinto AG, Wang YH, Chalasani N, et al. Inhibition of human intestinal wall metabolism by macrolide antibiotics: effect of clarithromycin on cytochrome P450 3A4/5 activity and expression. Clin Pharmacol Ther 2005; 77(3):178–188.
107. Gorski J, Wang Z, Hall S. Duration of CYP3A inhibition by clarithromycin. Clin Pharmacol Ther 2002; 71(2):105.
108. Pinto A, Horlander J, Chalasani N, et al. Diltiazem inhibits human intestinal cytochrome P450 3A (CYP3A) activity in vivo without altering the expression of intestinal mRNA or protein. Br J Clin Pharmacol 2005; 59(4):440–446.
109. Varhe A, Olkkola K, Neuvonen P. Oral triazolam is potentially hazardous to patients receiving systemic antimycotics ketoconazole or itraconazole. Clin Pharmacol Ther 1994; 56:601–607.
110. Backman J, Olkkola K, Aranko K, et al. Dose of midazolam should be reduced during diltiazem and verapamil treatments. Br J Clin Pharmacol 1994; 37:221–225.
111. Pichard L, Fabre I, Fabre G, et al. Cyclosporin A drug interactions. Screening for inducers and inhibitors of cytochrome P-450 (cyclosporin A oxidase) in primary cultures of human hepatocytes and in liver microsomes. Drug Metab Dispos 1990; 18(5):595–606.
112. Yeung PK, Buckley SJ, Hung OR, et al. Steady-state plasma concentrations of diltiazem and its metabolites in patients and healthy volunteers. Ther Drug Monit 1996; 18(1):40–45.
113. Grigoryan OR, Grodnitskaya EE, Andreeva EN, et al. Contraception in perimenopausal women with diabetes mellitus. Gynecol Endocrinol 2006; 22(4):198–206.
114. Belle DJ, Callaghan JT, Gorski JC, et al. The effects of an oral contraceptive containing ethinyloestradiol and norgestrel on CYP3A activity. Br J Clin Pharmacol 2002; 53(1):67–74.
115. Orme ML, Back DJ, Breckenridge AM. Clinical pharmacokinetics of oral contraceptive steroids. Clin Pharmacokinet 1983; 8:95–136.
116. Brody SA, Turkes A, Goldzieher JW. Pharmacokinetics of three bioequivalent norethindrone/mestranol-50 micrograms and three norethindrone/ethinyl estradiol-35 micrograms OC formulations: are 'low-dose' pills really lower? Contraception 1989; 40:269–284.
117. Faber MS, Fuhr U. Time response of cytochrome P450 1A2 activity on cessation of heavy smoking. Clin Pharmacol Ther 2004; 76(2):178–184.

118. Renwick AB, Watts PS, Edwards RJ, et al. Differential maintenance of cytochrome P450 enzymes in cultured precision-cut human liver slices. Drug Metab Dispos 2000; 28(10):1202–1209.
119. Fromm MF, Busse D, Kroemer HK, et al. Differential induction of prehepatic and hepatic metabolism of verapamil by rifampin. Hepatology 1996; 24(4):796–801.
120. Hsu A, Granneman GR, Witt G, et al. Multiple-dose pharmacokinetics of ritonavir in human immunodeficiency virus-infected subjects. Antimicrob Agents Chemother 1997; 41(5):898–905.
121. Lai AA, Levy RH, Cutler RE. Time-course of interaction between carbamazepine and clonazepam in normal man. Clin Pharmacol Ther 1978; 24(3):316–323.
122. Correia MA. Cytochrome P450 turnover. Methods Enzymol 1991; 206:315–325.
123. Ito K, Ogihara K, Kanamitsu S, et al. Prediction of the in vivo interaction between midazolam and macrolides based on in vitro studies using human liver microsomes. Drug Metab Dispos 2003; 31(7):945–954.
124. Kanamitsu SI, Ito K, Okuda H, et al. Prediction of in vivo drug-drug interactions based on mechanism-based inhibition from in vitro data: inhibition of 5-fluorouracil metabolism by (E)-5-(2-bromovinyl)uracil. Drug Metab Dispos 2000; 28(4):467–474.
125. Takanaga H, Ohnishi A, Matsuo H, et al. Pharmacokinetic analysis of felodipine-grapefruit juice interaction based on an irreversible enzyme inhibition model. Br J Clin Pharmacol 2000; 49(1):49–58.
126. Zhao P, Kunze KL, Lee CA. Evaluation of time-dependent inactivation of CYP3A in cryopreserved human hepatocytes. Drug Metab Dispos 2005; 33(6):853–861.
127. McGinnity DF, Berry AJ, Kenny JR, et al. Evaluation of time-dependent cytochrome P450 inhibition using cultured human hepatocytes. Drug Metab Dispos 2006; 34(8):1291–1300.

12

Transporter-Mediated Drug Interactions: Molecular Mechanisms and Clinical Implications

Jiunn H. Lin
Department of Drug Metabolism,
Merck Research Laboratories,
West Point, Pennsylvania, U.S.A.

I. INTRODUCTION

Most mammalian cell membranes contain various transporter systems, which are effective in taking up a great number of endogenous and exogenous molecules that the cells need, or in extruding endogenous and exogenous molecules that the cells want to get rid of. In humans, some of these transporters also play a significant role in the processes of drug absorption, distribution, metabolism, and excretion (ADME) of drugs. With the great advance in molecular biology and biotechnology, many drug transporters have been identified and characterized over the last 10 years. Although the importance of the role of drug transporters in the processes of ADME is becoming well recognized, our understanding of their physiological function is still nascent. In spite of a large body of information on the kinetic characterization of transporters in cell culture models and heterologous expression systems, there are still large gaps in our knowledge about how to utilize the information obtained from in vitro studies to interpret the in vivo kinetic behavior of drugs and for developing new drug candidates that are absorbed, distributed, or excreted by means of transporters (1,2).

The in vivo function of a given drug transporter is influenced by many factors, including the local drug concentration in the vicinity of the transporter, the substrate affinity, and regional (tissue) expression level of the transporter. In addition, the cellular localization and transport direction of the transporter are also important factors that must be considered. Another factor that adds to the complexity of in vivo function of drug transporters is that they interplay with drug-metabolizing enzymes (3,4). It is well known that there is a striking overlap of substrate specificity between drug transporters and drug-metabolizing enzymes (1,2). Therefore, it is important to dissect the function of drug transporters from that of drug metabolizing enzymes when assessing drug interactions. Experimentally, it is very difficult to accurately differentiate the relative role of a particular transporter from drug-metabolizing enzymes or other transporters. The relative role of a given transporter that interplays with other transporters is further complicated by the fact that many drug transporters may not have yet surfaced. It is expected that many new drug transporters will continue to emerge in the future.

Although less frequently compared with the reported cases of cytochrome P450 (CYP)-mediated drug interactions, transporter-mediated drug interactions have been reported in animals and humans. Unlike the CYP-mediated drug interactions, which can be readily defined by inhibition or induction of CYP enzymes, the examples of transporter-mediated drug interactions are often less conclusive (5–7). In many cases, transporter-mediated interactions are postulated on the basis of circumstantial evidence. Sometimes they are referred to as transporter-mediated drug interactions because they cannot be explained by CYP inhibition or induction. Owing to the broad overlap in substrate specificity, many inhibitors and inducers can simultaneously affect both drug transporters and CYP enzymes. Therefore, care should be exercised when exploring the underlying mechanisms of drug interactions. The main purpose of this chapter is to explore the molecular mechanisms of drug interactions involving drug transporters. While the discussion will be focused predominantly on human data, examples from animal studies will also be used to assist in our understanding of the transporter-mediated drug interactions.

II. MOLECULAR MECHANISMS FOR TRANSPORTER INHIBITION AND INDUCTION

Although many so-called transporter-mediated drug interactions have been reported, assessment of the underlying mechanisms for these interactions is not as straightforward as was generally believed. In fact, there are many conflicting reports with regard to the interpretation of the underlying mechanisms for the so-called transporter-mediated drug interactions. There are many reasons for the conflicts in the data interpretation. Limited knowledge about the inhibition and induction of transporters at the molecular level is one of the major reasons. The molecular mechanisms of inhibition and induction for most drug transporters are still not fully understood up to date.

A. Inhibition of Transporters

The inhibition of transporters does not always follow simple kinetics. Inhibition of drug transport by efflux transporter, P-glycoprotein (P-gp), an ATP-binding cassette (ABC) transporter, could potentially result from either the competition between drug and inhibitor on substrate-binding sites or ATP-binding sites or from the blockage of ATP hydrolysis. For example, verapamil inhibited the transport of other P-gp substrate in a competitive manner without interrupting the ATP hydrolysis process of P-gp, while vanadate interacted with the ATP-binding domains of P-gp without interacting with the substrate-binding sites. On the other hand, cyclosporin A inhibited transport function by interfering with both substrate recognition and ATP hydrolysis (8–10). These results suggest that P-gp contains multiple drug-binding sites. Recent studies by Loo et al. supported the notion of multiple drug-binding sites (11,12). They have shown that the drug-binding pocket of human P-gp is at the interface between transmembrane (TM) domains and can simultaneously bind two different drug substrates. In their so-called ''substrate induced fit'' model, they suggest that slight rotational and/or lateral movement in any TM segment could result in numerous permutations of residues contributing to the drug-binding site. A substrate would cause specific shifts in the different TM segments responsible for its binding (induced fit). Taken together, these results highlight the complexity of the competitive inhibition of P-gp by two drug substrates.

Although the competition of two substrates for the same P-gp normally results in an inhibitory effect on the P-gp-mediated transport of the substrates, stimulation of P-gp-mediated efflux transport has been reported in some cases. The P-gp-mediated doxorubicin efflux out of multidrug-resistant HCT-15 colon cells was significantly increased by some flavonoids (13). Similarly, rhodamine 123 and Hoechst 33342 stimulated the rate of P-gp-mediated transport of each other in P-gp-enriched plasma membrane vesicles isolated from Chinese hamster ovary CHRB30 cells (14). Interestingly, Hoechst 33342 transport was increased by daunorubicin and doxorubicin, while rhodamine 123 transport was inhibited by daunorubicin and doxorubicin (14). These results strongly suggest that molecular mechanisms of P-gp interaction are quite complex and cannot be predicted readily.

Complex inhibition and stimulation patterns have also been reported for another ABC efflux transporter, multidrug resistance-related protein 2 (MRP2). In vitro studies with MRP2-enriched membrane vesicles revealed that probenecid strongly inhibited the transport of methotrexate, but stimulated the transport of estradiol-17β-D-glucuronide ($E_2 17\beta G$) (15). Similarly, furosemide inhibited the MRP2 transport of *N*-ethylmaleimide glutathione conjugate, but stimulated the transport of $E_2 17\beta G$ (16). With these results, the investigators suggest that MRP2 contains two drug-binding sites—one site from which substrate is transported and a second site that allosterically regulates the affinity of the transport site for the substrate (16). Similar to efflux transporters, the inhibition of influx transporters also does not always follow simple kinetics. Transport of estrone-3-sulfate by organic anion transporting polypeptide 2B1 (OATP2B1) and organic anion transporting polypeptide 1B1 (OATP1B1) followed single- and biphasic-saturation kinetics, respectively. Uptake of estrone-3-sulfate by OATP2B1 was inhibited by sulfate conjugates but not by glucuronide conjugates, whereas its uptake by OATP1B1 was inhibited by both types of conjugates (17).

For the ABC efflux transporters, the transport from the cells can take place against a steep concentration gradient in an active and ATP-dependent manner. Therefore, depletion of ATP can result in a decrease in the functional activity of the ABC transporters. Pluronic block copolymer P85 (P85) inhibited P-gp, MRP1, and MRP2 transport activity by depleting intracellular ATP and inhibiting ATPase activity (18). Interestingly, the effect of P85 on P-gp ATPase activity was considerably greater compared with the effects on MRP1 and MRP2 ATPase activity. In parallel with the effect on ATPase activity, P85 also had a greater inhibitory effect on the P-gp transport compared with the MRP transport (18). It is also of interest to note that a striking difference between P-gp- and MRP-mediated ATP hydrolysis appears to be related to their structural difference in ATP-binding sites. Two ATP sites of P-gp are essentially symmetrical, while those of MRP transporters do not appear to be symmetrical (19,20). These results suggest that the symmetry of ATP-binding sites of P-gp appears to be more sensitive to the changes of the ATP levels as compared with the asymmetrical ATP-binding sites of MRP transporters.

Because of the complexity, it is difficult to predict the magnitude of drug interactions via transporter inhibition when transporter substrates and inhibitors are given simultaneously. This complexity can be further exacerbated by recent findings that inhibition of the transport of a substrate could result from alterations in the so-called transporter trafficking/sorting processes of endocytic retrieval and exocytic insertion of transporters between the apical membrane and intracellular pools of vesicles caused by a second substrate (21,22). For example, $E_2 17\beta G$ induced endocytic internalization of rat Mrp2, which occurred in parallel with decreased bile flow and Mrp2 transport activity (23). Confocal analysis demonstrated endocytic retrieval of Mrp2 from the canalicular membrane into pericanalicular domains after intravenous administration of $E_2 17\beta G$ (15 μmol/kg) to rats (23). Although drug interactions caused by alterations in transporter trafficking/sorting between membranes and intracellular pools have not been demonstrated, it is conceivable that this type of drug interaction could occur in vivo.

B. Induction of Transporters

Like some of the CYP isozymes, the expression of some transporters is inducible. Induction of the expression of transporters in response to chemical inducers has been primarily studied in the in vitro models using cell lines derived from animals and humans. Rifampicin significantly induced the mRNA and protein levels of pig P-gp and MRP2 in LLC-K1 cells derived from pig kidney (24). Consistent with the increase in the protein levels of P-gp and MRP2, the intracellular uptake of doxorubicin (a substrate for both P-gp and MRP2) was reduced in the rifampicin pretreated cells. Induction of human P-gp and MRP2 by rifampicin or other inducing agents in the human colon carcinoma cell lines has also been reported (25,26). The induction of P-gp and MRP2 by rifampicin is believed to be mediated by the orphan nuclear receptor, pregnane X receptor (PXR). It has been demonstrated that induction of human P-gp is mediated by a DR4 motif in the upstream enhancer of MDR1 gene at about −8 kbp (thousand base pairs), to which PXR binds (25). Similarly, an element at 440 bp upstream of the transcription initiation site of rat Mrp2 has been identified (27). PXR binds with high affinity to this element as heterodimers with the retinoid X receptor (RXR).

Induction of P-gp by rifampicin has also been reported in vivo in humans. In a clinical study, duodenal biopsies were obtained and the duodenal P-gp contents in healthy volunteers were determined before and after oral administration of rifampicin at 600 mg/day for nine days (28). Treatment with rifampicin resulted in a significantly increased expression of duodenal P-gp content by 4.2-fold. In another clinical study, treatment with rifampicin at 600 mg/day for 10 days resulted in a 3.5-fold increase in intestinal P-gp in health volunteers (29). Similarly, induction of intestinal MRP2 by rifampicin (600 mg/day for 9 days) has been reported in humans. In a clinical study with 16 healthy volunteers, rifampicin induced duodenal MRP2 mRNA in 14 out of 16 individuals, while MRP2 protein was significantly induced by rifampicin in 10 out of 16 subjects (30). Interestingly, St. John's wort, a known PXR activator, also induces the expression of P-gp and MRP2 in animals and humans (31,32).

In vitro and in vivo studies have demonstrated that dexamethasone, a ligand of the glucocorticoid receptor (GR), induces the expression of P-gp in vitro, suggesting the involvement of GR in P-gp induction. Dexamethasone significantly induced P-gp expression in rat hepatocytes (33). Consistent with in vitro observations, pretreatment of rats with dexamethasone (oral dose at 40 mg/kg/day for 3 days) resulted in significant increases in both intestinal and hepatic P-gp expression level by approximately two- to threefold (34). Dexamethasone is also a potent inducer of Mrp2 expression at mRNA and protein levels in rat hepatocytes (35). Induction of rat Mrp2 by dexamethasone suggests the involvement of GR, which is in line with the identification of several glucocorticoid-responding elements in the promoter region of the Mrp2 gene (36). Although it is generally believed that GR is involved in the regulation of P-gp and Mrp2, there are conflicting reports regarding the inductive effect of dexamethasone on these two transporters. In a study with rat hepatocytes, Courtois et al. have shown that dexamethasone downregulated the expression of P-gp mRNA, while it markedly upregulated the expression of Mrp2 mRNA (37). Because the dexamethasone-mediated induction of rat Mrp2 was not inhibited by the antiglucocorticoid RU486, the investigators concluded that dexamethasone induced Mrp2 expression in rat hepatocytes through a mechanism that seems not to involve the classical GR pathway (37). The reason for these discrepancies is not clear at the present time.

Induction has also been demonstrated for breast cancer resistance protein (BCRP) and OATP2. In a study, wild-type [PXR(+/+)] and PXR-null [PXR(−/−)] C57BL/6 mice were treated with 2-acetylaminofluorene (2-AAF), a PXR activator, up to 150 mg/kg/day for seven days. Hepatic expression of both Oatp2 and BCRP in PXR(+/+) mice was increased in a dose-dependent manner following the treatment with 2-AAF, but not PXR-null mice (38). These results suggest the involvement of PXR in the induction of Oatp2 and BCRP transporters. In addition to PXR, induction of BCRP may also involve the aryl hydrocarbon receptor (AhR). 2,3,7,8-Tetrachlorodibenzo-*p*-dioxin, a potent AhR agonist, significantly increased both mRNA and protein levels of BCRP in Caco-2 cells (39). On the other hand, phenobarbital treatment at 80 mg/kg per day for five days increased expression of Oatp2 more than twofold on both mRNA and protein in rats, suggesting the possible involvement of constitutive androstane receptor (CAR) in the induction of Oatp2 in rats (40).

III. TRANSPORTER-MEDIATED DRUG INTERACTIONS

As mentioned earlier, the underlying mechanisms for many of the transporter-mediated interactions are not fully understood and remain elusive at the present time. With the limited knowledge on the molecular mechanisms of transporter-mediated interaction and the fact that many inhibitors and inducers can simultaneously affect both drug transporters and CYP enzymes, it is difficult to quantitatively differentiate transporter-mediated interactions from CYP-mediated interactions. From the literature, it becomes clear that evidence of transporter-mediated drug interactions, with few exceptions, is often indirectly derived from in vitro transport studies with cellular culture models and heterologous expression systems.

A. Direct Evidence

Perhaps the most compelling clinical evidence of a transporter-mediated drug interaction is obtained from drugs that are eliminated predominately by drug transporters. The digoxin-valspodar interaction is a good example. As shown in Table 1, coadminstration of a single 400-mg dose of valspodar (PSC833) caused an average 75% increase in the steady-state digoxin AUC (area under the curve) and a 62% decrease in renal clearance in healthy volunteers after chronic dosing with digoxin for seven days (41). Because digoxin, a good P-gp substrate, is exclusively eliminated in humans by renal excretion with minimal metabolism and PSC833 is known to be a potent and specific inhibitor of P-gp, it is most likely that P-gp inhibition caused digoxin-valspodar interaction. In another clinical study, coadministration of verapamil caused an increase in the systemic

Table 1 Effect of Valspodar (400 mg PO) on the Absorption Kinetics of Digoxin (1 mg PO) in Healthy Volunteers

Parameters	Digoxin alone	Digoxin + single-dose valspodar
T_{max} (hr)	0.9	1.0
C_{max} (ng/mL)	1.01 ± 0.26	1.74 ± 0.51[a]
AUC (ng·hr/mL)	8.8 ± 2.3	15.3 ± 3.5[a]
CL_R (L/hr)	6.5 ± 2.5	2.3 ± 1.1[a]

Parameters are median values for T_{max} and mean values ± SD for all others.
[a]$p = 0.0001$ compared to control values.
Abbreviations: C_{max}, maximum measured drug concentration; T_{max}, time to reach C_{max}; AUC, area under the concentration-time curve; CL_R, renal clearance.
Source: Adapted from Ref. 41.

exposure of digoxin in humans in a dose-dependent manner. A daily dose of 160-mg verapamil caused a 40% increase in digoxin plasma concentrations, while a daily dose of 240-mg verapamil increased the digoxin plasma concentrations by 60–80% (42). Verapamil is also known to be a potent P-gp inhibitor. Therefore, the digoxin-verapamil interaction is highly likely due to P-gp inhibition. Interaction of digoxin with other P-gp inhibitors, such as quinidine and dipyridamole, has also been reported (43,44). The values of IC_{50} have been determined to be in the low micromolar range for PSC833, verapamil, quinidine, and dipyridamole, which are lower than the clinical concentrations of these inhibitors.

Talinolol, a good P-gp substrate, is eliminated from the body mainly by intestinal and renal excretion with minimal metabolism in humans. In a clinical study, a P-gp-mediated interaction between talinolol and verapamil has been reported (45). The inhibitory effect of verapamil on the intestinal secretion of talinolol was determined in six healthy volunteers by using the intestinal perfusion technique. While perfusing the small intestine with a verapamil-free solution, the mean intestinal secretion rate of talinolol was 4.0 μg/min after an intravenous dose of talinolol. The intestinal secretion rate decreased to 2.0 μg/min when a verapamil-containing solution was perfused (45). Similar to the clinical data, talinolol-verapamil interaction was also observed in rats. Coadministration of verapamil (4 mg/kg, PO) resulted in a 2.5- and 2.2-fold increase in the plasma oral AUC for *S*- and *R*-talinolol, respectively, after an oral dose of racemic talinolol in rats. On the other hand, after an intravenous dose of racemic talinolol, the inhibitory effect of verapamil on talinolol was less significant and there was only a 40% and 30% increase in AUC, respectively, for *S*- and *R*-talinolol (46). These results suggest that the larger increase in AUC of talinolol (~2- to 2.5-fold) after oral administration was likely due to the combination of an increase in oral absorption and a decrease in intestinal secretion, while the smaller increase in AUC (30–40%) after intravenous dosing was attributed mainly to a decrease in intestinal secretion.

A major challenge in the therapeutic treatment of cancer is the so-called multidrug resistance to anticancer drugs. Because over expression of P-gp has often been observed in tumor biopsies, it is believed that P-gp is one of the major factors responsible for the drug resistance, and inhibition of P-gp function may increase the sensitivity of cancer cells to anticancer drugs. The idea of reversing P-gp-mediated multidrug resistance has led to intensive efforts to develop potent P-gp inhibitors, such as PSC833 and GF120918. A significant increase in systemic exposures of paclitaxel (a P-gp substrate) in cancer patients was observed when used in combination with PSC833 or GF120918 (47,48). Similarly, both PSC833 and GF120918 caused an increase in systemic exposure of doxorubicin, a good P-gp substrate, associated with a decrease in doxorubicin clearance (49,50). It should be noted that although PSC833 and GF120918 are potent and specific P-gp inhibitors, their inhibitory activity against CYP enzymes is minimal. For example, the IC_{50} values for PSC833 to inhibit the functional activity of P-gp and CYP3A4 were 0.15 and 5 μM, respectively (51). Similarly, the IC_{50} values for LY-335979 (another potent and specific P-gp inhibitor) to inhibit the activity of P-gp transport and CYP3A4 were 0.024 and 5 μM, respectively (52). Taken together, it is clear that the drug interactions of paclitaxel and doxorubicin caused by PSC833 and GF120918 are likely due mainly to the inhibition of P-gp.

In addition to transporter inhibition, drug interactions caused by transporter induction have also been reported. Greiner et al. provided the most compelling evidence to date describing transporter induction as the cause of drug interactions (29). In a clinical study, the pharmacokinetics of digoxin before coadministration of rifampicin (600 mg/day for 10 days) was compared with those after rifampicin treatment in eight healthy volunteers. The plasma C_{max} and AUC of digoxin decreased from 5.4 ng/mL and 55 ng·hr/mL before rifampicin treatment to 2.6 ng/mL and 38 ng·hr/mL, respectively, after rifampicin treatment when the volunteers received a single dose of digoxin (Table 2). In this study, duodenal biopsies were obtained from each volunteer before and after administration of rifampicin. Rifampicin treatment increased intestinal P-gp content 3.5-fold, which correlated inversely with the oral AUC of digoxin. Taken together, these results strongly suggest that the digoxin-rifampicin interaction was mediated mainly by P-gp induction. This means that the decreased plasma concentration of digoxin during rifampicin treatment is caused by a combination of reduced bioavailability of digoxin as a result of P-gp induction. Similarly, administration of St. John's wort extract (3 doses of 300 mg/day for 14 days) resulted in a 1.2-fold decrease in the plasma AUC after a single digoxin dose (0.5 mg) in healthy volunteers (31). Consistent with the decrease in the AUC of digoxin, treatment with the St. John's wort extract resulted in a 1.4-fold increase

Table 2 Effect of Rifampicin (600 mg/day for 10 days) on the Absorption Kinetics of Digoxin (1 mg PO) in Healthy Volunteers

Parameters	Digoxin alone	Digoxin + rifampicin
T_{max} (hr)	0.7 ± 0.2	0.9 ± 0.3
C_{max} (ng/mL)	5.4 ± 1.9	2.6 ± 0.7^a
AUC (ng·hr/mL)	54.8 ± 11.6	38.2 ± 12.4^a
CL_R (L/hr)	9.5 ± 1.8	9.5 ± 2.3
F (%)	63 ± 11	44 ± 14^a

Mean values ± SD for all parameters.
[a] $p < 0.05$ compared to control.
Abbreviations: C_{max}, maximum measured drug concentration; T_{max}, time to reach C_{max}; AUC, area under the concentration-time curve; CL_R, renal clearance; F, bioavailability.
Source: Adapted from Ref. 29.

in the expression of duodenal P-gp. The decreased plasma AUC correlated reasonably well with an increased expression of intestinal P-gp.

The inductive effect of rifampicin on the pharmacokinetics of talinolol, which is eliminated from the body predominantly by renal and intestinal excretion with minimal metabolism (<1.5%), has been studied in normal volunteers (28). The bioavailability and oral plasma AUC of talinolol were decreased from 55% and 873 ng·hr/mL before rifampicin treatment to 35% and 565 ng·hr/mL, respectively, during rifampicin coadministration (600 mg/day for 9 days). On the other hand, the total clearance of talinolol was increased significantly by 30% after intravenous administration of the drug during rifampicin treatment. In addition, treatment with rifampicin resulted in a significant increase in the expression of duodenal P-gp content by about fourfold in these volunteers (28). The duodenal P-gp expression correlated significantly with the total clearance of talinolol. Since talinolol undergoes minimal metabolism, these results clearly demonstrated that the observed talinolol-rifampicin interaction was attributed mainly to a combination of a decrease in absorption and an increase in elimination via the induction of P-gp.

In conclusion, direct evidence of transporter-mediated drug interaction can be obtained relatively readily if a transporter substrate, such as digoxin or talinolol, undergoes minimal metabolism. In addition, transporter-mediated drug interactions can also be readily defined if potent and specific transporter inhibitors (or inducers), such as PSC833 and GF120918, are available.

B. Circumstantial Evidence

As aforementioned, the underlying mechanisms of many transporter-mediated drug interactions cannot be readily defined because of the interplay between drug transporters and CYP enzymes. In many cases, a transporter-mediated drug interaction was postulated simply on the basis of circumstantial evidence. The cerivastatin-cyclosporine interaction is a good example. In a clinical study, plasma concentrations of cerivastatin were determined after oral administration of 0.2-mg single dose of cerivastatin to 12 kidney transplant recipients who were receiving cyclosporine treatment. The AUC of cerivastatin (36.2 ng·hr/mL) in the kidney transplant recipients treated with cyclosporine was about fourfold higher than that in healthy volunteers (9.5 ng·hr/ml) who received the same oral dose level of cerivastatin without cyclosporine, while the elimination half-life (2.5 vs. 3.0 hours) was almost identical in both groups (53). Cerivastatin, a cholesterol lowering drug, is eliminated predominantly by CYP2C8- and CYP3A4-mediated metabolism in humans. Because cerivastatin is known to be almost completely absorbed after oral administration and because the elimination half-life was unaffected by cyclosporine, these investigators suggested that the increased AUC observed in transplant patients cannot be explained by CYP inhibition caused by cyclosporine. Together with the observation that the volume of distribution (V_c/F) appeared to be lower in the transplant patients compared with that in normal volunteers, they concluded that the cerivastatin-cyclosporine interaction is transporter mediated because of the inhibition of liver transport processes of cerivastatin by cyclosporine (53). Unfortunately, their conclusion is based on speculation without any supporting data.

To explore the underlying mechanisms for the cerivastatin-cyclosporine interaction, Shitara et al. (54) conducted a series of experiments. Their in vitro studies with human liver microsomes revealed that cyclosporine was not a potent inhibitor for cerivastatin metabolism with an IC_{50} greater than 50 μM. In contrast, studies with human hepatocytes suggested that cyclosporine was a potent inhibitor of cerivastatin (a substrate of OATP1B1) hepatic uptake with a K_i value of 0.3 μM. Similarly, cyclosporine was shown to inhibit the uptake of cerivastatin into the OATP1B1/MRP2 double-transfected MDCK cells with a K_i value of 0.2 μM (54). Taken together, Shitara et al. concluded that the cerivastatin-cyclosporine interaction was mainly due to inhibition of the hepatic uptake transporter, OATP1B1. However, when taking the peak plasma concentrations of cyclosporine (~1μM) at clinical doses and its unbound fraction in plasma (~5%) into consideration, the increase in the AUC of cerivastatin would be expected only to be moderate. Therefore, a fourfold increase in the AUC of cerivastatin cannot be explained by the inhibition of OATP1B1 alone. This argument is further supported by the observation that coadministration with cyclosporine only caused a modest increase in the plasma AUC of pravastatin (<1.5-fold), a good OATP1B1 substrate with minimal metabolism, in transplant patients (55).

In another clinical study, coadministration with gemfibrozil increased the plasma C_{max} and AUC of cerivastatin by three- and fivefold, respectively, in healthy volunteers (56). The C_{max} and AUC were 3.2 ng/mL and 20.9 ng·hr/mL, respectively, in the placebo phase and the corresponding values were 8 ng/mL and 91 ng·hr/mL in the gemfibrozil phase. Shitara et al. conducted in vitro metabolic and transport studies to explore the possible underlying mechanism for the cerivastatin-gemfibrozil interaction (57). Although OATP1B1 is known to be involved in the hepatic uptake of cerivastatin, gemfibrozil and its major *O*-glucuronide metabolite are not potent OATP1B1 inhibitors. The IC_{50} values for gemfibrozil and its *O*-glucuronide to inhibit OATP1B1-mediated uptake of cerivastatin were 72 and 24 μM, respectively. Because of the low inhibitory activity of gemfibrozil against OATP1B1, these investigators concluded that the interaction between gemfibrozil and cerivastatin was mainly due to the inhibition of CYP2C8-mediated metabolism, rather than the inhibition of OATP1B1-mediated transport. However, even for the CYP2C8-mediated metabolism of cerivastatin, gemfibrozil and its glucuronide are not potent inhibitors, the IC_{50} values for gemfibrozil and its *O*-glucuronide being 28 and 4 μM, respectively. Taking the plasma protein binding of gemfibrozil (>97%) into consideration, the unbound plasma concentrations of gemfibrozole and its glucuronide are expected lower than the IC_{50}. Therefore, the conclusion that the ceivastatin-gemfibrozil interaction is mainly due to the inhibition of CYP2C8-mediated metabolism is questionable.

A significant increase in plasma concentrations of pravastatin has also been reported when coadministered with gemfibrozil (58). Gemfibrozil increased the mean plasma AUC of pravastatin by twofold—from 139 ng·hr/mL in the absence of gemfibrozil to 281 ng·hr/mL in the presence of gemfibrozil. In parallel with a twofold increase in the AUC, there was a twofold decrease in the renal clearance of pravastatin from 417 mL/min to 233 mL/min. After intravenous administration of radiolabeled pravastatin to healthy volunteers, approximately 47% of total body clearance was via renal excretion and 53% by nonrenal routes, namely biliary excretion (59). Since gemfibrozil reduced the renal clearance of pravastatin by twofold, it is clear that the pravastatin-gemfibrozil interaction is due at least partly to the inhibition of renal transporters. However, the identities of the renal transporters have not been well characterized. In addition, gemfibrozil could also inhibit the hepatic transporters of pravastatin since biliary excretion is also a major route of pravastatin elimination. At least two transporters, OATP1B1 and MRP2, are known to be involved in the hepatobiliary elimination of pravastatin. Although gemfibrozil is not a potent inhibitor of OATP1B1 (57), there is no information regarding the inhibitory effect of gemfibrozil on MRP2 and other hepatic transporters. Therefore, the underlying mechanisms for the pravastatin-gemfibrozil interaction are still not fully understood.

Ambiguity also exists in the interpretation of the underlying mechanisms for the fexofenadine-rifampicin interaction. Pretreatment of rifampicin significantly decreased the systemic exposure of fexofenadine (a good P-gp substrate) in healthy volunteers (60). The C_{max} and AUC of fexofenadine were decreased by two- and threefold, respectively, in volunteers after rifampicin treatment (600 mg/day for 6 days). On the basis of the assumption that fexofenadine undergoes minimal metabolism in humans, the investigators concluded that the decreased plasma concentrations were the result of a reduced bioavailability caused by induction of intestinal P-gp. The assumption that fexofenadine is metabolized only to a minor extent in humans came originally from an abstract (61). In the abstract, it was stated that approximately 80% and 11% of an oral dose of [^{14}C]fexofenadine was recovered in the feces and urine, respectively, in a mass balance study in humans. Because MDL4829, the major metabolite of fexofenadine, accounted for about 20–30% of the radioactivity in urine (2–3% of the dose), it was concluded that fexofenadine is subject to minimal metabolism. However, it is unknown if the fecal component represents unabsorbed drug or the result of biliary and intestinal excretion. Moreover, information on the metabolite profiles in feces was not available. If the fraction of fexofenadine absorbed from the intestine is low (in the range of 10%) or if metabolites account for a significant fraction of radioactivity in the feces, the assumption that fexofenadine is subject to minimal metabolism in humans may not be valid. Therefore, it is possible that the observed fexofenadine-rifampicin interaction could be attributed to a combination of both CYP and P-gp induction. In addition, it has been demonstrated that organic anion transporting polypeptide (OATP1A2) is involved in the hepatic uptake of fexofenadine (62). It is possible that rifampicin is able to induce hepatic OATP1A2, resulting in an increased elimination clearance of fexofenadine. Thus, the involvement of OATP1A2 in the hepatobiliary excretion of fexofenadine may further complicate the interpretation of the observed fexofenadine-rifampicin interaction.

Similarly, the interpretation of the mechanism of the grapefruit juice–fexofenadine interaction may not necessarily be reasonable. In a clinical study, grapefruit juice or water at a volume of ‘‘1200 mL’’ was ingested within three hour after oral administration of 120-mg fexofenadine in a crossover study in 10 healthy subjects (63). Grapefruit juice decreased the plasma AUC of fexofenadine by approximately 2.5-fold compared with the corresponding volume of water. On the basis of the information that grapefruit juice inhibited the OATP1A2-mediated uptake in HeLa cells, these investigators concluded that the grapefruit juice–fexofenadine interaction was mainly due to the inhibition of OATP1A2 in the intestine (63). However, it should be noted that OATP1A2 is expressed at a very low level in the small intestine. In addition, it is still not known where OATP1A2 is located, either at the luminal or at the basolateral surface of intestinal epithelia cells (4). Recently, Satoh et al. (64) have shown that grapefruit juice significantly inhibited the uptake of substrates by human OATP2B1, which is highly expressed in the small intestine. In this regard, it is not known whether fexofenadine is a substrate for OATP2B1. Even though a direct inhibitory effect by grapefruit juice on intestinal OATP transporter(s) may partly explain the interaction, other mechanisms need to be considered. Ingestion of such an unusually large volume of grapefruit juice (1200 mL within 3 hour after drug dosing) may alter intestinal pH, osmolarity, gastric emptying time, and intestinal transit time of fexofenadine. Therefore, it is arguable that changes in gastrointestinal physiology may have indirect effects on the oral absorption of fexofenadine.

Because of the possible effects of the large volume of grapefruit juice on the gastrointestinal physiology, these investigators subsequently conducted a clinical study to evaluate the inhibitory effect of grapefruit juice on the absorption kinetics of fexofenadine at a more reasonable volume (300 mL) of grapefruit juice (65). The mean AUC value of fexofenadine decreased from 2167 ng·hr/mL after ingestion of 300 mL water to 1379 ng·hr/mL after 300 mL grapefruit juice. Interestingly, ingestion of grapefruit juice at 300 mL caused a much smaller decrease, approximately 30%, in the AUC of fexofenadine compared with a 2.5-fold decrease after ingestion of a large volume (1200 mL) of grapefruit juice (65). These results support the argument that a large volume of grapefruit juice could cause significant changes in gastrointestinal physiology and thereby complicate data interpretation.

Many HIV protease inhibitors are known to be good substrates for both CYP3A4 and P-glycoprotein (66). Ketoconazole is a potent inhibitor not only for CYP3A4 but also for human P-gp. The K_i value of ketoconazole to inhibit the metabolism of midazolam in human liver microsomes was determined to be 0.015 μM (67). The IC_{50} value of ketoconazole to inhibit digoxin transport in Caco-2 cells was determined to be 1.4 μM (52). An interaction between ketoconazole and HIV protease inhibitors, ritonavir and saquinavir, has been reported in AIDS patients (68). Coadministration of ketoconazole resulted in significant increases in the plasma and CSF concentration of ritonavir and saquinavir, while ketoconazole had little effect on the plasma protein binding of the drugs. These results suggest that ketoconazole inhibits the functional activity of both CYP3A4 and P-gp in humans. The ratio of CSF drug concentration to plasma unbound drug concentration (C_u) for saquinavir and ritonavir was increased from 0.06 and 0.09, respectively, in the absence of ketoconazole to 0.35 and 0.26 with ketoconazole treatment (68).

The CSF/C_u ratio of much less than unity is consistent with the notion that ritonavir and saquinavir are good P-gp substrates. If ketoconazole only inhibits CYP3A4 and does not alter the distribution of the drugs across the brain, the increase in plasma concentrations of ritonavir and saquinavir is expected to cause a proportional increase in CSF concentrations, meaning a relatively constant CSF/C_u concentration ratio of the drugs before and after treatment with ketoconazole. A three- to fivefold increase in the CSF/C_u concentration ratios indicated that the increase in CSF concentration was more than proportional to the increase in C_u. These results strongly suggest that ketoconazole may also have a significant effect on the function of P-gp. Similar observations were also observed for the interaction between ketoconazole and nelfinavir, an HIV protease inhibitor, which is eliminated predominantly by CYP3A enzymes, in mice. Coadministration of ketoconazole (50 mg/kg, IV) caused an eightfold increase in brain levels of nelfinavir and a threefold increase in plasma concentrations in mdr1a(+/+) mice (52). Since nelfinavir is a good P-gp substrate, the greater increase in brain concentration compared with plasma concentration strongly suggests the inhibition of P-gp in the BBB. However, it is important to emphasize that it is still quite difficult to quantitatively estimate the relative contribution of the inhibition of CYP3A and P-gp to overall interactions, even in animal studies.

The involvement of both CYP3A4 and P-gp in drug interactions has also been reported for the rifampicin-cyclosporine interaction (69). The pharmacokinetics of cyclosporine were studied in six healthy volunteers after oral and intravenous administration of the drug before and after rifampicin pretreatment (600 mg/day for 11 days). Blood clearance of cyclosporine increased from 5 mL/min/kg before rifampicin treatment to 7 mL/min/kg during rifampicin treatment, while the bioavailability decreased from 27% before rifampicin treatment to 10% after rifampicin treatment. Rifampicin not only increased the elimination clearance of cyclosporine but also decreased its bioavailability to a greater extent than would have been predicted by the increased clearance. Because cyclosporine is a good substrate for both CYP3A4 and P-gp and rifampicin can induce both the expression of CYP3A4 and P-gp, the increased clearance and decreased bioavailability of cyclosporine during rifampicin treatment is most likely due to a combination of CYP3A4 and P-gp induction. Although attempts have been made to estimate the relative contribution of CYP3A4 and P-gp to the overall interaction (70), there is still no simple way by which this can be quantified, because of the complexity of the interplay between intestinal and hepatic CYP3A4 and P-gp.

As the examples cited above indicate, many clinical drug interactions have been considered to be mediated by inhibition or induction of transporters based only on circumstantial evidence. Because of the lack of potent and specific inhibitors for each transporter, it is difficult to accurately assess the relative contributions CYP enzymes and transporters in drug absorption and excretion. The mechanisms become even more complex when multiple CYP enzymes and drug transporters are involved in the processes of drug absorption and excretion. Therefore, care should be taken when exploring the underlying mechanism of drug interactions.

IV. CLINICAL IMPLICATIONS OF TRANSPORTER-MEDIATED DRUG INTERACTIONS

An important question that is often asked is how significant the transporter-mediated drug interactions would likely be in clinical settings. It is of interest to note that the magnitude of changes in plasma concentrations of drugs caused by transporter-mediated interactions is generally much smaller than that by CYP-mediated drug interactions. As shown in Table 1, coadministration of valspodar (PSC833), a potent P-gp inhibitor, caused only a 75% increase in the plasma AUC of digoxin. In contrast, coadministration of ketoconazole, a potent CYP3A4 inhibitor, caused a 16-fold increase in the AUC of midazolam in healthy

Table 3 Effect of Ketoconazole (400 mg PO) on the Absorption Kinetics of Midazolam (7.5 mgPO) in Healthy Volunteers

Parameters	Midazolam alone	Midazolam + ketoconazole
T_{max} (hr)	1.3 ± 0.4	2.5 ± 0.5[a]
C_{max} (ng/mL)	22 ± 6	90 ± 7[b]
AUC (ng·hr/mL)	65.0 ± 10.0	1033 ± 100[c]
$t_{1/2}$ (hr)	2.8 ± 0.6	8.7 ± 1.0[b]

Mean values ± SD for all parameters.
[a] $p < 0.05$ compared to control.
[b] $p < 0.001$ compared to control.
[c] $p < 0.005$ compared to control.
Abbreviations: C_{max}, maximum measured drug concentration; T_{max}, time to reach C_{max}; AUC, area under the concentration-time curve; $t_{1/2}$, terminal half-life.
Source: Adapted from Ref. 71.

Table 4 Effect of Rifampicin (600 mg/day for 10 days) on the Absorption Kinetics of Midazolam (15 mg PO) in Healthy Volunteers

Parameters	Midazolam alone	Midazolam + rifampicin
T_{max} (hr)	1	1.25
C_{max} (ng/mL)	55 ± 4	3.5 ± 0.7[a]
AUC (ng·hr/mL)	170.0 ± 13.3	7.0 ± 0.8[a]
$t_{1/2}$ (hr)	3.1 ± 0.2	1.3 ± 0.2[a]

Parameters are median values for T_{max} and mean values ± SD for all others.
[a] $p < 0.001$ compared to control.
Abbreviations: C_{max}, maximum measured drug concentration; T_{max}, time to reach C_{max}; AUC, area under the concentration-time curve; $t_{1/2}$, terminal half-life.
Source: Adapted from Ref. 72.

volunteers (71). The AUC of midazolam increased from 65 ng·hr/mL in the absence of ketoconazole to 1033 ng·hr/mL in the presence of ketoconazole (Table 3). Similarly, pretreatment with rifampicin caused only a 30% decrease in the plasma AUC of digoxin (Table 2). However, pretreatment with rifampicin resulted in a 24-fold decrease in the AUC of midazolam in normal subjects (72). The AUC of midazolam decreased from 170 ng·hr/mL before rifampicin treatment to 7.0 μM·min/mL during rifampicin treatment (Table 4). It should be noted that midazolam is eliminated in humans exclusively by CYP3A4-mediated metabolism and that it is not a P-gp substrate. Conversely, digoxin is a good P-gp substrate, but is not a CYP substrate in humans.

However, from animal studies it becomes evident that the changes in tissue (intracellular) concentrations of drugs caused by inhibition of transporters are much greater than the corresponding changes in plasma concentrations. For example, coadministration of LY335979, a potent and specific P-gp inhibitor, caused a 37-fold increase in the brain concentrations of nelfinavir, but to only a twofold increase in plasma concentrations in wild-type mdr1a mice (52). Similarly, pretreatment with the potent P-gp inhibitor GF120918 led to a 13-fold increase in brain levels of amprenavir, but only a modest increase (2-fold) in plasma concentrations of amprenavir in mdr1a(+/+) mice (73). In another study, a 10-fold increase in brain concentrations of digoxin was observed, when PSC833 (50 mg/kg) was given orally to mdr1a(+/+) mice (74). Similar to the brain, the placenta is also very sensitive to P-gp inhibitors. Potent P-gp inhibitors, such as PSC833 and GF120918, were able to completely block the placental P-gp function in mice. The concentrations of digoxin and saquinavir in the wild-type [mdr1a/1b(+/+)] fetus were increased to levels that were comparable to that in the mdr1a/1b(−/−) fetus after oral coadministration of the P-gp inhibitors to heterozygous mothers (75). The notion that inhibition of transporters has a much greater impact on the distribution of drugs into tissues than on plasma concentrations is further supported by studies with transporter-deficient (transgenic) mice. For example, the digoxin concentration in brain were about 28-fold higher in mdr1a/1b(−/−) mice compared with those in wild-type mice, while there was only a 2.5-fold difference in plasma concentrations between mdr1a/1b(−/−) and wild-type mice (76). Similarly, the hepatic uptake of TEA was sixfold lower in Oct1 knockout mice (Oct1−/−) compared with the wild-type (Oct1+/+) mice, while there was only a slight difference in plasma concentration between (Oct1−/−) and wild-type mice following an intravenous administration of [^{14}C]TEA at a dose of 0.2 mg/kg (77).

One important lesson learned from these animal studies is that transporter inhibition has a much greater impact on the tissue distribution of drugs, particularly with regard to the brain, than on the systemic exposure of drugs. Hence, the potential risk of transporter-mediated drug interactions might be underestimated if only plasma concentration is monitored. Therefore, one should carefully assess the potential risk of transporter-mediated drug interactions when potent transporter inhibitors are administered together.

V. CONCLUSION

Although the importance of drug transporters in the processes of drug ADME has been widely recognized, our understanding of their functional role in drug ADME is still at a very early stage due to the complexity of polarized expression of multiplicity of transporters. With recent advances in molecular biology and biotechnology, the number of documented transporters continues to grow in an exponential manner, although the majority of the newly identified transporters are still not fully characterized. Therefore, there is a long way to go before we can fully understand the physiological function of all the drug transporters and their interplay in relation to drug absorption and disposition.

The molecular complexity of transporter inhibition and induction impairs our ability to predict the potential role of transporter-mediated drug-drug interactions, either in a quantitative or qualitative sense. In addition, the involvement of multiple transporters and CYP enzymes in drug absorption and disposition further complicates the interpretation of transporter-mediated drug interactions. Until the relative contribution of the role of transporters and CYP enzymes to overall drug interaction can be quantitatively estimated, care should be taken when exploring or interpreting the underlying mechanism of drug interactions. Another important lesson learned from animal studies is that transporter inhibition has a much greater impact on the tissue distribution of drugs, particularly with regard to the brain, than on the systemic exposure of drugs measured in plasma. The potential risk of transporter-mediated drug interactions might be underestimated if only plasma concentrations are monitored. Therefore, one should carefully assess the potential risk of transporter-mediated drug interactions when potent inhibitors of transporters are administrated.

REFERENCES

1. Lin JH, Yamazaki M. Role of P-glycoprotein in pharmacokinetic: clinical implications. Clin Pharmacokinet 2003; 42:59–98.
2. Lin JH, Yamazaki M. Clinical relevance of P-glycoprotein in drug therapy. Drug Metab Rev 2003; 35:417–454.
3. Kim RB. Transporters and xenobiotic disposition. Toxicology 2002; 181–182:291–297.
4. Ito K, Suzuki H, Horie T, et al. Apical/basolateral surface expression of drug transporters and its role in vectorial drug transport. Pharm Res 2005; 22:1559–1577.
5. Lin JH, Lu AYH. Inhibition and induction of cytochrome P450 and the clinical implications. Clin Pharmacokinet 1998; 35:361–390.
6. Lin JH. CYP induction-mediated drug interactions: In vitro assessment and clinical implications. Pharm Res 2006; 23:1089–1116.
7. Lin JH. Drug-drug interaction mediated by inhibition and induction of P-glycoprotein. Adv Drug Deli Rev 2003; 55:53–81.
8. Tamai I, Safa AR. Azidopine noncompetitively interacts with vinblastine and cyclosporin A binding to P-glycoprotein in multidrug resistant cells. J Biol Chem 1991; 266:16796–16800.
9. Ramachandra M, Ambudkar SV, Chen D, et al. Human P-glycoprotein exhibits reduced affinity for substrates during a catalytic transition state. Biochemistry 1998; 37:5010–5019.
10. Senior AE, Al-Shawi MK, Urbatsch IL. The catalytic cycle of P-glycoprotein. FEBS Lett 1995; 377:285–289.
11. Loo TW, Bartlett MC, Clarke DM. Substrate-induced conformational changes in the transmembrane segments of human P-glycoprotein. Direct evidence for the substrate-induced fit mechanism for drug binding. J Biol Chem 2003; 278:13603–13606.
12. Loo TW, Bartlett MC, Clarke DM. Simultaneous binding of two different drugs in the binding pocket of the human multidrug resistance P-glycoprotein. J Biol Chem 2003; 278:39706–39710.

13. Critchfield JW, Welsh CJ, Phang JM, et al. Modulation of adriamycin accumulation and efflux by flavonoids in HCT-15 colon cells. Biochem Pharmacol 1994; 48: 1437–1445.
14. Shapiro AB, Ling V. Positively cooperative sites for drug transport by P-glycoprotein with distinct drug specificities. Eur J Biochem 1997; 250:130–137.
15. Zelcer N, Huisman MT, Reid G, et al. Evidence for two interacting ligand binding sites in human multidrug resistance protein 2 (ATP binding cassette 2). J Biol Chem 2003; 278:23538–23544.
16. Bakos E, Evers R, Sinko E, et al. Interaction of human multidrug resistance proteins MRP1 and MRP2 with organic anions. Mol Pharmacol 2000; 57:760–768.
17. Tamai I, Nozawa T, Koshida M, et al. Functional characterization of human organic anion transporting polypeptide B (OATP-B) in comparison with liver-specific OATP-C. Pharm Res 2001; 18:1262–1269.
18. Batrakova EV, Li S, Li Y, et al. Effect of pluronic P85 on ATPase activity of drug efflux transporters. Pharm Res 2004; 21:2226–2233.
19. Yang R, Cui L, Hou Y, et al. ATP binding to the first nucleotide binding domain of multidrug resistance-associated protein plays a regulatory role at low nucleotide concentration, whereas ATP hydrolysis at the second plays a dominant role in ATP-dependent leukotriene C4 transport. J Biol Chem 2003; 278:30764–30771.
20. Sauna ZE, Ambudkar SV. Characterization of the catalytic cycle of ATP hydrolysis by human P-glycoprotein: the two ATP hydrolysis events in a single catalytic cycle are kinetically similar but affect different functional outcomes. J Biol Chem 2001; 276:11653–11661.
21. Trauner M, Boyer JL. Bile salt transporters: molecular characterization, function and regulation. Physiol Rev 2002; 83:633–671.
22. Kipp H, Pichetshote N, Arias IM. Transporters on demand: Intracellular pools of canalicular ATP binding cassette transporters in rat liver. J. Biol. Chem. 2001; 276:7218–7224.
23. Mottino AD, Cao J, Veggi LM, et al. Altered localization and activity of canalicular Mrp2b in estradiol-17-β-D-glucuronide-induced cholestasis. Hepatology 2002; 35:1409–1419.
24. Magnarin M, Morelli M, Rosati A, et al. Induction of proteins involved in multidrug resistance (P-glycoprotein, MRP1, MRP2, LRP) and of CYP3A4 by rifampicin in LLC-K1 cells. Eur J Pharmacol 2004; 483:19–28.
25. Geick A, Eichelbaum M, Burk O. Nuclear receptor response elements mediate induction of intestinal MDR1 by rifampin. J Biol Chem 2001; 276:14581–14587.
26. Bock KW, Eckle T, Ouzzine M, et al. Coordinate induction of antioxidants by UDP-glucuronosyltransferase UGT1A6 and the apical conjugate export pump MRP2 (multidrug resistance protein 2) in Caco-2 cells. Biochem Pharmacol 2000; 59:467–470.
27. Kast HR, Goodwin B, Tarr PT, et al. Regulation of multidrug resistance-associated protein 2 (ABCC2) by the nuclear receptors pregnane X receptor, farnesoid X-activated receptor, and constitutive androstane receptor. J Biol Chem 2002; 277:2908–2915.
28. Westphal K, Weinbrenner A, Zschiesche M, et al. Induction of P-glycoprotein by rifampin increases intestinal secretion of talinolol in human beings: a new type of drug-drug interaction. Clin Pharmacol Ther 2000; 68:345–355.
29. Greiner B, Eichelbaum M, Fritz P, et al. The role of intestinal P-glycoprotein in the interaction of digoxin and rifampin. J Clin Invest 1999; 104:147–153.

30. Fromm MF, Kaufmann HM, Fritz P, et al. The effect of rifampin treatment on intestinal expression of human MRP transporters. Am Pathol 2000; 157:1575–1580.
31. Durr D, Stieger B, Kullak-Ublick GA, et al. St John's wort induces intestinal P-glycoprotein/MDR1 and intestinal and hepatic CYP3A4. Clin Pharmarcol Ther 2000; 68:598–604.
32. Shibayama Y, Ikeda R, Motoya T, et al. St John's wort (Hypericum perforatum) induces overexpression of multidrug resistance protein 2 (MRP2) in rats: a 30-day ingestion study. Food Chem Toxicol 2004; 42:995–1002.
33. Fardel O, Lecureur V, Guillouzo A. Regulation by dexamethasone of P-glycoprotein expression in cultured rat hepatocytes. FEBS Lett 1993; 327:189–193.
34. Lin JH, Chiba M, Chen IW, et al. Effect of dexamethasone on the intestinal first-pass metabolism of indinavir in rats: evidence of cytochrome P-450 3A and P-glycoprotein induction. Drug Metab Dispos 1999; 27:1187–1193.
35. Kubitz R, Warskulat U, Schmitt M, et al. Dexamethasone- and osmolarity-dependent expression of the multidrug-resistance protein 2 in cultured rat hepatocytes. Biochem J 1999; 340:585–591.
36. Kauffmann HM, Schrenk D. Sequence analysis and functional characterization of the 5'-flanking region of the rat multidrug resistance protein 2 (MRP2) gene. Biochem Biophys Res Commun 1998; 245:325–331.
37. Courtois A, Payen L, Guillouzo A, et al. Up-regulation of multidrug resistance-associated protein 2 (MRP2) expression in rat hepatocytes by dexamethasone. FEBS Lett 1999; 459:381–385.
38. Anapolsky A, Teng S, Dixit S, et al. The role of pregnane X receptor in 2-acetylaminofluorene-mediated induction of drug transport and metabolizing enzymes in mice. Drug Metab Dispos 2006; 34:405–409.
39. Ebert B, Seidel A, Lampen A. Identification of BCRP as transporter of benzo[α] pyrene conjugates metabolically formed in Caco-2 cells and its induction by Ah-receptor agonists. Carcinogenesis 2005; 26:1754–1763.
40. Hagenbuch N, Reichel C, Stieger B, et al. Effect of Phenobarbital on the expression of bile salt and organic anion transporters of rat liver. J Hepatol 2001; 34:881–887.
41. Kovarik JM, Rigaudy L, Guerret M, et al. Longitudinal assessment of a P-glycoprotein-mediated interaction of valspodar on digoxin. Clin Pharmacol Ther 1999; 66:391–400.
42. Verschraagen M, Koks CHW, Schellens JHM, et al. P-glycoprotein system as a determinant of drug interactions: the case of digoxin-verapamil. Pharmacol Res 1999; 40:301–306.
43. Mordel A, Halkin H, Zulty L, et al. Quinidine enhancers digitalis toxicity at therapeutic serum digoxin levels. Clin Pharmacol Ther 1993; 53:457–462.
44. Verstuyft C, Strabach S, El Morabet H, et al. Dipyridamole enhances digoxin bioavailability via P-glycoprotein inhibition. Clin Pharmacol Ther 2003; 73:51–60.
45. Gramatte T, Oertel R. Intestinal secretion of intravenous talinolol is inhibited by luminal R-verapamil. Clin Pharmacol Ther 1999; 66:239–245.
46. Spahn-Langguth H, Baktir G, Radschuweit A, et al. P-glycoprotein transporters and the gastrointestinal tract: evaluation of the potential in vivo relevance of in vitro data employing talinolol as model compound. Int J Clin Pharmacol Ther 1998; 36:16–24.
47. Malingre MM, Beijnen JH, Rosing H, et al. Co-administration of GF120918 significantlyincreases the systemic exposure to oral paclitaxel in cancer patients. Br J Cancer 2001; 84:42–47.

48. Fracasso PM, Westerveldt P, Fears CA, et al. Phase I study of paclitaxel in combination with a multidrug resistance modulator, PSC833 (valspodar), in refractory malignancies. J Clin Oncol 2000; 18:1124–1134.
49. Advani R, Fisher GA, Lum BL, et al. A phase I trial of doxorubicin, paclitaxel, and valspodar (PSC833), a modulator of multidrug resistance. Clin Cancer Res 2001; 7:1221–1229.
50. Sparreboom A, Planting AS, Jewell RC, et al. Clinical pharmacokinetics of doxorubicin in combination with GF120918, a potent inhibitor of MDR1 P-glycoprotein. Anticancer Drugs 1999; 10:719–728.
51. Achira M, Suzuki H, Ito K, et al. Comparative studies to determine the selective inhibitors for P-glycoprotein and cytochrome P4503A4. AAPS PharmSci 1999; 1(4):E18.
52. Choo EF, Leake B, Wandel C, et al. Pharmacological inhibition of P-glycoprotein transport enhances the distribution of HIV-1 protease inhibitors into brain and testes. Drug Metab Dispos 2000; 28:655–660.
53. Muck W, Mai I, Fritsche L, et al. Increase in cerivastatin systemic exposure after single and multiple dosing in cyclosporine-treated kidney transplant recipients. Clin Pharmacol Ther 1999; 65:251–261.
54. Shitara Y, Itoh T, Sato H, et al. Inhibition of transporter-mediated hepatic uptake as a mechanism for drug-drug interaction between cerivastatin and cyclosporine A. J Pharmacol Exp Ther 2003; 304:61–616.
55. Regazzi MB, Iacona I, Campana C, et al. Altered disposition of pravastatin following concomitant drug therapy with cyclosporine A in transplant recipients. Transplant Proc 1993; 25:2732–2734.
56. Backman JT, Kyrklund C, Neuvonen , et al. Gemfibrozil greatly increases plasma concentrations of cerivastatin. Clin Pharmacol Ther 2002; 72:685–691.
57. Shitara Y, Hirano M, Sato H, et al. Gemfibrozil and its glucuronide inhibit the organic anion transporting polypeptide 2 (OATP2/OATP1B1: SLC21A6)-mediated hepatic uptake and CYP2C8-mediated metabolism of cerivastatin: analysis of the mechanism of the clinically relevant drug-drug interaction between cerivastatin and gemfibrozil. J Pharmacol Exp Ther 2004; 311:228–236.
58. Kyrklund C, Backman JT, Neuvonen M, et al. Gemfibrozil increases plasma pravastatin concentrations and reduces pravastatin renal clearance. Clin Pharmacol Ther 2003; 73:538–544.
59. Pravachol® package insert. New Jersey: Bristol-Myers Squibb Company, 2004.
60. Hamman MA, Bruce MA, Haehner-Daniels BD, et al. The effect of rifampin administration on the disposition of fexofenadine. Clin Pharmacol Ther 2001; 69:114–121.
61. Lippert C, Ling J, Brown P, et al. Mass balance and pharmacokinetics of MDL16,455A in healthy male volunteers. Pharm Res 1999; 12:S390 (abstr).
62. Cvetkovic M, Leake B, Fromm MF, et al. OATP and P-glycoprotein transporters mediate the cellular uptake and excretion of fexofenadine. Drug Metab Dispos 1999; 27:866–871.
63. Dresser GK, Bailey DG, Leake BF, et al. Fruit juices inhibit organic anion transporting polypeptide-mediated drug uptake to decrease the oral availability of fexofenadine. Clin Pharmacol Ther 2002; 71:11–20.
64. Satoh H, Yamashita F, Tsujimoto M, et al. Citrus juices inhibit the function of human organic anion-transporting polypeptide OATP-B. Drug Metab Dispos 2005; 33:518–523.

65. Dresser GK, Kim RB, Bailey DG. Effect of grapefruit juice volume on the reduction of fexofenadine bioavailability: possible role of organic anion transporting polypeptides. Clin Pharmacol Ther 2005; 77:170–177.
66. Lee CGL, Gottesman MM. HIV-1 protease inhibitors and the MDR1 multidrug transporter. J Clin Invest 1998; 101:287–288.
67. Gibbs MA, Thummel KE, Shen DD, et al. Inhibition of cytochrome P450 3A (CYP3A) in human intestinal and liver micromoses: comparison of K_i values and impact of CYP3A5 expression. Drug Metab Dispos 1999; 27:180–187.
68. Khaliq Y, Gallicano K, Venance S, et al. Effect of ketoconazole on ritonavir and saquinavir concentrations in plasma and cerebrospinal fluid from patients infected with human immunodeficiency virus. Clin Pharmacol Ther 2000; 68:637–646.
69. Hebert MF, Roberts JP, Prueksaritanont T, et al. Bioavailability of cyclosporine with concomitant rifampin administration is markedly less than predicted by hepatic enzyme induction. Clin Pharmcol Ther 1992; 52:453–457.
70. Wacher VJ, Silverman JA, Zhang Y, et al. Role of P-glycoprotein and cytochrome P450 3A in limiting oral absorption of peptides and peptidomimetics. J Pharm Sci 1998; 87:1322–1330.
71. Olkkola KT, Backman JT, Neuvonen PJ. Midazolam should be avoided in patients receving the systemic antimycotics ketoconazole or itraconazole. Clin Pharmacol Ther 1994; 55:481–485.
72. Backman JT, Olkkola KT, Neuvonen PJ. Rifampin drastically reduces plasma concentrations and effects of oral midazolam. Clin Pharmacol Ther 1996; 59:7–13.
73. Polli JW, Jarrett JL, Studenberg SD, et al. Role of P-glycoprotein on CNS disposition of amprenavir (141W94), an HIV protease inhibitor. Pharm Res 1999; 16: 1206–1212.
74. Mayer U, Wagenaar E, Dorobek B, et al. Full blockade of intestinal P-glycoprotein and extensive inhibition of blood-brain barrier P-glycoprotein by oral treatment of mice with PSC833. J Clin Invest 1997; 100:2430–2436.
75. Smit JW, Huisman MT, van Tellingen O. Absence or pharmacological blocking of placental P-glycoprotein profoundly increases fetal drug exposure. J Clin Invest 1999; 104:1441–1447.
76. Schinkel AH, Mayer U, Wagenaar E, Mol CAAM, van Deemter L, Smit JJM, van der Valk MA, Voordouw AC, Sprits H, van Tellingen O, Zijlmans JMJM, Fibbe WE, Borst P. Normal viability and altered pharmacokinetics in mice lacking mdr1-type (drug-transporting) P-glycoproteins. Proc Natl Acad Sci USA 1997; 94:4028–4033.
77. Jonker JW, Wagenaar F, Mol CA, Buitelaar M, Koepsell H, Smit JW, Schinkel AH. Reduced hepatic uptake and intestinal excretion of organic cations in mice with a targeted disruption of the organic cation transporter 1 (Oct1 [Slc22a1]) gene. Mol Cell Biol 2001; 21:5471–5477.

13

Metabolism and Transport Drug Interaction Database: A Web-Based Research and Analysis Tool

Houda Hachad, Isabelle Ragueneau-Majlessi, and René H. Levy

Department of Pharmaceutics, University of Washington, Seattle, Washington, U.S.A.

I. INTRODUCTION

It has become apparent that the most effective approach to decrease the interaction potential of a drug candidate is to address it during the earliest stages of drug development. Guidances were issued in 1997 and 1999 by regulatory agencies in the United States (1,2) and Europe (3), outlining the need for in vitro and in vivo studies for new chemical entities and the possibility of providing "class labeling." Recently, a Web site (4) was released by the FDA with the current agency's understanding of how to conduct drug-interaction studies and resulting labeling. Moreover, a new draft guidance, including additional discussions on emerging areas (transporters, nuclear receptors, etc.), is being finalized. For scientists in academia, regulatory agencies, and the pharmaceutical industry, building a drug interaction (DI) program requires rapid access to vast literature sets on metabolic isozymes, transporters, substrates, inducers, and inhibitors. That type of information is spread within a large body of literature that is relatively recent but expanding at a fast pace.

The Metabolic and Transport Drug Interaction Database (DIDB) (http://www.druginteractioninfo.org), developed by the Drug Interaction Prediction group at the University of Washington, includes one of the largest set of comprehensive data pertaining to DIs in humans. The DIDB was designed to serve as a tool for researchers and clinicians interested in correlating in vitro and in vivo findings on interactions associated with metabolic enzymes (phase I and II) and transporters. A description of the database, examples of queries, and sample outputs are presented in this chapter.

II. DATABASE DESIGN AND CONTENT

An earlier version of the DIDB was described in Chapter 14 of the previous edition of this book (5). The new DIDB application launched in 2005 has a typical multitier architecture in a Microsoft® .NET environment. The back end is a Microsoft SQL Server 2000 and the current application is deployed on a Web farm. Currently, the database has data extracted from more than 6280 published articles (1966 to 2007) related to drug metabolism and DIs and 260 product labels (1998 to 2007). The use of the Web facilitates worldwide access as well as upgrades and updates; the DIDB is updated daily.

The unit of information is the original research article. Detailed records are generated from each research article, highlighting study results as well as experimental conditions; the data extracted from each article are structured in the database according to a defined hierarchy. For example, relevant information collected from in vitro studies pertain to the role of particular metabolic enzymes in the various metabolic pathways of substrates and the inhibition and induction spectra of drugs toward metabolic enzymes. Particular attention is paid to experimental conditions used in the determination of enzyme kinetic parameters, including K_m, K_i, IC_{50}, K_I, and k_{inact}. In vivo studies include pharmacokinetic studies with blood level measurements, pharmacokinetic-pharmacodynamic studies, as well as case reports. In addition to research articles, the DIDB team has built original excerpts from product label of recently approved drugs (1998–2007) in the United States, available from the FDA Web site (6).

III. EXAMPLES OF QUERIES AND OUTPUT

The DIDB search interface utilizes a list of queries. Queries are structured along intuitive themes such as drug, enzyme, therapeutic class, transporter, and thus allow the user to quickly select the appropriate queries without the need for extensive training. The eight sets of queries can be categorized into qualitative or quantitative queries as shown below:
Qualitative:

Drug	Search by drug name, using generic names
Enzyme	Search by enzyme name
Therapeutic class	Search by therapeutic class
Transporters	Search by transporter name
Other	Search for articles using journal or author name. Search for side effects, pharmacodynamic effects, and more

Quantitative:

Area under the curve (AUC)/Clearance (CL) changes	Search for AUC and CL changes observed in DI studies
In vitro parameters	Search for parameters K_m and V_{max}, K_i, IC_{50}, and K_I and k_{inact}
Pharmacokinetics	Search for pharmacokinetic parameters of selected drugs

The only terminology specific to the DIDB pertains to drugs/compounds that appear as object or precipitant depending on their role in specific DI. *Object* refers to a compound that acts as the modified agent (i.e., substrate), and *precipitant* refers to a compound that acts as the causative agent. A precipitant can be an inhibitor, inducer, or activator, but may not have any effect.

The following section describes three examples of the use of DIDB, highlighting the three-step logic used to perform a search:

1. Defining the drug-drug interaction (DDI) issue (background and question)
2. Selecting the query (or queries) (data examined from different perspectives)
3. Analyzing the result output

A. Example 1: Inhibitors of CYP2B6

1. Background

Contrary to earlier beliefs, the role of CYP2B6 in drug metabolism has grown in recent years and progress has been achieved in characterizing in vivo and in vitro substrates and inhibitors, allowing further studies of this enzyme. CYP2B6 is expressed in 20–100% of the population and represents 3–5% of total hepatic CYP content (7). There is a wide (from 20- to 288-fold) individual variability in CYP2B6 protein expression and enzyme activity due to environmental factors and polymorphisms of the CYP2B6 gene (8,9). CYP2B6 is responsible for the metabolism of about 3% of drugs (10), and the interindividual difference in CYP2B6 catalytic capacity may result in variable systemic exposure of substrates, including the antineoplastics cyclophosphamide and ifosfamide (11), the antiretrovirals nevirapine (12) and efavirenz (13), the anesthetics propofol (14) and ketamine (15), and the anti-Parkinson agent selegiline (16). However, it remains difficult to evaluate the in vivo activity of CYP2B6 in humans because of the lack of specific substrates. When the in vitro biotransformation of bupropion to hydroxybupropion was studied in human liver microsomes and microsomes containing heterologously expressed human cytochromes P450 (CYP), hydroxybupropion formation was found to be mediated almost exclusively by CYP2B6 and was proposed as an index reaction to assess the activity of this isoform (17,18). In the September 2006 draft guidance on DIs, the FDA

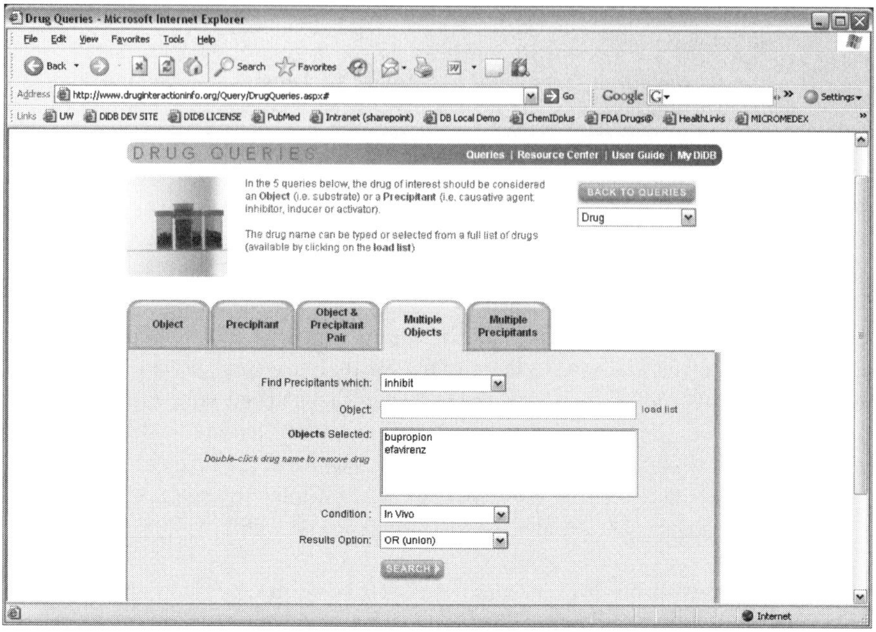

Figure 1 Drug query labeled "multiple objects" used to find in vivo inhibitors of CYP2B6. Display from the Metabolism and Transport Drug Interaction Database (http://www.druginteractioninfo.org, accessed October 2006).

recommended the use of efavirenz as a probe substrate to study CYP2B6 activity both in vivo and in vitro (4). In human liver microsomes, efavirenz undergoes hydroxylation to 8-hydroxyefavirenz (major in vivo) and 7-hydroxyefavirenz (minor) and secondary metabolism to 8,14-dihydroxyefavirenz. Correlation studies and incubation with specific inhibitors indicate that CYP2B6 is the principal catalyst of efavirenz sequential 8- and then 14-hydroxylation (13).

2. Question

Consider the case of a drug candidate for which in vitro studies indicate that it is a substrate of CYP2B6 and the user needs to identify usable in vivo inhibitors of this enzyme.

3. Appropriate Queries

A first query will address *effective inhibitors in vivo* of CYP2B6, using the specific probe substrate approach (probes substrates are drugs metabolized primarily by one enzyme that become useful tools for studying the inhibition/induction of that enzyme).

The query used is labeled "multiple objects" and is available under the section "Drug Queries" (Fig. 1). The objects used for the search are *bupropion* and *efavirenz*.

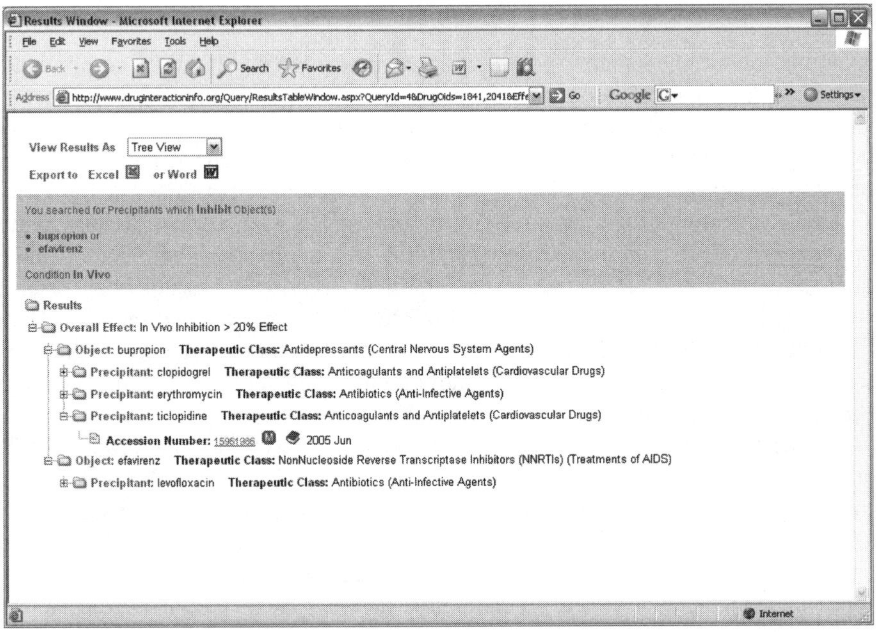

Figure 2 List of precipitants evaluated with the substrates bupropion and efavirenz and which have shown more than 20% effect in in vivo inhibition. Display from the Metabolism and Transport Drug Interaction Database (http://www.druginteractioninfo.org, accessed October 2006).

4. Result Output

The display shown on Figure 2 has an alphabetical list of four precipitants (clopidogrel, erythromycin, ticlopidine, and levofloxacin) that have been evaluated with the substrates bupropion and efavirenz and which have shown more than 20% effect in in vivo inhibition. Each precipitant (inhibitor) in the list has its own folder containing more detailed information:

Accession Number of the source article (matches PMID number); by clicking directly on this number, the full description of the article is retrieved (experimental conditions, results, design, etc.).

Abstract of the article is visualized with the M button.

Reference PK parameters for the drug are retrieved by clicking on the button.

There are several options of displaying the results in a table and performing filter operations as well as exporting capabilities into Microsoft Excel or Microsoft Word.

5. Analyzing the Result

Examination of the extent of inhibitors associated with the four compounds identified above (clopidogrel, erythromycin, ticlopidine, and levofloxacin) shows that only clopidogrel and ticlopidine significantly inhibited the CYP2B6-catalyzed bupropion hydroxylation in vivo. Clopidogrel reduced the AUC ratio of hydroxybupropion/bupropion by 68%, while ticlopidine reduced it by 90%. The corresponding increases in AUC of bupropion were 60% and 85% with clopidogrel and ticlopidine, respectively (19). As a side note, the results of this query also indicate that based on the literature, efavirenz has not been extensively studied as an in vivo marker of CYP2B6. The product label indicates that the compounds examined for their inhibitory effects on efavirenz are: indinavir, lopinavir/rit, nelfinavir, ritonavir, saquinavir, azithromycin, clarithromycin, fluconazole, paroxetine, sertraline, ethinyl estradiol, famotidine, voriconazole, and cetirizine (20). Only ritonavir and voriconazole had noticeable effects: 21% and 44% increase in AUC, respectively. At present, it appears that the list of inhibitors that could be used in an in vivo DI program with the candidate CYP2B6 substrate includes clopidogrel and ticlopidine.

To ascertain those results, the user can examine in vitro inhibitors using the following enzyme query: *Find precipitants which inhibit CYP2B6 in vitro* (Fig. 3).

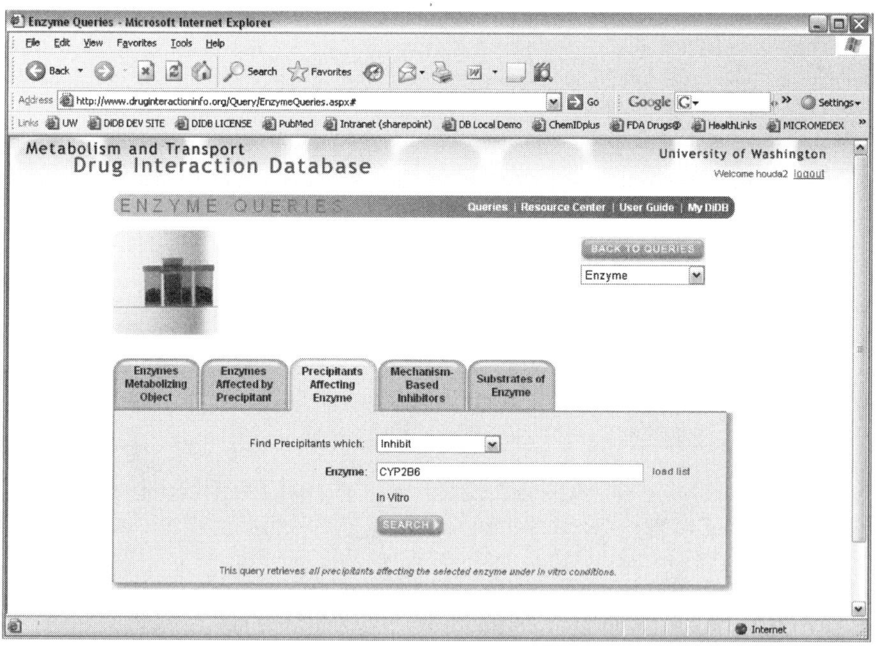

Figure 3 Enzyme query used to find precipitants that inhibit CYP2B6 in vitro. Display from the Metabolism and Transport Drug Interaction Database (http://www.druginteractioninfo.org, accessed October 2006).

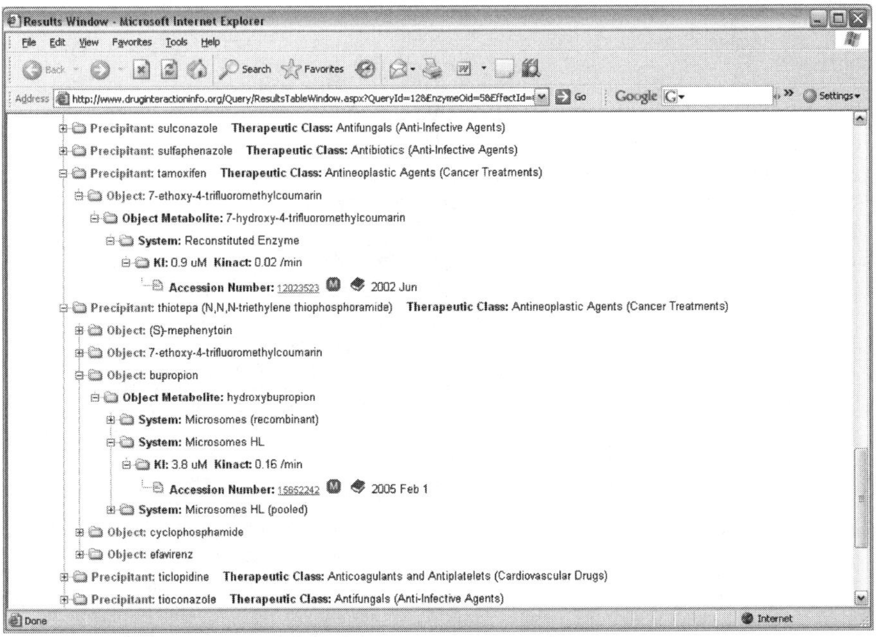

Figure 4 List of compounds that had some inhibitory effects toward CYP2B6 in vitro as measured by the parameters K_i, IC_{50}, K_I, and k_{inact} values. Display from the Metabolism and Transport Drug Interaction Database (http://www.druginteractioninfo.org, accessed October 2006).

6. Result Output

The result of this query, as shown in Figure 4, lists several compounds that had some inhibitory effects toward CYP2B6 in vitro as measured by the parameters K_i, IC_{50}, K_I, and k_{inact} values.

7. Interpretation

Inhibitors obtained from in vitro data include a number of compounds with different selectivities and specificity toward this enzyme. Most of them have not been tested for their in vivo effects and some may also inhibit other enzymes (ticlopidine, fluvoxamine, miconazole, nefazodone, paroxetine, etc.). This query did not yield any additional inhibitors. A summary of substrates (including partial ones) and inhibitors is shown in Table 1 and indicates that clopidogrel and ticlopidine, whose effects have been well documented in vivo and in vitro, are effective inhibitors of CYP2B6 and can be used in an in vivo DI program.

Table 1 Selected CYP2B6 Substrates and Inhibitors

Substrate[a]	Inhibitors[b]
S-Mephenytoin	**Clopidogrel**
S-Mephobarbital	Clotrimazole
3,4-Methylenedioxyamphetamine	Efavirenz
7-Benzyloxyresorufin	*Ethinyl estradiol*
7-Ethoxy-4-trifluoromethylcoumarin	Fluvoxamine
7-Pentoxyresorufin	Glabridin
All-*trans*-retinoic acid	Lasofoxifene
Artemisinin	Memantine
β-Arteether	Methoxsalen
Bupropion	Metyrapone
Clotiazepam	Miconazole
Cyclophosphamide	Montelukast
Dexloxiglumide	Nefazodone
Ecstasy	*Nelfinavir*
Efavirenz	Norfluoxetine
α-Endosulfan	Orphenadrine
Fenproporex	*Paroxetine*
Ifosfamide	Phencyclidine
Ketamine	Phenethyl isothiocyanate
Levomethadyl	**Ritonavir**
Meperidine	Selegiline
Methadone	*Sertraline*
Mexiletine	Sulconazole
Nevirapine	Tamoxifen
Nicotine	Thiotepa
Piclamilast	**Ticlopidine**
Propofol	Tioconazole
Selegiline	Tranylcypromine
Tamoxifen	ε-Viniferin
Testosterone	
Thiotepa	
Tramadol	

[a]Substrates shown to be at least partially metabolized by CYP2B6.
[b]*Bold*: Inhibitor tested in vivo and in vitro with bupropion or efavirenz.
Italic: In vitro Inhibitors with no effects in vivo on efavirenz.

B. Example 2: CYP3A Classification of Inhibitors: Evaluation of Substrate Independence

1. Background

A proposal of classification of CYP3A inhibitors was put forth, recently, in which the fold increase in AUC of a single oral dose of midazolam allows discrimination between strong, moderate, and weak inhibitors (21). A proposal for an extension of the midazolam-based classification of CYP3A inhibitors is included in the September 2006 draft of FDA guidance for drug-drug interaction studies (4) based on the assumption that the midazolam-based classification is independent of the *sensitive CYP3A substrate* used [i.e. sensitive CYP3A substrate refers to a drug whose AUC increased fivefold or higher when coadministered with known CYP3A inhibitor (4)].

2. Question

Is the classification of an inhibitor independent of the sensitive CYP3A substrate used?

3. Appropriate Queries

Using the "Changes in AUC or CL" query shown in Figure 5 and the ability to select (1) multiple objects (in our example, midazolam and simvastatin) and (2) the inhibitors tested with BOTH drugs (intersection), the AUC changes of the substrates observed with shared inhibitors can be easily displayed.

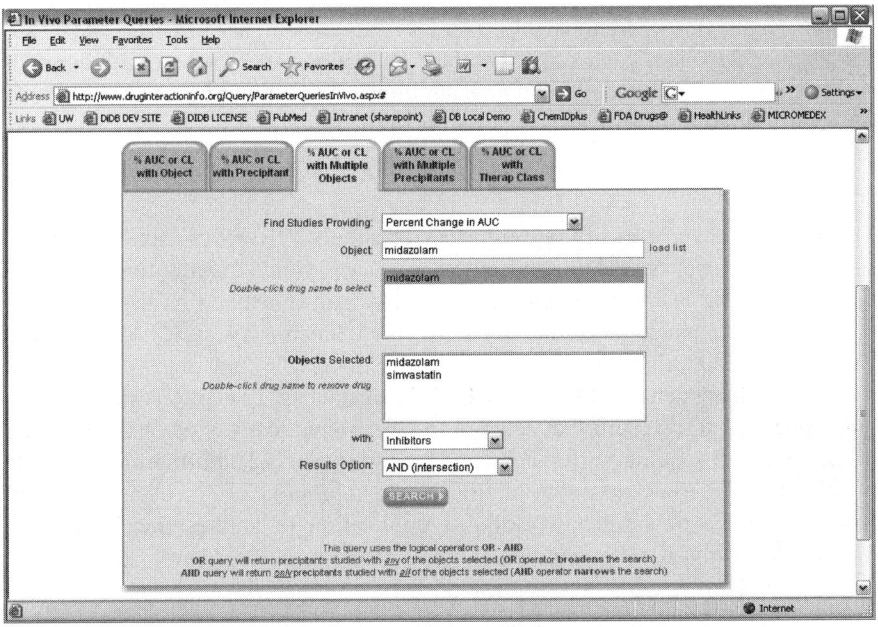

Figure 5 Query labeled "% AUC or CL with Multiple Objects" used to retrieve the change in AUC of simvastatin and midazolam with shared inhibitors (operator AND will return only precipitants studied with BOTH objects selected). Display from the Metabolism and Transport Drug Interaction Database (http://www.druginteractioninfo.org, accessed October 2006).

Table 2 Fold Increases in AUC of Simvastatin and Midazolam Observed with Shared CYP3A Inhibitors

	AUCi/AUCc[a] Simvastatin (SIM)	AUCi/AUCc Midazolam (MID)	Ratio SIM over MID
Diltiazem	4.81	3.75	1.28
Erythromycin	6.21	3.81	1.63
Clarithromycin	9.96	7.00	1.42
Itraconazole	11.00	6.64	1.66
Verapamil	4.65	2.92	1.59
Cyclosporine	2.60	1.00	2.60
Telithromycin	8.51	6.11	1.39
Ketoconazole	12.552	9.51	1.32
GFJ double strength	13.50	5.93	2.28
GFJ regular strength	3.60	2.40	1.5

[a]AUCi is the AUC in presence of inhibitor and AUCc is the AUC control.

4. Result Output

In the case of simvastatin and midazolam, nine shared inhibitors, tested at a comparable dose regimen, are available. The fold increases in the substrates' AUC are summarized in Table 2 and show that the AUC ratios are consistently higher with simvastatin than midazolam for all nine inhibitors, suggesting a higher sensitivity to CYP3A inhibition for simvastatin [mean of AUC ratios of 1.67 (1.28–2.60)].

5. Interpretation

This approach can be easily reproduced with other CYP3A-sensitive substrates, such as buspirone, triazolam, and eplerenone, and yields quantitative examination of the relationship between these sensitive substrates and midazolam. This question was pursued in some depth by Ragueneau-Majlessi et al. (22) and showed that classification of inhibitors (five moderate and eight potent) was substrate independent in 74% to 83% of the instances. Exceptions pertained to buspirone and simvastatin that seemed to be systematically more sensitive than midazolam and saquinavir that appeared less sensitive (22). Furthermore, results of this analysis allow an extrapolation of the inhibitory effect of a compound from one probe to another avoiding a duplication of studies (i.e., effects of inhibitor on simvastatin calculated from its effect on midazolam).

IV. ONGOING DEVELOPMENTS

The DIDB is a tool that is constantly being optimized as a result of feedback from a large base of users, including requests for specific searches. These can be in the format of new queries or special reports tailored by the DIDB team. New features currently being developed include the following.

A. Product Labels in Searchable Format

Until now, information on the DI profile of newly approved drugs (2000–2007), extracted from product labels was available as printable PDF summaries. The DIDB system has been recently modified to allow the inclusion of data from other sources such as DI studies described only in product labels and/or clinical reviews [both available from the FDA Web site (6)].

B. Information on Potential Risk for Drugs to Increase QTc Interval, Leading to Cardio Toxicity

In view of the clinical importance of the corrected Q–T interval (QTc) prolongation in drug development, the DIDB team has created a new set of data summarizing QTc effects of drugs. In addition, 150 publications already in the database were recently reviewed to highlight the articles with QTc measurement and cases of QTc prolongation. Case reports of torsades de pointes/ventricular fibrillation were also individualized for easy access by DIDB users.

C. In Vitro Data Related to Transporters

Transporter-based interactions have been increasingly documented. The DIDB currently includes a set of queries that allow retrieval of in vivo DI studies involving transporters such as P-glycoprotein (P-gp), organic anion transporting polypeptide (OATP), organic cation multidrug resistance-associated proteins (MRP), etc.; articles with in vitro basis supporting transporter-related DI will also be included in the next DIDB version.

REFERENCES

1. Food and Drug Administration. Guidance for Industry: Drug Metabolism/Drug Interaction Studies in the Drug Development Process, Studies In Vitro. Department of Health and Human Services U.S. Food and Drug Administration. Center for Drug Evaluation and Research. Center For Biologics Evaluation and Research. April 1997. Available at: http://www.fda.gov/cder/guidance.htm/. Accessed October 2006.
2. Food and Drug Administration. Guidance for Industry: In Vivo Drug Metabolism/Drug Interaction Studies—Study Design, Data Analysis, and Recommendations for Dosing and Labeling U.S. Department of Health and Human Services. Food and Drug Administration. Center for Drug Evaluation and Research (CDER). Center for Biologics Evaluation and Research (CBER). November 1999. Available at: http://www.fda.gov/cder/guidance.htm/. Accessed October 2006.
3. Guidance on the Investigation of Drug Interactions. Committee for Proprietary Medicinal Products. The European Agency for the Evaluation of Medicinal Products. Human Medicines Evaluation Unit. December 1997. Available at: http://www.emea.eu.int. Accessed October 2006.

4. Food and Drug Administration. Guidance for Industry: Drug Interaction Studies—Study Design, Data Analysis, and Implications for Dosing and Labeling. Draft Guidance (released September 2006). Available at: http://www.fda.gov/cder/guidance/6695dft.pdf. Accessed October 2006.
5. Carlson SP, Ragueneau-Majlessi I, Levy RH. Development of a metabolic drug interaction database at the University of Washington. In: Rodrigues AD, ed. Drugs and the Pharmaceutical Sciences, vol. 116. New York: Marcel Dekker Inc, 2002:549–563.
6. Food and Drug Administration. Available at: http://www.accessdata.fda.gov/scripts/cder/drugsatfda/. Accessed October 2006.
7. Ekins S, VandenBranden M, Ring BJ, et al. Examination of purported probes of human CYP2B6. Pharmacogenetics 1997; 7(3):165–179.
8. Ekins S, Vandenbranden M, Ring BJ, et al. Further characterization of the expression in liver and catalytic activity of CYP2B6. J Pharmacol Exp Ther 1998; 286(3):1253–1259.
9. Hesse LM, He P, Krishnaswamy S, et al. Pharmacogenetic determinants of interindividual variability in bupropion hydroxylation by cytochrome P450 2B6 in human liver microsomes. Pharmacogenetics 2004; 14(4):225–238.
10. Rendic S. Summary of information on human CYP enzymes: human P450 metabolism data. Drug Metab Rev 2002; 34(1–2):83–448.
11. Roy P, Yu LJ, Crespi CL, et al. Development of a substrate-activity based approach to identify the major human liver P-450 catalysts of cyclophosphamide and ifosfamide activation based on cDNA-expressed activities and liver microsomal P-450 profiles. Drug Metab Dispos 1999; 27(6):655–666.
12. Erickson DA, Mather G, Trager WF, et al. Characterization of the in vitro biotransformation of the HIV-1 reverse transcriptase inhibitor nevirapine by human hepatic cytochromes P-450. Drug Metab Dispos 1999; 27(12):1488–1495.
13. Ward BA, Gorski JC, Jones DR, et al. Cytochrome P450 2B6 (CYP2B6) is the main catalyst of efavirenz primary and secondary metabolism: implication for HIV/AIDS therapy and utility of efavirenz as a substrate marker of CYP2B6 catalytic activity. J Pharmacol Exp Ther 2003; 306(1):287–300.
14. Court MH, Duan SX, Hesse LM, et al. Cytochrome P-450 2B6 is responsible for interindividual variability of propofol hydroxylation by human liver microsomes. Anesthesiology 2001; 94(1):110–119.
15. Yanagihara Y, Kariya S, Ohtani M, et al. Involvement of CYP2B6 in N-demethylation of ketamine in human liver microsomes. Drug Metab Dispos 2001; 29(6):887–890.
16. Hidestrand M, Oscarson M, Salonen JS, et al. CYP2B6 and CYP2C19 as the major enzymes responsible for the metabolism of selegiline, a drug used in the treatment of Parkinson's disease, as revealed from experiments with recombinant enzymes. Drug Metab Dispos 2001; 29(11):1480–1484.
17. Hesse LM, Venkatakrishnan K, Court MH, et al. CYP2B6 mediates the in vitro hydroxylation of bupropion: potential drug interactions with other antidepressants. Drug Metab Dispos 2000; 28(10):1176–1183.
18. Faucette SR, Hawke RL, Lecluyse EL, et al. Validation of bupropion hydroxylation as a selective marker of human cytochrome P450 2B6 catalytic activity. Drug Metab Dispos 2000; 28(10):1222–1230.
19. Turpeinen M, Tolonen A, Uusitalo J, et al. Effect of clopidogrel and ticlopidine on cytochrome P450 2B6 activity as measured by bupropion hydroxylation. Clin Pharmacol Ther 2005; 77(6):553–559.

20. Sustiva® Package Insert. US Prescribing Information; Bristol-Myers Squibb Company; http://www.bms.com/. Accessed October 2006.
21. Bjornsson TD, Callaghan JT, Einolf HJ, et al. Pharmaceutical Research and Manufacturers of America (PhRMA) Drug Metabolism/Clinical Pharmacology Technical Working Group; FDA Center for Drug Evaluation and Research (CDER). The conduct of *in vitro* and *in vivo* drug-drug interaction studies: a Pharmaceutical Research and Manufacturers of America (PhRMA) perspective. Drug Metab Dispos 2003; 31(7):815–832.
22. Ragueneau-Majlessi I, Boulenc X, Rauch C, et al. Quantitative correlations among CYP3A sensitive substrates and inhibitors: Literature Analysis. Current Drug Metabolism 2007; 8 (in press).

14

In Vivo Probes for Studying Induction and Inhibition of Cytochrome P450 Enzymes in Humans

Grant R. Wilkinson[†]
Vanderbilt University School of Medicine, Nashville, Tennessee, U.S.A.

The discovery of cytochrome P450 (EC1.14.14.1, CYP) in the early 1960s resulted in an explosion of knowledge about the monooxygenase system that still continues today. However, even before this critical event, it was apparent that the oxidative metabolism of drugs often exhibits large interindividual variability; moreover, drug-metabolizing activity may be modulated by environmental, pathophysiological, and genetic factors (1). Research during the subsequent four decades has largely focused on determining the mechanisms involved in such variability and, in the case of drug metabolism in humans, its clinical significance and importance. In certain situations, genotyping with respect to the presence of allelic variants can be of some value in accounting for this interindividual variability, especially, if a strong genetic determinant is involved (2,3). However, even when genetic polymorphism is present, considerable variability is often present within a phenotypic group (2); moreover, genotyping cannot take into account the modulation of catalytic activity by environmental and disease-state factors. In vitro approaches using tissue preparations, e.g., liver microsomes and recombinant expressed enzymes, have considerable merit in this

[†]Deceased.

regard (see Chaps. 2, 3, and 7). However, the application of such invasive procedures to the clinical situation is obviously limited, especially when studying healthy subjects. Accordingly, so-called "noninvasive" procedures, utilizing readily available fluids, such as plasma and saliva, or excretions, such as urine and expired air, form the basis for measuring in vivo metabolizing ability. These measures are generally applied to two related types of experimental questions: What is the basal level of catalytic activity in an individual subject, i.e., phenotyping? What are the determinants of interindividual variability within or between populations, e.g., the effects of drug and environmental interactions, genetics, and disease states? The use of "model" compounds or, as currently termed, in vivo probes, has been extensively applied for these purposes since its conception some 30 years ago (4). This chapter considers the rationale, development, validation, and application of currently useful in vivo probes to assess the catalytic activity of specific human CYP isoforms in individual subjects.

I. ENDOGENOUS COMPOUNDS AS IN VIVO PROBES

By analogy to the use of creatinine clearance as an indicator of kidney function and the renal excretion of drugs, attempts have been made to identify an endogenous compound that could be used to assess drug-metabolizing activity. The plasma levels of γ-glutamyltransferase and bilirubin as markers of hepatic dysfunction and the urinary excretion of endogenous 6β-hydroxycortisol and D-glucaric acid have been sporadically investigated for this purpose over the years (5,6). With the exception of 6β-hydroxycortisol, these approaches have proven fruitless, but even the measurement of the hydroxysteroid's excretion has limitations. 6β-Hydroxycortisol is a minor metabolite of cortisol that is subsequently excreted unchanged in the urine; changes in adrenal corticoid generation, rather than formation of the metabolite, are accounted for by expressing 6β-hydroxycortisol excretion relative to that of cortisol. A number of drugs, e.g., rifampin and anticonvulsant agents, increase the excretion of this metabolite, consistent with the induction of a CYP-mediated pathway (5,6). CYP3A appears to be the major enzyme involved in the formation of 6β-hydroxycortisol (7,8); therefore, it has been inferred that changes in the metabolite's excretion reflect modulation of this isoform. Unfortunately, studies to investigate the relationship between 6β-hydroxycortisol excretion and the basal level of CYP3A activity using other in vivo probes, such as the erythromycin breath test and midazolam's hydroxylation (see sec. VIII), have been consistently unsuccessful (9–12). Furthermore, troleandomycin—a mechanism-based inhibitor of CYP3A—was not found to consistently affect the urinary excretion of endogenous 6β-hydroxycortisol, despite the fact that erythromycin breath test was markedly affected by such pretreatment (9). Collectively, these data raise serious questions regarding the nature and interpretation of any measured increase in urinary 6β-hydroxycortisol excretion over its basal level. It does not appear to reflect hepatic CYP3A alone, and possibly the localization of the isoform in other organs such as the kidney and intestinal epithelium may be contributory to 6β-hydroxycortisol's overall urinary excretion (9,13). Regardless, the current status of this endogenous probe would appear to be limited to its use as a relatively nonspecific indicator of enhanced oxidative metabolism following pretreatment with a putative inducing agent (14–16). As such, any change in the 6β-hydroxycortisol:cortisol ratio requires further investigation with respect to the specific CYP isoform(s) affected, the magnitude of induction, and in the case of CYP3A, whether this occurs in the liver and/or the intestinal epithelium and possibly other extrahepatic tissues.

II. EXOGENOUS COMPOUNDS AS IN VIVO PROBES

A. Desirable Phenotypic Trait Characteristics

Beginning with the use of antipyrine (4), administration of a model drug to quantitatively assess oxidative drug-metabolizing activity has been an important experimental tool. A number of compounds have been investigated for this purpose. However, with the recognition that CYP is a multigene super family of related heme-thiolate proteins with separate but potentially overlapping substrate specificities, the goal in recent years has been to develop and use specific probe drugs for the individual isoforms. The major effort has focused on isoforms of the CYP1, CYP2, and CYP3 families, since in humans these appear to be responsible for the metabolism of most drugs and other xenobiotics.

Although the liver is the major organ involved in CYP-mediated drug metabolism, other tissues, including the intestinal epithelium, kidney, and additional organs, all have a similar potential, depending on the individual isoform. However, available in vivo probes provide a collective estimate of the measured catalytic function within the body. That is, assessment of activity by an individual organ is usually not possible, despite the fact that this may be critical to interpreting the phenotyping result. For example, it is not unreasonable to suggest that CYP2E1 and CYP2A6 localization within the lung is more important than the isoforms' hepatic levels in lung carcinogenesis resulting from the metabolic activation of environmental chemicals. However, such a level of refinement is not currently possible.

Following administration of an in vivo probe, an experimental measure characterizing the enzyme's functional activity is obtained. Ideally, this phenotypic trait should exclusively reflect the catalytic activity of a single pathway of metabolism mediated by the isoform of interest. In practice, evidence of such absolute specificity is difficult, if not impossible, to obtain in vivo, so the trait measure should be considered a primary, rather than an exclusive, reflection of the isoform's activity. It is also desirable that the trait measure be sensitive to changes/differences in the enzyme's catalytic activity produced, for example, by a drug interaction or a genetic factor. Unless this characteristic is present, small changes/differences in activity will not be recognized. Additionally, differences in enzyme activity should ideally result in a linear change in the phenotypic value so that discrimination between values is readily interpretable. If other relationships are present, e.g., the rectangular hyperbola associated with debrisoquine's urinary metabolic ratio (MR) used for phenotyping CYP2D6 activity (sec. VI), large differences in the trait value do not necessarily reflect comparable differences in catalytic activity. It also makes common sense and is esthetically more satisfying if the value of the trait measure increases with an increase in catalytic activity. From a practical standpoint, the phenotyping procedure and associated chemical analyses should be robust so that reliable and reproducible results are obtained regardless of the particular circumstances under which the testing is performed. This is especially important in the case of field studies, where a relatively sophisticated research environment and facilities may not be present, especially with regard to analysis of the biological sample. Also, from the perspective of the individual being phenotyped, it is important that the involved procedure be simple, rapid, and as noninvasive as possible. A further practical issue relates to the availability of the in vivo probe and its measured metabolite in the specific locale where the study is being conducted. Ideally, the in vivo probe should be available and approved for clinical use worldwide, and the phenotypic trait should not be affected by the particular dosage form used. In practice, regulatory factors sometimes limit the use of a specific in vivo probe; e.g., sparteine is not approved by the Food and Drug Administration in the United States and, therefore, has not been used for investigating CYP2D6 activity in that country.

B. Pharmacokinetic Basis of Phenotypic Traits

1. Plasma Clearance Values

The most appropriate and closest in vivo measure of an enzyme's catalytic activity is a drug's intrinsic clearance ($CL_{int,u}$) in terms of its unbound concentration in the plasma (17). At low drug concentrations relative to the enzyme's K_m value, i.e., first-order conditions, this may be viewed as the ratio of the apparent V_{max} to the K_m value describing metabolism by an individual enzyme (see Chaps. 1 and 2). Intrinsic clearance associated with hepatic drug metabolism is closely reflected in a drug's clearance following oral administration (CL_o) rather than by intravenous or any other route; moreover, this applies regardless of the rate-limiting process of hepatic clearance, i.e., intrinsic clearance or hepatic blood flow (17). In addition, any CYP-mediated metabolism occurring in the intestinal epithelium, e.g., CYP3A, is also reflected in the first-pass effect following an oral dose. Accordingly, the "gold standard" approach to estimating the level of metabolic activity of a drug-metabolizing enzyme is to express the ratio of the total area under the plasma concentration–time curve (AUC_o) for unbound drug to the administered oral dose (D_o) in order to provide an estimate of oral clearance (Eq. (1)). In practice, the fraction of the drug unbound in plasma (f_u) and any changes/differences are rarely taken into account, but they may be a factor when this parameter is altered. Thus, the estimated intrinsic clearance (CL_o) is generally based on total drug concentrations.

$$\frac{D_o}{AUC_o} = CL_o \sim CL_{int} = f_u \cdot CL_{int,u} \tag{1}$$

If more than one metabolic pathway is present and several different enzymes are involved, this must be taken into account by "partitioning" the intrinsic clearance value according to the relative contributions of each enzyme/pathway (17,18). This is generally accomplished by correcting the total oral clearance by the fraction of the administered dose that is metabolized along the pathway of interest (f_{m1}) on the basis of the total amount of the individual metabolite and any associated secondary metabolites excreted in urine ($A_{e,ml}$, where m1 represents a single route of metabolism) (Eq. (2)):

$$CL_o = f_{m1} \cdot CL_o = \frac{A_{e\infty,m1}}{D_o} \cdot CL_o \tag{2}$$

The major advantage of (fractional) oral clearance as a phenotypic trait is that its value is linearly related to the enzyme's catalytic activity, provided that first-order conditions are present. This requirement, along with any safety considerations, is the main reason the dose of an in vivo probe should be as low as possible, consistent with analytical considerations. Furthermore, it is possible to directly extrapolate this type of trait measure to the disposition of other drugs whose metabolism is mediated by the measured enzyme and also to place the trait value within a therapeutic context. On the other hand, estimation of oral clearance requires multiple blood and urine collections, often over many hours, that are an inconvenience for the study subject and require considerable amounts of analytical time and effort. Because of this, simpler and less time-consuming approaches have often been used. However, it is not always appreciated that such phenotyping tests provide only an indirect measure of metabolizing activity and may be affected by factors other than the enzyme's intrinsic clearance. In addition, it is difficult to relate an indirect trait measure to parameters that are of clinical importance, such as the drug's clearance.

2. Urinary MRs

Because urinary MR requires only a single measure involving the relative excretion of unchanged drug and metabolite, it is the commonest indirect approach used for characterizing drug-metabolizing ability. It was first applied with respect to debrisoquine in order to identify individuals with absent or low CYP2D6 activity, i.e., poor metabolizers (PMs) (see sec. VI.B). This type of trait value expresses the ratio of the amount of unchanged drug, e.g., debrisoquine, to that of metabolite formed by the isoform of interest, e.g., 4-hydroxydebrisoquine, excreted in a fixed period, e.g., zero to eight hours, following administration of a single oral dose of the in vivo probe. Such a ratio provides the greatest discrimination between PMs and extensive metabolizers (EMs), since impaired metabolism results in a large numerator and a small denominator that provide a multiplier effect. As a result, the higher the metabolic activity, the smaller the trait value, and the relationship is a rectangular hyperbola. More important is the fact that a urinary MR is an entirely empirical trait measure that reflects pharmacokinetic factors besides that of the enzyme's intrinsic clearance, in particular, the renal clearance of the in vivo probe (18). In the case of debrisoquine, not only is this latter value relatively large but is time dependent, decreasing from a value approximating renal plasma flow to that of the glomerular filtration rate during the zero- to eight-hour urine collection period (19). An alternative approach for calculating a urinary MR is to express the amount of metabolite excreted over a fixed time period relative to the sum of this amount and that of unchanged drug (20,21). This relative recovery ratio (RR) approach, which is similar in form to the fractional urinary recovery of the metabolite, has the attraction that the trait value increases with increasing metabolic activity; however, its value also reflects the probe's renal clearance (18). In addition, interpretation of such approaches is critically dependent on the completeness of the urine collection; an incomplete specimen will result in underestimation of metabolizing ability.

With certain drugs that have been used as in vivo probes, e.g., coumarin, chlorzoxazone, and mephenytoin, essentially no drug is excreted unchanged into the urine; thus, an MR approach cannot be used. In the case of mephenytoin (see sec. V.B.1), a ratio expressing the administered oral dose of *S*-mephenytoin to the amount of 4′-hydroxymephenytoin formed by CYP2C19 over a fixed period after dosing—the hydroxylation index (HI)—has been used to discriminate between PMs and EMs (22,23). This relationship is similar to the inverse of the urinary recovery of metabolite; therefore, the trait value reflects not only the intrinsic clearance of the isoform of interest but, in addition, the intrinsic clearance values associated with other metabolic pathways and enzymes (18). Finally, it must be recognized that in the presence of renal dysfunction, the validity of a urinary MR is highly questionable because of the trait value's dependency on the renal clearance of the in vivo probe and/or its measured metabolite (24).

3. Plasma MRs

The urinary MR should provide similar information to that of the ratio of the plasma AUC values of the in vivo probe drug and metabolite over the collection time period, but it has the practical advantage of a single urine collection, compared with multiple plasma samples. In order to simplify a plasma-based trait measure, the plasma concentration ratio of metabolite to unchanged drug at a single time point after drug administration has also been used, e.g., plasma 6-hydroxychlorzoxazone:chlorzoxazone ratio as a measure of CYP2E1 activity

(25,26). In this situation, choice of an appropriate sampling time is critical. Sufficient time must elapse for a significant amount of metabolism to have occurred. But if the selected time is too soon after drug administration (1–2 hour), the MR will reflect the absorption characteristics of the probe drug in addition to its metabolism. This may vary not only between individuals but also between different manufacturers' products, especially for a generic drug. Sampling during the elimination phase obviates this potential problem; for metabolites whose elimination is formation limited, this represents the best approach, since the ratio should be a constant value over this time period (27). On the other hand, when removal from the body of the metabolite is elimination limited, e.g., 6-hydroxy-chlorzoxazone (28), then the plasma levels of drug and metabolite do not decline in parallel. As a result, the MR depends not only on the formation of the involved metabolite but also on its elimination relative to that of the parent drug (27). Unfortunately, such pharmacokinetic considerations rarely appear to be appreciated when such trait measures are developed.

4. CO_2 Breath Tests

N-Demethylation is a common CYP-mediated metabolic pathway, and the resulting formaldehyde subsequently enters the one-carbon pool and appears ultimately in the exhaled breath as CO_2. Labeling of the methyl group of an appropriate in vivo probe with either ^{14}C or ^{13}C and measurement of the expired radio- or stable-isotope, therefore, provides an index of the N-demethylation process. This is the basis of a number of CO_2 breath tests that have been used to evaluate in vivo drug metabolism. Experimental approaches include both the assessment of the complete time course of exhalation of labeled CO_2 in the breath as well as a more limited sampling including a single time point determination. Moreover, the involved quantitative collection procedure for expired CO_2 is now relatively simple; in certain instances, e.g., erythromycin breath test, a simple kit including all needed items is commercially available. Not only is a CO_2 breath test noninvasive, other than for drug administration, but by using a stable-labeled in vivo probe, studies may be performed in young infants and pregnant women, contrary to the situation when a radiolabel is used, with its associated radiation exposure. A breath test is generally sensitive and reproducible but suffers a major disadvantage in that it is an indirect measure of the responsible N-demethylating enzyme and a number of potentially limiting assumptions are involved (29).

The pathway from N-demethylation to exhaled CO_2 involves a number of steps many of which are also metabolic, but catalyzed by several enzyme systems distinct from the initiating CYP isoform. A critical assumption is that all steps prior to (e.g., absorption for a probe that is not administered parenterally) or subsequent to N-demethylation are not rate limiting. In general, the steps involved in intermediary metabolism of the one-carbon pool meet this criterion. However, a percentage, estimated to be between 37% and 57%, of labeled

N-demethylated groups is lost in transit through the one-carbon pool, and this exhibits interindividual variability (29), which is generally not accounted for. A further potential complication is the fact that endogenous CO_2 production is not constant and depends on a number of factors, including physical activity, food intake, body temperature, age, and body size. Accordingly, the use of a mean CO_2 production rate based on either body weight or surface area only partially normalizes for this factor, even if the in vivo probe is administered after a period of fasting and the subject under investigation remains recumbent for the period of the study (29). A final limitation is related to the fact that a CO_2 breath test is a measure of the rate of excretion of the labeled carbon. Accordingly, it is dependent on the volume of distribution of the in vivo probe; i.e., a large volume of distribution will result in a reduction in the breath test result, even though the enzyme's intrinsic activity is the same (30,31). However, interindividual variability in such a parameter is common with all drugs; therefore, the assumption that a constant value is applicable to all subjects results in some error in evaluating the involved enzyme's activity.

C. Validation of a Phenotypic Trait Measure

Following the in vitro discovery that a particular metabolic pathway is mediated by a single CYP isoform and that such metabolism is a major route of elimination, it is not uncommon to speculate that appropriate assessment of the formation of the metabolite in vivo could serve as a phenotypic trait measure (see Chaps. 3 and 7). Moreover, knowledge of the disposition of the drug and metabolite may indicate a putative quantitative trait for this purpose, e.g., urinary MR. However, considerably greater effort and information is, in fact, required before such a trait value can be accepted as a valid measure of the enzyme's metabolic activity. Unfortunately, several of the earlier-developed in vivo probes were not rigorously evaluated prior to their application, and interpretation of differences/changes in their trait values is therefore not easy.

Ideally, the trait measure should be correlated directly with the target enzyme's intrinsic clearance as measured, for example, in a tissue biopsy, e.g., liver from the same subjects. However, from a practical standpoint this gold standard approach is difficult, especially in health subjects. Moreover, even if applied, it does not address the issue of any extrahepatic metabolism. On the other hand, if the target enzyme exhibits genetic polymorphism so that a null phenotype exists, e.g., CYP2C19 and CYP2D6 (secs. V.B and VI), advantage of an experiment of this nature can be taken. However, in general, the best practical validation approach would appear to be the demonstration of a close and meaningful correlation between the putative trait value and the in vivo probe's fractional oral clearance determined by a conventional pharmacokinetic study. Care must be taken in such a correlation to ensure that a large enough population size, perhaps 50 or more subjects, be investigated for this purpose and that the data are not inappropriately weighted by individuals with metabolic activities at the extremes of the distribution curve, i.e., subjects having inhibited or induced metabolizing activity. A false and overly positive impression of the trait measure may be obtained if these two factors are not adequately considered; in this regard, it is important to recognize that the appropriate statistical measure of the potential usefulness of any correlation is the coefficient determination (r^2) and not the regression coefficient (r). As a corollary, it is also important to demonstrate that the phenotypic trait also correlates with the fractional clearances of other substrates metabolized by the same target enzyme. Such correlations are particularly critical in establishing that factors other than metabolism are not rate limiting. Additional validation steps generally focus on modulating the target enzyme's activity and its effect on the trait value. Enzyme induction and inhibition, especially involving mechanism-based inhibitors, are usually used for this purpose, with the trait measure appropriately increasing or decreasing. Finally, if metabolism is limited to the liver or if a liver-specific test is required, then changes in the trait value would be expected in the presence of severe liver disease. In this regard, advantage may be taken of the anhepatic period during a liver transplant operation, when no functioning liver is present.

Importantly, no single criterion is itself sufficient to validate a particular phenotypic trait value; rather, several of the described approaches must provide collective and consistent evidence. Finally, it should be recognized that to some extent validation depends on the purpose to which the in vivo probe is to be applied. For example, if evidence is required to demonstrate the presence or absence of a drug interaction, a less rigorous level of validation might be acceptable than if a quantitative measure of the extent of modulation of metabolic activity is necessary. Thus, the erythromycin breath test (see sec. VIII.B) is a useful in vivo probe for answering such a semiquantitative question, despite the fact that it only reflects hepatic CYP3A4 and not that localized in the intestinal epithelium, which importantly contributes to first-pass metabolism after oral drug administration. Similarly, a trait measure that discriminates between PMs and EMs associated with a genetic polymorphism may not necessarily be suitable for quantifying smaller within-phenotype differences in metabolism.

D. In Vivo Probe Cocktails

Determination of the activity of individual CYP isoforms by use of a single in vivo probe has the advantage that only the enzyme of interest is targeted, and metabolic or other types of interaction occurring because of coadministration of other drugs can be disregarded. On the other hand, such a narrow focus has disadvantages, given that multiple CYP isoforms are present and more than one of these may be of interest. In this case, a single-probe strategy would require multiple sequential studies using different in vivo probes to assess each individual enzyme. Not only is this time consuming but also results in an inefficient use of resources. To overcome these disadvantages, a "cocktail" approach has been applied based on the simultaneous administration of more than one in vivo probe, each of which assesses the metabolic activity of a different enzyme. This concept was originally developed using nonspecific model drugs, such as antipyrine and hexobarbital (32), but more recently it has been applied with cocktails of several ($n = 2$–6) different selective in vivo probes. For example, various combinations of debrisoquine or dextromethorphan (DTM) with mephenytoin, caffeine, nifedipine, coumarin, and chlorzoxazone have been used to simultaneously assess CYP2D6, CYP2C19, CYP1A2, CYP3A, CYP2A6, and CYP2E1 activities, respectively (33–36).

The cocktail approach appears to be particularly suited to indirect phenotypic trait values based on a single time point determination, e.g., MRs, where only one or two samples are sufficient for the experimental objective. For example, it is an attractive strategy for investigating the CYP isoform selectivity of enzyme inhibition or induction of a drug prior to more focused studies (34,37). However, a critical issue in the application of any cocktail study is whether one or more of the individual drugs interacts with another in the mixture. This may occur within the body through a metabolic interaction such as inhibition of another CYP isoform's catalytic activity, or a pharmacodynamic interaction may occur that either affects the phenotypic trait value of another drug or results in an undesirable clinical effect. Accordingly, it is important to establish prior to application that combining two or more in vivo probes has no effect on any of the individual phenotypic trait values and that the combination is safe. A second potential complicating factor is that multiple drugs and their metabolites are present and must be analyzed in the same biological sample. Accordingly, the involved analytical methodologies must not only be sufficiently sensitive but also specific so that no analytical interference is present.

A related approach to using a cocktail strategy that has been suggested is based on measuring the metabolism of a single in vivo probe in which two or more metabolites are formed involving different CYP isoforms. For example, the O-demethylation of DTM to dextrorphan (DT) is mediated by CYP2D6, whereas CYP3A is importantly involved in the N-demethylation pathway leading to the formation of 3-methoxymorphinan (3MM) (38). Accordingly, it has been proposed that the urinary MRs DTM:DT and DTM:3MM ratios could be used to separately assess the metabolic activities of the two involved isoforms (39,40).

However, the demonstration that enzymes other than CYP3A may also significantly contribute to N-demethylation raises questions as to the reliability and validity of the DTM:3MM MR as a measure of CYP3A activity (41). In principle, the stereo- and regioselective metabolism of warfarin could also serve to assess the activity of multiple CYP isoforms. For example, *S*-warfarin is primarily metabolized to 7-hydroxywarfarin by CYP2C9, whereas CYP1A2 is responsible for R-enantiomer's 6- and 8-hydroxylation; also, CYP3A is importantly involved in the formation of 10-hydroxy-*R*-warfarin, and hydroxylation of *R*-warfarin at the 8-position is mediated mainly by CYP2C19 (42,43). However, this approach has yet to be extended to the in vivo situation, probably because of analytical considerations, which would be further complicated by the small dose of warfarin that would need to be used for safety purposes in healthy subjects undergoing such phenotyping. Nevertheless, the overall strategy of a single probe and multiple enzymes appears to be worthwhile pursuing further.

III. CYTOCHROME P450 1A2

The cytochrome P450 1A (CYP1A) subfamily consists of two members, CYPlAl and CYP1A2, which are both important in the metabolic activation of chemical procarcinogens. However, unlike its related isoform, CYPlAl is not constitutively expressed in the liver and is primarily an extrahepatic enzyme; moreover, it is not involved in the metabolism of therapeutically useful drugs. Accordingly, the in vivo probes investigated for this subfamily have been directed almost exclusively toward CYP1A2. Drugs whose metabolism importantly involves CYP1A2 include phenacetin (O-deethylation), caffeine (N1-, N3-, and N7-demethylation), theophylline (N1- and N3-demethylation), tacrine (1- and 7-hydroxylation), tamoxifen (N-demethylation), R-warfarin (6- and 8-hydroxylation), and acetaminophen (ring oxidation) (44). A number of these substrates have been studied and applied as in vivo probes in humans.

A. Caffeine

Caffeine (1,3,7-trimethylxanthine) is one of the most widely and frequently consumed xenobiotics throughout the world. The diet is the principal source of such intake, with estimates of per capita daily consumption in Europe and North America exceeding 200 mg/day. Following oral administration, caffeine is rapidly and completely absorbed, and it is then eliminated essentially completely (>95%) by metabolism. Such metabolism is complex, with at least 17 metabolites being formed and excreted in the urine. However, these arise from three primary pathways that contribute to over 95% of the drug's overall metabolic clearance; N-demethylation to form paraxanthine (80%), N1-demethylation to form theobromine (11%), and N7-demethylation to form theophylline (4%). C8-Hydroxylation and C8–N9 bond scission together account for the remaining 5% or so of caffeine's metabolism. Subsequently, these primary metabolites are extensively further metabolized. Importantly, CYP1A2 is almost exclusively responsible for all three of the initial N-demethylation steps, especially after small doses. At high in vitro substrate concentrations, CYP2E1 and possibly other CYP isoforms are contributory, but this is of little relevance to caffeine exposure in vivo. Not surprising, therefore, caffeine has become the most widely used in vivo probe for assessing CYP1A2 activity (45,46).

1. Caffeine Oral Clearance

Over 95% of caffeine's plasma clearance can be accounted for by the three N-demethylation pathways, all of which are mediated exclusively by CYP1A2 (45,46). Accordingly, measurement of the in vivo probe's oral clearance is the gold standard by which this isoform's activity can be evaluated. This notion is supported by reasonably strong correlation ($r = 0.74$, $p < 0.01$) between this parameter and caffeine's intrinsic clearance with regard to N3-demethylation, as estimated in vitro from liver microsomes obtained from the same subjects studied in vivo (47). In practice, phenotyping is a simple procedure involving the oral administration of a single 100- to 200-mg dose of caffeine followed by serial blood sampling over 12 to 24 hours and appropriate HPLC measurement of caffeine's plasma level and pharmacokinetic analysis. The caffeine may be given in the form of an available over-the-counter formulation or as a measured amount of coffee of known caffeine content or similarly as a cola drink. Some investigators have administered caffeine intravenously and used systemic clearance as the phenotypic trait measure (48,49). However, since caffeine is a low-clearance drug, the potential value of this approach is not particularly great and is outweighed by the disadvantage of administering the caffeine by intravenous injection. Because caffeine is ubiquitous in the diet, phenotyping involves a caffeine-free period of one to three days prior to and also during the study period. However, if measurable caffeine is present in the plasma prior to administration of the in vivo probe, a pharmacokinetically based correction can take this into account (47). Several studies have demonstrated that caffeine's oral clearance is appropriately altered by factors that have been shown to modulate CYP1A2 activity either in vitro or in animals, e.g., tobacco usage and administration of other inducers, oral contraceptives, and mechanism-based inhibitors (46). In addition, this phenotypic trait value is robust and reproducible when studied in the same subjects over a four-month period (50). In order to minimize the number of required blood samples, it has been suggested that estimation of caffeine's elimination half-life based on three or four postabsorption plasma levels could provide an alternative phenotypic trait measure (51,52). Since caffeine's volume of distribution is similar to total body water, such an estimate can also be used to obtain an approximate value of the probe's clearance (48). However, such approaches still require that the blood sampling period be sufficiently long to accurately define caffeine's elimination half-life, which ranges from about 2 to 12 hours in healthy subjects but can be considerably longer in patients with liver disease or when CYP1A2 is inhibited because of a drug interaction. This factor probably accounts for the lower accuracy and higher intrasubject variability found with foreshortened sampling protocols (49). Recently, a Bayesian estimation of caffeine clearance based on a single plasma level obtained at either 12 or 24 hours after intravenous administration of the probe was shown to be well correlated with the directly determined value (49). Further study of this simple approach would appear warranted.

Caffeine is not extensively bound to plasma proteins; therefore, it readily distributes into saliva with a saliva:plasma concentration ratio of total drug between 0.74 and 0.94 (52,53). Not surprisingly, therefore, a close correlation exists between saliva and plasma caffeine levels and derived pharmacokinetic parameters. Accordingly, an alternative approach for estimating caffeine's oral clearance following administration of the in vivo probe is to measure salivary levels of the drug (47,52–55). Because of its noninvasiveness, even when salivary flow rate is stimulated by chewing Parafilm® or by the application of citric acid to the tongue, a large number of samples may be obtained for pharmacokinetic analysis, and collection can be extended beyond the clinical setting. Indeed, comparisons of caffeine's oral clearance values independently determined from plasma and saliva concentrations are essentially the same (47,53–55).

In an attempt to further simplify the caffeine phenotyping test, a trait measure based on the plasma or salivary paraxanthine:caffeine concentration ratio between three hours and seven hours after administration of the probe has been suggested (56). High linear correlations (>0.89) have been observed between this trait value and caffeine's oral clearance, and if necessary, a predicted caffeine clearance value may be calculated from the ratio (56). Currently, this phenotyping approach appears to be the simplest and most noninvasive means of readily assessing CYP1A2 activity using caffeine as a probe; in addition, the method is reproducible and appears to be robust (56), despite the theoretical dependency of the trait value on the urine flow rate (51).

2. Caffeine and Phenacetin Breath Tests

The use of radiolabeled caffeine to determine drug-metabolizing ability based on a $^{14}CO_2$ breath test was an early example of the applicability of this general approach (57,58), even before it was recognized as a measure of CYPIA2 activity. Generally, labeling has been at the *N*-3 methyl position, since this is the site of the major pathway of caffeine metabolism; however, labeling of all three *N*-methyl groups (58) and N-7 labeling has also been investigated (59). Additionally, both radio-(^{14}C) and stable-(^{13}C) labeling has been successfully used (57–64). The need for mass spectrometry-based analytical methodology in the case of stable-labeled caffeine is in most instances outweighed by the safety issue related to exposure to radioactivity associated with the use of radiolabeled carbon. Other than equipment requirements, the caffeine breath test is simple to perform and for [^{13}C]-(*N*-3-methyl) caffeine, a commercial kit is available for this purpose. Typically, exhaled breath is collected at several intervals up to one to eight hours following an oral dose of labeled caffeine. Either the cumulative amount (57–63) or the hourly rate (64) of labeled CO_2 excretion is used as the phenotypic trait value. The caffeine breath test appears to be reproducible, although extensive testing of this characteristic has not been reported. However, excellent correlations (r = 0.84–0.90) have been found between the two-hour cumulative caffeine breath test and the drug's oral clearance (58,62,63). In addition, the breath test has been shown to alter in response to modulating conditions that either decrease or increase CYP1A2 activity (45,60,61).

The O-deethylation of phenacetin is CYPlA2-mediated and results in the liberation of acetaldehyde that is subsequently metabolized to acetate and then CO_2. Thus, a breath test based on the use of phenacetin labeled with ^{14}C in the 1-position of the ethyl side chain could function to assess CYP1A2 activity. Early studies demonstrated the feasibility of this approach and its potential application to evaluating hepatic function (65,66). No extensive validation was attempted, so it is difficult to determine how well this test reflects the enzyme's intrinsic clearance, rather than perhaps some other determinant, such as liver blood flow. However, the situation appears to be moot since phenacetin is no longer an approved drug worldwide because of its renal side effects following chronic dosing; accordingly, further studies of this approach are unlikely.

3. Urinary MRs

Following initial *N*-demethylation, caffeine's primary metabolites undergo extensive further metabolism. For example, the major metabolite paraxanthine (17X) is demethylated to form 1-methylxanthine (IX), 1-methylurate (1U), and 5-acetylamino-6-formylamino-3-methyluracil (AFMU), which spontaneously degrades, especially under basic assay conditions, to 5-acetylamino-6-amino-3-methyluracil (AAMU) (45). These metabolites account for about 20%, 40%, and 15%, respectively, of the urinary recovery of caffeine-derived products. In addition, approximately 10% of 17X is excreted unchanged and another 20% is hydroxylated to form 1,7-dimethylurate (17U). Theobromine (37X) is in part excreted unchanged (10%), and about 20% is metabolized to 3-methylurate (37U) and approximately 50% to 7-methylxanthine (7X). About 10 to 15% of theophylline (13X) is excreted into urine, with about 50% of this primary metabolite being metabolized to 1,3-demethylurate (13U) and some 23% to 1U. Finally, a small amount of caffeine is excreted unchanged in urine, and some additional minor metabolites are formed (45,51).

CYP1A2 is primarily responsible for the N1-, N3-, and N7-monodemethylations of 17X, 37X, and 13X, respectively, and also the 8-hydroxylation of these primary metabolites. However, other CYP isoforms, e.g., CYP2A6, CYP2E1, and CYP3A, are also importantly involved in the formation of the various dimethyluric acids. Polymorphic *N*-acetyltransferase-2 (NAT2) mediates the conversion of an unstable 17X intermediate to AFMU, and xanthine oxidase is responsible for the oxidation of IX to 1U (45,51). Thus, the metabolism of caffeine results in a complex urinary recovery profile involving multiple primary and secondary metabolites as well as unchanged drug.

Because CYP1A2 is predominantly involved in caffeine's primary and secondary metabolism, a urinary MR approach has been applied with an expectation that this would provide a simple and convenient noninvasive means of assessing the isoform's metabolic activity. Over 12 different urinary molar MRs have been suggested as putative trait measures of CYP1A2 activity, generally based on a 0- to 24-hour urine collection following an oral dose of caffeine, although "spot"

Table 1 Common Caffeine Urinary Metabolite Ratios Used as Phenotypic Measures of CYP1A2 Activity

		Refs.
Ratio 1	$\frac{(17X)}{(137X)}$	47,56,61,83
Ratio 2	$\frac{(17U)+(17X)}{(137X)}$	47,56,61,69–73,81
Ratio 3	$\frac{(AFMU)+(1X)+(1U)}{(17X)}$	69,74
Ratio 4	$\frac{(AFMU)+(1X)+(1U)}{(17U)}$	45,47,54,56, 69,70,75–80
Ratio 5	$\frac{(AFMU)+(1X)+(1U)+(17X)}{(137X)}$	68
Other ratios	Various	51,69,78

For simplicity AFMU is used in all equations when the actual analyte is AAMU after converting all AFMU to AAMU.

sampling within a shorter defined time period (2–6 hours) has also been used (45,51,67–69). The most common of these are shown in Table 1. A major difficulty in the application of these phenotypic trait measures is that they are essentially all empirical, and until recently their limitations were not understood or, more importantly, appreciated. It is clear that CYP1A2 activity affects both the numerator and the denominator of the MRs to a varying extent. Also, CYP isoforms other than CYP1A2 and urine flow rate affect the excretion of caffeine and several of its metabolites (51,70). All of these factors exhibit intra- and interindividual variability, so it is not surprising that in a study involving 237 healthy subjects, the common MRs appeared to reflect three different parameters, and no one ratio correlated particularly well with any other (68). Ratios 1, 2, and 5 (Table 1) were the best correlated ratios ($r = 0.73$–0.88), but, even so, considerable variability was present within the relationship. A rigorous sensitivity analysis based on a pharmacokinetic model of caffeine's metabolism and urinary excretion profile identified a number of confounding variables that contributed to this situation (51). Moreover, this analysis concluded that none of the caffeine urinary MRs is specific for CYP1A2, although Ratio 4 may be useful when studying the modulation of CYP1A2 activity within the same subject. It was also suggested that the plasma-saliva ratio of 17X:137X measured a short time after caffeine administration might be a robust CYP1A2 marker.

Experimental investigations have subsequently confirmed these theoretical findings. For example, significant correlations were obtained between Ratio 4 and caffeine's oral clearance ($r = 0.66$–0.77, $p < 0.002$) by several different investigators (47,69,70). By contrast, the correlations between Ratios 1 and 2 and caffeine clearance were generally much poorer (47,69,70); other, less common MRs were also found to be poor measures of caffeine clearance (69). Thus, with the exception of Ratio 4, the validity of other urinary MRs would appear to be questionable; however, considerable experimental data indicate that this urinary MR is a robust and reproducible trait measure that is sensitive to the modulation of CYP1A2 activity (45).

Unfortunately, many investigators have used various of the caffeine urinary MRs without recognizing their potential limitations. As a result, conclusions drawn from the interpretation of such flawed data may be inaccurate. One area where this may exist and remains a controversial issue concerns the population distribution of inferred CYP1A2 activity. For example, several studies using Ratios 1 or 2 have concluded that CYP1A2 activity is either bi- or trimodally distributed within the population, and ''slow,'' ''intermediate,'' and ''rapid'' phenotypes can be identified; additionally, interethnic, and racial differences in the frequency of these putative phenotypes exist (71–73). These observations are in strong contrast to similar studies by other investigators, using Ratio 4 as an index of CYP1A2 activity, where a log-normal distribution has been found (70,75–80). Moreover, numerous studies based on the determination of caffeine's plasma clearance in large numbers of subjects have not provided any evidence of discrete subgroups with either low or high values within a log-normal distribution. Modeling analysis also indicates the likelihood that the polymodal distribution could be an artifact (51); this is supported by the observations that despite the fact that the frequency distributions of Ratio 4 and caffeine clearance were unimodal, the distribution for Ratio 2 in the same subjects was bimodal (70). Also, the effects of cigarette smoking on CYP1A2 activity, as measured by Ratio 2, was noted to depend on racial background (71,81), an observation that again is inconsistent with data based on clearance or Ratio 4 measures and one that is difficult to mechanistically explain. These discordances might be considered trivial except for the potential value of identifying the level of CYP1A2 activity as a possible risk factor for the development of certain types of cancer (81). Use of a valid in vivo probe would appear to be critical for such studies.

Although the validity of the urinary MR approach for assessing CYP1A2 activity is debatable, there is substantial evidence indicating that its use for determining NAT2 activity is appropriate. This application is based on the involvement of NAT2 in the formation of AFMU, and both the molar ratios of urinary AFMU:1MX and AFMU:(AFMU + IX + 1U) have been demonstrated to be reliable phenotypic indicators that categorize populations into three subgroups according to genotype (45–47,67,80,82–85). In practice, pretreatment of the urine to convert all of the AFMU to AAMU is advisable to avoid misclassification (45,85). Similarly, the molar ratios 1U:1X and 1U:(IX + 1U) have been used to determine xanthine oxidase activity (74,76,80).

The metabolism of theophylline (1,3-dimethylxanthine) is similar to that of caffeine but less complex (vide supra). Consequently, some consideration has been given to using it as a CYP1A2 probe. However, potential analytical sensitivity problems and, more importantly, safety considerations do not suggest that theophylline has any advantage over caffeine for this purpose (86).

In summary, caffeine is an acceptable and validated in vivo probe for assessing CYP1A2 activity in humans. The gold standard approach depends on determination of the drug's oral clearance following a single phenotyping dose under dietary caffeine-free conditions. Both plasma and saliva concentrations may be used for this purpose. Comparisons of caffeine's elimination half-life may be an alternative approach when within-subject changes in CYP1A2 activity are being investigated. However, such temporal monitoring applicable to drug interactions is probably best accomplished by the 17X:IX plasma/saliva concentration ratio determined at a single time point after caffeine administration. Alternatively, a caffeine breath test can similarly provide such within-subject information. Interpretation of caffeine urinary MRs is more difficult than with other approaches, and that based on the molar ratio (AFMU + IX + 1U):17U (Ratio 4) is probably the best of those developed. However, given the comparable simplicity and noninvasiveness of the salivary 17X:IX ratio, it is difficult to justify why even urinary Ratio 4 should continue to be used.

IV. CYTOCHROME P450 2A6

CYP2A6 appears to be the only catalytically active isoform of the CYP2A subfamily that is expressed in humans. Activity is localized mainly in the liver; however, extrahepatic distribution is also present, especially in the nasal epithelium and lung. Such localization is likely to be critically important, since CYP2A6 mediates the metabolism and activation of nicotine, cotinine, and tobacco-smoke-related nitrosamines like 4-(methylnitrosamino)-l-(3-pyridyl)-l-butanone (NNK), 4-(methylnitrosamino)-l-(3-pyridyl)-l-butanol (NNAL), and *N*-nitrosonornicotine (NNN), which are among the most potent of known lung carcinogens. CYP2A6 also activates a number of other established procarcinogens, many of which are also metabolized by CYP2E1 (44). Only a small number of drugs are currently known to be importantly metabolized by CYP2A6; these include coumarin (87,88), methoxyflurane (89), halothane (90), valproic acid (91), disulfiram (92), losigamone (93), letrozole (94), and (+)-*cis*-3,5-dimethyl-2(3-pyridyl) thiazolidine-4-one (SM-12502) (95). Importantly, genetic polymorphisms are present in the CYP2A6 gene, and, although current data are relatively limited, these appear to have functional consequences. At least three defective alleles have been reported, the most prevalent of which (CYP2A6*2) appears to be a Leu160His substitution that yields an inactive enzyme (96–99). By contrast, a deletion mutation (CYP2A6*4) is the most common variant (15–20%) in Asian populations (100). Duplication of the CYP2A6 gene also occurs and appears to be associated with increased catalytic activity (101).

The 7-hydroxylation of coumarin (1,2-benzopyrone) is a major urinary metabolic pathway that accounts for about 60% of an orally administered dose (102). This pathway is almost exclusively mediated by CYP2A6 (87,88) and forms the basis of the "coumarin index" or "2-hr coumarin test" used to measure in vivo CYP2A6 activity. The phenotypic trait measure is simply the percentage of a 5-mg dose of coumarin excreted in urine as 7-hydroxycoumarin over the zero- to two-hours period following oral administration in the fasted state (102). Because the 7-hydroxy metabolite is excreted mainly as a conjugate, urine is pretreated with β-glucuronidase prior to analysis, and a methodology based on chromatographic separation would appear to be preferable to one using solvent extraction (103). Application of this phenotyping procedure to various population groups has shown that the trait measure exhibits considerable interindividual variability, and it is unimodally distributed in a normal fashion (102–104). While a few individuals with lower values could be identified, no clear polymorphism was apparent, consistent with the low frequency of the CYP2A6*2 allele in Caucasians. On the other hand, a gene dose effect has been reported with respect to individual CYP2A6*1 homozygotes, CYP2A6*1/CYP2A6*2 heterozygotes, and CYP2A6*2 homozygotes (105). Accordingly, it would be expected that in the general population all three phenotypes (extensive, intermediate, and poor) would be present.

The "coumarin index," however, has several problems that make it less than ideal as a CYP2A6 trait measure. First, is the fact that the trait value is entirely empirical and has been validated and characterized to only a very limited extent. Accordingly, its sensitivity to determinants of CYP2A6 activity are largely unknown, other than that its value is reduced by age (106) and the administration of grapefruit juice (107) but not disulfiram (108), and that it is increased by antiepileptic drugs (109). As expected, severe but not mild liver disease reduces the urinary recovery of 7-hydroxycoumarin, but, not unexpectedly, renal dysfunction has also been found to affect the trait value (109). A second difficulty is that the coumarin index is an indirect measure of CYP2A6 activity, and perhaps, more importantly, it is unlikely that it can ever be validated against a gold standard such as the formation clearance of 7-hydroxycoumarin. This is because of the extreme analytical difficulties associated with measuring plasma coumarin levels because of its relatively high volatility, and this problem is further compounded by the low dose used for phenotyping (5 mg). Coumarin is also excreted in the urine as a result of dietary and environmental exposure through fragrances and other sources. Such daily exposure may be as high as 25 mg (110), which probably accounts for the finding that in certain subjects the urinary molar recovery of 7-hydroxycoumarin exceeds the molar dose of coumarin administered to determine the trait value (103,110). An additional consideration, especially in North America, is the absence of an available approved formulation containing coumarin, which was removed from the market 45 years ago because of its hepatotoxicity and carcinogenic properties in animals (111). More recently, limited use of coumarin in certain types of cancer has been investigated (112), but the strength of the available tablet is 100 mg, i.e., 20-fold greater than the dose used to determine the coumarin index. Because of these problems, attempts have been made to develop an alternative phenotyping method to assess in vivo CYP2A6 activity.

A potential method that was recently reported is based on a MR approach, namely, the 2-hydroxyphenylacetic acid:7-hydroxycoumarin ratio in a zero- to eight-hour urine sample following oral administration of 2-mg coumarin (110). 2-Hydroxyphenylacetic acid is the terminal metabolite of an alternative pathway of coumarin metabolism besides 7-hydroxylation, and usually accounts for about 2.5–30% of the administered dose. The frequency distribution of this trait measured in 103 subjects identified two individuals with values markedly greater than the remainder of the population. Moreover, within the major subgroup, there was evidence of overlapping bimodality. Unfortunately, the CYP2A6 genotypes of the subjects were not determined to show that the apparent phenotypes reflected the genetic polymorphism. Future studies will undoubtedly investigate this possibility and demonstrate the value of this urinary MR as an indicator of CYP2A6 activity.

The major human urinary metabolites of nicotine are cotinine, nicotine N'-oxide, and *trans*-3′-hydroxycotinine (113). CYP2A6 appears to be the major enzyme responsible for formation of an iminium ion that is the first step in the C-5 oxidation of nicotine to cotinine and also the subsequent 3-hydroxylation of this metabolite (114,115). These facts, coupled with the known acute safety profile and wide use of nicotine through tobacco smoking, suggest that appropriate measurement of the drug's metabolism might provide a means to assess in vivo CYP2A6 activity. One reported approach is based on the 30-minute intravenous infusion of a 50:50 mixture (2 μg base/kg) of 3′,3′-dideuterium-labeled nicotine and 2,4,5,6-tetradeutero cotinine followed by serial blood sampling over the following 96 hours and a 0- to 8-hour urine collection (116,117). Using gas chromatography–mass spectrometric–based assays, the levels of nicotine and cotinine derived from each stable-labeled form are measured. Appropriate pharmacokinetic analysis then allows estimation of nicotine's formation clearance to cotinine and also the latter's clearance. To date, this methodology has been applied primarily to investigating nicotine's metabolism within the context of cigarette smoking and addiction (116,117). It should be noted, however, that an individual deficient in the CYP2A6-mediated conversion of nicotine to nicotine was identified using this approach (118). Clearly, substantial further research is required before a simple and routine nicotine/cotinine-based phenotyping procedure for CYP2A6 is established. Despite the need for stable-labeled drugs and the associated sophisticated instrumentation for their measurement, such an approach would provide a gold-standard against which alternative trait measures such as the coumarin index or others could be evaluated and validated. Possibly, a simpler, single-point plasma- or urine-based measure could be developed using nicotine/cotinine. Finally, if the new drug candidate SM-12502 ever becomes clinically available, it is possible that an appropriate phenotyping measure could be developed using this drug since a genotype:phenotype relationship appears to be present (95,100).

V. CYTOCHROME P450 2C

Four closely related CYP2C genes have been definitively identified in humans: CYP2C8, CYP2C9, CYP2C18, and CYP2C19. However, additional genes/gene-like sequences are present. At one time, CYP2C10 was thought to be a discrete protein but is now considered to be an allelic variant of CYP2C9 or possibly a cloning artifact. Little information is currently available regarding CYP2C8, especially substrates that are preferentially metabolized by this isoform. One such reaction is the 6α-hydroxylation of taxol, and the 4-hydroxylations of retinol and retinoic acid also appear to be CYP2C8-mediated. CYP2C18 appears to be a minor member of the CYP2C subfamily; to date, no metabolic reactions have been demonstrated to be selectively catalyzed by this enzyme (44). On the other hand, many drugs are substrates of CYP2C9 and CYP2C19, and in many instances the involved metabolic reactions are highly specific. Accordingly, these two isoforms have received the most attention with regard to the development and application of in vivo probes.

A. Cytochrome P450 2C9

CYP2C9 is importantly involved in the oxidation of a large number of drugs, many of them widely used in clinical practice. Such drugs for which the isoform catalyzes the formation of a principal metabolite include phenytoin, tolbutamide, fluoxetine, losartan, *S*-warfarin, torsemide, valproic acid, and many nonsteroidal anti-inflammatory agents (diclofenac, ibuprofen, naproxen, piroxicam, suprofen, and tenoxicam). Thus, most, but not all, CYP2C9 substrates are weak acids with pKa values between 3.8 and 8.1 (119). Increasing evidence indicates that genetic polymorphisms are present in the CYP2C9 gene that have functional consequences. Three alleles resulting from Arg → Cys and Ile → Leu substitutions at amino acids 144 and 359 have been noted. CYP2C9*1 (Arg^{144}/Ile^{359}) represents the wild-type protein, whereas CYP2C9*2 (Cys^{144}/Ile^{339}) and CYP2C9*3 (Arg^{144}/ Leu^{359}) appear to be relatively rare variants. As with many other genetic polymorphisms, population frequency distributions differ according to racial ancestry; for example, CYP2C9*2 has not yet been found in Asian groups (Chinese and Japanese), and its prevalence is also very low in African-Americans (119).

In vitro studies with human liver microsomes and expressed CYP2C9 allelic variants have found markedly impaired catalytic activity of CYP2C9*3 compared with CYP2C9*1 (119). Furthermore, this difference has also been noted to be present in patients receiving warfarin therapy, where a gene-dose effect leads to reduced clearance of the anticoagulant's *S*-enantiomer (120–122). As a result, a dangerously exacerbated therapeutic response to normal doses of racemic warfarin is produced in CYP2C9*3/CYP2C9*3 homozygotes (120). Similarly, an individual identified as a PM of tolbutamide was also subsequently found to be homozygous for CYP29*3 (123); in the clinical trials associated with the development of losartan, two subjects, corresponding to <1% of the study population, had markedly impaired conversion of the drug to an active metabolite and both were CYP2C9*3/CYP2C9*3 homozygotes (124). Comparable information regarding CY2C9*2 is currently less definitive, since in vitro studies provide conflicting data on the effect of the Arg 144 Cys substitution (119). And, while the median warfarin maintenance dose was 20% lower in CYP2C9*1/CYP2C9*2 heterozygotes compared with homozygous wild-type patients, there was considerable overlap in the dosage requirements (125).

Thus, valuable information on CYP2C9 activity in vivo has been obtained through studies of warfarin's metabolism, and it has even been suggested that the drug's *S:R* enantiomeric concentration ratio in plasma could be used to identify homozygous CYP2C9*3 patients (121). However, for safety and analytical reasons, it is unlikely that the anticoagulant could be used as an in vivo probe in healthy subjects. A similar safety issue also would appear to apply to phenytoin, despite the fact that its major route of metabolism, viz, 4-hydroxylation, is mediated by CYP2C9. Moreover, the plasma clearance of phenytoin exhibits nonlinearity due to saturable metabolism, and a urinary 4-hydroxyphenytoin: phenytoin MR has been shown to be overly variable and of limited usefulness as a phenotypic trait measure of CYP2C9 activity (126). Accordingly, efforts have been directed toward other potential drugs.

1. Tolbutamide

The metabolism of tolbutamide (l-butyl-3-*p*-tolysulfonylurea) in humans involves a single pathway, with the initial and rate-limiting step being tolyl methyl-hydroxylation to form hydroxytolbutamide, which is further oxidized to carboxy-tolbutamide by alcohol and aldehyde dehydrogenases. Overall, this pathway of metabolism accounts for up to 85% of tolbutamide's clearance and is exclusively mediated by CYP2C9 (119). Accordingly, determination of tolbutamide's (fractional) clearance following a single oral dose (500 mg) has been used as a phenotypic trait value for assessing in vivo CYP2C9 activity (127). Since the drug's half-life ranges between 4 and 12 hours, this approach requires not only multiple blood samples but collection over a considerable time period (24–36 hr). An alternative and simpler MR approach has also been developed based on the relative recovery of metabolites and unchanged drug excreted into urine over the 6- to 12-hour period following a 500-mg oral dose of tolbutamide. Initially, this ratio—(hydroxytolbutamide + carboxytolbutamide):tolbutamide—was shown to be sensitive to the inhibition of CYP2C9 activity by pretreatment with sulfaphenazole (128). Subsequently, the trait measure was shown to be unimodally distributed in an Australian population of 106 healthy subjects (129), which is not surprising given the low frequency of the CYP2C9*3 allele (vide supra). Also, a previously identified PM, who was later shown to be a CYP2C9*3 homozygote (123), was found to have a markedly lower value than the reference population (129). It therefore appears that tolbutamide is a useful in vivo probe for CYP2C9. However, such use is not without problems, in particular, the safety issue associated with the hypoglycemic response produced by tolbutamide administration. This is not usually a problem in subjects who have been fed. However, in fasted individuals, blood glucose levels may be significantly reduced by tolbutamide and require reversal using glucose supplementation (130); use of a lower dose (250 mg) may obviate this problem. Clearly, this limits application of the test to a controlled clinical situation.

2. Nonsteroidal Anti-Inflammatory Agents

The nonsteroidal anti-inflammatory agents (NSAIDs) are relatively safe drugs, and since CYP2C9 is a major determinant in the metabolism of many of these agents, measurement of the involved pathway in principle could serve as an indicator of the isoform's activity. For example, the oral clearance of diclofenac, which reflects primarily CYP2C9-mediated 4′-hydroxylation, was found to be reduced following pretreatment with fluvastatin, an established in vitro inhibitor of the enzyme (131). A decrease was also noted in a urinary MR (4′-hydroxydiclofenac:diclofenac) based on a zero- to four-hour collection period but not over four to eight hours. Unfortunately, validation of neither of these putative trait measures using diclofenac has been reported. More recently, the 4′-hydroxylation of flurbiprofen has been investigated in a similar fashion, with the intent to develop this drug as a CYP2C9 in vivo probe (132). Such preliminary information will obviously require appropriate substantiation before the described trait measures will be widely accepted.

B. Cytochrome P450 2C19 (CYP2C19)

The enzyme now known as CYP2C19 (133,134) was first identified because of its major involvement in the 4′-hydroxylation of the *S*-enantiomer of mephenytoin (22,23). Although, its substrate specificity was originally thought to be limited to related anticonvulsant agents, the in vivo metabolism of an increasing number of structurally unrelated drugs appear to be mediated by this isoform. These include *R*-mephobarbital (4′-hydroxylation) (135), hexobarbital (3′-hydroxylation) (136), proguanil (ring cyclization) (137), omeprazole, and related proton pump inhibitors (5′-methylhydroxylation) (138–140), diazepam (N-demethylation) (141), certain tricyclic antidepressants (N-demethylation) (142–144), carisoprodil (N-dealkylation) (145), citalopram (N-demethylation) (146), moclobemide (C-hydroxylation) (147), propranolol (side-chain oxidation) (148), and nelfinavir (methylhydroxylation) (149). A major characteristic of CYP2C19 is the presence of a genetic polymorphism that subdivides populations into PMs and EMs (150). The molecular genetic basis of such phenotypes is now well recognized. The two most common defects in volve null alleles arising from G → A base pair mutations in exon 5 (CYP2C19*2) and exon 4 (CYP2C19*3), respectively (3), and they account for over 99% of defective alleles in populations with Asian ancestry but only in about 87% in individuals of Caucasian origin (151–153). A transition mutation in the initiation codon (CYP2C19*4) accounts for an additional 3% of defective alleles in Caucasian PMs (154), and three rare mutations (CYP2C19*5, CYP2C19*6, and CYP2C19*7) have also been identified (155–158). Not unexpectedly, the population allelic frequencies (159) and phenotypes, i.e., EMs, representative of both wild-type homozygotes and heterozygotes, and PMs (homozygous mutants), varies according to racial/geographical origin. For example, in populations of European descent, the frequency of the PM phenotypes ranges from 1.3% to 6.1%, with a mean value of about 3.5% (160). A similar low prevalence rate is also present in Africans and African Americans (161–163). By contrast, a much higher frequency (13–23%) is found in indigenous populations living in Southeast Asia, such as Chinese, Japanese, and Koreans (138,164–168). Importantly, the impaired metabolism present in PMs is often associated with marked differences in a substrate's disposition and pharmacokinetics (150). Moreover, drug interactions involving CYP2C19 can occur only in individuals with the EM phenotype, since no enzyme is present in PMs (169). These factors have led to the development of in vivo probes to classify individuals according to phenotype.

1. Mephenytoin

The genetic polymorphism in CYP2C19 was first discovered serendipitously during clinical studies to investigate the enantioselective metabolism of racemic mephenytoin at Vanderbilt University (150). Subsequently, two alternative phenotyping procedures were developed that have been widely used throughout the world by numerous investigators. Both of these are based on the fact that CYP2C19 metabolizes racemic mephenytoin to its 4′-hydroxy metabolite, and this is essentially complete and stereospecific for the *S*-enantiomer.

a. Urinary S:R enantiomeric ratio. The 4′-hydroxylation of *S*-mephenytoin is not only extensive but also rapid, and this is in contrast to metabolism of the *R*-enantiomer, which involves mainly N-demethylation (150). Accordingly, only a small fraction of an administered oral dose of racemic mephenytoin (50–100 mg) is excreted into urine over zero to eight hours as the unchanged *S*-enantiomer in EMs, whereas a much larger amount of *R*-mephenytoin is excreted. In the absence of CYP2C19 activity, i.e., PMs, impaired metabolism of *S*-mephenytoin by 4′-hydroxylation results in increased excretion of this enantiomer; because the metabolism of *R*-mephenytoin is not different between EMs and PMs, the *S*:*R* enantiomeric concentration ratio in a zero- to eight-hour urine sample is increased (23). Thus, the *S*:*R* ratio ranges from less than 0.03 to 0.8 in EMs and 0.9 to 1.2 in PMs (150). The pharmacokinetic basis of the trait measure has been described, and studies have confirmed that the urinary *S*:*R* ratio is the same as the comparable ratio of the areas under the plasma concentration–time profiles of the enantiomers during the collection period, which in turn reflects the relative intrinsic clearances of the two isomers (169). Also, this trait value has been found to be reproducible in individuals over a long period of time (170). Although the *S*:*R* ratio is widely used, it has a reciprocal and, therefore, rectangular hyperbolic relationship with CYP2C19 activity. For this reason, there is some merit in using the *R*:*S* ratio, which is linearly related to such activity, so the smaller its value, the lower the 4′-hydroxylating activity (21,171).

A minor route of metabolism of *S*-mephenytoin results in the formation and urinary excretion of an acid-labile conjugate that is probably a cysteinyl derivative (172). This pathway appears to be associated with CYP2C19, since it is present only in EMs (173). The significance of this urinary metabolite is that it is easily hydrolyzed back to *S*-mephenytoin, and this can occur to an unpredictable extent during sample storage, even at −20°C (173). The resulting artifactual *S*:*R* ratio value may, therefore, misclassify an individual's phenotype. Several approaches have been used to obviate or minimize this problem. The most widely used procedure is to obtain a repeat S:R value but following acid hydrolysis of the urine sample prior to analysis. The enantiomeric ratio is significantly increased by such pretreatment in EMs but remains essentially unchanged in PMs (173–175). Another approach (176) includes measurement of the *S*:*R* ratio in a urine sample collected 24 to 32 hours after drug administration, since little or no acid-labile metabolite is present at this time. However, the *S*:*R* ratio in EMs at this later time is smaller (0.16–0.5) than that in a zero- to eight-hour sample, whereas the value in PMs is still close to unity. Also, it is possible to extract the collected urine immediately after collection and store the dried extract at −20°C until subsequent analysis (176).

The major advantage of using the urinary *S*:*R* ratio as a phenotypic trait for assessing CYP2C19 activity is that the method is fairly robust with regard to any incompleteness of urine collection or noncompliance with respect to dose administration. This is because the *R*-enantiomer serves as an in vivo "internal standard." On the other hand, the required enantiospecific assay uses chiral capillary column gas chromatography with a nitrogen-specific detector, and such instrumentation is not commonly available. For this reason, an alternative phenotypic trait measure based on the formation and urinary elimination of 4′-hydroxymephenytoin has also been frequently used.

b. Urinary 4′-hydroxymephenytoin excretion. After its formation, 4′-hydroxymephenytoin is glucuronidated and excreted in this form into the urine. Accordingly, the aglycone must be liberated prior to determination of the amount of CYP2C19-mediated metabolite formed, using either β-glucuronidase pretreatment or acid hydrolysis (176). In EMs, the zero- to eight-hour urinary recovery of 4′-hydroxy-mephenytoin is between 5% and 52% (25–240 μmol) of the administered 100-mg (460 μmol) phenotyping dose of racemic mephenytoin (mean 18–20%) (150). By contrast, in PMs from undetectable to 3% of the dose (mean 0.5–0.8%) is excreted (150). An alternative phenotypic trait measurement based on excretion of the 4′-hydroxy metabolite is the mephenytoin HI:

$$\mathrm{HI} = \frac{\text{molar doses of } S\text{-mephenytoin(230 } \mu\text{mol in 100-mg racemic dose)}}{\mu\text{mol 4}'\text{-hydroxymephenytoin excreted in } 0 - 8 \text{ hr}} \tag{3}$$

In EM individuals, this value ranges from about 0.6 to 20, whereas a much higher value (30–2500) is observed in PMs (22,23,150). Importantly, the metabolite's concentration in urine samples from PM subjects is often close to the lower limit of sensitivity of the HPLC-based assay. Thus, an antimodal value that discriminates between the two phenotypes cannot be defined with absolute precision and varies between laboratories. As a result, phenotypic misclassification may occur when the zero- to eight-hour urinary recovery is in the range of about 15 to 25 μmol. A further interpretive difficulty is the trait's dependency on a complete zero- to eight-hour urine collection, since a low recovery of the 4′-hydroxy metabolite can also reflect poor subject compliance. Some investigators, therefore, confirm the completeness of urine collection in putative PMs by measurement of the amount of creatinine in the sample (>50 mg in 0–8 hour). Alternatively, the phenotype is determined by combining the information provided by both the excretion of 4′-hydroxymephenytoin and the urinary *S*:*R* ratio.

Despite the described approaches and precautions, the trait values of a small number of individuals may not be consistent with genotypic information or are uninterpretable based on the assumption that only two phenotypes are present. One reason for this is the very low frequency of "intermediate metabolizers" who have an *S*:*R* enantiomeric ratio consistent with the PM phenotype (>0.9) but who excrete more (20–60 μmol) of the 4′-hydroxy metabolite than would be expected if this was the case but at a rate less than that in EMs (176). It is likely that such rare individuals have an as-yet-unidentified allelic variant of CYP2C19 with reduced catalytic activity compared with the wild-type enzyme, as occurs with CYP2D6 (see sec. VI). It is possible that the urinary *S*:*R* ratio measured at 24 to 32 hours may also identify such individuals, since occasional subjects have been noted to have values between 0.2 and 0.5, whereas in most EMs this trait value is about 0.1 (175,177).

Mephenytoin has been extensively used for phenotyping purposes; however, such use is not without practical problems. For example, sedation is often observed in PMs, especially those of small body size, e.g., children and Southeast Asians, following administration of a 100-mg dose usually used for phenotyping (178). Accordingly, a dose of 50-mg mephenytoin is often used to phenotype such individuals. A further complicating factor is that racemic mephenytoin (Mesantoin®, Sandoz/Novartis, Basal, Switzerland) is not available in many parts of the world. For these reasons, other in vivo probes have been investigated.

2. Omeprazole

The 5′-hydroxylation of omeprazole cosegregates with CYP2C19 activity both in vitro using human liver microsomes (179–181) and in vivo in several different populations (138,182). In addition, the drug has a short elimination half-life (1–2 hour) and has a far wider therapeutic ratio than mephenytoin, resulting in better tolerability by subjects. Accordingly, trait measures reflective of in vivo CYP2C19 activity have been investigated.

In general, phenotyping has been performed following oral administration of a single 20-mg oral dose of omeprazole and obtaining a single blood sample for determination of the plasma concentrations of omeprazole and its 5′-hydroxy metabolite by HPLC. However, different sampling times and trait values have been used by different investigators. A MR (omeprazole: 5′-hydroxyomeprazole) determined three hours after administration of the probe was able to identify EMs who had trait values between 0.05 and 5.6, whereas individuals with the PM phenotype had larger ratios (183). In fact, the presence of metabolite in the plasma was not always detectable in PMs. A MR of about 7 appeared to be the antimode of the population distribution curve in 160 Swedish Caucasians (184). A similar approach, but based on sampling at two hours after omeprazole administration and using the logarithm of the MR—termed the "omeprazole hydroxylation index"—was also able to discriminate, using an antimode of one, between the two CYP2C19 phenotypes in 85 healthy subjects of various racial backgrounds (185,186) and 77 African Americans (163). A third trait value, using a single four-hour blood sample, has also been used: (omeprazole + omeprazole sulfone):5′-hydroxyomeprazole (187). Again, phenotypic classification was possible based on an antimode value of 12. Regardless of the precise trait value used, concordance was noted in the various studies between the measured CYP2C19-mediated omeprazole metabolism and the urinary *S*:*R* ratio for mephenytoin and also the genotype of the individual. This suggests that measurement of omeprazole's 5′-hydroxylation using a single-time-point plasma MR is a valid approach for CYP2C19 phenotyping. However, wider application of the methods will probably be necessary before omeprazole replaces mephenytoin as the gold standard in vivo probe for assessing CYP219 activity. For example, analytical sensitivity issues may limit omeprazole's wider application, since in one population study 23 of 100 subjects could not be phenotyped because of unmeasurable concentration of omeprazole and/or its 5′-hydroxy metabolite (185,186). Also, omeprazole sulfone formation is mediated primarily by CYP3A (180), which exhibits considerable interindividual variability in its activity (sec. VIII). Accordingly, the MR incorporating this metabolite's concentration into its estimation would be affected by this factor and presumably altered if CYP3A was inhibited or induced by drug administration in addition to the in vivo probe (187).

3. Proguanil

The antimalarial effects of proguanil (chloroguanide) are dependent on its metabolic activation to cycloguanil. This pathway is mediated in large part by CYP2C19, and the formation clearance of the metabolite cosegregates with the mephenytoin phenotype (188). Furthermore, the relative amounts of cycloguanil and unchanged drug excreted in urine collected 0 to 8 hours after an oral dose of 200-mg proguanil has been shown to be dependent on the CYP2C19 phenotype (137,189–194). Also, a good correlation ($r = 0.96$) was found between the urinary proguanil:cycloguanil MR and the formation clearance of 4-proguanil to cycloguanil (195), and an antimode of 10 was able to discriminate between EMs and PMs (190,192). However, in other studies this value did not necessarily separate the two phenotypes (191,193), and a better antimode was suggested to be 15 (196). More recent studies based on genotype:phenotype relationships have found a gene-dose effect in the proguanil MR (197,198). However, there was substantial overlap in the MR in subjects of different CYP2C19 genotypes (198), and it was difficult to define an exact antimode (197). Additional reservations about this phenotyping approach have also been expressed on the basis of the observation that no correlation was found between the proguanil MR and mephenytoin's HI (199,200). This may reflect the fact that the cyclization of proguanil to cycloguanil is mediated not only by CYP2C19 but also by CYP3A (201). Such considerations indicate that the proguanil urinary MR approach may have limitations as the method of choice for CYP2C19 phenotyping.

VI. CYTOCHROME P450 2D6

CYP2D6 is the most thoroughly investigated human CYP isoform because of its genetic polymorphism and involvement in the metabolism of many drugs of clinical importance. Accordingly, a plethora of reports and reviews addressing various aspects of the enzyme have been published (202,203). The genetic polymorphism was discovered independently by three groups of investigators, two studying the metabolism of debrisoquine (204,205) and another interested in sparteine (206). With both drugs, a bimodal frequency distribution was observed in the urinary excretion of measured metabolites and unchanged drug, and the subgroups were termed EMs and PMs. More recently, based on genetic analysis, an "intermediate" metabolizer (IM) phenotype has been defined (207), and in addition, "ultrarapid" metabolizers (UMs) have been described (208). The molecular genetic bases of the CYP2D6 polymorphism are now well established. Currently, some 48 mutations resulting in 53 alleles are known and additional rare ones continue to be identified (203,207,209). However, the five most common alleles represent over 95% of the variants (209), and a formal nomenclature scheme has been adopted (210). Many types of null mutations result in impaired CYP2D6 activity, and homozygosity is associated with the PM phenotype, e.g., CYP2D6*3, CYP2D6*4, CYP2D6*5, and CYP2D*6. Other variant alleles, such as CYP2D6*9, CYP2D6*10, and CYP2D6*17, lead to an enzyme with reduced catalytic activity compared with the wild-type allele (CYP2D6*1). Finally, gene duplication and amplification up to 13 copies (CYP2D6*2XN, where *N* indicates the number of genes) is associated with ultrarapid metabolism (210).

Considerable heterogeneity is present in the frequencies of these various alleles in different worldwide populations, dependent on racial/geographic factors (203,207). Consequently, the frequencies of CYP2D6 phenotypes differ among these groups. Most information is available on the PM phenotype, which is present in about 4–10% (mean 7.4%) of populations of European descent (160). By contrast, a lower frequency, 0.6–1.5%, has been observed in Southeast Asians, such as Japanese (165), Chinese (166), and Koreans (167). Also, the population distribution curve of CYP2D6 activity is shifted to the right in Chinese and Koreans, compared with Caucasians, as a consequence of the lower frequency of CYP2D6*4 and the increased prevalence of CYP2D6*10 alleles. In general, a similar low phenotypic frequency also appears to be the case with populations of African descent, due to the fact that CYP2D6*17 is more common than CYP2D6*4 (211–213). However, considerable heterogeneity also appears to be present among various African populations (sec. VI.E).

A large number of drugs, estimated to be over 50, have been shown to be metabolized by CYP2D6. These include, for example, β-adrenoceptor blockers (metoprolol, propranolol, timolol), antiarrhythmic agents (sparteine, propafenone, mexilitene, encainide, flecainide), antidepressants (tricyclics, selective serotonin reuptake inhibitors), neuroleptics (haloperidol, perphenazine, thioridazine, zuclopenthixol), opioids (codeine, dihydrocodeine, dextromethorphan), amphetamines (methamphetamine, methylenedioxymethylamphetamine—''ectasy,'' fenfluramine), and various other drugs (202). Although such drugs exhibit diverse chemical structures, a critical characteristic appears to be a basic nitrogen atom, which is ionized at physiological pH and interacts with an aspartic acid residue in the active site of CYP2D6. In the cases where CYP2D6-mediated metabolism is of major importance in the overall elimination of a drug, differences in drug disposition and pharmacokinetics is present between the two phenotypes, and this may be quite marked (202). The clinical consequences in PMs of such differences (202) include a higher propensity to develop adverse drug reactions following a conventional dose or a reduced clinical effect because an active metabolite is not formed, e.g., the conversion of codeine to morphine (214). In addition, therapeutic failure may occur in UMs as a result of inefficacious drug levels (215). These clinical considerations and general interest in the CYP2D6 genetic polymorphism have resulted in the development and application of a number of in vivo probes for assessing the enzyme's activity.

A. Sparteine

A polymorphism in sparteine metabolism was initially identified because of the involvement of CYP2D6 in the formation of 2- and 5-dehydrosparteine (206). This led to the development of a urinary MR type of phenotypic trait measure (sparteine:dehydrosparteines) based on a 0- to 8-hour or 0- to 12-hour urine collection following oral administration of 100-mg sparteine. Such a phenotyping approach is robust and has been applied to several thousand individuals. An antimode value greater than 20 appears to reliably distinguish PMs from EMs. More recently, subphenotyping within the conventional EM phenotype has been suggested (2). For example: a sparteine MR of 2 to 20 reflects the IM phenotype, who are generally CYP2D6*2/CYP2D6 null heterozygotes; a value from 0.2 to 2 is indicative of wild-type homozygotes and some heterozygotes; and values below 0.2 are usually associated with UMs who have duplicated/amplified genes. However, overlap in the trait values between the various subphenotypic groups does not permit any useful predictability of genotype from the sparteine MR, or vice versa.

Unfortunately, sparteine is not marketed and approved for clinical use in many countries, including North America and the United Kingdom, and analytical reference compounds required for the GLC analytical procedure are not readily available. Accordingly, the use of sparteine for CYP2D6 phenotyping has had limited use outside a small number of mainly European investigators.

B. Debrisoquine

CYP2D6 mediates the alicyclic hydroxylation of debrisoquine at the 1-, 3-, and 4-positions (216). However, formation of 4-hydroxydebrisoquine is quantitatively the most important of these pathways, accounting for between 1% and 30% of an administered dose, depending on genotype, and the metabolite is almost exclusively of the *S*-enantiomer configuration (19). This urinary metabolic profile led to the empirical development of the commonly used CYP2D6 trait measure of the zero- to eight-hour urinary MR (debrisoquine:4-hydroxydebrisoquine) determined following an oral 10-mg dose of debrisoquine (204,205). An alternative "urinary recovery ratio" [4-hydroxydebrisoquine: (debrisoquine + 4-hydroxy-debrisoquine)] has also been used, but to a far lesser extent (21,35). In most populations, a MR above 12.6 has generally been used to discriminate PMs from EMs. Also, within the latter phenotype, there is a general trend for the MR to decrease as the number of functional alleles increases, and IM (2.0–12.6), EM (0.1–2.0), and UM (<0.1) subphenotypes may be defined (19,217). However, there is considerable overlap in the phenotypic trait value between the various subgenotypes, so the debrisoquine MR cannot be used to identify a particular genotype, and vice versa.

Thousands of individuals worldwide have been phenotyped with debrisoquine. This extensive experience by numerous investigators has established that it is safe even in PMs, and it provides a reproducible and robust method for determining in vivo CYP2D6 activity (170). Moreover, the MR appears to be sufficiently sensitive to modulation of CYP2D6 activity for it to be used to identify determining factors (202). However, a significant practical problem associated with debrisoquine (Declinax®, Hoffman La-Roche, Nutley, New Jersey, U.S.) is its lack of availability and regulatory approval in several countries, including the United States. As a result, alternative drugs suitable for in vivo CYP2D6 phenotyping in such places have been investigated.

C. Metoprolol

Over 95% of an administered dose of metoprolol is metabolized in humans to a number of metabolites, including α-hydroxymetoprolol, which accounts for up to 10% of the eliminated dose (218). Considerable interindividual variability is present in the β-adrenoceptor blocker's oral clearance, and this was found to be determined by the CYP2D6 polymorphism as assessed by debrisoquine (219) and sparteine MRs (220). These observations led to the development and application of a zero- to eight-hour urinary metoprolol:α-hydroxymetoprolol MR approach for CYP2D6 phenotyping following an oral dose of 100-mg metoprololtartrate (221,222). A good correlation was observed between this ratio and the debrisoquine MR, with an antimode value of about 12.5, clearly separating EMs from PMs. Although metoprolol's urinary excretion may be affected by urinary pH (218) and the drug's metabolism exhibits stereoselectivity (223), neither of these factors appears to be an important variable in the trait measure, which has been shown to be reproducible with respect to phenotypic assignment (220). A single-point, 3-hr postdose metabolic ratio approach has also been suggested based on its good agreement with the equivalent urinary trait measure (224); however, its use has been limited.

Despite the relative safety of metoprolol and its general availability for clinical studies, the metoprolol metabolic ratio has been used to only a limited extent as an in vivo probe for assessing CYP2D6 activity. Such use has been directed mainly toward the investigation of racial/geographic differences in CYP2D6 and the dissociation between measures of such activity provided by different in vivo probes (Sec. VI.E). One reason for this limited use has been the application of an even safer and more widely used drug for phenotyping purposes, namely, dextromethorphan.

D. Dextromethorphan

Dextromethorphan (3-methoxy-17-methylmorphinan) is a widely used and effective non-narcotic antitussive. After oral administration, it is rapidly and extensively metabolized in humans by O- and N-demethylation to form dextrorphan and 3-methoxymorphinan; a small amount of secondary metabolite, 3-hydroxymorphinan, is also formed. N-Demethylation is mediated mainly by

CYP3A, whereas the formation of the major metabolite, dextrorphan, is determined by CYP2D6 (38). This dependency led to the suggestion that the conversion of dextromethorphan to dextrorphan could be used to assess in vivo CYP2D6 activity (225). Moreover, its high safety profile and global availability would permit its universal application, even in subjects where use of unapproved drugs like sparteine and debrisoquine was not possible (e.g., children, pregnant women).

A 0–8-hr urinary metabolic ratio (dextromethorphan:dextrorphan) is usually used as the trait measure following an oral dose of 15-40 mg dextromethorphan hydrobromide in a solid dosage form or as cough syrup. Because dextrorphan is conjugated prior to excretion, hydrolysis of the urine by pretreatment with (β-glucuronidase is usually performed, although some investigators have suggested that this may not be necessary (226). Using the conventional approach, an antimode of 0.3 is able to discriminate between EMs and PMs (225). Pharmacokinetic studies have substantiated that the urinary metabolic ratio reflects the plasma levels of the unchanged drug to its metabolite (227), and possible factors affecting the trait measure have been investigated (228). It is also possible to determine a single time-point, dextromethorphan metabolic ratio in plasma within 2–5 hr after an oral dose or in a 6-hr saliva sample (229,230); however, neither of these alternative approaches has been widely used.

In general, excellent agreement has been obtained between CYP2D6 phenotypic assignment based on the use of dextromethorphan compared to debrisoquine, both in vitro (231,232) and in vivo (38,225,233). Furthermore, concordance between EM and PM CYP2D6 genotypes and dextromethorphan's metabolic ratio has more recently been established (217,233–235). On the other hand, this trait measure appears less suitable for defining subphenotypes within the EM phenotype, since there is considerable overlap in the trait value between the groups (217). Accordingly, it is not possible to identify ultrarapid or intermediate metabolizers. However, the dextromethorphan urinary metabolic ratio is modulated by factors that alter CYP2D6 activity, such as inhibition by quinidine and substrates like selective serotonin-reuptake inhibitors (236,237).

E. Correlations Between Different CYP2D6 Probes

In populations of European descent, good agreement in the CYP2D6 phenotype has generally been found regardless of the particular in vivo probe that has been used. For example, a high correlation ($r = 0.81$) was observed in a white British population between the MRs of debrisoquine and metoprolol (221). In Japanese, similar good relationships ($r = 0.78–0.83$) were obtained between alternative probes (238). However, discordances have been noted in other populations, especially African. For example, in Ghanaians and Zambians much lower correlations ($r = 0.41$ and 0.60) were found between the urinary MRs of sparteine, metoprolol, and debrisoquine (239,240). Also, in another study the correlation between the MRs of debrisoquine, sparteine, and dextromethorphan were considerably weaker in Ghanaians compared with Caucasians and Chinese (241). Moreover, no polymorphism was apparent in the population distribution curve in Nigerians using either debrisoquine or sparteine (212,242), and no significant relationship ($r = 0.31$) was found between debrisoquine and sparteine MRs (243). Studies in black Africans of the Venda have also indicated dissociations between the various in vivo CYP2D6 probes, namely, a log-normal distribution of the sparteine MR with no evidence of a PM subgroup, whereas phenotyping with debrisoquine identified a 4% prevalence of this phenotype (244,245). However, the metoprolol MR indicated a PM incidence of 7.4% (246). Similarly, only 1 of 18 PMs in a San Bushmen population in South Africa phenotyped with debrisoquine (247) were subsequently found to have impaired metabolism of metoprolol (246), and a similar discordance was noted in Zimbabweans (211). It has been speculated that such dissociations might reflect poor patient compliance with the clinical protocol or that the antimode value established in populations of European descent does not always apply to other populations. However, a more likely explanation, which is also consistent with the shift to the right of the frequency distribution curve of the phenotypic trait measure in African populations compared with those of European descent, involves differences in allele frequency, especially CYP2D6*17. This variant allele, which expresses a protein with reduced catalytic activity compared with the wild-type gene (248,249), is not present in Europeans but is common among black African populations (203,207,248). Significantly, such decreased CYP2D6 activity is associated with reduced affinity for some CYP2D6 substrates and altered substrate specificity (249), which could also account for the insensitivity of Africans to the effect of quinidine on CYP2D6 (241). Alternatively, a presently unidentified CYP2D6 variant may be present in such populations.

In summary, any of the described in vivo probes and associated trait measures provide safe, simple, and practical approaches for identifying CYP2D6 EMs and PMs, and these have been validated and successfully applied by numerous investigators, especially in non-African populations. The choice of probe depends mainly on the local regulatory situation and the availability of the particular drug and its metabolites. Debrisoquine and sparteine appear to be more suitable than dextromethorphan for subphenotyping within the EM population. However, in neither case does it appear possible to identify an individual's genotype on the basis of the subphenotypic trait value. Just as important is the reverse situation, namely, the CYP2D6 genotype does not predict the level of CYP2D6 catalytic activity within EMs, which can be established only by phenotyping. Likewise, modulation of such activity requires the use of an in vivo probe, and all three of the widely used approaches appear to be sufficiently sensitive for this purpose. The situation in various African and possibly other populations (250,251) is less clear because of the high prevalence of CYP2D6*17, which appears to have different substrate and inhibitor specificities than other allelic variants. As a result, identification of

PMs in such groups by phenotyping is not as clear-cut as in European- and Southeast Asian-derived populations; this problem is compounded by the low prevalence of this subgroup in these populations. This would appear to be one situation in which putative classification as a PM requires confirmation by genotyping.

VII. CYTOCHROME P450 2E1

CYP2E1 is importantly involved in the metabolic activation of a large number of environmental xenobiotics many of which have carcinogenic or toxic effects, for example, *N*-nitrosamines, benzene, styrene, and halogenated hydrocarbons. Thus, along with CYP2A6 (sec. IV) this isoform is considered to be important in the etiology of and individual susceptibility to disease states associated with exposure to such agents. CYP2E1 also mediates the metabolism of a small number of drugs, which include ethanol, acetaminophen, chlorzoxazone, and certain fluorinated anesthetic agents (44,252). The regulation of CYP2E1 is complex and can be affected by both physiological and exogenous factors, such as drug-induced inhibition and induction. Thus, in vivo CYP2E1 activity varies quite markedly within a population, with the basal variability being about four- to fivefold; however, this can be greater if obese individuals and chronic alcoholics are included (25,28). In addition, CYP2E1 activity is normally distributed; to date, no evidence of polymorphism or rare individuals with unexplained impairment have been described (28).

Several genetic polymorphisms have been identified in the CYP2E1 gene (253–256). This has led to considerable speculation regarding their possible involvement as risk factors in, for example, alcoholic liver disease and cancer. Efforts to address such issues are complicated by the fact that the frequencies of the polymorphisms vary substantially according to the racial/geographic characteristics of the study population. As a result, such molecular epidemiological findings have become controversial, since the resulting data are often conflicting and suffer from low statistical power. The latter problem also applies to population studies attempting to relate genotype to phenotype. In general, CYP2E1 activity as measured both in vitro (256–259) and in vivo (28,259,260) by the 6-hydroxylation of chlorzoxazone does not appear to be under genetic regulation by the various identified allelic variants. However, a recent study investigating the RsaI restriction fragment length polymorphism in the 5′-regulatory sequence of the CYP2E1 gene (C1019T) found that chlorzoxazone's oral clearance was greatest in homozygous wild-type individuals (c_1/c_1) and decreased with the number of variant c_2 alleles (261). However, the difference in CYP2E1 activity between the two homozygous groups was less than twofold. In addition, a recent report has described a further mutation in the 5′-regulatory region that appears to be involved in the inducibility of CYP2E1 activity (255). It is also not clear what role, if any, such genetic factors may have in the 30–40% lower CYP2E1 activity noted in populations of Japanese ancestry and possibly other Southeast Asians compared with those of European descent (259,261).

The only in vivo probe developed and used for assessing human CYP2E1 activity has been chlorzoxazone, based on the finding that this isoform selectively mediates the drug's 6-hydroxylation (262). Subsequent studies also found that CYP1A1 was able to catalyze this reaction (263,264); however, this is not a constitutive isoform, and its affinity for chlorzoxazone is much less than that of CYP2E1. Accordingly, its overall role in vivo is likely to be very minor (264). This probability is supported by the observation that chlorzoxazone's metabolism is similar in nonsmokers and smokers in whom CYP1A1 would be expected to be induced (25). Preliminary in vitro evidence has been presented showing that recombinant CYP1A2 also mediates chlorzoxazone's 6-hydroxylation (265), but this has not been confirmed (264). Similarly, it is not clear whether CYP3A has a role in the in vivo formation of chlorzoxazone's 6-hydroxy metabolite as has been suggested by studies in vitro (266). Overall, the current available evidence indicates that the in vivo 6-hydroxylation of chlorzoxazone in humans is predominantly, if not exclusively, mediated by CYP2E1, and its measurement provides a valid estimate of the enzyme's activity.

Chlorzoxazone [5-chloro-2(3H)-benzoxazolone] is a skeletal muscle relaxant that has been approved for clinical use for over 40 years. Single-dose (250 mg) oral administration is safe and well tolerated, and 6-hydroxylation is the major pathway of elimination, accounting for about 50–80% of the dose (25,28). This metabolite is rapidly conjugated; therefore, determination of its concentration in either plasma or urine requires pretreatment with either β-glucuronidase or acid, respectively (25,267,268). The gold standard for assessing CYP2E1 activity is estimation of chlorzoxazone's fractional oral clearance to its 6-hydroxy metabolite, following an overnight fast, based on determination of the drug's plasma level-time profile over eight hours and measurement of the metabolite's 0- to 8-hour or 0- to 12-hour urinary recovery (28,259,269). Because ethanol is an inhibitor of CYP2E1, while present in the body and an inducer after chronic use, it is important that alcoholic beverages not be consumed for about 72 hours prior to phenotyping. Such a clearance approach has been validated in humans by a number of studies demonstrating that factors known to alter CYP2E1 activity in animals or in vitro also modulate chlorzoxazone's clearance in an appropriate fashion. These include mechanism-based inhibition by disulfiram (269), chlormethiazole (270), and phytochemicals in watercress (271), as well as induction by obesity (267) and pre treatment with isoniazid (272,273). Importantly, the last interaction is masked by the antitubercular agent's inhibition of CYP2E1 while it is present in the body. In contrast to findings in animals, fasting over 36 hours reduces CYP2E1 activity compared with a standard 8- to 12-hour overnight fast (267). The main practical disadvantage of this clearance approach is the need to obtain multiple blood samples over eight hours to define the plasma concentration–time profile along with collection of urine for 8 to 12 hours. Because 6-hydroxylation is the major route of elimination of chlorzoxazone, use of the probe's oral clearance is probably as informative as the formation clearance estimate. Besides simplifying the phenotyping procedure, another advantage of this latter trait measure is that it reduces the error of the value's estimation associated, for example, with incomplete collection of urine. However, extensive blood sampling is still required; for this reason, a simpler MR approach has been advocated. Chlorzoxazone is generally rapidly absorbed, with a peak value approximately two hours after oral administration, and plasma levels of the 6-hydroxy metabolite are detectable prior to this time (25,28). Thus, a single-time-point, plasma MR (6-hydroxychlorzoxazone:chlorzoxazone) is readily determinable. Different investigators have used different times to determine this trait measure. Many employ a value determined two hours after oral chlorzoxazone administration, based on a high correlation ($r = 0.88$) between it and the clearance of chlorzoxazone to its 6-hydroxy metabolite (25,268). Also, the ratio of the areas under the plasma concentration–time curves for the chlorzoxazone and its 6-hydroxy metabolite has been found to correlate with the MR (25). Others have suggested that a four-hour sampling time point is more optimal and similarly have reported a high correlation (0.89) between the two trait measures (26). However, in larger population studies such correlations have been found to be much lower; for example, $r = 0.42$ to 0.53 (28). Nevertheless, chlorzoxazone's plasma MR undoubtedly reflects CYP2E1 activity and is reduced by inhibition, e.g., chlormethiazole (274) and liver disease (275,276), and increased by induction, e.g., ethanol (25,277). Moreover, it has been demonstrated to provide a simple method for investigating temporal aspects associated with changes in CYP2E1 activity, as, for example, following the effects of ethanol withdrawal in chronic alcoholics (277).

Phenotypic trait measures based solely on the urinary excretion of 6-hydroxychlorzoxazone have also been reported based on the amount of metabolite excreted in zero to eight hours (278), or a "hydroxylation index" approach (279), or estimating the elimination half-life of the metabolite (280). However, the validity of these approaches is highly questionable, so they have not been widely applied.

VIII. CYTOCHROME P450 3A

The human CYP3A subfamily includes at least three functional proteins, CYP3A4, CYP3A5, and CYP3A7; however, the last one is not found in significant amounts in adults. The substrate specificities of CYP3A4 and CYP3A5 are in general similar, although some important distinctions do exist; for example, erythromycin and quinidine do not appear to be metabolized by CYP3A5, although they are good CYP3A4 substrates (44,281). Because of this and the current difficulty in distinguishing between the individual CYP3A isoforms, which may be present in different amounts in the same tissue, they are collectively referred to as CYP3A. Importantly, CYP3A is the most abundant of all of the human CYP isoforms and constitutes, on average, about 30% of total CYP protein in the liver. CYP3A is also present in the small intestinal epithelium, particularly in the apical region of mature enterocytes at the tip of the microvillus, where it accounts for about 70% of total CYP protein and is present at about 50% of the hepatic level. Catalytic activity, mainly associated with CYP3A5, is also present in the kidney (44,281).

The substrate specificity of CYP3A is very broad; accordingly, an extremely large number of structurally divergent chemicals are metabolized by a variety of different pathways, often in a regio- and stereoselective fashion (44). Estimates, based primarily on in vitro studies, suggest that the metabolism of about 40–50% of drugs used in humans involves CYP3A, and frequently such biotransformation is extensive. Moreover, because of its localization in the intestinal epithelium and liver, CYP3A is an important factor in the first pass metabolism of drugs following their oral administration (282). A further characteristic of CYP3A is the large interindividual variability in activity, which reflects both genetic and environmental factors. Basal variability appears to be about fivefold, but this range can be significantly increased by inhibition or induction. In this respect, CYP3A is extremely prone to both types of interactions (281). Furthermore, the distribution of CYP3A activity is unimodal, and to date no evidence of a functional polymorphism has been reported. Recently, it has been recognized that the substrate/inhibitor specificity of CYP3A overlaps with that of the membrane efflux transporter termed P-glycoprotein (283). However, this overlap is not complete, and it is probably a fortuitous one that reflects the broad substrate specificities of the individual proteins (284). In fact, several important CYP3A substrates, such as nifedipine and midazolam, are not transported by P-glycoprotein (284). Nevertheless, the colocalization of the two proteins at important sites for drug disposition, such as the enterocyte and hepatocyte, results in an interrelationship that functions in a concerted fashion to reduce the intracellular drug concentration. Thus, the hepatic elimination, for example, of erythromycin—a substrate for both P-glycoprotein and CYP3A—is determined by both proteins. In addition, the coadministration of two CYP3A substrates can result in drug interactions that reflect inhibition of metabolism alone, reduced P-glycoprotein efflux only, or a combination of both effects (281).

The overall importance of CYP3A in human drug metabolism and the propensity of this isoform's activity to be readily modulated by drug interactions have resulted in considerable effort to identify a suitable in vivo probe. In principle, any of the many drugs metabolized by CYP3A could be used to determine the enzyme's activity in the body. This, of course, is the approach used to determine whether a significant interaction occurs between a specific CYP3A substrate and a known inhibitor or inducer. On the other hand, investigation of whether a new drug candidate interacts with CYP3A and studying determinants of the enzyme's activity require a more focused approach. It is now recognized that measurement of the endogenous 6-β-hydroxylation of cortisol does not fulfill this requirement (sec. I), and therefore alternative approaches have been investigated (285).

A. Dapsone

Dapsone (4,4′-diaminodiphenylsulfone) has been widely used for phenotyping with respect to acetylation by NAT-2; however, the drug is also N-hydroxylated. Formation of the hydroxylamine metabolite by human liver microsomes was found to be selectively mediated by CYP3A (286); this led to the development of a zero- to eight-hour urinary metabolic recovery ratio approach [dapsone hydroxylamine:(dapsone + dapsone hydroxylamine)] to quantitatively assess this pathway of metabolism (287,288). Subsequently, the trait measure has been applied as part of a cocktail approach (35) in a number of studies investigating the putative role of CYP3A as a risk factor in cancer (289–291) and other disease states (288,292,293).

Attempts to correlate the dapsone recovery ratio to other measures of CYP3A activity have, however, been disappointing (11,294). This lack of success is not surprising, since it is now recognized that several CYP isoforms besides CYP3A contribute to dapsone's N-hydroxylation, including CYP2E1 (295) and CYP2C9 (296). Furthermore, CYP3A is involved to a major extent only at high concentrations that are not achieved in vivo (295). Additionally, it has been found that the dapsone urinary recovery ratio is not altered by pretreatment with ketoconazole, even though this markedly affects CYP3A activity as measured by the erythromycin breath test (297). Collectively, this evidence strongly suggests that the dapsone recovery ratio is not a useful trait measure to selectively measure CYP3A activity. Nevertheless, there may be some merit in continuing its use as an in vivo probe in certain types of investigations, since a high trait value appears to be associated with a reduced risk of developing aggressive bladder cancer (289–291).

B. Erythromycin

CYP3A selectively N-demethylates erythromycin, and if [^{14}C]-N-methyl drug is used, the carbon of the methyl group is eventually excreted as $^{14}CO_2$. Measurement of radioactivity present in a breath sample collected after intravenous administration and expressing this as a fraction of the dose excreted per hour is the basis of the erythromycin breath test (285). Since the first introduction of this phenotyping procedure, the sampling schedule has changed from multiple collections over one to two hours (31,298,299) to a single sample 20 minutes after drug administration (300–303). Considerable validation and application of this simple and rapid phenotyping approach have been reported (31,285,298–304), and there is no question that the erythromycin breath test provides a measure of CYP3A activity under certain circumstances and for some types of investigation.

A limiting factor of this breath test is that it appears to measure only CYP3A4-mediated metabolism and not that involving CYP3A5; thus, overall CYP3A activity is underestimated, especially in the 25–30% of individuals with significant hepatic levels of this isoform (281). More important, however, is the fact that the erythromycin breath test reflects predominantly CYP3A4 activity only in the liver and, therefore, metabolism in the intestinal tract is not measured. This is an obvious limitation with respect to an orally administered drug and may account, in part, for the relatively poor correlation between this trait measure and the oral clearance of several other CYP3A substrates (10,11,294). Similar poor relationships have also been observed with regard to the breath test and the systemic clearance of drugs that are CYP3A substrates (12,305), which is somewhat unexpected, since the intravenous route of administration is common to both these estimates of drug metabolism. It is, therefore, unclear what the erythromycin breath test is exactly measuring. Clearly this is not erythromycin's clearance or the CYP3A-mediated formation clearance to the N-demethylated metabolite, since the antibiotic's half-life is one to two hours and breath sampling is at 20 minutes. The recent findings that erythromycin is a substrate for P-glycoprotein (284,306) and that this probably contributes to the significant biliary excretion of erythromycin (307) further complicate the situation.

Despite these questions and limitations, the erythromycin breath test provides a practical and useful means of assessing CYP3A activity, especially in studies involving its potential modulation. For example, serial monitoring of relative CYP3A activity has proven of importance in the drug development process for identifying whether a drug candidate or a new drug affects CYP3A activity and would, therefore, potentially produce a drug interaction with other CYP3A substrates. Even though the findings reflect only hepatic CYP3A4, there should be sufficient evidence that an interaction does (302,304,308) or does not (309,310) occur. Only in the former situation would further targeted studies with additional CYP3A substrates be necessary. However, the need for an in vivo probe that takes into account the CYP3A activity involved in presystemic metabolism related to the oral first-pass effect has led to the investigation of approaches other than the erythromycin breath test.

C. Midazolam

In humans, midazolam is rapidly and almost completely metabolized to its primary 1′-hydroxy metabolite and, to a much lesser extent, to 4-hydroxymidazolam. Both of these pathways are selectively mediated by CYP3A (311,312). In addition, both intestinal and hepatic microsomes exhibit high midazolam hydroxylation activity, which in the case of the liver is significantly correlated with the drug's systemic clearance (313). Moreover, scale-up of such in vitro measures (282) was found to provide an excellent prediction of the in vivo extraction ratios of the two organs (313,314). Liver dysfunction markedly impairs midazolam's elimination (315,316), and plasma levels during the anhepatic phase of liver transplantation are elevated (317). A further characteristic of midazolam's metabolism is that it is readily altered by administration of known CYP3A inhibitors (e.g., azole antifungal agents, certain macrolide antibiotics, HIV protease inhibitors, and grapefruit juice) and inducing agents (e.g., anticonvulsants and rifampin) (281). Collectively, these characteristics fulfill most of the criteria usually accepted for validation of an in vivo probe (sec. II.C).

A plasma-metabolic-ratio (l′-hydroxymidazolam:midazolam) approach based on a single-time-point value at 30 minutes following an intravenous dose of the probe has been considered as a possible simple phenotypic trait value (313,318). However, this trait measure now appears to be less valid and useful than originally suggested (314). Accordingly, a clearance approach is currently the only available way to assess CYP3A activity with this in vivo probe. Ideally, a formation clearance, involving measurement of the amount of l′-hydroxy metabolite formed and eliminated as conjugate in the urine, provides the gold standard. However, simply using the drug's plasma clearance without partitioning this into its individual pathways would also appear to be a valid approach, since midazolam's metabolism and systemic elimination appear to be predominantly, if not exclusively, mediated by CYP3A.

Since midazolam's elimination half-life is only one to two hours, phenotyping consists of determining the drug's plasma concentration-time curve by obtaining multiple samples for 6 to 8 hours following administration of a suitable dose and, if required, a 0- to 8-hour or 0- to 12-hour urine collection. If midazolam is given by the oral route, then the measure of CYP3A activity reflects both intestinal and hepatic CYP3A activity. Thus, the phenotypic trait allows quantification of CYP3A function in a way that can be related to therapeutic situations involving oral drug administration. So long as the oral midazolam dose is 5 to 7.5 mg, adverse effects associated with the drug's sedative effect are generally minimal, unless CYP3A activity is impaired, when a lower dose should be used. Significantly, midazolam's oral clearance is very sensitive to modulation of CYP3A and, therefore, is capable of detecting small changes/differences in the level of activity. For example, rifampin pretreatment was found to reduce the area under the drug's plasma concentration–time curve by 96% (319), whereas ketoconazole increased this parameter by 15-fold (320). Accordingly, the full range of CYP3A activity that is possible in vivo is several 100-fold. Moreover, more modest changes (281), including no effect at all (321–323), can be readily detected.

Midazolam may also be safely administered in doses below about 2 mg by the intravenous route, but in some subjects this is accompanied by mild sedation, which requires appropriate clinical monitoring. Thus, estimation of the drug's systemic clearance resulting from CYP3A-mediated metabolism is possible, and in contrast to the drug's oral clearance, this measure appears to reflect predominantly hepatic elimination; i.e., the contribution of extrahepatic CYP3A is relatively small (317). This probably reflects the fact that intestinal CYP3A is localized in the apical region of the enterocytes and, therefore, access of drug in the systemic blood is limited. Moreover, midazolam is extensively bound to plasma proteins (>98%), which further impairs midazolam's distribution to this site. Accordingly, a second CYP3A trait measure is available that focuses mainly on the enzyme's activity in the liver. By appropriate analysis of both of these clearance values determined in the same individual, it is possible to further characterize CYP3A activity by estimating the separate contributions of the intestine and the liver to the overall metabolism of midazolam (314). These two clearance estimates may be obtained by serially administering midazolam by the two routes of administration on separate occasions. However, a more desirable approach, which removes any error associated with intrasubject variability in CYP3A activity, employs the simultaneous administration of a stable-labeled [$^{15}N_3$] form of the drug along with unlabeled drug by the other route. Recent applications of this elegant approach have found that CYP3A in the intestine contributes almost equally to midazolam's first-pass metabolism after an oral dose (314). Also, the inhibition of CYP3A activity by pretreatment with clarithromycin affects mainly the intestinal enzyme, which in turn is the major determinant of the drug's interindividual variability and not hepatic CYP3A activity (324). The potential to obtain further mechanistic understanding about drug-drug interaction involving CYP3A using this approach would appear to be considerable.

Both the erythromycin breath test and estimation of the oral and/or intravenous clearances of midazolam appear to provide practical and useful information of the in vivo level of CYP3A activity. However, it is clear that these different phenotypic trait values reflect different aspects of such activity. Accordingly, selection of the more appropriate approach depends to a large extent on the particular question being addressed. In general, the erythromycin breath test provides a simple but relative measure that is well suited for temporal monitoring of hepatic CYP3A4 activity, as would be applicable to detect alterations in this enzyme's activity associated with inhibition or induction caused by a drug interaction. On the other hand, the use of midazolam provides a more absolute measure, and, in the case of the drug's oral clearance, this finding can be quantitatively extrapolated to other CYP3A substrates when these are used in a therapeutic situation. Attempts have been successfully made to combine the use of the two in vivo probe drugs by administering them concurrently (325). However, the simultaneous administration of midazolam by the oral and intravenous routes would appear to be a more rigorous and informative approach (314,324), providing the necessary analytical instrumentation is available.

IX. PERSPECTIVES

Over the past decade, considerable progress has occurred in the development and application of in vivo probes suitable for assessing the catalytic activity of individual CYP isoforms in human subjects. CYP2C19 and CYP2D6 have benefited most from this effort, which has focused mainly on phenotypic classification because of the extremely large interindividual variability in enzyme activity across the population distribution curve, i.e., from ultrarapid/extensive to PMs. It is likely that a similar level of investigation will occur with the other isoforms that have more recently been shown to have rare or polymorphic variant alleles, e.g., CYP2A6, CYP2C9, and CYP2E1. Undoubtedly, molecular genetic approaches will continue to identify new genomic polymorphisms for these and possibly other CYP isoforms, similar to the situation that has been found with the more established genetic polymorphisms. A future critical issue, therefore, will be to establish any functional significance of these mutant alleles by appropriate in vitro approaches and, importantly, to confirm that a genotype:phenotype relationship exists and is important in the in vivo setting. Such studies will be facilitated by the availability of new, simple, reliable, and valid phenotypic trait measures for the isoforms of interest.

In general, the in vivo activity of most of the characterized and expressed CYP isoforms that are important in the metabolism of drugs and other xenobiotics of toxicologic interest can currently be assessed, at least at the level of the whole organism. The exception is CYP2B6; an isoform whose role and importance in the metabolism of xenobiotics has yet to be adequately defined. Recent studies suggest that CYP2B6 may be more important than previously considered, despite the fact that it is a minor hepatic CYP (326,327). However, identification of a selective probe, even for in vitro studies, has been problematic (326,328), although the N-demethylation of S-mephenytoin to nirvanol shows promise in this regard (328,329). Unfortunately, the substrate concentrations used with this putative probe make it unsuitable for in vivo phenotyping, since they are rarely encountered in humans (329). CYP2B6 was the only one of 10 recombinant expressed human CYP isoforms able to *trans*-hydroxylate the investigational drug RP73401 (330). Since this is the primary route of metabolism of this compound in vivo, it is possible that a suitable phenotypic trait measure for CYP2B6 could be developed using RP73401, if the drug becomes clinically available.

All of the available in vivo probes and associated trait measures appear to be sufficiently sensitive and suitable for evaluating changes/differences in the particular isoform's level of activity. Accordingly, they may be applied to investigating the presence or absence of a drug interaction and provide insight into its mechanism. Selection of the most appropriate approach, when several in vivo probes or trait measures are available for a particular isoform, depends to a large extent on the purpose of the study. Indirect trait values based on a single point determination such as a saliva/plasma/urine MR are well suited to screening studies designed to answer the question of whether an interaction occurs or not. Incorporating several in vivo probes into a cocktail strategy further facilitates this goal. On the other hand, more quantitative questions related to the extent to which metabolism is inhibited or induced and to sites of interaction (intestine versus liver) may require the use of trait values based on more direct measures, such as clearance approaches. Regardless, interpretation of any change in the trait measure is critically dependent on an understanding of its basis and limitations. Finally, it should be appreciated that the in vivo evaluation of enzyme activity is in most cases complementary to information obtained by applying the approaches of molecular genetics. However, it has the added advantage that it also reflects the contributions of other determinants, including the effects of environmental factors and disease states; moreover, in many instances, phenotyping has direct therapeutic relevance.

ACKNOWLEDGMENT

This work was supported, in part, by grants GM31304 and CA76020 from the United States Public Health Service.

REFERENCES

1. Gillette JR. Keynote address: man, mice, microsomes, metabolites, and mathematics—40 years after the revolution. Drug Metab Rev 1995; 27:1–44.
2. Griese EU, Zanger UM, Bradermanns U, et al. Assessment of the predictive power of genotypes for the in vivo catalytic function of CYP2D6 in a German population. Pharmacogenetics 1998; 8:15–26.
3. Goldstein JA, Blaisdell J. Genetic tests which identify the principal defects in CYP2C19 responsible for the polymorphism in mephenytoin metabolism. In: Johnson EF, Waterman MR, eds. Cytochrome P450, Part B: Methods in Enzymology. San Diego, CA: Academic Press, 1996; 272:210–218.
4. Vesell ES, Page JG. Genetic control of drug levels in man: antipyrine. Science 1968; 161:72–73.
5. Park BK. Assessment of urinary 6(3-hydroxycortisol as an in vivo index of mixed-function oxygenase activity. Br J Clin Pharmacol 1981; 12:97–102.
6. Wilkinson GR. Prediction of interpatient variability of drug metabolizing ability. In: Wilkinson GR, Rawlins MD, eds. Drug Metabolism and Disposition: Considerations in Clinical Pharmacology. Lancaster, PA: MTP Press, 1985, 183–209.
7. Waxman DJ, Attisano C, Guengerich FP, et al. Human liver microsomal steroid metabolism: identification of the major microsomal steroid hormone 6β-hydroxylase cytochrome P-450 enzyme. Arch Biochem Biophys 1988; 263:424–436.
8. Ged C, Rouillon JM, Pichard L, et al. The increase in urinary excretion of 6β-hydroxycortisol as a marker of human hepatic cytochrome P450IIIA induction. Br J Clin Pharmacol 1989; 28:373–387.
9. Watkins PB, Turgeon DK, Saenger P, et al. Comparison of urinary 6-beta-cortisol and the erythromycin breath test as measures of hepatic P450IHA (CYP3A) activity. Clin Pharmacol Ther 1992; 52:265–273.
10. Hunt CM, Watkins PB, Saenger P, et al. Heterogeneity of CYP3A isoforms metabolizing erythromycin and cortisol. Clin Pharmacol Ther 1992; 51:18–23.
11. Kinirons MT, O'Shea D, Downing TE, et al. Absence of correlations among three putative in vivo probes of human cytochrome P4503A activity in young healthy men. Clin Pharmacol Ther 1993; 54:621–629.

12. Kinirons MT, O'Shea D, Groopman JD, et al. Route of administration does not explain the lack of correlation between putative in vivo probes of cytochrome P4503A. Br J Clin Pharmacol 1994; 37:501P–502P.
13. Seidegard J, Dahlstrom K, Kullberg A. Effect of grapefruit juice on urinary 6β-hydroxycortisol/cortisoi excretion. Clin Exp Pharmacol Physiol 1998; 25:379–381.
14. Kovacs SJ, Martin DE, Everitt DE, et al. Urinary excretion of 6-beta-hydroxycortisol as an in vivo marker for CYP3A induction: applications and recommendations. Clin Pharmacol Ther 1998; 63:617–622.
15. Fleishaker JC, Pearson LK, Peters GR. Induction of tirilazad clearance by phenytoin. Biopharm Drug Dispos 1998; 19:91–96.
16. Monsarrat B, Chatelut E, Royer I, et al. Modification of paclitaxel metabolism in a cancer patient by induction of cytochrome P450 3A4. Drug Metab Dispos 1998; 26:229–233.
17. Wilkinson GR. Clearance approaches in pharmacology. Pharmacol Rev 1987; 39: 1–47.
18. Jackson PR, Tucker GT, Lennard MS, et al. Polymorphic drug oxidation: pharmacokinetic basis and comparison of experimental indices. Br J Clin Pharmacol 1986; 22:541–550.
19. Dalen P, Dahl ML, Eichelbaum M, et al. Disposition of debrisoquine in Caucasians with different CYP2D6-genotypes including those with multiple genes. Pharmacogenetics 1999; 9:697–706.
20. Inaba T, Otton SV, Kalow W. Debrisoquine hydroxylation capacity: problems of assessment in two populations. Clin Pharmacol Ther 1981; 29:218–223.
21. Kaisary A, Smith P, Jacqz E, et al. Genetic predisposition to bladder cancer: ability to hydroxylate debrisoquine and mephenytoin as risk factors. Cancer Res 1987; 47:5488–5493.
22. Kiipfer A, Preisig R. Pharmacogenetics of mephenytoin: a new drug hydroxylation polymorphism in man. Eur J Clin Pharmacol 1984; 26:753–759.
23. Wedlund PJ, Aslanian WS, McAllister CB, et al. Mephenytoin hydroxylation deficiency in Caucasians: frequency of a new oxidative drug metabolism polymorphism. Clin Pharmacol Ther 1984; 36:773–780.
24. Kevorkian JP, Michel C, Hofmann U, et al. Assessment of individual CYP2D6 activity in extensive metabolizers with renal failure: comparison of sparteine and dextromethorphan. Clin Pharmacol Ther 1996; 59:583–592.
25. Girre C, Lucas D, Hispand E, et al. Assessment of cytochrome P4502E1 induction in alcoholic patients by chlorzoxazone pharmacokinetics. Biochem Pharmacol 1994; 47:1503–1508.
26. Frye RF, Adedoyin A, Mauro K, et al. Use of chlorzoxazone as an in vivo probe of cytochrome P450 2E1: choice of dose and phenotypic trait measure. J Clin Pharmacol 1998; 38:82–89.
27. Houston JB. Drug metabolite kinetics. Pharmacol Ther 1982; 15:521–552.
28. Kim RB, O'Shea D, Wilkinson GR. Interindividual variability of chlorzoxazone 6-hydroxylation in men and women and its relationship to CYP2E1 genetic polymorphisms. Clin Pharmacol Ther 1995; 57:645–655.
29. Lambert GH, Kotake AN, Schoeller D. The CO_2 breath tests as monitors of the cytochrome P450 dependent mixed function monooxygenase system. Prog Clin Biol Res 1983; 135:119–145.

30. Lane EA, Parashos I. Drug pharmacokinetics and the carbon dioxide breath test. J Pharmacokinet Biopharm 1986; 14:29–49.
31. Lown KS, Thummel KE, Benedict PE, et al. The erythromycin breath test predicts the clearance of midazolam. Clin Pharmacol Ther 1995; 57:16–24.
32. Breimer DD, Schellens JHM. A ''cocktail'' strategy to assess in vivo oxidative drug metabolism in humans. Trends Pharmacol Sci 1990; 11:223–225.
33. Sanz EJ, Villen T, Alm C, et al. S-Mephenytoin hydroxylation phenotypes in a Swedish population determined after coadministration with debrisoquin. Clin Pharmacol Ther 1989; 45:495–499.
34. Schellens JHM, Ghabrial H, van der Wart HHF, et al. Differential effects of quinidine on the disposition of nifedipine, sparteine, and mephenytoin in humans. Clin Pharmacol Ther 1991; 50:520–528.
35. Frye RF, Matzke GR, Adedoyin A, et al. Validation of the five-drug ''Pittsburgh cocktail'' approach for assessment of selective regulation of drug-metabolizing enzymes. Clin Pharmacol Ther 1997; 62:365–376.
36. Endres HGE, Henschel L, Merkel U, et al. Lack of pharmacokinetic interaction between dextromethorphan, coumarin and mephenytoin in man after simultaneous administration. Pharmazie 1996; 51:46–51.
37. Adedoyin A, Frye RF, Mauro K, et al. Chloroquine modulation of specific metabolizing enzymes activities: investigation with selective five drug cocktail. Br J Clin Pharmacol 1998; 46:215–219.
38. Jacqz-Aigrain E, Funck-Brentano C, Cresteil T. CYP2D6- and CYP3A-dependent metabolism of dextromethorphan in humans. Pharmacogenetics 1993; 3:197–204.
39. Jones DR, Gorski JC, Hamman MA, et al. Quantification of dextromethorphan and metabolites: a dual phenotypic marker for cytochrome P450 3A4/5 and 2D6 activity. J Chromatogr B Biomed Appl 1996; 678:105–111.
40. Ducharme J, Abdullah S, Wainer IW. Dextromethorphan as an in vivo probe for the simultaneous determination of CYP2D6 and CYP3A activity. J Chromatogr B Biomed Appl 1996; 678:113–128.
41. Schmider J, Greenblatt DJ, Fogelman SM, et al. Metabolism of dextromethorphan in vitro: involvement of cytochromes P450 2D6 and 3A3/4, with a possible role of 2E1. Biopharm Drug Dispos 1997; 18:227–240.
42. Kaminsky LS, Zhang ZY. Human P450 metabolism of warfarin. Pharmacol Ther 1997; 73:67–74.
43. Wienkers LC, Wurden CJ, Storch E, et al. Formation of (R)-8-hydroxywarfarin in human liver microsomes—a new metabolic marker for the (S)-mephenytoin hydroxylase, P4502C19. Drug Metab Dispos 1996; 24:610–614.
44. Guengerich FP. Human cytochrome P450 enzymes. In: Ortiz de Montellano PR, ed. Cytochrome P450: Structure, Mechanism, and Biochemistry. New York: Plenum, 1995; 473–535.
45. Kalow W, Tang BK. The use of caffeine for enzyme assays: a critical appraisal. Clin Pharmacol Ther 1993; 53:503–514.
46. Miners JO, Birkett DJ. The use of caffeine as a metabolic probe for human drug metabolizing enzymes. Gen Pharmacol 1996; 27:245–249.
47. Fuhr U, Rost KL, Engelhardt R, et al. Evaluation of caffeine as a test drug for CYPTA2, NAT2 and CYP2E1 phenotyping in man by in vivo versus in vitro correlations. Pharmacogenetics 1996; 6:159–176.

48. Nagel RA, Dirix LY, Hayllar KM, et al. Use of quantitative liver function tests—caffeine clearance and galactose elimination capacity—after orthotopic liver transplantation. J Hepatol 1990; 10:149–157.
49. Denaro CP, Jacob P III, Benowitz NL. Evaluation of pharmacokinetic methods used to estimate caffeine clearance and comparison with a Bayesian forecasting method. Ther Drug Monit 1998; 20:78–87.
50. Balogh A, Harder S, Vollandt R, et al. Intra-individual variability of caffeine elimination in healthy subjects. Int J Clin Pharmacol Ther Toxicol 1992; 30: 383–387.
51. Rostami-Hodjegan A, Nurminen S, Jackson PR, et al. Caffeine urinary metabolite ratios as markers of enzyme activity: a theoretical assessment. Pharmacogenetics 1996; 6:121–149.
52. Jost G, Wahllander A, von Mandach U, et al. Overnight salivary caffeine clearance: a liver function test suitable for routine use. Hepatology 1987; 7:338–344.
53. Fuhr U, Klittich K, Staib AH. Inhibitory effect of grapefruit juice and its bitter principal, naringenin, on CYP1A2 dependent metabolism of caffeine in man. Br J Clin Pharmacol 1993; 35:431–436.
54. Campbell ME, Spielberg SP, Kalow W. A urinary metabolite ratio that reflects systemic caffeine clearance. Clin Pharmacol Ther 1987; 42:157–165.
55. Wahllander A, Mohr S, Paumgartner G. Assessment of hepatic function—comparison of caffeine clearance in serum and saliva during the day and at night. J Hepatol 1990; 10:129–137.
56. Fuhr U, Rost KL. Simple and reliable CYP1A2 phenotyping by the paraxanthine/caffeine ratio in plasma and in saliva. Pharmacogenetics 1994; 4:109–116.
57. Wietholtz H, Voegelin M, Arnaud MJ, et al. Assessment of the cytochrome P-448 dependent liver enzyme system by a caffeine breath test. Eur J Clin Pharmacol 1981; 21:53–59.
58. Kotake AN, Schoeller DA, Lambert GH, et al. The caffeine CO_2 breath test: dose response and route of N-demethylation in smokers and nonsmokers. Clin Pharmacol Ther 1982; 32:261–269.
59. Horsmans Y, De Koninck X, Geubel AP, et al. Microsomal function in hepatitis B surface antigen healthy carriers: assessment of cytochrome P450 1A2 activity by the ^{14}C-caffeine breath test. Pharmacol Toxicol 1995; 77:247–249.
60. Rost KL, Brosicke H, Brockmoller J, et al. Increase of cytochrome P450IA2 activity by omeprazole: evidence by the ^{13}C-(N-3-methyl)-caffeine breath test in poor and extensive metabolizers of S-mephenytoin. Clin Pharmacol Ther 1992; 52:170–180.
61. Rost KL, Roots I. Accelerated caffeine metabolism after omeprazole treatment is indicated by urinary metabolite ratios: coincidence with plasma clearance and breath test. Clin Pharmacol Ther 1994; 55:402–411.
62. Renner E, Wietholtz H, Huguenin P, et al. Caffeine: a model compound for measuring liver function. Hepatology 1984; 4:38–46.
63. Pons G, Blais JC, Rey E, et al. Maturation of caffeine N-demethylation in infancy: a study using the $^{13}CO_2$ breath test. Pediatr Res 1988; 23:632–636.
64. Fontana RJ, Turgeon DK, Woolf TF, et al. The caffeine breath test does not identify patients susceptible to tacrine hepatotoxicity. Hepatology 1996; 23:1429–1435.
65. Breen KJ, Bury RW, Calder IV, et al. A (^{14}C) phenacetin breath test to measure hepatic function in man. Hepatology 1984; 4:47–52.

66. Schoeller DA, Kotake AN, Lambert GH, et al. Comparison of the phenacetin and aminopyrine breath tests: effect of liver disease, inducers and cobaltous chloride. Hepatology 1985; 5:276–281.
67. Tang BK, Kadar D, Quian L, et al. Caffeine as a metabolic probe: validation of its use for acetylator phenotyping. Clin Pharmacol Ther 1991; 49:648–657.
68. Notarianni LJ, Oliver SE, Dobrocky P, et al. Caffeine as a metabolic probe: a comparison of the metabolic ratios used to assess CYP1A2 activity. Br J Clin Pharmacol 1995; 39:65–69.
69. Denaro CP, Wilson M, Jacob P III, et al. Validation of urine caffeine metabolite ratios with use of stable isotope-labeled caffeine clearance. Clin Pharmacol Ther 1996; 59:284–296.
70. Tang BK, Zhou Y, Kadar D, et al. Caffeine as a probe for CYP1A2 activity: potential influence of renal factors on urinary phenotypic trait measurements. Pharmacogenetics 1994; 4:117–124.
71. Butler MA, Lange NP, Young JF, et al. Determination of CYP1A2 and NAT2 phenotypes in human populations by analysis of caffeine urinary metabolites. Pharmacogenetics 1992; 2:116–127.
72. Nakajima M, Yokoi T, Mizutani M, et al. Phenotyping of CYP1A2 in Japanese population by analysis of caffeine urinary metabolites: absence of mutation prescribing the phenotype in the CYP1A2 gene. Cancer Epidemiol Biomarkers Prev 1994; 3:413–421.
73. Yokoi T, Sawada M, Kamataki T. Polymorphic drug metabolism: studies with recombinant Chinese hamster cells and analyses in human populations. Pharmacogenetics 1995; 5:S65–S69.
74. Relling MV, Lin JS, Ayers GD, et al. Racial and gender differences in *N*-acetyltransferase, xanthine oxidase, and CYP1A2 activities. Clin Pharmacol Ther 1992; 52:643–658.
75. Kalow W, BK Tang. Caffeine as a metabolic probe: exploration of the enzyme-inducing effect of cigarette smoking. Clin Pharmacol Ther 1991; 49:44–48.
76. Kalow W, Tang BK. Use of caffeine metabolite ratios to explore CYP1A2 and xanthine oxidase activities. Clin Pharmacol Ther 1991; 50:508–519.
77. Vistisen K, Poulsen HE, Loft S. Foreign compound metabolism capacity in man measured from metabolites of dietary caffeine. Carcinogenesis 1992; 13:1561–1568.
78. Carrillo JA, Benftez J. Caffeine metabolism in a healthy Spanish population: *N*-acetylator phenotype and oxidation pathways. Clin Pharmacol Ther 1994; 55:293–304.
79. Catteau A, Bechtel YC, Poisson N, et al. A population and family study of CYP1A2 using caffeine urinary metabolites. Eur J Clin Pharmacol 1995; 47:423–430.
80. Rasmussen BB, Brosen K. Determination of urinary metabolites of caffeine for the assessment of cytochrome P4501A2, xanthine oxidase, and N-acetyltransferase activity in humans. Ther Drug Monit 1996; 18:254–262.
81. Lang NP, Butler MA, Massengill J, et al. Rapid metabolic phenotypes of acetyltransferase and cytochrome P4501A2 and putative exposure to food-borne heterocyclic amines increase the risk for colorectal cancer or polyps. Cancer Epidemiol Biomarkers Prev 1994; 3:675–682.
82. Kilbane AJ, Silbart LK, Manis M, et al. Human N-acetylation genotype determination with urinary caffeine metabolites. Clin Pharmacol Ther 1990; 47:470–477.

83. McQuilkin SH, Nierenberg DW, Bresnick E. Analysis of within-subject variation of caffeine metabolism when used to determine cytochrome P4501A2 and N-acetyltransferase-2 activities. Cancer Epidemiol Biomarkers Prev 1995; 4:139–146.
84. Hardy BG, Lemieux C, Walker SE, et al. Interindividual and intraindividualvariability in acetylation: characterization with caffeine. Clin Pharmacol Ther 1988; 44:152–157.
85. Tang BK, Kadar D, Kalow W. An alternative test for acetylator phenotyping with caffeine. Clin Pharmacol Ther 1987; 42:509–513.
86. Rasmussen BB, Brosen K. Theophylline has no advantages over caffeine as a putative model drug for assessing CYP1A2 activity in humans. Br J Clin Pharmacol 1997; 43:253–258.
87. Raunio H, Syngelma T, Pasanen M, et al. Immunochemical and catalytical studies on hepatic coumarin 7-hydroxylase in man, rat, and mouse. Biochem Pharmacol 1988; 37:3889–3895.
88. Miles JS, McLaren AW, Forrester LM, et al. Identification of the human liver cytochrome P-450 responsible for coumarin 7-hydroxy-lase activity. Biochem J 1990; 267:365–371.
89. Kharasch ED, Hankins DC, Thummel KE. Human kidney methoxyflurane and sevoflurane metabolism. Intrarenal fluoride production as a possible mechanism of methoxyflurane nephrotoxicity. Anesthesiology 1995; 82:689–699.
90. Spracklin DK, Hankins DC, Fisher JM, et al. Cytochrome P4502E1 is the principal catalyst of human oxidative halothane metabolism in vitro. J Pharmacol Exp Ther 1997; 281:400–411.
91. Sadeque AJM, Fisher MB, Korzekwa KR, et al. Human CYP2C9 and CYP2A6 mediate formation of the hepatotoxin 4-ene-valproic acid. J Pharmacol Exp Ther 1997; 283:698–703.
92. Madan A, Parkinson A, Faiman MD. Identification of the human and rat P450 enzymes responsible for the sulfoxidation of S-methyl N,N-diethylthiolcarbamate (DETC-ME). The terminal step in the bioactivation of disulfiram. Drug Metab Dispos 1995; 23:1153–1163.
93. Torchin CD, McNeilly PJ, Kapetanovic IM, et al. Stereoselective metabolism of a new anticonvulsant drug candidate, losigamone, by human liver microsomes. Drug Metab Dispos 1996; 24:1002–1008.
94. Wirz B, Valles B, Parkinson A, et al. CYP3A4 and CYP2A6 are involved in the biotransformation of letrozole. ISSX Proc 1996; 10:359.
95. Nunoya KI, Yokoi T, Kimura K, et al. (+)-Cis-3,5-dimethyl-2-(3-pyridyl) thiazolidin-4-one hydrochloride (SM-12502) as a novel substrate for cytochrome P450 2A6 in human liver microsomes. J Pharmacol Exp Ther 1996; 277:768–774.
96. Yamano S, Tatsuno J, Gonzalez FJ. The CYP2A3 gene product catalyzes coumarin 7-hydroxylation in human liver microsomes. Biochemistry 1990; 29: 1322–1329.
97. Hadidi H, Zahlsen K, Idle JR, et al. A single amino acid substitution (Leul60His) in cytochrome P450 CYP2A6 causes switching from 7-hydroxylation to 3-hydroxylation of coumarin. Food Chem Toxicol 1997; 35:903–907.
98. Oscarson M. Genetic polymorphisms in the cytochrome P450 2A6 (CYP2A6) gene: implications for interindividual differences in nicotine metabolism. Drug Metab Dispos 2001; 29:91–95.

99. Fernandez-Salguero P, Hoffman SM, Cholerton S, et al. A genetic polymorphism in coumarin 7-hydroxylation: sequence of the human CYP2A genes and identification of variant CYP2A6 alleles. Am J Hum Genet 1995; 57:651–660.
100. Nunoya K, Yokoi T, Kimura K, et al. A new deleted allele in the human cytochrome P4502A6 (CYP2A6) gene found in individuals showing poor metabolic capacity to coumarin and (+)-cis-3,5-dimethyl-2-(3-pyridyl)thiazolidin-4-one hydrochloride (SM-12502). Pharmacogenetics 1998; 8:239–249.
101. Rao Y, Hoffmann E, Zia M, et al. Duplications and defects in the CYP2A6 gene: identification, genotyping and in vivo effectson smoking. Mol Pharmacol 2000; 58:747–755.
102. Rautio A, Kraul H, Kojo A, et al. Interindividual variability of coumarin 7-hydroxylation in healthy volunteers. Pharmacogenetics 1992; 2:227–233.
103. Cholerton S, Idle ME, Vas A, et al. Comparison of a novel thin-layer chromatographic-fluorescence detection method with a spectrofluorometric method for the determination of 7-hydroxycoumarin in human urine. J Chromatogr 1992; 575: 325–330.
104. Iscan M, Rostami H, Iscan M, et al. Interindividual variability of coumarin 7-hydroxylation in a Turkish population. Eur J Clin Pharmacol 1994; 47:315–318.
105. Oscarson M, Gullsten H, Rautio A, et al. Genotyping of human cytochrome P4502A6 (CYP2A6), a nicotine C-oxidase. FEBS Letters 1998; 438:201–205.
106. Sotaniemi EA, Lumme P, Arvela P, et al. Age and CYP3A4 and CYP2A6 activities marked by the metabolism of lignocaine and coumarin in man. Therapie 1996; 51:363–366.
107. Merkel U, Sigusch H, Hoffmann A. Grapefruit juice inhibits 7-hydroxylation of coumarin in healthy volunteers. Eur J Clin Pharmacol 1994; 46:175–177.
108. Kharasch ED, Hankins DC, Baxter PJ, et al. Single-dose disulfiram does not inhibit CYP2A6 activity. Clin Pharmacol Ther 1998; 64:39–45.
109. Sotaniemi EA, Rautio A, Backstrom M, et al. CYP3A4 and CYP2A6 activities marked by the metabolism of lignocaine and coumarin in patients with liver and kidney diseases and epileptic patients. Br J Clin Pharmacol 1995; 39:71–76.
110. Hadidi H, Irshaid Y, Vagbo CB, et al. Variability of coumarin 7- and 3-hydroxylation in a Jordanian population is suggestive of a functional polymorphism in cytochrome P450 CYP2A6. Eur J Clin Pharmacol 1998; 54:437–441.
111. Egan D, O'Kennedy R, Moran E, et al. The pharmacology, metabolism, analysis, and applications of coumarin and coumarin-related compounds. Drug Metab Rev 1990; 22:503–529.
112. Marshall ME, Mohler JL, Edmonds K, et al. An updated review of the clinical development of coumarin (1,2-benzopyrene) and 7-hydroxycoumarin. J Cancer Res Clin Oncol 1994; 120:S39–S42.
113. Jacob P III, Benowitz NL, Shulgin AT. Recent studies of nicotine metabolism in humans. Pharmacol Biochem Behav 1988; 30:249–253.
114. Messina ES, Tyndale RF, Sellers EM. A major role for CYP2A6 in nicotine C-oxidation by human liver microsomes. J Pharmacol Exp Ther 1997; 282: 1608–1614.
115. Nakajima M, Yamamoto T, Nunoya KI, et al. Characterization of CYP2A6 involved in 3′-hydroxylation of cotinine in human liver microsomes. J Pharmacol Exp Ther 1996; 277:1010–1015.

116. Benowitz NL, Jacob P III. Metabolism of nicotine to cotinine studied by a dual stable isotope method. Clin Pharmacol Ther 1994; 56:483–493.
117. Perez-Stable EJ, Herrera B, Jacob P III, et al. Nicotine metabolism and intake in Black and White smokers. JAMA 1998; 280:152–156.
118. Benowitz NL, Jacob P III, Sachs DPL. Deficient C-oxidation of nicotine. Clin Pharmacol Ther 1995; 57:590–594.
119. Miners JO, Birkett DJ. Cytochrome P4502C9: an enzyme of major importance in human drug metabolism. Br J Clin Pharmacol 1998; 45:525–538.
120. Steward DJ, Haining RL, Henne KR, et al. Genetic association between sensitivity to warfarin and expression of CYP2C9*3. Pharmacogenetics 1997; 7:361–376.
121. Henne KR, Gaedigk A, Gupta G, et al. Chiral phase analysis of warfarin enantiomers in patient plasma in relation to CYP2C9 genotype. J Chromatogr B Biomed Sci Appl 1998; 710:143–148.
122. Takahashi H, Kashima T, Nomizo Y, et al. Metabolism of warfarin enantiomers in Japanese patients with heart disease having different CYP2C9 and CYP2C19 genotypes. Clin Pharmacol Ther 1998; 63:519–528.
123. Bhasker CR, Miners JO, Coulter S, et al. Allelic and functional variability of cytochrome P4502C9. Pharmacogenetics 1997; 7:51–58.
124. McCrea JB, Cribb A, Rushmore T, et al. Phenotypic and genotypic investigations of a healthy volunteer deficiency in the conversion of losartan to its active metabolite E-3174. Clin Pharmacol Ther 1999; 65:348–352.
125. Furuya H, Fernandez-Salguero P, Gregory W, et al. Genetic polymorphism of CYP2C9 and its effect on warfarin maintenance dose requirement in patients undergoing anticoagulation therapy. Pharmacogenetics 1995; 5:389–392.
126. Tassaneeyakul W, Birkett DJ, Pass MC, et al. Limited value of the urinary phenytoin metabolic ratio for the assessment of cytochrome P4502C9 activity in vivo. Br J Clin Pharmacol 1996; 42:774–778.
127. Knodell RG, Hall SD, Wilkinson GR, et al. Hepatic metabolism of tolbutamide: characterization of the form of cytochrome P-450 involved in methyl hydroxylation and relationship in vivo disposition. J Pharmacol Exp Ther 1987; 241:1112–1119.
128. Veronese ME, Miners JO, Randies D, et al. Validation of the tolbutamide metabolic ratio for population screening with use of sulfaphenazole to produce model phenotypic poor metabolizers. Clin Pharmacol Ther 1990; 47:403–411.
129. Veronese ME, Miners JO, Rees DLP, et al. Tolbutamide hydroxylation in humans: lack of bimodality in 106 healthy subjects. Pharmacogenetics 1993; 3:86–93.
130. Krishnaiah YSR, Satyanarayana S, Visweswaram D. Interaction between tolbutamide and ketoconazole hi healthy subjects. Br J Clin Pharmacol 1994; 37: 205–207.
131. Transon C, Leemann T, Vogt N, et al. In vivo inhibition profile of cytochrome P450TB (CYP2C9) by (±)-fluvastatin. Clin Pharmacol Ther 1995; 58:412–417.
132. Frye RF, Tracy TS, Hutzler JM, et al. Flurbiprofen as a selective in vivo probe of CYP2C9 activity. Clin Pharmacol Ther 2000; 67:109.
133. Wrighton SA, Stevens JC, Becker GW, et al. Isolation and characterization of human liver cytochrome P4502C19: correlation between 2C19 and *S*-mephenytoin 4′-hydroxylation. Arch Biochem Biophys 1993; 306:240–245.
134. Goldstein JA, Faletto MB, Romkes-Sparks M, et al. Evidence that CYP2C19 is the major (S′)-mephenytoin 4′-hydroxylase in humans. Biochemistry 1994; 33:1743–1752.

135. Kiipfer A, Branch RA. Stereoselective mephobarbital hydroxylation cosegregates with mephenytoin hydroxylation. Clin Pharmacol Ther 1985; 38:414–418.
136. Adedoyin A, Prakash C, O'Shea D, et al. Stereoselective disposition of hexobarbital and its metabolites: relationship to the S-mephenytoin polymorphism in Caucasian and Chinese subjects. Pharmacogenetics 1994; 4:27–38.
137. Ward SA, Helsby NA, Skjelbo E, et al. The activation of the biguanide antimalarial proguanil co-segregates with the mephenytoin oxidation polymorphism—a panel study. Br J Clin Pharmacol 1991; 31:689–692.
138. Andersson T, Regardh CG, Lou YC, et al. Polymorphic hydroxylation of S′-mephenytoin and omeprazole metabolism in Caucasian and Chinese subjects. Pharmacogenetics 1992; 2:25–31.
139. Sohn DR, Kwon JT, Kim HK, et al. Metabolic disposition of lansoprazole in relation to the S-mephenytoin 4′-hydroxylation phenotype status. Clin Pharmacol Ther 1997; 61:574–582.
140. Tanaka M, Ohkubo T, Otani K, et al. Metabolic disposition of pantoprazole, a proton pump inhibitor, in relation to S-mephenytoin 4′-hydroxylation phenotype and genotype. Clin Pharmacol Ther 1997; 62:619–628.
141. Bertilsson L, Henthorn TK, Sanz E, et al. Importance of genetic factors in the regulation of diazepam metabolism: relationship to *S*-mephenytoin, but not debrisoquin hydroxylation phenotype. Clin Pharmacol Ther 1989; 45:348–355.
142. Skjelbo E, Brosen K, Hallas J, et al. The mephenytoin oxidation polymorphism is partially responsible for the N-demethylation of imipramine. Clin Pharmacol Ther 1991; 49:18–23.
143. Nielsen KK, Brosen K, Hansen MGJ, et al. Single-dose kinetics of clomipramine: relationship to the sparteine and 5-mephenytoin oxidation polymorphisms. Clin Pharmacol Ther 1994; 55:518–527.
144. Breyer-Pfaff U, Pfandl B, Nill K, et al. Enantioselective amitriptyline metabolism in patients phenotyped for two cytochrome P450 isozymes. Clin Pharmacol Ther 1992; 52:350–358.
145. Dalen P, Alvan G, Wakelkamp M, et al. Formation of meprobamate from carisoprodol is catalyzed by CYP2C19. Pharmacogenetics 1996; 6:387–394.
146. Sindrup SH, Brosen K, Hansen MGJ, et al. Pharmacokinetics of citalopram in relation to the sparteine and the mephenytoin oxidation polymorphisms. Ther Drug Monit 1993; 15:11–17.
147. Gram LF, Guentert TW, Grange S, et al. Moclobemide, a substrate of CYP2C19 and an inhibitor of CYP2C19, CYP2D6, and CYP1A2: a panel study. Clin Pharmacol Ther 1995; 57:670–677.
148. Ward SA, Walle T, Walle UK, et al. Propranolol's metabolism is determined by both mephenytoin and debrisoquine hydroxylase activities. Clin Pharmacol Ther 1989; 45:72–79.
149. Lillibridge JH, Lee CA, Pithavala YK, et al. The role of CYP2C19 in the formation of nelfmavir hydroxy-f-butylamide (M8): in vitro/in vivo correlation. ISSX Proc 1998; 13:55.
150. Wilkinson GR, Guengerich FP, Branch RA. Genetic polymorphism of S-mephenytoin hydroxylation. Pharmacol Ther 1989; 43:53–76.
151. de Morais SMF, Wilkinson GR, Blaisdell J, et al. The major genetic defect responsible for the polymorphism of S-mephenytoin in humans. J Biol Chem 1994; 269:15419–15422.

152. de Morais SMF, Wilkinson GR, Blaisdell J, et al. Identification of a new genetic defect responsible for the polymorphism of *S*-mephenytoin metabolism in Japanese. Mol Pharmacol 1994; 46:594–598.
153. Brosen K, de Morais SMF, Meyer UA, et al. A multifamily study on the relationship between CYP2C19 genotype and S-mephenytom oxidation phenotype. Pharmacogenetics 1995; 5:312–317.
154. Ferguson RJ, deMorais SMF, Benhamou S, et al. A novel defect in human CYP2C19: mutation of the initiation codon is responsible for poor metabolism of S-mephenytoin. J Pharmacol Exp Ther 1998; 284:356–361.
155. Ibeanu GC, Blaisdell J, Ghanayem BI, et al. An additional defective allele, CYP2C19*5, contributes to the S-mephenytoin poor metabolizer phenotype in Caucasians. Pharmacogenetics 1998; 8:129–135.
156. Xiao ZS, Goldstein JA, Xie HG, et al. Differences in the incidence of the CYP2C19 polymorphism affecting the S-mephenytoin phenotype in Chinese Han and Bai populations and identification of a new rare CYP2C19 mutant allele. J Pharmacol Exp Ther 1997; 281:604–609.
157. Ibeanu GC, Goldstein JA, Meyer U, et al. Identification of new human CYP2C19 alleles (CYP2C19*6 and CYP2C19*2B) in a Caucasian poor metabolizer of mephenytoin. J Pharmacol Exp Ther 1998; 286:1490–1495.
158. Ibeanu GC, Blaisdell J, Ferguson RJ, et al. A novel transversion in intron5 donor splice junction of CYP2C19 and a sequence polymorphism in exon 3 contribute to the poor metabolism of the anticonvulsant drug S-mephenytoin. J Pharmacol Exp Ther 1999; 290:635–640.
159. Goldstein JA, Ishizaki T, Chiba K, et al. Frequencies of the defective CYP2C19 alleles responsible for the mephenytoin poor metabolizer phenotype in various Oriental, Caucasian, Saudi Arabian and American black populations. Pharmacogenetics 1997; 7:59–64.
160. Alvan G, Bechtel P, Iselius L, et al. Hydroxylation polymorphisms of debrisoquine and mephenytoin in European populations. Eur J Clin Pharmacol 1990; 39:533–537.
161. Masimirembwa C, Bertilsson L, Johnansson I, et al. Phenotyping and genotyping of S-mephenytoin hydroxylase (cytochrome P450 2C19) in a Shona population of Zimbabwe. Clin Pharmacol Ther 1995; 57:656–661.
162. Edeki TI, Goldstein JA, de Morais SMF, et al. Genetic polymorphism of S-mephenytoin 4′-hydroxylation in African-Americans. Pharmacogenetics 1996; 6: 357–360.
163. Marinac JS, Balian JD, Foxworth JW, et al. Determination of CYP2C19 in phenotype in black Americans with omeprazole: correlation with genotype. Clin Pharmacol Ther 1996; 60:138–144.
164. Jurima M, Inaba T, Kadar D, et al. Genetic polymorphism of mephenytoin p(4′)-hydroxylation: difference between Orientals and Caucasians. Br J Clin Pharmacol 1985; 19:483–487.
165. Nakasaura K, Goto F, Ray WA, et al. Interethnic differences in genetic polymorphism of debrisoquin and mephenytoin hydroxylation between Japanese and Caucasian populations. Clin Pharmacol Ther 1985; 38:402–408.
166. Bertilsson L, Lou YQ, Du YL, et al. Pronounced differences between native Chinese and Swedish populations in the polymorphic hydroxylations of debrisoquin and *S*-mephenytoin. Clin Pharmacol Ther 1992; 51:388–397.

167. Roh HK, Dahl ML, Johansson I, et al. Debrisoquine and S-mephenytoin hydroxylation phenotypes and genotypes in a Korean population. Pharmacogenetics 1996; 6:441–447.
168. Sohn DR, Kusaka M, Ishizaki T, et al. Incidence of S-mephenytom hydroxylation deficiency in a Korean population and the inter-phenotypic differences in diazepam pharmacokinetics. Clin Pharmacol Ther 1992; 52:160–169.
169. Jacqz E, Hall SD, Branch RA, et al. Polymorphic metabolism of mephenytoin in man: pharmacokinetic interaction with a co-regulated substrate, mephobarbital. Clin Pharmacol Ther 1986; 39:646–653.
170. Lerena A, Valdivielso MJ, Benitez J, et al. Reproducibility over time of mephenytoin and debrisoquine hydroxylation phenotypes. Pharmacol Toxicol 1993; 73:46–48.
171. Bluhm RE, Wilkinson GR, Shelton R, et al. Genetically determined drug-metabolizing activity and desipramine-associated cardiotoxicity: a case report. Clin Pharmacol Ther 1993; 53:89–95.
172. Tybring G, Nordin J, Bergman T, et al. An S-mephenytoin cysteine conjugate identified in urine of extensive but not of poor metabolizers of S-mephenytoin. Pharmacogenetics 1997; 7:355–360.
173. Wedlund PJ, Sweetman BJ, Wilkinson GR, et al. Pharmacogenetic association between the formation of 4-hydroxymephenytoin and a new metabolite of *S*-mephenytoin in man. Drug Metab Dispos 1987; 15:277–279.
174. Zhang Y, Blouin RA, McNamara PJ, et al. Limitation to the use of the urinary S/R-mephenytoin ratio in pharmacogenetic studies. Br J Clin Pharmacol 1991; 31: 350–352.
175. Tybring G, Bertilsson L. A methodological investigation on the estimation of the S-mephenytoin hydroxylation phenotype using the urinary *S/R* ratio. Pharmacogenetics 1992; 2:241–243.
176. Wedlund PJ, Wilkinson GR. In vivo and in vitro measurement of CYP2C19 activity. In: Johnson EF, Waterman MR, eds. Cytochrome P450, Part B: Methods in Enzymology. San Diego: Academic Press, 1996; 272:105–114.
177. Sanz EJ, Villen T, Alm C, et al. S-mephenytoin hydroxylation phenotypes in a Swedish population determined after coadministration with debrisoquin. Clin Pharmacol Ther 1989; 45:495–499.
178. Setiabudy R, Chiba K, Kusaka M, et al. Caution in the use of a 100 mg dose of racemic mephenytoin for phenotyping Southeastern Oriental subjects. Br J Clin Pharmacol 1992; 33:665–666.
179. Chiba K, Kobayashi K, Manabe K, et al. Oxidative metabolism of omeprazole in human liver microsomes: cosegregation with S-mephenytoin 4′-hydroxylation. J Pharmacol Exp Ther 1993; 266:52–59.
180. Andersson T, Miners JO, Veronese ME, et al. Identification of human liver cytochrome P450 isoforms mediating omeprazole metabolism. Br J Clin Pharmacol 1993; 36:521–530.
181. Karam WG, Goldstein JA, Lasker JM, et al. Human CYP2C19 is a major omeprazole 5-hydroxylase, as demonstrated with recombinant cytochrome P450 enzymes. Drug Metab Dispos 1996; 24:1081–1087.
182. Sohn DR, Kobayashi K, Chiba K, et al. Disposition kinetics and metabolism of omeprazole in extensive and poor metabolizers of S-mephenytoin 4′-hydroxylation recruited from an Oriental population. J Pharmacol Exp Ther 1992; 262:1195–1202.

183. Chang M, Tybring G, Dahl ML, et al. Interphenotype differences in disposition and effect on gastrin levels of omeprazole—suitability of omeprazole as a probe for CYP2C19. Br J Clin Pharmacol 1995; 39:511–518.
184. Chang M, Dahl ML, Tybring G, et al. Use of omeprazole as a probe drug for CYP2C19 phenotype in Swedish Caucasians: comparison with S-mephenytoin hydroxylation phenotype and CYP2C19 genotype. Pharmacogenetics 1995; 5: 358–368.
185. Ieiri I, Kubota T, Urae A, et al. Pharmacokinetics of omeprazole (a substrate of CYP2C19) and comparison with two mutant alleles, $CYP2C19_{ml}$ inexon 5 and $CYP2C19_{m2}$ in exon 4, in Japanese subjects. Clin Pharmacol Ther 1996; 59: 647–653.
186. Balian JD, Sukhova N, Harris JW, et al. The hydroxylation of omeprazole correlates with S-mephenytoin metabolism: a population study. Clin Pharmacol Ther 1995; 57:662–669.
187. Rost KL, Brockmoller J, Esdorn F, et al. Phenocopies of poor metabolizers of omeprazole caused by liver disease and drug treatment. J Hepatol 1995; 23: 268–277.
188. Setiabudy R, Kusaka M, Chiba K, et al. Metabolic disposition of proguanil in extensive and poor metabolisers of S-mephenytoin 4′-hydroxylation recruited from an Indonesian population. Br J Clin Pharmacol 1995; 39:297–303.
189. Ward SA, Watkins WM, Mberu E, et al. Inter-subject variability in the metabolism of proguanil to the active metabolite cycloguanil in man. Br J Clin Pharmacol 1989; 27:781–787.
190. Helsby NA, Ward SA, Edwards G, et al. The pharmaco-kinetics and activation of proguanil in man: consequences of variability in drug metabolism. Br J Clin Pharmacol 1990; 30:593–598.
191. Brosen K, Skjelbo E, Flachs H. Proguanil metabolism is determined by the mephenytoin oxidation polymorphism in Vietnamese living in Denmark. Br J Clin Pharmacol 1993; 36:105–108.
192. Wanwimolruk S, Pratt EL, Denton JR, et al. Evidence for the polymorphic oxidation of debrisoquine and proguanil in a New Zealand Maori population. Pharmacogenetics 1995; 5:193–198.
193. Wanwimolruk S, Thou MR, Woods DJ. Evidence for the polymorphic oxidation of debrisoquine and proguanil in a Khmer (Cambodian) population. Br J Clin Pharmacol 1995; 40:166–169.
194. Skjelbo E, Mutabingwa TK, Bygbjerg I, et al. Chloro-guanide metabolism in relation to the efficacy in malaria prophylaxis and the *S*-mephenytoin oxidation in Tanzanians. Clin Pharmacol Ther 1996; 59:304–311.
195. Somogyi AA, Reinhard HA, Bochner F. Pharmacokinetic evaluation of proguanil: a probe phenotyping drug for the mephenytoin hydroxylase polymorphism. Br J Clin Pharmacol 1996; 41:175–179.
196. Basci NE, Bozkurt A, Kortunay S, et al. Proguanil metabolism in relation to S-mephenytoin oxidation in a Turkish population. Br J Clin Pharmacol 1996; 42: 771–773.
197. Coller JK, Somogyi AA, Bochner F. Association between CYP2C19 genotype and proguanil oxidative polymorphism. Br J Clin Pharmacol 1997; 43:659–660.

198. Hoskins JM, Shenfield GM, Gross AS. Relationship between proguanil metabolic ratio and CYP2C19 genotype in a Caucasian population. Br J Clin Pharmacol 1998; 46:499–504.
199. Funck-Brentano C, Bosco O, Jacqz-Aigrain E, et al. Relation between chlorguanide bioactivation to cycloguanil and the genetically determined metabolism of mephenytoin in humans. Clin Pharmacol Ther 1992; 51:507–512.
200. Partovian C, Jacqz-Aigrain E, Keundjian A, et al. Comparison of chlorguanide and mephenytoin for the in vivo assessment of genetically determined CYP2C19 activity in humans. Clin Pharmacol Ther 1995; 58:257–263.
201. Birkett DJ, Rees D, Andersson T, et al. In vitro proguanil activation to cycloguanil by human liver microsomes is mediated by CYP3A isoforms as well as by S-mephenytoin hydroxylase. Br J Clin Pharmacol 1994; 37:413–420.
202. Eichelbaum M, Gross AS. The genetic polymorphism of debrisoquine/sparteine metabolism-clinical aspects. Pharmacol Ther 1990; 46:377–394.
203. Meyer UA, Zanger UM. Molecular mechanisms of genetic polymorphisms of drug metabolism. Ann Rev Pharmacol Toxicol 1997; 37:269–296.
204. Mahgoub A, Idle JR, Dring LG, et al. Polymorphic hydroxylation of debrisoquine in man. Lancet 1977; 2:584–586.
205. Tucker GT, Silas JH, Iyun AO, et al. Polymorphic hydroxylation of debrisoquine. Lancet 1977; 2:718.
206. Eichelbaum M, Spannbrucker N, Steincke B, et al. Defective N-oxidation of sparteine in man: a new pharmacogenetic defect. Eur J Clin Pharmacol 1979; 16: 183–187.
207. Daly AK. Molecular basis of polymorphic drug metabolism. J Mol Med 1995; 73: 539–553.
208. Johansson I, Lundqvist E, Bertilsson L, et al. Inherited amplification of an active gene in the cytochrome P450 2D-locus as a cause of ultrarapid metabolism of debrisoquine. Proc Natl Acad Sci U S A 1993; 90:11825–11829.
209. Marez D, Legrand M, Sabbagh N, et al. Polymorphism of the cytochrome P450 C YP2D6 gene in a European population: characterization of 48 mutations and 53 alleles, their frequencies and evolution. Pharmacogenetics 1997; 7:193–202.
210. Daly AK, Brockmoller J, Broly F, et al. Nomenclature of human CYP2D6 alleles. Pharmacogenetics 1996; 6:193–201.
211. Masimirembwa C, Hasler J, Bertilsson L, et al. Phenotype and genotype analysis of debrisoquine hydroxylase (CYP2D6) in a black Zimbabwean population—reduced enzyme activity and evaluation of metabolic correlation of CYP2D6 probe drugs. Eur J Clin Pharmacol 1996; 51:117–122.
212. Iyun AO, Lennard MS, Tucker GT, et al. Metoprolol and debrisoquin metabolism in Nigerians: lack of evidence for polymorphic oxidation. Clin Pharmacol Ther 1986; 40:387–394.
213. Relling MV, Cherrie J, Schell MJ, et al. Lower prevalence of the debrisoquin oxidative poor metabolizer phenotype in American black versus white subjects. Clin Pharmacol Ther 1991; 50:308–313.
214. Sindrup SH, Brosen K. The pharmacogenetics of codeine hypoalgesia. Pharmacogenetics 1995; 5:335–346.
215. Bertilsson L, Dahl ML, Sjoqvist F, et al. Molecular basis for rational megaprescribing in ultrarapid hydroxylators of debrisoquine. Lancet 1993; 341:63.

216. Eiermann B, Edlund PO, Tjernberg A, et al. 1- and 3-hydroxylations, in addition to 4-hydroxylation, of debrisoquine are catalyzed by cytochrome P450 2D6 in humans. Drug Metab Dispos 1998; 26:1096–1101.
217. Sachse C, Brockmoller J, Bauer S, et al. Cytochrome P450 2D6 variants in a Caucasian population: allele frequencies and phenotypic consequences. Am J Hum Genet 1997; 60:284–295.
218. Regardh CG, Johnsson G. Clinical pharmacokinetics of metoprolol. Clin Pharmacokinet 1980; 5:557–569.
219. Lennard MS, Silas JH, Freestone S, et al. Oxidation phenotype—a major determinant of metoprololol metabolism and response. New Engl J Med 1982; 307: 1558–1560.
220. Clark DWJ. Genetically determined variability in acetylation and oxidation. Therapeutic implications. Drugs 1985; 29:342–375.
221. McGourty JC, Silas JH, Lennard MS, et al. Metoprolol metabolism and debrisoquine oxidation polymorphism—population and family studies. Br J Clin Pharmacol 1985; 20:555–566.
222. Sohn DR, Shin SG, Park CW, et al. Metoprolol oxidation polymorphism in a Korean population: comparison with native Japanese and Chinese populations. Br J Clin Pharmacol 1991; 32:504–507.
223. Lennard MS, Tucker GT, Silas JH, et al. Differential stereoselective of metoprolol in extensive and poor debrisoquin metabolisers. Clin Pharmacol Ther 1983; 34: 732–737.
224. Sohn DR, Kusaka M, Shin SG, et al. Utility of a one-point (3-hour postdose) plasma metabolic ratio as a phenotyping test using metoprolol in two East Asian populations. Ther Drug Monit 1992; 14:184–189.
225. Schmid B, Bircher J, Preisig R, et al. Polymorphic dextromethorphan metabolism: co-segregation of oxidative O-demethylation with debrisoquin hydroxylation. Clin Pharmacol Ther 1985; 38:618–624.
226. Basci NE, Bozkurt A, Kayaalp SO, et al. Omission of the deconjugation step in urine analysis and the unaltered outcome of CYP2D6 phenotyping with dextromethorphan. Eur J Drug Metab Pharmacokinet 1998; 23:1–5.
227. Capon DA, Bochner F, Kerry N, et al. The influence of CYP2D6 polymorphism and quinidine on the disposition and antitussive effect of dextromethorphan in humans. Clin Pharmacol Ther 1996; 60:295–307.
228. Kiipfer A, Schmid B, Pfaff G. Pharmacogenetics of dextromethorphan O-demethylation in man. Xenobiotica 1986; 16:421–433.
229. Hou ZY, Pickle LW, Meyer PS, et al. Salivary analysis of determination of dextromethorphan metabolic phenotype. Clin Pharmacol Ther 1991; 49:410–419.
230. Hu YP, Tang HS, Lane HY, et al. Novel single-point plasma or saliva dextromethorphan method for determining CYP2D6 activity. J Pharmacol Exp Ther 1998; 285:955–960.
231. Dayer P, Leemann T, Striberni R. Dextromethorphan O-demethylation in liver microsomes as a prototype reaction to monitor cytochrome P-450 db 1 activity. Clin Pharmacol Ther 1989; 45:34–40.
232. Kerry NL, Somogyi AA, Bochner F, et al. The role of CYP2D6 in primary and secondary oxidative metabolism of dextromethorphan: in vitro studies using human liver microsomes. Br J Clin Pharmacol 1994; 38:243–248.

233. Evans WE, Relling MV. Concordance of P450 2D6 (debrisoquine hydroxylase) phenotype and genotype: inability of dextromethorphan metabolic ratio to discriminate reliably heterozygous and homozygous extensive metabolizers. Pharmacogenetics 1991; 1:143–148.
234. Funck-Brentano C, Thomas G, Jacqz-Algrain E, et al. Polymorphism of dextromethorphan metabolism: relationships between phenotype, genotype and response to the administration of encainide in humans. J Pharmacol Exp Ther 1992; 263:780–786.
235. Zimmermann T, Schlenk R, Pfaff P, et al. Prediction of phenotype for dextromethorphan O-demethylation by using polymerase chain reaction in healthy volunteers. Arzneimittelforschung 1995; 45:41–43.
236. Ereshefsky L, Riesenman C, Lam YW. Antidepressant drug interactions and the cytochrome P450 system—the role of cytochrome P4502D6. Clin Pharmacokinet 1995; 29(suppl 1):10–19.
237. Richelson E. Pharmacokinetic drug interactions of new antidepressants: a review of the effects of the metabolism of other drugs. Mayo Clin Proc 1997; 72:835–847.
238. Horai Y, Taga J, Ishizaki T, et al. Correlations among the metabolic ratios of three test probes (metoprolol, debrisoquine and sparteine) for genetically determined oxidation polymorphism in a Japanese population. Br J Clin Pharmacol 1990; 29: 111–115.
239. Woolhouse NM, Eichelbaum M, Oates NS, et al. Dissociation of co-regulatory control of debrisoquin/phenformin and sparteine oxidation in Ghanaians. Clin Pharmacol Ther 1985; 37:512–521.
240. Simooya OO, Njunju E, Hodjegan AR, et al. Debrisoquine and metoprolol oxidation in Zambians: a population study. Pharmacogenetics 1993; 3:205–208.
241. Droll K, Bruce-Mensah K, Otton SV, et al. Comparison of three CYP2D6 probe substrates and genotype in Ghanaians, Chinese and Caucasians. Pharmacogenetics 1998; 8:325–333.
242. Ritchie JC, Mitchell SC, Smith RL. Sparteine metabolism in a Nigerian population. Drug Metabol Drug Interact 1996; 13:129–135.
243. Lennard MS, Iyun AO, Jackson PR, et al. Evidence for a dissociation in the control of sparteine, debrisoquine and metoprolol metabolism in Nigerians. Pharmacogenetics 1992; 2:89–92.
244. Sommers DK, Moncrieff J, Avenant J. Non-correlation between debrisoquine and metoprolol polymorphisms in the Venda. Hum Toxicol 1989; 8:365–368.
245. Sommers DK, Moncrieff J, Avena JC. Absence of polymorphism of sparteine oxidation in the South African Venda. Hum Exp Toxicol 1991; 10:175–178.
246. Sommers DK, Moncrieff J, Avenant J. Metoprolol alpha-hydroxylation polymorphism in the San Bushmen of Southern Africa. Hum Toxicol 1989; 8: 39–43.
247. Sommers DK, Moncrieff J, Avenant J. Polymorphism of the 4-hydroxylation of debrisoquine in the San Bushmen of Southern Africa. Human Toxicol 1988; 7:273–276.
248. Masimirembwa D, Persson I, Bertilsson L, et al. A novel mutant variant of the CYP2D6 gene (CYP2D6*17) common in a black African population: association with diminished debrisoquine hydroxylase activity. Br J Clin Pharmacol 1996; 42: 713–719.

249. Oscarson M, Hidestrand M, Johansson I, et al. A combination of mutations in the CYP2D6*17 (CYP2D6Z) allele causes alterations in enzyme function. Mol Pharmacol 1997; 52:1034–1040.
250. Al-Hadidi HF, Irshaid YM, Rawashdeh NM. Metoprolol alpha-hydroxylation is a poor probe for debrisoquine oxidation (CYP2D6) polymorphism in Jordanians. Eur J Clin Pharmacol 1994; 47:311–314.
251. Irshaid YM, Al-Hadidi HF, Latif A, et al. Dextromethorphan metabolism in Jordanians: dissociation of dextromethorphan *O*-demethylation from debrisoquine 4-hydroxylation. Eur J Clin Metab Pharmacokinet 1996; 21:301–307.
252. Ronis MJJ, Lindros KO, Ingelman-Sundberg M. The CYP2E1 subfamily. In: Ioannides C, ed. Cytochromes P450: Metabolic and Toxicological Aspects. Boca Raton, FL: CRC Press, 1996:211–239.
253. Uematsu F, Kikuchi H, Ohmachi T, et al. Two common RFLPs of the human CYP2E gene. Nucleic Acids Res 1991; 19:2803.
254. Hu Y, Oscarson M, Johansson I, et al. Genetic polymorphism of human CYP2E1: characterization of two variant alleles. Mol Pharmacol 1997; 51:370–376.
255. McCarver DG, Byun R, Hines RN, et al. A genetic polymorphism in the regulatory sequence of human CYP2E1: association with increased chlorzoxazone hydroxylation in the presence of obesity and ethanol intake. Toxicol Appl Pharmacol 1998; 152:276–281.
256. Fairbrother KS, Grove J, de Waziers I, et al. Detection and characterization of novel polymorphisms in the CYP2E1 gene. Pharmacogenetics 1998; 8:543–552.
257. Carriere V, Berthou F, Baird S, et al. Human cytochrome P450 2E1 (CYP2E1): from genotype of phenotype. Pharmacogenetics 1996; 6:203–211.
258. Powell H, Kitteringham NR, Pirmohamed M, et al. Expression of cytochrome P4502E1 (CYP2E1) in human liver: assessment of mRNA, genotype and phenotype. Pharmacogenetics 1998; 8:411–421.
259. Kim RB, Yamazaki H, Chiba K, et al. In vivo and in vitro characterization of CYP2E1 activity in Japanese and Caucasians. J Pharmacol Exp Ther 1996; 279: 4–11.
260. Lucas D, Menez C, Girre C, et al. Cytochrome P450 2E1 genotype and chlorzoxazone metabolisminhealthy and alcoholic Caucasian subjects. Pharmacogenetics 1995; 5:298–304.
261. Le Marchand L, Wilkinson GR, Wilkens LR. Genetic and dietary predictors of CYP2E1 activity: a phenotyping study in Hawaii Japanese using chlorzoxazone. Cancer Epidemiol Biomarkers Prev 1999; 8:495–500.
262. Peter R, Bocker R, Beaune PH, et al. Hydroxylation of chlorzoxazone as a specific probe for human liver cytochrome P450IIE1. Chem Res Toxicol 1990; 3:566–573.
263. Carriere V, Goasduff T, Ratanasavanh D, et al. Both cytochromes P450 2E1 and 1A1 are involved in the metabolism of chlorzoxazone. Chem Res Toxicol 1994; 6:852–857.
264. Yamazaki H, Guo Z, Guengerich FP. Selectivity of cytochrome P4502E1 in chlorzoxazone 6-hydroxylation. Drug Metab Dispos 1995; 23:438–440.
265. Ono S, Hatanaka T, Hotta H, et al. Chlorzoxazone is metabolized by human CYP1A2 as well as by human CYP2E1. Pharmacogenetics 1995; 5:141–148.
266. Gorski JC, Jones DR, Wrighton SA, SD Hall. Contribution of human CYP3A subfamily members to the 6-hydroxylation of chlorzoxazone. Xenobiotica 1997; 27: 243–256.

267. O'Shea D, Davis SN, Kim RB, et al. Effect of fasting and obesity in humans on the 6-hydroxylation of chlorzoxazone: a putative probe of CYP2E1 activity. Clin Pharmacol Ther 1994; 56:359–367.
268. Lucas D, Menez JF, Berthou F. Chlorzoxazone: an in vitro and in vivo substrate probe for liver CYP2E1. In: Johnson EF, Waterman MR, eds. Cytochrome P450, Part B: Methods in Enzymology. San Diego, CA: Academic Press, 1996; 272: 115–123.
269. Kharasch ED, Thummel KE, Mhyre J, et al. Single-dose disulfiram inhibition of chlorzoxazone metabolism: a clinical probe for P4502E1. Clin Pharmacol Ther 1993; 53:643–650.
270. Eap CB, Schnyder C, Besson J, et al. Inhibition of CYP2E1 by chlormethiazole as measured by chlorzoxazone pharmacokinetics in patients with alcoholism and in healthy volunteers. Clin Pharmacol Ther 1998; 64:52–57.
271. Leclercq I, Desager JP, Horsmans Y. Inhibition of chlorzoxazone metabolism, a clinical probe for CYP2E1, by a single ingestion of watercress. Clin Pharmacol Ther 1998; 64:144–149.
272. Zand R, Nelson SD, Slattery JT, et al. Inhibition and induction of cytochrome P4502El-catalyzed oxidation by isoniazid in humans. Clin Pharmacol Ther 1993; 54:142–149.
273. O'Shea D, Kim RB, Wilkinson GR. Modulation of CYP2E1 activity by isoniazid in rapid and slow N-acetylators. Br J Clin Pharmacol 1997; 43:99–103.
274. Gebhardt AC, Lucas D, Menez JF, et al. Chlormethiazole inhibition of cytochrome P450 2E1 as assessed by chlorzoxazone hydroxylation in humans. Hepatology 1997; 26:957–961.
275. Dilger K, Metzler J, Bode JC, et al. CYP2E1 activity in patients with alcoholic liver disease. J Hepatol 1997; 27:1009–1014.
276. Dupont I, Lucas D, Clot P, et al. Cytochrome P4052E1 inducibility and hydroxyethyl radical formation among alcoholics. J Hepatol 1998; 28:564–572.
277. Lucas D, Menez C, Girre C, et al. Decrease in cytochrome P4502E1 as assessed by the rate of chlorzoxazone hydroxylation in alcoholics during the withdrawal phase. Alcohol Clin Exp Res 1995; 19:362–366.
278. Kim RB, O'Shea D, Wilkinson GR. Relationship in healthy subjects between CYP2E1 genetic polymorphisms and the 6-hydroxylation of chlorzoxazone—a putative measure of CYP2E1 activity. Pharmacogenetics 1994; 4:162–165.
279. Dreisbach AW, Ferencz N, Hopkins NE, et al. Urinary excretion of 6-hydroxychlorzoxazone as an index of CYP2E1 activity. Clin Pharmacol Ther 1995; 58: 498–505.
280. Vesell ES, DeAngelo Seaton T, A-Rahim YI. Studies on interindividual variations of CYP2E1 using chlorzoxazone as an in vivo probe. Pharmacogenetics 1995; 5: 53–57.
281. Thummel KE, Wilkinson GR. In vitro and in vivo drug interactions involving human CYP3A. Annu Rev Pharmacol Toxicol 1998; 38:389–430.
282. Thummel KE, Kunze KL, Shen DD. Enzyme-catalyzed processes of first-pass hepatic and intestinal drug extraction. Adv Drug Deliv Rev 1997; 27:99–127.
283. Wacher VJ, Wu CY, Benet LZ. Overlapping substrate specificities and tissue distribution of cytochrome P4503A and P-glycoprotein: implications for drug delivery and activity in cancer chemotherapy. Mol Carcinog 1995; 13:129–134.

284. Kim RB, Wandel C, Leake B, et al. Interrelationship between substrates and inhibitors of human CYP3A and P-glycoprotein. Pharmacol Res 1999; 16:408–414.
285. Watkins PB. Noninvasive tests of CYP3A enzymes. Pharmacogenetics 1994; 4: 171–184.
286. Fleming CM, Branch RA, Wilkinson GR, et al. Human liver microsomal N-hydroxylation of dapsone by cytochrome P-4503A4. Mol Pharmacol 1992; 41: 975–980.
287. May DG, Porter J, Wilkinson GR, et al. Frequency distribution of dapsone N-hydroxylase, a putative probe for P4503A4 activity, in a white population. Clin Pharmacol Ther 1994; 55:492–500.
288. May DG, Arns PA, Richards WO, et al. The disposition of dapsone in cirrhosis. Clin Pharmacol Ther 1992; 51:689–700.
289. Fleming CM, Kaisary A, Wilkinson GR, et al. The ability to 4-hydroxylate debrisoquine is related to recurrence of bladder cancer. Pharmacogenetics 1992; 2: 128–134.
290. Fleming CM, Persad R, Kaisary A, et al. Low activity of dapsone N-hydroxylation as a susceptibility risk factor in aggressive bladder cancer. Pharmacogenetics 1994; 4:199–207.
291. Branch RA, Chern HD, Adedoyin A, et al. The procarcinogen hypothesis for bladder cancer: activities of individual drug metabolizing enzymes as risk factors. Pharmacogenetics 1995; 5:S97–S102.
292. May DG, Black CM, Olsen NJ, et al. Scleroderma is associated with differences in individual routes of metabolism: a study with dapsone, debrisoquin, and mephenytoin. Clin Pharmacol Ther 1990; 48:286–295.
293. Black C, May G, Csuka ME, et al. Activity of oxidative routes of metabolism of debrisoquine, mephenytoin, and dapsone is unrelated to the pathogenesis of vinyl chloride-induced disease. Clin Pharmacol Ther 1992; 52:659–667.
294. Stein CM, Kinirons MT, Pincus T, et al. Comparison of the dapsone recovery ratio and the erythromycin breath test as in vivo probes of CYP3A activity in patients with rheumatoid arthritis receiving cyclosporine. Clin Pharmacol Ther 1996; 59: 47–51.
295. Mitra AK, Thummel KE, Kalhorn TF, et al. Metabolism of dapsone to its hydroxylamine by cytochrome P-450 2E1 in vitro and in vivo. Clin Pharmacol Ther 1995; 58:556–566.
296. Gill HJ, Tingle MD, Park BK. N-Hydroxylation of dapsone by multiple enzymes of cytochrome P450: implications for inhibition of haemotoxicity. Br J Clin Pharmacol 1995; 40:531–538.
297. Kinirons MT, Krivoruk Y, Wilkinson GR, et al. Effects of ketoconazole on the erythromycin breath test and the dapsone recovery ratio. Br J Clin Pharmacol 1999; 47:223–224.
298. Watkins PG, Hamilton TA, Annesley TM, et al. The erythromycin breath test as a predictor of cyclosporine blood levels. Clin Pharmacol Ther 1990; 48:120–129.
299. Turgeon DK, Normolle DP, Leichtman AB, et al. Erythromycin breath test predicts oral clearance of cyclosporine in kidney transplant recipients. Clin Pharmacol Ther 1992; 52:471–478.
300. Turgeon DK, Leichtman AB, Lown KS, et al. P4503A activity and cyclosporine dosing in kidney and heart transplant recipients. Clin Pharmacol Ther 1994; 56: 253–260.

301. Turgeon DK, Leichtman AB, Blake DS, et al. Prediction of interpatient and intrapatient variation in OG 37-325 dosing requirements by the erythromycin breath test: a prospective study in renal transplant recipients. Transplantation 1994; 57: 1736–1741.
302. Cheng CL, Smith DE, Carver PL, et al. Steady-state pharmacokinetics of delavirdine in HIV-positive patients: effect on erythromycin breath test. Clin Pharmacol Ther 1997; 61:531–543.
303. Wagner D. CYP3A4 and the erythromycin breath test. Clin Pharmacol Ther 1998; 64:129–130.
304. Watkins PB, Murray SA, Winkelman LG, et al. Erythromycin breath test as an assay of glucocorticoid-inducible liver cytochromes P-450. Studies in rats and patients. J Clin Invest 1989; 83:688–697.
305. Krivoruk Y, Kinirons MT, Wood AJJ, et al. Metabolism of cytochrome P4503A substrates in vivo administered by the same route: lack of correlation between alfentanil clearance and erythromycin breath test. Clin Pharmacol Ther 1994; 56: 608–614.
306. Schuetz EG, Yasuda K, Arimori K, et al. Human MDR1 and mouse mdrla P-glycoprotein alter the cellular retention and disposition of erythromycin, but not of retinoic acid or benzo(a)pyrene. Arch Biochem Biophys 1998; 350:340–347.
307. Chelvan P, Hamilton-Miller JMT, Brumfitt W. Biliary excretion of erythromycin after parenteral administration. Br J Clin Pharmacol 1979; 8:233–235.
308. Jamis-Dow CA, Pearl ML, Watkins PB, et al. Predicting drug interactions in vivo from experiments in vitro: human studies with paclitaxel and ketoconazole. Am J Clin Oncol 1997; 20:592–599.
309. Tateishi T, Graham SG, Krivoruk Y, et al. Omeprazole does not affect measured CYP3A4 activity using the erythromycin breath test. Br J Clin Pharmacol 1995; 40: 411–412.
310. Craig PI, Tapner M, Farrell GC. Interferon suppresses erythromycin metabolism in rats and human subjects. Hepatology 1993; 17:230–235.
311. Kronbach T, Mathys D, Umeno M, et al. Oxidation of midazolam and triazolam by human liver cytochrome P450DIA4. Mol Pharmacol 1989; 36:89–96.
312. Gorski JC, Hall SD, Jones DR, et al. Regioselective biotransformation of midazolam by members of the human cytochrome P4503A (CYP3A) subfamily. Biochem Pharmacol 1994; 47:1643–1653.
313. Thummel KE, Shen DD, Podoll TD, et al. Use of midazolam as a human cytochrome P450 3A probe: I. In vitro-in vivo correlations in liver transplant patients. J Pharmacol Exp Ther 1994; 271:549–556.
314. Thummel KE, O'Shea D, Paine MF, et al. Oral first-pass elimination of midazolam involves both gastro-intestinal and hepatic CYP3A-mediated metabolism. Clin Pharmacol Ther 1996; 59:491–502.
315. MacGiichrist AJ, Birnie GG, Cook A, et al. Pharmacokinetics and pharmacodynamics of intravenous midazolam in patients with severe alcoholic cirrhosis. Gut 1986; 27:190–195.
316. Pentikainen PJ, Valisalmi L, Himberg JJ, et al. Pharmacokinetics of midazolam following intravenous and oral administration in patients with chronic liver disease and in healthy subjects. J Clin Pharmacol 1989; 29:272–277.
317. Paine MF, Shen DD, Kunze KL, et al. First-pass metabolism of midazolam by the human intestine. Clin Pharmacol Ther 1996; 60:14–24.

318. Thummel KE, Shen DD, Podoll TD, et al. Use of midazolam as a human cytochrome P450 3A probe: II. Characterization of inter- and intra-individual hepatic P4503A variability after liver transplantation. J Pharmacol Exp Ther 1994; 271: 557–566.
319. Backman JT, Olkkola KT, Neuvonen PJ. Rifampin drastically reduces plasma concentrations and effects of oral midazolam. Clin Pharmacol Ther 1996; 59:7–13.
320. Olkkola KT, Backman JT, Neuvonen PJ. Midazolam should be avoided in patients receiving the systemic antimycotics ketoconazole or itraconazole. Clin Pharmacol Ther 1994; 55:481–485.
321. Backman JT, Olkkola KT, Neuvonen PJ. Azithromycin does not increase plasma concentrations of oral midazolam. Int J Clin Pharmacol Ther 1995; 33:356–359.
322. Mattila MJ, Vanakoski J, Idanpaan-Heikkla JJ. Azithromycin does not alter the effects of oral midazolam on human performance. Eur J Clin Pharmacol 1994; 47:49–52.
323. Ahonen J, Olkkola KT, Neuvonen PJ. Effect of itraconazole and terbinafine on the pharmacokinetics and pharmacodynamics of midazolam in healthy volunteers. Br J Clin Pharmacol 1995; 40:270–272.
324. Gorski JC, Jones DR, Haehner-Daniels BD, et al. The contribution of intestinal and hepatic CYP3A to the interaction between midazolam and clarithromycin. Clin Pharmacol Ther 1998; 64:133–143.
325. McCrea J, Prueksaritanont T, Gertz BJ, et al. Concurrent administration of the erythromycin breath test (EBT) and oral midazolam (MDZ) as in vivo probes for CYP3A activity. J Clin Pharmacol Ther 1999; 39:1212–1220.
326. Ekins S, VandenBraden M, Ring BJ, et al. Examination of purported probes of human CYP2B6. Pharmacogenetics 1997; 7:165–179.
327. Code EL, Crespi CL, Penman BW, et al. Human cytochrome P4502B6: interindividual hepatic expression, substrate specificity, and role in procarcinogen activation. Drug Metab Dispos 1997; 25:985–993.
328. Ekins S, VandenBraden M, Ring BJ, et al. Further characterization of the expression in liver and catalytic activity of CYP2B6. J Pharmacol Exp Ther 1998; 286: 1253–1259.
329. Ko JW, Desta Z, Flockhart DA. Human N-demethylation of (S)-mephenytoin by cytochrome P450s 2C9 and 2B6. Drug Metab Dispos 1998; 26:775–778.
330. Stevens JC, White RB, Hsu SH, et al. Human liver CYP2B6-catalyzed hydroxylation of RP 73401. J Pharmacol Exp Ther 1997; 282:1389–1395.

15

Drug-Drug Interactions: Clinical Perspective

David J. Greenblatt and Lisa L. von Moltke
Tufts University School of Medicine and
Tufts-New England Medical Center,
Boston, Massachusetts, U.S.A.

I. INTRODUCTION

During the last 20 years, the general problem of pharmacokinetic drug interactions has received increasing attention. Over this period a number of new and unique classes of medications have been introduced into clinical practice. These include the selective serotonin reuptake inhibitor (SSRI) and related mixed-mechanism antidepressants, novel azole antifungal agents, newer macrolide antimicrobial agents, and the highly active antiretroviral therapies (HAART) used against human immunodeficiency virus (HIV) infection and the acquired immunodeficiency syndrome (AIDS). While these and other classes of agents have had a major beneficial impact of the therapy on some serious and life-threatening illnesses, many of the agents have the secondary pharmacologic property of inducing or inhibiting the human cytochrome P450 (CYP) enzymes responsible for oxidative metabolism of most drugs used in clinical practice (Table 1) (1–14). As such, pharmacokinetic drug interactions have become a clinical issue of increasing concern.

One principal objective of the drug development process is the generation of scientific information on drug interactions so that treating physicians will have the data necessary to proceed with safe clinical treatment involving more than

Table 1 Representative Drugs Having Large and Clinically Important Effects on the Human CYP Enzymes

	Inhibition of	Induction of
Azole antifungals		
Ketoconazole	CYP3A	
Itraconazole	CYP3A	
Fluconazole	CYP3A, CYP2C9	
Voriconazole	CYP3A	
Terbinafine	CYP2D6	
Antidepressants		
Fluoxetine	CYP2D6	
Paroxetine	CYP2D6	
Fluvoxamine	CYP1A2, CYP2C19	
Nefazodone	CYP3A	
Bupropion	CYP2D6	
Anticonvulsants		
Carbamazepine		CYP3A
Anti-infectives		
Erythromycin	CYP3A	
Clarithromycin	CYP3A	
Ciprofloxacin	CYP1A2	
Rifampin		CYP3A
Viral protease inhibitors		
Ritonavir	CYP3A	
Saquinavir	CYP3A	
Amprenavir		CYP3A
Nonnucleoside reverse transcriptase inhibitors		
Delavirdine	CYP3A	
Nevirapine		CYP3A
Efavirenz		CYP3A
Cardiovascular agents		
Quinidine	CYP2D6	
Diltiazem	CYP3A	
Verapamil	CYP3A	

Abbreviation: CYP, cytochrome P450.

one medication. However, complete attainment of this objective is seldom possible, because the number of possible drug interactions is very large, and time and resources available for implementation of controlled clinical pharmacokinetic studies are inevitably limited. Some needed drug interaction studies will therefore be postponed until after a new drug is marketed, and some studies may be bypassed altogether. As such, in vitro data are becoming increasingly important as an approach to identifying which drug interactions are probable, possible, or unlikely, and thereby allow more informed planning of actual clinical studies (1,2,4,6,15–28).

II. CLINICAL CONSIDERATIONS IN EVALUATION OF DRUG INTERACTIONS

A. Nomenclature

A useful although legalistic nomenclature system refers to the agent causing the drug interaction as the "perpetrator," while the drug being affected by the interaction is the "victim" (Fig. 1). A pharmacokinetic drug interaction implies that the perpetrator causes a change in the metabolic clearance of the victim, in turn either decreasing or increasing concentrations of the victim drug in plasma and presumably also at the site of action (29). This change may or may not alter the clinical activity of the victim drug. A pharmacokinetic interaction "variant" is one in which the perpetrator does not change the systemic clearance or plasma

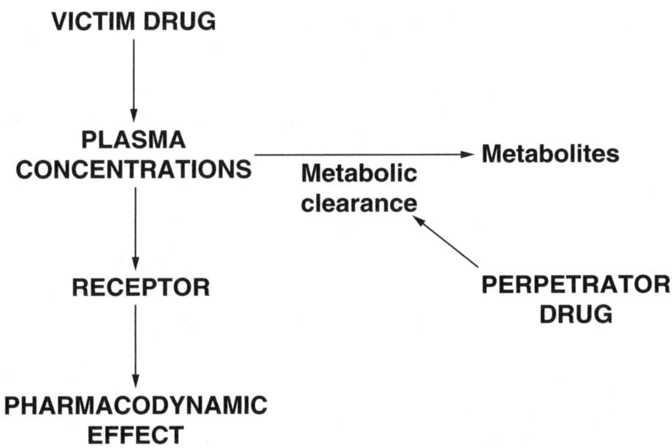

Figure 1 Schematic representation of the mechanism of pharmacokinetic drug interactions. Plasma concentrations of the "victim" drug are determined by its dosing rate and metabolic clearance. Plasma levels, in turn, determine drug concentrations at the receptor site and ultimately its pharmacodynamic effect. A pharmacokinetic drug interaction involves the effect of the "perpetrator" on the metabolic clearance of the victim. When the perpetrator is an inducer, clearance of the victim is increased, plasma levels are diminished, and pharmacological effect is reduced. Conversely, when the perpetrator is an inhibitor, clearance of the victim is reduced, plasma levels are increased, and pharmacodynamic effect is enhanced.

levels of the victim, but rather modifies the access of the victim to its pharmacologic receptor site. A familiar example is the antagonism of benzodiazepine activity by flumazenil; a less familiar example is benzodiazepine receptor antagonism by ketoconazole (30).

A pharmacodynamic interaction involves either inhibition or enhancement of the clinical effects of the victim drug as a consequence of similar or identical end-organ actions. Examples are the increase or decrease of the sedative-hypnotic actions of benzodiazepine agonist drugs due to coadministration of ethanol or caffeine, respectively (31,32).

B. Inhibition vs. Induction of Metabolism

Mechanistically different processes are involved in drug interactions involving inhibition as opposed to induction of metabolism mediated by CYP enzymes (Table 2) (15–28,33–37). Chemical inhibition is an immediate phenomenon. The effect becomes evident as soon as the inhibitor comes in contact with the enzyme and is in principle reversible when the inhibitor is no longer present (an exception is ''mechanism-based'' inhibition, see chap. 11) (38–42). The magnitude of inhibition—that is, the size of the interaction—depends on the concentration of the inhibitor at the intrahepatic site of enzyme activity relative to the intrinsic potency of the inhibitor. Inhibition potency can be measured using in vitro systems, yielding quantitative estimates such as the inhibition constant (K_i) or the 50% inhibitory concentration (IC_{50}). Methods of calculating K_i and IC_{50}, including the limitations and drawbacks inherent in the calculations, are reviewed elsewhere (15–28). Our current technological capacity to determine K_i or IC_{50} using in vitro systems is more advanced than our understanding of how to apply the numbers to quantitative predictions of drug interaction in vivo, which require knowledge of the effective concentration of inhibitor that is available to the enzyme. A generally applicable scheme to relate total or unbound plasma concentrations of inhibitor in the systemic or portal circulation to effective enzyme-available concentrations has not been

Table 2 Mechanistic Comparison of CYP Inhibition and Induction

	Inhibition	Induction
Mechanism	Direct chemical effect on enzyme	Indirect effect through enhanced production of CYP protein
Onset and reversibility	Rapid	Slow
Immediate exposure	Needed	Not needed
Prior exposure	Not needed	Needed
In vitro study	Straightforward	Difficult

Abbreviation: CYP, cytochrome P450.

established. It is now abundantly clear based on numerous examples that the theoretical assumption of equality of unbound systemic plasma concentrations and enzyme-available intrahepatic concentrations is incorrect in reality and will frequently yield underestimates of observed in vivo drug interactions by as much as an order of magnitude or even more (15–17,23,25,43–45). Modified scaling models have recently been proposed in which the inhibitor concentration available to the enzyme is postulated to be the estimated maximum unbound inhibitor concentration at the inlet to the liver—that is, in the portal vein (23,25). Although this is reported to yield some improvement in the predictive validity of the model, the overall predictive accuracy continues to be unsatisfactory.

Induction of CYP-mediated metabolism requires prior exposure of the hepatocyte's CYP-synthesis mechanism to a chemical inducer. The inducer signals the synthetic mechanisms to upregulate the production of one or more CYP isoforms, a process that takes time. Consequently, evidence of increased CYP activity is of slow onset following initiation of exposure to the inducer, and conversely slowly reverts to baseline after the inducer is removed (46,47). Enhancement of CYP expression/activity due to chemical induction therefore reflects prior, but not necessarily current, exposure to the inducer. Nuclear receptors, such as the pregnane X receptor (PXR) and the constitutive androstane receptor (CAR), have been identified as key elements in the process of transcriptional activation (48–56).

The quantitative extent of CYP induction depends on the dosage (concentration) of the inducer and on the duration of exposure. However, the induction process, in contrast to inhibition, is not as straightforward to study in vitro, since induction requires intact cellular protein synthesis mechanisms as available in cell culture models (57–62).

Inducers and inhibitors of CYP3A can be expected to influence both hepatic and gastrointestinal CYP3A (see chap. 10), although not necessarily to the same extent (2,63). Nonetheless, profound changes in both hepatic and gastrointestinal CYP3A will be caused by very strong inhibitors (such as ketoconazole or ritonavir) (2,63–65) or very strong inducers (such as rifampin) (3,12). A potentially complex situation arises for drugs such as ritonavir that are both inhibitors and inducers of CYP3A. In principle, interactions of ritonavir with CYP3A substrate drugs could be time dependent, with initial exposure producing CYP3A inhibition (64,65), but with CYP3A induction during extended exposure counterbalancing the inhibitory effects of acute exposure. However, it is now established that with extended exposure to ritonavir, the inhibitory effect predominates over any induction that may have occurred (66–69).

C. Clinical Importance of Drug Interactions

Given the prevalence of polypharmacy in clinical practice, noninteractions of drugs are far more common than interactions (70). The usual outcome of coadministration of two drugs is no detectable pharmacokinetic or pharmacodynamic interaction. That is, the pharmacokinetic disposition and clinical activity of each drug proceed independent of each other. Less common is the occurrence of a kinetic interaction. That is detectable using controlled study design methods but is of no clinical importance under usual therapeutic circumstances because (1) the interaction, while statistically significant, is not large enough in magnitude to produce a clinically important change in dynamics of the victim drug; (2) the therapeutic index of the victim drug is large enough that even a substantial change in plasma levels of the victim will not alter therapeutic effects or toxicity; or (3) kinetics and response to the victim drug is so variable that changes in plasma levels due to the drug interaction are far less important than inherent variability. Even less common are clinically important interactions that require modification in dosage of the perpetrator, the victim, or both. The most unusual consequence of a drug interaction is a situation in which the drug combination is so hazardous as to be contraindicated, as in the case of ketoconazole and terfenadine (71). These situations are rare, but unfortunately they receive disproportionate attention in the public media.

Many secondary sources are available to clinicians as guidelines to anticipate and avoid drug interactions. These compendia often serve as excellent and comprehensive collections of published data on drug interactions, but they generally are less helpful to clinicians in critically sorting out the literature and deciding what interactions are actually of real concern in the course of drug therapy. A useful general guideline for clinicians is that drug interactions are more likely to be important when (1) the perpetrator drug produces a very large change in the kinetics and plasma levels of the victim drug, that is, the perpetrator is a powerful inducer or inhibitor and (2) the therapeutic index of the victim is narrow. The first is exemplified by powerful inducers or inhibitors of CYP3A (ketoconazole, ritonavir, rifampin) coadministered with CYP3A substrates (72), or powerful inhibitors of CYP2D6 (quinidine, fluoxetine, paroxetine) coadministered with CYP2D6 substrates. The second case is exemplified by victim drugs such as phenytoin, warfarin, and digoxin, for which small changes in plasma levels could have important clinical consequences.

The intrinsic kinetic properties of the victim drug also influence the potential clinical consequences of an interaction. For orally administered medications that undergo significant presystemic extraction, impairment of clearance by a CYP inhibitor may produce increases in bioavailability (reduced presystemic extraction) as a consequence of reduced clearance. The effects may be particularly dramatic for CYP3A substrates (such as triazolam, midazolam, or buspirone) that undergo both hepatic and enteric presystemic extraction. As an example, coadministration of the CYP3A inhibitor ketoconazole with triazolam produced very large increases in area under the plasma concentration–time curve

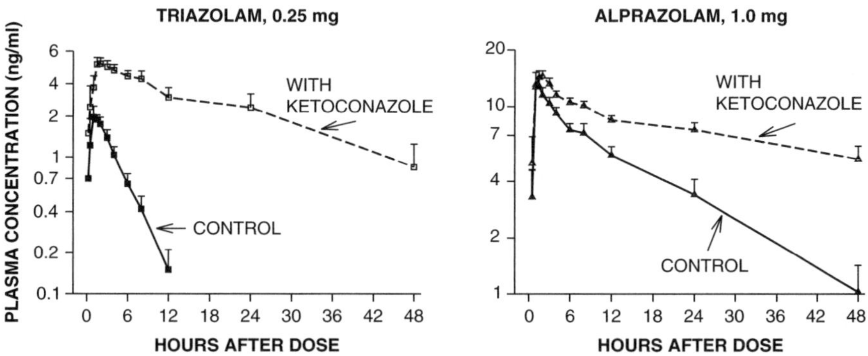

Figure 2 Mean (±SE) plasma concentrations of triazolam (*left*) or alprazolam (*right*) in a series of healthy individuals who participated in a clinical pharmacokinetic study. In one phase of the study, they ingested a single 0.25-mg oral dose of triazolam with ketoconazole, 200 mg twice daily, or with placebo to match ketoconazole (control). In the second phase of the study, they took 1.0 mg of alprazolam orally, either with the same dosage of ketoconazole or with placebo to match ketoconazole (control). Note that ketoconazole increases AUC and reduces clearance of both triazolam and alprazolam. For triazolam (a high-extraction compound), the effect is evident as reduced presystemic extraction, increased $C_{\max}$, and prolonged half-life. However, for alprazolam (a low-extraction compound), the effect of ketoconazole is evident only as a prolongation of half-life. *Abbreviation*: AUC, the plasma concentration–time curve. *Source*: Adapted, in part, from Ref. 74.

(AUC) and increases in peak plasma concentration ($C_{\max}$) of triazolam (73,74) (Fig. 2). Cotreatment of ketoconazole with alprazolam, also a CYP3A substrate, produced a large increase in AUC for alprazolam (74). However, alprazolam is a low-extraction compound with bioavailability ordinarily in the range of 90% (75). As such, the reduction in alprazolam clearance caused by ketoconazole was evident mainly as prolonged elimination half-life but without a significant change in $C_{\max}$.

III. INTEGRATING KINETIC AND DYNAMIC STUDY OBJECTIVES

A. Background

Clinical pharmacokinetic drug interaction protocols increasingly incorporate pharmacodynamic endpoints into the study design such that the dynamic

Table 3 Options for Pharmacodynamic Endpoints in Kinetic-Dynamic Studies of GABA-Benzodiazepine Agonist Drugs

Classification (with examples)	Relevance to primary therapeutic action	Influence of placebo	Influence of adaptation/ practice	''Blind'' conditions needed	Quantitative approach
Subjective					
Global assessments, rating scales	Close	Yes	Yes	Yes	Transformation of ratings into numbers
Semi-objective					
Psychomotor function tests, memory tests	Linked to adverse effect profile	Yes	Yes	Yes	Test outcomes are quantitative
Objective					
Electro-encephalography	Not established	No	No	Yes	Fully objective computer-determined quantitation

consequences of drug interactions may be estimated concurrently with the usual pharmacokinetic outcome measures. The level of complexity of an integrated kinetic-dynamic study depends on the nature of the pharmacodynamic actions of the drug under study as well as the type of pharmacodynamic outcome measures that are required. A number of methodological principles and dilemmas are illustrated by kinetic-dynamic design options for drug interaction studies involving sedative-hypnotic and anxiolytic drugs acting on the γ-aminobutyric acid (GABA)-benzodiazepine receptor system (76). For this category of drugs, a variety of outcome measures is available, but the approaches may differ substantially in their relevance to the principal therapeutic actions of the drug, the stability of the measure in terms of response to placebo or changes caused by practice or adaptation, the objective or subjective nature of the quantitative assessment, and the comparability of results across different investigators and different laboratories (Table 3). The extent to which the various pharmacodynamic measures provide unique information, as opposed to being overlapping or redundant, is not clearly established. For this reason, most kinetic-dynamic studies of GABA-benzodiazepine agonists utilize multiple parallel dynamic outcome measures.

B. Clinical Application

The kinetic and dynamic interaction of the triazolobenzodiazepine triazolam with various macrolide antimicrobial agents illustrates a number of these principles (77). The ''victim'' drug in this model, triazolam, is established as a relatively ''pure'' substrate for human CYP3A isoforms (73) and has been extensively prescribed in clinical practice as a hypnotic agent. The principal biotransformation pathways for triazolam involve parallel hydroxylation at two sites on the molecule, yielding α-OH-triazolam and 4-OH-triazolam as principal metabolites. Previous studies have established that triazolam biotransformation is strongly inhibited in vivo and in vitro by CYP3A inhibitors such as ketoconazole, itraconazole, ritonavir, and nefazodone (2,65,73,74,78,79). Some, but not all, of the macrolide antimicrobial agents also are CYP3A inhibitors (11,13,14). The mode of inhibition by this class of compounds is described as ''mechanism based,'' in that the parent compound binds to the metabolically active site on the CYP3A enzyme yielding a metabolic intermediate that irreversibly inactivates the enzyme (39–42). Mechanism-based inhibition of enteric CYP3A by natural substances contained in grapefruit juice also explains pharmacokinetic drug interactions involving grapefruit juice and certain CYP3A substrate drugs that ordinarily undergo presystemic extraction by CYP3A isoforms present in the mucosa of the small bowel (80–82). These inhibitors, chemically classified as furanocoumarins (of which the most important is 6′,7′-dihydroxybergamottin), irreversibly inactivate enteric CYP3A (80–89). Recovery from inhibition depends on the normal process of enzyme turnover and regeneration (83). It is of interest that other fruit beverages, such as pomegranate juice, that do not contain irreversible furanocoumarin inhibitors also do not cause clinically important interactions with CYP3A substrate drugs, even though the beverage may cause reversible inhibition of CYP3A based on in vitro models (90).

The following study of a drug interaction with macrolide antimicrobial agents illustrates the link between in vitro and in vivo findings as well as methods to define the pharmacodynamic consequences of a pharmacokinetic interaction (77). For the in vitro model, varying concentrations of four macrolide antimicrobial agents [troleandomycin (TAO), erythromycin, clarithromycin, azithromycin] were preincubated with liver microsomes and appropriate cofactors, followed by addition of a fixed concentration (250 μM) of triazolam, a substrate used to profile CYP3A activity. At the completion of the 20-minute incubation period, samples were processed and concentrations of α-OH- and 4-OH-triazolam determined by high-performance liquid chromatography (63,73,77). Rates of formation of the metabolites with coaddition of inhibitor were expressed as a percentage of the control velocity with no inhibitor present. The reaction velocity ratio versus inhibitor concentration

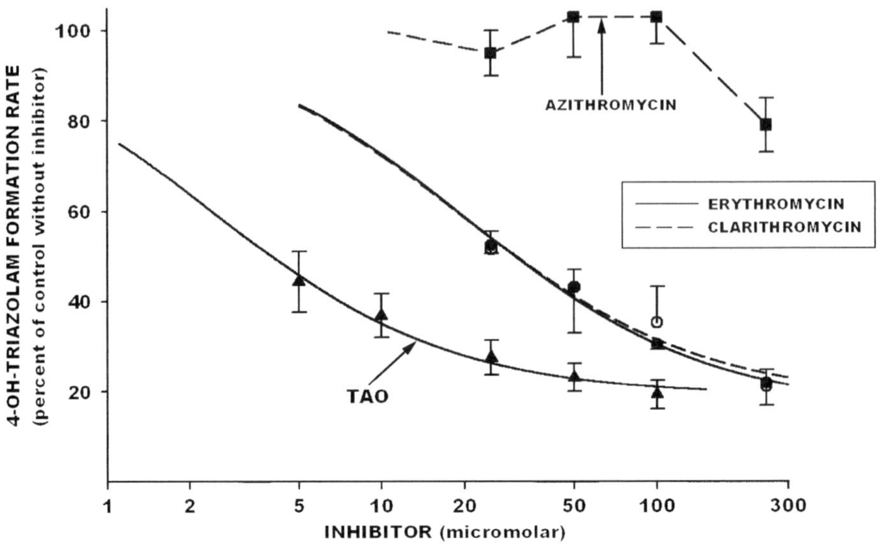

Figure 3 Rates of formation of 4-OH-triazolam from triazolam (250 μM) by human liver microsomes in vitro. Each point is the mean (±SE) of four microsomal preparations. Reaction velocities when preparations were preincubated with the macrolide agents are expressed as a percentage of the control velocity with no inhibitor present (inhibitor = 0). Mean IC_{50} were TAO, 3.3 μM; erythromycin, 27.3 μM; clarithromycin, 25.2 μM; azithromycin, >250 μM. *Abbreviations*: IC_{50}, 50% inhibitory concentrations; TAO, troleandomycin. *Source*: Adapted, in part, from Ref. 77.

relationship was used to determine a 50% inhibitory concentration (IC_{50}) (63). TAO, erythromycin, and clarithromycin all produced significant in vitro inhibition of triazolam hydroxylation and would appear to have the potential to produce a significant interaction with triazolam in vivo (Fig. 3). However, azithromycin was a very weak inhibitor of triazolam in vitro and is anticipated to produce no significant interaction in vivo.

The clinical pharmacokinetic-pharmacodynamic study had a double blind, randomized, five-way crossover design, with at least seven days elapsing between trials. Twelve healthy volunteers participated and the treatment conditions were:

1. Triazolam placebo plus macrolide placebo
2. Triazolam (0.125 mg) plus macrolide placebo
3. Triazolam (0.125 mg) plus azithromycin

	DAY 1		DAY 2		
TRIAL	AM	PM	8AM	9AM	PM
A	PL	PL	PL	PL	PL
B	PL	PL	PL	TRZ	PL
C	AZ-500	PL	AZ-250	TRZ	PL
D	ERY	ERY	ERY	TRZ	ERY
E	CLAR	CLAR	CLAR	TRZ	CLAR

PL = PLACEBO

TRZ = TRIAZOLAM, 0.125 mg

AZ-500 = AZITHROMYCIN, 500 mg

AZ-250 = AZITHROMYCIN, 250 mg

ERY = ERYTHROMYCIN, 500 mg

CLAR = CLARITHROMYCIN, 500 mg

Figure 4 Schematic representation of the design of the clinical study.

4. Triazolam (0.125 mg) plus erythromycin
5. Triazolam (0.125 mg) plus clarithromycin

Dosage schedules of the coadministered macrolides were chosen to be consistent with usual dosage recommendations (Fig. 4).

Following each dose of triazolam (or placebo to match triazolam), multiple venous blood samples were drawn over a period of 24 hours and multiple pharmacodynamic testing procedures were performed. These included subjective measures (observer's ratings and subjects' self-ratings of sedation and mood), a semiobjective measure of psychomotor performance (the digit-symbol substitution test, or DSST), and the fully objective quantitative of electroencephalographic (EEG) amplitude falling in the ''beta'' (13–30 cycles per second) frequency range (66,73,74,76,91–97). Note that for purposes of a kinetic-dynamic study of a CNS-active agent, Trial A is necessary, whereas for a purely pharmacokinetic study, Trial A would not be necessary. Even so, the five-way crossover design still does not rule out the possibility, although very unlikely, of CNS pharmacodynamic effects attributable to the macrolide agents alone. This would have required three additional trials—triazolam placebo plus azithromycin, triazolam placebo plus erythromycin, and triazolam placebo plus clarithromycin.

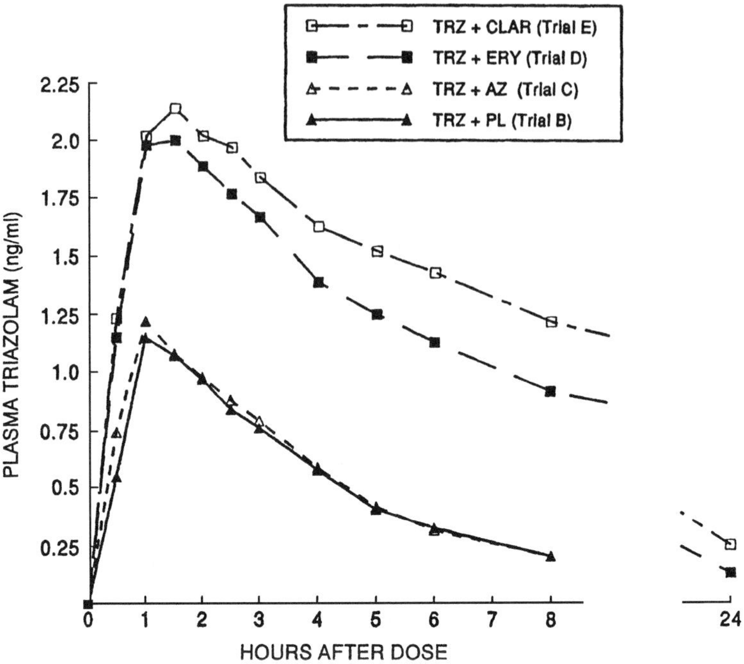

Figure 5 Mean plasma TRZ concentrations following a single 0.125-mg oral dose of TRZ administered in Trials B, C, D, and E, as described in the text and in Fig. 4. *Abbreviations*: TRZ, triazolam; CLAR, clarithromycin; ERY, erythromycin; AZ, azithromycin; PL, placebo. *Source*: From Ref. 77.

Triazolam plasma concentrations were determined by gas chromatography with electron capture detection (73,95). The pharmacokinetic results demonstrated that mean clearance during Trials B and C were nearly identical (413 and 416 mL/min, respectively); that is, coadministration of azithromycin had no effect on the pharmacokinetics of triazolam (Fig. 5). However, triazolam clearance was significantly reduced to 146 mL/min by erythromycin (Trial D) and to 95 mL/min by clarithromycin (Trial E). Thus, the in vivo kinetic results are highly consistent with the in vitro data.

The pharmacodynamic data indicated that the benzodiazepine agonist effects of triazolam plus placebo (Trial B) and of triazolam plus azithromycin (Trial C) were similar to each other and greater than the effects of placebo plus placebo (Trial A). However, coadministration of erythromycin (Trial D) or

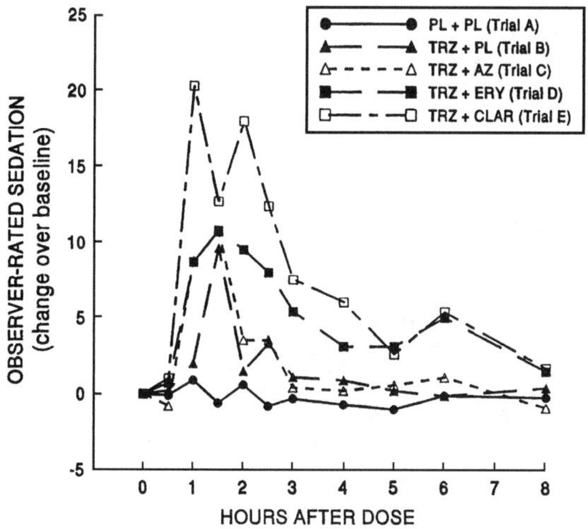

Figure 6 Mean changes over baseline in observer-rated sedation during each of the five trials, as described in the text and in Fig. 4. *Abbreviations*: TRZ, triazolam; CLAR, clarithromycin; ERY, erythromycin; AZ, azithromycin; PL, placebo. *Source*: From Ref. 77.

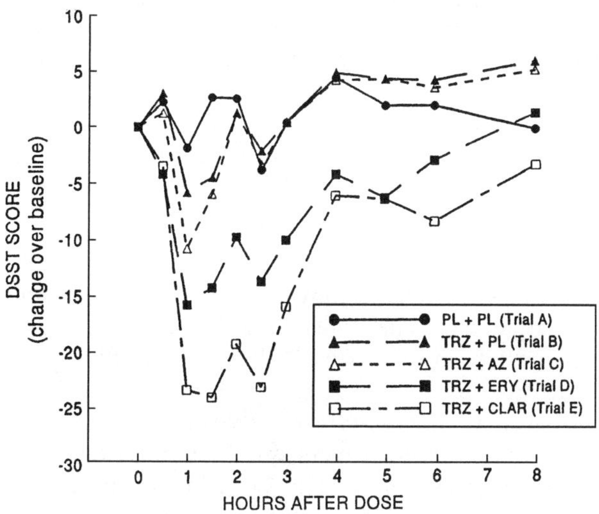

Figure 7 Mean changes over baseline in scores on the DSST during each of the five trials, as discussed in the text and in Fig. 4. *Abbreviations*: DSST, digit-symbol substitution test; TRZ, triazolam; CLAR, clarithromycin; ERY, erythromycin; AZ, azithromycin; PL, placebo. *Source*: From Ref. 77.

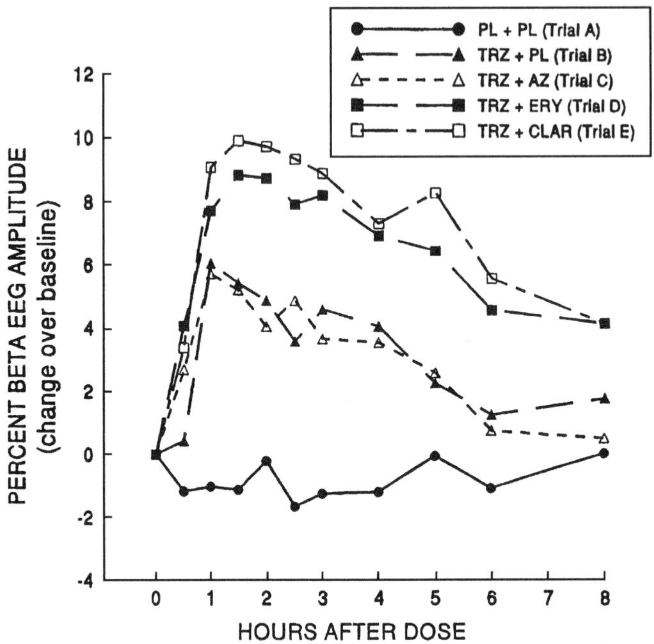

Figure 8 Mean changes over baseline in percentage of EEG amplitude falling in the beta frequency range (13–30 cycles/sec) during each of the five trials, as discussed in the text and in Fig. 4. *Abbreviations*: EEG, electroencephalographic; TRZ, triazolam; CLAR, clarithromycin; ERY, erythromycin; AZ, azithromycin; PL, placebo. *Source*: From Ref. 77.

clarithromycin (Trial E) augmented the pharmacodynamic effects of triazolam when compared with Trials B and C. The outcome was similar based on subjective measures, a semiobjective measure (the DSST), or the fully objective measure (the EEG) (Figs. 6–8). Kinetic-dynamic modeling indicated that the augmentation in benzodiazepine agonist effects of triazolam caused by coadministration of erythromycin or clarithromycin was fully consistent with the increase in triazolam plasma concentrations (Fig. 9). As anticipated, there was some redundancy among the various pharmacodynamic measures, in that the changes in these outcome measures at corresponding times were significantly intercorrelated (Fig. 10).

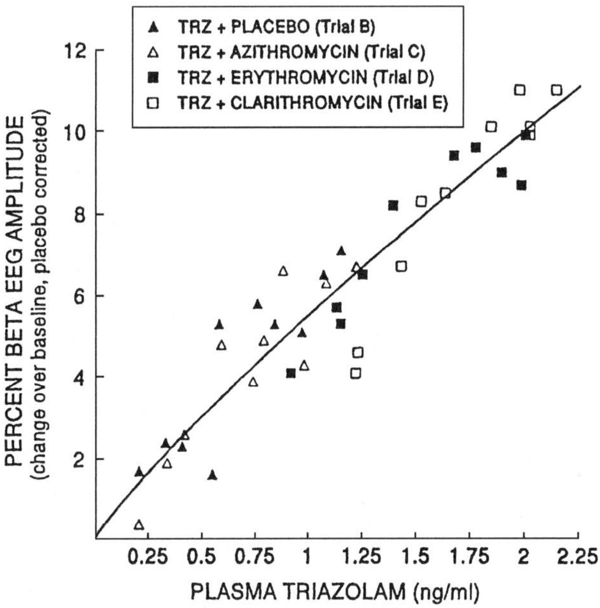

Figure 9 Mean changes over baseline in percentage of EEG amplitude falling in the beta frequency range (normalized for placebo-associated changes) in relation to mean plasma TRZ concentrations at corresponding times. The line represents a function of the form $y = Bx^A$ determined by nonlinear regression. *Abbreviations*: EEG, electroencephalographic; TRZ, triazolam. *Source*: From Ref. 77.

IV. COMMENT

The basic and clinical scientific issues underlying pharmacokinetic drug interactions are becoming increasingly complex as polypharmacy becomes more common and more drugs with enzyme-inducing or enzyme-inhibiting properties are introduced into clinical practice. A well-planned, integrated approach is needed to address the clinical problems. Ideally, the approach should incorporate the collaborative participation of individuals with expertise in molecular pharmacology, cytochrome biochemistry, in vitro metabolism, clinical pharmacokinetics-pharmacodynamics, and clinical therapeutics. The ultimate goal should be the informed and safe use of drug combinations in clinical practice.

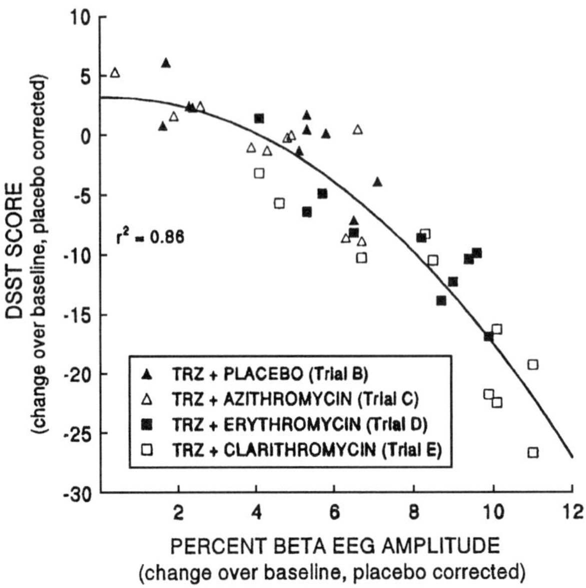

Figure 10 Mean changes over baseline in percentage of EEG amplitude falling in the β-frequency range in relation to mean changes over baseline in DSST score at corresponding times. Both measures were normalized for placebo-associated changes. The line represents a function of the form $y = Bx^A$ determined by nonlinear regression. *Abbreviations*: EEG, electroencephalographic; DSST, digit-symbol substitution test. *Source*: From Ref. 77.

ACKNOWLEDGMENTS

We are grateful for the collaboration and assistance of Dr. Richard I. Shader and Jerold S. Harmatz. The work is supported by Grants AG-017880, AT-003540, DA-005258, AI-058784, AT-001381, DA-013834, DK-058496, MH-058435, and RR-000054 from the Department of Health and Human Services.

REFERENCES

1. Greenblatt DJ, von Moltke LL, Harmatz JS, et al. Human cytochromes and some newer antidepressants: kinetics, metabolism, and drug interactions. J Clin Psychopharmacol 1999; 19(suppl 1):23S–35S.
2. Venkatakrishnan K, von Moltke LL, Greenblatt DJ. Effects of the antifungal agents on oxidative drug metabolism in humans: clinical relevance. Clin Pharmacokinet 2000; 38:111–180.

3. Niemi M, Backman JT, Fromm MF, et al. Pharmacokinetic interactions with rifampicin: clinical relevance. Clin Pharmacokinet 2003; 42:819–850.
4. Bertz RJ, Granneman GR. Use of in vitro and in vivo data to estimate the likelihood of metabolic pharmacokinetic interactions. Clin Pharmacokinet 1997; 32:210–258.
5. Antoniou T, Tseng AL. Interactions between antiretrovirals and antineoplastic drug therapy. Clin Pharmacokinet 2005; 44:111–145.
6. Hemeryck A, Belpaire FM. Selective serotonin reuptake inhibitors and cytochrome P-450 mediated drug-drug interactions: an update. Curr Drug Metab 2002; 3:13–37.
7. de Maat MM, Ekhart GC, Huitema AD, et al. Drug interactions between antiretroviral drugs and comedicated agents. Clin Pharmacokinet 2003; 42:223–282.
8. Piscitelli SC, Gallicano KD. Interactions among drugs for HIV and opportunistic infections. N Engl J Med 2001; 344:984–996.
9. Fichtenbaum CJ, Gerber JG. Interactions between antiretroviral drugs and drugs used for the therapy of the metabolic complications encountered during HIV infection. Clin Pharmacokinet 2002; 41:1195–1211.
10. Smith PF, DiCenzo R, Morse GD. Clinical pharmacokinetics of non-nucleoside reverse transcriptase inhibitors. Clin Pharmacokinet 2001; 40:893–905.
11. Gillum JG, Israel DS, Polk RE. Pharmacokinetic drug interactions with antimicrobial agents. Clin Pharmacokinet 1993; 25:450–482.
12. Finch CK, Chrisman CR, Baciewicz AM, et al. Rifampin and rifabutin drug interactions: an update. Arch Intern Med 2002; 162:985–992.
13. von Rosensteil NA, Adam D. Macrolide antibacterials. Drug interactions of clinical significance. Drug Saf 1995; 13:105–122.
14. Westphal JF. Macrolide-induced clinically relevant drug interactions with cytochrome P-450A (CYP) 3A4: an update focused on clarithromycin, azithromycin and dirithromycin. Br J Clin Pharmacol 2000; 50:285–295.
15. von Moltke LL, Greenblatt DJ, Schmider J, et al. In vitro approaches to predicting drug interactions in vivo. Biochem Pharmacol 1998; 55:113–122.
16. Venkatakrishnan K, von Moltke LL, Greenblatt DJ. Human drug metabolism and the cytochromes P450: application and relevance of in vitro models. J Clin Pharmacol 2001; 41:1149–1179.
17. Venkatakrishnan K, von Moltke LL, Obach RS, et al. Drug metabolism and drug interactions: application and clinical value of in vitro models. Curr Drug Metab 2003; 4:423–459.
18. Lin JH. Sense and nonsense in the prediction of drug-drug interactions. Curr Drug Metab 2000; 1:305–331.
19. Brown HS, Galetin A, Hallifax D, et al. Prediction of in vivo drug-drug interactions from in vitro data: factors affecting prototypic drug-drug interactions involving CYP2C9, CYP2D6 and CYP3A4. Clin Pharmacokinet 2006; 45:1035–1050.
20. Obach RS, Walsky RL, Venkatakrishnan K, et al. In vitro cytochrome P450 inhibition data and the prediction of drug-drug interactions: qualitative relationships, quantitative predictions, and the rank-order approach. Clin Pharmacol Ther 2005; 78:582–592.
21. Brown HS, Ito K, Galetin A, et al. Prediction of in vivo drug-drug interactions from in vitro data: impact of incorporating parallel pathways of drug elimination and inhibitor absorption rate constant. Br J Clin Pharmacol 2005; 60:508–518.
22. Yao C, Levy RH. Inhibition-based metabolic drug-drug interactions: predictions from in vitro data. J Pharm Sci 2002; 91:1923–1935.

23. Obach RS, Walsky RL, Venkatakrishnan K, et al. The utility of in vitro cytochrome P450 inhibition data in the prediction of drug-drug interactions. J Pharmacol Exp Ther 2006; 316:336–348.
24. Galetin A, Burt H, Gibbons L, et al. Prediction of time-dependent CYP3A4 drug-drug interactions: impact of enzyme degradation, parallel elimination pathways, and intestinal inhibition. Drug Metab Dispos 2006; 34:166–175.
25. Ito K, Brown HS, Houston JB. Database analyses for the prediction of in vivo drug-drug interactions from in vitro data. Br J Clin Pharmacol 2004; 57:473–486.
26. Bachmann KA, Lewis JD. Predicting inhibitory drug-drug interactions and evaluating drug interaction reports using inhibition constants. Ann Pharmacother 2005; 39:1064–1072.
27. Bachmann KA. Inhibition constants, inhibitor concentrations and the prediction of inhibitory drug drug interactions: pitfalls, progress and promise. Curr Drug Metab 2006; 7:1–14.
28. Bjornsson TD, Callaghan JT, Einolf HJ, et al. The conduct of in vitro and in vivo drug-drug interaction studies: a Pharmaceutical Research and Manufacturers of America (PhRMA) perspective. Drug Metab Dispos 2003; 31:815–832.
29. Greenblatt DJ, von Moltke LL. Sedative-hypnotic and anxiolytic agents. In: Levy RH, Thummel KE, Trager WF, et al. Metabolic Drug Interactions. Philadelphia: Lippincott Williams and Wilkins, 2000:259–270.
30. Fahey JM, Pritchard GA, von Moltke LL, et al. The effects of ketoconazole on triazolam pharmacokinetics, pharmacodynamics and benzodiazepine receptor binding in mice. J Pharmacol Exp Ther 1998; 285:271–276.
31. Chan AW. Effects of combined alcohol and benzodiazepine: a review. Drug Alcohol Depend 1984; 13:315–341.
32. Cysneiros R, Farkas D, Harmatz JS, et al. Pharmacokinetic and pharmacodynamic interactions between zolpidem and caffeine. Clin Pharmacol Ther 2007; 82:54–62.
33. Lin JH, Lu AY. Interindividual variability in inhibition and induction of cytochrome P450 enzymes. Annu Rev Pharmacol Toxicol 2001; 41:535–567.
34. Pelkonen O, Mäenpää J, Taavitsainen P, et al. Inhibition and induction of human cytochrome P450 (CYP) enzymes. Xenobiotica 1998; 28:1203–1253.
35. Lin JH, Lu AY. Inhibition and induction of cytochrome P450 and the clinical implications. Clin Pharmacokinet 1998; 35:361–390.
36. Park BK, Kitteringham NR, Piromohamed M, et al. Relevance of induction of human drug-metabolizing enzymes: pharmacological and toxicological implications. Br J Clin Pharmacol 1996; 41:477–491.
37. Waxman DJ, Azaroff L. Phenobarbital induction of cytochrome P-450 gene expression. Biochem J 1992; 281:577–592.
38. Zhou S, Yung Chan S, Cher Goh B, et al. Mechanism-based inhibition of cytochrome P450 3A4 by therapeutic drugs. Clin Pharmacokinet 2005; 44:279–304.
39. Silverman R. Mechanism-based enzyme inactivators. Methods Enzymol 1995; 249:240–283.
40. Blobaum AL. Mechanism-based inactivation and reversibility: is there a new trend in the inactivation of cytochrome P450 enzymes? Drug Metab Dispos 2006; 34:1–7.
41. Bertelsen KM, Venkatakrishnan K, von Moltke LL, et al. Apparent mechanism-based inhibition of human CYP2D6 in vitro by paroxetine: comparison with fluoxetine and quinidine. Drug Metab Dispos 2003; 31:289–293.

42. Perloff ES, Duan SX, Skolnik PR, et al. Atazanavir: effects on P-glycoprotein transport and CYP3A metabolism in vitro. Drug Metab Dispos 2005; 33:764–770.
43. Yao C, Kunze KL, Kharasch ED, et al. Fluvoxamine-theophylline interaction: gap between in vitro and in vivo inhibition constants toward cytochrome P4501A2. Clin Pharmacol Ther 2001; 70:415–424.
44. Culm-Merdek KE, von Moltke LL, Harmatz JS, et al. Fluvoxamine impairs single-dose caffeine clearance without altering caffeine pharmacodynamics. Br J Clin Pharmacol 2005; 60:486–493.
45. Yao C, Kunze KL, Trager WF, et al. Comparison of in vitro and in vivo inhibition potencies of fluvoxamine toward CYP2C19. Drug Metab Dispos 2003; 31:565–571.
46. Faber MS, Fuhr U. Time response of cytochrome P450 1A2 activity on cessation of heavy smoking. Clin Pharmacol Ther 2004; 76:178–184.
47. Ohnhaus EE, Breckenridge AM, Park BK. Urinary excretion of 6 beta-hydroxycortisol and the time course measurement of enzyme induction in man. Eur J Clin Pharmacol 1989; 36:39–46.
48. Gibson GG, Plant NJ, Swales KE, et al. Receptor-dependent transcriptional activation of cytochrome P4503A genes: induction mechanisms, species differences and interindividual variation in man. Xenobiotica 2002; 32:165–206.
49. Wang H, LeCluyse EL. Role of orphan nuclear receptors in the regulation of drug-metabolising enzymes. Clin Pharmacokinet 2003; 42:1331–1357.
50. Willson TM, Kliewer SA. PXR, CAR and drug metabolism. Nat Rev Drug Discov 2002; 1:259–266.
51. Waxman DJ. P450 gene induction by structurally diverse xenochemicals: central role of nuclear receptors CAR, PXR, and PPAR. Arch Biochem Biophys 1999; 369: 11–23.
52. Burk O, Koch I, Raucy J, et al. The induction of cytochrome P450 3A5 (CYP3A5) in the human liver and intestine is mediated by the xenobiotic sensors pregnane X receptor (PXR) and constitutively activated receptor (CAR). J Biol Chem 2004; 279:38379–38385.
53. Smirlis D, Muangmoonchai R, Edwards M, et al. Orphan receptor promiscuity in the induction of cytochromes P450 by xenobiotics. J Biol Chem 2001; 276: 12822–12826.
54. El-Sankary W, Plant NJ, Gibson GG, et al. Regulation of the CYP3A4 gene by hydrocortisone and xenobiotics: role of the glucocorticoid and pregnane X receptors. Drug Metab Dispos 2000; 28:493–496.
55. Honkakoski P, Negishi M. Regulation of cytochrome P450 (CYP) genes by nuclear receptors. Biochem J 2000; 347:321–337.
56. Savas U, Griffin KJ, Johnson EF. Molecular mechanisms of cytochrome P-450 induction by xenobiotics: an expanded role for nuclear hormone receptors. Mol Pharmacol 1999; 56:851–857.
57. Li AP, Reith MK, Rasmussen A, et al. Primary human hepatocytes as a tool for the evaluation of structure-activity relationship in cytochrome P450 induction potential of xenobiotics: evaluation of rifampin, rifapentine and rifabutin. Chem Biol Interact 1997; 107:17–30.
58. Kostrubsky VE, Ramachandran V, Venkataramanan R, et al. The use of human hepatocyte cultures to study the induction of cytochrome P-450. Drug Metab Dispos 1999; 27:887–894.

59. LeCluyse E, Madan A, Hamilton G, et al. Expression and regulation of cytochrome P450 enzymes in primary cultures of human hepatocytes. J Biochem Mol Toxicol 2000; 14:177–188.
60. Maurel P. The use of adult human hepatocytes in primary culture and other in vitro systems to investigate drug metabolism in man. Adv Drug Deliv Rev 1996; 22:105–132.
61. Madan A, Graham RA, Carroll KM, et al. Effects of prototypical microsomal enzyme inducers on cytochrome P450 expression in cultured human hepatocytes. Drug Metab Dispos 2003; 31:421–431.
62. McGinnity DF, Berry AJ, Kenny JR, et al. Evaluation of time-dependent cytochrome P450 inhibition using cultured human hepatocytes. Drug Metab Dispos 2006; 34:1291–1300.
63. von Moltke LL, Greenblatt DJ, Grassi JM, et al. Protease inhibitors as inhibitors of human cytochromes P450: high risk associated with ritonavir. J Clin Pharmacol 1998; 38:106–111.
64. Greenblatt DJ, von Moltke LL, Harmatz JS, et al. Alprazolam-ritonavir interaction: implications for product labeling. Clin Pharmacol Ther 2000; 67:335–341.
65. Greenblatt DJ, von Moltke LL, Harmatz JS, et al. Differential impairment of triazolam and zolpidem clearance by ritonavir. J Acquir Immune Defic Syndr 2000; 24:129–136.
66. Culm-Merdek KE, von Moltke LL, Gan L, et al. Effect of extended exposure to grapefruit juice on cytochrome P450 3A activity in humans: comparison with ritonavir. Clin Pharm Ther 2006; 79:243–254.
67. Yeh RF, Gaver VE, Patterson KB, et al. Lopinavir/ritonavir induces the hepatic activity of cytochrome P450 enzymes CYP2C9, CYP2C19, and CYP1A2 but inhibits the hepatic and intestinal activity of CYP3A as measured by a phenotyping drug cocktail in healthy volunteers. J Acquir Immune Defic Syndr 2006; 42:52–60.
68. Fellay J, Marzolini C, Decosterd L, et al. Variations of CYP3A activity induced by antiretroviral treatment in HIV-1 infected patients. Eur J Clin Pharmacol 2005; 60:865–873.
69. Mouly S, Rizzo-Padoin N, Simoneau G, et al. Effect of widely used combinations of antiretroviral therapy on liver CYP3A4 activity in HIV-infected patients. Br J Clin Pharmacol 2006; 62:200–209.
70. Bergk V, Gasse C, Rothenbacher D, et al. Drug interactions in primary care: impact of a new algorithm on risk determination. Clin Pharmacol Ther 2004; 76:85–96.
71. von Moltke LL, Greenblatt DJ, Duan SX, et al. In vitro prediction of the terfenadine-ketoconazole pharmacokinetic interaction. J Clin Pharmacol 1994; 34:1222–1227.
72. Tsunoda SM, Velez RL, von Moltke LL, et al. Differentiation of intestinal and hepatic cytochrome P450 3A activity with use of midazolam as an in vivo probe: effect of ketoconazole. Clin Pharmacol Ther 1999; 66:461–471.
73. von Moltke LL, Greenblatt DJ, Harmatz JS, et al. Triazolam biotransformation by human liver microsomes in vitro: effects of metabolic inhibitors, and clinical confirmation of a predicted interaction with ketoconazole. J Pharmacol Exp Ther 1996; 276:370–379.
74. Greenblatt DJ, Wright CE, von Moltke LL, et al. Ketoconazole inhibition of triazolam and alprazolam clearance: differential kinetic and dynamic consequences. Clin Pharmacol Ther 1998; 64:237–247.

75. Greenblatt DJ, Wright CE. Clinical pharmacokinetics of alprazolam: therapeutic implications. Clin Pharmacokinet 1993; 24:453–471.
76. Laurijssens BE, Greenblatt DJ. Pharmacokinetic-pharmacodynamic relationships for benzodiazepines. Clin Pharmacokinet 1996; 30:52–76.
77. Greenblatt DJ, von Moltke LL, Harmatz JS, et al. Inhibition of triazolam clearance by macrolide antimicrobial agents: in vitro correlates and dynamic consequences. Clin Pharmacol Ther 1998; 64:278–285.
78. Barbhaiya RH, Shukla UA, Kroboth PD, et al. Coadministration of nefazodone and benzodiazepines: II. A pharmacokinetic interaction study with triazolam. J Clin Psychopharmacol 1995; 15:320–326.
79. Varhe A, Olkkola KT, Neuvonen PJ. Oral triazolam is potentially hazardous to patients receiving systemic antimycotics ketoconazole or itraconazole. Clin Pharmacol Ther 1994; 56:601–607.
80. Mertens-Talcott SU, Zadezensky I, De Castro WV, et al. Grapefruit-drug interactions: can interactions with drugs be avoided? J Clin Pharmacol 2006; 46: 1390–1416.
81. Greenblatt DJ, Patki KC, von Moltke LL, et al. Drug interactions with grapefruit juice: an update. J Clin Psychopharmacol 2001; 21:357–359.
82. Bailey DG, Malcolm JAO, Spence JD. Grapefruit juice—drug interactions. Br J Clin Pharmacol 1998; 46:101–110.
83. Greenblatt DJ, von Moltke LL, Harmatz JS, et al. Time-course of recovery of cytochrome P450 3A function after single doses of grapefruit juice. Clin Pharmacol Ther 2003; 74:121–129.
84. Paine MF, Widmer WW, Hart HL, et al. A furanocoumarin-free grapefruit juice establishes furanocoumarins as the mediators of the grapefruit juice-felodipine interaction. Am J Clin Nutr 2006; 83:1097–1105.
85. Widmer W, Haun C. Variation in furanocoumarin content and new furanocoumarin dimers in commercial grapefruit (Citrus paradisi Macf.) juices. J Food Sci 2005; 70: c307–c312.
86. Schmiedlin-Ren P, Edwards DJ, Fitzsimmons ME, et al. Mechanisms of enhanced oral availability of CYP3A4 substrates by grapefruit constituents. Decreased enterocyte CYP3A4 concentration and mechanism-based inactivation by furanocoumarins. Drug Metab Dispos 1997; 25:1228–1233.
87. Edwards DJ, Bellevue FH, Woster PM. Identification of 6′,7′-dihydroxybergamottin, a cytochrome P450 inhibitor, in grapefruit juice. Drug Metab Dispos 1996; 24:1287–1290.
88. Fujita K, Hidaka M, Takamura N, et al. Inhibitory effects of citrus fruits on cytochrome P450 3A (CYP3A) activity in humans. Biol Pharm Bull 2003; 26: 1371–1373.
89. Girennavar B, Poulose SM, Jayaprakasha GK, et al. Furocoumarins from grapefruit juice and their effect on human CYP 3A4 and CYP 1B1 isoenzymes. Bioorg Med Chem 2006; 14:2606–2612.
90. Farkas D, Oleson LE, Zhou Y, et al. Pomegranate juice does not impair clearance of oral or intravenous midazolam, a probe for cytochrome P450-3A activity: comparison with grapefruit juice. J Clin Pharmacol 2007; 47:286–294.
91. Greenblatt DJ, Harmatz JS, von Moltke LL, et al. Comparative kinetics and response to the benzodiazepine agonists triazolam and zolpidem: evaluation of sex-dependent differences. J Pharmacol Exp Ther 2000; 293:435–443.

92. Greenblatt DJ, von Moltke LL, Ehrenberg BL, et al. Kinetics and dynamics of lorazepam during and after continuous intravenous infusion. Crit Care Med 2000; 28:2750–2757.
93. Greenblatt DJ, Ehrenberg BL, Culm KE, et al. Kinetics and EEG effects of midazolam during and after one-minute, one-hour, and three-hour intravenous infusions. J Clin Pharmacol 2004; 44:605–611.
94. Greenblatt DJ, Harmatz JS, von Moltke LL, et al. Age and gender effects on the pharmacokinetics and pharmacodynamics of triazolam, a cytochrome P450 3A substrate. Clin Pharmacol Ther 2004; 76:467–479.
95. Greenblatt DJ, Gan L, Harmatz JS, et al. Kinetics and dynamics of single-dose triazolam: electroencephalography compared to the digit-symbol substitution test. Br J Clin Pharmacol 2005; 60:244–248.
96. Greenblatt DJ, Legangneux E, Harmatz JS, et al. Dynamics and kinetics of a modified-release formulation of zolpidem: comparison with immediate-release standard zolpidem and placebo. J Clin Pharmacol 2006; 46:1469–1480.
97. Greenblatt DJ, Harmatz JS, Karim A. Age and gender effects on the pharmacokinetics and pharmacodynamics of ramelteon, a hypnotic agent acting via melatonin receptors MT_1 and MT_2. J Clin Pharmacol 2007; 47:485–496.

16

An Integrated Approach to Assessing Drug-Drug Interactions: A Regulatory Perspective

Shiew-Mei Huang, Lawrence J. Lesko, and Robert Temple
U.S. Food and Drug Administration,
Silver Spring, Maryland, U.S.A.

I. INTRODUCTION

Change in one or more safety and efficacy outcomes of a drug by concomitant administration of another drug that interacts with it has become of increasing interest in the last decade. This increased interest has arisen in part because of many documented adverse clinical consequences of drug-drug interactions, coupled with improved understanding as to their cause. Interest in drug-drug interactions has also increased because of the rise in polypharmacy, where patients may take many drugs in the course of a day. The following regimen would not be uncommon today in patients over 50 years of age—one or more antihypertensive agents, aspirin, a lipid-lowering drug, a hypnotic, an antihistamine, an antidepressant, one or more oral hypoglycemics, a proton pump inhibitor, an NSAID, and, if the patient is female, a drug to prevent osteoporosis and hormone replacement therapy might be used. Depending on various short-term conditions, an antibiotic or antifungal might be used.

To avoid serious harm, health care practitioners must be aware of and manage potential important interactions. To provide optimum information in product labeling for practitioners and patients, drug development and regulatory scientists should work cooperatively to ensure that each approved new drug is well characterized with respect to its metabolic or transport vulnerability as a substrate and its action as an inhibitor or inducer (1). Further, product labeling for older drugs should be updated as additional information about their potential for being a part of important drug-drug interactions becomes available.

Pharmacokinetic drug-drug interactions result from alteration in the dose/systemic exposure relationship, as reflected in a blood or plasma concentration–time curve, when an interacting drug induces or inhibits one or more routes of elimination or transport of a substrate drug. Inhibition of metabolism may be associated with increased blood levels and pharmacological activity of the substrate, but if the substrate is a prodrug, pharmacological activity may be reduced; in some cases, when the parent drug and its metabolite have equal effects, there may be no change in pharmacological activity despite large changes in blood levels of parent and metabolite (Chaps. 1 and 15). The magnitude of clinical effect of an inhibitor depends on the magnitude of the effect of the inhibitor on clearance of the substrate, which in turn depends on the extent of inhibition and the extent to which the substrate is cleared by the affected pathway. A well-known interacting drug is ketoconazole, a powerful inhibitor of cytochrome P450 3A4 (CYP3A4) metabolism that has recently also been shown to inhibit transport mechanisms (2). Coadministration of ketoconazole or a similar drug, itraconazole, increases the blood levels of many drugs, including dihydropyridine calcium channel blockers, short-acting benzodiazepines, some HMG-CoA reductase inhibitors, astemizole, and cisapride (3–5). Drugs that induce metabolic pathways and reduce systemic exposure may result in loss of effectiveness (Chaps. 1 and 6).

While most literature reports have focused on metabolic drug-drug interactions involving the CYP enzyme systems, every possible clearance pathway for a substrate may be altered by an interacting drug, including active renal secretion, drug degradation in the gut, biliary excretion, and secretion based on cellular transport mechanisms (Chaps. 5, 8, and 12). Examples include the inhibition of the renal tubular secretion of penicillins by probenecid, which results in major increases in penicillin blood levels (6) or the increase in digoxin blood levels by the coadministration of quinidine, presumably by the inhibition of digoxin renal tubular secretion through inhibition of the P-glycoprotein (P-gp) transporter (7). While these nonmetabolic examples of drug-drug interactions may be common, clearance pathways by the CYP enzymes of the liver and gut have proved especially vulnerable. Many of the drug-drug interactions that result in large (>50%) changes in exposure do so through inhibition or induction of CYP enzymes in the liver or gut.

Less commonly recognized than pharmacokinetic interactions—perhaps because fewer studies have been performed to detect them—are pharmacodynamic drug-drug interactions, changes in response to a drug caused by alteration in exposure/response relationships. This type of drug-drug interaction may arise when the substrate and interacting drug affect the same physiological system or when one drug prevents an appropriate response to the other. As an example of the former, both organic nitrates and sildenafil inhibit NO-mediated vasoconstriction and together cause marked hypotension (8). As an example of the latter, marked hypotension was observed in patients switched from the calcium channel blocker mibefradil to a dihydropyridine calcium channel blocker, apparently because residual mibefradil inhibited the usual compensatory tachycardia caused by the dihydropyridine. The effect may have been exaggerated by the increased levels of the dihydropyridine resulting from mibefradil's inhibition of the CYP3A4 route of elimination (9). Both pharmacokinetic and pharmacodynamic drug-drug interactions should be considered when two or more drugs are administered concurrently.

The critical question in considering drug interactions is: Does the dose of a substrate drug need to be adjusted in the presence of the interacting drug? More specifically, is the pharmacokinetic and/or pharmacodynamic change in the substrate drug in the presence of the interacting drug of sufficient magnitude require adjustment of the substrate dose (or avoidance of the interacting drug)? If we are willing (as we generally are) to assume that exposure-response relationships for the substrate drug are undisturbed in the presence of a pharmacokinetically interacting drug and have a reasonable idea at what the exposure-response relationship is, then we can rely on the effect of the interacting drug on systemic exposure pharmacokinetic measures such as area under the plasma concentration–time curve (AUC) and $C_{\max}$ and on whether other effect would lead to a clinical problem to determine whether the dose of the substrate needs to be altered in the presence of the interacting drug. Finally, how confident we need to be in the answer depends on the nature of interaction and the consequences of error. The larger question about what to do about a clinically important interaction has led to removal of substrates from the market (terfenadine, cisapride, astemizole), removal of a strong CYP3A4 inhibitor from the market, and many warnings and boxed warnings about the interactions and contraindication to concomitant use.

II. METHODS TO ASSESS DRUG-DRUG INTERACTIONS

Assessment of a potential drug-drug interaction begins with an understanding of the absorption, distribution, and elimination processes for both substrate and interacting drugs. On the basis of this information, the potential importance of one or more routes of elimination in contributing to a clinically important drug-drug interaction can be estimated. Even when a metabolic route is important for the elimination of a substrate and is affected by an interacting drug, additional studies may be needed to understand whether a metabolic drug-drug interaction has clinical impact. Various methods may be used to develop the requisite information, including in vitro studies, in vivo pharmacokinetic and pharmacodynamic studies, population pharmacokinetic studies, clinical safety and efficacy studies, and postmarketing observational studies. All of these approaches can generate useful information about potentially important drug-drug interactions and each has special strengths and limitations. Many of these approaches are described in FDA guidance for industry and updated recommendation in a recently established FDA Web site (10–12). Metabolic drug-drug interactions involving CYP3A4 may require special consideration because they may occur in the wall of the gastrointestinal tract and/or the liver. Interactions in the gastrointestinal tract can increase bioavailability, as reflected in C_{max} and AUC, but may cause little or no effect on half-life. Interactions in the liver may have only a small effect on single-dose C_{max}, but may alter half-life and accumulation index. Interpretation of drug-drug interaction data is sometimes complicated when a substrate drug is actively transported from the serosal to the mucosal side of the gastrointestinal tract by transporters such as P-gp. Like CYP3A4, these transporters are subject to inhibition/induction.

III. GENERAL APPROACHES

As discussed in section II, early in vitro and in vivo investigations can enhance the quality and efficiency of drug development, in some cases fully addressing a question of interest, and in others providing information to guide further studies (Chaps. 6 and 7). The early elucidation of drug metabolism, for example, permits in vitro investigations of drug-drug interaction that in turn provide information useful in guiding the clinical program and possibly avoiding some clinical studies. Metabolism data can also provide information on the relevance of preclinical metabolism and toxicological data and permit early identification of drugs that are likely to have large interindividual pharmacokinetic variability due to genetically determined polymorphisms in drug-metabolizing enzymes or drug-drug interactions. An integrated approach is most useful, one in which evidence for and against a drug-drug interaction is examined at all stages of drug development, including (1) preclinical in vitro human tissue studies of drug metabolism and drug-drug interactions to determine which in vivo studies should be conducted, (2) early-phase in vivo studies to assess the most important potential drug-drug interactions suggested by in vitro data, (3) late-phase drug development population pharmacokinetic studies to expand the range of potential interactions studied, including unexpected ones, and to allow examination of pharmacodynamic drug-drug interactions. The further sections of this chapter provide more specific information about these approaches.

A. In Vitro Methodologies

Pharmaceutical sponsors now frequently conduct in vitro studies in the preclinical phase of drug development programs to assess the contribution of CYP or other enzymes to the metabolic elimination of an investigational drug and the ability of an investigational drug to inhibit specific metabolic pathways. The utility of these studies has been enhanced by the availability of specific enzyme preparations, microsomal preparations, and liver cell preparations, together with standard substrates and inhibitors/inducers. Information from in vitro metabolic studies can suggest not only that a substrate drug is or is not likely to be a candidate for certain metabolic drug-drug interactions but also whether a drug's metabolism will be affected by genetic polymorphisms.

If a drug is not metabolized or its metabolism is not mediated by CYP enzymes, in vivo metabolic drug interactions with CYP enzyme inhibitors and inducers are not needed. If a drug does not inhibit any of the CYP enzymes, in vivo interaction studies with CYP substrate are not needed. Detection of the involvement of certain metabolic pathways, notably CYP1A2, CYP3A4, CYP2D6, CYP2B6, CYP2C8, CYP2C9, and CYP2C19, from in vitro studies suggests the possibility of important drug-drug interactions and usually results in significant effort to detect and define them. Thus, in vivo studies may be avoided through an in vitro study showing that an investigational drug's metabolism is not affected by furafylline (CYP1A2), ketoconazole (CYP3A4), quinidine (CYP2D6), clopidogrel (CYP2B6), gemfibrozil (CYP2C8), sulfaphenazole (CYP2C9) (no potential effect on the substrate), and that the drug does not affect the metabolism of caffeine (CYP1A2), midazolam (CYP3A4), dextromethorphan (CYP2D6), efavirenz (CYP2B6), repaglinide (CYP2C8), *S*-warfarin (CYP2C9), or *S*′-mephenytoin (CYP2C19) (no potential inhibition/induction). The recently published FDA guidance for industry entitled *Drug-Drug Interactions—Study Design, Data Analysis, and Implications for Dosing and Labeling* (10) describes the techniques and approaches to in vitro study of metabolic-based drug-drug interactions, in vitro and in vivo correlations, the timing of these studies, and the labeling of drug products based on in vitro metabolism and drug-drug interaction data. This guidance emphasizes the value of in vitro studies in human biomaterials in ruling out important metabolic pathways in a drug's metabolism or the possibility of the drug's ability to affect certain enzyme systems.

Previous chapters have detailed the relative advantages and disadvantages of various in vitro techniques in providing information pertinent to drug-drug interactions. These include the following preparations:

1. Cellular-based in vitro models, such as isolated hepatocytes and precision-cut liver preparations
2. Subcellular elements, such as microsomes or S9 (cytosolic) fractions
3. Expressed human drug-metabolizing enzymes

These systems can be used to define a drug's metabolic pathway, to assess its potential to inhibit the metabolism of other drugs, and to determine whether other drugs influence its metabolism.

The complex interrelationship of cellular transport mechanisms and drug-metabolizing enzymes, particularly CYP3A, in mediating systemic drug availability and drug-drug interactions is under increasing study. P-gp is the best-understood cellular transporter. It is abundantly present in the intestinal epithelium and serves as an efflux pump for a variety of drugs and xenobiotics. It is also highly expressed in bile canaliculi, the apical membrane of the renal tubule epithelium and other tissues. Many inhibitors of P-gp also inhibit CYP3A metabolism, and many, although not all, substrates for CYP3A are also actively transported by P-gp. For this reason, the relative contribution of CYP and transporter effects to a drug-drug interaction may be difficult to quantify. In vitro models currently available allow investigation of transporter-mediated drug-drug interactions, including a human colon carcinoma cell line, Caco-2 (10).

B. In Vitro–In Vivo Correlation

A complete understanding of the relationship between in vitro findings and in vivo results of metabolism/drug-drug interaction studies is still emerging. Quantitative prediction of the magnitude of clinical drug-drug interactions based on in vitro methodologies has been the topic of numerous publications and is described in earlier chapters (Chaps. 5, 7, 10, and 11). Although excellent quantitative concordance of in vitro and in vivo results has been shown, in some cases in vitro data may also under- or overestimate the clinical effect (13), and at present an observed in vitro effect needs further elucidation in in vivo studies. The bases for in vitro/in vivo disassociations have been described and include (1) irrelevant substrate concentrations and inappropriate in vitro model systems, (2) mechanism-based inhibition, (3) activation/induction phenomena, (4) physical-chemical effects on absorption, (5) parallel elimination pathways that decrease the importance of the in vitro–assessed pathway, and (6) modulation of an important cellular transport mechanism.

C. Specific Clinical Investigations

If metabolism is an important mechanism of clearance and in vitro studies suggest that metabolic routes can be inhibited or that the drug may inhibit important clearance pathways of other drugs, in vivo studies are needed to evaluate the extent of these potential interactions. A recently published FDA guidance for industry entitled *Drug-Drug Interaction Studies—Study Design, Data Analysis, and Implications for Dosing and Labeling* (10) provides recommendations on study design, study population, choice of interacting drugs, route of administration, dose selection, and statistical considerations for clinical drug-drug interaction studies. As with in vitro studies, in vivo studies can often use a screening approach involving probe drugs. For example, if ketoconazole, a powerful CYP3A4 inhibitor, does not have a significant effect on the pharmacokinetics of a drug with some evidence of CYP3A4 metabolism in vitro, further interaction studies with other CYP3A4 inhibitors are not necessary. Similarly, if the drug does not affect the pharmacokinetics of a sensitive CYP3A4 substrate, such as midazolam, it will not pose problems with other CYP3A4-metabolized drugs. Where interactions are found, the studies of probe drugs and other drugs will provide a basis for specific recommendations on product labeling as to what concomitant uses should be avoided or what dosage adjustments to make. A critical determination for substrate effects is the size of the effect, measured in the in vivo interaction study, and the importance of the effect. Thus, a 50% increase in blood levels of a well-tolerated drug with little dose-related toxicity may require no dosage adjustment. The same degree of increase for a drug with a narrow therapeutic range might require careful adjustment in dose or avoidance of coadministration. The issues in the areas of study design and data analysis are discussed in more detail in the following section.

If in vitro studies and other information suggest a need for in vivo metabolic drug-drug interaction studies, the following general issues and approaches should be considered. Depending on the study objectives, the substrate and interacting drug may be investigational agents or approved products.

1. Study Design

In general, interaction studies compare substrate levels with and without the interacting drug. Several different study designs have been used to study drug-drug interactions. Any may be suitable, depending on the specific objectives of the study and the desired outcome. The study may use a randomized crossover, a one-way (fixed sequence) crossover, or a parallel design. Depending on circumstances, the studies can use various durations of exposure for substrate and interacting drug: single dose/single dose, single dose/multiple dose, multiple dose/single dose, and multiple dose/multiple dose. The details of the study design depend on a number of factors for both the substrate and interacting drug, including (1) pharmacokinetic and pharmacodynamic characteristics of the substrate and interacting drugs; (2) the need to assess induction as well as inhibition (induction generally needs longer study duration); and (3) safety considerations, including whether the substrate is a narrow therapeutic range (NTR) or non-NTR drug. In general, the inhibiting/inducing drugs and the substrates should be dosed so that the exposure of both drugs is relevant to their clinical use. The following specific examples may be useful in choosing among study designs.

1. A substrate drug intended for chronic administration should generally be given until steady state is attained, with assessment of pharmacokinetics over one or more dosing intervals followed by administration of the interacting drug, which is also given until steady-state concentration is reached, again with collection of pharmacokinetic data on the substrate. The studies of erythromycin-terfenadine and ketoconazole-terfenadine interactions in healthy volunteers (14,15) are examples of this one-way, or fixed-sequence, crossover design.
2. If the substrate drug has a long half-life and accumulates, the probability of seeing an effect may be enhanced by giving the substrate drug as a single dose and the interacting drug as multiple doses. One example of this design

is the study of the effect of terfenadine on the pharmacokinetics of buspirone, a CYP3A4 substrate, where a randomized two-way crossover design was utilized (16). Note that although sensitivity to detecting inhibitory effect may be increased, it could be argued that effect on steady-state $C_{\max}$ and AUC is more relevant.

3. When the substrate has complex metabolism (e.g., a long-acting active metabolite) or the interacting drug has a long half-life or active metabolite, attainment of steady state may pose problems. Multiple-dose studies would generally be necessary to ensure that relevant metabolites can be assessed and that the relevant dose of the interacting drug is used, but special approaches may also be useful. For example, a loading dose of the potential inhibitor may allow relevant levels to be obtained more rapidly and selection of a one-way (fixed-sequence) crossover or a parallel design, rather than a randomized crossover study design, may also help. Using a one-way crossover design, a recent study (17) showed that multiple-dose administration of sertraline inhibited the clearance of desipramine to a considerably greater extent than did a single-dose administration. The long half-lives and the nonlinear accumulation of sertraline and its desethyl metabolite, both of which are CYP2D6 inhibitors, appeared to have contributed to the higher exposure of these two components and thus greater inhibition effects after multiple dosing.
4. The dosing duration depends on whether inhibition or induction is to be studied. Inducers may take several days or longer to exert their effects, while inhibitors generally exert their effects more rapidly. For this reason, a more extended period of exposure to interacting drug may be necessary if induction is to be assessed. The study design should also allow assessment of how long the inhibition or induction effect will last after an interacting drug has been removed from the dosing regimen. This effect can be observed in the randomized crossover design and in the one-sequence or parallel designs by adding an additional period in which the interacting drug is withdrawn. A recent publication (9) describing serious adverse events (including one death) observed when dihydropyridine calcium channel blockers (CYP3A4 substrates) were given to patients immediately after mibefradil (a CYP3A4 inhibitor) was withdrawn illustrates the importance of this consideration. In this case, mibefradil both increased blood levels of the dihydropyridine and inhibited the increased heart rate needed to overcome the lowered blood pressure.
5. For an inhibitor drug that induces its own metabolism, a multiple-dose study design should be used so that the extent of interaction is not overestimated. Multiple doses of ritonavir have been shown to have smaller inhibitory effects on other CYP3A substrates (18,19) than a single dose. This inhibition may be partially explained by the lower exposure to ritonavir after multiple doses than after a single dose.
6. When a pharmacodynamic effect is also being measured, attainment of steady state for the parent or metabolite whose pharmacodynamic effects

are being measured is important. In addition, inclusion of a period of the interacting drug alone in the sequence is often advisable so that its contribution to the pharmacodynamic effects can be assessed. For example, erythromycin is known to prolong QT intervals at some doses. The assessment of QT interval change due to substrate accumulation resulting from erythromycin inhibition of CYP3A4 metabolism cannot be evaluated without an erythromycin-alone group to examine the effect of erythromycin in the population.

7. Studies can usually be open label (unblinded), unless pharmacodynamic endpoints (e.g., adverse events whose interpretation is potentially subject to bias) are part of the assessment of the interaction.

2. Study Population

Clinical drug-drug interaction studies can generally be performed in healthy volunteers unless safety considerations preclude their participation. Sometimes, use of subjects/patients for whom the substrate drug is intended offers advantages, including the opportunity to study pharmacodynamic endpoints not present in healthy subjects. If metabolic polymorphisms for a pathway being studied exist, the availability of genotype or phenotype information may be important, as inhibitors or inducers may have no effect or little effect in poor metabolizers, an observation that is particularly important for substrates eliminated by the CYP2D6, CYP2C9, and CYP2C19 pathways.

3. Choice of Substrates and Interacting Drugs

While past experience (20,21) revealed a reasonable number of interaction studies with such drugs as digoxin and warfarin, the drugs used in these interaction studies generally did not reflect a clear understanding of interaction potential related to CYP enzyme inhibition. Improved understanding of the metabolic basis of drug-drug interactions allows the use of more informative approaches to choosing substrates and potential interacting drugs. Figure 1 describes a decision-making process (20) for the conduct of in vivo drug interaction studies once a new drug is characterized as a substrate for a particular metabolic pathway or an inhibitor of that pathway.

a. Investigational drug as an inhibitor or an inducer of CYP enzymes. In contrast to earlier approaches that focused mainly on a specific group of approved drugs (e.g., digoxin, hydrochlorothiazide) where coadministration was likely or the clinical consequences of an interaction were of concern, improved understanding of the mechanistic basis of metabolic drug-drug interactions allows more general approaches to, and conclusions from, specific drug-drug interaction studies. In studying an investigational drug as the interacting (inhibiting or inducing) drug, the choice of substrates (approved drugs) for initial

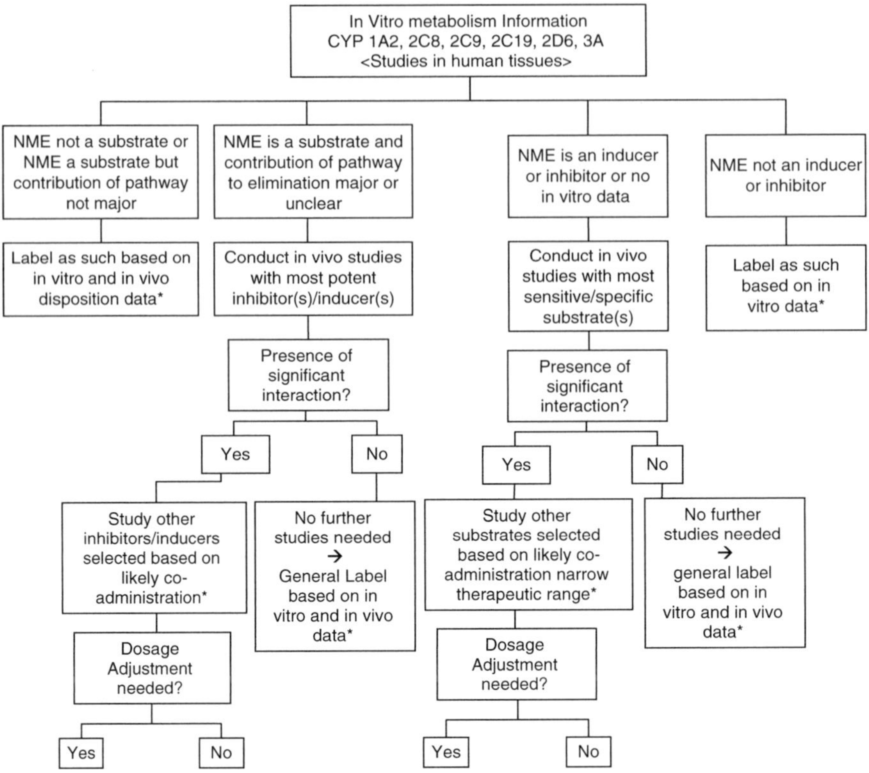

NME: New molecular entity

* Additional population pharmacokinetic analysis may assist the overall evaluation

Figure 1 CYP-based drug-drug interaction studies—decision tree. *Source*: From Ref. 20.

in vivo studies depends on the CYP enzymes affected by the interacting drug. In testing inhibition, the substrate selected should generally be one whose pharmacokinetics are markedly altered by coadministration of known specific inhibitors of the enzyme systems to assess the impact of the interacting investigational drug. Examples of substrates include (1) midazolam for CYP3A inhibition, (2) theophylline for CYP1A2 inhibition, (3) repaglinide for CYP2C8 inhibition, (4) warfarin for CYP2C9 inhibition (with the evaluation of *S*-warfarin), (5) omeprazole for CYP2C19 inhibition, and (6) desipramine for CYP2D6 inhibition (Table 1). Additional examples of substrates, along with inhibitors and inducers of specific CYP enzymes, are available on the FDA Web site (12). If the initial study shows that an investigation drug either inhibits or induces metabolism, further studies using less sensitive substrates, based on the likelihood of coadministration, may be useful. If the initial study is negative with

Table 1 Examples of In Vivo Substrate, Inhibitor, and Inducer for Specific CYP Enzymes Recommended for Study (Oral Administration)

CYP	Substrate	Inhibitor	Inducer
1A2	Theophylline, caffeine	Fluvoxamine	Smokers vs. nonsmokers[a]
2B6	Efavirenz	Clopidogrel	Rifampin
2C8	Repaglinide, rosiglitazone	Gemfibrozil	Rifampin
2C9	Warfarin, tolbutamide	Fluconazole, amiodarone (use of PM vs. EM subjects)[b]	Rifampin
2C19	Omeprazole, esoprazole, lansoprazole, pantoprazole	Omeprazole, fluvoxamine, moclobemide (use of PM vs. EM subjects)[b]	Rifampin
2D6	Desipramine, dextromethorphan, atomoxetine	Paroxetine, quinidine, fluoxetine (use of PM vs. EM subjects)[b]	None identified
2E1	Chlorzoxazone	Disulfirum	Ethanol
3A4/3A5	Midazolam, buspirone, felodipine, lovastatin, eletriptan, sildenafil, simvastatin, triazolam	Atazanavir, clarithromycin, indinavir, itraconazole, ketoconazole, nefazodone, nelfinavir, ritonavir, saquinavir, telithromycin	Rifampin, carbamazepine

This is not an exhaustive list. For an update list, see the following: http://www.fda.gov/CDER/drug/drugInteractions/default.htm

Substrates for any particular CYP enzyme listed in this table are those with plasma AUC values increased by twofold or higher when coadministered with inhibitors of that CYP enzyme; for CYP3A, only those with plasma AUC increased by fivefold or higher are listed. Inhibitors listed are those that increase plasma AUC values of substrates for that CYP enzyme by twofold or higher. For CYP3A inhibitors, only those that increase AUC of CYP3A substrates by fivefold or higher are listed. Inducers listed are those that decrease plasma AUC values of substrates for that CYP enzyme by 30% or higher.

[a]A clinical study can be conducted in smokers compared with nonsmokers (in lieu of an interaction study with an inducer), when appropriate.

[b]A clinical study can be conducted in PM compared with EM for the specific CYP enzyme (in lieu of an interaction study with an inhibitor), when appropriate.

Abbreviations: CYP, cytochrome P450; PM, poor metabolizers; EM, extensive metabolizers; AUC, area under the plasma concentration–time curve.

the most sensitive substrate, it can be presumed that less sensitive substrates will also be unaffected.

The FDA guidance proposed a classification of CYP inhibitors by magnitude of inhibition (1,10,12). As shown in Table 2, if an investigational drug increases the AUC of oral midazolam or other CYP3A substrates by fivefold or more, it can be considered a strong CYP3A inhibitor. If an investigational drug,

Table 2 Classification of CYP Inhibitors

CYP	Strong inhibitors (≥5-fold increase in AUC)	Moderate inhibitors (≥2 but <5-fold increase in AUC)	Weak inhibitors (≥1.25 but <2-fold increase in AUC)
3A	Atazanavir, clarithromycin, indinavir, itraconazole, ketoconazole, nefazodone, nelfinavir, ritonavir, saquinavir, telithromycin	Amprenavir, aprepitant, diltiazem, erythromycin, fluconazole, fosamprenavir, grapefruit juice[a], verapamil	Cimetidine
1A2	Fluvoxamine	Ciprofloxacin, mexiletine, propafenone, zileuton	Acyclovir, cimetidine, famotidine, norfloxacin, verapamil
2C8	Gemfibrozil		Trimethoprim
2C9		Amiodarone, fluconazole, oxandrolone	Sulfinpyrazone
2C19	Omeprazole		
2D6	Fluoxetine, paroxetine, quinidine	Duloxetine, terbinafine	Amiodarone, sertraline

Please note the following:

- A strong inhibitor is one that causes a ≥fivefold increase in the plasma AUC values or more than 80% decrease in clearance of *CYP substrates (not limited to sensitive CYP substrate)* in clinical evaluations
- A moderate inhibitor is one that causes a ≥two- but <fivefold increase in the AUC values or 50–80% decrease in clearance of *sensitive CYP substrates when the inhibitor was given at the highest approved dose and the shortest dosing interval* in clinical evaluations.
- A weak inhibitor is one that causes a ≥1.25- but <2-fold increase in the AUC values or 20–50% decrease in clearance of *sensitive CYP substrates when the inhibitor was given at the highest approved dose and the shortest dosing interval* in clinical evaluations
- This is not an exhaustive list. For an updated list, see the following link http://www.fda.gov/cder/drug/drugInteractions/default.htm

[a]The effect of grapefruit juice varies widely.
Abbreviations: CYP, cytochrome P450; AUC, area under the plasma concentration–time curve.

given at its highest dose and shortest dosing interval, increases the AUC of oral midazolam or other sensitive CYP3A substrates by between two- and fivefold, it can be considered a moderate CYP3A inhibitor. Finally, if an investigational drug, given at the highest dose and shortest dosing interval, increases the AUC of oral midazolam or other sensitive CYP3A substrates by between 1.25- and 2-fold, it can be considered a weak CYP3A inhibitor. When an investigational drug is determined to be an inhibitor of CYP3A, its interaction with sensitive

CYP3A substrates or CYP3A substrates with a narrow therapeutic range will need to be described in various sections of the labeling, as appropriate. Similar classifications of inhibitors of other CYP enzymes are also discussed in the guidance and the FDA Web site (10,12) and listed in Table 2.

When an in vitro evaluation cannot rule out the possibility that an investigational drug is an inducer of a CYP enzyme, an in vivo evaluation can be conducted using the most sensitive substrate (e.g., oral midazolam for CYP3A).

Simultaneous administration of a mixture of substrates of CYP enzymes in one study (i.e., a ''cocktail approach'') in human volunteers is another way to valuate a drug's inhibition or induction potential (35), provided that the study is designed properly and the following factors are present: (1) the substrates are specific for individual CYP enzymes, (2) there are no interactions among these substrates, and (3) the study is conducted in a sufficient number of subjects. Negative results from a cocktail study can eliminate the need for further evaluation of particular CYP enzymes. However, positive results can indicate the need for further in vivo evaluation to provide quantitative exposure changes (such as AUC and $C_{\max}$), if the initial evaluation only assessed the changes in the urinary parent to metabolite ratios.

b. Investigational drug as a substrate of CYP enzymes. In testing an investigational drug for the possibility that its metabolism is inhibited or induced (i.e., as a substrate), selection of the interacting drugs should be based on in vitro or in vivo studies identifying the enzyme systems that metabolize the drug. The choice of interacting drug can then be based on known, important inhibitors of the pathway under investigation. For example, if the investigational drug is shown to be importantly metabolized by CYP3A (the contribution of this enzyme to the overall elimination of this drug is >25% of the clearance pathway or is unknown), the choice of inhibitor and inducer could be ketoconazole and rifampin, respectively, because they are the strongest inhibitor and inducer, respectively, of the CYP3A pathway. If the study results are negative, then absence of a clinically important drug-drug interaction for the metabolic pathway would have been demonstrated. If the clinical study of the strong, specific inhibitor/inducer is positive, it should generally be determined in further clinical studies whether there is an interaction between the test drug and less potent specific inhibitors or inducers. If a drug is metabolized by CYP3A and its plasma AUC is increased fivefold or more by a CYP3A inhibitor, it is considered a sensitive substrate of CYP3A. Labeling would indicate that it is a ''sensitive CYP3A substrate.'' Its use with strong or moderate inhibitors might call for caution, depending on the drug's exposure-response relationship. If a drug is metabolized by CYP3A and its exposure-response relationship indicates that increases in the exposure levels by the concomitant use of CYP3A inhibitors may lead to serious safety concerns (e.g., torsades de pointes), it is considered as a ''CYP3A substrate with narrow therapeutic range.'' Similar classifications of substrates of other CYP enzymes are also discussed in the guidance (10,12).

If an orally administered drug is a substrate of CYP3A and has low oral bioavailability because of extensive presystemic extraction contributed by enteric CYP3A, grapefruit juice may have a significant effect on its systemic exposure (21,22). Use of the drug with grapefruit juice may call for caution depending on the drug's exposure-response relationship (23).

If a drug is a substrate of CYP3A, coadministration with St. John's wort, a CYP3A inducer, can decrease the systemic exposure and effectiveness. St. John's wort may be listed in the labeling along with other known inducers, such as rifampin, rifabutin, rifapentin, dexamethasone, phenytoin, carbamazepine, or phenobarbital, as possibly decreasing plasma levels.

If a drug is metabolized by a polymorphic enzyme (such as CYP2D6, CYP2C9, or CYP2C19), the comparison of pharmacokinetic parameters of this drug in poor metabolizers versus extensive metabolizers may indicate the extent of interaction of this drug with strong inhibitors of these enzymes, and make interaction studies with such inhibitors unnecessary. When the above study shows significant interaction, further evaluation with weaker inhibitors may be necessary.

There are situations when multiple-inhibitor studies may be appropriate (10,24). For example, the AUC of repaglinide increased 19-fold when taken with itraconazole and gemfibrozil together, while the AUC of repaglinide plasma levels only increased 1.4- and 8.1-fold, respectively, when taken with itraconazole or gemfibrozil separately. The large effect of coadministration of itraconazole and gemfibrozil on the systemic exposure (AUC) of repaglinide may be attributed to collective effects on both enzyme and transporters (10,24–26).

c. Investigational drug as an inhibitor or an inducer of P-gp transporter. In testing an investigational drug for the possibility that it may be an inhibitor/inducer of P-gp in vivo, digoxin or other known substrates of P-gp should be used.

d. Investigational drug as a substrate of P-gp transporter. In testing an investigational drug for the possibility that its transport may be inhibited or induced in vivo (as a substrate of P-gp), an inhibitor of P-gp should be studied. In cases where the drug is also a CYP3A substrate, inhibition should be studied by using a strong inhibitor of both P-gp and CYP3A.

e. Investigational drug as a substrate of other transporters. In testing an investigational drug for the possibility that its disposition may be inhibited or induced (i.e., as a substrate of transporters other than or in addition to P-gp), it may be appropriate to use an inhibitor of multiple transporters. Recent interactions involving drugs that are substrates for transporters other than or in addition to P-gp include some HMG-CoA reductase inhibitors, such as rosuvastatin and pravastatin.

4. Route of Administration

For an investigational agent used as either an interacting drug or substrate, the route of administration should generally be the one being studied in trials. If only oral dosage forms will be marketed, studies with an intravenous formulation are not usually necessary, although information from oral and intravenous dosings may be useful in discerning the relative contributions of alterations in absorption and/or presystemic clearance to the overall effect observed for a drug interaction. For example, the interaction studies of clarithromycin and intravenous or oral doses of midazolam enabled Gorski et al. to estimate the changes in the intestinal and hepatic availability of midazolam in the presence of clarithromycin (27). Sometimes the use of certain routes of administration may reduce the utility of information from a study. For example, intravenous administration of a substrate or inhibitor would not reveal an effect on intestinal CYP3A activity that markedly altered oral bioavailability.

5. Dose Selection

For both substrate and interacting drug, testing should maximize the possibility of finding an interaction. In general, the maximum doses of the interacting drug should be used. Doses smaller than those to be used clinically may be needed for substrates on safety grounds and should provide an adequate assessment of an interaction. The differential effects of different doses of ritonavir on the plasma levels of saquinavir (18) demonstrate the dose effect of an interacting drug.

6. Endpoints

The following measures and parameters are recommended to assess changes in substrate pharmacokinetic endpoints: (1) systemic exposure measures, such as AUC, C_{max}, time to C_{max} (T_{max}), and others as appropriate; and (2) pharmacokinetic parameters, such as clearance, volumes of distribution, and half-lives. In some cases, these measures may be of interest for the inhibitor or inducer as well, notably where the study is intended to assess possible interactions between both study drugs. Additional measures may help in steady-state studies (e.g., trough concentration, C_{min}) to demonstrate that dosing strategies were adequate to achieve steady state before and during the interaction. In certain instances, an understanding of the relationship between dose, blood levels, and response may lead to a special interest in particular pharmacokinetic measures/parameters. For example, if a clinical outcome is most closely related to peak concentration (e.g., tachycardia with sympathomimetics), C_{max} or another early exposure measure might be most appropriate. Conversely, if the clinical outcome is related more to extent of absorption, AUC would be preferred. A CDER/CBER guidance for industry (34) provides considerations in the evaluation of exposure-response relationships. In certain instances, reliance on endpoints in addition to pharmacokinetic measures/parameters may be useful. Examples include international normalized ratio (INR) measurement (when studying warfarin interactions) or QT interval measurements.

IV. REGULATORY CONSIDERATIONS

Information gathered in properly conducted in vitro and clinical studies of drug metabolism, drug absorption, and drug-drug interactions provides critical data for drug development decisions. For example, early knowledge that a candidate drug with a relatively narrow therapeutic range has high interindividual variability in pharmacokinetics because of oxidative metabolism by a polymorphically distributed CYP enzyme might influence the decision to invest in further development. Increasingly, also, these factors can affect the regulatory decision to approve such a drug and/or how it is labeled. Section 505 of the Food Drug and Cosmetic Act requires that, for approval, a drug must be demonstrated to be both effective and safe when used as labeled. Safety is not an absolute measure but rather reflects a conclusion that the drug's benefits outweigh its risks. Among the risks that must be considered is the presence of individuals who are at particular risk because of individual characteristics (e.g., poor metabolizers) or concomitant drug administration. It is striking that several important drugs—terfenadine, mibefradil, astemizole, and cisapride—have been removed from the market, at least partly, because of drug-drug interaction problems (28–31). The importance of both mean and between- and within-individual variability must be assessed in light of many factors. These include the toxicity of the drug (wide therapeutic range drugs may not be harmful even if their pharmacokinetics are very variable, e.g., propranolol, loratidine), the disease being treated, the availability of alternative therapy, the value of treatment, and the consequences of treatment failure resulting from inadequate drug concentrations. Thus, development of a drug to treat seasonal allergic rhinitis that shows significant cardiac toxicity when taken with a CYP3A4 inhibitor would not be prudent. In the context of a non-life-threatening condition for which numerous safe and effective alternative therapies exist, such a drug would be unlikely to be approved for marketing today. In contrast, the potential for serious toxicities due to drug interactions is not an insurmountable impediment for drugs intended to treat severe or life-threatening conditions, particularly when alternative treatments are not available. In these instances, close attention to labeling and other aspects of risk management will be needed to inform practitioners and patients about the likelihood and consequences of interactions and the ways to avoid them.

Labeling for drug products in the United States must be in the format specified in the Code of Federal Regulations (21 CFR 201.56). Drug absorption, metabolism, and excretion, and drug-drug interaction information appears, as appropriate, in some or all of the following sections of the approved product label—Clinical Pharmacology, Contraindications, Warnings and Precautions, Adverse Reactions, or Dosage and Administration (32). Certain basic pharmacokinetic information is almost always included (e.g., bioavailability, food effects, clearance, and half-life), as is all available information about drug-drug interactions, often including negative results. Clinically important interactions are emphasized and discussed in more detail. Potential interactions based on metabolic pathways may also be included. Recently approved product labels have reflected the increased understanding of the pathways and consequences of drug metabolism by health care practitioners. Newer labels almost always include mention of the drug's effect on specific CYP enzymes as well as the clinical consequences of their perturbation on coadministered drugs and the influence of concomitantly administered drugs on the drug itself. The following section describes the appropriate location for drug metabolism and drug-drug interaction information. The role of P-gp and other transporter mechanisms and their relationship to drug-metabolizing enzymes remain to be fully understood, and the effects on P-gp-mediated transport are only beginning to be reflected in labeling at this time. It is easy to envision, however, that the role of transporters and the clinical consequences of their modulation will soon be better understood and studied so that information on these systems will appear regularly in labeling.

V. INFORMATION APPROPRIATE FOR THE APPROVED PRODUCT LABEL

It is important that all relevant information on the metabolic pathways and metabolites and pharmacokinetic interactions be included in the pharmacokinetics subsection of the clinical pharmacology section of the labeling. The clinical consequences of metabolism and interactions should be placed in drug interactions, warnings and precautions, boxed warnings, contraindications, or dosage and administration sections, as appropriate. Information related to clinical consequences should not be included in detail in more than one section, but rather reported fully in one section and then referenced in other sections, as appropriate. When the metabolic pathway or interaction data results in recommendations for dosage adjustments, contraindications, or warnings (e.g., coadministration should be avoided) that are included in the boxed warnings, contraindications, warnings and precautions, or dosage and administration sections, these recommendations should also be included in highlights. Refer to the guidance for industry on labeling (32) for more information on presenting drug interaction information in labeling.

In certain cases, information based on clinical studies not using the labeled drug can be described with an explanation that similar results may be expected for that drug. For example, if a drug has been determined to be a strong inhibitor of CYP3A, it does not need to be tested with all CYP3A substrates to warn about an interaction with sensitive CYP3A substrates and CYP3A substrates with narrow therapeutic range. An actual test involving a single substrate would lead to labeling concerning use with all sensitive and NTR substrates.

If a drug has been determined to be a sensitive CYP3A substrate or a CYP3A substrate with a narrow therapeutic range, it does not need to be tested with all strong or moderate inhibitors of CYP3A to warn about an interaction with strong or moderate CYP3A inhibitors, and it might be labeled in the absence of any actual study if its metabolism is predominantly by the CYP3A route. Similarly, if a drug has been determined to be a sensitive CYP3A substrate or a CYP3A substrate with a narrow therapeutic range, it does not need to be tested with all CYP3A inducers to warn about an interaction with CYP3A inducers. Examples of CYP3A inducers include rifampin, rifabutin, rifapentin, dexamethasone, phenytoin, carbamazepine, phenobarbital, and St. John's wort.

VI. SUMMARY

In vitro and in vivo metabolism and drug-drug interaction data are critical for the complete evaluation and labeling of a drug. The information provided by these studies needs to be appreciated and understood by prescribers and utilized in individualizing pharmacotherapy. An integrated approach to studying and evaluating drug-drug interactions during the drug development and regulatory review process and incorporating language into labeling has been described. This integrated approach should be based on good understanding and utilization of the primary question, our willingness to rely on in vitro and in vivo pharmacokinetic and pharmacodynamic data, and our understanding of the degree to which an observed change in substrate measures caused by an interacting drug is or is not clinically important. In recent years, understanding the metabolic disposition and identifying the potential for metabolic drug-drug interactions such as inhibition and induction of enzymes has become an integral part of the drug development process. Improved understanding of the mechanistic basis of metabolic drug-drug interactions has enabled standardized and focused approaches to evaluating interactions with generalizable conclusions. Similar progress is anticipated in the transporter area (33). As science progresses and new tools become available, the FDA updates its recommendations. The recently published guidance (10) reflects the agency's current view in the evaluation of drug-drug interactions during drug development and includes the following principles. Future efforts in assessing, managing, and communicating the risks of drug-drug interactions may focus on (1) improved uses of in vitro tests to evaluate transporter-based interactions, (2) better use of in vitro data as a surrogate for in vivo findings, e.g., through in vitro/in vivo correlations, (3) better evaluation and prediction of the clinical consequences of multiple metabolic/transporter interactions, (4) better use of pharmacokinetic and pharmacodynamic data in understanding the clinical consequences of drug-drug interactions, and (5) better ways to communicate information about important drug-drug interactions to patients and practitioners.

ACKNOWLEDGMENTS

The authors would like to acknowledge the hard work of the following FDA Drug Interaction Working Group members who contributed to the development of the 2006 CDER/CBER guidance entitled *Drug Interaction Studies—Study Design, Data Analysis, and Implications for Dosing and Labeling*: Sophia Abraham, Sayed Al-Habet, Ray Baweja, Gilbert Burckart, Sang Chung, Phil Colangelo, Jerry Collins, David Frucht, Martin Green, Paul Hepp, Shiew-Mei Huang, Ron Kavanagh, Ho Sum Ko, Lawrence J. Lesko, Patrick Marroum, Srikanth Nallani, Janet Norden, Wei Qiu, Atik Rahman, Kellie Reynolds, Soloman Sobel, Toni Stifano, John Strong, Robert Temple, Kenneth Thummel, Douglas C. Throckmorton, Xiaoxiong Wei, Sally Yasuda, Lei K Zhang, and Jenny H. Zheng. Drs. Gilbert Burckart, and Kenneth Thummel contributed to the guidance preparation while on sabbatical at the FDA. The authors would also like to acknowledge the critical contributions of Drs. Peter Honig and Roger Williams, who were at the FDA when the previous version of this chapter was prepared.

REFERENCES

1. Huang SM, Temple R, Throckmorton DC, et al. Drug interaction studies—study design, data analysis and implications for dosing and labeling. Clin Pharmacol Ther 2007 Feb; 81(2):298–304.
2. Kim RB, Cvetkovic M, Fromm MF, et al. OATP and P-glycoprotein mediate the uptake and excretion of fexofenadine. Clin Pharmacol Ther 1999; 65:111.
3. Bedford TA, Rowbotham DJ. Cisapride. Drug interactions of clinical significance. Drug Saf 1996; 15:167–175; published erratum appears in Drug Saf 1997; 17:196.
4. Albengres E, Le Louet H, Tillement JP. Systemic antifungal agents. Drug interactions of clinical significance. Drug Saf 1998; 18:83–97.
5. Neuvonen PJ, Kantola T, Kivisto KT. Simvastatin but not pravastatin is very susceptible to interaction with the CYP3A4 inhibitor itraconazole. Clin Pharmacol Ther 1998; 63:332–341.
6. Barza M, Weinstein L. Pharmacokinetics of the penicillins in man. Clin Pharmacokinet 1976; 1:297–308.
7. Fromm MF, Kim RB, Stein CM, et al. Inhibition of P-glycoprotein-mediated drug transport: a unifying mechanism to explain the interaction between digoxin and quinidine. Circulation 1999; 99:552–557.
8. Webb DJ, Freestone S, Allen MJ, et al. Sildenafil citrate and blood-pressure-lowering drugs: results of drug interaction studies with an organic nitrate and a calcium antagonist. Am J Cardiol 1999; 83:21C–28C.
9. Mullins ME, Horowitz BZ, Linden DH, et al. Life-threatening interaction of mibefradil and beta-blockers with dihydropyridine calcium channel blockers. JAMA 1998; 280:157–158.
10. Food and drug administration. Draft guidance for industry: Drug-Drug interactions—study design, data analysis and implications for dosing and labeling. http://www.fda.gov/cder/guidance/6695dft.htm; published September 2006; accessed November 10, 2007; a final guidance will be available at: http://www.fda.gov/cder/.

11. Food and drug administration. Guidance for industry: Population pharmacokinetics, available at: http://www.fda.gov/cder/guidance/1852fnl.pdf; accessed November 10, 2007.
12. FDA Internet Web site: http://www.fda.gov/cder/drug/drugInteractions/default.htm/. Established May 2006.
13. Davit B, Reynolds K, Yuan R, et al. FDA evaluations using in vitro metabolism to predict and interpret in vivo metabolic drug-drug interactions: impact on labeling. J Clin Pharmacol 1999; 39:899–910.
14. Honig PK, Woosley RL, Zamani K, et al. Changes in the pharmacokinetics and electrocardiographic pharmacodynamics of terfenadine with concomitant administrations of erythromycin. Clin Pharmacol Ther 1992; 52:231–238.
15. Honig PK, Wortham DC, Sarmani K, et al. Terfenadine-ketoconazole interactions: pharmacokinetic and electrocardiographic consequences. JAMA 1993; 269:1513–1518.
16. Lamberg TS, Kivisto KT, Neuvonen PJ. Lack of effect of terfenadine on the pharmacokinetics of the CYP3A4 substrate buspirone. Pharmacol Toxicol 1999; 84: 165–169.
17. Kurtz DL, Bergstrom RF, Goldberg MJ, et al. The effect of sertraline on the pharmacokinetics of desipramine and imipramine. Clin Pharmacol Ther 1997; 62: 145–156.
18. Hsu A, Granneman G, Cao G, et al. Pharmacokinetic interactions between two HIV-protease inhibitors, ritonavir and saquinavir. Clin Pharmacol Ther 1998; 63:453–464.
19. Hsu A, Granneman GR, Sun E, et al. Assessment of single- and multiple-dose interactions between ritonavir and saquinavir. XI International Conference on AIDS. Vancouver, B.C., Canada, 1996 July, 7–12.
20. Huang SM, Lesko LJ, Williams RL. Assessment of the quality and quantity of drug-drug interaction studies in recent NDA submissions: study design, data analysis issues. J Clin Pharmacol 1999; 39:1006–1014.
21. Huang SM, Lesko LJ, Drug-drug, drug-dietary supplement, and drug-citrus fruit and other food interactions—what have we learned? J Clin Pharmacol 2004; 44: 559–569.
22. Huang SM, Hall SD, Watkins P, et al. Drug interactions with herbal products & grapefruit juice: a conference report. Clin Pharmacol Ther 2004; 75:1–12.
23. Huang SM, Temple R, Lesko LJ. Drug-drug, drug-dietary supplement, and drug-citrus fruit and other food interactions—labeling implications. In: Lam F, Huang SM, Hall S, eds. Herbal Supplements and Drug Interactions. New York: Taylor & Francis, 2006.
24. Thummel K, Chung S, Nallani S, et al. and the CDER/CBER working group. When is a multiple-inhibitor study necessary? To be presented at the ASCPT annual meeting, San Diego, California, March 2007.
25. Huang S-M, Lesko LJ, Temple R. Adverse drug reactions and drug interactions. In: Part IV. Pharmacology and Therapeutics: Principles to Practice. Elsevier (in press).
26. Niemi M, Backman JT, Neuvonen M, et al. Effects of gemfibrozil, itraconazole, and their combination on the pharmacokinetics and pharmacodynamics of repaglinide: potentially hazardous interaction between gemfibrozil and repaglinide. Diabetologia 2003; 46(3):347–351; [Epub 2003 Feb 27].
27. Gorski JC, Jones DR, Haehner-Daniels BD, et al. The contribution of intestinal and hepatic CYP3A to the interaction between midazolam and clarithromycin. Clin Pharmacol Ther 1998; 64:133–143.

28. Friedman MA, Woodcock J, Lumpkin MM, et al. The safety of newly approved medicines: do recent market removals mean there is a problem? JAMA 1999; 281: 1728–1734.
29. Huang SM, Booth B, Fadiran E, et al. What have we learned from the recent market withdrawal of terfenadine and mibefradil? Presentation at the 101st Annual Meeting of American Society of Clinical Pharmacology and Therapeutics, March 15–17, 2000, Beverly Hills, California. Clin Pharmacol Ther 2000; 67(2):148 (abstr).
30. Food and Drug Administration. Talk Paper: Janssen Pharmaceutica Stops Marketing Cisapride in the US. Available at: http://www.fda.gov/bbs/topics/ANSWERS/ans01007.html.
31. Huang SM, Miller M, Toigo T, et al. Drug metabolism/clinical pharmacology, (Schwartz J, section ed.). In: Legato M, ed. Principles of Gender-Specific Medicine. New York: Elsevier Acad. Press, 2004.
32. Clinical Studies Section of Labeling for Human Prescription Drug and Biological Products—Content and Format [HTML] or [PDF] (Issued 1/18/2006; Posted 1/18/2006). Available at: http://www.fda.gov/CDER/guidance/5534fnl.pdf. Accessed October 6, 2006.
33. Zhang L, Strong JM, Qiu W, et al. Scientific perspectives on drug transporters and their role in drug interactions. Mol Pharm 2006; 3(1):62–69. [Epub 2006 Jan 4].
34. Food and Drug Administration. Guidance for Industry: Exposure-Response Relationship, Study Design, Data Analysis, and Regulatory Applications, April 2003 (posted May 2003). Available at: http://www.fda.gov/cder/. Accessed October 6, 2006.
35. Fuhr U, Jetter A, Kirchheiner J. Appropriate phenotyping procedures for drug metabolizing enzymes and transporters in humans and their simultaneous use in the ''Cocktail'' Approach. Clin Pharmacol Ther 2007; 81:270–283.

17

Drug-Drug Interactions: Toxicological Perspectives

Sidney D. Nelson
Department of Medicinal Chemistry, School of Pharmacy, University of Washington, Seattle, Washington, U.S.A.

I. INTRODUCTION

Major problems facing the pharmaceutical industry include loss of drug candidates due to preclinical toxicology and unanticipated adverse drug reactions after the introduction of new drugs into clinical practice (1). Most adverse drug reactions occur in only a small percentage of patients and are termed idiosyncratic, and many of these reactions are caused by reactive metabolites formed from drugs (2–4). Reactions of reactive metabolites with tissue macromolecules can lead to direct or intrinsic toxic effects and/or cause toxicity by forming haptens that lead to immunotoxic effects. Although new animal models are being developed that provide insights into factors that play a role in these idiosyncratic toxicities (5–7), no generally useful models are yet available.

In some cases a new drug may be the precipitator or perpetrator of toxicity of another drug by altering its metabolism and/or disposition, or the new drug may be the object or victim of altered metabolism and/or disposition caused by a drug already on the market. In many instances, the object or victim is a drug with a narrow therapeutic index, window, or ratio (for a discussion, see Ref. 8). Several definitions have been applied to this terminology, including the qualitatively simple one of a drug ''for which relatively small changes in systemic concentrations lead to marked changes in pharmacodynamic response'' (9). The FDA has defined narrow therapeutic ratio to include those drugs for which there is less than a twofold difference in median lethal dose (LD_{50}) and median effective dose (ED_{50}), or for which there is less than a twofold difference in the minimum toxic concentrations and minimum effective concentrations in the blood, or for which safe and effective use of the drug requires careful titration and patient monitoring (10).

This chapter will focus on those metabolic drug-drug interactions that have led or can lead to serious toxicological consequences in humans. Most of the chapter will describe examples of metabolic drug-drug interactions that have caused serious toxicities. As discussed in chapter 15, the majority of drug-drug interactions of clinical significance have occurred through interactions at the level of cytochromes P450. Since substantial information is now either available or readily obtainable about induction and inhibition of these enzymes as well as the kinetic parameters associated with the metabolism of drugs and other probe substrates, many metabolic drug-drug interactions can be predicted prior to clinical trials. However, because the situation in vivo is complicated by a variety of genetic and environmental factors that affect drug absorption, distribution, and metabolism and because the physiological response to a toxic insult may vary from one individual to another, it is often difficult to predict that a particular drug-drug interaction will lead to a toxic insult. Nonetheless, the results of preclinical studies should provide the basis for more informed planning of clinical studies.

II. DRUGS AND CLASSES OF DRUGS AS OBJECTS (VICTIMS) OF CYTOCHROMES P450-MEDIATED DRUG-DRUG INTERACTIONS THAT LEAD TO TOXICITIES

A. Warfarin

Because of its narrow therapeutic window and extensive oxidation to inactive metabolites by cytochromes P450, warfarin (and the closely related drug acenocoumarol) is subject to many metabolic drug-drug interactions that can place patients at severe risk of either hyper- or hypocoagulability. Drug interactions with warfarin have been reviewed (11–13), and it is clear that most interactions occur through either induction or inhibition of CYP2C9, which forms the major 7-hydroxylation metabolite of the most active (*S*)-warfarin enantiomer (14).

Several inducers of cytochromes P450, including rifampin, several barbiturates, aminoglutethimide, primidone, phenytoin, and carbamazepine increase requirements for warfarin dosing, although mechanisms for most of these interactions have not been thoroughly investigated (11–13). Clinically, this effect becomes manifest either when a patient stabilized on warfarin adds one or more of these drugs to his or her therapy or, more commonly, when the patient removes one of these drugs from his or her therapy after stabilization on the combination therapy. Rifampin induces several P450s, including CYP2C9, and has been shown to increase the formation clearance of the major hydroxylated metabolites of (*S*)-warfarin (15). Substantial clinical and other indirect data implicate enhanced clearance of (*S*)-warfarin by CYP2C9 as one mechanism of the interaction (16), although increased glucuronidation may also play a role.

Several inhibitors of cytochromes P450 can substantially decrease requirements for warfarin dosage that, if not attended to, can lead to life-threatening bleeding episodes. Some drugs, such as sulfaphenazole, metronidazole, danazol, cotrimoxazole (trimethoprim-sulfamethoxazole), miconazole, and fluconazole, contain heterocyclic rings with sp^2-hybridized nitrogen, a structural unit known to bind to the heme iron of P450s, and investigations implicate inhibition of CYP2C9 oxidation of (*S*)-warfarin as the mechanism for the drug-drug interactions caused by several of these drugs (12).

However, the presence of a nitrogen-containing heterocyclic ring in a drug is not sufficient for potent inhibition of CYP2C9. Cimetidine contains an imidazole moiety, but it is a much better inhibitor of the metabolism of (*R*)-warfarin (17), the least potent enantiomer, so that an effect on warfarin therapy is observed only at high doses of cimetidine (18). Also, other potent inhibitors of CYP2C9 that inhibit (*S*)-warfarin metabolism and thereby increase the hypoprothrombinemic response to warfarin, such as phenylbutazone, sulfinpyrazone, and amiodarone (12), do not contain such structures. Although many case reports have appeared of interactions between warfarin and a variety of other drugs with many different drug structures (19), only a few of these have resulted in serious toxic effects, and mechanisms are largely unknown. Because of their increased use, further investigations with some of these drugs, such as tamoxifen (20,21), seems warranted.

B. Theophylline

General aspects of the drug-drug interactions involving theophylline are similar to those described for warfarin, because it too is a drug with a narrow therapeutic index. Increases in its rate of metabolism, either by some inducers of cytochromes P450 or by removal of an inhibitor of those P450s given concomitantly with theophylline, lead to diminution of therapeutic effect, resulting in increased dyspnea. Conversely, decreases in its rate of metabolism either by inhibitors of P450s involved in the metabolism of theophylline or by removal of an inducer given concomitantly can lead to serious toxicities, including convulsions and heart arrhythmias that can be serious enough to cause death.

The major P450 involved in the oxidation of theophylline to inactive metabolites is CYP1A2 (see Ref. 22 for a review). Interestingly, there are no reports of serious toxicity resulting from interactions of CYP1A2 inducers, such as cigarette smoking, even though theophylline clearance is increased (23). Several case reports have appeared of increased theophylline clearance by barbiturates, carbamazepine, phenytoin, and rifampin, which are thought to induce CYP3A4 with little effect on CYP1A2, and adjustments to theophylline dosage are often required for optimal therapeutic effect.

In contrast, decreases in theophylline metabolism by selective inhibitors of CYP1A2, such as fluvoxamine and some quinolone antibiotics, or by selective and potent inhibitors of CYP3A4, such as the macrolide antibiotics, have resulted in serious theophylline toxicity (22). It is postulated that taken over time, the macrolide antibiotics act as mechanism-based inhibitors of CYP isoforms other than just CYP3A4. Some nonselective inhibitors of P450s, such as cimetidine, some β-blockers and calcium channel blockers, and others (19,22), also appear to inhibit the metabolism of theophylline enough to cause toxicity.

C. Nonsedating Antihistamine Drugs

Terfenadine and astemizole were removed from the market in 1997 and 1999, respectively, because of drug interactions that led to QT interval prolongation (24). Both of these drugs are prodrugs that are metabolized primarily by CYP3A4 to their therapeutically active metabolites (for a review, see Ref. 25). Inhibition of CYP3A4 by azole antifungal agents (25,26) and most macrolide antibiotics (25,27) can lead to sufficient increases in terfenadine concentrations to cause torsades de pointes as a result of the prodrug's ability to inhibit delayed rectifier potassium currents (28). Similar interactions occur with astemizole (25,29). Neither loratadine nor cetirizine, nor the active metabolites of terfenadine (fexofenadine) and astemizole (norastemizole), cause this cardiotoxic effect to any significant extent (25).

D. Cisapride

The promotility agent cisapride was removed from the market because of over 300 reports of heart rhythm abnormalities similar to those caused by terfenadine, including 80 deaths (30). Cisapride also is metabolized extensively by CYP3A4, and the same macrolide antibiotics, azole antifungal agents, and other inhibitors of this enzyme, such as grapefruit juice (31,32), sustain high enough concentrations of the parent drug to cause heart problems such as torsades de pointes.

E. Cyclosporine

The widely used immunosuppressive agent cyclosporine is extensively metabolized by CYP3A4 (and to a lesser extent by CYP3A5) in human intestine and liver (33), and therefore it is subject to similar metabolic drug-drug interactions as described for terfenadine. However, in the case of cyclosporine, induction of its metabolism can lead to loss of its immunosuppressive activity to the point of transplant organ rejection, and inhibition of its metabolism can lead to kidney damage as a major toxicity (33,34). Cyclosporine also is pumped out of intestinal epithelial cells by P-glycoprotein, and many of the drugs that inhibit or induce CYP3A4 also inhibit or induce this transporter (33,35). Thus, this effect also contributes, in part, to many of the observed drug-drug interactions with cyclosporine. It is noteworthy that ketoconazole, an azole antifungal agent that increases cyclosporine blood concentrations by its inhibition of CYP3A4 and P-glycoprotein, can be used concurrently to decrease the high cost of cyclosporine therapy in transplant recipients (36). Tacrolimus, a newer immunosuppressive agent related to cyclosporine, apparently, is subject to similar drug interactions as cyclosporine, though it has not been in use for long, and limited data are available (37).

F. The Statins

Of the statin HMG-CoA reductase inhibitors on the market in the United States, lovastatin, simvastatin, atorvastatin, and rosuvastatin are metabolized mainly by CYP3A4, whereas fluvastatin is metabolized by CYP2C9 and pravastatin by phase II pathways (38–40). Consistent with an important role for CYP3A4 in the metabolism of simvastatin, induction of CYP3A4 by rifampin decreases the area under the plasma concentration–time curve (AUC) of simvastatin by approximately 90% (41). Most toxic drug interactions caused by the statins (myopathies and rhabdomyolysis) are related to supratherapeutic concentrations achieved as a result of inhibition of CYP3A4 by macrolide antibiotics, the azole antifungal agents, and cyclosporine (38–40), though some have resulted from combined therapy with other lipid-lowering agents, such as gemfibrozil, apparently due to inhibition of the glucuronidation of some statin metabolites by UGT1A1 and UGT1A3 (42). Another statin drug, cerivastatin, was withdrawn from the market in 2001 because of a five- to sevenfold higher incidence of myopathies and rhabdomyolysis mostly associated with drug interactions (43–45). Mibefradil, a unique benzimidazole-containing calcium channel–blocking drug, was removed from the market because of its potent inhibition of the metabolism of several drugs and resultant toxicities, including life-threatening rhabdomyolysis in patients on lovastatin and simvastatin (46).

Thus, safety issues related to statin therapy are often related to drug-drug or drug-food interactions that in many cases are clinically manageable (19,47–49). A task force on statin safety concluded that benefits of statin therapy far outweigh their risks in most individuals, even those who have concomitant drug therapy (50). However, it is unwise to treat statins as over-the-counter drugs because of the potential for drug interactions that can lead to serious toxicities.

G. Calcium Channel Blockers

The dihydropyridine class of calcium channel blockers undergoes extensive first-pass oxidation by CYP3A isoforms to their pyridine metabolites, and several studies have shown that inducers and inhibitors of these P450s decrease and increase the blood concentrations of the active dihydropyridine structures, respectively (51). The calcium channel blockers verapamil and diltiazem are unrelated structures that also undergo significant metabolism by cytochromes

P450 of the CYP3A family (51). However, apparently in only very few cases has this metabolism caused significant enough loss of antihypertensive activity (in the case of concomitant administration of inducers of CYP3A isoforms) or hypotension and edema (in the case of concomitant administration of inhibitors of CYP3A isoforms) to cause toxic drug reactions (19,51). More commonly, it is the ability of these drugs to inhibit CYP3A isoforms that leads to toxicities caused by some other object drug.

H. Sedative-Hypnotic and Anxiolytic Agents

Benzodiazepine and azapirone derivatives are widely used drugs in this class, and most are metabolized extensively by enzymes of the CYP3A family, except oxazepam, lorazepam, and temazepam, which are mostly glucuronidated (52). Again, several studies have shown that inducers and inhibitors of CYP3A can markedly alter plasma concentrations of many of these drugs, but in only a few cases have toxic effects, such as deep unconsciousness, been reported (19,52,53). Nonetheless, patients on these drugs should probably be monitored carefully, particularly the elderly, who may suffer severe physical injury as a result of falls from impairment of psychomotor function.

I. Antidepressants

Toxicities associated with antidepressant drugs have been most commonly reported for the tricyclic antidepressants as a result of inhibition of cytochromes P450, particularly CYP2D6 (54). They include bradycardia, seizures, and delirium. Inhibitors of CYP2D6, such as paroxetine, fluoxetine, perfenazine, quinidine, and β-blockers, have all been shown to significantly increase plasma concentrations of tricyclic antidepressants, such as imipramine and desipramine, in some cases with overt signs of toxicity (54–56). However, in other cases, toxicity is minimized either because other pathways of metabolism involving CYP3A4, CYP2C19, and CYP1A2 are not affected or because of genetic polymorphisms of CYP2D6 (57–59) that decrease its activity. Selective serotonin reuptake inhibitors (SSRIs), like paroxetine, are susceptible to CYP2D6 polymorphisms and drug interactions, but appear to be safe drugs even in very high doses (60,61).

J. β-Blockers

The β-adrenoceptor antagonists (β-blockers) are widely used drugs that are metabolized by cytochromes P450, particularly CYP2D6 (62). Fortunately, these drugs have a rather large therapeutic index and only a few instances of severe toxicity have been reported, which, in part, may be related to CYP2D6 polymorphisms (63). Reports of cardiac effects ranging from significant decreases in heart rate to orthostatic hypotension to cardiac arrest have occurred with β-blockers when combined with inhibitors of P450 metabolism, such as amiodarone (64), quinidine (65), propafenone (66), and fluoxetine (67). The over-the-counter antihistamine diphenhydramine has been shown to cause bradycardia and other hemodynamic changes in subjects on metoprolol who were CYP2D6 extensive metabolizers but not poor metabolizers (68). The COX-2 inhibitor, celecoxib, also inhibits CYP2D6, and has been shown to significantly increase plasma concentrations of metoprolol (69).

K. Anesthetics

The volatile ''flurane'' anesthetics are metabolized primarily by CYP2E1, with lesser involvement by CYP2A6 and CYP3A4 (70). A severe idiosyncratic immune-mediated toxic effect of most of these agents is liver necrosis as a result of oxidative dehalogenation of the anesthetics by CYP2E1 to form acyl halides that acylate hepatic proteins yielding antigens (71–73). Thus, it might be anticipated that inducers of CYP2E1 would increase the risk of hepatotoxicity caused by the flurane anesthetics and inhibitors of CYP2E1 would decrease the risk. The only evidence to support this observation is that obesity induces CYP2E1 activity (74) and is an increased risk factor for halothane hepatitis (75). Obesity also leads to increased halothane oxidation in humans (76). An interesting suggestion has been put forth to use disulfiram, a CYP2E1 inhibitor, in patients administered fluranes, because it markedly decreases the oxidation of halothane in humans to the proposed toxic metabolite (77).

The only other anesthetic to cause serious toxicity for which a metabolic drug interaction has been reasonably well characterized is the local anesthetic and antiarrhythmic agent lidocaine. Amiodarone decreased lidocaine systemic clearance in a patient (primarily by inhibition of CYP3A4 N-dealkylation of lidocaine) and yielded concentrations of lidocaine that led to seizures (78,79).

L. Antiepileptics

1. Carbamazepine

Carbamazepine is considered a relatively safe antiepileptic drug that is subject to dose-related neurologic toxicities (e.g., drowsiness, vertigo, loss of coordination) in adults and children (80). Since a major route of elimination of carbamazepine is via epoxidation catalyzed by CYP3A4 (81), there are several reports and studies that demonstrate CNS toxic effects of carbamazepine in individuals who also take CYP3A4 inhibitors (82).

The most serious toxicities associated with carbamazepine use are idiosyncratic skin rashes, hematological disorders, hepatotoxicity, and teratogenicity (80). On the basis of studies with mice, teratogenicity is most likely related to formation of arene oxide and/or quinone-like metabolites of carbamazepine (83), and studies in humans suggest that a reactive iminoquinone of 2-hydroxystilbene is formed (84). It is known that coadministration of cytochrome P450 inducers (e.g., phenobarbital) with carbamazepine increases the risk of serious toxic effects (85–87). Felbamate apparently increases the metabolism of carbamazepine by heteroactivation of CYP3A4 (88), but toxicological consequences of this activation have not been documented.

2. Phenytoin

Phenytoin, like carbamazepine, causes dose-related neurological toxicities (89). Since phenytoin is cleared mostly via CYP2C9 and CYP2C19 aromatic oxidation to *p*-hydroxyphenytoin (90), inhibitors of CYP2C9 (e.g., pyrazole nonsteroidal anti-inflammatory agents, some azole antifungal agents, amiodarone, isoniazid, and sulfa drugs) and inhibitors of CYP2C19 (e.g., cimetidine, felbamate, omeprazole, ticlopidine, and fluvoxamine) can increase concentrations of phenytoin and increase the incidence of CNS-related toxicities (82,91).

As with carbamazepine, phenytoin also causes idiosyncratic toxic effects, including hematological and connective tissue toxicities, hepatotoxicity, and teratogenicity (89). Although some of these toxicities have been hypothesized to be caused by P450 oxidative metabolism (92,93) or peroxidase-mediated reactions (94,95), mechanisms for these toxic effects in humans are unknown.

3. Valproic Acid

The two most serious toxic effects of valproic acid are hepatocellular injury (96) and teratogenesis (97). Since CYP2A6 and CYP2C9 are known to oxidize valproic acid to a 4-ene metabolite that is hepatotoxic, inducers of these isoforms, including other antiepileptic agents, are likely to increase the risk of hepatotoxicity (98). However, valproic acid also is metabolized by several other pathways that may be involved in causing its toxicities (99).

M. Antineoplastic Agents

Several drugs used to treat cancer are metabolized by cytochromes P450, and it would be anticipated that if the parent drug were the cytotoxic species, inhibition of its metabolism would enhance cytotoxicity, which could either be beneficial if controlled or cause severe toxicity to bone marrow, the nervous system, etc., if concentrations of the parent drug became too high (100).

Alternatively, P450 inducers may decrease therapeutic effectiveness of the drugs (101). Interestingly, only a few cases of toxicities to patients due to such drug-drug interactions have been reported, probably because most chemotherapy regimens are administered until some undesired toxic effect (e.g., leukopenia) limits the dosing. CYP3A isoforms appear to play the most significant role in the metabolism of many of the drugs (including paclitaxel, docetaxel, vincristine, vinblastine, vinorelbine, etoposide, teniposide, cyclophosphamide, and tamoxifen), and in many cases P-glycoprotein transport is also affected (102). For example, (*R*-verapamil is an inhibitor both of CYP3A isoforms and of P-glycoprotein, and it significantly reduces the clearance and increases the hematological toxicity of paclitaxel (103). The same reasoning applies to cases of severe neurotoxicity when itraconazole is administered with vincristine (104,105).

The relatively new class of selective tyrosine kinase inhibitors (imatinib, gefitinib, and erlotinib) are both substrates and inhibitors of CYP3A4 (106). Inducers of CYP3A4 (e.g., rifampin and St. John's wort) can markedly decrease the plasma concentrations of these drugs leading to decreased clinical efficacy, and inhibitors (e.g., azole antifungals) can markedly increase plasma concentrations leading to toxicities (106–108). An example of their ability to inhibit CYP3A4 is imatinib inhibition of simvastatin metabolism (109), thereby increasing the risk for development of myopathies.

N. HIV Drugs

The advent of highly active antiretroviral therapy (HAART) to minimize the rapid development of viral resistance in the treatment of HIV infection may result in multiple drug interactions (110–113). Both the nonnucleoside reverse transcriptase inhibitors and the protease inhibitors are substrates and inhibitors of some CYP enzymes, and some act as inducers as well (110,111). The major effects are on the CYP3A isoforms, and this has been used to advantage to increase concentrations of some HIV drugs. For example, delavirdine is a mechanism-based irreversible inhibitor of CYP3A4, and thereby is used to increase exposure to protease inhibitors (114). Ritonavir is a protease inhibitor, but it is used primarily for its ability as a potent inhibitor of CYP3A4 to increase concentrations of other protease inhibitors (115).

Interestingly, only a few cases of toxicities have been reported that are related to interactions with HIV drugs. The use of St. John's wort in a patient taking indinavir and lamivudine led to an increase in HIV RNA load (116), and St. John's wort has been shown to significantly increase the clearance of indinavir (117). Similarly, rifampin markedly increases the clearance of delavirdine (114). Thus, it is likely that most inducers of CYP3A4 will decrease plasma concentrations of those HIV drugs metabolized by CYP3A4, leading to decreased efficacy and increased resistance (118).

O. Ergot Alkaloids

Serious, life-threatening peripheral ischemia (ergotism) has resulted from the use of CYP3A4 inhibitors (e.g., macrolide antibiotics and HIV protease inhibitors) with Cafergot, and a black box warning has been added to the drug information literature about this drug (106,119).

III. DRUGS AND CLASSES OF DRUGS AS OBJECTS (VICTIMS) OF NONCYTOCHROME P450-MEDIATED PHASE I DRUG-DRUG INTERACTIONS THAT LEAD TO TOXICITIES

A. Antidepressant Serotonergic Drugs and Sympathomimetics

These two classes of drugs are subject to life-threatening interactions (e.g., mania, convulsions, hypertension, heart arrythmias) with monoamine oxidase (MAO) inhibitors, such as isocarboxazide, phenelzine, selegiline, and tranylcypromine, because they inhibit the metabolism of serotonin and sympathomimetic amines (19,120). This interaction is one of the earliest toxic drug-drug interactions to be recognized; however, these interactions are not often observed because the MAO inhibitors are now used sparingly.

B. Digoxin

Digoxin is a narrow therapeutic index drug whose primary drug-drug interactions appear to involve the P-glycoprotein transporter (121). An additional drug-drug interaction may occur at the level of reduction of the lactone ring double bond by intestinal microbial reductases that yields an inactive metabolite. Some antibiotic drugs can kill these microbes and lead to increases in digoxin concentrations (122).

C. Arylamine Sulfonamides and Hydrazine Drugs

Several of these drugs can cause immune-mediated idiosyncratic toxicities, such as immune hemolysis, agranulocytosis, aplastic anemia, drug-induced lupus, and severe skin rashes (123,124). It is well known that for most drugs in these classes acetylation of the amine or hydrazine group protects against the toxic effects based on significantly higher incidences of toxicity in individuals that genetically are slow acetylators (125). However, there are no reported drug-drug interactions with the *N*-acetyltranferases. Oxidation products of the arylamino or hydrazine groups are implicated as the haptenic reactive metabolites (123–131), and both cytochromes P450 and peroxidases have been implicated in the oxidation process (124,129,130). Although it might be anticipated that inducers and inhibitors of these enzymes would affect toxicities associated with the drugs, no reports of such drug-drug interactions on toxicity have appeared. The idiosyncratic nature of the toxicities makes them very difficult to study.

D. 6-Mercaptopurine and Azathioprine

Both of these drugs are metabolized by xanthine oxidase, and concomitant administration of the xanthine oxidase inhibitor allopurinol leads to elevated plasma concentrations of 6-mercaptopurine that can cause significant bone marrow depression (132,133).

IV. DRUG-DRUG INTERACTIONS WITH PHASE II METABOLIC ENZYMES THAT LEAD TO TOXICITIES

There are only a few published reports of serious toxicities caused by drugs as a result of drug-drug interactions with phase II metabolic enzymes. This in part may reflect less attention given to these enzymes and/or lesser extents of induction and inhibition of these enzymes by drugs (chap. 4).

In addition to being cleared by xanthine oxidase (see sec. III.D), 6-mercaptopurine is cleared by S-methylation catalyzed by the genetically polymorphic thiopurine methyltransferase (134). This enzyme is inhibited by the drug sulfasalazine, leading to bone marrow suppression as a result of increased 6-mercaptopurine concentrations (135,136).

Valproic acid is extensively glucuronidated, and the coadministration of valproate with other drugs eliminated extensively by glucuronidation, such as lamotrigine (137) and zidovudine (138), can significantly decrease the clearance of these latter two drugs with resultant toxicities. Sertraline has been found to cause a similar effect with lamotrigine (139) and fluconazole with zidovudine (138). Interestingly, increased incidences of convulsions observed when carbapenem antibiotics are administered to patients on valproic acid may be caused by carbapenem inhibition of glycolytic enzymes that hydrolyze valproic acid glucuronide back to free valproic acid (140).

V. DRUG-DRUG INTERACTIONS THAT AFFECT HEPATOTOXICITY CAUSED BY ACETAMINOPHEN: A COMPLEX EXAMPLE

Acetaminophen is a widely used analgesic-antipyretic agent, and several instances of drug interactions have been reported (141). However, in only a few cases have these interactions apparently increased the risk of hepatotoxicity, the major serious toxicity observed in humans who ingest this drug (142). In part, this may be a consequence of multiple pathways of metabolism for acetaminophen and, in part, because relatively high concentrations of the drug (>1 mM) are usually required to cause hepatotoxicity, which is an order of magnitude greater than therapeutic concentrations.

The major toxic metabolite of acetaminophen is *N*-acetyl-*p*-benzoquinone imine (NAPQI), which is an oxidation product formed by several human cytochromes P450 (for a review see Ref. 142). Therefore, it would be anticipated that inducers of cytochromes P450 would increase the rate of formation of NAPQI and thereby increase the risk for hepatotoxicity. Surprisingly, only a few cases of hepatotoxicity, caused by the use of normal doses of acetaminophen in patients on anticonvulsant drugs that are inducers of cytochromes P450, have been reported (143–150). This may be due to the ability of these same drugs to induce glucuronosyl transferases, which would increase the formation of acetaminophen glucuronide, a nontoxic metabolite (151,152). However, recent reports also show that some anticonvulsants can inhibit some glucuronosyl transferases involved in acetaminophen glucuronidation (153,154). Thus, the end effect on toxicity is not easy to predict. Furthermore, many inducers of drug metabolizing enzymes affect animals and humans differently, likely due to species differences in their orphan nuclear receptors (155,156).

Human CYP2E1 is one of the most efficient P450s to catalyze the oxidation of acetaminophen to NAPQI (157–159). Ethanol and isoniazid cause a time-dependent inhibition and induction of acetaminophen oxidation to NAPQI in humans (160,161) that can decrease risk for hepatotoxicity over the interval of concurrent administration and increase risk for hepatotoxicity a few hours after removal of ethanol or isoniazid. The latter induction phase of CYP2E1 may, in part, be responsible for cases of acetaminophen hepatotoxicity associated with the use of ethanol (162–165) or isoniazid (166–168). However, the induction is modest (2- to 3-fold); therefore, other susceptibility factors, genetic and others such as decreased glutathione stores and nutritional status, are likely to play an important role in some individuals (169–174).

VI. SUMMARY

Toxicities caused by drugs often limit their usefulness, and drug-drug interactions can cause enough of a change in tissue concentrations of some drugs, particularly those with a narrow therapeutic index, to cause serious toxic effects. Most of these interactions occur at the level of metabolism, though interactions with transporters, such as P-glycoprotein, are also becoming better recognized.

Unfortunately, we still do not have a good enough understanding either of the metabolism of some drugs or of mechanisms of toxicity (particularly idiosyncratic toxicities) to be able to predict whether or not a drug will cause toxic effects and under what conditions. For example, several nonsteroidal anti-inflammatory drugs have caused idiosyncratic toxicities that may be related to acyl glucuronide formation and/or cytochrome P450 activation (175–177). Therefore, it would be anticipated that other drugs that affect these pathways might either increase or decrease the risk of toxicities, but almost no data are available because of the idiosyncratic nature of the toxicities and lack of knowledge about susceptibility factors and/or immune system involvement in mechanisms leading to drug-induced toxicity. The same reasoning applies to hepatic injury caused by the drugs trovofloxacin (178) and troglitazone (179) that were removed from the market. There is very little published information about whether it was the drugs themselves or their metabolites that were responsible for the observed toxicities.

However, new methods are beginning to provide useful information on structure/toxicity relationships that can be applied to safer drug design (180–182). In cases like that of mibefradil, the basic science of drug-drug interactions has progressed enough to make informed benefit/risk decisions. Thus, it is important to continue basic and clinical investigations of drug-drug interactions as well as studies of mechanisms of toxicity to effect safer drug therapy.

REFERENCES

1. Kola I, Landis J. Can the pharmaceutical industry reduce attrition rates? Nat Rev Drug Discov 2004; 3:711–715.
2. Nelson SD. Molecular mechanisms of adverse drug reactions. Curr Ther Res 2001; 62:885–899.
3. Liebler DC, Guengerich FP. Elucidating mechanisms of drug-induced toxicity. Nature Rev Drug Discov 2005; 4:410–420.
4. Park BK, Kitteringham NR, Maggs JL, et al. The role of metabolic activation in drug-induced hepatotoxicity. Annu Rev Pharmacol Toxicol 2005; 45:177–202.
5. Roth RA, Luyendyk JP, Maddox JF, et al. Inflammation and drug idiosyncracy - is there a connection? J Pharmacol Exp Ther 2003; 307:1–8.
6. Shenton JM, Chen J, Uetrecht JP. Animal models of idiosyncratic drug reactions. Chem Biol Interact 2004; 150:53–70.
7. Welch KD, Reilly TP, Bourdi M, et al. Genomic identification of potential risk factors during acetaminophen-induced liver disease in susceptible and resistant strains of mice. Chem Res Toxicol 2006; 19:223–233.
8. Levy G. What are narrow therapeutic index drugs? Clin Pharmacol Ther 1998; 63: 501–505.
9. Benet LZ, Goyan JE. Bioequivalence and narrow therapeutic index drugs. Pharmacotherapy 1995; 15:433–440.
10. Williams RL. FDA position on product selection for ''narrow therapeutic index'' drugs. Am J Health Syst Pharm 1997; 54:1630–1632.
11. Ansell J, Hirsh J, Poller L, et al. The pharmacology and management of the vitamin K antagonists. Chest 2004; 26(suppl 3):S204–S233.
12. Trager WF. Oral anticoagulants. In: Levy RH, Thummel KE, Trager WF, et al., eds. Metabolic Drug Interactions. Philadelphia: Lippincott Williams and Wilkins, 2000:403–413.
13. Tirona RG, Bailey DG. Herbal product - drug interactions mediated by induction. Br J Clin Pharmacol 2006; 61:677–681.
14. Rettie AE, Korzekwa KR, Kunze KL, et al. Hydroxylation of warfarin by human cDNA-expressed cytochrome P450: a role for P-4502C9 in the etiology of (*S*)-warfarin-drug interactions. Chem Res Toxicol 1992; 5:54–59.
15. Heimark LD, Gibaldi M, Trager WF, et al. The mechanism of the warfarin-rifampin drug interaction in humans. Clin Pharmacol Ther 1987; 42:388–394.
16. Cropp JS, Bussey HI. A review of enzyme induction of warfarin metabolism with recommendations for patient management. Pharmacotherapy 1997; 17:917–928.
17. Niopas I, Toon S, Rowland M. Further insight into the stereoselective interaction between warfarin and cimetidine in man. Br J Clin Pharmacol 1991; 32:508–511.
18. Toon S, Hopkins KJ, Garstang FM, et al. Comparative effects of rantidine and cimetidine on the pharmacokinetics and pharmacodynamics of warfarin in man. Eur J Clin Pharmacol 1987; 32:165–172.
19. Hansten PD, Horn JR. Hansten and Horn's Drug Interactions Analysis and Management. St. Louis: Facts and Comparisons, 2006.
20. Lodwick R, McConkey B, Brown AM. Life-threatening interaction between tamoxifen and warfarin. Br Med J 1987; 295:1141.
21. Tenni P, Lalich DL, Byrne MJ. Life-threatening interaction between tamoxifen and warfarin. Br Med J 1989; 298:93.

22. Birkett DJ, Miners JO. Methylxanthines. In: Levy RH, Thummel KE, Trager WF, et al., eds. Metabolic Drug Interactions. Philadelphia: Lippincott Williams and Wilkins, 2000:469–482.
23. Grygiel JJ, Birkett DJ. Cigarette smoking and theophylline clearance and metabolism. Clin Pharmacol Ther 1981; 30:491–496.
24. Gibaldi M. Gibaldi's Drug Therapy 2000. New York: McGraw-Hill, 2000:120.
25. Shen DD, Madani S, Banfield C, et al. H2-Receptor antagonists. In: Levy RH, Thummel KE, Trager WF, et al., eds. Metabolic Drug Interactions. Philadelphia: Lippincott Williams and Wilkins, 2000:435–446.
26. Monahan BP, Ferguson CL, Killeavy ES, et al. Tor-sades des pointes occurring in association with terfenadine use. J Am Med Assoc 1990; 264:2788–2790.
27. Honig PK, Wortham DC, Zamani K, et al. Comparison of the effect of macrolide antibiotics erythromycin, clarithromycin and azithromycin on terfenadine steady-state pharmacokinetics and electrocardiographic parameters. Drug Invest 1994; 7: 148–156.
28. Woosley RL, Chen Y, Freiman JP, et al. Mechanism of the cardiotoxic actions of terfenadine. J Am Med Assoc 1993; 269:1532–1536.
29. Woosley RL. Cardiac actions of antihistamines. Annu Rev Pharmacol Toxicol 1996; 36:233–252.
30. Henney JE. Withdrawal of troglitazone and cisapride. J Am Med Assoc 2000; 283:2228.
31. Bedford TA, Rowbotham DJ. Cisapride: drug interactions of clinical significance. Drug Safety 1996; 15:167–175.
32. Gross AS, Goh YD, Addison RS, et al. Influence of grapefruit juice on cisapride pharmacokinetics. Clin Pharmacol Ther 1999; 65:395–401.
33. Hebert MF. Immunosuppressive agents. In: Levy RH, Thummel KE, Trager WF, et al., eds. Metabolic Drug Interactions. Philadelphia: Lippincott Williams and Wilkins, 2000:499–510.
34. Henderson L, Yue Y, Bergquist C, et al. St John's wort (*Hypericum perforatum*): drug interactions and clinical outcomes. Br J Clin Pharmacol 2002; 54:349–356.
35. Lown KS, Mayo RR, Leichtman AB. Role of intestinal P-glycoprotein (mdrl) in interpatient variation in the oral bioavailability of cyclosporine. Clin Pharmacol Ther 1997; 62:1–13.
36. First MR, Schroeder TJ, Weiskittel P, et al. Concomitant administration of cyclosporin and ketoconazole in renal transplant recipients. Lancet 1989; 2:1198–1201.
37. Christians U, Jacobsen W, Benet LZ, et al. Mechanisms of clinically relevant drug interactions associated with tacrolimus. Clin Pharmacokinet 2002; 41:813–851.
38. Hersmans Y. Cholesterol-lowering agents and cardiac glycosides. In: Levy RH, Thummel KE, Trager WF, et al., eds. Metabolic Drug Interactions. Philadelphia: Lippincott Williams and Wilkins, 2000:379–390.
39. Herman RJ. Drug interactions and the statins. Can Med Assoc J 1999; 161: 1281–1286.
40. Williams D, Feely J. Pharmacokinetic-pharmacodynamic drug interactions with HMG-CoA reductase inhibitors. Clin Pharmacokinet 2002; 41:343–370.
41. Kyrklund C, Backman JT, Kivisto KT, et al. Rifampin greatly reduces plasma simvastatin and simvastatin acid concentrations. Clin Pharmacol Ther 2000; 68: 592–597.

42. Prueksaritanont T, Zhao JJ, Ma B, et al. Mechanistic studies on metabolic interactions between gemfibrozil and statins. J Pharmacol Exp Ther 2002; 301: 1042–1051.
43. Staffa JA, Chang J, Green L. Cerivastatin and reports of fatal rhabdomyolysis. N Engl J Med 2002; 346:539–540.
44. Gotto AM. Statins, cardiovascular disease, and drug safety. Am J Cardiol 2006; 97 (suppl):3C–5C.
45. Bags H. Statin safety: an overview and assessment of the data - 2005. Am J Cardiol 2006; 97(suppl):6C–26C.
46. Krayenbuhl JC, Vozeh S, Kando-Ostreicher M, et al. Drug-drug interactions of new active substances: mibefradil example. Eur J Clin Pharmacol 1999; 55:559–565.
47. Thompson PD, Clarkson P, Karas RH. Statin-associated myopathy. J Am Med Assoc 2003; 289:1681–1690.
48. Talbert RL. Safety issues with statin therapy. J Am Pharm Assoc 2006; 46:479–490.
49. Bottorff MB. Statin safety and drug interactions: clinical implications. Am J Cardiol 2006; 97(suppl):27C–31C.
50. McKenney JM, Davidson MH, Jacobsen TA, et al. Final conclusions of the National Lipid Association Statin Safety Assessment Task Force. Am J Cardiol 2006; 97 (suppl):89C–94C.
51. Jones DR, Hall SD. Calcium channel blockers. In: Levy RH, Thummel KE, Trager WF, et al., eds. Metabolic Drug Interactions. Philadelphia: Lippincott Williams and Wilkins, 2000:333–345.
52. Greenblatt DJ, von Moltke LL. Sedative-hypnotic and anxiolytic agents. In: Levy RH, Thummel KE, Trager WF, et al., eds. Metabolic Drug Interactions. Philadelphia: Lippincott Williams and Wilkins, 2000:259–270.
53. Hiller A, Olkkola KT, Isohanni P, et al. Unconsciousness associated with midazolam and erythromycin. Br J Anaesth 1990; 65:826–828.
54. Chiba K, Kobayashi K. Antidepressants. In: Levy RH, Thummel KE, Trager WF, et al., eds. Metabolic Drug Interactions. Philadelphia: Lippincott Williams and Wilkins, 2000:233–243.
55. Bergstrom RF, Peyton AL, Lemberger L. Quantification and mechanism of the fluoxetine and tricyclic antidepressant interaction. Clin Pharmacol Ther 1992; 51:239–248.
56. Preskorn SH, Alderman J, Chung M, et al. Pharmacokinetics of desipramine coadministered with sertraline and fluoxetine. J Clin Psychopharmacol 1994; 14:90–98.
57. Cohen LJ, DeVane CL. Clinical implications of antidepressant pharmacokinetics and pharmacogenetics. Ann Pharmacother 1996; 3:1471–1480.
58. Brosen K, Hansen JG, Nielsen KK, et al. Inhibition by paroxetine of desipramine metabolism in extensive but not poor metabolizers of sparteine. Eur J Clin Pharmacol 11993; 44:349–355.
59. Eichelbam M, Gross AS. The genetic polymorphism of debrisoquin/sparteine metabolism—clinical aspects. Pharmacol Ther 1990; 46:377–394.
60. Barbery JT, Roose SP. SSRI safety in overdose. J Clin Psychiatry 1998; 59(suppl 15): 42–48.
61. Hilleret H, Voirol P, Bovier P, et al. Very long half-life of paroxetine following intoxication in an extensive cytochrome P4502D6 metabolizer. Ther Drug Monit 2002; 24:567–569.

62. Lennard MS. 3-Adrenoceptor antagonists. In: Levy RH, Thummel KE, Trager WF, et al., eds. Metabolic Drug Interactions. Philadelphia: Lippincott Williams and Wilkins, 2000:347–358.
63. Lennard MS, Tucker GT, Silas JH, et al. Debrisoquine polymorphism and the metabolism and action of metoprolol, timolol, propranolol and atenolol. Xenobiotica 1986; 16:435–437.
64. Lesko LJ. Pharmcokinetic drug interactions with amiodarone. Clin Pharmacokinet 1989; 17:130–140.
65. Loon NR, Wilcox CS, Folger W. Orthostatic hypotension due to quinidine and propranolol. Am J Med 1986; 81:1101–1104.
66. Wagner F, Kalushce D, Trenk D, et al. Drug interaction between propafenone and metoprolol. Br J Clin Pharmacol 1987; 24:213–220.
67. Drake WM, Gordon GD. Heart block in a patient on propranolol and fluoxetine. Lancet 1994; 343:425–426.
68. Hamelin BA, Bouayad A, Methot J, et al. Significant interaction between the non-prescription antihista-mine diphenhydramine and the CYP2D6 substrate metoprolol in healthy men with low or high CYP2D6 activity. Clin Pharmacol Ther 2000; 67: 466–477.
69. Werner U, Werner D, Rau T, et al. Celecoxib inhibits metabolism of cytochrome P4502D6 substrate metoprolol in humans. Clin Pharmacol Ther 2003; 74:130–137.
70. Kharasch ED, Ibrahim AE. Volatile, intravenous, and local anesthetics. In: Levy RH, Thummel KE, Trager WF, et al., eds. Metabolic Drug Interactions. Philadelphia: Lippincott Williams and Wilkins, 2000:271–295.
71. Pohl LR, Satoh H, Christ DD, et al. The immunologic and metabolic basis of drug hypersensitivities. Annu Rev Pharmacol Toxicol 1988; 28:367–387.
72. Pohl LR. An immunochemical approach to identifying and characterizing protein targets of toxic reactive metabolites. Chem Res Toxicol 1993; 6:786–793.
73. Bourdi M, Chen W, Peter R, et al. Human cytochrome P450 2E1 is a major autoantigen associated with halothane hepatitis. Chem Res Toxicol 1996; 9: 1159–1166.
74. O'Shea D, Davis SN, Kim RB, et al. Effect of fasting and obesity in humans on the 6-hydroxylation of chlorzoxazone, a putitive probe of CYP2E1 activity. Clin Pharmacol Ther 1994; 56:359–367.
75. Cousins MJ, Plummer JL, Hall PM. Risk factors for halothane hepatitis. Aust NZ J Surg 1989; 59:5–14.
76. Bentley JB, Vaughan RW, Gandolfi AJ, et al. Halothane biotransformation in obese and nonobese patients. Anesthesiology 1982; 57:94–97.
77. Kharasch ED, Hankins D, Mautz D, et al. Identification of the enzyme responsible for oxidative halothane metabolism: implications for prevention of halothane hepatitis. Lancet 1996; 347:1367–1371.
78. Siegmund JB, Wilson JH, Imhoff TE. Amiodarone interaction with lidocaine. J Cardiovasc Pharmacol 1993; 21:513–515.
79. Ha HR, Candinas R, Steiger B, et al. Interaction between amiodar one and lidocaine. J Cardiovasc Pharmacol 1996; 21:533–539.
80. Holmes GL. Carbamazepine toxicity. In: Levy RH, Mattson RH, Meldrum BS, eds. Antiepileptic Drugs. 4th ed. New York: Raven Press, 1995:567–579.

81. Kerr BM, Thummel KE, Wurden CJ. Human liver carbamazepine metabolism. Role of CYP3A4 and CYP2C8 in 10,11-epoxide formation. Biochem Pharmacol 1994; 47:1969–1979.
82. Mather GG, Levy RH. Anticonvulsants. In: Levy RH, Thummel KE, Trager WF, et al., eds. Metabolic Drug Interactions. Philadelphia: Lippincott Williams and Wilkins, 2000:217–232.
83. Amore BM, Kalhorn TF, Skiles GL, et al. Characterization of carbamazepine metabolism in a mouse model of carbamazepine teratogenicity. Drug Metab Dispos 1997; 25:953–962.
84. Ju C, Uetreeht JP. Detection of-2-hydroxy irninostilbene in the urine of patients taking carbamazepine and its oxidation to a reactive iminoquinone intermediate. J Pharmacol Exp Ther 1999; 288:51–56.
85. Lindhout D, Hoppener RJEA, Meinardi H. Teratogenicity of antiepileptic drug combinations with special emphasis on epoxidation of carbamazepine. Epilepsia 1984; 25:77–83.
86. Kaneko S, Otani K, Fukushima Y, et al. Teratogenicity of antiepileptic drugs: analysis of possible risk factors. Epilepsia 1988; 29:459–467.
87. Omzigt JGC, Los FJ, Meijer JWA, et al. The 10,ll-epoxide-10,ll-diol pathway of carbamazepine in early pregnancy in maternal serum, urine, and amniotic fluid: effect of dose, comedication, and relation to outcome of pregnancy. Ther Drug Monit 1993; 15:1–10.
88. Egnell A-C, Houston B, Boyer S. In vivo CYP3A4 heteroactivation is a possible mechanism for the drug interaction between felbamate and carbamazepine. J Pharmacol Exp Ther 2003; 305:1251–1262.
89. Bruni J. Phenytoin toxicity. In: Levy RH, Mattson RH, Meldrum BS, eds. Antiepileptic Drugs. 4th ed. New York: Raven Press, 1995:345–350.
90. Bajpai M, Roskos LK, Shen DD, et al. Roles of cytochrome P4502C9 and cytochrome P4502C19 in the stereoselective metabolism of phenytoin to its major metabolite. Drug Metab Dispos 1996; 24:1401–1403.
91. Mamiya K, Kojima K, Yukawa E, et al. Phenytoin intoxication induced by fluvoxamine. Ther Drug Monit 2001; 23:75–77.
92. Leeder JS, Riley RJ, Cook J, et al. Human anti-cytochrome P450 antibodies in aromatic anticonvulsant induced hypersensitivity reactions. J Pharmacol Exp Ther 1992; 263:360–366.
93. Zhou LX, Pihlstrom B, Hardwick JP, et al. Metabolism of phenytoin by the gingiva of normal humans: the possible role of reactive metabolites of phenytoin in the initiation of gingival hyperplasia. Clin Pharmacol Ther 1996; 60:191–198.
94. Kubow S, Wells P. In vitro bioactivation of phenytoin to a reactive free radical intermediate by prostaglandin synthase, horseradish peroxidase and thyroid peroxidase. Mol Pharmacol 1989; 35:1–8.
95. Uetrecht J. Drug metabolism by leukocytes and its role in drug-induced lupus and other idiosyncratic drug reactions. Toxicology 1990; 20:213–235.
96. Porubek DJ, Grillo MP, Olsen RK, et al. Toxic metabolites of valproic acid: inhibition of rat liver acetoacetyl-CoA thiolase by 2-*n*-propyl-4-pentenoic acid (A^4-VPA) and related branched chain carboxylic acids. In: Levy RH, Penry JK, eds. Idiosyncratic Reactions to Valproate: Clinical Risk Patterns and Mechanisms of Toxicity. New York: Raven Press, 1991:53–58.

97. Bojic U, Elmazaar MMA, Hauck R-S, et al. Further branching of valproate-related carboxylic acids reduces the teratogenic activity, but not anticonvulsant effect. Chem Res Toxicol 1996; 9:866–870.
98. Sadeque AJM, Fisher MB, Korzekwa KR, et al. Human CYP2C9 and CYP2A6 mediate formation of the hepatotoxin 4-ene-valproic acid. J Pharmacol Exp Ther 1997; 283:698–703.
99. Baillie TA, Sheffels PR. Valproic acid chemistry and biotransformation. In: Levy RH, Mattson RH, Meldrum BS, eds. Antiepileptic Drugs. 4th ed. New York: Raven Press, 1995:589–604.
100. McLeod HL. Clinically relevant drug-drug interactions in oncology. Br J Clin Pharmacol 1998; 45:539–544.
101. Relling MV, Pui C-H, Sandlund JT, et al. Adverse effect of anticonvulsants on efficacy of chemotherapy for acute lymphoblastic leukemia. Lancet 2000; 356: 285–290.
102. Mangold JB, Fischer V. Antineoplastic agents. In: Levy RH, Thummel KE, Trager WF, et al., eds. Metabolic Drug Interactions. Philadelphia: Lippincott Williams and Wilkins, 2000:545–554.
103. Tolcher AW, Cowan KH, Solomon D. Phase I crossover study of paclitaxel with R-verapamil in patients with metastatic breast cancer. J Clin Oncol 1996; 14: 1173–1184.
104. Bohme A, Ganser A, Hoelzer D. Aggravation of vincristine-induced neurotoxicity by itraconazole in the treatment of adult ALL. Ann Hematol 1995; 71:311–312.
105. Gillies J, Hung KA, Fitzsimons E, et al. Severe vincristine toxicity in combination with itraconazole. Clin Lab Haematol 1998; 20:123–124.
106. Physician's Desk Reference, 60th ed. Montvale, NJ: Thompson PDR, 2006.
107. Smith P. The influence of St John's wort on the pharmacokinetics and protein binding of imatinib mesylate. Pharmacotherapy 2004; 24:1508–1514.
108. Frye RF, Fitzgerald SM, Lagattuta TF, et al. Effect of St John's wort on imatinib mesylate pharmacokinetics. Clin Pharmacol Ther 2004; 76:323–329.
109. O'Brien SG, Peng B, Dutrix C, et al. A pharmacokinetic interaction of Glivec and simvastatin, a cytochrome 3A substrate, in a patient with chronic myeloid leukemia. Blood 2001; 98(suppl):141a, Abstract 593.
110. Unadkat JD, Wang Y. Antivirals. In: Levy RH, Thummel KE, Trager WF, et al., eds. Metabolic Drug Interactions. Philadelphia: Lippincott Williams and Wilkins, 2000:403–413.
111. Rainey PM. HIV drug interactions: the good, the bad and the other. Ther Drug Monit 2002; 24:26–31.
112. Tirona RG, Bailey DG. Herbal product-drug interactions mediated by induction. Br J Clin Pharmacol 2006; 61:677–681.
113. van den Bout-van den Beukel CJP, Koopmans PP, van der Ven AJAM, De Smet PAGM, Burger DM. Possible drug metabolism interactions of medicinal herbs with antiretroviral agents. Drug Metab Rev 2006; 38:477–514.
114. Tran JQ, Gerber JG, Kerr BM. Delavirdine: clinical pharmacokinetics and drug interactions. Clin Pharmacokinet 2001; 40:207–226.
115. Flexner C. Dual protease inhibitor therapy in HIV-infected patients: pharmacologic rationale and clinical benefits. Annu Rev Pharmacol Toxicol 2000; 40:649–674.
116. Hu Z, Yang X, Ho PC, et al. Herb-drug interactions: a literature review. Drugs 2005; 65:1239–1282.

117. Piscitelli SC, Burstein AH, Chaitt D, et al. Indinavir concentrations and St John's wort. Lancet 2000; 355:547–548.
118. deMaat MM, Ekhart GC, Huitema AD, et al. Drug interactions between antiretroviral drugs and comedicated agents. Clin Pharmacokinet 2003; 42:223–282.
119. Horowitz RS, Hart RC, Gomez JF. Clinical ergotism with lingual ischemia induced by clarithromycin-ergotamine interaction. Arch Intern Med 1996; 156:456–458.
120. Shad MU, Preskorn SH. Antidepressants. In: Levy RH, Thummel KE, Trager WF, et al., eds. Metabolic Drug Interactions. Philadelphia: Lippincott Williams and Wilkins, 2000:563–577.
121. Silverman JA. P-Glycoprotein. In: Levy RH, Thummel KE, Trager WF, et al., eds. Metabolic Drug Interactions. Philadelphia: Lippincott Williams and Wilkins, 2000:135–144.
122. Lindenbaum J, Rund DG, Butler VP Jr., et al. Inactivation of digoxin by the gut flora: reversal by antibiotic therapy. N Engl J Med 1981; 305:789–794.
123. Park BK, Kitteringham NR. Drug-protein conjugation and its immunological consequences. Drug Metab Rev 1990; 22:87–144.
124. Uetrecht J. Drug metabolism by leukocytes and its role in drug-induced lupus and other idiosyncratic drug reactions. Crit Rev Toxicol 1990; 20:213–235.
125. Meyer UA. Molecular mechanisms of genetic polymorphisms. Annu Rev Pharmcol Toxicol 1997; 37:269–296.
126. Bourdi M, Tinel M, Beaune PH, et al. Interactions of dihydralazine with cytochromes P4501A: a possible explanation for the appearance of anti-cytochrome P4501A2 autoantibodies. Mol Pharmacol 1994; 45:1287–1295.
127. Cribb AE, Lee BL, Trepanier LA, et al. Adverse reactions to sulphonamide and sulphonamide-trimethoprim antimicrobials: clinical syndromes and pathogenesis. Adverse Drug React Toxicol Rev 1996; 15:9–50.
128. Reilly TP, Woster PM, Svensson CK. Methemoglobin formation by hydroxylamine metabolites of sulfamethoxazole and dapsone: implications for differences in adverse drug reactions. J Pharmacol Exp Ther 1999; 288:951–959.
129. Bluhm RE, Adedoyin A, McCarver DG, et al. Development of dapsone toxicity in patients with inflammatory dermatoses: activity of acetylation and hydroxyl-ation as risk factors. Clin Pharmacol Ther 1999; 65:598–605.
130. Gill HJ, Tingle MD, Park BK. N-Hydroxylation of dapsone by multiple enzymes of cytochrome P450: implications for inhibition of haemotoxicity. Br J Clin Pharmacol 1995; 40:531–538.
131. Pirmohamed M, Madden S, Park BK. Idiosyncratic drug reactions: metabolic bioactivation as a pathogenic mechanism. Clin Pharmacokinet 1996; 31:215–230.
132. Poplack DG, Balis FM, Zimm S. The pharmacology of orally administered chemotherapy. A reappraisal. Cancer 1986; 58:473–480.
133. Kennedy DT, Hayney MS, Lake KD. Azathioprine and allopurinol: the price of an avoidable drag interaction. Ann Pharmacother 1996; 30:951–954.
134. Winshelboum RM, Sladek SL. Mercaptopurine pharmacogenetics: monogenic inheritance of erythrocyte thiopurine methyltransferase activity. Am J Hum Genet 1980; 32:651–662.
135. Szumlanski C, Winshelboum RM. Sulphasalazine inhibition of thiopurine methyltransferase: possible mechanism for interaction with 6-mercaptopurine and azathioprine. Br J Clin Pharmacol 1995; 39:456–459.

136. Lewis LD, Benin A, Szumlanski C. Olsalazine and 6-mercaptopurine-related hematologic suppression: a possible drug-drug interaction. Clin Pharmacol Ther 1997; 62:464–475.
137. Yuen AW, Land G, Weatherley BC, et al. Sodium valproate acutely inhibitslamotrigine metabolism. Br J Clin Pharmacol 1992; 33:511–513.
138. Trapnell CB, Klecker RW, Jamis-Dow C, et al. Glucuronidation of 3′-azido-3′-deoxythymidine (zidovudine) by human liver microsomes: relevance to clinical pharmacokinetic interactions with atovaquone, fluconazole, methadone, and val-proic acid. Antimicrob Agents Chemother 1998; 42:1592–1596.
139. Kaufman KR, Gerner R. Lamotrigine toxicity secondary to sertraline. Seizure 1998; 7:163–165.
140. Nakajima Y, Mizobuchi M, Nakamura M, et al. Mechanism of the drug interaction between valproic acid and carbapenem antibiotics in monkeys and rats. Drug Metab Dispos 2004; 32:1383–1391.
141. Prescott LF. Paracetamol (Acetaminophen): A Critical Bibliographic Review. London: Taylor and Francis, 1996.
142. Nelson SD. Analgesic-antipyretics. In: Levy RH, Thummel KE, Trager WF, et al., eds. Metabolic Drug Interactions. Philadelphia: Lippincott Williams and Wilkins, 2000:447–455.
143. Wright JN, Prescott LF. Potentiation by previous drag therapy of hepatotoxicity following paracetamol overdosage. Scott Med J 1973; 18:56–58.
144. Wilson JT, Kasantikul V, Harbison R, et al. Death in an adolescent following an overdose of acetaminophen and phenobarbital. Am J Dis Child 1978; 132:466–473.
145. Minton NA, Henry JA, Frankel RJ. Fatal paracetamol poisoning in an epileptic. Hum Toxicol 1988; 7:33–34.
146. Bray GP, Harrison PM, O'Grady JG, et al. Long-term anticonvulsant therapy worsens outcome in paracetamol-induced fulminant hepatic failure. Hum Exp Toxicol 1992; 11:265–270.
147. Lystback BB, Norregaard P. A case of paracetamol retard poisoning with fatal outcome. Ugeskr Laeger 1995; 157:899–900.
148. Pirotte JH. Apparent potentiation of hepatotoxicity from small doses of acetaminophen by phenobarbital. Ann Intern Med 1984; 101:403.
149. Parikh S, Dillon LC, Scharf SL. Hepatotoxicity possibly due to paracetamol with carbamazepine. Intern Med J 2004; 34:441–442.
150. Young CR, Mazure CM. Fulminant hepatic failure from acetaminophen in an anorexic patient treated with carbamazepine. J Clin Psychiatry 1998; 59:622.
151. Miners JO, Atwood J, Birkett DJ. Determinants of acetaminophen metabolism: effects of inducers and inhibitors of drug metabolism on acetaminophen's metabolic pathways. Clin Pharmacol Ther 1984; 35:480–486.
152. Prescott LF, Critchley JAJH, Bulali-Mood M, et al. Effects of microsomal enzyme induction on paracetamol metabolism in man. Br J Clin Pharmacol 1981; 12:149–153.
153. Kostrubsky SE, Sinclair JF, Strom SC, et al. Phenobarbital and phenytoin increased acetaminophen hepatotoxicity due to inhibition of UDP-glucuronosyl transferases in cultured human hepatocytes. Toxicol Sci 2005; 87:146–155.
154. Mutlib AE, Goosen TC, Bauman JN, et al. Kinetics of acetaminophen glucuronidation by UDP-glucuronosyl transferases 1A1, 1A6, 1A9 and 2B15. Potential implications in acetaminophen-induced toxicity. Chem Res Toxicol 2006; 19:701–709.

155. Nelson SD, Slattery JT, Thummel KE, et al. CAR unlikely to significantly modulate acetaminophen hepatotoxicity in most humans. Hepatology 2003; 38:254–257.
156. Nelson SD, Bruschi SA. Mechanisms of acetaminophen-induced liver disease. In: Kaplowitz N, DeLeve LD, eds. Drug Induced Liver Disease. 2nd ed. New York: Informa Healthcare, 2007:353–388.
157. Raucy JL, Lasker JM, Lieber CS, et al. Acetaminophen activation by human liver cytochromes P450IIE1 and P450IA2. Arch Biochem Biophys 1989; 271:270–283.
158. Patten CJ, Thomas PE, Guy RL, et al. Cytochrome P450 enzymes involved in acetaminophen activation by rat and human liver microsomes and their kinetics. Chem Res Toxicol 1993; 6:511–518.
159. Chen W, Koenigs LL, Thompson SJ, et al. Oxidation of acetaminophen to its toxic quinone imine and nontoxic catechol metabolites by baculovirus-expressed and purified human cytochromes P4502E1 and 2A6. Chem Res Toxicol 1998; 11: 295–301.
160. Thummel KE, Slattery JT, Ro H, et al. Ethanol and production of the hepatotoxic metabolite of acetaminophen in healthy adults. Clin Pharmacol Ther 2000; 67:591–599.
161. Zand R, Nelson SD, Slattery JT, et al. Inhibition and induction of cytochrome P450 2E1 catalyzed acetaminophen oxidation by isoniazid in humans. Clin Pharmacol Ther 1993; 54:142–149.
162. McClain CJ, Kromhaut JP, Peterson FJ, et al. Potentiation of acetaminophen hepatotoxicity by alcohol. J Am Med Assoc 1980; 244:251–253.
163. Seeff LB, Cuccherini BA, Zimmerman HJ, et al. Acetaminophen toxicity in the alcoholic: a therapeutic misadventure. Ann Intern Med 1986; 104:399–404.
164. Zimmerman HJ, Maddrey WC. Acetaminophen (paracetamol) hepatotoxicity with regular intake of alcohol: analysis of instances of therapeutic misadventure. Hepatology 1995; 22:762–777.
165. Johnston SC, Pelletier LL. Enhanced hepatotoxicity of acetaminophen in the alcoholic patient: two case reports and a review of the literature. Medicine 1997; 76:185–191.
166. Murphy R, Scartz R, Watkins PB. Severe acetaminophen toxicity in a patient receiving isoniazid. Ann Intern Med 1990; 113:799–800.
167. Moulding TS, Redeker AG, Kanel GC. Acetaminophen, isoniazid, and hepatic toxicity. Ann Intern Med 1991; 114:431.
168. Nolan CM, Sandblom RE, Thummel KE, et al. Hepatotoxicity associated with acetaminophen usage in patients receiving multiple drug therapy for tuberculosis. Chest 1994; 105:408–411.
169. Lauterburg BH, Velez ME. Glutathione deficiency in alcoholics: risk factor for paracetamol hepatotoxicity. Gut 1988; 29:1153–1157.
170. Whitcomb DC, Block GD. Association of acetaminophen hepatotoxicity with fasting and ethanol use. J Am Med Assoc 1994; 272:1845–1850.
171. Slattery JT, Nelson SD, Thummel KE. The complex interaction between ethanol and acetaminophen. Clin Pharmacol Ther 1996; 60:241–246.
172. Lee TD, Sadda MR, Mendler MJ, et al. Abnormal hepatic methionine and glutathione metabolism in patients with alcoholic hepatitis. Alcohol Clin Exp Res 2004; 28:173–181.
173. Rumack BH. Acetaminophen misconceptions. Hepatology 2004; 40:10–15.

174. Kaplowitz N. Acetaminophen hepatotoxicity: what do we know, what don't we know, and what do we do next? Hepatology 2004; 40:23–26.
175. Spahn-Langguth H, Benet LZ. Acyl glucuronides revisited: Is the glucuronidation process a toxification as well as detoxification mechanism? Drug Metab Rev 1992; 24:5–48.
176. Shen S, Davis MR, Doss GA, et al. Metabolic activation of diclofenac by human cytochrome P4503A4: role of 5-hydroxydiclofenac. Chem Res Toxicol 1999; 12:214–222.
177. Tang W, Stearns RA, Wang RW, et al. Roles of human hepatic cytochrome P450s 2C9 and 3A4 in the metabolic activation of diclofenac. Chem Res Toxicol 1999; 12:192–199.
178. Chen HJL, Boch KJ, MacLean JA. Acute eosinophilic hepatitis from trovofloxacin. N Eng J Med 2000; 342:359–360.
179. Kassahun K, Pearson PG, Tang W, et al. Studies on the metabolism of troglitazone to reactive metabolites *in vitro* and *in vivo*. Evidence for novel biotransformation pathways involving quinone methide formation and thiazolidine ring scission. Chem Res Toxicol 2001; 14:62–70.
180. Evans DC, Watt AP, Nicoll-Griffith DA, et al. Drug-protein covalent adducts: an industry perspective on minimizing the potential for drug bioactivation in drug discovery and development. Chem Res Toxicol 2004; 17:3–16.
181. Walgren JL, Mitchell MD, Thompson DC. Role of metabolism in drug-induced idiosyncratic hepatotoxicity. Crit Rev Toxicol 2005; 35:325–361.
182. Baillie TA. Future of toxicology - metabolic activation and drug design: challenges and opportunities in chemical toxicology. Chem Res Toxicol 2006; 19:889–893.

18

Drug-Drug Interactions: Marketing Perspectives

Kevin J. Petty
Johnson and Johnson, Raritan, New Jersey, U.S.A.
Jose M. Vega
Amgen, Thousand Oaks, California, U.S.A.

I. INTRODUCTION

The number of drugs available to treat patient illness is steadily increasing as drug development benefits from advances in molecular biology and from increasing automation of drug screening through the use of robotics and combinatorial chemistry. This ever-expanding pharmaceutical arsenal is available to physicians to treat a large number of diseases (both human and veterinary). As the mean age of industrialized nations increases, in part due to advances in medical care, the need to treat multiple disease processes simultaneously increases the probability that large numbers of people will receive concomitant therapy with multiple drugs. Consequently, there is an increased risk of adverse drug-drug interactions as more drugs are used to treat a variety of conditions in the same patient.

Identifying the potential for adverse drug-drug interactions is increasingly difficult when patients are cared for by multiple specialists, each primarily concerned with one organ system, without overall coordination of the patient's management by one person. In many situations, the potential for drug-drug interactions can be minimized by appropriate choices of agents, particularly when options exist with different mechanisms of action, sites of metabolism, and routes of excretion. There are, unfortunately, situations where interactions may not be avoidable and the risks and benefits must be carefully assessed. For instance, the treatment of cancer, AIDS, or other life-threatening diseases might require treatment with a drug known to inhibit enzymes that metabolize other drugs [e.g., HIV protease inhibitor inhibition of cytochrome P450 3A4 (CYP3A4)]. Thus, drug-drug interactions must be evaluated in light of the therapeutic class and the risk/benefit ratio. These interactions can have a significant impact on the marketing of drugs. This chapter will focus on drug-drug interactions and their effect on the market place.

II. MARKET SIZE

The prescription drug market is large and continuously growing. Prescription drug sales (retail pharmacies) in the United States in 1997 totaled \$81.2 billion (1). Each of the 10 top-selling prescription drugs in 1997 had U.S. sales of over \$800 million and ranged from \$804.8 million for Augmenting® to \$2.28 billion for Prilosec®. Prilosec became the first prescription drug to exceed \$5 billion, with worldwide sales in 1998 of \$5.14 billion (2). The six top-selling drugs in 1997 (Prilosec, Prozac®, Zocor®, Epogen®, Zoloft®, and Zantac®) each had U.S. sales greater than \$1 billion. New products launched in 1997 produced \$3.28 billion in U.S. sales, led by Lipitor® (\$587 million) and Rezulin (\$325 million). In 1998, U.S. retail pharmaceutical sales were \$86.6 billion, and 25 drugs achieved worldwide annual sales of at least \$1 billion. Worldwide (North America, Europe, Japan, Latin America, Australia) retail pharmacy sales totaled more than \$185 billion in 1998 (3).

This enormous market is influenced by a complex array of factors. Among the most significant of these factors are the efficacy and safety of a given drug. Those two factors are the most important considerations in the process by which drugs receive approval from regulatory agencies to allow their marketing. Those factors are also important for drugs after they gain entry to the market, together with additional factors, such as dosing convenience and cost when more than one drug is available to treat the same condition.

There are many ways in which drugs can interact to produce adverse events. Perhaps the most common type of interaction is where one drug alters the pharmacokinetics of a second drug. Alteration of pharmacokinetics can include inhibition by one drug of the metabolism of a second drug (e.g., erythromycin inhibition of warfarin metabolism), leading to accumulation of the second drug with its resultant toxicity (4). Conversely, induction of metabolism of one drug by another can also produce untoward effects if plasma levels of the second drug become subtherapeutic. An example of such an interaction was the reported interaction of rifampin with oral contraceptives containing ethinyl estradiol, where concomitant use of rifampin accelerated the metabolism of ethinyl estradiol, resulting in decreased efficacy as contraceptive and unwanted pregnancies (4).

Since 1964, ~60 drug products have been withdrawn from the U.S. market because they were found to be ineffective or unsafe (5). Most of the compounds withdrawn primarily for safety had toxicities directly attributable to the compound. Only two of these drugs (terfenadine and mibefradil) were withdrawn primarily for their high incidence of adverse drug-drug interactions. The discussion that follows will describe the experience with these two drugs and the experience with cimetidine, where drug-drug interactions have had a significant impact on its market position.

A. Terfenadine

Terfenadine was introduced into the marketplace as the first nonsedating histamine-1 (H_1) receptor antagonist. It was launched in the United States in 1985 under the brand name Seldane®. Its patent protection was near expiration when the drug was voluntarily withdrawn from the antihistamine market in 1997.

1. Clinical Background

During its early development, terfenadine was found to act as a competitive antagonist for histamine binding to the H_1 receptor with a 50% inhibitory concentration (IC_{50}) of 0.7 μM. It was thought that the antihistaminic effect of terfenadine was due to direct interaction with the H_1 receptor. Subsequent studies revealed that terfenadine was completely metabolized in vivo to fexofenadine, a metabolite entirely responsible for the antihistaminic effect (6). The unique property of fexofenadine compared to first-generation antihistamines (diphenhydramine, chlorpheniramine) was its inability to cross the blood-brain barrier, thereby avoiding the sedation seen with the first-generation antihistamines. Terfenadine was indicated for use in allergic rhinitis (both seasonal and perennial), and the recommended dose was 60 mg twice daily.

2. Clinical Pharmacology

Terfenadine is at least 70% absorbed after oral administration but is rapidly metabolized by first-pass metabolism to fexofenadine (terfenadine carboxylate) and an inactive dealkylated product. Metabolism appears to be mediated entirely by the CYP (CYP3A4). Fexofenadine is about 70% protein bound and exhibits biphasic elimination with an initial plasma half-life of 3.5 hours and a terminal plasma half-life of 6 to 12 hours. Fexofenadine is excreted mostly unchanged (80% in feces, 12% in urine), with <10% converted to inactive metabolites (7). Fexofenadine excretion can be affected by compounds (e.g., ketoconazole) that interact with the P-glycoprotein transporter because fexofenadine is a substrate for this transporter (8).

3. Drug-Drug Interactions

Terfenadine itself has no effect on CYP activity and thus does not affect metabolism of other CYP substrates. The drug-drug interactions of significance were due to inhibition of CYP3A4 by other drugs, leading to toxic accumulation of terfenadine in plasma, where it normally would only be found in trace amounts (9,10).

4. Adverse Experiences

The first published report of a serious adverse event due to an interaction of terfenadine with another drug was that of a young woman who was taking terfenadine and subsequently began taking ketoconazole. Within a few days after beginning ketoconazole therapy, she experienced syncopal episodes and was found to have torsade de pointes (polymorphic ventricular tachycardia) (11). Torsade de pointes was also seen in patients with liver failure who took terfenadine and in patients who simultaneously received erythromycin and terfenadine (12). On the basis of the initial reports of torsade with terfenadine, a ''Dear Doctor'' letter was issued by the manufacturer of Seldane in 1990. A retrospective analysis of concomitant drug use within a large cohort of Medicaid patients revealed that there was a significant correlation between terfenadine toxicity and concomitant use of either erythromycin or ketoconazole (both potent CYP3A4 inhibitors) (13). Additional reports of torsade de pointes in 1992 prompted the need for the manufacturer to incorporate a black box warning in the Seldane label that contraindicated concomitant use of terfenadine with CYP3A4 inhibitors, including the azole antifungals (ketoconazole, itraconazole) and macrolide antibiotics (erythromycin, clarithromycin, troleandomycin), and use in patients with hepatic insufficiency.

The occurrence of cardiac toxicity was closely correlated with terfenadine use, and subsequent in vitro studies confirmed that terfenadine (but not fexofenadine) efficiently blocks cardiac potassium channels (14). A study in healthy volunteers treated concomitantly with terfenadine and ketoconazole found a linear relationship between trough terfenadine concentrations and QT_C intervals. The QT_C interval lengthened up to 110 millisecond at the highest plasma concentrations of ~45 ng/mL (9). Thus, the direct inhibitory effect of terfenadine on cardiac potassium channels results in prolongation of cardiac repolarization, which is a well-known cause of ventricular arrhythmias. In one death in which terfenadine was implicated, plasma level of the drug was 55 ng/mL several hours after the last ingestion of the drug (when it normally should be undetectable).

While fexofenadine is also metabolized primarily by CYP3A4 and its levels can be elevated in the presence of potent inhibitors of CYP3A4, its lack of effect on cardiac potassium channels allows for higher fexofenadine levels to be safely tolerated without QTc interval prolongation and without an increased risk of ventricular arrhythmias.

5. Market Dynamics

Seldane held market exclusivity as a nonsedating antihistamine from its launch in 1985 until 1989, when astemizole (Hismanal®) entered the market. Hismanal did not penetrate significantly into the market because of perceived inferior efficacy (longer onset of action) and cardiac toxicities similar to Seldane (7). On the basis of its nonsedating property, efficacy, and convenient dosing, Seldane maintained market leadership, with a ranking of the fifth most commonly prescribed drug in the United States in 1990. In 1991, 17 million prescriptions were issued, and there were more than 100 million users of Seldane worldwide. It had peak U.S. retail pharmacy sales of $540 million and 54% market share in 1992 (combined antihistamine and cold markets). Despite the black box warning, Seldane had worldwide sales of $700 million in 1994 and held up to 29% market share in the United States in 1995. With the launch of Zyrtec® in 1996 (promoted as safer and equally effective), market share of Seldane plummeted to 2.5% in 1997, when it was withdrawn from the market.

When it was recognized that fexofenadine was the active metabolite, efforts were begun to register it as a separate entity. Because of existing patent coverage of fexofenadine by Sepracor, Inc. (Marlborough, Massachusetts, U.S.), Hoechst (Frankfurt, Germany) obtained the licensing rights for its development. Fexofenadine was approved (as Allegra®) for marketing in the United States in July 1996.

B. Mibefradil

Mibefradil (Posicor®) was launched in the United States in June 1997 by Roche, Switzerland. It was promoted as a unique calcium channel blocker (CCB) that affected both transient (T) and long (L) calcium channels. At the time of launch, it was projected to eventually provide 1–3% of Roche's sales. It was withdrawn from the market in June 1998.

1. Clinical Background

Mibefradil is a tetralol derivative developed as a unique CCB. Its efficacy as an antihypertensive was demonstrated in phase III trials, where doses of 50 to 100 mg were compared to other CCBs (nifedipine SR, diltiazem CD, nifedipine GITS, amlodipine). Mibefradil was shown to be equally effective as or more effective than nifedipine SR, diltiazem CD, nifedipine GITS, or amlodipine in reducing blood pressure in mild to moderate hypertension. Average reductions of diastolic blood pressure of as much as 15 mmHg were seen with the 100-mg dose. It was also found to be effective in the treatment of chronic stable angina. Thus, it was indicated for use in hypertension and stable angina at doses of 50 or 100 mg once daily (15).

Studies to support its registration included 2636 patients. It was reported to be well tolerated, with the most common adverse experiences being headache, leg edema, dizziness, and fatigue at incidences similar to those with placebo. The incidence of leg edema with mibefradil was found to be lower than with other CCBs, which was an attribute important in its registration and marketing.

Compared with other CCBs, mibefradil was found to have more negative chronotropic effects, with a significant incidence of dose-dependent first-degree AV block and sinus bradycardia. No effect on QT intervals was detected in the phase III studies.

2. Clinical Pharmacology

Oral bioavailability of mibefradil is dose dependent and ranges from 37% to over 90% with doses of 10 mg or 160 mg, respectively. The plasma half-life is 17 to 25 hours after multiple doses, and it is more than 99% protein bound (15). The metabolism of mibefradil is mediated by two pathways: esterase-catalyzed hydrolysis of the ester side chain to yield an alcohol metabolite and CYP3A4-catalyzed oxidation. After chronic dosing, the oxidative pathway becomes less important and the plasma level of the alcohol metabolite of mibefradil increases. In animal models, the pharmacological effect of the alcohol metabolite is about 10% compared to that of the parent compound. After metabolic inactivation, mibefradil is excreted into the bile (75%) and urine (25%), with less than 3% excreted unchanged in the urine.

Studies in human liver microsomal preparations have demonstrated that mibefradil is a powerful inhibitor of liver CYP3A4, with both competitive and mechanism-based effects on this enzyme at therapeutically relevant concentrations (16). In particular, the potency of competitive inhibition of CYP3A4 is such that the IC_{50} value (<1 μM) falls within the therapeutic plasma concentrations of mibefradil and is comparable to that of ketoconazole. However, the inhibition of CYP3A4 by mibefradil, unlike that of ketoconazole, is at least partially irreversible. On the basis of the in vitro results, mibefradil is one of the most potent mechanism-based inhibitors of CYP3A4 reported to date. Therefore, it should have been anticipated that clinically significant drug-drug interactions would likely ensue when mibefradil was coadministered with the large number of agents metabolized primarily by CYP3A4. In addition, in vitro studies by the manufacturer demonstrated an inhibitory effect of mibefradil on CYP2D6 and CYP1A2, thus suggesting the possibility of additional potential drug-drug interactions.

3. Clinical Drug-Drug Interactions

Prior to registration, several clinical drug-drug interaction studies were done. In those studies, mibefradil or its metabolites were found to inhibit CYP3A4 and CYP2D6 but not CYP1A2. Coadministration of mibefradil with metoprolol (a substrate for CYP2D6) in healthy subjects resulted in a twofold increase in the peak plasma concentrations of total (*R*- and *S*-enantiomeric) metoprolol and a four- to fivefold increase in AUC. Coadministration of terfenadine (a substrate for CYP3A4) and mibefradil in healthy subjects resulted in elevated plasma concentrations of terfenadine up to 40 ng/mL.

Twice-daily dosing of 60 mg terfenadine increased the mean QT_C interval by 12%. The levels of cyclosporine A (another CYP3A4 substrate) increased about twofold with concomitant treatment with 50-mg mibefradil for eight days. In healthy volunteers, elevations in peak quinidine (a CYP3A4 substrate) plasma levels (15–19%) and in AUC (50%) were found during coadministration of single doses of mibefradil at doses of 50 mg and 100 mg. Despite in vitro evidence of inhibition of CYP1A2, no pharmacokinetic interaction was observed with theophylline, a CYP1A2 substrate. It was also reported that no clinically important interaction was seen between mibefradil and cimetidine, digoxin, angiotensin-converting enzyme (ACE) inhibitors, nonsteroidal anti-inflammatory agents, long-acting nitrates, or warfarin (15).

4. Adverse Experiences

Within several weeks of its launch, numerous serious adverse events involving mibefradil were being reported on a regular basis. These included severe bradycardia when used concomitantly with β-blockers, cardiogenic shock when switching from Posicor to dihydropyridine CCBs (17), and rhabdomyolysis when used with HMG-CoA reductase inhibitors (18).

On the basis of these postmarketing adverse experiences, the manufacturer issued a Dear Doctor letter in which it was mentioned that mibefradil was found to interfere with the metabolism of 26 other medicines and that concomitant use of mibefradil with several of these medications was contraindicated.

In a second Dear Doctor letter in June 1998, mibefradil was voluntarily withdrawn from the market by Roche due to ''complexities'' of drug interactions. The withdrawal apparently was precipitated by the analysis of the results of a study of ~2500 patients with congestive heart failure (Mortality Assessment in Congestive Heart Failure or MACH-1 trial). This was a three-year study in which patients were treated with mibefradil, an ACE inhibitor, or placebo. The study reportedly found no difference between mibefradil and placebo in treating congestive heart failure, but it ''provided further information on drug interactions'' (19). It is likely that most of the interactions involved drugs metabolized by CYP3A4 and CYP2D6, but details of the study and its results have not yet been published.

5. Market Dynamics

Although the CCB market is sizable (more than $4 billion in U.S. retail sales in 1998), it is occupied by a variety of compounds, both branded and generic. Thus, Posicor was launched in a very competitive market, and it had virtually no impact on that market. Before its withdrawal, it had been prescribed to ~200,000 patients worldwide and generated <$27 million in U.S. retail sales during its brief time on the market.

C. Cimetidine

Cimetidine was the first histamine-2 (H_2) receptor antagonist in the antiulcerant market. It was introduced to the U.S. market as Tagamet® in 1977, and its patent expired in May 1994. It was approved for over-the-counter (OTC) marketing in September 1995, and it currently remains available on both the prescription and OTC markets.

1. Clinical Background

Cimetidine is a specific antagonist of the H_2 receptor. It binds to the H_2 receptor on gastric parietal cells and competitively blocks the action of histamine in the signaling pathway that regulates gastric acid secretion (both basal and stimulated). It is indicated for the treatment of duodenal and gastric ulcers, gastroesophageal reflux disease (GERD), and hypersecretory conditions such as Zollinger-Ellison syndrome and systemic mastocytosis. It is also indicated for prevention of upper gastrointestinal bleeding in critically ill patients.

Controlled clinical trials supporting the registration of cimetidine found that it was significantly better than placebo in achieving healing of duodenal ulcers. Subsequent studies also found cimetidine was effective in controlling symptoms of peptic ulcer disease and GERD. In addition, it significantly decreased the duration of gastrointestinal bleeding in patients with peptic ulcers (20). During phase III studies, cimetidine was generally well tolerated. A variety of adverse experiences were reported with cimetidine but not an incidence greater than placebo. Drug-drug interactions were not significant when cimetidine was used concomitantly with sedatives, analgesics, thiazide diuretics, bronchodilators, digoxin, or propranolol in the phase III studies.

More extensive experience with cimetidine showed that it was reasonably well tolerated. Postmarketing adverse experiences included dizziness and somnolence, reversible confusional states (in severely ill patients), mild diarrhea, gynecomastia, impotence, neutropenia, thrombocytopenia, increased hepatic transaminases, and increased serum creatinine (due to competition by the drug of renal tubular secretion of creatinine). Most of these adverse experiences are uncommon (occurring in $<1\%$ of patients) and dose related. Gynecomastia has been reported in as many as 4% of patients taking prolonged high doses of cimetidine for hypersecretory conditions (Zollinger-Ellison syndrome).

2. Clinical Pharmacology

Cimetidine is over 90% absorbed with $\sim$50% bioavailability after oral administration. The plasma half-life is two hours with a volume of distribution of 1 L/kg. In patients with renal failure, plasma half-life is prolonged to $\sim$5 hours. It has relatively low plasma protein binding (13–25%) and is readily removed by hemodialysis. Nearly 50% of the drug is excreted unchanged in urine after an oral dose and 75% is excreted unchanged in the urine after an intravenous dose. The remaining drug after oral dosing is metabolized primarily to cimetidine sulfoxide and to a lesser extent to 5-hydroxymethyl cimetidine and guanylurea cimetidine, which are excreted in urine. Less than 2% of the drug is excreted unchanged in bile.

Although cimetidine itself does not appear to be a significant substrate for CYP enzymes, it has been shown to inhibit several enzymes to varying degrees, including CYP1A2, CYP2C19, and CYP3A4. The inhibitory effect of cimetidine on these enzymes is the basis for its drug-drug interaction profile.

3. Drug-Drug Interactions

Drugs metabolized by CYP that interact with cimetidine include, but are not limited to, the following: lidocaine, quinidine, midazolam, triazolam, nifedipine, verapamil, and fentanyl (4). In each instance, inhibition of CYP by cimetidine results in reduced metabolic clearance and increases in serum concentrations of the other drug, which can lead to the expected toxicity and adverse experiences characteristic of the other drug.

4. Market Dynamics

Cimetidine was launched in the United States as a first-in-class compound in 1977. It held virtually 100% of the U.S. prescription antiulcerant market from 1977 to 1981. In 1982, Carafate® captured 4% of the U.S. market, and it was not until 1983 that the second H_2 antagonist (Zantac) became available. The market share of Tagamet decreased steadily from 1983 until it went off patent in 1994 (Fig. 1). The U.S. retail pharmacy sales of Tagamet peaked at $534 million in 1986, when it comprised 56% of the dispensed prescriptions in the U.S. market. Zantac held the greatest U.S. market share of dispensed prescriptions from 1988 (44%) through 1996 (33%), with peak sales of $1.95 billion in 1994. The U.S. prescription antiulcerant market is currently led by Prilosec, which has been the largest-selling drug in the world from 1996 through 1998.

Several factors contributed to Tagamet's market decline. Dosing convenience (twice a day) made Zantac much more attractive than Tagamet (four times a day) to physicians and patients. Tagamet was eventually approved for twice-daily dosing, but not in time to prevent the market uptake of Zantac. However, the adverse drug-drug interactions attributed to Tagamet also played a prominent role in the marketing strategy and success of Zantac and other competitors. Despite being the first H_2 blocker in the prescription market and the first to apply for approval for OTC marketing, the approval for OTC marketing for Tagamet did not occur until September 1995 (more than one year after patent expiration), and another H_2 receptor antagonist (Pepcid®) actually launched first in the OTC market. The concern of drug-drug interactions continues to be a factor in marketing efforts against cimetidine in both the prescription and OTC markets.

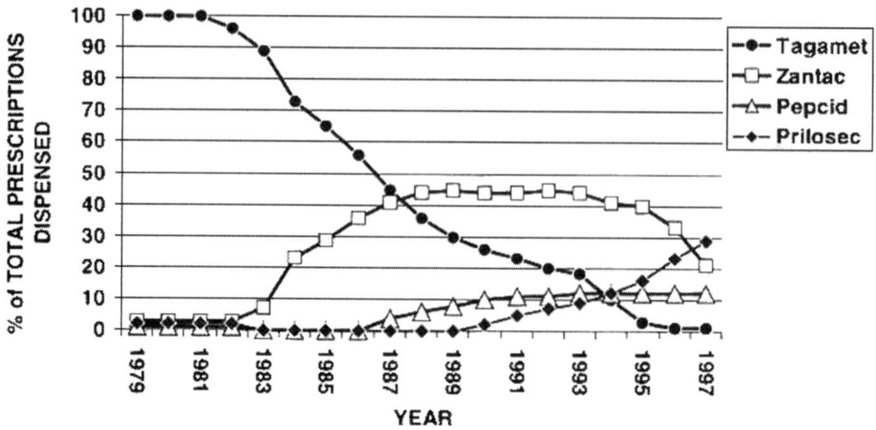

Figure 1 Antiulcerant market share, 1979–1997. The market share (expressed as percent of total prescriptions dispensed) of the antiulcerant market is shown over the period from 1979 to 1997. Only four products (Tagamet, Zantac, Pepcid, and Prilosec) are shown. Other products in this market not shown here are Carafate, Axid®, Cytotec, Propulsid®, Prevacid®, and generic cimetidine and ranitidine. Note that the initial increase in market share for Zantac (1983–1986) came exclusively at the expense of Tagamet and that further erosion of Tagamet market share occurred as additional products were introduced into the market. Patent expiration for Tagamet was in 1994 when it held 10% of the market. *Source*: From the National Prescription Audit, IMS America.

III. CONCLUSIONS

Drug-drug interactions have always been important in the development of safe and effective drug therapies. The increasingly competitive marketplace requires that every possible advantage be highlighted, particularly with the more extensive use of direct-to-consumer advertising. The potential impact of drug-drug interactions must be considered at all phases of drug development.

REFERENCES

1. Mirasol F. Pharma industry is poised for double-digit growth. Chem Market Report 1998; 253:14–16.
2. Galewitz P. Ulcer drug is world's top seller. Associated Press. February 16, 1999.
3. IMS Health Drug Monitor, 12 months to December 1998.
4. Michalets EL. Update: clinically significant cytochrome P-450 drug interactions. Pharmacotherapy 1998; 18:84–112.
5. Food and Drug Administration. Federal Register, vol. 63, no. 195, October 8, 1998.
6. Markham A, Wagstaff AJ. Fexofenadine. Drugs 1998; 55:269–274.
7. Gonzalez MA, Estes KS. Pharmacokinetic overview of oral second-generation H1 antihistamines. Int J Clin Pharmacol Ther 1998; 36:292–300.

8. Cvetkovic M, Leake B, Fromm MF, et al. OATP and P-glycoprotein transporters mediate the cellular uptake and excretion of fexofenadine. Drug Metab Dispos 1999; 27(8):866–871.
9. Honig PK, Wortham DC, Zamani K, et al. Terfena-dine-ketoconazole: pharmacokinetic and electrocardiographic consequences. JAMA 1993; 269:1513–1519.
10. Honig PK, Woosley RL, Zamani K, et al. Changes in the pharmacokinetics and electrocardiographic pharmacodynamics of terfenadine with concomitant administration of erythromycin. Clin Pharm Ther 1992; 52:231–238.
11. Monahan BP, Ferguson CL, Killeavy ES, et al. Torsades de pointes occurring in association with terfenadine use. JAMA 1990; 264:2788–2790.
12. Slater JW, Zechnich AD, Haxby DG. Second-generation antihistamines, a comparative review. Drugs 1999; 57:31–47.
13. Pratt CM, Hertz RP, Ellis BE, et al. Risk of developing life-threatening ventricular arrhythmia associated with terfenadine in comparison with over-the-counter antihistamines, ibuprofen and clemastine. Am J Cardiol 1994; 73:346–352.
14. Woolsley RL, Chen Y, Freiman JP, et al. Mechanism of the cardiotoxic actions of terfenadine. JAMA 1993; 269:1532–1537.
15. Brogden RN, Markham A. Mibefradil, a review of its pharmacodynamic and pharmacokinetic properties, and therapeutic efficacy in the management of hypertension and angina pectoris. Drugs 1997; 54:774–793.
16. Praeksaritanont T, Ma B, Tang C, et al. Metabolic interactions between mibefradil and HMG-CoA reductase inhibitors: an in vitro investigation with human liver preparations. Br J Clin Pharmacol 1999; 47:291–298.
17. Mullins ME, Horowitz BZ, Linden DHJ, et al. Life-threatening interaction of mibefradil and [beta]-blockers with dihydropyridine calcium channel blockers. JAMA 1998; 280:157–158.
18. Schmassman D, Bullingham R, Gasser R, et al. Rhabdomyolysis due to interaction of simvastatin with mibefradil. Lancet 1998; 351:1929–1230.
19. F-D-C Reports, Inc. Roche Posicor withdrawn due to ''complexities'' of drug interactions, Roche says; FDA cites 25 drugs ''dangerous'' in combination with the calcium channel blocker, June 9, 1998.
20. Brogden RN, Heel RC, Speight RM, et al. Cimetidine: a review of its pharmacological properties and therapeutic efficacy in peptic ulcer disease. Drugs 1978; 15: 93–131.

Index